# optimal birth

## what, why & how

# Comments about this book:

"I bought this book as a midwife and a very skeptical first-time mom-to-be. After reading it, I feel quite empowered and am keen to give all the suggestions for optimal birth a try. I have already informed my caregivers (to be!) of what to expect. Thank you."

**A midwife and mom-to-be**

"I thought I would peruse the book prior to a session to 3rd yr students on control. I must admit I was easily captivated. I found the narrative style of the book easy to read and the stories compelling. In fact, I found it difficult to put down. This is a useful book for all midwives and students…"

**Principal lecturer in midwifery**

"This book reaffirms my professional and personal experiences. Thank you so much for sending me a copy."

**Senior lecturer in midwifery at another university**

"I have read the whole book and I am impressed. The content is practical, well organized and the overall impression is of a very personal, warm, respectful and humane approach of a very complex, highly personal and sometimes difficult subject."

**Midwife and project manager for clinical quality & public health**

I think the book is fantastic. I love your clear non-interventionist approach, and it's also great to have something so fresh, as I feel that a lot of the books out there say the same things. I also really appreciate it being written for practitioners. Many books written for mothers vilify staff somehow, so it's a welcome relief to be the intended audience and for that tension not to be there.

**A student midwife**

"I think the content is very moving and insightful. I think your strategy of publishing in this area is very useful in moving the normal/optimal birth agenda forward. All the very best with your work for now and the future."

**Professor at another university and director of a midwifery research group**

*A warm thank you to all the midwives and lecturers in midwifery who provided detailed feedback and suggestions for improvements.*

# optimal birth

## what, why & how

A reflective, narrative approach
based on research evidence

Sylvie Donna

with forewords by Rebecca Wright and Professor Soo Downe, and contributions from numerous other people, both lay and professional, including midwives, obstetricians and researchers

Fresh ♥ Heart
PUBLISHING

## 1st American edition

First published in the USA in 2010 by
FRESH HEART PUBLISHING, a division of Fresh Heart
PO Box 225, Chester le Street, DH3 9BQ, United Kingdom
www.freshheartpublishing.com

ISBN: 978 1 906619 13 8

Library of Congress Cataloging-in-Publication Data
   Donna, Sylvie, 1959—. Optimal Birth: Why, Why & How. Includes bibliographical references and index. (Pbk : alk. paper) 1. Childbirth—United States. 2. Midwifery—United States. 3. Maternal health services—United States. 4. Obstetrics—United States. I. Title.

An earlier, British edition of this book was published as *Optimal Birth: The What, The Why & The How* in 2008 and reprinted twice. This edition is the American version of the 2nd British edition of the same book.

Set in different fonts to reflect different 'voices' as follows:
• Franklin Gothic Book—for all the author's commentary
• Bookman Old Style—for all commentary from experts
• Comic Sans MS—for all other contributors' commentary

Printed in the USA by Lightning Source Inc
Cover photo of Nuala OSullivan and Ciara © Jill Furmanovsky
Designed and typeset by Fresh Heart Publishing

## Disclaimer

# Contents

Acknowledgments viii

A few words from a mother and doula... xii

... and a few words from Soo Downe, Professor of Midwifery xiii

**Introduction: Have you ever been swimming? 1**

## W h a t are we talking about exactly? 11

**The physiological processes 14**

Pregnancy ♦ Labor ♦ Birth ♦ Postpartum ♦ The natural norm ♦ Life-saving intervention ♦ Grey areas

**Cesareans 96**

Optimal or not? The facts... ♦ The operation itself ♦ The fashion ♦ The feeling ♦ Are cesareans really necessary?

## W h y is maximally physiological birth best? 145

**1 The physiological processes are complex and efficient 146**

The hormonal cocktail of pregnancy ♦ The hormonally-triggered switchover to 'instinctive' ♦ The hormones of labor and birth ♦ Hormones produced shortly after the birth

**2 Intervention is an uncontrolled experiment 150**

From the birth of time... ♦ 1800-1900 ♦ The 1900s and 1910s ♦ The 1920s and 1930s ♦ The 1940s ♦ The 1950s ♦ The 1960s ♦ The 1970s ♦ The 1980s and 1990s ♦ How much intervention takes place now? ♦ How much intervention is needed?

**3 Research suggests it's best not to disturb the processes 162**

Pregnancy ♦ Labor ♦ Birth and beyond ♦ Disturbance as a result of pain relief ♦ Other disturbance which gets too little press

**4 Pain relief seems to cause more problems than it solves 170**

Pain relief offered ♦ Key reasons for inadequacy ♦ What about complementary approaches? ♦ Where does it all leave us?

**5 Babies are disadvantaged, perhaps far into the future 192**

The baby's perspective ♦ What about a baby's capacity to love?

## Dedication
*For all the professionals working in health care who are fighting to give families the right to safer, more effective and more compassionate care*

Photo © Colin Smith

## 6 Women end up disempowered, upset, depressed  199

Negative emotions postpartum... ✦ Do you want an empowered woman?

## 7 Hidden problems are truly horrendous  210

Perineal damage, urinary, fecal and flatus incontinence ✦ Sexual dysfunction ✦ ineffective anesthesia

## 8 Optimal birth is more beautiful...  218

The dignity and experience of optimal birth ✦ A summary of reasons why...

# HOW can optimal birth be facilitated?  263

## The prenatal period  270

What's the purpose of prenatal care? ✦ What are the potential problems? ✦ What are the potential opportunities? ✦ What are the potential risks? ✦ The importance of communication ✦ The chance to develop even better relationships ✦ Risk assessment ✦ The importance of an accurate due date ✦ Women whose birth went wrong last time ✦ Prenatal testing ✦ Monitoring procedures ✦ Ultrasound ✦ Prenatal checks ✦ Grey areas of risk assessment ✦ Providing psychological support ✦ Building rapport and open communication ✦ Freebirth ✦ The importance and role of birth plans ✦ Requests for home birth

## Intrapartum care  446

The importance of non-disturbance ✦ Disturbance vs negligence ✦ The prevalence of accidental disturbance ✦ The reality of non-disturbance ✦ Using water ✦ Prolonged labor / failure to progress ✦ What about doulas? ✦ What about the baby's father? ✦ Children ✦ Positions for labor and birth ✦ Cord-cutting ✦ Unexpected events ✦ PPH ✦ Prematurity ✦ What if the worse really comes to the worst?

## Conclusion: How can we move toward optimality?  572

Ground rules for ensuring optimality ✦ Research and the need to evaluate it ✦ How can we make decisions with inadequate data?

Week-by-week easy-use guide to pregnancy  598

Bibliography  634

Recommended reading  636

Useful websites 638

Birthframes index  642

Reflection index  645

General index  648

Who is Sylvie Donna? Who are the contributors? What inspired this book?  664

# Acknowledgments

I would like to thank everyone who helped me and contributed to this project on behalf of all the women, men and babies who might be helped by it. I am particularly grateful to five professionals who provided support or reviewed and commented on the first edition of this book before it was published: a Head of Midwifery, a former Head of Midwifery, a midwife working in a top-end of the market London hospital, an independent midwife and a midwife who is a consultant for a legal firm, which deals with litigation and dispute management, amongst other things. For detailed feedback on the book after it was published, I am particularly grateful to Prof Shirley R Jones, Tracey Hodgson, Inge Baader, Selvamalar Ratnasingam, Pat Donovan, Gill Smith, Julie Dawid and Hildur Kristjansdottir. Any remaining errors remain my own.

As far as the content is concerned, I consulted with everybody on how material by other people would be used and even the rather critical or analytical introductory blurbs and commentaries did get approved. My reason for writing blurbs, incidentally, was to place each birth story in the context of the book and make the overall message of the book consistent and clear. My consultation with contributors was useful in many cases because it allowed me to identify misunderstandings, which might otherwise have been left unresolved. Of course, I also took the opportunity to request many explanations and additions, which is why some accounts, which had initially appeared elsewhere, appear in this book in much longer form. Most of the accounts and comments are original, though. I did, incidentally, decide to adapt the spelling (and occasionally the wording) of some accounts to American English for the sake of easier reading but only vocabulary or spelling changes were made.

As you will see, many of the contributors were very happy to be named. Initially, I thought almost everybody would want to be anonymous. Then, one day, very early on in the process, I received an email from a woman who said: "I don't want or need to be anonymous!" Other contributors said "I'd love to be named" or similar, and explained why they wanted to contribute towards this project. In the words of one contributor: "I would be really pleased for you to use my story in the book. I do think that it's important for other mothers, who may be in the same situation, to feel that they can be in control and make their own decisions about what they want for their birth." It soon became clear that this was a subject on which people wanted to 'stand up and be counted'. Perhaps, also, they wanted it to be clear to readers that their contributions were totally authentic.

In cases where contributors did request anonymity this was sometimes at the request of husbands or children and sometimes because the content of what had been written was sensitive or potentially embarrassing in some way. I must admit, there were cases where I suggested to people that they should be anonymous where they simply refused. Of course, I generally respected people's decisions on this, although I have taken two names out for legal reasons. Please forgive me for making this compromise. Note that in cases where I felt it necessary to use a false name, I have indicated that it's false by putting it in quotation marks (e.g. "Clare").

Birth stories were contributed by Bhavna Amlani, Phil Anderton, Helen Arundell, Janet Balaskas, Elaine Batchelor, Debbie Brindley, Sarah Buckley, 'Tina C', Sarah Cave, 'Céline', Ruth Clark, Krisanne Collard, Esther Culpin, Mave Denyer, Beth Dubois, Pauline Farrance, Sarah-Jane Forder, Jill Furmanovsky, Elise Hansen, Janet Hanton, Sarah Hobart, Angela Horn, Deborah Jackson, Jennifer Jacoby, Kathy Kleere, Nina Klose, Katya Korochantseva, 'Katya S', Liliana Lammers, Karen, Alan and Cara Low, 'Marie-Claude', Christina Mansi, Ashley Marshall, Nathalie Meddings, Steve and Olga Mellor, David Newbound, Michel Odent, Clare O'Ryan, Nuala OSullivan, Sue Pakes, Anne Phillips, Gaia Pollini, Monica Reid, Justine Rowan, Jenny Sanderson, Maria Shanahan, Gemma Shepherd, Jo Siebert (and Dr Lawrence Impey), Fiona Lucy Stoppard, 'Pat T', Jan Tritten, Caroline Turner, Rachel Urbach, Carol Walton, Joanne Whistler, Michael White, Clare Winter, Sonia Winter, Liz Woolley, June Worcester, Rebecca Wright and Heba Zaphiriou-Zarifi.

Extracts from birth stories, diaries, birth plans, letters or emails, interviews, self-standing comments or information texts (some of which were used in these pages, all in the companion volume for pregnant women, *Preparing for a Healthy Birth*) were contributed by Helen Arundell, Celina Barton, Elaine Batchelor, Paula Bays, Wendy Blumfield, Debbie Brindley, Bill Bryson (with CARE International), Sarah Buckley, Emma Cameron, Luisa Cescutti-Butler, Amanda Chalfen, Ruth Clark, Kathryn Clarke, Kathy Cleere, Rachel Cockburn, Janine DeBaise, Beth Dubois, Sarah-Jane Forder, Mary Frankland, Jill Harradine, Jenny Hodge, Kris Holloway, Angela Horn, Eleanor Jackson, Jennifer Jacoby, Libby Kelly, Eliza Klose, Nina Klose, Liliana Lammers, Dr Nicolette Lawson, Julia Lockwood, (and Melanie Milan), Karen Low, Liz Perry, Anne Phillips, Ashley Marshall, Nancy Redheffer, Shari Henry Rife, Justine Rowan, Hazel Rymell, Katya S from Moscow, Jenny Sanderson, Kay Sawford, Claire Saxby, Amanda Sealy, Joanne Searle, Gemma Shepherd, Teri Small, Sarah Stanley, Fiona Lucy Stoppard, Clare Swain, Fiona Taylor, Georgina Taylor, Jan Tritten, Jill Unwin, Rachel Urbach, Juliana van Olphen-Fehr, Ulrike von Moltke, Carol Walton, Julie White, Clare Winter, Sonia Winter, Rebecca Wright and by numerous anonymous contributors. A very special thank you to everyone who asked to remain anonymous!

Thank you, too, to everybody who contributed photographs. Particular thanks (again, in alphabetical order) to Elaine Batchelor, Jill Furmanovsky, John Huson, Nina Klose, Ashley Marshall, Jenny Matthews (of CARE International), Aly Neely, Nuala OSullivan, Jenny Sanderson, Colin Smith and Rebecca Wright, as well as to my pregnant models Sarah Morris and 'Linda' (who asked me not to reveal her surname) and to one anonymous, naked, pregnant woman. Thank you, too, to the other anonymous contributors of photos.

I would also like to thank Sarah Buckley, Esther Culpin, Angela Horn, Sheila Kitzinger, Nina Klose, Liliana Lammers, Nancy Radford and Dr Claire Robson for practical support. I am particularly grateful to Dr Sarah Buckley for allowing me to use adapted extracts from her excellent research articles. Thank you too, to Dani Zur of *Mother and Baby Magazine* for letting me use some of the results from the survey conducted in association with Persil in 2002. An enormous thank you to my various anonymous advisers in the realms of midwifery and obstetrics.

A very special thank you to my original reviewers and editors (in alphabetical order) Nina Klose, Liliana Lammers, Michel Odent, Clare O'Ryan, Theresa Prentice, Nancy Radford, Jenny Sanderson, Jenna Shaw-Battista, Liz Woolley and Rebecca Wright. (There was also another editor, who asked to remain anonymous... Thank you, mystery mother of four!) Thank you also to Nancy Radford for her invaluable advice and to Clare O'Ryan for acting as a perfect sounding board to my ideas. My thanks are also due to my husband and to our daughters for encouraging me over the years and for allowing me to spend many, many hours at my computer. My mother has also been a wonderful source of support and I really appreciate the interest my mother-in-law and sister have taken in this project too—not to mention my friends.

A heartfelt thank you to Soo Downe and Rebecca Wright for their forewords and to Michel Odent for all the material he contributed or checked for me. A very special thank you too, to my various anonymous contacts at the British Royal College of Midwives, the Nursing & Midwifery Council, Doula UK, the National Childbirth Trust (another UK-based organization) and Midwifery Today, whose informal advice and encouragement at conferences, via email or snail mail has been invaluable; thank you too, to the midwives, obstetricians, professors and senior lecturers of midwifery who have been in touch with me by email or letter to offer support or encouragement. Thank you, too, to Sarah Buckley, Soo Downe, Hélène Vadeboncoeur and Denis Walsh for reviewing research evidence and therefore providing me with a great starting point for the notes and references at the back. Particular thanks to Hélène Vadeboncoeur and Christine Morton for additional support with research, and to Robin Russell, anesthesiologist at the John Radcliffe Hospital in Oxford, England, for helping me understand pain relief issues. Any errors in this text or in the references remain my own, of course.

Finally, I would like to mention the obstetrician who attended the birth of my first child in Sri Lanka and all the staff who supported me. I realize that despite all the research supporting physiological birth, your decision was considered somewhat dangerous at that time. Thank you! I am also grateful to Elaine Batchelor for helping me to work out a backup program for my second labor for the period when Michel wasn't available, and to the British NHS (the state-run health system) for providing my routine prenatal and postpartum care. Thank you to Jo Farrington for that wonderful, long, reassuring pre-birth chat we had one afternoon, and to my family doctor for not striking me off his list! Thank you to the NHS midwife, who helped smooth the path to my third and final labor by not hassling me and at the same time offering me invaluable support. (It's a great shame you weren't able to attend the actual birth.) My thanks to the midwives who did arrive for the last few minutes and went along with my requests. A big thank you to Lawrence Impey in Oxford and Donald Gibb in London, both consultants who appeared out of the blue in birth stories. Along with all the other supportive professionals I came into contact with while researching this book, you really do seem to be working to optimize conditions for mother, father, baby, family and society at large each and every time a new child is born. Perhaps my first obstetrician was right when he said he was convinced that violent births result in a violent society, while gentle births result in gentle, loving societies. Let's hope we find out the outcome of a mass move towards gentle birth over the next few decades. Our world needs a bit more peace and harmony.

## From the publisher...

The publisher would also like to thank the following for the use of previously published material:

- Clairview Books for allowing the reproduction of extracts from *Birth and Breastfeeding* by Michel Odent (Forest Row, 2007)
- Care International and Bill Bryson for the extract from the *African Diary* (Doubleday 2002) (p373)
- *LLL GB News* for allowing the publication of adapted and extended versions of accounts which appeared in the magazine (Birthframes 19 and 86)
- *New Beginnings* for allowing the publication of a longer version of an account which first appeared in the May/June 2003 issue (Birthframe 100)
- *Reader's Digest* for information and ideas which appeared in an article in the July 2003 edition of *Reader's Digest* magazine (p434)
- Shufu no Tomosya for the photographs of Liliana Lammers which were originally published in the Japanese-language magazine *Balloon* (pp 197 and 633)
- *The Practising Midwife* for Esther Culpin's account (Birthframe 48), which first appeared in the January 2003 edition (2003 Jan 6(1):10-1)

If any material has accidentally been used without appropriate acknowledgment, or if any details are incorrect, please contact the publisher so that amendments can be made in future editions. Every possible effort has been made to ensure that all details of contributions are correct.

## Your input:

If you would like to write to the author to comment on any aspect of this book, please write to her c/o Fresh Heart Publishing, PO Box 225, Chester-le-Street, DH3 9BQ, United Kingdom or click on 'Contact us' link on the website at www.freshheartpublishing.com. Fresh Heart is interested in receiving:
- feedback and opinions based on your experience
- corrections, additional information or photographs
- additional birthframes or comments from women or professionals
- additional advice you would give to other midwives or caregivers

All contributors to future editions will be acknowledged, as preferred.

# A few words from a mother and doula...

During my first labor and birth my midwife surprised me. Contrary to all I had read (and I had read a lot in preparation for this birth), she did not tell me when I had reached 10cm and was ready to push. She didn't ask to examine me internally at all. She was simply a quiet figure, present, but unobtrusive. She respectfully asked if she could check the baby's heart rate with her hand-held Doppler and waited (without asking unnecessary questions) until I was in between contractions. She seemed to know what I needed without my saying. She could see when the contractions were coming fast and furious, with no break in between, that I was losing my concentration. She gently brought me back to my breath and those moments of lesser intensity, as one contraction tailed off and the next one began. When she could see I was getting fed up, she encouraged me gently back into the birthing pool, where nearly immediately my pushing intensified and the baby was born. She didn't direct me unnecessarily—she trusted my body and the process. When my baby was born, she handed her to me directly and let me rest with her in the pool. There was no sense of worry or fear, no hurry, no pressure. I learned about birth from her and this experience. Only later did I learn how fortunate I had been and how different my experience was to so many other women.

Sadly, too often birthing women are not respected—they are violated by unnecessary medical procedures, many of which are described to them in euphemisms that disguise their true impact on women and babies. Women are left disempowered, hurt and angry but with little recourse. As a doula, I've heard from women who reported bullying into procedures they did not want, from women who expressly denied permission for procedures that were then carried out without their consent, women who asked for support and were given drugs and left alone, women who were separated from their babies unnecessarily. It is as if these violations are unimportant because they are carried out in the name of medical necessity for the woman's or her baby's 'own good', as if she herself were incapable of judging her own or her baby's best interest. Often, if she complains afterwards, she is considered mentally unstable or a troublemaker. Little wonder many women don't bother, although they and their babies carry this pain with them.

Birth is important work, perhaps the most important work. It remains a deeply contested space, where women's bodies are the battleground. It takes a special kind of person to take on the mantle of midwife, step into the fray to be truly 'with woman' —to trust her, empower her and protect her and her baby, all the while letting her be the star of the show. Respect, honesty, holding the space, knowing how a labor is progressing by the sounds a woman makes... these are some of the components of the art of midwifery. Midwives who practice this art in the face of pressure to do otherwise are true heroes, celebrated by mothers and babies with deep gratitude and love.

Sylvie Donna' s book is a timely call to revive this art. Her work brings welcome scientific collaboration of the safety of normal, undisturbed birth even as it demonstrates a healthy respect for the potentially life-saving wonders of modern medical science. It should prove thought-provoking and inspirational to midwives working in the current climate.

*Rebecca Wright*

This book should prove thought-provoking and inspirational
to midwives working in the current climate

# ... and a few words from Soo Downe, Professor of Midwifery at UCLAN*

Evidence-based medicine (and, by extension, evidence-based maternity care) has become a talisman for health services across the world. Where the care and treatment of populations is concerned and where the interventions under scrutiny are drugs or treatments which have a straightforward operation that works in the same way for most people, this kind of evidence is invaluable. Magnesium sulphate for pre-eclamsia, placing babies on their backs to sleep and continuous support in labor—all of these effective treatments have been reinforced by good quality randomized controlled trials and systematic reviews.[1]

However, there are many situations in health care where the issues are complex, ill-defined and highly personal. In these cases, trials would be too expensive to run, too complicated to make sense of or, conversely, too simple to take account of the relevant issues for each individual involved. This gap in knowledge is increasingly taken seriously in the medical and health care literature. Indeed, a journal has recently been launched to present and discuss case studies called, appropriately, the *Cases Journal*[2]. In parallel with this development, narrative medicine is gaining prominence[3]. This usually takes the form of (auto)biographical accounts of people's lives, which provide a context for their illnesses. Such approaches to understanding health mirror the way in which many health care practitioners actually work, especially those who have the opportunity to build long-term relationships with those they are working with. This includes family doctors, CNMs and, of course, many midwives. However, while individual case reports have been included as a matter of routine in highly academic journals, paradoxically this has not been the norm in academic midwifery journals.

This book takes a narrative approach to understanding the accounts of women, and, in some cases, their partners, who have experienced what Sylvie Donna has termed an 'optimal' birth. There are a number of definitions of optimal birth that are currently in the literature, including a birth that is as physiological as possible for any particular woman[4]—so, in some cases, this might mean making a necessary cesarean section a positive experience, for example. This book follows the same definition but focuses mainly on optimal physiological birth and on the many and varied ways in which a woman and her partner might encounter this. Using what she has termed 'birthframes' (personal accounts or commentaries on birth), Sylvie has set out to provide a counterpoint to some of the more standardized ways of dealing with childbirth, which tend to dominate in many large maternity units across the world. This offers the possibility of integrating more personalized approaches for women who might be laboring normally for them, but differently from the average or the 'standard' established at any one point in time. The aim of the book is to facilitate reflection on possibilities for labor and birth and to suggest that moving outside boundaries considered standard might not only be safe, but also optimal for individual women and babies—and, based on some of the accounts by midwives that are included, optimal for staff as well.

Given that client satisfaction and litigation are such everyday concerns for health practitioners today, focusing on how to optimize experiences for individuals, in whatever way, can only be a very positive move into the future. I hope, therefore, that this book will provide inspiration and support for midwives and others who want to maximize the possibility of optimal birth for pregnant and laboring women.

*Soo Downe*

*University of Central Lancashire in the United Kingdom, www.uclan.ac.uk

# NOTES & REFERENCES

1   See the following studies:
- Magpie Trial Follow-Up Study Collaborative Group. The Magpie Trial: a randomised trial comparing magnesium sulphate with placebo for pre-eclampsia. Outcome for women at 2 years. *BJOG.* 114(3):300-9.
- Mitchell EA.(2009) SIDS: past, present and future. *Acta Paediatr.* 98(11):1712-9.
- Hodnett ED, Gates S, Hofmeyr GJ, Sakala C.(2007) Continuous support for women during childbirth. *Cochrane Database Syst Rev.* (3):CD003766.

2   Cases journal: Available at: http://casesjournal.com/ [Accessed 6 January 2010]

3   Greenhalgh T, Hurwitz B. (1991) Narrative based medicine: Why study narrative? *British Medical Journal* 318:48-50.

4   Kennedy HP (2006) A concept analysis of optimality in perinatal health. *Journal of Obstetric, Gynecologic & Neonatal Nursing.* 35(6):763-9.

*"A ship in harbor is safe... but that's not what ships are for." (Grace Hopper)*

# Introduction:

## Have you ever been swimming?

Years ago, when I was an English teacher, teaching working adults or students business, technical or academic English, I was sometimes able to slip away for an exotic holiday. (Things have changed a lot since then.) Once, while I was based in Singapore I spent an idyllic week on the tropical island that was used to film *South Pacific.* (You know, that one with the song 'I'm gonna wash that man right out of my hair.') And while I was there, I met a sailor... Say no more.

Actually, I am going to say quite a lot more, as you've probably realized from the thickness of this book. (Hope you've got a cup of coffee by your side.) This American sailor was extremely interesting to talk to, mainly because he was an ex-Marine, about halfway through sailing round the world and very knowledgeable on all kinds of relevant maritime topics.

In an email conversation with an obstetrician the other day, I suddenly remembered this man, for some strange reason. I was reminded of a story I heard, even further back into my long, complicated past. It was the tale of a teenager who got taken on as crew on a large ocean liner. He soon became friends with the Captain, who would regale the boy for literally hours with all kinds of maritime tales and fascinating trivia. (A bit like the ex-Marine and me, really.) Often, the Captain—being rather an arrogant man—would gently mock the boy and ask him, "Don't you know anything at all about astronomy?" or "Haven't you ever studied marine biology?" or even "Can't you even *explain* the word oceanography?" Tedious as it was, since these two characters were out at sea, the situation continued for some time and I suppose both enjoyed these long conversations for different reasons. The boy drank up information about all kinds of new subjects (which seemed much more interesting than school had ever been) and the Captain enjoyed gloating over this ignorant boy. Finally, one day, the boy came running into the Captain's cabin, gasping for breath, shouting, "Captain! Captain! Did you ever study swimmology?" Of course, the Captain was merely bemused and barely paid the poor boy any attention. But the lad persisted and eventually, in total exasperation, shouted, "Captain! Captain! *Can you swim?*" Well, it just so happened that this particular captain couldn't. (All right! All right! It's just a story.) So he died. And the stupid, naïve, poorly educated, inexperienced boy swam to the shore and lived on...

In case you haven't already guessed, this book is about both swimming and swimmology, as it were. You might be able to swim or you might be a swimmologist—I really don't know. Whatever your experience of birth I hope you find something of interest in these pages. If you've never swum in real water (i.e. experienced physiological birth for yourself[1]), I hope that your expertise as a swimmologist will not put you off reading about that experience and perhaps thinking about it too. Recent developments in research certainly justify that...[2]

After all, as you may agree, recent research has confirmed many things that some very ordinary women have been trying to talk about for decades—the endorphins of labor, the emotional postpartum 'high', the all-round benefits of giving birth without pain relief. And my own personal research has convinced me that both swimmers and swimmologists need each other, if birth is going to be optimized from both safety and experiential points of view. While birthing women need midwives, CNMs or obstetricians for the sake of safety, caregivers also need to witness the process of birth (and possibly even experience it for themselves) so they can gain a real insight into the processes and their particular role within them.

In case you're wondering how I fit in, I'm not really sure if I'm primarily a swimmer or a swimmologist. I certainly started out as a swimmer, flailing around and gasping for breath, then I became a little better at swimming and I finally took up the science of swimmology myself. However, as you probably realize, there are all kinds of swimmologists... Midwives and CNMs are hardly the same as obstetricians. And obstetricians themselves may also have a completely different perspective to researchers, whose whole focus is on a systematic collection of data. Then there are journalists, partners of women who've given birth physiologically, husbands of women who've had a traumatic birth, and mothers of mothers, who have their own perspective, based perhaps on an entirely different experience of birth. And what about prenatal teachers, doulas, maternity care assistants, anesthesiologists, neonatal nurses and pediatricians? Family doctors fit in there somewhere too, in a very significant position.

Anyway, my reason for beginning and continuing my own studies of swimmology were the numerous conversations I had with other women after giving birth myself. I was disturbed by what I heard, or read in facial expressions or other body language. And comments some women made, or experiences they described, left me upset and puzzled for days or weeks at a time. The first consequence was that I started working on a book, which was eventually published as BIRTH: Countdown to Optimal. (This has now been updated and improved, notes and academic references have been added, and it's been republished under the title Preparing for a Healthy Birth: Inspiration and Information for Pregnant Women. I hope it's a book you might find useful for use with women who say they want to 'do without drugs' if they can, but who are not entirely confident that they'll be able to.) After the first book came out I was encouraged to adapt it for caregivers... so I did, at great speed, in time for a conference (six weeks later!) on optimal childbirth. This book is a much extended version of that original edition, with references added in.

There seemed to be room for yet another book on childbirth because although numerous books about maternity care are available, there doesn't seem to be one which has a clear focus on ways of supporting women who opt for safest or most humanitarian approaches, according to what most research has indicated so far, and/or who embrace the physiological model of birth perhaps for philosophical or ideological reasons, or simply because of a belief in a healthy body's ability to grow and birth a baby effectively.

In most books on midwifery and obstetrics somehow, the advice which would be relevant when caring for at least 95% of pregnant women (if they chose a physiological birth) has got lost amongst the advice which relates to the other 5% or less, who are likely to need drugs or interventions from the point of view of safety. The focus on pathology or drug-based 'pain relief' (a misnomer, in my view) has meant that fewer and fewer caregivers feel confident about supporting normal labor—although thankfully this is now a major focus of much midwifery training, in some countries at least. Needless to say, perhaps, it really is vital that all caregivers know how to support normal labor and birth (following the physiological processes) because women who choose this safest route need and deserve excellent care.[3] So this book has been written for midwives and other caregivers who want to consider, or reconsider, best ways of supporting women who want to have a physiological birth, if at all possible.

In the recent past the emphasis on pathology over normality, on medical management instead of watchful waiting, has meant that 'normal' almost came to mean 'abnormal', because women who were hoping for a straightforward, but safe, physiological labor and birth were being treated as unusual. Unnecessary interventions had become so commonplace in the labor ward and delivery suite that a woman who insisted on having none (unless they were needed for safety) was branded an extremist, even though research was very definitely on her side. Hopefully, this situation has changed radically, thanks to the publication of many books which are promoting normal birth (of which 'optimal' is definitely a subset), and thanks to ongoing inspiration and recommendations which are coming from reports or surveys, such as Listening to Mothers conducted in 2002 (at www. maternitywise.org.).

Nevertheless, in some places there may still be caregivers who are shocked or cynical about requests for a 'natural birth'. Perhaps there are some very understandable reasons for this. Firstly, they may have come to associate this phrase with unrealistic idealists, whose births are actually not very 'natural'. Secondly, they may never actually have witnessed a truly 'normal' birth, i.e. a physiological one with no complications. (Of course, this is because many women are requesting pain relief, or are consenting to it when it's offered. And while some caregivers may have witnessed a normal birth—without any anesthesia or analgesia—they may not have witnessed one which has proceeded in optimal conditions, according to the true physiological processes.) Another reason why certain professionals may react negatively to any request for a 'natural birth' is perhaps their worry about things going wrong. Although even the most superficial study of statistics will make it clear that there is *always* a maternal and fetal mortality rate, the reality of an individual death is horrifying, particularly given the high risk of litigation. In fact, the risk of a criminal investigation might be what puts many caregivers off supporting other women. They may be concerned that their superiors may fail to support them if there are not sufficient records, which will in turn lead to excessive (and probably intrusive or invasive) monitoring. Perhaps it's time a 'normal birth' consent form was introduced, requesting non-disturbance!

Whatever your own views, feelings or experience, the aim of this book is to support you if you want to help improve the safety of the births you facilitate[4] and if you want to help women *actually have* the natural or normal births they request. While I have tried to avoid giving patronizing advice, I have made some tentative suggestions which are all based on research, feedback and suggestions I've received and/or my own experience as a mother who has given birth three times, entirely physiologically. Of course, I have at all times tried to take typical protocols into account. In any cases where I suggest changes or extensions to these I hope you will view these in the spirit they were intended—as a springboard for thought, discussion and possibly even action! Most often, I have presented questions to aid reflection.

If, as you embark on reading this book, you doubt the value of supporting physiological birth (perhaps because you have seen so many women request pain relief after all), please suspend your disbelief for as long as it takes you to read the whole book. After all, we have to recognize that while the English Queen Victoria set in motion a hope for painfree birth in 1853 (when she agreed to try out chloroform for the birth of her eighth child), this hope has not yet become a reality. Even women who have the best of anesthesia complain of pain beforehand or afterwards, and side effects are unfortunately all too common.[5] As we all know, but maybe sometimes prefer to forget, the use of any unnecessary drugs or interventions during labor and birth compromises safety. For this reason alone it's well worth knowing how to make labor and birth as good as possible for women who opt out of this artificial approach, perhaps after reading research evidence or simply because they trust that the natural, physiological processes will be pretty effective, thank you very much! From personal experience and from the extensive research I've carried out (over a period of over 12 years now), I'm convinced that the physiological processes are much more effective than we often assume. I'm also sure that they're worth 'enduring' because outcomes really do seem to be better for both mother and baby, not to mention families and the caregivers who provide care.

Before I sign off—in case you're still wondering who on earth I am—I'll give you a little more information. After teaching for around 20 years and finally meeting the man who is now my husband, I had three babies. I mention my own experience of birth here and there because I think it's important to convey how difficult it was to arrange my own 'optimal' births. I had to fight long and hard to exercise my right to give birth as I wanted, perhaps because the trend for evidence-based care hadn't taken hold back then, and perhaps also because society hadn't yet validated a woman's right to choose. (Even now, I wonder how much real choice it's possible to have, for practical, financial and logistical reasons, apart from anything else.) Anyway, when I started writing, since I wasn't a medical expert myself, I wanted to add 'weight' and authority to my own words, so I also conducted a great deal of research—both academic and personal—and made contact with many professionals, some of whom have joined me on these pages. Comments or accounts from mothers are included so as to provide a window into the world of increased safety and satisfaction... and its alternative, help you gain insight into the processes and stimulate thought and reflection. After all, there are still many practical issues to resolve.

Photo © Colin Smith

*Perspectives on birth are changing and the world is beginning to look very different...*

## NOTES & REFERENCES

To find abstracts for any studies, go to www.pubmed.com and search using year and key words (e.g. any author's name and main words in the title). If you work for a hospital or are studying at an educational institution, you may well also have access to the full articles. By the way, if you notice errors of any kind in the Notes & references here, email info@freshheartpublishing.co.uk. Amendments will be made in future editions. Also, please make contact if there's anything in this section which you disagree with, or would like to add to.

1 In case you're wondering why I don't say here that swimming is like giving birth, period, i.e. in *any* way, the reason is quite simple... Giving birth with drugs in your system, with EFM and other interventions such as ARM and forceps, would be the equivalent to being 'swum' across the Atlantic under sedation, in a harness, pulled along by someone else in a boat. We're talking here about active and aware *swimming*—the only kind of swimming, I would suggest, which will really give you an insight into what birth can be like. If you've only been 'swum' how can you understand real swimming? How can you know how to make it as good as possible?

2 There are no references here as I provide plenty—with commentary—later, as topics come up. You are probably familiar with the research which generally shows that 'natural' is generally better, in terms of safety, than any kind of unnecessarily managed approach. If you haven't done so already, see:

- The book written and researched by former WHO director Marsden Wagner: *Born in the USA: How a broken maternity system must be fixed to put women and children first.* University of California Press, 2006
- The book written by Jennifer Block, former journalist and co-editor of the revised classic *Our Bodies Ourselves*: *Pushed.* Da Capo Lifelong, 2008
- The book co-authored by seven researchers (Enkin, *et al*) *Guide to Effective Care in Pregnancy and Childbirth.* Oxford University Press, 2000 and the Cochrane database at www.cochrane.org
- Enkin's book, written with another eminent researcher Jadad: *Randomised Control Trials: Questions, Answers and Musings.* Blackwell, 2007
- The book edited by professor of midwifery and director of a research group, Soo Downe: *Normal Childbirth: Evidence and Debate.* Churchill Livingstone, 2004
- Denis Walsh's discursive overview of evidence which points us back to physiological labor and birth as best 'first' options: *Evidence-based Care for Normal Labour and Birth.* Routledge, 2007

Beyond this, it's interesting to observe that as studies are increasingly published, initial conclusions about the superiority of 'natural' approaches are being confirmed.

To take breastfeeding as an example, if you read any book on the topic, the advantages will be well documented, for both mothers and babies. Nevertheless, newer studies are still revealing still more advantages. For example, the following study found that breastfeeding decreases a baby's chance of getting rheumatoid arthritis later on in life:

- Pikwer M, Bergström U, Nilsson JA, Jacobsson L, Berglund G, Turesson C. Breast feeding, but not use of oral contraceptives, is associated with a reduced risk of rheumatoid arthritis. *Annals of the Rheumatic Diseases,* 2009 Apr;68(4):526-30. Epub 2008 May 13

3   If you don't believe you really can be a woman's advocate, whatever her choices, consider why you are doing the work you are doing. What are your general aims in life? How do you feel about negative or positive consequences you personally create? If you have no direct, first-hand experience of birth, this book is intended to help, using of narrative accounts and reflective questions.

4   If you are an experienced caregiver, or even if you're a student, I'm sure you'll have your own views on how best to support and facilitate physiological labor. Nevertheless, there may be factors you haven't considered or behaviors or procedures you often engage in, which you consider innocuous, which are actually very disturbing for the laboring and birthing woman. In my own pregnancies, labors and births I was aware of a great range of caregiving approaches and I have since read about many more variations. I have also been alarmed by many accounts I've received from new mothers. For these reasons, I would like to ask you to be open-minded about how physiological labor is best facilitated. Even though he's a man, who you might say cannot possibility understand birth as a woman does, the French obstetrician, Michel Odent, has considered this issue in depth, and after attending over 15,000 births, he has drawn certain conclusions. In the following article, he outlines what he considers important in terms of facilitating the natural processes:

•   Odent M. New reasons and new ways to study birth physiology. *International Journal of Gynecology & Obstetrics* 2001, 75:S39-S45

(He expands further on these ideas in his book *Birth & Breastfeeding.*)

In case you feel wary of facilitating the physiological processes of labor and birth *without pain relief* you may be reassured to note (in case it's a surprise) that not all women who have normal labors find them painful and outrageously unpleasant—even if they are making all kinds of strange noises! Quite a few people who've contacted me have mentioned their surprise at the *lack* of pain in their labors or the surprising nature of the sensations. Others—including myself—experienced intense pain but found they were able to travel through it, thanks to the strange hormonal processes which were taking place, which inevitably have an effect on the mind as well as the body. Motivation to face the pain comes mainly from thoughts about the baby, who may be affected by any drugs, and for me the knowledge I have now would convince me that facing the pain of labor ultimately means less pain overall, if postpartum pain (both physical and emotional) is also taken into account.

5   Actually, research into the causative associations between drug use in labor and things such as alertness postpartum and breastfeeding is in its infancy, but results are already suggesting that drugs do have side- or after-effects; these are constantly discussed in the literature on anesthesiology. Possible (or probable) side- or after-effects are widely accepted as including nausea and vomiting, feelings of confusion, lowering of the blood pressure, sedation, urinary retention, slower emptying of stomach contents and itching (pruritis)—and the pain relief is not always effective. Of course, in all cases, it's not easy to establish what causes what.

Looking only at the issue of alertness for now—we'll come back to breastfeeding later—studies do not provide a clear overall picture. Years ago, some practitioners were apparently concerned about the observable depressive effects on newborns of analgesia used in labor. I deduce this because researchers (Bonta, *et al,* 1979) discovered that another drug (naloxone) provided an effective 'antidote' and restored what apparently seemed to be an acceptable level of alertness in newborns. However, is naloxone (or any equivalents) really a solution to reduced alertness or is the mother-baby dyad losing out when natural alertness is not present? (Would you prefer natural sexual arousal when you meet your life partner, or passion produced by sedatives, counteracted by Viagra? Of course, we need to remember that these early interactions can never be repeated, and also that they might have a significant effect on later interactions too.) Mothers, incidentally, are sometimes assessed by anesthesiologists for alertness using a four-point scale (1. awake/alert, 2. drowsy, but readily responsive, 3. drowsy and requires shaking to rouse, 4. unconscious). But how awake and alert do people expect women to be just after they've had a baby? Physiological labor usually results in extreme alertness, which is, of course, beneficial for the bonding process.

In a more recent study (Volikas, *et al*, 2005), which looked at potential side-effects of patient-controlled opioid analgesia (remifentanil) postpartum, 22 out of 50 women were reported to have experienced some drowsiness. (44% seems rather a high percentage...) The researchers reported that at the dose used in the study, remifentanil had 'an acceptable level of maternal side-effects and minimal effect on the neonate' (i.e. the newborn baby). Personally, I question whether any level of drowsiness is acceptable during this one-time encounter between new mother and baby. And I wonder how it could be established that there was only a 'minimal' effect on the neonate if there was no control group, i.e. if researchers did not compare these 50 neonates with 50 others, who were born entirely physiologically. What seems a normal level of alertness in a neonate might change if researchers were to document the extreme alertness many people have anecdotally reported when babies have been born without any drugs in their systems. There is an enormous difference, I would suggest, between a dull-eyed look and a vibrant gaze, in terms of bonding and simple joy in new motherhood. And when a newborn looks at his or her mother it's helpful, perhaps, if the mother isn't one of the 22 out of 50 women who were reported in this study 'to have experienced some drowsiness'. Hill (2008) later strongly recommended remifentanil, although Van de Velde disagreed in a follow-up article. Although remifentanil is more effective than pethidine and diamorphine (which perhaps explains why it is popular in Belfast, N Ireland), what price are women paying for their reduced alertness in terms of effective early bonding?

There is an enormous difference, I would suggest, between a dull-eyed look and a vibrant gaze, in terms of bonding

Another study (Wittels, *et al*, 1997) compared the alertness (amongst other things) of newborns exposed to either epidural morphine or intravenous patient-controlled analgesia. Of course, because the focus was on newborns of mothers who'd had a cesarean it was impossible to compare the alertness of babies born with drugs in their systems and that of babies who'd been born with absolutely no drugs in their system, so only 'relative' alertness could be tracked. Yet another study, back in 1981 (Rosenblatt, *et al*) looked at the influence of maternal analgesia (epidural bupivacaine) on the newborn. Significant effects were found: "Immediately after delivery, infants with greater exposure to bupivacaine in utero were more likely to be cyanotic [blue-skinned] and unresponsive to their surroundings. Visual skills and alertness decreased significantly with increases in the cord blood concentration of bupivacaine, particularly on the first day of life but also throughout the next six weeks. Adverse effects of bupivacaine levels on the infant's motor organization, his ability to control his own state of consciousness and his physiological response to stress were also observed." A recent study by Henrichs, et al (2009) considered whether alertness could be affected by a factor such as fetal size in mid- or late pregnancy. (The conclusion was that it could.) In a study such as this, I would imagine there could be numerous confounding factors, the principal one being the use of anesthesia or analgesia (or not) during labor. Personally, I would only trust the results of this study if all fetuses measured in utero had been born without any drugs in their systems. After all, while the motivation of these researchers appears to have been a desire to investigate behavioral problems in newborns (e.g. infant irritability), they do not appear to have taken into account the fact that one of the primary characteristics of narcotics-addicted neonates is that they are 'substantially more irritable' (Strauss, et al, 1975).

Given the vital importance of good bonding in the sensitive one-hour period following birth (from the point of view of later mothering behavior), I very much hope that other researchers will look further into the issues of alertness and breastfeeding success (or lack thereof), particularly in relation to drug-use in labor.

See:

- Bonta BW, Gagliardi JV, Williams V, Warshaw JB. Nalaxone reversal of mild neurobehavioral depression in normal newborn infants after routine obstetric analgesia. *Journal of Pediatrics*, 1979. Jan;94(1):102-5
- Volikas I, Butwick A, Wilkinson C, Pleming A, Nicholson G. Maternal and neonatal side-effects of remifentanil patient-controlled analgesia in labour. *British Journal of Anaesthesia*, 2005, Oct;95(4):504-9. Epub 2005 Aug 19
- Wittels B, Glosten B, Faure EA, Moawad AH, Ismail M, Hibbard J, Senal JA, Cox SM, Blackman SC, Karl L, Thisted RA. Postcesarean analgesia with both epidural morphine and intravenous patient-controlled analgesia: neurobehavioral outcomes among nursing neonates. *Anesthesia & Analgesia*, 1997. Sep;8 (3):600-
- Rosenblatt DB, Belsey EM, Lieberman BA, Redshaw M, Caldwell J, Notarianni L, Smith RL, Beard RW. The influence of maternal analgesia on neonatal behaviour: II. Epidural bupivacaine. *British Journal of Obstetric Gynaecology*, 1981. Apr;88(4): 407-13
- Henrichs J, Schenk JJ, Schmidt HG, Arends LR, Steegers EA, Hofman A, Jaddoe VW, Verhulst FC, Tiemeier H. Fetal size in mid- and late pregnancy is related to infant alertness: the generation R study. *Developmental Psychobiology*, 2009, Mar; 51(2):119-30
- Strauss ME, Lessen-Firestone JK, Starr RH Jr, Ostrea EM Jr. Behavior of narcotics-addicted newborns. *Child Development*, 1975. Dec;46(4):887-93
- Hill D. Remifentanil patient-controlled analgesia should be routinely available for use in labour. *International Journal of Obstetric Anesthesia*, 2008, 17(4),336-339.
- Van de Velde M. Controversy. Remifentanil patient-controlled analgesia should be routinely available for use in labour. *International Journal of Obstetric Anesthesia*, 2008 Oct;17(4):339-42. Epub 2008 Jul 9

Let's dare to really
explore the art
and science of
giving birth...

What are
the
consequences

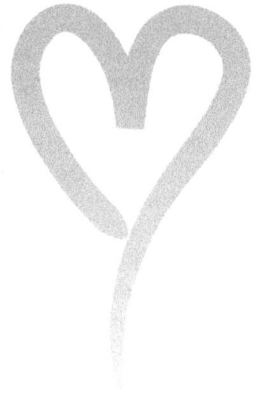

if we don't?

# Optimal Birth: Wh a t
## What are we talking about exactly?

The simplest way of defining an optimal birth is to say it's a birth which is as good as it possibly can be, for both mother and baby—and hopefully for caregivers too! In my view, for most women this means a birth without any drugs or unnecessary interventions or disturbance, i.e. a physiological birth, taking place spontaneously in the context of a healthy body and mind. For women who are not healthy it means only intervention which facilitates and optimizes the processes in terms of both safety and experience, for mother and baby.

Optimal birth as I'm describing it in this book is actually different from *natural* birth as it's been understood over the last few decades. I like to use the phrase 'old natural' to describe that... It's often meant a great departure from the real physiological processes because interventions have merely changed from being of the pharmacologic type to being 'New Age'. In other words, instead of IVs and shots, women in the 1970s, 80s and 90s typically had whale music playing in the background while they allowed themselves to be prodded by their acupuncturist's needles (or shiatsu therapist's fingers) and shouted at (or 'coached') by a committed birthing partner.[1] Occasionally, a beautiful birth occurred after all this 'natural' intervention, but most of the time it resulted in huge amounts of disappointment and feelings of failure. More often than not, women would give up on the idea of 'natural' after a few hours of excruciating contractions and would willingly accept Demerol or an epidural. Others would be diagnosed with dystocia (or 'failure to progress') and, after being put on an IV to augment contractions, would end up having either forceps or an emergency cesarean. Afterwards, these mothers were forced to admit they'd been wrong to assume it was all so possible and many of them felt disillusioned, disappointed, angry and even guilty. As you know, this is not just an historical scenario, it's still happening every day in certain maternity wards all over the country and around the world. What a shame so many mothers and caregivers have no first-hand experience of real 'optimal birth'.

As you may have guessed, or as you may already know, 'optimal birth' means allowing the physiological processes to proceed without *any* kind of interventions, whether pharmacologic or New Age. It means natural, as in giving birth as a healthy, active, *powerful* woman, with full midwifery or obstetric back-up and ready access to life-saving intervention in case it's needed. In a nutshell, optimal means an ultra-natural approach, backed-up by all the best our society has to offer. Your expertise is crucial to the optimality of this kind of birth because obviously, without it, maternal and fetal risks are much higher. In an optimal birth, according to this definition—one which is accepted elsewhere—no pain relief or intervention is used, unless absolutely necessary, because any interventions are likely to disturb the physiological processes and consequently the level of risk to both the laboring woman and her baby.[2]

In case you question this definition, consider for a moment how an epidural immediately puts a woman at risk because it inevitably results in a lowering of blood pressure; and consider how the use of artificial oxytocin (pitocin) for augmentation immediately necessitates higher levels of monitoring because no one can be sure precisely which dose is ideal (or dangerous) for any one particular woman. And—to make sure we're clear about this—let me say that 'no interventions' means no induction, electronic fetal monitoring, TENS, Entonox (gas and air, which is used in some places), Demerol or epidurals. It even means no complementary therapies such as shiatsu, acupuncture and homeopathy, precisely because they can all be so incredibly powerful.

Unlike 'old natural', which didn't facilitate the natural processes, 'optimal' (or 'new natural') really is possible for most women, as long as a few basic principles are respected. It's not only possible, it's desirable...

♥ My own research has made me conclude that an optimal birth usually means far less pain is experienced overall by the mother, if postpartum pain is taken into account too.[3] When a woman is fully alert and undistracted she is likely to have a shorter labor, a fast second stage[4] and she's far less likely to sustain a bad tear, or indeed any tear at all.[5] Any pain experienced is also likely to be more manageable because of the endorphins which are naturally produced and the subsequent unusual state of mind which results when the physiological processes remain truly undisturbed.[6] Postpartum psychological pain is likely to be less too, which brings us to the next point...

♥ After optimal births, bonding can proceed more smoothly not least because both mother and baby are extremely alert.[7] Higher rates of breastfeeding are also likely to result—sometimes even when the mother has planned to formula-feed—because many drugs used for pain relief (in particular opiates such as Demerol) weaken a baby's suck. With a normal suck, a newborn can easily get the colostrum and milk he or she needs, without hurting the mother through ineffectual sucking.[8]

♥ Babies face far fewer risks, both perinatally and postpartum—into adulthood and old age. (We'll come back to this in more detail later.)[9]

♥ Babies experience high levels of alertness at birth,[10] which may have enormous repercussions in terms of bonding and formative experiences. If these two types of early experience are at all important, experiencing an optimal birth may have an enormous impact on a baby's later psychological development.

♥ Higher levels of satisfaction are experienced by mothers because they experience the endorphins which are spontaneously produced postpartum.[11]

♥ Caregivers have greater job satisfaction because fully natural births have a strangely compelling atmosphere...

In short, 'optimal' means a less painful experience for the mother, both physically and psychologically, fewer health risks and a better deal for the baby,[12] not to mention satisfaction for you.

If you question this last point, let me ask you to consider for a moment why it is that midwives who have witnessed many completely physiological births are often prepared to go to great lengths to safeguard their right to attend women as they feel is right (e.g. by practicing in difficult or even illegal circumstances... Unfortunately, the situation in the USA is fairly unsupportive of a woman's right to choose to have a midwife, and of a midwife's right to support women, and it has now become common for the 'average woman' to automatically register with an obstetrician. This is a shame because it is midwives who can best safeguard the normality of birth for the vast majority of women, quite simply because obstetricians are usually trained to look out for pathology, not health!It's time we reminded ourselves what 'normal' really means.[13]

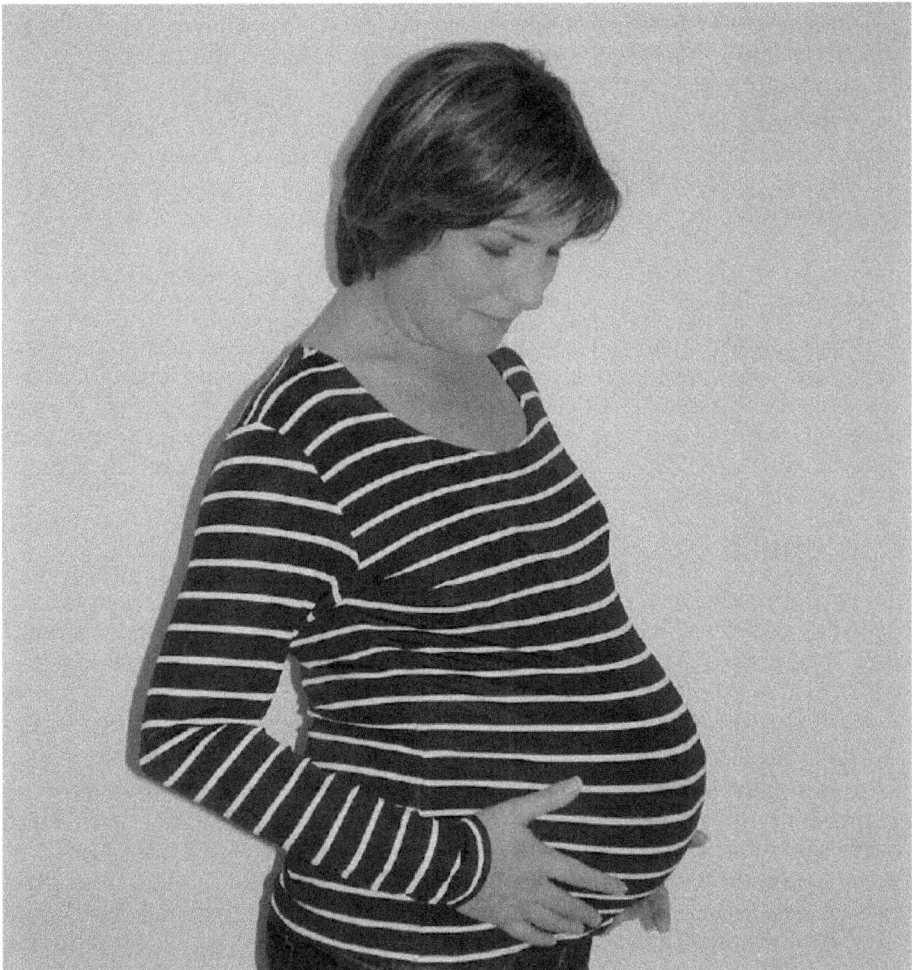

*Why are midwives who've witnessed physiological births keen to promote them?*

# The physiological processes

Of course, you know what happens in conception, pregnancy and birth... But do you know what constitutes optimal from a couple's point of view? More importantly, can you clearly differentiate between processes which are natural and those which are typical when analgesia, anesthesia or complementary therapies are used, or when non-essential interventions or even disturbance take place? If you're used to supporting women who are using anesthesia, analgesia or complementary therapies you may have become so used to dealing with risks caused by unnecessary interventions and may have forgotten the characteristics of a truly undisturbed physiological birth. Research over the last few decades has needed to relate to the use of drugs and interventions for legal reasons. As a result, relatively little attention has been paid to documenting the undisturbed processes and some caregivers have never even witnessed them. That's why we're focusing on the 'what' in this book.[14]

The effect of emotions on the natural processes of conception, pregnancy, labor and birth have been particularly neglected in recent decades. As we know from the simple effect fear has on the production of a hormone such as adrenaline, a person's emotional state can have an enormous impact on physiological states. The effect of emotions is also obvious if we remember that adrenaline usually inhibits the production of oxytocin—the hormone which stimulates contractions—which in turn is produced by the simple act of sharing a meal with a friend or loved one.

Neglecting the physiological processes or even underestimating the importance of something like emotions is dangerous if an appropriate decision tree for watchful waiting is to be developed and understood. In order to provide safe and helpful support for women laboring naturally—so as to produce optimal outcomes—it is therefore vital that the processes of an entirely physiological pregnancy, labor and birth are well understood.

## Conception

If you're an experienced midwife or other caregiver, who is very used to the everyday and rather messy processes of birth, I think it is just possible you may have forgotten the romance in people's hearts when they first consider the idea of having babies. To help jog your memory—or introduce you to a new world, if you're not remotely romantic yourself!—here's an account of a couple's attempts to get pregnant.

This account—the first of many birthframes in this book—is only partly included to amuse you. I'm also hoping that remembering it when you meet each new pregnant woman might help to remind you of the specialness of every individual woman's experience of pregnancy and birth. And I'm hoping that you will accept that even if a conception might not be optimal, it does not mean the subsequent pregnancy and birth needn't be optimal. (Birthframes included later should give you even more insight into other people's worlds.)

## Birthframe 1: Maria Shanahan and her partner—Conception

This account may seem unlikely, but I assure you it's entirely authentic...

I suppose my birth stories begin with where I was conceived, at the top of a mountain in France, St Paul de Vence. Knowing where I was conceived made me feel really special, and I was determined that when I decided to have a child, I would ask for it in the most wonderful place I could.

Knowing where I was conceived made me feel really special

Other details that I was determined about were that the birth would take place in private, with only my husband present. I never had any doubt that I would have a child, or that I was totally capable of the task myself. The only person I wanted present was a beloved husband, and the concession I made to all those shocked by my plans—I was about 10 years old—was that I would allow a doctor to wait outside the door, just in case.

These convictions may have arisen out of my mother's horrific stories. She was shaved, called 'disgusting' when she inadvertently peed during labor, on her back with her legs in the air, white coats all about, and there was a categorical refusal to her request to have my father with her—he had to wait outside. All her births went uncomplicatedly, despite the hospital, and her best experience was with a home birth which she was finally allowed after my sister and I. She knew of my passion to give birth naturally and one day invited me to attend a waterbirth seminar given by Michel Odent in London, in the UK. We both found him so inspiring and gentle, revolutionary really, and investigated more about waterbirths. I had always loved water and the ideas he presented seemed totally logical. A waterbirth seemed the best possible way to ease a child into the world outside the womb.

I finally met the man of my dreams when I was 27. We shared a passion for adventure and a zest for life and together determined to make our time as amazing as possible. He taught me to dive and told me of this amazing place in Oman, a place called the Tiwi Hole. It is a collapsed limestone cavern in the middle of the desert, 20 miles from the nearest village. After he had described it to me I knew it was the place—the place where we would conceive our special child: The Eye of the World.

### Over to the father...

Between the mountains and the sea there is a place, a very special place. When I went there the track became tortuous as it wound steeply up off the plain, rising to a plateau riven by ancient rivers, now seldom running. The engine roared in low gear, straining to reach the top. Far away the high mountains of the Eastern Hajar rose table upon table to the clouds drawing ever closer. At last the track dropped down one last hairpinned slope to a lower plain. To the left the sea shimmered, the light of the late morning sun reflecting from rays leaping from its surface like bright diamonds of polished steel. To the right the plain rose gradually to meet the foothills preceding the majesty of the soaring heights of the Jebel.

The turn-off to the right was unmarked, just the merest depressions of tires, found only by a stirring instinct. Nothing to be seen ahead except the scrub of the plain stretching away to the hills.

Between the mountains and the sea,
there is a place, a very special place...

And then the ground opened. A huge, roughly circular hole straight ahead. I brought my vehicle to a stop some way from the edge, turned the engine off and stepped out of the air-conditioned environment and into the humid silence of this land. The heat blasted stones crunched under my feet as I approached the rim and I saw a sight that would never leave me. This was the start of an adventure into more than my mind could perceive. The start to a story that ultimately would never end. I stood at the edge, my breath taken away. The ground before me long ago had collapsed into a subterranean cavern some 30m deep by 60m in diameter linked by deep channels from the sea half a mile distant to underground fresh water streams from the Jebel to the West. At the bottom, a crescent of pure turquoise water, undisturbed, reflecting in its calmness the sides of the hole.

I entered the crystal waters and dropped to a depth of 20m looking down into an impenetrable darkness to unguessed depths. Exhausted air bubbles collected in the roof of the overhang like pools of mercury. I turned away from the darkness towards the light and looked up. The rim of the hole and the water's edge formed the shape of an eye, bordered with yellow and blue. I was motionless inside this eye at the very heart of the magic, suspended in a vision looking out on the world above, through silver shafts of sunlight, which were cutting through this ether to greet me.

I broke the surface and left. I took the vision with me then and the vision held part of me there, waiting until it could show me the two people sitting at the water's edge in silence asking for their child to appear in their world.

This place really calls to me. It is an ancient subterranean cavern, the roof of which collapsed many hundreds or even thousands of years ago. This caused a sink hole some 150 to 200 feet in diameter and about 90 feet deep. It is mainly sheer-sided and undercut on a three-quarter crescent where fresh water from beneath the mountains of the Hajar meets and mixes with salt water entering deep below via some opening to the sea which is at least quarter to a half mile away. The water is crystal clear and the most brilliant bright turquoise. In 1985, a diver recorded a depth of 55m, and he still hadn't found the bottom!

The reason I called it 'The Eye' is that at one point if you dive down to around 10 to 15m and turn back to look up to the surface, the appearance is that you are looking out through an eye which is rimmed with light blue and yellow out into another world of earth and sky. It is a unique and amazing experience once you've overcome the apprehension of being suspended within a bottomless pit!

It is a unique and amazing experience

I feel a total isolation from our everyday life

There is such a silence of the earth here. I feel a total isolation from our everyday life of possessions and feel part of the surrounding rock, the water, the breathing wind and the silent skies. Somehow the reality of these elements runs far deeper than my normal existence as if there is some primeval relationship far beyond the reach of language. The wind from the plain above breathes around the rim with a hardly perceptible hiss dislodging the odd loose pebble which drops 'plop!' into the water sending out ripples across the calm surface to the fringes of the hole. I suddenly realize what we are witnessing—the timeless erosion of the earth. At that moment my existence is thrown into a deeper perspective. Immaterial, insignificant, a newborn child in a grown-up's game. The whole span of my life might be less than the time it took the wind to excavate and dislodge the last pebble that fell into the water.

I suddenly realize what we are witnessing—
the timeless erosion of the earth

The Eye will always remain in our minds and the magic of the Tiwi Hole will remain with us forever between the mountains and the sea, the surface and the air.

## Maria takes up the story again...

When my partner told me about this place, I visualized it so clearly... I said that this was the place where we should conceive our first child. Our dream was to ask for our child here. However, the actual practicalities of this plan were a little different from the theory!

When he told me about this place I visualized it so clearly...
I said this was the place we should conceive our first child

Amassing the camping and diving equipment, as well as the four-wheel drive to get there—to be lent to us at the drop of a hat when the ovulation test strips showed positive—required a fair quantity of people to be enrolled in our plan. At last the time came and we were ready. We kitted up, wearing less than usual, and entered the water from the small shoreline. Looking past the 5 to 10m shallows, the bottom recedes into darkness under the overhanging cliff, dropping at a sharp angle to unguessed depths. I was by this time somewhat apprehensive about our planned conception. For not only was our plan to conceive here, but to actually make love at 10m whilst diving in 'The Eye'. My thoughts at this point were something along the lines of: "Oh my god, what's all that green stuff??!!", "Why are there so many little fish about??", "How far does that go down and what's lurking down there?" and "Can we seriously be considering sex down there???!!!" And although we achieved it, amazingly, the end result was not conception... probably just as well since it might have turned out to be a fish! Never had so many people been interested in my next period! Fiohann was conceived nine months later in north west London.

*A Japanese couple, enjoying their wedding day.*
*How much do couples prepare before they embark on parenthood?*
*What, I wonder, could be done to influence them positively?*

# Pause to reflect...

## ...about conception and prenatal care

1 Do you think most couples prepare in any way before they try and conceive?

2 In your particular job is there anything you could do to influence people's behavior before they get pregnant?

3 What advice would you give a woman who came to you before getting pregnant?

4 What information and support can you provide to women who experience one or more miscarriages?

5 How do you think women deal with miscarriage?

6 What do you think women expect when they present for their first prenatal appointment with you?

7 What emotions do you want them to feel during the appointment and afterwards?

8 Are there any key issues you need to explore in a first appointment?

9 What can you do or say in very early prenatal appointments to help women optimize their pregnancy?

10 What can you do or say during prenatal appointments to help women optimize their experience of birth?

Do you have any other thoughts or questions?

For material relevant to these questions, see Birthframes 2, 4, 14, 15, 27, 33, 34, 35, 36, 37, 38, 40, 41, 47, 57, 58, 59, 62, 70, 97 and 99.

## Pregnancy

Of course, a woman's experience of pregnancy is likely to vary according to her health, personality, emotional responses, social environment, geographic location, prenatal care and culture.[15] This means not only that each woman starts from a different point than any other woman on the planet, but also that she can influence the course of her own pregnancy according to what she does, and what she allows other people to do to her mind or body. And we need to remember in all this that whatever happens or is done to her is also likely to affect her baby too. Of course, this has enormous implications for care.

Each woman can influence the course of her own pregnancy

Whether the woman perceives pregnancy as something positive or negative—or as a neutrally interesting change in her circumstances, or even as a life-threatening situation—may well depend on her experience of birth so far and even on how the professionals she encounters speak to her. (See Birthframes 33 and 50 if you would like to reflect more on this.)

How the woman perceives pregnancy may be affected by you

### THE FIRST TRIMESTER (WEEKS 1-12)

The first three months are often a time of heightened emotions and physical unease due to occasional bouts of nausea and tiredness, and unfamiliar bodily changes. Of course, the upheaval is caused by the different hormones circulating and the effect of the growing placenta. Emotions are often triggered by thoughts about pregnancy and birth and they might even stimulate shockingly vivid dreams. Women might experience an overwhelming need to sleep at any time of the day and they usually also need to go to the bathroom more often because of pressure on the bladder. It's in this period that the baby is going through its most crucial development—all the major body organs are being formed.

### THE SECOND TRIMESTER (WEEKS 13-28)

Most women experience fewer discomforts during this second three-month period. Any nausea usually disappears completely by Week 13 or 14 because the placenta should now be fully functioning. Hormones settle down and fewer trips are needed to the bathroom. Most women perceive the slowly emerging baby belly positively because it's tangible proof of the pregnancy, which is usually considered a good thing. Many, though, will still be grappling with odd thoughts and feelings and some will not have come to terms with the fact of being pregnant.

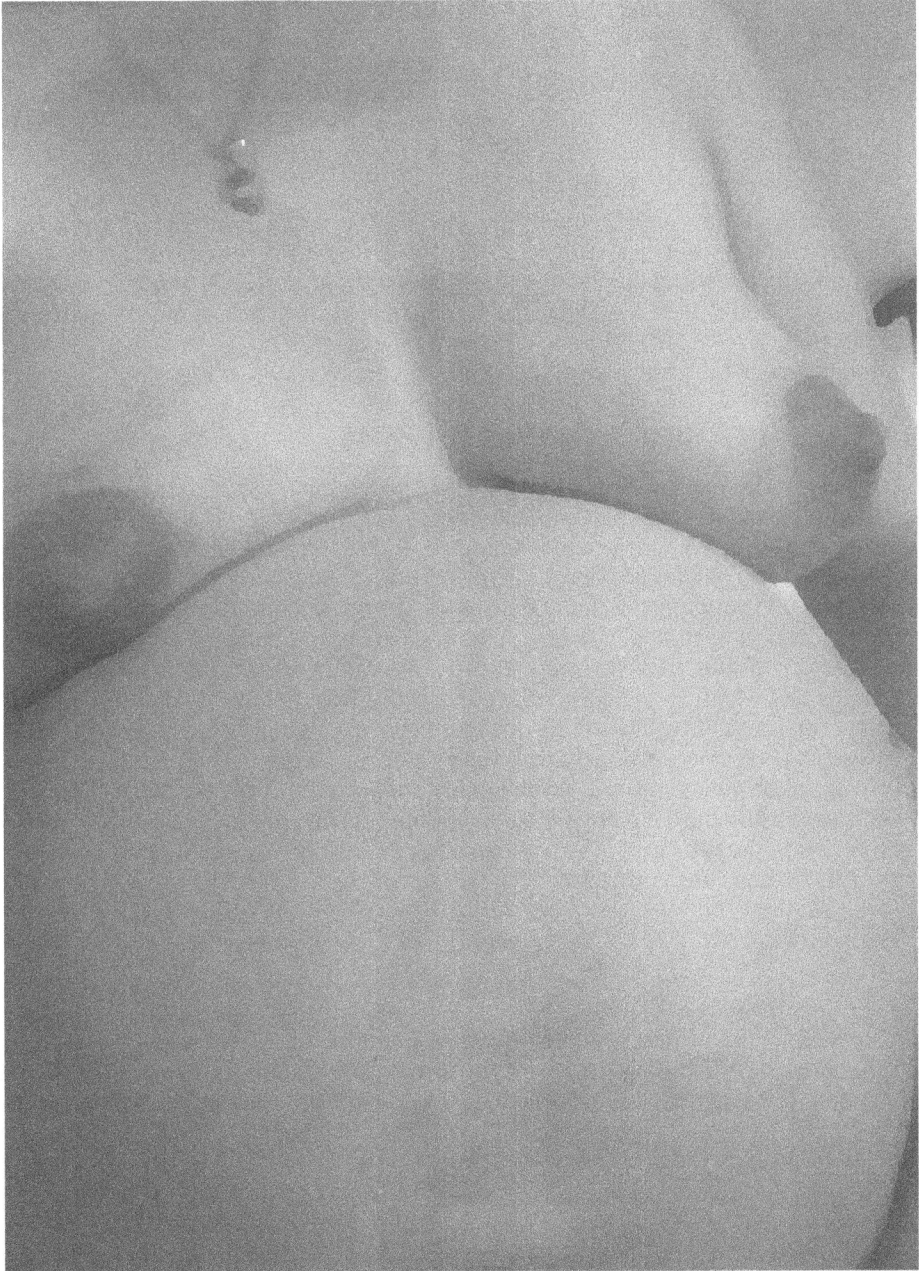

*In ages past, before celebrity Size 0 bodies, people used to make fertility dolls which looked a bit similar. Confetti is a reminder that in the past when people got married well-wishers would be hoping they would be very fertile and have lots of bellies like this!*

More blood is now circulating around the woman's body (meaning more work for her lungs, kidneys and heart), thanks to the work of the placenta. This will sometimes mean that women are mistakenly diagnosed as suffering from anemia, when really the lower percentage of hemoglobin in their blood is simply an indication that the placenta is functioning normally. (There will be more information and commentary on this later.)

Typically, women feel much more alert and energetic in this trimester but they sometimes worry about the idea of giving birth or about impending motherhood. They may have all sorts of fears about the welfare of the baby at this stage. All this is perfectly normal because it's a sign the woman is tuning in to her baby and his or her needs. From the 18th or 20th week onwards, most women can feel the baby kicking and moving about now and then.

## THE THIRD TRIMESTER (WEEKS 29-40+)

Often women experience no particular changes as they enter the third trimester, except for a dramatically increasing waistline. The fact of getting bigger may bother some women, especially if poor posture is causing backache or if they are afraid of giving birth. Tiredness may also be a big problem because of the woman's increased weight and uncomfortable or disrupted nights, caused by difficulty finding a good position and a bladder which is being pushed out of the way! The woman may well feel a need to drink more water than usual, which is logical considering that the amniotic fluid is constantly being replaced. Stretch marks may appear and nipples are likely to look different. Most women find themselves thinking a lot about the upcoming birth. Changing hormones in the woman's body are gradually preparing both her and her baby for the birth. Labor will begin spontaneously, the timing being affected by genetics, health, diet, lifestyle, psychology, circumstances... as well as an incorrectly calculated due date.

### Week by week...

It's useful to consider pregnancy week by week because this information may easily be forgotten or changed by an overly interventionist view of birth. Therefore, you can find a quick-reference guide at the back of the book. This appendix is intended to help you say something useful and encouraging to women each week of their pregnancies, whenever they happen to visit you.

## Labor

As you know, it's not always easy to know when a woman is really in active labor because the latent phase may seem very similar. To make matters worse, many women experience 'false' labor, or stop-start labor. And there also seems to be little agreement in the literature as to when precisely labor begins or even how fast it should progress, so as to be considered 'normal'.

The Friedman curve, once so closely followed, is now widely disregarded—and even the originator, Emanuel Friedman, apparently expressed horror at the way in which it came to be used. Some contemporary commentators (e.g. Denis Walsh, a male midwife and associate professor in the UK) are suggesting that ½cm per hour may be more appropriate and that even this should be a guideline only, as 'safe normal' varies considerably.

Michel Odent, who has observed 15,000 labors, many of which must have been entirely physiological (because of his eventual preference for this), joked to me that he only ever diagnosed labor in retrospect—when a baby had been born! Other practitioners who have been over-keen to diagnose have often ended up feeling the need to augment labor in one of the usual ways, because it has often not progressed as expected. Obstetricians or midwives who have not done this have later described a labor as lasting for 'five days', or similar. In Michel's view, these were not actually cases of five-day labors, but cases where a woman simply had a long latent phase, or where she had a stop-start labor. In any case, he prefers never to confirm to a woman that she is actually *in labor* because he says that doing so can put her under pressure to 'deliver'.

Despite this uncertainty about how or when labor begins, in very broad terms we can make some observations about the onset of labor and of course it's important for pregnant women to understand these too. When there are no attempts at induction, there might be various signs that labor is beginning.

♥ A woman will become aware that something is changing within her body. Perhaps she suddenly has either a bigger or a smaller appetite; she may become restless and have a sudden spurt of energy; she might experience a bout of diarrhea, which will indicate that her body is 'clearing the way' for the new baby.

♥ Sometimes, the pregnant woman might notice a 'show', a discharge of a jelly-like substance, which may be smeared with blood. Seeing this jelly would be a sign that the cervix is beginning to open so as to eventually release the baby into the outside world.

♥ Sometimes, but less frequently, a woman might experience a gush of 'water' from between her legs.

♥ Of course, in all these cases, eventually the woman will notice that her uterus is gradually—or suddenly—becoming active as it flexes so as to fully open up the cervix and push the baby down through her pelvis, out into the world.

Of course, the rhythmic 'flexes' of the uterine muscles are usually—rather unhelpfully, it must be said—called 'contractions' in most books on childbirth, or 'rushes' in some others, which is again sometimes an undescriptive, unhelpful term because at some points during her labor a woman may not feel that things are 'rushing' anywhere. Having said this, if a woman has not interfered with the physiological processes taking place within her at any point during her pregnancy or labor, she is certainly likely to feel that an enormously powerful, absorbing, irreversible process is taking place, and 'sweeping' her along towards a predetermined biological end. That end is, of course, the birth of the baby... or babies, in the case of a multiple birth.

Are the sensations of labor and birth painful? As you no doubt know they are to most women, but not all. Some only experience pain for a short part of their labor and some don't experience any pain at all during the birth itself. A few women experience some or all of the process as being interesting or absorbing, rather than painful. Some even report labor and birth as 'orgasmic'... (This makes sense when you consider that the natural hormone which opens up the cervix in labor—oxytocin—is the same one as is responsible for orgasm).[16] Even when women experience contractions as outrageously painful they usually go into an interesting state of mind if they're completely undisturbed. This means they become totally absorbed by the labor and birth and find they are able to cope as long as they take the contractions one by one.

If women become totally absorbed, they can cope

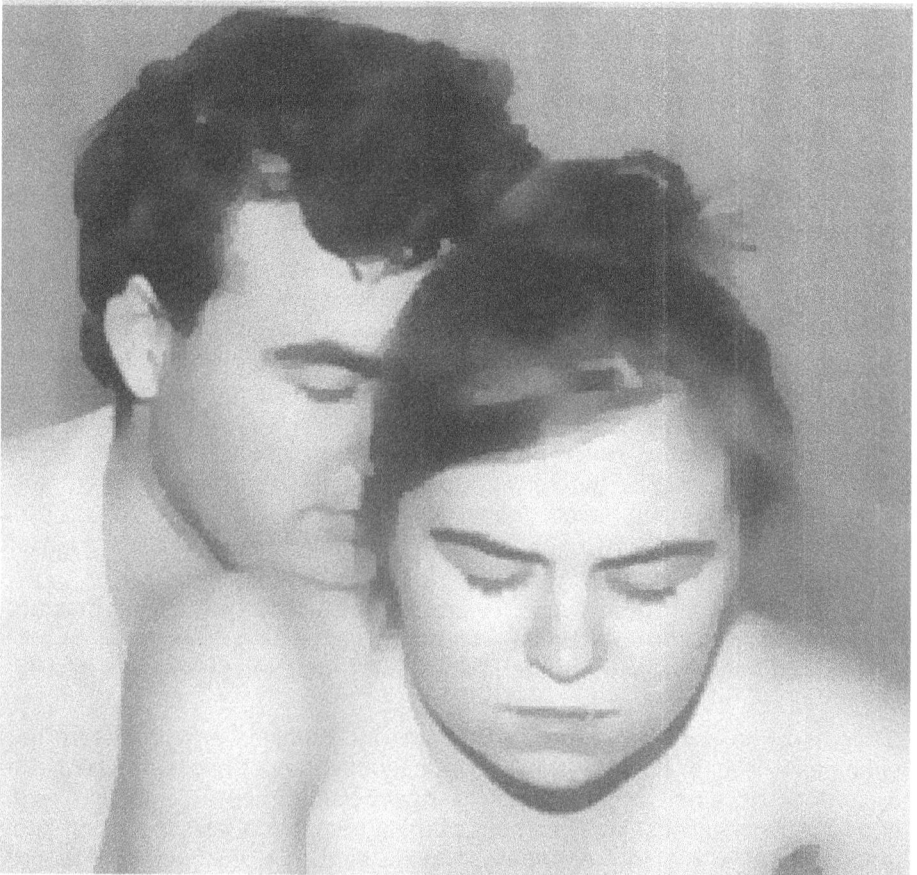

*Ashley Marshall, laboring normally. Because of the importance of not disturbing women in labor (which is easily done with a camera) I refused all photos of women in labor and actually giving birth except this one, and one other, from Ashley.*

Most caregivers, it must be said, have never witnessed this because their behavior or speech (either with the best intentions, or as a result of hospital-imposed protocols) cause a disturbance. Instead of providing a safety net, the caregiver then becomes a disruption to the smooth proceeding of the labor (and birth), which is rather tragic, really. When, on the other hand, caregivers give the laboring woman the feeling of being secure and when they monitor so unobtrusively that the laboring woman isn't even aware of it happening, they provide the perfect conditions... because of course they can intervene at any point if it is necessary, for safety, with the woman's consent. In most cases, though, women experiencing a fully physiological labor in a safe, but undisturbed environment find the resources within themselves to withstand all difficult sensations. Relatively fast they have dilated to the magical 10cm, ready to proceed to the second stage.

Since, as we've already noted, many caregivers have never witnessed a completely physiological labor and birth or experienced them first-hand (while giving birth to their own babies) I am including some comments from contributors. I hope these will give you some insight into how the physiological processes are typically experienced in an optimal labor and birth.

66 I had a whole night of very strong, regular contractions 24 hours before Asya was born. I was sure this was the real thing, but at 6:00AM they petered out. Apparently this kind of 'false labor' is very common, because the body produces the most oxytocin at night. It didn't feel false! The good news is that false labor does help to open the cervix, so your labor will probably be that much shorter when it does get going for real.

66 The night before Arion was born I had an innate sense that it would be the following day. I woke a few minutes to 6:00PM and had to go up to go to the bathroom. Soon afterwards, I had a show and things started to happen. I felt a tightening in my stomach that felt like a contraction. I had been unaware of anything up until this point. Very soon, I was having contractions every five minutes, lasting approximately 30 seconds. [Arion was in fact born a few hours later.]

66 I woke up needing to pee, and made it to the toilet with clear liquid coming down my legs. The odd thing was that I had no control over this liquid, it just sort of fell out when I stood up. [The baby was born the same day.]

66 Well, time went on. Due to finish work soon. On Sunday, June 25 I was a little off, and I slept like a baby. Must have needed it, as Darren could not believe I had slept so well due to the noise going off outside—neighbors. Thought nothing of it. On the Monday, I took my son to school, as normal, and waddled off to work for my last week. I was up and down like a yoyo, going back and forth to the bathroom. Well, something was happening and I was just not ready. Called the midwife, but by the time I had finished talking I was huffing and puffing... [Twins were born just after 3:00PM.]

*This is Ashley Marshall again. Incidentally, when I explained my policy of not using photos of women in labor to Ashley, she said: "I'm sorry to say that I was in very active labor when this photo was taken. It was at my request, though, as I wanted the event to be documented. I just love birth! I think a simple statement about children at birth would justify using the photo. Sometimes it is in fact reassuring for older children if they can be there when the baby's born—and this may be okay for the mother as long as she's not disturbed by having them around.*

66 Perhaps what I experienced was false labor, or perhaps it was interrupted labor. All I know is that my strong, regular contractions suddenly stopped after I'd been talking to my mother on the phone for five minutes. She sounded horrified that I was in labor when I answered the phone. She then immediately switched to a totally boring, irrelevant topic for the rest of the conversation... er, monolog. No amount of relaxation exercises could restart the contractions when I came off the phone. However, the next day, when my contractions started again, they were even more purposeful and my 'real' labor lasted less than two hours.

66 I did not want to leave the top floor of our house, I did not want the drapes open, and I did not want the lights on. Almost as soon as labor had begun I had started to withdraw into what I see as a sort of animal protective state.

66 Baby turned from posterior position to anterior a short time after I adopted an all-fours, butt-up position.

" At 32 weeks I had a prenatal check-up and was told the babies were roughly 6lb each in weight, so you can imagine my surprise and shock. The next few weeks I got bigger and bigger and everyone was hoping for an early delivery... even the doctors at the hospital couldn't understand after 36 weeks how I was still carrying them. Then 40 weeks approached and my due date. I attended another prenatal clinic, where I saw the obstetrician. He told me there was still plenty of room for them to grow so he would not induce me! 40 weeks + 3 days, my contractions started and I went to hospital, where they told me it was the first time they had ever seen a twin pregnancy overdue. By 6:00AM the first baby was ready to arrive: I pushed for about 15 minutes and Megan Victoria was born at 6:19AM, weighing 7lb 1oz, then six minutes later after pushing came Thomas William, weighing 7lb 12oz, both perfectly healthy.
*Clare Swain*

" At the beginning my temperature perception was going up and down, so during a contraction all my clothes came off, then I was freezing in between, and they all had to go on.

" In the middle of the night my water broke. I was asleep. It was about one o'clock in the morning and I woke up and I was just lying in bed and it was all wet. I called the nurse and she said, "No, you've probably just wet yourself," which made me feel really bad—it was really horrible. But I was certain it was my water that had broken and so I insisted and she came and checked then called the doctor who was on call that night. He examined me and I was already 7cm dilated. But I hadn't felt any pain or anything—just pressure. I'd been feeling that for some time when I was standing up.

" Before my labor I ate voraciously. Part of my first stage I then spent chucking up into the toilet! This felt fine, actually. Afterwards, I rationalized that I must have digested most of the food I'd had beforehand anyway.

" You asked how I felt about shouting but to be honest it was more like a cow 'mooing'. It felt good to me and I reached a point where I stopped even caring what anyone else may have thought of it.

" Found standing up and leaning over the bed on a beanbag, rocking and having someone massage my lower back excellent for pain relief.

" When they checked me over, I was 9cm dilated. The twins wanted out. There was nothing I could do but go with the flow.

" I had now been on my feet for about 9 hours with nothing to eat since the night before (my breakfast having hit the stairs some time before), and I really wanted to lie down. I did not care if labor stopped for half an hour; I just wanted a rest.

❝ Instead of putting on the Julian Cope compilation tape I had made specially for the labor, I knew I had to listen to music from my ancestors, from my soul, and put on some Welsh male voice choir albums... which then played for the next five hours. An email sent by a friend after the birth said, "After five hours of Welsh singing, the birth must have seemed a breeze."

❝ My midwife used a Doppler (a waterproof one bought especially for my water birth), but I can hardly recall it. I think she only listened once to the heartbeat at the same time as she examined me internally.

❝ I went from Stage 1 to Stage 2 in about five minutes. From being able to hold a normal phone conversation one minute, I was suddenly only capable of screaming blue murder the next. My husband has never shown a cooler nerve as he drove me to the hospital. By this time I was screaming so loudly I was sure the whole of the city could hear me. Incapable of sitting in the car I was raging around the back seat like a werewolf. I was in the grip of elemental forces that were telling me to go find a lair, a den, a quiet place to do this really important job. An overwhelmingly strong instinct was telling me to get my pants off so that this being could come out. I would have done it in the elevator. I would have done it in front of a million people. As it was, they found me a room and a couch and I was told the very encouraging news that I was already fully dilated.

*Infuriatingly, perhaps, most women labor mostly during the night...*
*Could this possibly be a hint that they want a little privacy?!*

# Pause to reflect...

## ...about labor itself and intrapartum care

1   How do you feel about labor yourself? Are your feelings influenced by what you've heard, what you've seen, what you guess it must be like or what you've experienced?

2   How do you imagine women typically feel when they first go into labor?

3   What is it best to say to a woman when you first meet her?

4   Do you think labor is always painful? Have you ever seen a woman appear to enjoy labor? Does sex look good or painful from the outside?

5   Can you see labor and birth as a psycho-sexual experience? What are the possible similarities between labor and birth and making love with someone you love?

6   If you are fairly sure a woman you're observing is in pain, does that make you feel very uncomfortable?

7   What kind of behavior (from a caregiver) do you think might best support a woman?

8   Why do some women opt to refuse drug-based pain relief?

9   Why do you think some caregivers recommend non-medicated labors and normal, physiological birth?

10  What non-pharmacologic methods of pain relief can be used to help women get through labor and birth?

Do you have any other thoughts or questions?

For material relevant to these questions, see Birthframes 3, 6, 7, 14, 21, 24, 25, 26, 27, 31, 32, 38, 39, 42, 43, 53, 54, 61, 71, 73, 74, 75, 76, 77, 78, 79, 81, 82, 83, 84 and 85. for other relevant material.

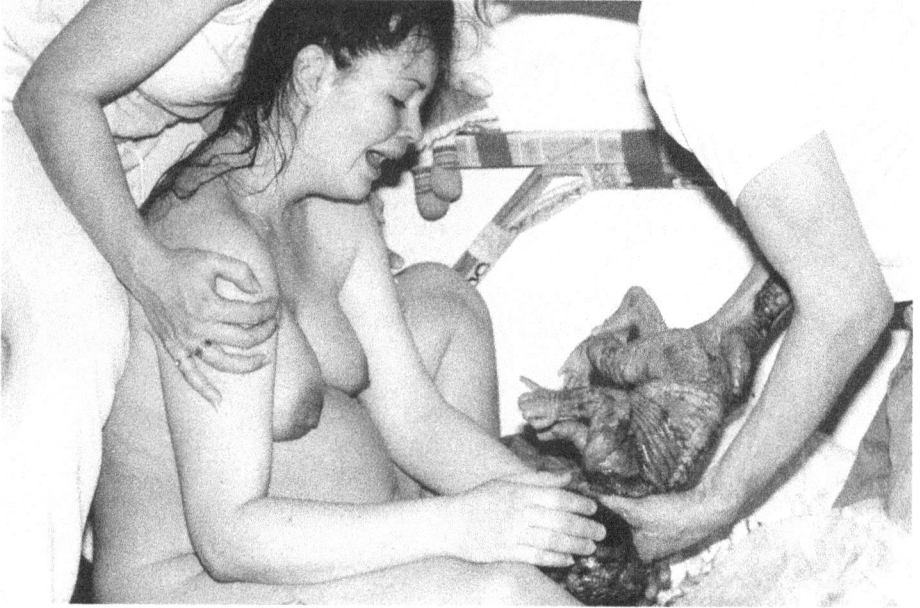

*The woman will either spontaneously pick up her newborn baby,*
*or have it handed to her silently by a birth attendant*

Photo © Jill Furmanovsky

# Birth

Towards the end of a physiological labor, without any intervention at all, the cervix will have opened up fully to about 10cm dilation, so as to allow the passage of the baby's head. Then, thanks to the powerful rhythmic muscular contractions which have been taking place and which still continue, the baby descends through the woman's pelvis and down through the soft, fleshy folds of her vagina. (There is sometimes a break of 20 minutes or longer between the cervix becoming fully dilated and the baby descending. During this period—or a shorter period—the woman experiences no contractions whatsoever. Other women continue on without a moment's break.) One of the bones which would normally block the baby's downwards passage—the tailbone (or coccyx)—moves out of the way when the woman is in an upright position, thereby increasing the diameter of the opening.[17] (It is much more likely that this will happen during an optimal birth because the woman will be entirely conscious and responding to her body's cues.)[18] When the baby's head stretches the perineum the woman usually experiences a sudden burning sensation, often called the 'ring of fire'. Then, suddenly, first the baby's head, then its body emerges through the woman's vagina. Of course, if it is a breech birth, the baby's bottom or feet will come first, followed by the rest of the body.

If the woman has not been disturbed in any way during her labor, she is likely to be in a pleasant 'submerged' state of mind and just before the moment of birth it is possible she may experience what Michel Odent calls a 'fetus ejection reflex', i.e. a sudden and compelling rush of energy which automatically makes birth simple, safe, active and intuitive. (This is when—mysteriously—adrenaline and oxytocin are successfully and only momentarily produced simultaneously. Instead of being antagonistic, as they normally are, they are both facilitative at this particular and very specific moment of birth.) The woman will typically pull herself to an upright position and spontaneously flick her hips forwards so as to facilitate the birth of the baby's head. As with other labors, there may be a time (during transition) when a woman feels intensely unsure of her ability to give birth, when she is fearful or angry or confused. This is normal... because of hormonal changes, the woman will spontaneously find her own way optimally without any speech or intervention.[19]

Again, since many professionals have not seen or experienced this kind of entirely physiological birth, here are some comments from women who have...

66 I have done some exciting stuff like parachute jumps, wing-walking, trips to war zones, but giving birth was definitely the most intense thing ever. My body felt totally out of control.

66 For me, the most important thing was to take things moment by moment. If I hadn't done I would almost certainly have been overwhelmed... Actually, I was overwhelmed but I always kept going for another minute. When I experienced the fetus ejection reflex my body seemed to act on its own, although thoughts were still going through my head. It was very sudden and spontaneous and came after great feelings of 'not being able to do this thing of birth' successfully. But I could actually.

66 In second stage my body seemed to know when I should push, to pant or hold back, and I didn't need the midwife's help to control the delivery.

66 At the next contraction I said that I could feel the head crowning; I don't think anyone believed me. My next comment was "Ah that's better, the head's out." I then remember someone saying that they needed to get me off the toilet; once again I had my own ideas. I also had another contraction and the body was born. With this contraction I lifted myself off the toilet seat and caught our baby. I brought her up between my legs and sat back down to cuddle her for the first time. It was 10.30am, and she immediately went to my breast.

66 I had gone from no baby to baby in about three minutes. This was a truly shocking experience, and I remember screaming as this thing fell out of my body.

66 As I lifted the baby up, just after he'd been born he didn't cry, just opened his eyes and looked around curiously. I felt like superwoman—amazed, empowered and overjoyed that birth could be such a wonderful experience!

" My daughter Isabel was born on December 9 at 8:43PM. She was born in my bedroom just by the door. I had told the midwives I would make it to the end of the bed but when push came to shove I didn't want to get any further into the room.

" I asked for an enema during my first labor because I felt blocked up. I was given a pessary and just a little poop came out a short time later. At the time, I didn't feel I'd been disturbed by this procedure but looking back I wonder whether I was in fact because I had a very long and difficult second stage. In my second labor, I again felt constipated, but my request for an enema was ignored. A few minutes later, I suddenly decided I'd stand up and do my 'poop' directly onto the floor! As I pulled myself up I also knew I was going to have a baby—not do a poop at all—so I think I wasn't really the least bit confused. The baby came out easily—she was obviously ready. It didn't even hurt!

" From what I'd read about births, I had expected pushing to feel good. It didn't feel good to me. But it was really weird and interesting. The best part was definitely the moment after the pushing ended, when here she was at last, a whole new person!

" Rebecca was delivered in a supported squatting position, Ros in a leaning forward crouching position, Juliet and Natasha in a kneeling position. Third stages were all in the kneeling-up position except for Rebecca, when I was lying down (and it was the least comfortable).

" This time I wasn't lying down when I gave birth. I was squatting. Michel asked David to hold me under the armpits while I squatted—which was, in fact, difficult for him to do. I think Michel wanted me to get into that position because he knew she was going to be a big baby. But I didn't have to do any pushing at all. There was never any need to push... it was all just coming slowly but surely—and then fast.

### A midwife's account:

" As the last few barriers to Jenny's relaxing disappear (the last of the children are taken out), contractions become more painful, impinging on her concentration. Contracting strongly, every four minutes for 30-40 seconds. Jenny feels she needs to move away from things happening around her. Darkened bathroom, candles. I remove all offensive smells (cooking in the kitchen). She's out of the bathtub, kneeling now, leaning over. Tim's applying pressure to Jenny's back. Things have hotted up. 15 minutes later: Pressure increases, intensity increases. Jenny thinks it's a boy—only a man would muck her about like this! Tim looks on, not making any comment. 30 minutes later: Shoulders and baby and all. Together Tim and Jenny investigate what their new baby looks like. "A girl," says Tim. "I'm glad you're a girl". Tears! Placenta born into a plastic bowl. All present and correct. Jenny has a bath with herb infusion. Baby checked and all there— not a *little* girl at 9lb 5oz (4.22 kg)!

" Keeping upright keeps the tailbone out of the way making the pelvic area wider so that the baby has a little more room to get out.

66 I felt no pain at all during the second stage, just an incredible surging strength. With two clear pushes the baby came down the birth canal. I did this in about two seconds, between two contractions. I'd say it was like doing a very easy, but big poop—not diarrhea exactly, but very easy and even more satisfying than producing two long, plump turds. What I produced, after I'd then felt the 'ring of fire' and had flicked my hips forward two times, with the next two contractions, was a beautiful, baby, who now lay at my feet on a crumpled up sheet, looking up at me with rapt attention. When she suddenly burst into tears, I picked her up and put her to my breast. She seemed relieved and nursed with gusto.

66 As I pushed the baby out, my whole body was rising up in the water. That took me by surprise, but it did not make the birth difficult. The midwife said, "Julia, get ready to catch the baby," and I said, "but what should I do with it?"

66 I found the whole experience of a natural birth very exciting and satisfying. It hurt like hell, but it was bearable (just). I would be happy to go through the whole thing again next week if I could. It was a real peak experience for me.

I found the whole experience very exciting and satisfying

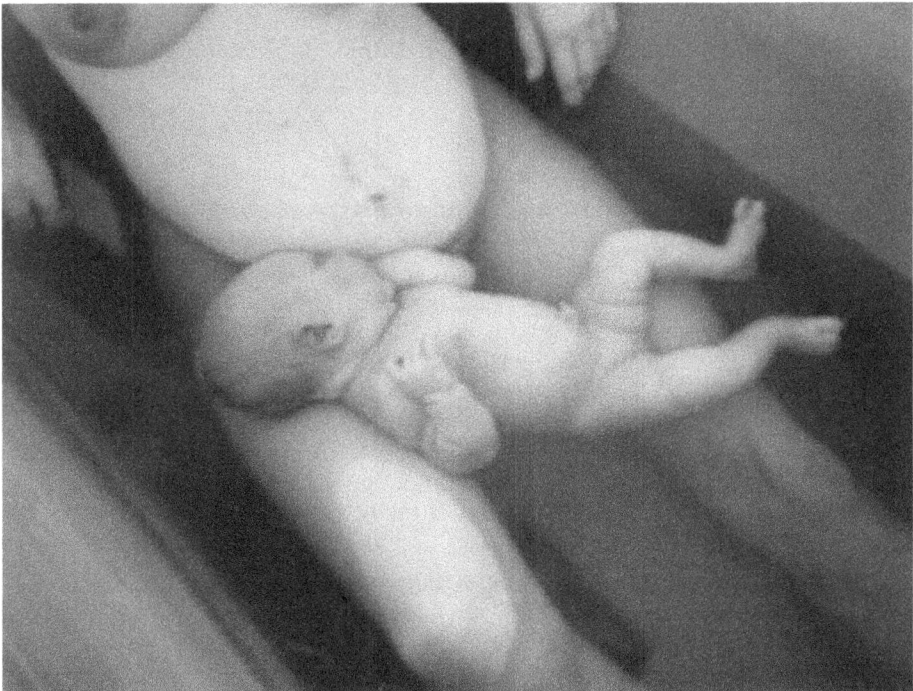

*Here, a new mom marvels at her new baby, soon after giving birth*

Note that no commands are necessary to tell the woman how to push. There is no straining, no fear, no control and no management.[20] I must mention at this point that one midwife disputed this point with me, saying that women often require instructions, guidance or encouragement during the second stage. (I hate mentioning specific conversations like this, but it is extremely relevant in this case, so I will just hope that the midwife concerned will forgive me for mentioning her anonymously like this.) Anyway, after talking to this midwife for a while it became clear that she'd only ever observed *one* physiological birth, and that particular one was accidental. (It was a case of a woman arriving in active labor and being virtually ready to push on admission.) My midwife friend had been about to embark on the usual admission protocols when a more experienced colleague stopped her, saying the mother was fine and she should be left undisturbed.

I think there are many cases in life when we *assume* that some kind of guidance or action is vital to the success of some endeavor, even though we might find Zen-like non-action (i.e. watchful waiting) to be more effective in actual fact. I don't know if I am the only one in the world who has had a row with her husband because of a comment made, which has prompted the response, "Yes, I know! What do you think I'm going to do?!" And am I the only person who's taken the advice of the childcare books on sibling rivalry? Sometimes, as advised, I've just sat and watched while my children squabble— only to find that 99% of the time they resolve their disputes on their own, without any 'help' from me. In fact, I'm often surprised by how harmonious and fair their own approach is and dismayed by the way in which my own well-intended interruptions often just make things much worse.

You may think I'm digressing terribly in mentioning these very different scenarios, but actually I think the principle of 'things going better when you leave well alone' applies *even more* in the case of childbirth because of the brain processes which are taking place. During an undisturbed physiological labor and birth a woman moves into a completely new state of mind... a kind of 'submerged' but very alert and intuitive state. She appears to switch off from her day-to-day reality and responds very dramatically to anything going on around her, even if *to some extent, in some circumstances* she is able to 'cut herself off' from the outside world. Any kind of speech, however brief, can jolt a woman out of an instinctive frame of mind back to cold reality, which naturally usually disturbs birth more than it facilitates it, not only because of the shift in hormones which the speech may initiate, but also because she may be switching from a 'right-brain' mode, back to a language-based 'left-brain' mode.

What, I think, is particularly interesting is that the laboring woman often *doesn't* feel she knows how to give birth. It is only by being left entirely undisturbed that she manages to reconnect with that deep, instinctive knowledge within herself, which typically emerges very suddenly.

Anyway, getting back to our documenting of the physiological processes... After the baby shoots out of the woman's body, hopefully onto something soft or into your hands, the baby will spontaneously take its first breath just a few moments later, sometimes crying as it does so, and sometimes not. Usually, the woman will pick the baby up or take it from you, if you've caught it.

At this point, both mother and child are usually in a state of heightened alertness, provided her birth attendants (including you!) do not disturb her in any way, either by talking, or attempting to communicate nonverbally.[21] She will continue in this state for some time (again, provided she is not disturbed in any way), and will typically instinctively put the baby to her breast... In many cases, the baby will slowly make its way there on its own, if placed on the mother's abdomen. The baby's suck is very strong and effective because it will have no opiates in its system. Of course, during this first feed the mother will usually be gazing at her new baby and as part of this first examination, she will spontaneously check the baby's gender. Needless to say, *telling* her the baby's gender would be a disturbance which would stop the birth from being entirely physiological. If this seems extreme at this stage, please bear with me here... As we will see later on, any disturbance changes the hormonal environment and therefore the smooth-running of the physiological processes—and speech is particularly powerful. If you doubt this, remember the link between fear and adrenaline, which is antagonistic to oxytocin, the main hormone which is facilitating this whole process. (We'll analyze these processes more later.)

The continued smooth production of hormones then ensures that the placenta is released safely and speedily—probably within two hours of the birth, but often within seconds or minutes of the baby's birth.[22] The umbilical cord is cut when it stops pulsing or can remain intact until it spontaneously breaks away from the baby's body a few days later—unless it snaps before this![23]

In an undisturbed physiological birth, the third stage happens very soon after the birth of the baby—even two seconds later

Suddenly, provided there is no disturbance, the placenta is born too—usually in one smooth, slippery movement. In an undisturbed physiological birth, this happens very soon after the birth of the baby (even one or two seconds later). Breastfeeding also stimulates the production of the hormone (oxytocin) which makes the birth of the placenta possible.

As we've already noted, if the woman was upright (squatting, semi-squatting, or on her hands and knees) at the moment of birth, it's very likely that her vagina, and indeed her entire pelvic floor will be undamaged after all these processes have taken place because the perineum is naturally stretchy and the woman's complete awareness of every sensation will help her to make sure the birth does not cause any injury. This means she'll have no pain after the birth, beyond a little discomfort the first time she passes urine and a bowel motion, as well as the experience of her womb contracting back to its normal size. She will feel completely normal when sitting or walking around. Obviously, this is in stark contrast to the postpartum experience of many women who have analgesia or anesthesia in whatever form because of the increased risk of tearing or other side-effects (such as nausea or postpartum drowsiness), not to mention the likelihood of a 'cascade of interventions' which might result in abdominal surgery too (a cesarean) and all that it implies.[24]

# Pause to reflect...

## ...about physiological birth and ways of facilitating it

1 Has anything shocked you so far?

2 Does the description of birth which you've just read tie in with your own experience? If not, how does it differ?

3 If you will accept—for the time being at least—that what has been described is a typical physiological birth, how does it differ from managed births?

4 What do you think are the likely differences—for the laboring and birthing woman—between a managed and a physiological birth (which takes place in optimal conditions, for example in a hospital birthing suite)—in terms of emotions, sensations and feelings of control?

5 If this model of birth is optimal, what kinds of conditions would best facilitate it, in your opinion?

6 If you were designing a midwifery-led birthing center, what kinds of architectural details would you stipulate as being essential and which details would you consider desirable?

7 Why are there comments on whether or not speech is required or ideal in this description of 'optimal'?

8 How could you optimize a woman's sense of privacy?

9 How can you support women, while at the same time giving them the feeling of being undisturbed?

10 How could you teach pregnant women about this kind of birth? And how can you learn more about it yourself?

For material relevant to these questions, see Birthframes 3, 5, 7, 20, 21, 28, 31, 32, 39, 45, 48, 53, 61, 69, 71, 72, 73, 76, 77, 78, 79 and 82.

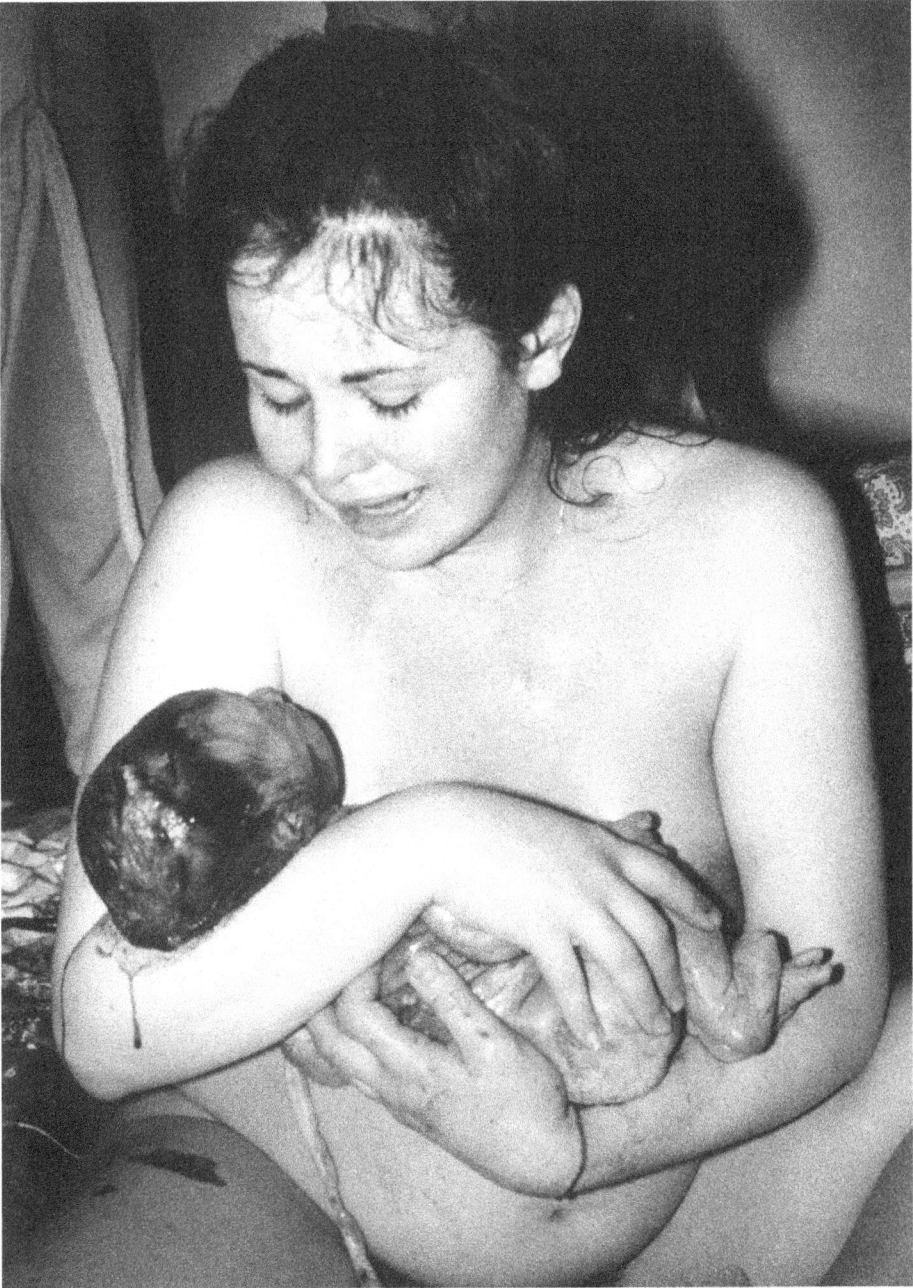

*Nuala OSullivan, a mother who you will read about soon, who was fully conscious
and actively participating during her entirely physiological birth*

Photo © Jill Furmanovsky

One very interesting aspect of all this, from a midwifery point of view, is that a truly physiological birth usually involves far less mess than occurs during a managed third stage. When the placenta detaches itself quickly and extremely efficiently (as happens when the woman and baby are completely undisturbed at the time of the birth and afterwards) there can be virtually no blood loss. (This is something midwives have told me they have observed many times.) And, of course, this also has enormous safety implications because an efficient third stage means a much lower chance of any life-threatening PPH.

The reduced blood loss after a true fetus ejection reflex also means that the woman is extremely strong after the birth, which puts her at an enormous advantage.

If this or any other aspect of this description seems unlikely or exaggerated to you, let me finish by saying that I've experienced all these things first-hand and many other women who've contributed accounts to this book have reported the same or similar experiences. I have been repeatedly distressed when listening to women describe their experience after opting for 'pain relief'. It really does seem a bit of a misnomer...[25]

I've been repeatedly distressed when listening to women describing their experience after opting for 'pain relief'.

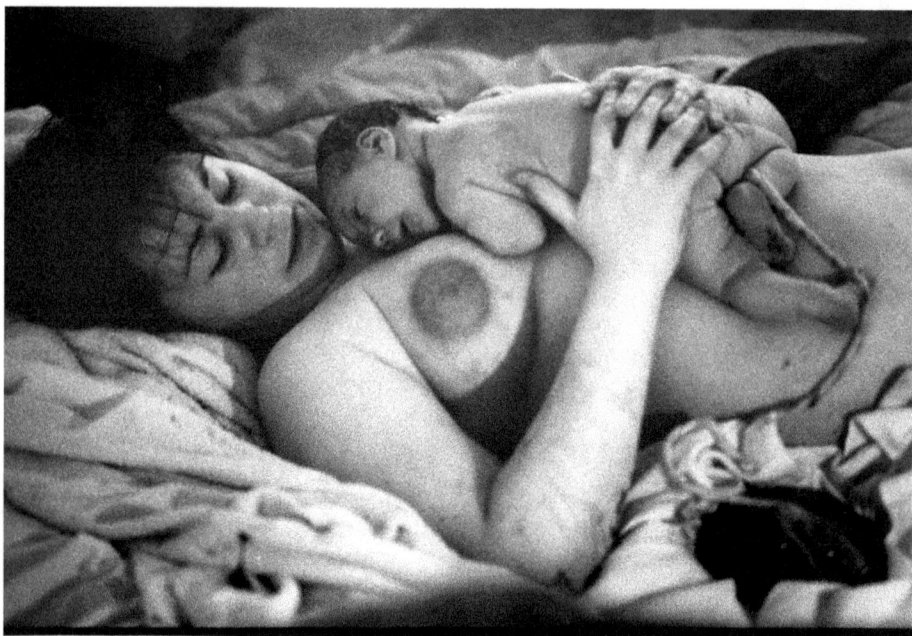

*Of course, skin-to-skin contact is important after the birth*
Photo © Jill Furmanovsky

*The mother we saw on pp 30, 37 and 38, Nuala OSullivan, with her newborn baby*
Photo © Jill Furmanovsky

A number of pleasant mind-body states are typical.
Both mother and baby will be full conscious and active.

Even if there is some mess after the birth (feces or blood), as is likely, both mother and baby will certainly be fully conscious and active during and after a completely physiological birth, and a number of pleasant mind-body states are typical. These include wonder, alertness and euphoria—and it seems, from both the babies' facial expressions and research evidence, that the baby experiences these too.[26] Many midwives have also reported greatly increased job satisfaction when attending physiological births because some of this euphoria seems to rub off on them! Perhaps it's just nice being around people who are clearly happy and fulfilled.

The hormones which the mother and baby will be benefitting from will continue to affect their behavior not just in the hour or so after birth, but even for a week or so afterwards. Of course, this all means that women have a much easier start to motherhood because they are behaving instinctively at each stage, rather than having to be directed by outsiders or prompted from within.

The impact of natural hormones on the processes really is quite remarkable. A woman who is usually self-conscious about being naked will happily strip off her clothes and get into unusual positions when she's completely involved in her labor. In a similar way, even if she has planned not to nurse her baby she is likely to do so anyway after a completely physiological birth. Obviously, the implications are enormous. Simply thanks to a hormonal cascade, which takes place spontaneously when the mom-to-be is feeling undisturbed and safe, all a woman's previous fears and worries about birth are overridden and her instinctive knowledge about birth and mothering emerges.

One particularly interesting thing about physiological birth is that the baby does not need to be 'caught'. The mom-to-be can simply prepare something soft for her baby to land on and can position herself above it. This means the woman can also be alone, while her midwife (or midwives) wait in the next room, or in the corridor, monitoring by listening and observing through a crack in a door. (Perhaps other designs could facilitate this non-disturbing watchful waiting?) An alternative to waiting behind the door, if this prospect makes you nervous, would be for only one very unobtrusive and sensitive midwife to stay with the woman giving birth, while another waits outside—if a second needs to be present. Whatever the arrangement, the processes seem to work best when the smallest possible number of people are present. (Many people have observed that the more people there are, the longer the labor turns out to be and this can be seen even in the birthframes in this book.) With only one or two other people around (or in an adjoining room), the laboring and birthing woman truly has a feeling of privacy and security (because she knows her attendants are there), so she relaxes completely into the birthing process.

The processes seem to work best with fewer people around

Finally, to state the obvious, I hope this description has made it clear that literally *nothing* needs to be done to facilitate a healthy birth: it all happens without any prompting or intervention. Although a back-up system is essential for optimality, no help of any kind is needed after most completely physiological births—except clearing up! After all, the baby's reflexes help him or her to adjust to life outside the womb. Coughing, sneezing, rooting and sucking all take place completely spontaneously, when necessary, and the newborn will almost always breathe without assistance. If something does need to be done by an outsider—for example, if the umbilical cord is wrapped around the baby's neck—the mother herself (who is fully conscious and active) will often spontaneously take action, so true sensitivity is required by midwives.

If you're used to something quite different, I hope this description of birth will stimulate thought and an openness to exploring new possibilities. This kind of birth is optimal because it's both safe and beautiful and also because it gives the woman a feeling of control. This seems to be crucial for her satisfaction with her care retrospectively and helpful in terms of the transition she needs to make from being a 'woman' to being a 'mother'. The alertness both mother and baby experience are also very helpful for effective bonding and breastfeeding. Having said all that, 'optimal birth' also includes the idea that intervention is fine when it's necessary—so the art and science of midwifery and obstetrics lie in identifying when any intervention is necessary, i.e. when it's necessary to depart from this norm. No intervention is needed for most mothers and babies, simply because of the advantages that physiological birth offers in terms of safety, satisfaction and longer-term outcomes.

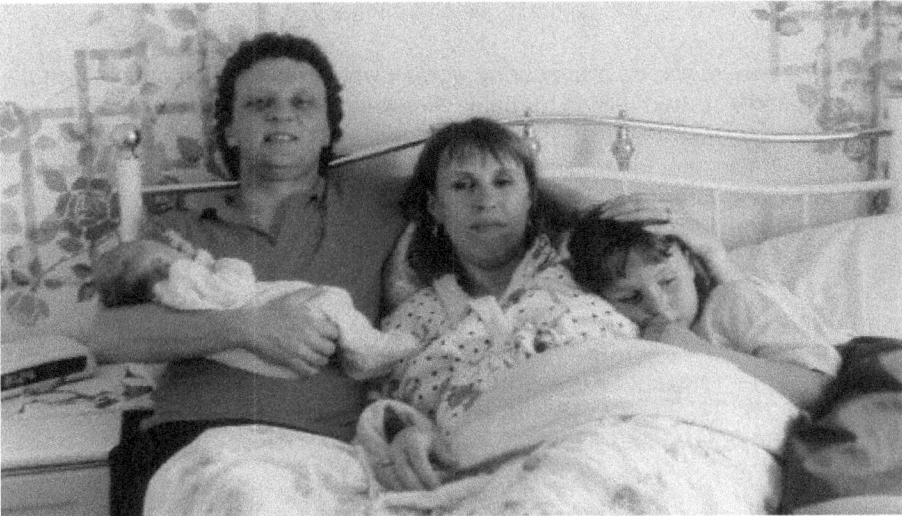

*The Low family, who you will read about later on. They experienced an optimal birth after initially having great difficulty finding caregivers who would support them in their wish for an undisturbed birth at home. All proceeded very smoothly.*

# Pause to reflect...

### ...about key issues in intrapartum care

1   To what extent is it good, in your opinion, for a woman to be conscious while she's giving birth?

2   Given this description of an optimal birth, how do you think a woman can most effectively be monitored?

3   What does 'watchful waiting' mean in practice?

4   Which signs would tell you that a laboring woman was having problems and needed intervention of some kind?

5   Which signs would make you extremely worried?

6   In what circumstances would you want to carry out a vaginal examination?

7   What are the pros and cons of using a) a Doppler and b) a Pinard during labor? How often would you want to use them so as to maintain the 'optimality' of a birth? When do you need to consult a client about using each one? Which positions can a Pinard be used in?

8   What would you do if a laboring woman who'd planned an optimal birth suddenly started asking for pain relief while she was in transition?

9   To what extent do you think you need to 'hand over' the process of labor and birth to a woman and to what extent do you feel you want to take control? What do you think laboring women feel about these issues?

10  What are your experiences of third stage?

For material relevant to these questions, see Birthframes 3, 4, 6, 7, 12, 14, 22, 26, 27, 31, 32, 48, 49, 52, 53, 54, 59, 61, 73, 74, 75, 76, 77, 78, 82, 84, 85, 94 and 98.

*Right: An alert newborn baby after a physiological birth.*
*Of course, babies who have been born without any drugs in their system*
*are extremely alert and keen to take in all that they find around them—*
*especially people, and in particular their mom!*

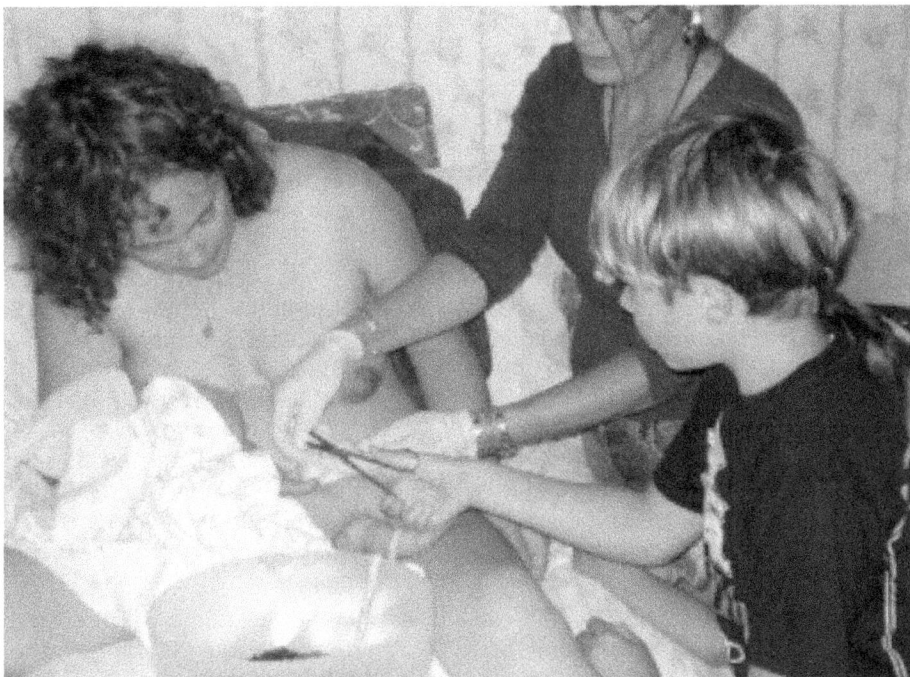

Cutting the cord

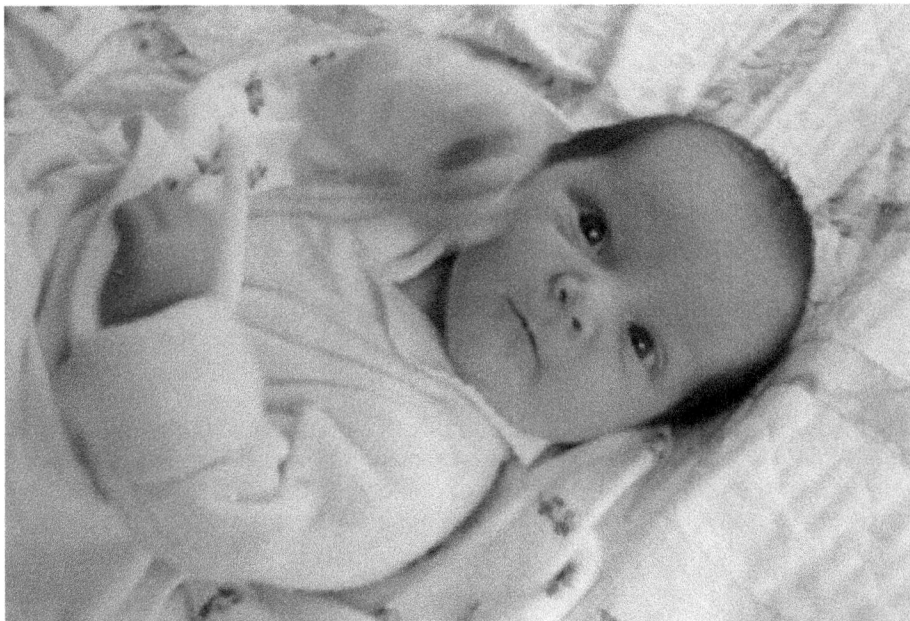

## Postpartum

After-birth scenarios obviously vary enormously. Mother and baby are usually both very alert for about an hour after the birth. In that time, as we've already noted, the baby usually starts nursing to get the creamy, health-sustaining colostrum. Mother and baby—and often the father too—start bonding, as the baby gazes, cuddles up, sucks and eventually drifts off into a contented sleep.[27]

If you are the birth attendant, you will carry out your normal checks at this point (Apgar scores, etc.). If you are making the birth entirely physiological you will avoid any practices which disturb the baby (or mother), such as suctioning (unless necessary to establish the baby's breathing), the administration of Vitamin K (by any means), or any 'routine' diagnostic tests. You will also not give any instructions so as to 'facilitate' breastfeeding or mothering behavior because this should all take place smoothly, triggered simply by the appropriate hormones being naturally produced by the mother's (and baby's) body. Having said all this, you would of course intervene if there were any life-threatening problem and your client may also accept or reject each of these interventions, depending on her outlook and evidence she has collected.

You will carry out your usual checks on the mother at a time which seems suitable, so as to cause minimal (or zero) disturbance in early bonding. If the baby's father is around, this might be a good time for him to say his first hello—meaning, of course, that the baby need never be separated from his or her parents in the first few hours (or days) of life.

If you're not used to this kind of birth, you may be surprised at this point to find you have an extremely active mother on your hands! She may want your support though while she has her first shower soon after the birth (if she feels messy). And she's sure to appreciate any help you can give her cleaning up. (The joys of providing care.) She may well also suffer from strong afterpains.

The baby's extreme alertness may also come as a surprise to you, if you're used to seeing babies born with drugs in their systems. The baby's personal priority at this stage is probably to make use of this alertness to take in as much as possible of his or her new world, outside the womb. He or she will definitely not be worried about having a first wash... so don't worry about arranging this. In fact, so as to facilitate bonding between the mother and newborn, it's best to leave the baby with his or her beautiful newborn aroma, because this can never be restored, once washed off. The baby can simply be cleaned up by wiping with a clean cotton cloth, put in some kind of diaper, and dressed or wrapped appropriately, depending on the weather. In any climate, at any time of the year, it's essential of course to wrap the baby so that he or she can keep warm, and the mother may well need extra warmth at this stage too.

The baby's extreme alertness may come as a surprise to you

Quite soon after the birth, perhaps after eating and drinking, both mother and baby—although initially in a heightened state of alertness—may well want to sleep. Obviously, they will have been through quite a lot together. Of course, there is no reason why they shouldn't sleep together... As is well reported, mothers who nurse are very sensitive to the needs of their babies and—incredibly, but reassuringly—this is true from the very moment of birth. Afterpains will continue postpartum, particularly during early breastfeeding, as the uterus returns to its pre-pregnancy size. After a physiological birth these afterpains may well be *more* painful, because the process is likely to take place more efficiently than after a 'managed' birth. Nevertheless, the process is good for ensuring the future health of the mother because the uterus will quickly involute to its usual size. No doubt, you will also have to complete your usual record-keeping after the birth...

**Here are few people explain how it happened for them...**

❝❝ I was surprised when the placenta shot out from between my legs as I reached down to pick up my new baby. So fast and easy!

❝❝ My physiological third stage was surprisingly painful, but very fast. Then suddenly, it was all over.

**A midwife's notes:**

10.24   Baby delivers, brought straight out of water by Georgina and G. Baby cries instantly.

10.35   Baby lying in Georgina's arms.

10.40   Baby still attached to placenta but cord ceased pulsating now. Georgina has period-like pains and feels the placenta beginning to come. Water temperature increased after the birth (temperature now 37.3 degrees centigrade). Baby feels warm. Baby is lovely pink now.

10.50   Baby is feeding from Georgina now.

10.55   Placenta membranes delivered spontaneously in the water. Baby still feels OK but has been covered up with a towel now. Cord clamped now and cut by his daddy.

11.00   Georgina out of the water. Bleeding minimal, estimated all in all at approx. 100 ml. Perineum observed, no tears at all. Georgina just feels sore and bruised but it all looks healthy and well.

11.05   Placenta looks complete. Membranes x 2 present and 3 vessels in cord.

❝❝ The placenta had a strange smell and didn't look as I imagined it would.

❝❝ I got out of the birthing pool for the third stage and had a small tear, which didn't require suturing.

❝❝ There was no need for sutures afterwards.

““ It might be of interest to note that although I had what seemed like a lot of blood loss when the placenta was expulsed (without the use of syntometrine), this large loss of blood is considered a good thing in many so-called undeveloped countries (according to Sheila Kitzinger's research) and this seemingly large quantity of blood was followed by very little lochia. On Day 1 I seemed to be soaking pads fairly consistently but from Day 2 onwards I have had very little discharge.

““ Postpartum, I just needed 'mini' pads.

““ Postpartum problems? I simply didn't have any. I think problems can sometimes be triggered by drugs given in labor, catching infections, unnecessary interventions, etc.

In case you haven't ever had a baby yourself, or need reminding of what it's like emotionally, practically and socially, here are a few comments from mothers who did the whole thing entirely physiologically...

““ I can't tell you what an amazing feeling it is to hold him, touch him, look at him making funny, suspicious faces and wrinkling his brow. I can see Mom was right when she said romantic love is nothing compared to this feeling of utter astonishment and prostrated adoration of this minute creature. My husband and I both feel it. My husband went home to get a few things and run some errands the next day. "I got home to the empty house and it all seemed so pointless. All I wanted was to be back with the two of you." He is utterly taken by the baby. We haven't needed any name for our son in these timeless days at the hospital, with no one but the three of us, a tiny universe.

““ Somehow the house feels entirely different now. We have a home now where we're sheltering somebody infinitely precious. Before it was just a post-college pad for two of us to hang out in and amuse ourselves. Now it all has a point.

““ I was a bit confused after the birth for a while. I almost felt like I'd failed because it wasn't totally enjoyable. I kept having to remind myself of what I'd achieved, which wasn't hard—I just had to look at my baby and I knew I'd done something wonderful! Being just me and my husband at the birth made it very special and intimate, we didn't need anyone else. After the birth it was all very peaceful and beautiful.

““ Babies are such a wonderful way to start people!

““ Mom left on Sunday. That afternoon we took him for the first walk of his life, in the park. All the other parents with children looked calm and collected. We got hot around the collar trying to put the stroller together with him in it.

Babies are such a wonderful way to start people!

*Here a mother who's just had her second VBAC relaxes, shortly after an optimal birth*

66 My husband and I argued terribly in the first weeks after our baby was born, partly because she was so incredibly interested in everything all the time and so disinterested in sleep! We constantly criticized each other's attempts to care for her and I felt tremendously alone and unloved. Fortunately, things eventually settled down and our second baby was incredibly easy and obliging, although also clearly as smart as can be.

66 I really threw myself into the childrearing and I remember at least one person said to me "What about you? You've got to give some time to yourself" and I remember saying: "This is me. This is important." Not to say, there were times when I didn't get exhausted.

**Words from a mother of six, expecting her seventh child:**

66 If there was one piece of advice I'd give to new mothers, it'd be this... Forget about the idea of having routines! Just try to meet your baby's needs and go with the flow.

With optimal outcomes, things usually settle down after a few months, provided there are no postpartum problems, of course

*A few months after the birth...*

**More on birthframes...**

Here are a few more 'birthframes'. To give you more of an idea of why I've adopted this narrative approach, let me say that these birth stories are intended to help you visualize real life scenarios and see where my own personal views have come from. As you probably already realize, there are parts of human experience which randomized controlled trials simply cannot access so a case study approach really is appropriate at times. Incidentally, I've created the word 'birthframe' to refer to any true account of a birth experience, part of it, or even only part of the labor or pregnancy—or it may cover more than one birth. The word 'birthframe' reminds me of 'window frame', through which you can get a glimpse into someone else's world, and 'frame of reference', which helps put other things in perspective.

## Birthframe 2: Sylvie Donna—Thoughts on first birth

Here's some background to my own first daughter's birth.

My first experience of birth was in Sri Lanka, with a Sri Lankan obstetrician in attendance. Although there was a great deal of disturbance, since a very large number of people came to 'ogle' me while I was in labor and giving birth, it was a good birth, which involved no interventions or drugs. This turned out to be an achievement not only because this kind of birth never usually took place at that hospital, but also because I'd been told my baby was in posterior position at the onset of labor and she didn't turn—so I experienced the full pain of a face-to-pubis birth. Nevertheless, I was delighted to have given birth entirely actively, without drugs, because the research I'd done during my pregnancy had convinced me that this kind of birth was best in all possible respects, but particularly for the baby.

When I first started reading about pregnancy and birth I was open-minded. I couldn't really say that anyone in my past had influenced my views on birth or that I'd formulated a clear idea of what I wanted for myself. I was interested to prepare for this new life experience, though, in the same way as I always prepared for new jobs or life in different countries.

The choices I'd made for the upcoming birth had been made in an unself-conscious way, after reading various books and research articles. We were lucky to have access not only to a good medical library in Colombo, but also to have an outstanding bookstore, which stocked all kinds of interesting and unusual books. We also had full Internet access (at a time when that was a little unusual) and an arrangement with an enormous bookstore in London, so I could buy literally any books with an ISBN number by mail order. In case you're wondering why I wanted to read so much about pregnancy and birth, all I can say is that it was my inclination to read up on literally any topic that affected my life... I suppose I'm just the kind of person who likes to be well-prepared!

Anyway, I was rather taken aback by my obstetrician's reaction to my birth plan. Nothing seemed to be acceptable about it—particularly my requests not to have EFM on admission,[28] an episiotomy,[29] commanded pushing[30] and a physiological third stage of labor[31]. I also heard from a couple of colleagues (who'd given birth at the same hospital) that the supine position was imposed, even if it meant that nurses had to forcibly hold the woman down, at the instruction of the obstetrician. All women labored with either Demerol or an epidural, so my request not to have any drug-based analgesia was also met with concern.

Reading some of the more 'angry' books about childbirth (e.g. *Open Season* by Nancy Wainer Cohen, or *Immaculate Deception II* by Suzanne Arms) made me realize that my chances of having what seemed to me to be a 'good birth' were slight. The fact that one colleague was repeatedly telling me about the trauma of her emergency cesarean (at the same hospital) a few years earlier did not help me to feel confident about the upcoming birth.

I started to look for a different caregiver, who would accept my requests. This proved extremely difficult... As well as searching locally, I also considered flying a midwife out from the UK or flying home myself—but my full-time job and limited finances made these options unrealistic in practice. Eventually, to my enormous relief, when I was 30 weeks pregnant I heard by chance that a 'hands-off' obstetrician had just returned to Colombo after a two-year secondment in Germany. The fact that he was already fully booked up by locals and—according to the secretary—would not accept any more 'patients' didn't daunt me. I pestered her until she allowed me 'only' a five-minute consultation with him, just to have a chat about pregnancy.

In my allotted five minutes I introduced myself extremely cautiously and politely and presented my birth plan. While this new obstetrician read through it, I commented on how my requests seemed to me to be evidence-based or at least extremely easily justifiable in terms of the research, particularly as it related to maternal and neonatal outcomes. Although this poor man was rather taken aback he seemed interested and—after a pause—announced that he'd be happy to take me on as a patient. (As with many other issues, I thought it best at that point not to launch into an explanation of why I preferred the term 'client' to 'patient'!)

Things went well over the next few weeks until one day my new obstetrician mentioned that he would be away (in Germany again) during the two weeks leading up to my due date. At around 35 weeks by then, I felt completely at a loss... but he did reassure me by saying he would try and think of someone who could 'cover' for that period. The next week he told me he'd asked his former obstetrics teacher (who was now retired) and that he had agreed in principle. He gave me a letter of introduction which I then took with me when I met this man.

By the way, I would tell you the names of these wonderful men, but I don't have written permission from them to do so and, given the situation in Sri Lanka (a place where home birth is illegal), I think it's best for their own safety if I don't reveal these details.

The 'obstetrics teacher' accepted everything in my birth plan except one thing: he said he would insist on inducing me if I went even one day past my due date. I was about to argue the point when suddenly I realized that the other obstetrician would be in charge of me again by that time!

I don't claim to have any superhuman pain threshold. I did indeed experience the pain in my labor as outrageous and I did ask for an epidural at one point. However, when I heard that having this would only be possible if I transfered to the local (bigger) hospital, I immediately renewed my resolve to keep going. I did also ask for Entonox—see the Index if you don't know what this is!—and I was extremely lucky that the obstetrician and nurses could not get the canisters going. I still to this day wonder whether my obstetrician 'pretended' they wouldn't work so as to support my wish for a drug-free labor, for the baby's sake. If this is what he did, I am enormously appreciative!

I find it interesting and rather perplexing that the resident doctor was disappointed by my SVD (which he felt should have been a cesarean). I was shocked when the obstetrician later wrote to tell me that he hadn't seen a 'birth like that' for 14 years, either in Sri Lanka or Germany, particularly since obstetricians oversee *all* births in Sri Lanka, even low risk ones. I was also completely shocked by everyone's reaction to my baby. Nobody, it seemed, could believe that a baby could be so alert; they were all just so used to seeing droopy, sleepy babies because they were always born with drugs in their systems. In short, I immediately became a local celebrity.

In my opinion, keys to the 'success' of this birth include the facts that:

- I had found out about all aspects of birth, so I was well-prepared.
- My husband was prepared to act as ongoing masseur and 'assistant'!
- I had a caregiver who had agreed to support me in my wishes.
- I'd also prepared the obstetric nurses by miming an 'active birth' and insisting that they agree not to hold me down on a bed to give birth!
- My obstetrician had obtained permission from the hospital manager for me *not* to be transfered to the delivery room before I gave birth
- I was allocated a very small room, with easy access to a bathroom
- The bed was taken out of the room—so I just had a mat on the floor, a wicker chair (to lean on) and a plastic bench for my husband to lie on
- I had full freedom of movement and was 'allowed' to be noisy!
- My caregiver understood the importance on non-disturbance (in terms of verbal communication) and quiet, reassuring support. He was wonderful.

*Our eldest daughter, in a playful mood (as usual!)—eating (or not eating) spaghetti*

## Birthframe 3: Sylvie Donna—Fetus ejection reflex

My second experience of birth exceeded my expectations.

The second time I gave birth I experienced an authentic fetus ejection reflex. I had merely hoped to *avoid* drugs and interventions—i.e. I'd had a negative mental framework for the experience—so I was completely unprepared for the notion that birth could in fact be so extremely positive and empowering. This birthing experience made me understand how far we've moved away from our innate physiological functioning in the birthing environments we've created in most parts of the world now. And it also made me understand why so many women consequently experience birth as a negative, disempowering event. This, I realized, is truly tragic, given the incredibly empowering possibilities of the instinctive process. I feel very grateful to Michel Odent, my birth attendant, for that birth, for making this secure, non-disturbed, exhilarating birth possible.

Throughout this book I will talk about the principles which made a true fetus ejection possible. It's worth aiming for because it is much safer for both mother and baby than any kind of alternative. It needs to be *rediscovered*. Here's an account of the birth, including any details which seem relevant...

In the three weeks before the evening of the birth I had experienced numerous painless or low-pain contractions. The contractions which started at 9:30PM on her 'birth day', while I was doing the dishes, came on very suddenly. They were much stronger than any I had felt in the days and weeks before. In fact, they were completely absorbing. I was immediately on my hands and knees on the kitchen floor.

Fortunately, Michel Odent was already in the house when these contractions started so there was no problem with this being a short labor. The day before I thought I'd gone into labor because I started having regular, fairly strong contractions. I called him immediately because we were living quite a way from his apartment and I knew the journey would take at least an hour, if not more. However, just after I called Michel my mother phoned me and, as usual, asked how I was. When I said (exhilarated and excited): "I'm in labor!" her reaction completely took me aback. After saying, "Oh dear" and seeming put out, she changed the subject and launched into an anecdote about someone I hardly knew. When I eventually managed to get off the phone I realized my contractions were much weaker and within minutes they'd stopped altogether. No amount of trying to 'psyche myself out' of my mother's nervousness would make any difference. When Michel arrived very soon after this he said, as soon as I opened the front door: "You don't look like a woman in labor." (I found this very annoying at the time, although I've since learned that it's a strategy of his for keeping the birthing environment calm and for avoiding misdiagnoses of labor. It's not one I entirely agree with, but I can see his point of view.) Anyway, the net result was a rather embarrassing evening in which I cooked a meal (to the highest standards possible at the time, since I was very aware that Michel is French!) and we entertained Michel as best we could. Michel then said he would stay overnight because he thought it was possible I might go into labor again during the night (which I didn't!) The next morning, he also said he would return in the evening, just in case the baby decided to come then. (Oh no, I thought. I'll be cooking and entertaining this man every day for the next two weeks!)

Anyway, my labor really did start the next evening. Michel had arrived at around 4:00PM and we'd had dinner again. (I ate huge quantities of everything.) Then Phil, my husband left Michel and me to talk at the dinner table afterwards and something did relax deep inside me during that conversation. We talked about all my worries, mainly about money. Michel's comments were reassuring—even though it turned out he was totally wrong!— and I was soon to go into labor again, this time with a vengeance. Michel then went up to bed and I was left to do the dishes. Halfway through, an enormous contraction ripped across my abdomen and I immediately fell to my knees, in an attempt to cope with the pain. More, extremely powerful contractions continued and I soon felt the need to stagger upstairs and start doing some sonorous groaning.

Hearing the noise, Michel emerged from the room we'd assigned him (next door to the bathroom, where I was) and asked if he could check the baby's heartbeat and position. His brief examination reassured him that conditions were ideal for me to labor undisturbed. Firstly, palpation of my belly had told him I had a full bag of water. This of course meant there was little danger of the umbilical cord prolapsing or becoming compressed.[32] Secondly, he established that the position of my baby was fine—it was LOA. (Later Michel told me his 'treatment'—to labor undisturbed—would not have been different if she had also been in a posterior position, like her big sister. He said he has found that women find better positions for helping the baby turn if left alone, undisturbed.) Thirdly, Michel confirmed with a Pinard that the fetal heartbeat was strong. He was also reassured to see that on my return to the bathroom I was spontaneously laboring in favourable positions— leaning forward, either standing, on my hands and knees, or on my knees in front of the toilet when I was chucking up that wonderful dinner! He knew this would mean avoiding compressing the vena cava, which would happen lying down, which is important for the baby's oxygen supply. While I labored alone, in peace, Michel waited and listened from his bedroom. He even waited outside the room while I gave birth.[33]

Phil, my husband, helped me throughout my brief two-hour labor. As each contraction ripped into me I fully understood why some women choose drug-based pain relief or a cesarean—but I knew I must not even mention any of these thoughts out loud. A few contractions into my labor I said to Phil: "Two children are enough! We really don't need to have a third one, you know." Wisely ignoring my comment, he ran up and down the stairs fetching things for me—candles for candlelight, something to tie my hair back... (We'd agreed he'd do this in advance.) Meanwhile, Michel continued to keep watch from the next room, without making himself seen. He knew from long experience that the best support would be to leave me completely undisturbed, with a feeling of being unobserved.[34] The sounds I made would be enough to tell him if all was going well. Of course, he administered no drugs and insisted on no routines. Nature had her own routines in mind...

Then, within two hours of feeling that first contraction, I gave birth. Moments before, Anjula had woken up so Phil had had to go and console her. After sitting on the toilet for a few minutes, straining to do a non-existent poop, I suddenly stood up, lifted up my arms, flexed my knees slightly and gave birth! Between my feet, bundled up in a large ball, was a plastic sheet which I'd been 'relocating' around the house in a preoccupied manner for the few days before I went into labor. So much strength was coursing through my body when I was ready to give birth that I didn't need anyone's help or support. I thought about various things I'd read... I thought of Inuit supporting their wives from behind and pushing down on the baby belly so as to help the baby, and I focused in on my own baby within me.

Then I thought, "Baby, be born!"—calm, joyful words which passed through my head quite spontaneously. Then suddenly, I gave two wonderful strong, clear, purposeful painfree pushes. The ring of fire I then felt made me consciously realize my new baby's head must have crowned. I momentarily felt worried I would tear. Then—thinking, "Oh, I don't care if I rip in two"—I flicked my hips forward. Reaching down to feel what was going on I was shocked to feel a head. "Michel!" I shouted, anxious that I should be giving birth without his help. Knowing it would be dangerous to disturb me at this point (because it would disturb the flow of hormones), he ignored me and carried on simply silently watching through the crack in the door. There was a moment's pause, then again I flicked my hips forward. In a sudden gush my baby was born and, confused, I felt another gush seconds later.

I looked down, transfixed for what must have been only a few seconds, admiring my new baby's features. Suddenly she let out a cry and saying, "Don't cry," I took her up in my arms and instinctively put her to my breast.

There was none of the hesitation I'd expected to feel. Having read about so many problem situations, I had wondered whether breastfeeding would be so easy the second time around. My new baby sucked as if for the thousandth time, not the first. As I looked down it dawned on me that the second gush had been the placenta... It was all so fast! Beautifully alert, my baby was gazing quizzically into my eyes as she nursed. Michel came in at this point and was reassured to see the placenta lying by my side, born and whole, a sign that all was properly finished and safe. My husband was sad to have missed the birth and 2-year-old Anjula, who'd just woken up, looked amazed at her mother holding a new baby in her arms. Michel let them come in to see me as soon as he was reassured that the placenta had been born.

It was wonderful not having someone else 'catch' the baby for me. It was also wonderful being able to discover her sex myself and pick her up for the first time without anyone observing or 'checking' on me. The most wonderful aspect of this birth, though, was the feeling of strength I had. I felt so strong, both physically and mentally. I didn't need anyone to support me as I suddenly stood up, swung round, raised my arms in the air, elbows bent, feet planted firmly on the floor some distance apart, knees flexed. I certainly didn't need anyone to tell me I was fully dilated. Somewhere deep inside, I knew it was time to push. The feeling of pushing—two long, clear, happy pushes—was very positive and also completely painless, as I said before. The ring of fire which I immediately became aware of at the end of these pushes was also not painful or 'weak-feeling'. There was never any need to 'pant' or control my breathing—I just felt poised, focused, very alert and decisive. And I was strong physically in the sense that I had no problems with low blood pressure and no feelings of weakness after the birth. This had been a truly authentic fetus ejection reflex.

To me, it's particularly interesting that this incredible feeling of strength came directly after a long phase (lasting at least 20 minutes!) in which I felt very uncertain and nervous. While I was sitting on the toilet trying to do that 'poop' I certainly didn't feel like I knew what I was doing or that the birth was proceeding along any known, instinctive curve. But somehow the cocktail of hormones must have been building up within me and prompting me into action at various points.

Of course, my sudden small envelope of complete privacy is also interesting because it was only when Phil got called away (by our first daughter, Anjula) that my 'instinct' to give birth kicked in decisively. It's almost as if I needed this extreme privacy in order to give birth.

As far as the safety of the birth is concerned I always felt confident that Michel knew what he was doing even though I hadn't been expecting him to actually be *outside the room* while I was in labor and giving birth. (It would have helped if I'd known this in advance, I'm sure.) My confidence in Michel came from the reading I'd done about him, which had reassured me that a) he was well-qualified and experienced, b) he cared passionately about honoring women and c) his approach of non-disturbance was logical (given what we know about the hormonal cascade) and grounded in research and/or his clinical experience. I would ideally have prefered him to be female, I must say, because some part of my mind still has a problem dealing with the idea of a man attending a woman in labor. After all, how can anyone truly be an expert in something if they've never done it themselves? Watching processes just isn't the same as being involved in them. I know this from long hours watching my mother doing things which she's still expert at, which I still fail at miserably—e.g. keeping plants alive or doing housework! The experience of doing something oneself provides insights which can never be gained from any amount of observation or academic study... And besides, I do think there's something extremely intimate about birth which maybe even means that the husband should be excluded. In any case, at points during my labor I certainly did feel extremely lonely at being a woman giving birth alone, only in the presence of two men. (Of course, I wasn't really expecting 2-year-old Anjula to be able to help.) Early on in my labor especially, I was longing to have an experienced woman with me. Really, I suppose, what I wanted was a midwife—of the archetypal variety.

Anyway, after the birth, and my third a couple of years later too, I realized again the enormous advantages postpartum of giving birth entirely physiologically—with no drugs or unnecessary interventions. I understood on a very deep level how *healthy* this type of birth is, physically and psychologically and how little pain there is overall. Most importantly, I saw the advantages for my babies, who were alert, nursed easily, and *thrived*. And whenever I met other women who described less exhilarating births (which was most of the time) I so wished they could have a better experience...

*Nina-Jay, aged 3, born by fetus ejection reflex. Like her big sister (and younger sister), she was alert and breathing immediately after the birth and nursed very easily*

## Birthframe 4: Sylvie Donna—Dealing with disturbance
Would it be three times a charm... or not?

By the time I conceived our third daughter, I was 43 and still working on a first book about pregnancy. Would I be able to give birth naturally again? Did those natural processes really work reliably when a body and a baby were left undisturbed? We'd moved to the Middle East, to Oman, some time before but we returned to the UK partly because of my pregnancy. I discovered that not only was home birth illegal in Oman (as well as Sri Lanka!), the hospitals were also extremely interventionist. Anyway, my husband managed to get a short-term job in the UK, so I had a lucky escape from my dilemma. I was 30 weeks pregnant when we moved into our new rented house, a short drive away from a large maternity hospital.

This time, I didn't have too much time finding a caregiver who accepted my birth plan. However, I did have to check out several doctors' offices and midwives before I found a supportive caregiver! (I did a lot of asking around in playgrounds and even in the street, whenever I saw a woman with a baby.) My midwife was extremely easy-going, and very supportive of my choices, although unfortunately she was not on duty on the day I went into labor.

I became aware that I might be going into labor when a powerful contraction woke me one day, at 3 o'clock in the morning. After getting up for awhile, I soon went back to bed and to sleep but gentle contractions kept coming here and there throughout the morning. Between-times I managed to get my two little ones to their playgroup and home again and while they were out I pottered round the house. When I'd brought them home again I sat on the floor, listening to a CD, rocking to and fro, while 2-year-old Nina-Jay and 4-year-old Anjula played in front of me in the living room.

Eventually I phoned my husband at work. As soon as he got home I went up to the bathroom alone. After two hours of labor there (spent mostly kneeling forward in the bathtub) I shouted downstairs to ask Phil to call the midwives. "Are you sure?" he said. "Yes. Call them now."

Five minutes after two unknown midwives arrived I gave birth standing up, completely unassisted. I'd told these government midwives to wait outside and had locked the bathroom door so I could give birth in peace. This may seem extreme to you, so please let me explain... When the main midwife had arrived I'd felt really disturbed by her presence because she spoke loudly and really seemed to be 'invading' my space. Her big black bag (with equipment and supplies) seemed ominous in the tiny bathroom where I was laboring and her presence made me feel far from calm. Intuitively, judging from what she said—"Oh, we've heard all about YOU!"—I felt that she did not really support my birth choices in practice, even though she might have agreed to support me in theory.[35]

I felt that she did not really support my birth choices in practice

The best approach really did seem to be to 'protect my birthspace'. I knew that by asking the midwives to wait outside the bathroom I would still have full access to their expertise. They could monitor by listening to the sounds I was making (or not) and in extremis could always have broken the flimsy lock to the bathroom door.

As any midwife knows, PPH has not normally been known to kill a woman within five minutes or render her unconscious—and there were no signs of any other pathology at the time. I agree that my actions would have been irresponsible if I had had any medication, but since I was entirely alert and active and participating in a healthy natural process, with a history of straightforward SVDs, I do not feel regretful of my actions. I do, however, feel sad that I was so uncertain that my midwives would act as true advocates, which of course it is their remit to do.

When I opened the door, a few minutes after the birth, the lead midwife advised me to have the cord clamped at the placental side of the snapped umbilical cord. I immediately agreed and this is obviously an indication of my intention to work *with* my midwives and take their advice when matters of safety were in question.

Maybe the experience of attending me prompted them to consider a few important issues all caregivers should consider

I realize I must have shocked these poor midwives enormously... but maybe the experience of attending me in labor did prompt them to consider a few important issues. Questions which all caregivers could consider, I think, based on this experience, include the following...

♥ How should a caregiver greet a woman in labor? Does she need to say anything at all?

♥ Is a vaginal examination really necessary when you first meet a woman in labor? In what ways might a vaginal examination disturb the woman in labor? What could the alternatives be?[36]

♥ What does 'watchful waiting' mean in practice? If you have a set of protocols for normal, vaginal birth in your particular working context, to what extent do you agree with each protocol? Could or should anything be changed in cases where a woman wants not just a 'normal' birth, but an entirely physiological one, involving a fetus ejection reflex?

♥ How much is it necessary to talk? Who is it legitimate to talk to? Does it matter whether or not the mom-to-be can hear what you are saying? What effect might speech have on the her? What benefit do you as a caregiver gain when you speak? Does the laboring woman obtain equal benefit?

♥ Where should you be *physically* when you are attending a woman laboring without drugs? Should you stand or sit, or what? Where should you put your equipment and supplies? How can you make the laboring woman feel supported, without also disturbing her in any way?

Anyway, coming back to this third birth of mine, another interesting thing to consider is how my body worked, given my experience of giving birth... This time I didn't experience a fetus ejection reflex for the simple reason, I think, that there was an enormous amount of disturbance around me when I gave birth. The fact that I didn't feel as safe and confident about my midwives was probably also a factor... These women weren't known to me, I was aware that one of them was only a student and the other one didn't seem sympathetic to me, as I've already mentioned.

The disturbance mainly took the form of bustling and talking.[37] Repeatedly, I asked the midwives to be quiet, but they continued to talk, ignoring my requests. (I realize they were probably nervous or even panicky at being asked to wait outside the bathroom while I gave birth.) Although this was a source of great annoyance at the time and probably had a real effect on the birth, I am at least grateful that it happened this way because this experience prompted a great deal of thought about this issue. Could speech really affect the birthing process? Being an activity governed by the left hemisphere of the brain (in a right-handed person), it seemed possible it could because birth did, after all, seem to be a process which was controlled by the right-hand hemisphere (being essentially instinctive and 'psycho-sexual', as Sheila Kitzinger has explained). Being a language teacher and linguist (in my previous work) these questions were of particular interest—so I can't honestly say I'm particularly angry or annoyed with my midwives!

Anyway, the atmosphere was certainly not one of calm, confident competence. Also, it would have *really* helped if I'd met and talked to these midwives on several occasions before. Oh, how I believe in continuity of care![38] I think it would make so much difference to women during their pregnancies, labors and births if they had the same midwife attending them right the way through. If it has to be a *team* of midwives, it would at least be great if women could meet the team and get to know the individual members.

Finally, you may be curious about the issue of 'catching the baby'. Actually, nobody caught the baby during this birth. She was simply born onto the soft carpet, which was further cushioned by a thick plastic tablecloth and towel. I picked her up immediately after she was born and put her to my breast. She nursed like a professional, gazing up into my eyes.

The midwives supervised while I had a quick shower (while my husband looked after newborn Jumeira) and left very soon afterwards. As usual, I experienced no problems postpartum, except for strong afterpains with vomiting too! I was up and about immediately after the birth and busy with housework and the usual errands the next day. This time I didn't have the 'problem' of being pregnant again—I found my breasts became too sensitive when I was pregnant—so I nursed Jumeira for over three years. Seven years' breastfeeding experience in all! My body had served its biological purpose...

My third 'baby' Jumeira aged 6, with curled hair, after 10-year-old
big sister Anjula had been experimenting with her new curling tongs!

## The natural norm

The last few accounts, describing my own births, are what I like to call 'the natural norm'. In the light of your own experience you may find this very surprising. Think about it though, please... Isn't this how human beings should be giving birth? (Yes, I realize I'm pushing the definition of the word 'norm' here, but I mean that physiological birth *should* be the norm, even if it isn't at the moment in most, or even many, parts of the world. I also mean it's the norm in a healthy body which has no unnecessary intervention.) In our efforts to improve birth experiences for mothers and get babies born quickly, we've moved a long, long way from what should be normal. To make matters worse, whenever births are featured in the media, they are made to look *abnormal*.

Media coverage is particularly unbalanced because editors are looking for unusual things to report. I was reminded of this when I contacted a Features Editor of a well-known women's magazine, when I was researching this book. At the time I was looking for a vaginal breech birth story. The editor initially seemed enthusiastic to pass material on to me, but completely changed her tone when I said I was also interested in any births which had involved life-saving intervention. "Ah no," she said. "You wouldn't be able to have any of that. That's precisely the kind of story *we're* interested in." In other words, it was clear that the magazine's objective was not to accurately portray reality, to present a balanced picture of what childbirth is for most women, or to help women consider what it might be. It was to catch people's interest through the sensational.

Perhaps as a result of our constant media exposure, perhaps because of our fear of failure (or litigation!) and our desire to control things which seem *out of control*, the present-day climate in hospitals is one of worry, habitual intervention and insensitivity to a laboring woman's needs.[39] Or is it oversensitivity to laboring women, who appear to be in an unbearable amount of pain, which must be stopped at all costs? In any case, it's not surprising that many women don't experience 'optimal'.

Of those who do, most feel no need to talk postpartum precisely because they haven't been traumatized in any way. Michel told me:

"Often women who talk a lot about the birth of their babies are those who had a difficult birth, problems and so on. And women who had a very easy birth tend not to talk about that. ... So I think it might be good to say somewhere that this book is special. It's full of accounts we might learn from."

When I chanced upon somebody who had given birth naturally, it was usually rather difficult to get an actual account. The women concerned were quite open about their experiences, but they didn't make a 'big deal' of them. This generally meant that if I asked people to write anything down they usually said, "But all I did was... There was nothing special about it." Sometimes my 'potential contributors' did indeed realize there was something special about their experience, but they clearly felt no driving need to talk about it. As a result I had to do quite a bit of hassling to get hold of many of the natural birth stories in this book. For many accounts I exchanged scores of emails over some years.

In the end though, after 12 years of perseverance, I managed to obtain accounts from all kinds of women (and men!), from vastly contrasting walks of life—making it clear that it wasn't just 'hippies' who give birth physiologically!

What was particularly interesting (if not obvious?) was that many of my contributors had a fully physiological birth the *second* time they gave birth. This reminds me of the viewpoint of the !Kung San tribe in Botswana: their ideal is to manage to give birth alone by their third child. Before that they are only 'practicing'. Clearly, many women in the developed world do the same. Another factor seemed to come into play though in many of the births... Things went much more successfully second (or third) time round when the laboring and birthing women weren't disturbed. To give you an idea of what I mean, below are some comments, followed by another account from Maria Shanahan.

" For my first child I'd planned a water birth but ended up with a four-and-a-half hour labor that was very traumatic. I was rushed into the hospital and into a very clinical environment, with bright lights and at least six people standing around me. Jamie was a forceps baby and I ended up having to be sutured. It was almost enough to put me off having any more children. So when I got pregnant again I knew I wanted things to be quite different and I was prepared more. After Arion's birth I felt so empowered that it had gone right the second time.

" After having forceps and a postpartum hemorrhage the first time round, my second birth was totally straightforward with no time for the TENS, birthing pool or gas and air, which I had planned. I stood up for delivery, supported by my husband and pushed the baby out myself, the midwife 'catching' her well. I had no shot [of syntometrine] to speed up the placenta and control the blood loss as it was just not necessary and this third stage was completed in about 20 minutes.

" This time round I was in control the whole time and I knew what was going on, I was more prepared mentally and physically.

## Birthframe 5: Maria Shanahan—Second optimal birth

This is the story of a woman full of ideals, who became disillusioned after a bad first initial experience of birth when all kinds of intervention had seemed inappropriate. Would she be able to rediscover her innate, instinctive ability to give birth if left to her own devices?

I was very frightened about my ability to give birth properly, and questioned all my ideas about home and water birth. I called the Active Birth Centre in London and explained my experiences at the hospital when I had my first baby, Fiohann. I asked if there was anyone I could talk it through with. They gave me a few names, and then suggested Michel Odent. I couldn't believe that I actually had my revered Michel Odent's phone number.

I was very frightened about my ability to give birth properly

Before I could think about it, I forced myself to phone him. I explained to him what had happened, and although, understandably, he wasn't prepared to comment on the hospital treatment, he did explain that not many midwives had experience of water births, and could panic, thinking the baby would drown. I told him it was a dream of mine that he would deliver our baby, and after hearing when the conception date was, he found that it did fit in with his schedule of international conferences, and that yes he would. I had to pinch myself!

He came around to meet us and I cannot describe what a gentle, intuitive and passionate man he is. He filled me with a quiet confidence and reassurance, and was so different to our doctor, who had originally told me that if I wanted a home birth I would have to find another doctor. Another difference was that he had absolutely no problem with calculating the due date from the conception date: it was a simple nine months later. The midwife could not cope with anything more than the date of my last period, which I did not know exactly, since I never paid any attention to my cycle. Her date was two weeks before Michel's estimate. The pregnancy went well in that I was fit and healthy, but my husband and I started to have problems and when I was six months pregnant he left home for a month. I felt desperately insecure and cried through most of my pregnancy. Another pregnancy filled with grief, but so different to the first. We went on vacation and decided to try again.

I rented the birthing pool two days before the earliest due date and waited. The midwife got increasingly agitated and by the July 21 was telling me that I would be causing the fetus brain damage because my placenta was past its sell-by date. Michel calmed her by telling her it was due on the 26th, which she could not comprehend. Sure enough, on the 25th, which happily was a Saturday, I went into labor. I called Michel and he promised to be over in a couple of hours.

We filled the pool, put the low music on and had candles ready, and I tried to relax, but found it difficult because of my doubts in myself. Michel arrived and asked me how I was, felt my belly, and said he would go to sleep in the next room, because I had a few hours to go yet. My husband went to sleep in our bed, and I spent most of the night sitting on the toilet having contractions; it felt the most comfortable place to be, looking at the stars through the bathroom roof window, and going back to bed trying to doze before the next one.

Early next morning, things started hotting up and I decided to get into the birthing pool to ease the pains. It did ease them immediately, and as soon as he heard the change in my noises, Michel got up and came in. He can tell where a woman is in her labor by the noises she makes, and just that small change had indicated to him a change in me. I told him I didn't know what to do, asked him what I should be doing, so different from my first birth, and he told me just to listen to my body, just as I had done in the first. I yelled that I didn't know what it was saying, and was very fearful. How different to the first time, with my confidence now in tatters.

He told me just to listen to my body... I yelled that I didn't know what it was saying, and was very fearful.

Michel calmly created an environment where I would have to listen to myself, by leaving me alone and getting my husband to go and have some breakfast. Our Brazilian au pair was beside herself that I could be moaning and wailing alone with no doctor, and took it into her own hands to go and tell Michel in no uncertain terms to go in to me and induce me! In Brazil, she said, over 40% of women have cesarean sections because it keeps their passages honeymoon fresh! He handled her sweetly and reassured her.

Things really got going at about 8:30AM. I was gnawing on the side of the birthing pool, thinking about the benefits of knives and drugs, half hoping that she wouldn't come out at all, would go back and stay safe inside me.

But that doesn't happen, and finally I got that huge push urge, when all you can do is that colossal push and your whole body is intent on turning itself inside out.

## My body did know what to do after all...

It did know after all what to do, I just had to get my doubts out of the way. Her head came out in the water, just like Fiohann—my son. Michel reached down and checked the umbilical cord was not in the way, and then said that because she was so big, we needed gravity to help us. With her head still between my legs, he lifted my legs, and my husband lifted my torso out of the pool, and I hung from his arms, with Michel ready to catch her.

Two more big pushes and out she slid, our beautiful girl. I flopped to the floor and Michel gave her to me immediately. I held her close and within a few minutes she was nuzzling at my breast ready to nurse, still attached to the umbilical cord.

Michel was delighted at the perfection of it, but said that before she settled into it he would tie off her cord with string rather than the metal clamps because it was more comfortable for the baby. I lay on our bed with our wonderful baby and as she suckled I realized that it would have been no different for Fiohann. If only.

Michel left me to deliver Eowyn's placenta naturally, which came easily soon afterwards. What I hadn't expected was the sharper pains as my uterus contracted back again, but apparently this is a normal feature of a second childbirth.

Michel wrote to our doctor to inform him of the birth and our baby's 'top' scores, knowing that we would be left undisturbed until Monday. Once again, I had no tears and was perfectly fit. The day following Eowyn's birth we had a celebration barbecue, and I walked around Tesco shopping for it, feeling so proud and happy, as if everyone must be able to tell that I had just had a baby! Michel came with his son, and was thrilled to see us so clearly well and happy. It was perfect.

Our doctor and his doctor wife arrived Monday morning, demanding to know how long she nursed on each breast and making appointments for pediatricians to see her. I had no idea how long on each breast and I declined the pediatrician offer, which they seemed a bit put out about. Finally they left, leaving the midwives and health visitors to do their checks, etc. and eventually we were left in peace again.

Eowyn is now a very fit and healthy 4-year-old.

Some midwives and researchers have rightly expressed a
concern over maternal dissatisfaction after labors in which the
mother has felt 'abandoned' or 'neglected'...

*Maria with Eowyn and Fiohann*

Research could usefully investigate best ways of preparing
women so that support and lack of disturbance are features

# Pause to reflect...

## ...about women's hopes, expectations and realities

1 Why do so many first births not go as women want?

2 Imagine you meet a primigravida early on in her pregnancy and she is worried that things are likely to go wrong... What would you say to her to prepare and reassure her?

3 What, in fact, do you think are the key elements in preparing a woman successfully for the birth she wants?

4 When a woman mentions that she wants to avoid 'drugs' or 'interventions' if she possibly can, how can you best prepare her for an optimal physiological birth?

5 If a multigravida comes to you, talking about bad previous experiences of birth, how do you respond?

6 In Maria Shanahan's second birth, what were the key elements influencing the successful outcomes?

7 If you'd been the caregiver, would you have done anything differently?

8 If a multigravida comes to you, talking about wonderful previous births, how do you respond?

9 In the case of a multigravida, how can you talk about risks without making the woman fearful—i.e. how can you avoid what Michel Odent calls the 'nocebo effect'?

10 To what extent do you transfer your own fears and hopes?

Do you have any other thoughts or questions?

For material relevant to these questions, see Birthframes 4, 21, 28, 29, 32, 33, 36, 38, 42, 43, 44, 45, 46, 48, 49, 55, 58, 59, 67, 69, 71, 73, 76, 79, 84, 85, 88, 89, 90 and 99.

Here's a quick overview of a few more reasons why many women might not experience the 'natural norm' or at least some form of optimality:

♥ Pregnant women may be overly cautious and fearful about childbirth, so request interventions which aren't really needed. (For example, they may ask to have their water broken if they're overdue.)

♥ They might feel convinced they can't face the pain of labor, so request pain relief which necessitates other interventions.[40]

♥ Perhaps they've never met a woman who's had a completely physiological birth and can't imagine it.

♥ They've met, read about and watched (on television) numerous women who've had interventionist births for various reasons.

♥ Their caregivers may be cautious because of a lack of experience dealing with women who are unmedicated. Mostly, they might fear litigation.

♥ Pregnant women might be classified as 'high risk' out of over-cautiousness or for the wrong reasons.[41]

♥ Lack of one-on-one care during labor or birth might mean women are disturbed and lose confidence.

♥ Pregnant women may not know their midwife or obstetrician well.

♥ Without good support, pregnant women are likely to lack confidence in their ability to give birth.[42]

## More questions for reflection...

### In the prenatal period:

1 How do you feel when women tell you they're going to need lots of pain relief?

2 How often do you feel, during prenatal appointments, that you're dealing with a woman who is confident about giving birth?

3 How often are you skeptical about a woman's ability to give birth?

4 How often do you find yourself expecting things to go wrong?

5 How often do you make negative comments to women in response to their comments, questions or assertions, and how often are you encouraging?

### During labor and birth:

1 What kind of comments do you make to women when they phone you up, reporting they're in labor?

2 Which kinds of comments are helpful and which are not, in your opinion?

3 How do you usually first greet a woman when you first see her in labor?

4 How much do you speak when a woman is in labor—to the woman herself, or to other people, when she can hear?

5 When do you think it's legitimate or helpful to ignore a woman's questions, requests or comments?

6 How do you feel about your own silence and a woman's noisiness?

This is Beth Dubois, a mother who experienced a physiological birth for the birth of her first child, despite a history of sexual abuse—and all the psychological baggage which that usually entails. Through careful preparation and support from her midwives, she managed to give birth naturally and also breastfeed. You'll read her account soon.

## Birthframe 6: Beth Dubois—Considering her midwife's care

Beth emailed me with some comments about her midwifery care during her second birth. Do her comments and reflections change any of your earlier responses to questions? Could they prompt a change in behavior in any way?

The experience of Zoe's birth was really profound. During the most difficult part of it, I traversed an inner territory so intense that the words that came to me to describe it were 'walking through the valley of the shadow of death'. From a medical standpoint, everything went perfectly for me and Zoe and both of us were always safe. I consider the care I received to be the gold standard. Here's what I found helpful...

♥ My midwife came to my home for all prenatal visits and often stayed for several hours. I had plenty of relaxed time to ask questions and get used to relating to her. She told me that her midwifery group purposely goes to the woman's house for the prenatal visits because the woman is then used to having them come over, so when it's time for the birth, it feels comfortable to the woman to have them there. It really was like this for me. When she arrived for the birth, it seemed much like one of her regular visits, although much more exciting since the birth was impending!

♥ My midwife had everything needed to resuscitate me and/or the baby if needed, medication to stop postpartum hemorrhage, a regular Doppler, a Doppler that could be used underwater, a birthing pool, materials for suturing. She had antibiotics available for women who are GBS positive and wish to receive antibiotics during labor. I felt very comfortable that she had everything in terms of skill and medical equipment needed to manage most complications that could occur. We also had a hospital less than two miles away that we could have gone to, if needed. Having said all this, another great thing was that I don't remember seeing any of it (except the Doppler at the time she used it and the suturing materials at the time she used them) in my house during the labor and birth. She was very discreet with placing the equipment so that I didn't at all feel the birth was 'medical' or that she was expecting an 'emergency'. However, in the back of my mind I knew it was there and could be used if we needed it, which gave me peace of mind.

♥ Although my midwife usually works in a team with both of the midwives attending the prenatal visits she allowed me to have only one consistent midwife at my prenatal visits so I could focus on relating to and developing rapport with only one person. I did want a second midwife at the birth to be on hand in case there were any complications. However, I did not want her to even be in my home until the birth was very close. My midwife respected my wishes on all counts. The second midwife came to my house about an hour before my daughter was born and both midwives waited outside the room I was in, listening. Occasionally, the primary midwife would check the baby's heartbeat. She only came in the room I was laboring in when I asked her to come in as the contractions became very intense.

`She only came in the room when I asked her to come in`

♥ Our midwife spent time getting to know our older child and developing a rapport with him. She made it clear she encourages siblings to be at the birth. However, our child and I both felt that it would be best for him to be at a friend's house during the birth. My son attended all the prenatal visits and often got to help our midwife take my blood pressure, listen to the baby with the fetoscope, etc. He got to help the midwife weigh his sister when she was 1 day old. She also got to know my husband well and he felt well included in all visits, the labor and birth. She gave her care in the context of our unique family.

♥ Our midwife did a great job of suturing my tear. She used sterile techniques and the same materials that would be used in the hospital.

♥ The postpartum care was incredible. Women in the US who give birth at hospital must leave their home for postpartum care, which I know from doing this three days after the birth of my first child, is absolutely exhausting. And visits for the baby are separate and involve additional exhausting trips outside the home. Also, women who birth in the hospital typically only have one visit at six weeks postpartum. In contrast, my midwife made home visits on Days 1, 3, 7, 14, 28, and at six weeks, as well as an extra visit for me to check my perineum since I had a second degree tear (which was very rare in her practice). She also made an extra visit to check on our baby since her respiratory rate was fast. And at about nine months postpartum she came to see me when I had mastitis. Each of these postpartum visits included attention to the physical well-being of me and Zoe, as well as the emotional well-being of our family, especially me. I wish this level of care was standard for all postpartum women. There would be much higher breastfeeding rates and much lower postpartum depression rates if all women received this level of in-home care.

Having said that, during a particularly difficult part of my labor, there was one particular aspect of the midwifery care that I didn't respond to well...

During my first labor, when things became intense I asked my husband to get the midwife, Valerie. I remember saying to Valerie "What's happening to me?!!" and she replied "You're having a baby!" I know this sounds obvious, but when she said it, I relaxed and felt assured that everything I was going through was totally normal. I continued on with my contractions, yelling my way through them while I sat on the toilet. When each contraction subsided, I would collapse onto Valerie who was sitting on a stool in front of me. This was all reassuring and it was how I made it triumphantly through the most difficult part of my first labor. So, in my second labor when I felt I was at that same difficult point, I tried what had worked for me with the first labor—I asked my husband to get the midwife to come and help me!

Sandi came in to our bedroom where I was laboring in the birthing pool. I started telling her how hard it was. She said: "You're SOOO close" and I countered: "How do you know?" She said: "I just know." I felt frustrated. She 'just knew' but I wasn't sure for myself. (Looking back, I wonder if there is something she could have said that would have put the power and knowing back with me.)

Seeing that I wanted more proof, she offered to check my dilation and said it was a 'stretchy 9cm' and that there was a little lip of the cervix left. She offered to massage that area during a contraction. I agreed and then she told me I was fully dilated. However, it seemed as though once she came into my space and checked my dilation, she became personally involved with my labor. Seeing how much pain I was in, she then started strongly encouraging me to push my baby out so that I would be out of pain. She offered to talk me through a contraction which I agreed to. She was saying something about having gratitude, which just didn't go with how I was feeling at the time! I waved her away and she immediately stopped.

At one point I screamed—yelling is a great pain-coping mechanism for me—and Sandi told me to make the sound deeper. I was furious! I was thinking: "You're not in labor. I am in labor and this is MY cervix yelling. I will yell as my cervix tells me to!" Her interventions continued. I felt my 'space' was being invaded. I couldn't follow my body since I was trying to do what Sandi was suggesting. But her suggestions weren't what I needed and with all the pain I was already facing, her suggestions felt like a real imposition and broke my concentration. She said: "Push, push, push out through your bottom," which I did, but I didn't like being told what to do. With Theodore's birth the pushing had been no problem for me and I was certain that this time, when I was ready to push and felt the urge to push, I'd be able to do that. She again told me to bring my voice lower.

Finally I was totally frustrated with her suggestions and hurt that she didn't seem to think I could figure out on my own how to birth my baby, but still not wanting to hurt her feelings I said: "Sandi, do you think I'm not doing it right?" (Afterwards, my husband said the tone was: "Sandi, shut the hell up!" and she backed off.) I was in the birthing pool at that time. In response, my husband leaned over it and said quietly: "You're doing everything *exactly* right." With that encouragement I suddenly summoned all my strength and determination, pulled myself up to a standing/squatting position, leaning somewhat forward with my hands on my thighs. My husband helped hold me up from behind. Then, I could feel the baby there ready to come out. (In retrospect, on all fours in the pool there just wasn't enough gravity for me to feel the baby and push her out.) With my eyes squeezed shut in concentration, I yelled and pushed. I was determined to get her whole head out on that one push. It was a long yell and a long push and then I was sure I had got it out. I was certain I could push the rest of the baby out with one more push. Yet, right then, when that big push was over, I was startled to hear a baby cry. I opened my eyes and my midwife was holding a baby! It had just taken the one push. I remember saying to Zoe: "Thank you, baby." I was so grateful to her for having come out!

My experience illustrates that even in the case of a wonderfully skilled midwife who has a great rapport with a family and strong intentions of respecting the birthing mother's privacy and innate power, the tendency to 'help' can run so deep in caregivers that it can pop up even when the midwife's truest intention is to empower the woman. Sandi apologized the next day for insisting on me pushing out the baby, saying that she knew, based on our prenatal conversations, that I didn't want that.

The tendency to 'help' can run so deep in caregivers...

She acknowledged that the baby was never in any distress and she didn't know why she had been so insistent on me pushing. I think that it came from her wanting me to be out of pain and her concept that she knew what I could do to get out of pain. I think she made a mistake in trying to 'rescue' the birthing woman by directing me, rather than pointing me back to what I could know in my own body.

[To protect her privacy and to respect her openness, the name 'Sandi' is not the real name of the midwife. Beth and I respect her postpartum honesty and humility. Everything else is unchanged.]

## Questions for reflection...

1 What could the midwife have done to 'help' Beth without invading her territory and distracting her from what her body was 'telling her' what to do?
2 What other things do caregivers sometimes do which might upset women who ask for help, but then feel that what is done is not appropriate?
3 What was the significance of the speech in this account and how could it have been improved? What would you have said yourself at each stage?

### Birthframe 7: Rachel Urbach—Going to another planet

Rachel, a first-time mom, was amazed to find her labor really did proceed as Michel had predicted and the support she received seemed appropriate...

I gave birth at home, at age 38, in water without using any drugs. My [identical] twin sister had had a baby the previous year, with the usual story of failed home delivery due to minor complications followed by the cascade of interventions at hospital ending in a cesarean.

I read Michel Odent's *Birth Reborn* and was inspired at the stories of natural birth—the fact that he finally avoided the use of chemical pain relief altogether because it seemed to interfere with the natural pain-relieving processes of the body.

I searched widely for a sympathetic midwife and was so lucky to eventually find one, who visited me at home and helped my confidence. I think state of mind is so important—I also did some hypnosis, or visualization of the birthing process, which I think helped.

My baby was born after a 9-hour labor. To my amazement, it was like Michel described—I became almost like an animal, or went into another level of consciousness beyond pain. The part I remember as the most difficult was the beginning—moments of fear—then gaining confidence—more moments of fear—reassurance from the midwife—then feeling like a fish in water being thrown about by the contractions and finally roaring as she came out, not with pain (as is so often the stereotype we see of laboring women) but with power.

It left me feeling so proud of my body.

It left me feeling so proud of my body

# Life-saving intervention

## Birthframe 8: Anonymous—IVF and placenta previa 4+

Sometimes, of course, medical intervention is necessary to ensure optimal outcomes. In the next account we see how it made it possible for a couple to have another child against the odds. This is a clear case of medical support being life-saving for both mother and child.

Having had one fairly straightforward birth, it came as a shock when I realized things can go wrong in pregnancy. My second pregnancy was ectopic so at nine weeks I had emergency surgery. Luckily, I did not lose the fallopian tubes but they did end up damaged. So, in order to have any future children the only option was to go through a lengthy, costly IVF program. We were extremely lucky to be successful on the first attempt, where I became pregnant with twins. However, fairly early on I developed hyper-stimulation (painful swelling) and one of the twins died by the 10-week ultrasound. (There are a lot of early ultrasounds with IVF to detect ectopic pregnancies and check that everything is developing normally.) As I had lost one of the babies I was told to expect some minor bleeding. They also mentioned that bleeding can be a response to all the vaginal ultrasounds... So when I had some small bleeds before the 13-week ultrasound I was not surprised, but a little worried. At the 13-week ultrasound I was told I had placenta previa, which was fairly low down, but was told that for most women by the mid-term ultrasound and certainly by 28 weeks, the placenta moves sideways and away from the cervix, along with the growth of the baby. I continued to get small bleeds and was up and down to the hospital like a yoyo. At the mid-term ultrasound (20 weeks) I was told that I had major placenta previa, being classed as Grade 4+ (1 being minor, and 4+ being serious). This meant that the placenta would not be able to move away from the cervix as it was completely covering the os, and in some places was adhered to the cervix. I was told I would not be able to give birth naturally and would have to have a cesarean.

At 25 weeks I started to have a fairly heavy bleed and was admitted for the duration of my pregnancy. Luckily, I stopped bleeding but had to keep a cannula in just in case I started to hemorrhage. I was allowed home occasionally as long as someone was with me 24/7 and I was within 10 minutes of the hospital. This is because I would have to be on the operating table within 20 minutes if I started to hemorrhage, as the worst-case scenario is death of both mother and child. At my 28-week ultrasound the earlier diagnosis was confirmed—no way was I getting away with a natural birth.

I started to have contractions at 32 weeks, which caused problems as I was not allowed to go into labor. (This is because the cervix dilates in labor and this would mean the placenta would rip, causing a major hemorrhage.) After being monitored on the labor ward to see the strength and frequency of contractions I had an emergency cesarean under general anesthetic. [As you no doubt realize, it was necessary to have a general anesthetic in this case for the sake of speed. An epidural takes much longer to set up and sometimes doesn't become effective the first time it is administered—which then means needing to try again.]

I was extremely poorly after the cesarean because I did end up having a hemorrhage, as well as some sort of reaction to the anesthetic. My daughter spent a short time on the special baby ward before being whisked into the Neonatal Unit as she was having breathing difficulties. Mother and baby are now both fine—and the 'baby' is now 3 years old!

In general, I would definitely be on the pro 'natural birth' side of things, wherever possible. My pregnancy was clinical and from the outset not only was it a traumatic experience, I was unable to hold my baby for nearly two weeks after she was born. I was not aware I had had a baby until four hours after the birth.

Cesareans as a birth option are ridiculous. Mothers 'opting' to have an elective cesarean when it is not a necessity are totally nuts! I did not recover properly until four months after the birth. It is major abdominal surgery.

## Birthframe 9: Tina C from the UK—Hyperemesis

In this case, the problem was ongoing nausea and vomiting in pregnancy.

You've probably heard of the drug 'thalidomide', which eventually became known as a teratogen (i.e. harmful to the growing baby). It was originally prescribed to pregnant women in the late 1950s to treat morning sickness, but it eventually became clear that it caused gross deformities when taken within a certain period during the first trimester.

Health care professionals are now more careful about any interventions or prescriptions during the first trimester because the after-effects in pregnancy or during birth are still unknown in most cases, or unconfirmed. However, experimental intervention does sometimes seem necessary or desirable in certain cases, such as the one described in this case...

It was December 2001 and I was delighted to find out that I was expecting our second child, if a little apprehensive following a miscarriage earlier in the year. With our first child I had started feeling sick at about Week 8. This sickness had then rapidly intensified until I was vomiting up to six times a day.

A sympathetic doctor diagnosed hyperemesis and allowed me time off from my full-time teaching job. After three months of this sickness level and some anti-emetic drugs, I gradually felt better and the pregnancy continued uneventfully until the birth of our healthy first son in the May of 1999.

The memories of this were still fresh two and a half years later when we decided to try for our second child. I had not really researched the condition of hyperemesis and naively believed the many comments I had heard about every pregnancy being different and the unlikely event of this level of sickness returning.

I had heard about every pregnancy being different...

In the fifth week of this pregnancy I was understandably anxious when I started to bleed again. After two ultrasounds, however, the hospital were able to reassure me as far as they could and told me to come back in Week 8 for a further ultrasound to see if they could detect a heartbeat.

*I was understandably anxious when I started to bleed again*

Then, on the Saturday commencing my sixth week of pregnancy I woke up feeling weak, battered and extremely sick. The vomiting started on the Sunday, and by the Monday—which was Christmas Eve—I could hardly get off the couch and was wondering how I was going to manage to get through the celebrations of Christmas Day. At first I felt mad with myself and kept repeating that it was only morning sickness, even if it did last all day. But by the following week I was having difficulty keeping even sips of water down and my New Year's was spent lying dehydrated on the couch until my husband arrived home smelling of wine, which immediately sent me rushing for the toilet again. I could not tolerate the smell of anything and started to sleep on my own. Even the smell of the sheets on my bed would make me retch uncontrollably.

Two days later and despite having a hospital appointment for an ultrasound the following day I felt I could not wait and went to see the doctor. By this stage my husband had to practically carry me in and out of the car. The doctor asked me to give her a urine sample but, as I explained, I couldn't because I had not really had anything properly to drink for days. She took some blood and explained that although it was unlikely, if the results were abnormal, she would contact me the following day. She did contact me the following day and left a message on the answering machine. I never heard it as I had already been admitted to the hospital.

*I never heard her message as I'd already been admitted...*

My hospital appointment on the next day appeared to go well. I had been having some bleeding and I was reassured to see a healthy heartbeat. After the ultrasound, my husband and I were unclear about whether we should wait again to see a doctor or go straight home. I wanted to go home because of the terrible way that I was feeling, but we waited anyway to have the results of the ultrasound confirmed by a doctor. I was sick again while we were waiting and heard a nurse say, "What's the matter with her?" "Oh, just morning sickness", came the reply.

The doctor confirmed the pleasing results of the ultrasound but said that it was obvious that I was suffering. She explained that high levels of ketones had been found in my urine, that I would need to come into the hospital to have some anti-emetic drugs and an IV, and that I would feel a lot better after about 24 hours. I had not known about the existence of ketones, or that it was possible to be hospitalized for hyperemesis. I also felt frightened as, apart from the birth of my son, I had been lucky enough not to be hospitalized before. Little did I know then this was to be the first of seven hospitalizations and the beginning of six really difficult months for our small family.

The attitudes of the nurses in the hospital varied. Some were sympathetic, others appeared to think that I was wasting a valuable bed—despite repeated vomiting and fainting every morning when I was forced to get up to allow the bed to be changed. (I later found out that low blood pressure is one of the symptoms of this condition.) There were other difficult aspects to being in the hospital with this condition—for example, sometimes I was placed in wards with women suffering from miscarriages, and I was forced to smell the food at mealtimes.

There were other side effects of the illness that I did not know, or that nobody explained to me. As I previously mentioned, my blood pressure was very low and I passed out daily, which was both unpleasant and frightening. After a few months of the sickness, apart from the weight loss, my skin started to look yellow, I was constantly cold and shivering, my hair started to fall out and I started to vomit small amounts of bile and blood. The latter, I was reassured, was normal and simply caused by small tears in the stomach. It was intensely painful and started to create a fear of being sick any more. I was constantly cajoled in the hospital to drink more if I wanted to get better, when every sip made my sore stomach retch. I found the cajoling patronizing and felt that it showed a lack of understanding, with the implication that I was not helping myself. I was also given a variety of anti-emetic drugs which did not work for me, and this was echoed by other women that I talked to with this condition. The pattern followed that after being on an IV for two or three days, the ketones would be gone, I would be sent home feeling only marginally better (because of not being dehydrated anymore) and within 24 hours I would be back in the same state again. After two or three hospital visits, I was given some ketone sticks with which I could measure the levels of ketones myself at home. For the first four months of the pregnancy I had constant ketones and would only go back to the hospital when I really felt that I could not stand it at home any longer.

Being in the hospital did provide the invaluable opportunity to chat with other women suffering from this condition, and we talked about how even the smell of our children and husbands would make us retch. I also discovered some Internet sites where other women who had similarly suffered had written their stories and I found this really supportive in the times when I thought I must be going crazy.

*Being in the hospital did provide the invaluable opportunity to chat to other women suffering from this condition*

My son, who was 2½ at the time, appeared to take it all in his stride—that his mommy had simply stopped looking after him and spent the whole time sleeping and chucking up. My husband and I had a bedroom in the attic and when he left for work I would be too weak to get out of bed and climb down the ladder, so I simply stayed in bed the whole day, vomiting and retching into a bucket. He would arrive home and I would attempt to get up, but even this effort would send me into more retching and vomiting fits. Without his unceasing sympathetic support I would not have got through this pregnancy.

We moved house when I was 12 weeks' pregnant as the sale had already gone so far along. Three days later, I was back in the hospital again, dehydrated. My husband had a wonderful boss who was supportive whenever he took time off to take me to the hospital or to take care of our small son on the days when the kindergarten had no space. Before the pregnancy, I had been studying and working part-time as a supply teacher. I suspended the study at 20 weeks and, of course, received no pay for the supply teaching as I was only casually employed. That was the financial effect of the hyperemesis.

At about four months, in desperation I started to research the condition and I found one particular study linking the use of antibiotics to curing hyperemesis, so I went to persuade my doctor to give this a try. This was my own doctor, who I had not seen since the start of the pregnancy. She was shocked at my condition and despite having no ketones in my urine, she sent me in an ambulance straight to the hospital. There, after breaking down in tears, they readmitted me and I persuaded a sympathetic doctor to try out the antibiotic theory. He did, but it didn't work and the pain of taking the antibiotics was difficult with such a sore stomach. He suggested, however, that there was one last solution: a high dose of steroids with about a 1 in 5 chance that they would affect the adrenal glands of the baby. I was desperate so immediately agreed to take them.

After about three days, the effect was remarkable. The constant nausea and retching had gone and I was finally able to slowly start living a more normal life again. The high dosage made me agitated and unable to sit still, but this seemed a small price to pay. I was ecstatic and could not stop moving about and eating, after months of enforced starvation. A follow-up appointment a week later with the obstetrician led to the immediate question as to who had given me these and why. The obstetrician explained that I had to come off them as soon as possible, which has to be done gradually with steroids. This took about five weeks and as the dose started to diminish I could feel the nauseous sensation starting to return and then gradually I started to actually be sick again. After coming off them completely, I ended up in the hospital again, but this time for the last time. After about 25 weeks, the sickness was no longer so severe that I needed to be hospitalized, but it did continue 5-6 times daily until the last 6 weeks, when it was just once or twice a day.

The controversy of the steroid treatment led to a close monitoring of the rest of the pregnancy, with numerous growth ultrasounds and blood tests, etc. The heart of this controversy appeared to be that this was not a conventional treatment for this condition because of the risk to the baby's adrenal glands. I did feel that the steroids relieved the symptoms of hyperemesis, but also that the hyperemesis probably reduced in severity on its own by about seven months.

At almost two weeks' overdue and following an induced birth, I had our second whopping 9lb 8oz son. He is 11 weeks old now and completely gorgeous. When I hold him, I cannot believe how placid and gentle he is, and how lucky I am to have him.

I cannot believe how placid and gentle he is and how lucky I am

## Birthframe 10: Jo Siebert—Obstetric cholestasis with twins

Here, we have a clear case of obstetric knowledge being used to ensure an optimal result. Although the woman giving birth used a TENS machine for pain relief, she did manage to avoid all other drugs and interventions while she was in labor. Clearly, both she and her care providers realized the advantages—from the babies' point of view—of minimizing interference in the natural, physiological processes.

I found out I was expecting twins when I was only 10 weeks pregnant and was immediately put under the excellent care of Lawrence Impey and his team at the Feto-Maternal Medicine Unit at the John Radcliffe Hospital in Oxford, England. Due to the twins being identical—sharing a placenta and monochorionic—I was immediately informed of the complications that could occur, particularly of the risk of twin-to-twin transfusion syndrome. I was told I would be monitored closely and have regular ultrasounds.

Throughout my pregnancy I was healthy and reasonably comfortable until at 32 weeks I developed obstetric cholestasis. I was given medication to control this [ursodeoxycholic acid] and my pregnancy continued as normal. At 36 weeks I went into labor and I gave birth to healthy twin boys, weighing 4lb 6oz and 4lb 8oz, with no cesarean, no epidural and only the use of a TENS machine for pain relief.

I was very lucky to receive such outstanding care at the John Radcliffe Hospital. I believe that it was the positive attitude of the medical staff towards vaginal twin births and natural pain relief that enabled me to have such a wonderful natural birth experience.

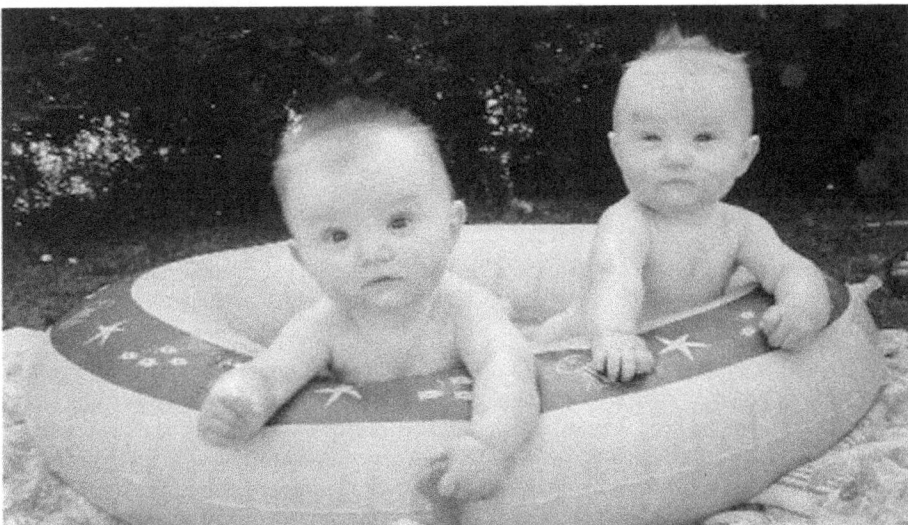

*Frederick John and Robert Joseph, Jo Siebert's twin boys*

## Birthframe 11: Jill Furmanovsky—Lack of fetal descent

Here's an account where medical support became necessary during labor. As you will see, the patience and sensitivity of her caregivers optimized this woman's experience, leaving her feeling very positive after the birth. Nevertheless, Jill makes a postpartum evaluation... Do you agree with her comments? If you do, how could protocols be changed (in your area) so as to make outcomes even better for women in a similar situation?

I'm a professional photographer and in July 1986 I met Michel Odent when I took a portrait of him for London-based *Time Out* magazine, with his son Pascal. He was a very influential figure in the mid 1980s. I went on to work with him on a few projects, including shooting pictures of two home births he attended but, in between, I was to have my own child, though as it happened, not with him in attendance.

After two miscarriages in my early 30s I was thrilled to finally become a mother at the age of 34. During my pregnancy I was very much aware of Michel Odent's pioneering work. I had read his books, seen the groundbreaking documentary on his work in Pithiviers and heard him give a talk to a gathering of people in London with another pioneer of natural childbirth, Janet Balaskas. I was fascinated by what he was saying, and wanted, when my turn came, to have the sort of birth he described—a birth in privacy and, if possible, without medical intervention.

I chose a local family doctor, Dr Smith, now retired, who was prepared to attend and supervise home births and for the next nine months was in the hands of a series of great homebirth midwives who I got to know well. They were a very cheerful lot and I loved seeing them. The fact that I was choosing to have a home birth made my actual pregnancy feel a lot more special, and even though I ended up giving birth in a hospital I am sure that the planning of a home birth made the experience of being pregnant, as well as the actual birth and recovery, much more straightforward than it would have been if I'd opted for a hospital birth from the start.

My baby was due around Christmas—any time from the December 23 onwards. Here's a transcript of the journal I wrote at the time...

**8:15AM December 25, i.e. Christmas morning!**

I've been having contractions since about 3:15AM. At 6:15AM Ron (husband) phoned Frances (one of the home birth midwifery team) and told her he'd taken the chocolate cake out of the fridge—a signal that labor had begun. She groaned but was very sweet about it and said she would come over in a couple of hours. He also phoned my mom so that she would be with me in thought.

Now it's 8:15AM and I've just had another contraction, this time in the kitchen. I wiggled my hips hula-hoop style to meet the pain. Ron has gone back to bed to get a bit of sleep... I hope he does. I'm quite happy to walk about in the kitchen. My bowels want to empty and I just had a little shit. In between the contractions I feel perfectly okay. It's almost as though I dream them. However the feeling is real enough when it comes—a gripping period-type pain that goes from the front of the vagina through to the small of the back. I try to visualize the cervix opening up.

**2:00PM, December 25**

Frances came, but it was a false alarm. The pains became less and less after that last entry and then faded away. In fact, I just ate Christmas dinner! Frances said it was the baby's head trying to move down the pelvis although it hasn't done so, so heaven knows what it actually was... a dress rehearsal perhaps. Useful though. It showed me that it was going to be painful, that is PAINFUL. I managed okay with that breathing technique I learned in prenatal classes and Ron was delightful, so here's to January 1, recommended by Frances as a good day to produce!

**December 29**

How can I begin to describe the birth? Leah Anne was born at 4:59AM on December 27 1987, a Sunday. There's the announcement, for the record, and a happier mom would be hard to find. But I want to try and recall the birth because it was the most extraordinary physical experience of my life. It was a journey through untold pain, discomfort, fear and loss of control. It was dignity for me in the face of adversity—something to be proud of for the rest of my life, something that took my spirit to another place entirely. It was also an experience of love of the highest quality from the people who helped to deliver my beautiful daughter.

After that last entry on December 25, I rested and tried to catch up on my missed sleep. I did. Later that evening, still December 25, at around 8PM, the contractions started again. This time Ron and I were less freaked out and decided to wait before calling anyone out. Michelle (a friend) came round to visit at around 11PM and I chatted to her in between contractions. When they came—every seven minutes or so and lasting about 30 seconds—I had to concentrate on my breathing. Michelle left at around midnight and Ron went to bed. About this time I had a 'show' and phoned Frances, to let her know. She said to call her again when the contractions were every five minutes for a minute each. That didn't happen.

I stayed on my own downstairs and tried to sleep in between contractions—sometimes they were 15 minutes apart but never as regular as Frances' description. At 3:00AM I'd had enough of being on my own and I woke Ron. He brought a quilt downstairs and slept on the floor while I tried to relax on the couch.

It wasn't until daylight that we called the home birth midwifery service. Frances was now off duty but her replacement, a delightful Nigerian lady named Mary, with a warm and generous presence, came over pretty promptly. She told us it could be a long labor because the head hadn't dropped much and I was barely dilated, but she said she would open me up further (by hand) to get things moving a bit. She spent the rest of the day going backwards and forwards to our house while she did other errands in the area, checking on me every hour and half or so. Meanwhile, I was in quite a lot of pain but coping well with Ron's help, and trying to stay calm. We went for a short walk outside, stopping for each contraction. Not a lot was happening but I was gradually dilating. At around 4PM Mary gave me an enema, which helped clear out my bowels and I felt a lot better inside, although the pain got worse.

My midwife had a warm and generous presence

At 4:30PM I moved to the bed and my water bag suddenly popped. More contractions followed, some very painful, but still the baby wasn't really engaged. Dr Smith (the home birth doctor) was called. Then a long period followed where Mary or Ron held my hand while I sat propped up on cushions doing deep breathing, concentrating on each contraction. Some contractions lasted nearly two minutes. Dr Smith arrived. Then things got surreal. I remember Mary going home for about 45 minutes and Ron taking me for a pee. In midstream I got a contraction so powerful I screamed. I hung round his neck while Dr Smith rubbed my back. I was close to hysterics. Mary returned. I still wasn't fully dilated but I was close—apparently the baby was higher up than expected. Dr Smith's partner, Steve, a trainee home birth doctor, arrived. Now there were four people in the room. Ron walked me up and down, not sure if I asked him to or whether the doctors had suggested it. I was naked. During the contractions I was in agony. I remember groaning and Dr Smith saying 'Oh Jilly' in a very sympathetic way. At the request of the doctors I started pushing at some point, although I felt it wasn't right to push. A further examination revealed that the baby was stuck at nine and half centimetres. They tried to prize the head through but it didn't work. I'd been having the shakes for about eight hours now and I was exhausted. Suddenly Dr Smith said we had to go to the hospital. I was bitterly disappointed but just wanted to be free of the intense pain. Ron drove Mary and I and the others followed. It was a very painful journey but I remember walking into the maternity unit myself.

We went straight to the maternity suite. A nice hospital doctor said it might have to be forceps or even a cesarean but he would give me two hours with an epidural, so I could have a rest and see what happened. That epidural was a tremendous relief and I did manage to rest or even sleep for a short time. As the epidural wore off I gradually felt the baby move down the birth canal because of pressure on my anus. I was now in the hands of a hospital midwife called Anne, who was also very kind. When she examined me, she told me she could see the baby's head and directed my hand so I could feel it too—a beautiful moment that renewed my energy. Even so, it took me another two hours to push my daughter out, with Ron directing in a plastic apron. It took the last of my strength but in the end it was a normal delivery and I didn't even need sutures.

Holding my baby Leah for the first time, seeing her calm, wise eyes, checking her 10 perfect fingers and toes, knowing she was a healthy weight—6lb 2ozs— I felt such joy!

In the years that followed Leah's birth I attended a few of Michel Odent's home births as a photographer and learned a few things that would have helped with my own birth, had I known them at the time. Firstly, Michel would never have allowed so many people in the birth room. In particular he would not have allowed a trainee doctor to attend a first-time mother's home birth. I remember him saying to me that, ideally, trainee midwives and doctors should only be allowed to observe women having their third or fourth child. There is a sound physiological reason behind this protection of the new mother's privacy: it is so that the primitive instincts needed for an optimal birth—to crouch or howl or do whatever—are given a chance to arise. This is more likely to happen if the mother is unobserved. Had he attended my birth, I cannot imagine he would have told me to push. The last stage of labor is an involuntary process that does not require instruction, only patience.

The birth attendants (or fathers of the child) in the cases I saw were encouraged to sit in the kitchen unless the mother specifically wanted them to be with her. Michel himself dozed outside the door, ignoring most of the moans and groans from within. Occasionally, he checked on the woman with a reassuring, hand-held fetoscope.

It was obvious, even to me, that there was a distinct change in the sounds when it was time for the woman to push. At that point Michel quickly but calmly entered the room with the birth friend or husband and instructed that person to support the woman, sometimes under the arms, while he caught the baby. Those were the brief moments I photographed without the mother even being aware of my presence.

After the birth the woman was gently lowered down and handed the warm and gooey baby still attached to the cord and placenta. She was then left alone in peace for a few minutes to bond with her baby. The ecstatic new mother inevitably put the baby to her breast, thereby stimulating the correct hormones to expel the placenta, which occured naturally in both the cases I saw.

My labor was truly hard and I ended up needing hospital facilities, for which I was very grateful, after more than 24 hours. However, the whole experience of being a homebirth mom-to-be was wonderful. Since I had only one dose of an epidural and no sutures I recovered quickly and was home within 12 hours of giving birth.

Leah is now a stunning 22-year-old, but having transcribed that account from my original notes, more or less verbatim, I feel once again as though I have just given birth—one of the greatest normal experiences on earth, if one has the courage to let nature do its own thing.

*Jill with her daughter, when she was 13 months old.* Photo © Matt Anker

## Birthframe 12: Liliana Lammers—In-labor cesarean

Here's an account where medical support became necessary even when the mother and all those around her were hoping for a completely physiological birth. Perhaps it's never possible to know the real reasons for undesired events in our lives. I find it interesting that there is a clue in the following account for what follows even in the first line—in the phrase 'all being well'—the feelings of the mother's *mother* are intriguing...

My daughter had planned, all being well, to give birth at home.

The very strange thing was that months in advance I couldn't see it happening. I never said a word to anyone, but on a few occasions when it was quiet, I sat down and wondered... Why? What could happen? Was my daughter to change her mind? Would something go very wrong? Impossible to tell, I just could not see her home birth a reality. But don't think I was full of fear or negative thoughts, it was more like a fact, a reality that I could not comprehend.

Three weeks before her due date she went into labor. Michel Odent came at 10PM and we all went to rest. At 2AM Marisa woke me up, she was fine, she wanted to be by herself, but thought Michel had better know she was still having contractions. Michel, after listening to the baby, asked if he could do a vaginal exam. Marisa had no problems with that. I wondered why, but I didn't ask questions. No words, but Michel looked different, a bit tense. Marisa decided to go back to her room, where her boyfriend was asleep.

At 5AM I saw Michel was nervous. I questioned him and his reply was: "Footling breech. I can't wait any longer." I had no time to think, he was pushing us into a cab. Marisa said she was having contractions every 20 minutes. She looked lovely. What to do? Michel could not take it anymore. You probably know, Sylvie, that for footling breech the policy is elective cesarean. For a breech birth Michel will only accept a home birth if it is quick, between four and five hours. They can be very quick. Some of these babies are born on the way to the hospital. The worst thing to do is touch them when you see the lower part of the body appearing—any stimulation can get them 'stuck', so there are many deaths or brain damaged babies. I suppose you know all this. But quick undisturbed, undiagnosed breech births can go very well.

So I talked to Marisa and she said, "That's fine. Let's go". Baby Ryo was born two hours later by cesarean. To me it was a shock. First time in my family.

Marisa, with her 18 years, took it quite well. She is fine, the baby is adorable, and she had plenty of milk from Day 1. They all sleep together. Even in the hospital after a C-section, they were happy for mom and baby to be together all the time.

We will never know what could have happened if we had stayed at home. One of the risks is cord prolapse. With their feet they kick so much that if the membranes rupture, cord prolapse can follow (although not necessarily, of course).

Sometimes, it is a problem to know all this.

*Liliana with her grandson*

My daughter had planned, all being well, to give birth at home...
The very strange thing was, I couldn't see it happening.

## Grey areas

In some cases it's not easy for professionals to decide on the best course of action. Different midwives have different approaches and obstetricians may disagree with them. The mom-to-be may also have her own opinion, deduced not from years of training and experience, but from an evening spent researching the Internet. Other women may be better informed, having researched their own situation and medical condition thoroughly over a course of weeks and months, their concentration having been made sharper by their very personal need to 'get things right'. What can be done?

Many professionals are loathe to challenge the protocols within their institution. Others stick their necks out... but why? What could their motivation be? Why do some midwives go so far as to endanger their jobs, their bank balances and even their freedom (in the case of midwives in some states who go to prison for the sake of their principles)? Why do some midwives support women who seem to be making outrageous requests, which quite frankly come across as demands?

The answer is twofold. Firstly, research is on their side. Isn't it odd that so many protocols fly in the face of research, especially since much of the research was published in respectable medical journals quite a while ago? For example, the protocols in some units stipulate that women need to have EFM for 20 minutes on admission, despite the well-known fact that EFM has not been associated with any improvement in outcomes; the only confirmed outcome is an increase in the cesarean rate, which is not seen as being a good thing.[43] Secondly, women have the right to make decisions about their own care and the care of their babies. Their only problem is to find professionals (especially midwives) who will agree to support them. After all, most women who make seemingly bold choices as regards their care do so after extensive research and do not want to give birth unassisted. They *want* the support and expertise of midwives who will act in their interests with love and sensitivity.

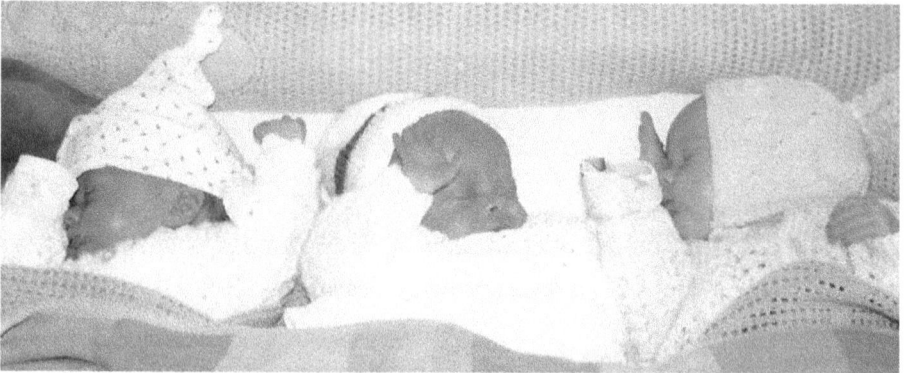

*Triplet pregnancies may seem to some an obvious indication for a cesarean, but some mothers disagree. These babies, Kate, Sophie and Louisa (7 weeks old here) were born vaginally in London in 1999 with an obstetrician and midwife in attendance*

## Birthframe 13: Anonymous—Vaginal twin birth

I am a 40-year-old woman who gave birth to twins naturally five months ago in a government hospital. It is possible!

I did it with a lot of persistence and research, after several meetings with various midwives, obstetricians, and hospital managers, and by finding independent midwives and an obstetrician who supported me in trying for natural deliveries for our babies. With some luck, a very supportive partner and midwives, and an obstetrician who went beyond the call of duty by being available for me in the labor ward (though by my request, not in the delivery room) when he was not normally on call, I gave birth to the first twin in a birth pool, and the second on land, without any drugs or intervention at 39 weeks and 2 days. Both babies were well and weighed in at just over and under 6lb. I would like to encourage other expectant twin moms or those with a high risk pregnancy to inform themselves of what the risks associated with their pregnancy are, what the birth options are with their various pros and cons and to persist in trying to lay the conditions and plans for a natural delivery, if that is what they want. Giving birth to our babies is one of the most satisfying, joyous, proud experiences I have had.

Here's another case of a woman avoiding intervention, while still ensuring she received care to help ensure safe outcomes.

## Birthframe 14: Debbie Brindley—Midwife's vaginal twin birth

I first read about the British midwife Debbie Brindley in *The Complete Book of Pregnancy* (NCT Publishing 2000). Thanks to careful preparation, she managed to arrange a very natural water birth for the birth of her twins. (For her first baby, she'd had a water birth at home.) When I wrote her to ask a few questions, I discovered that even during her pregnancy, Debbie really did care about the care she received...

### What's your view on the use of ultrasound in pregnancy?

Our unit policy is for regular serial ultrasounds. However, it only appears to be safe, doesn't it? Research is not conclusive so I feel it should be used as with all medical intervention and technology—when it's needed and not routinely. Following this philosophy, I chose to have two ultrasounds: an anomaly ultrasound at 20 weeks and one further ultrasound at 34 weeks to check presentation (they were both cephalic)—so I could then plan how to manage my labor. Growth was diagnosed clinically on palpation.

### But the midwives who looked after you during pregnancy used a Doppler to check the fetal heartbeats at each appointment?

Carole, my midwife, did all my prenatal care (at home—I'm so lucky) and she agreed to listen to the fetal hearts with a fetoscope instead of a Doppler. They always had good movements so we were both happy with this.

### And how were the fetal heartbeats checked while you were in labor?

I was happy to have the heart rates checked with the Doppler, as the fetoscope's awkward in labor because the laboring woman may have to lie down for it to be used effectively. I had a 10-minute CTG [cardiotocograph], i.e. electronic fetal monitoring—the student midwife held the monitor on—while the pool was running. Beforehand I had agreed to a short CTG of the second twin when the first had been born but in the event, it turned out there was no time!

*I was happy to have the heart rates checked with the Doppler during labor because the fetoscope's awkward in labor*

### Did you make any special requests of your birth attendants for your labor. I mean, did you want them to behave in any particular way?

I wanted as much privacy as I would have had if I'd been at home and I didn't want people 'dropping in' or coming for a look. (Not that any of my colleagues would dare!) I believe in needing a quiet, relaxing environment so as to raise endorphins and lower adrenaline. I agreed to Carole's student witnessing, though, as Carole had been so kind to me. Otherwise, I wanted midwives only. I wrote a detailed birth plan and then made an appointment with my extremely supportive obstetrician to discuss it. (I met him two times in all, once early on in my pregnancy and then at 34 weeks in order to discuss my birth plan.) He agreed with most of it, we discussed one or two points regarding monitoring and physiological third stage and then he signed it. In the birth plan, I requested privacy and he and the pediatrician kindly agreed to wait outside in the corridor during the birth. My obstetrician was there in case I needed any help—even though he wasn't officially on duty. He's a very special man.

*I requested privacy and they kindly agreed to wait outside*

### What was the room like where you gave birth in hospital? Did you make any requests about this in advance?

It is common to deliver twins in a 'theater' delivery room—large and clinical, usually with an IV, epidural and in the lithotomy position [lying on your back]. I definitely didn't want this, so asked to use the small standard delivery room connected to the pool room. The room was reasonably but not fantastically comfortable, but at 9cm dilated on admission I didn't care. It was night, so it was dark and warm and it was down the quiet end of the corridor.

### Under what circumstances would you have accepted intervention?

I trusted my midwife implicitly and if she had voiced any concerns I would have accepted any interventions necessary. I strongly believe in the importance of a positive birth experience but my babies' welfare was always more important. I wouldn't have accepted any interventions unless I had thought they were for the direct benefit of my twins. I wouldn't have accepted a routine intervention.

*I trusted my midwife implicitly and if she had voiced any concerns I would have accepted any interventions necessary*

**Presumably, you had syntometrine for the third stage—this is pretty much routine in the UK, isn't it?**

No, I didn't have syntometrine for my third stage. I'd had a physiological third stage with my first baby and had wanted to keep this labor natural too. I discussed this with my midwife and we agreed to have something ready in case I needed it, in which case I would have happily accepted it. Actually, this was pitocin, rather than syntometrine because it has fewer side effects because it includes no ergometrine. Syntometrine is a combination of ergometrine (which is rapid-acting) and pitocin—i.e. synthetic oxytocin (which has sustained action). Ergometrine's side effects can be raised blood pressure and vomiting, dizziness, headaches... so pitocin (synthetic oxytocin) alone can help avoid these.

Anyway, I felt if I had strong labor contractions then the chances are that my uterus would effectively expel the placenta. And I had no complications. The placenta was expelled within a few minutes and blood loss was minimal. Of course, having a physiological third stage means that the cord can be cut later and I think the transition to extra-uterine life is more gentle with delayed cord-cutting. Cutting the cord later, babies receive their intended extra bonus of blood.

*Here are some other twins who were born entirely naturally, which the parents believed was optimal. Actually, in this case it was an undiagnosed twin delivery at home! The next birthframe gives you more information about how this all came about and provides an interesting insight into differing viewpoints in different countries ...*

## Birthframe 15: Steve and Olga Mellor—Surprise twin birth

Here Steve Mellor, an American, explains how he and his Russian wife un-expectedly came to give birth to twins at home. While you're reading consider...
- Does the Mellor's experience affect your view of prenatal education at all?
- In the light of this viewpoint, to what extent do women have a choice when they are offered drugs during labor, even though they opt out in advance?
- How would you have felt if you had been the caregiver for these clients?

Toward the end of April I was out shopping for some new shoes when my cell phone buzzed. It was Olga (my Russian fiancée) calling with a big surprise. It seems our last visit together had been more productive than either of us thought. Yes, she was pregnant and we were going to have our first baby. So within one year I went from being single to finding my life partner and expecting my first child.

Because of the visa processing time I was not around for the first few months of the pregnancy as Olga was still in Moscow. It was hard for both of us as we did not get to celebrate like most couples, but what we did do was start the process of preparing for our child coming into the world. This is where it got a little interesting.

It was by no means anything like the Cold War of our two nations but, because of our different life experiences, our ideas about how to have a baby differed. Myself, having been born and raised in America, knew exactly how to have a baby: you get pregnant, wait nine months and then go to the hospital where the doctor can give drugs to get through the pain and then he delivers it.

Olga had a completely different plan. Having been raised in Russia and not always having faith in the medical system, she didn't want to have a hospital birth. She wanted to have a home birth, as did many of her friends, with no medication, as she understood that any drug she took, the baby would receive 10 times as strong.[4]

Now I had neighbors years earlier who had had a home birth and at the time I had thought, "How interesting", but I had never thought to myself, "Yes, that is what I want." So Olga wanted to have a home birth and I knew nothing about it. One of my good friends, who was a chiropractor and who had studied a lot about natural ways to heal the body, gave me a book on natural childbirth. So I read the book cover to cover and learned all I could. The book had me questioning a lot of the traditional things I had heard about giving birth or had seen in movies or on TV, when they were showing a woman having a baby. At the same time Olga was attending a series of classes in Moscow, where the main idea was to do everything as naturally as possible, without any interference.

Olga's classes were great as she was not only learning how to deliver, but also talking about birth on a physical, emotional and spiritual level—something I think is missing in the typical American system. So as she would tell me about the classes I would think to myself that some of the stuff she was learning was crazy. But then I would do some more research and find out for myself that it wasn't really. I remember our first conversation about having a home birth. I was scared of the possibility of things going wrong.

*Olga with her two little girls*

And then I would find statistics stating that most C-sections were done on Fridays—wouldn't want to ruin the doctor's weekend!—and that women were actually scheduling their C-sections before they even gave a regular birth a try. It started to appear to me that the medical system seemed just as crazy.

There were many other questions that I found myself asking. Olga didn't want to have any ultrasounds and that again concerned me. Then I started to ask around to find out the purpose of doing them. No one could really give me a good reason. I heard things like: "To know the baby's sex", "To help predict the due date" or "To find out if the baby is healthy".[5]

But then I would also find out that even though they could find some things out, in most cases there was nothing that could be done. I further learned that there have been cases of couples who, learning about some defect, chose to abort the pregnancy, only to find out afterwards that the baby was perfectly healthy. And I loved the way Olga always told people that if she were meant to know the sex of the baby or know how it was going, God would have installed a window on her belly. Also, I read in one book that 94% of women could deliver naturally without any problem. So I decided to trust the process and be responsible for creating an environment where everything would live up to the vision I had created.

Finally Olga's visa was approved and we got married in a small ceremony in Oregon. We didn't have time to arrange a big ceremony as we were given 90 days to get married from the day she arrived, and we had no idea when the 90 days would start.

One night, when we were out shopping there just happened to be a farmers' market going on. Right outside the shop we were in was a woman advertising a home birth collective, run by a group of midwives. We got a lot of information that night and ended up contacting them later in the week to interview our potential midwives. Ellen and Kenna came over the next week and we talked through what we wanted to create in having our baby and what they could do for us. It really helped me more than anything to see that they had lots of experience and training and could handle the most common problems, should any occur. So Ellen would be our primary midwife and Kenna would be her back-up. They would come by the house every other week to do check-ups. Also, at that time Ellen was training Julie in midwifery, which would later prove to be a blessing.

For the next three and a half months we went through the basics of daily living. I worked, and Olga worked hard at being pregnant. And I mean this in the most positive way. Every day she took the time to walk a mile to the gym and swim for an hour in the pool. She took the time to eat really healthy food that would provide our child with a great source of natural nutrition. Olga took this time because she didn't want to take man-made, prenatal vitamins, which to her were not as good a source of nutrition as real healthy food could provide. She wanted our child to have only food that was from nature and full of life.

Every day Olga walked a mile to the gym and swam for an hour

We took Bradley natural birth classes and this became one of my biggest learning experiences. We were in a class with five other couples who wanted to have their babies born naturally, but they were having hospital births. It was amazing for Olga and me to experience this class, where we spent half of each meeting learning how to defend ourselves from a hospital and their possible interventions when not needed or wanted. I remember in one class Olga was actually in tears hearing about all the problems with hospitals and we were not even going to one. It was an eye-opener to hear how they would tempt women to have some drugs regardless of what their wishes were. I was never more thankful for Olga's strength and determination to have our child at home in a loving environment.

*We spent half the classes learning to defend ourselves*

Then came the big day. We were actually about two weeks from the projected due date, sitting in a movie theater, watching *Star Wars I*. The big battle at the end was just starting when Olga got this very funny look on her face. I'm sure I would look like that too if my water had just broken and my pants were all wet. Well, needless to say, Olga never saw the end of the movie and wouldn't do so for a couple more years. We headed home, just a couple of blocks away, and called the midwives...

Olga gave birth the next day to her first baby in a birthing pool, supported by her midwives and husband. Back to Steve for the rest of the story...

It was 3:40AM in the morning, 29 hours from when Olga's water first broke and Olga gave one more push and out came our beautiful baby girl right into my hands. We slowly brought her out of the water of the birthing pool and onto Olga's chest so mother and baby could start the bonding process.

*It was 3:40AM in the morning, 29 hours from when Olga's water first broke and Olga gave one more push...*

It was great to have this time. No one carried her away, cleaned her up, probed or weighed her. We all just held each other and experienced the miracle we had just been given. I remember at one point about 13 minutes into working on the placenta Olga jumped and asked Ellen if she had long fingernails. She said, "No," and we went on. Then about a minute later Ellen was checking the progress of the placenta when she uttered those now famous words to us: "This is too hard for a placenta. Oh my god, you're having another baby!" And just as those words were leaving Ellen's lips Olga gave a push and out came another baby girl. Everyone in that room was completely amazed, and moved in quickly to deal with the extra child. For nine months no one knew there were two babies. I have to say the excitement in the room was pretty wonderful. Truly, we had a small miracle on our hands.

*Everyone in that room was completely amazed*

After some time, Olga delivered the placenta and we eventually brought the girls out of the birthing pool. Having Julie as an apprentice was again just perfect, as we needed three people to take the girls and the placenta out of the pool, while Ellen attended to Olga. We also had to remember which child came out first.

We had happened to wrap up the first girl in a yellow blanket and the second one in a blue one, so we created the rhyme 'blue two' which to this day still holds true. Olga had originally wanted to keep the placenta attached for 24 hours. This was to allow all the blood that is pushed into the placenta during the birth process to return to the child and also because of an old Russian spiritual tradition that recognizes the placenta's importance in the development of the child. Well, with having just one placenta and two babies connected to it, this was not a good option. But we managed to go a couple of hours before I cut the cords. We then wrapped them up in blankets and brought Olga and the girls into the bedroom and put them in our bed for a well-deserved rest. Also, during this time the midwives took the time to get all the statistics like weight, height, and do all the other checks they do on newborn babies.

We spent the morning talking about the shock and beauty of what had just occurred, about the signs that had been there but that no one had picked up on, and for good reason. Ellen had, of course, never delivered twins at home and told us had she known that we were having them she would have sent us to a hospital. Knowing Olga like I do, I know she would have refused to go, as she wanted the birth experience she had and really was fearful of what would have happened to her in the hospital.

We don't think hospitals are bad places, they are just not for us as we understand they have two goals: do the best they can to help women deliver babies, and make sure they avoid lawsuits. We had our girls 29 hours after the water broke and if we had been in a hospital we would have had only 18 hours to deliver before they would have started to intervene and force the girls out.[6] In reality, we had twins born perfectly, head first, with no complications. We didn't spend the nine months worrying about all that could go wrong, we focused on how it would go just right, and it did.

> We focused on how it would go just right, and it did

A lot of people ask us how we could not have known we were having twins, and all I can say is it was a series of events that helped us to create the birth experience we wanted. Throughout the birth Olga only gained 15 pounds and was actually six pounds lighter after the birth than the day she found out she was pregnant. So she had about 21 pounds total for the pregnancy and the girls weighed $4\frac{1}{2}$ and 5 pounds. This was because she ate right, exercised and just took excellent care of herself—something most working women are not given the time to do these days. Also, Olga had really strong stomach muscles from many years of riding horses, which made the midwives' job of examining her belly very difficult. The midwives didn't ever think to look for multiple heartbeats and, since we had chosen not to have an ultrasound, the girls were able to support our goal by hiding from us all, so that they could be born at home in a beautiful, loving, peaceful and drug-free environment.

## Olga's comments...

I had a beautiful and wonderful birth experience. I was very grateful for all the workshops I attended in Russia and for the Bradley classes in the USA. I was very conscious and confident about what I was doing. I think what helped me most on this journey was the deep belief that everything would be fine. I was taking care of my body—exercising and feeding it well—and it did a great job of getting through pregnancy and giving birth with all the wisdom of nature that it has inside it. In the classes in Moscow we discussed the process of birth through the experience of a child, which makes you understand that normal, healthy childbirth is the best way to bring a child into the world.

*I think what helped me most on this journey was the deep belief that everything would be fine*

I am very grateful to Steven because he gave me a lot of love and support, which helped me to gain strength and confidence. During the birth we worked together like a great team. I am sure that creating a loving relationship in the family is very important for having a healthy, easy pregnancy and birth. It is not a rule but I believe that a happy woman with a positive attitude has a better chance to have an easy birth. Creating good relationships depends on the people themselves.

In my opinion, these are the most important things, if you want to have a great, healthy birth:

♥ Have a deep knowledge of the benefits of healthy childbirth for both the child and the mother and make a conscious choice to do it 100% healthily. It is very important to select people (especially midwives or doctors) who will support and encourage you in your choices. It will never work when the woman says: "I will go to the hospital and try to do it." There is a big chance that there will be a nurse who, after several hours of labor and pain, will say: "And now this is your last chance to get an epidural. Would you like it?" It's terribly hard to answer "No" when you're in pain.

♥ Have a deep belief in yourself that you can do it, trusting your own body with all its wisdom. Have a very positive attitude and a lot of gratitude for all the processes that are taking place within you.

♥ Take good care of your body. Have a very healthy, natural diet and do a lot of exercise (like swimming, walking, stretching).

♥ Consciously accept who you are and work to develop positive thoughts and a peaceful state of mind, and nurture a deep feeling of love for your child, yourself and the world. Work through the problems in your relationships with others (especially in your family) and within yourself.

*Have a deep knowledge of the benefits of healthy childbirth for both the child and the mother*

# Cesareans

## Optimal or not?

Despite the fashion for elective cesareans in some places, amongst certain groups of people, the cesarean certainly doesn't seem to be an optimal choice in cases where it is not medically indicated. In fact, in the light of research it's quite absurd to suggest that the modern cesarean surgery represents an ideal.[44] Instead, it is simply an extremely effective and now mostly safe rescue surgery for cases where things would otherwise go wrong.

In the light of research it's absurd to suggest a section is best

Most midwives would not disagree, but since a few obstetricians still view the cesarean overly positively in cases where a woman could otherwise give birth vaginally, it is worth considering what's 'wrong' with it. The main problem is that it represents a great departure from the natural processes and is less pleasant, less safe and less healthy for both mother and baby, unless of course the surgery really is lifesaving.[45] It's possible that being born by cesarean has long-term effects on the psychological and physical development of the baby firstly because of the enormously different hormonal environment present when a woman has this kind of major abdominal surgery, secondly because of the experience itself and thirdly because of the disruption to the normal bonding processes after the surgery.[46] The absence of oxytocin (dubbed by a researcher, Niles Newton, 'the hormone of love') is the key concern here and may interfere with the baby's development of the ability to love.[47] Michel Odent has discussed this at length in his book *The Scientification of Love* (Free Association Books 1999). Of course, efforts can be made to counteract this potential problem but the point is that a cesarean birth is not ideal for the baby, unless of course it saves his or her life, or that of his or her mother. It is also not ideal for a mother in terms of minimizing damage to her sexual anatomy firstly because damage to the pelvic floor (which includes the muscles around a woman's vagina) usually occurs during pregnancy (particularly when the woman is overweight)[48] and secondly because of the enormously important psychological aspect of arousal, response and orgasm, which may be affected by artificial birth experiences.[49]

## The facts

Research was actually summarized in 1998 in the *British Medical Journal*. According to this summary, women who have a cesarean (rather than a vaginal birth) open themselves up to all kinds of dangers.[50] In the short term, they have a far greater risk of hemorrhage, infection, ileus, pulmonary embolism and Mendelson's syndrome. Long term health problems faced by cesarean mothers apparently include the formation of adhesions, intestinal obstruction, and bladder injury.

Uterine rupture is also a possibility in a subsequent pregnancy, especially if labor is induced or augmented.[51] Research also shows that women who've had cesareans are more likely to have fertility problems, an ectopic pregnancy or a placenta previa later on.[52] It's also more likely that there will be a need for a hysterectomy, which would also have an obvious impact on fertility. There is even evidence that the health of cesarean mothers' future children may be affected by having an older cesarean-born sibling. Psychological problems, such as postpartum depression, are also more likely for cesarean mothers, especially if the surgery was carried out under general anesthetic, and these psychological problems are likely to affect the mother's ability to bond effectively with her baby (or babies).[53]

Postpartum depression is more likely for cesarean mothers

Research has also shown a significantly higher death rate for cesarean mothers, but it's impossible to determine how high the risk is.[54] This is partly because it would never be possible to carry out a randomized controlled trial and partly because the level of risk would depend on the reason for the cesarean and the woman's general level of health. Some writers and researchers claim that a woman having a cesarean is two times as likely to die as a woman giving birth vaginally, and some even say she is 16 times more likely to die. The imprecision of these estimates is obvious if you consider any other risk in life—such as the risk of crossing the road, which is affected by numerous personal and situational factors. The estimates are also absurd in a sense because without the surgery any one woman might have died anyway, or her baby might have. For a comprehensive review of research relating to the effects of cesareans, see *Pushed* (Da Capo 2007).

In conclusion, we can see that in cases where either mother or baby is in serious danger it is obviously best for this life-saving intervention to take place because it is currently the only means we have of dealing with certain obstetric emergencies, so it is indeed 'optimal' in these cases. The risk of the surgery then becomes less than the risk of the alternative (i.e. potential death of either the mother or baby). So the small minority of women who really do need to have this surgery are best advised to focus on the life-saving properties of the modern C-section than on any negative aspects and to be thankful that they can be helped in a difficult situation. If, on the other hand, any particular woman is likely to be capable of a vaginal birth, my research has certainly led me to conclude it is better for her to take this route.

In other words, the cesarean is real rescue surgery in high risk cases but for low-risk women capable of having a safe vaginal birth (even a VBAC) the evidence shows that it's far from ideal. This is not really surprising, when we think that it constitutes major surgery and a lengthy disruption to the normal mother and baby bonding processes.[55]

## THE SURGERY ITSELF

We have already considered vaginal births in great detail, describing their less pleasant aspects as well as the exhilarating moments. Therefore, the equivalent level of detail is provided below for a cesarean birth. Seeing this in print will hopefully give you a chance to reflect on what women are letting themselves in for when they commit themselves to a date for an elective cesarean, particularly in cases where they could just as easily opt for a vaginal birth. And of course, you may want to share these details with some pregnant women, if you feel this is appropriate in any particular cases.

Of course, given the difference in protocols in different institutions, not to mention changes based on new evidence, the precise procedure in the institution(s) you work in may be different.[56]

## BEFORE THE SURGERY...

The mom-to-be (or her husband) is required to sign a consent form.
- In some countries, in the case of an elective cesarean, the evening before the surgery some glycerine suppositories may be given. These will help the woman clear her rectum. The woman's pubic prep is also carried out.
- An antacid may be given so as to prevent the contents of the woman's stomach from becoming acidic.
- Drugs to prevent thrombosis are sometimes given as well as antibiotics, in case infection develops. An IV is inserted for the further administration of drugs.
- In many countries a catheter is installed so that urine can be continually drained from the bladder.
- A blood pressure cuff is placed around the woman's arm.
- Jewelry, nail polish and make-up are removed, so that small changes in coloring can be observed. Contact lenses are also removed and teeth checked so no loose ones are inadvertently swallowed during the surgery!
- The woman is asked to change into a clean surgical gown.
- The woman is asked to lie down on her left side and a wedge is placed under her right buttock.
- She is then asked to curl up into the fetal position if an epidural is to be inserted or topped up, or a general anesthetic is injected via the IV. (In the case of an epidural, she may also be allowed to sit up while this is inserted.)

## DURING THE SURGERY...

- The surgeon wears a double layer of plastic gloves and a clear plastic shield around his or her face. This is to protect him or her from exposure to the woman's bodily fluids.

- The woman is turned onto her back so that her abdomen can be easily accessed. At the same time, the operating table is tilted slightly so that the woman's head is lower than her feet. This is to help establish a level of anesthesia which will be high enough to make the woman's abdomen numb, but not so high that it will affect her ability to breathe.
- A nurse then paints the skin on the woman's abdomen with an antiseptic solution. Someone on the surgical team will then cover the abdomen with drapes and sterile plastic—leaving exposed only the area which is to be cut. The drapes are put over a bar above the woman's chest so that she cannot see the surgery itself... er, except in the reflection of the light above in some cases!
- In one clear motion, the obstetrician then makes a horizontal, crescent-shaped incision, just above the usual site of the pubic hair. In the standard technique, the incision is made deep into the skin, so as to cut through all the superficial layers. In a more recently proposed technique (developed by Michael Stark in the 1990s), which Michel considers to be optimal, the scalpel only sinks one inch into the layer of fat and the so-called lateral tissue underneath is torn apart using two fingers. (Techniques involving tearing, rather than cutting, are generally recommended, where possible.)[57]
- Continuing to use fingers and a thumb, or using a scalpel and tiny forceps (which are like tweezers), or even electrocautery (if the 'Pelosi' technique is used), the obstetrician then tears or cuts deeper into the subcutaneous tissue. He or she then reaches a thick layer of fibrous tissue, called the fascia. After that the mucous membrane called the peritoneum, which is the last layer of the abdominal wall, is opened. This is done either by cutting or tearing in order to gain access to the abdominal cavity—which, of course, contains the uterus.
- After the bladder is pushed away, another peritoneum (which loosely lies over the 'low segment' of the uterus) is cut. The so-called 'low segment' of the uterus is then also cut open using short, careful strokes of a knife.
- The small original cut is then extended to a length of at least 15cm with the use of fingers or blunt-ended scissors.
- If the membranes around the baby are still intact—they may have burst open by this time—they are punctured and opened.
- The obstetrician then places one hand inside the uterus, under the baby's head and exerts pressure on the upper end of the uterus with the other hand so as to push the baby through the incision. (An assistant usually helps with this.)
- In some places, the baby's mouth is immediately suctioned with a small ear syringe and then the shoulders and the rest of the baby are eased out.
- Held up in the air, the baby usually begins to cry. When this happens the new baby is frequently held over the mother (if she is conscious) so that she can see the baby's genitals! The assumption is that her first priority will be to know whether she's got a boy or a girl.

- The cord is clamped and the baby handed over to a nurse holding a warmed towel. Mothers and babies may also have some skin-to-skin contact at this stage because modern cesareans rarely require general anesthesia.
- There have been reports of some women complaining of pain during the next few stages of the surgery, or of having difficulty breathing, although both are probably rare occurrences nowadays. Some women also vomit while their organs are being handled and the damage repaired. If any of these occur in the next stages of the surgery, the woman will be immediately sedated.
- The placenta is then pulled out and passed to an attendant, who gives it a thorough check. He or she might be soaked in blood and amniotic fluid by this stage.
- Next, the uterus is sutured (sewn together), either while still inside the abdomen (as advised by *Obstetrics for Ten Teachers*) or after being taken out to rest on the outside of the abdomen (as advised by Michael Stark).
- The loose peritoneum covering the uterus is sutured closed, or left open for spontaneous healing if the obstetrician considers this to be preferable. Then the fascia (the thick fibrous layer) is brought together with heavy thread stitches because it is this layer which holds all the abdominal organs inside and keeps them from coming through the incision. The subcutaneous tissue (which is mostly fat) is sometimes also closed using loose stitches. The skin, the final layer, is sewn up with silk or synthetic thread, or joined with metal staples.
- A dry dressing is placed over the woman's incision and taped to her skin.
- Finally, the drapes are removed.

## AFTER THE SURGERY...

- Both the woman and her baby are transfered to the postpartum ward together as soon as possible.
- Wherever she is, the woman's blood pressure and pulse may be recorded every now and then and her temperature noted every two hours or so. Whether or not this is done, and the frequency with which it is done will depend on the woman's apparent condition. If she is talking and alert, no checks may be considered necessary.
- The woman's wound is inspected every 30 minutes to check for blood loss and her lochia is also inspected. If the woman has had a general anesthetic, she is left in the recovery position until she is fully conscious because she is still at risk of airway obstruction or regurgitation and silent aspiration of her stomach contents.
- Analgesia is prescribed and given. This may take the form of an epidural opioid, rectal analgesia or intramuscular analgesia (i.e. an injection). If the epidural route is chosen, the woman's breathing also needs to be recorded.

- The baby is put on the mother, skin-to-skin if possible. He or she may also be put to the mother's breast in such a way that the mother is disturbed as little as possible.[58]
- Even if the woman feels very hungry, she is only allowed to have fluids at first because there is a small risk of paralytic ileus. The woman still needs to have an intravenous infusion for a while after the surgery.
- The woman is helped out of bed and encouraged to walk around as soon as possible so as to reduce the risk of deep vein thrombosis. She is also encouraged to perform breathing exercises. A physiotherapist usually teaches these and may also give chest physiotherapy. Low-dose heparin (an anticoagulant drug) and TED anti-embolism stockings are often prescribed to prevent thrombosis.
- Blood pressure, temperature and pulse continue to be checked periodically. The intravenous infusion continues and the urinary catheter remains in the bladder until the woman can get up and go to the bathroom. The wound and lochia are observed hourly.
- When the urinary catheter has been removed, urinary output is monitored carefully because some women have difficulty passing urine at first or emptying the bladder completely.
- A woman is reassured if—as frequently happens after a general anesthetic— she feels tired and drowsy for hours or even days after the surgery. If, as often happens, the woman complains of feelings of detachment and unreality or feels she cannot relate well to her new baby, she is encouraged to talk freely and is again reassured.
- Painkillers are given as often as the woman asks for them. This usually means giving intramuscular opiates for up to 48 hours (i.e. injections) and oral analgesics (tablets) after that. Alternatively, some women have their epidural topped up and some women don't feel they need any painkillers at all.
- The woman is encouraged to rest as much as possible and visitors may be discouraged. Because she is likely to tire quickly, the woman will probably need help looking after her new baby.
- Finally, the woman is not allowed to drive a vehicle or carry other children for six weeks after the surgery—or for the length of time determined by the insurance company.

Clearly, a cesarean surgery is a complicated surgical procedure, with all kinds of risks, indignities, discomforts and inconveniences. So why on earth has it become fashionable in some cultures and subcultures, to have a cesarean section instead of a vaginal birth?

A cesarean is a complicated surgical procedure with all kinds of risks, indignities, discomforts and inconveniences

## The fashion

Cesarean rates have been steadily increasing around the world. In 2001 in the US, the rate was 22.9% and in the same year in Australia it was 21.9%. In Brazil the overall rate for the country is well above 50% and in private hospitals in big cities such as Sao Paulo and Rio de Janeiro the rate has been estimated at a staggering 90%. According to private consultation with a professor of midwifery in China, the trend there is the same, with most urban women being sectioned the one-and-only time they give birth (given the ongoing one-child policy there). In England and Wales in 1960 only 2.8% of births were achieved by cesarean section but the figure had risen to 18.0% by 1999—and 25% of births in England in 2008 took place by cesarean section.

This modern fashion for cesareans is actually quite easy to understand.[59] Choosing a cesarean section appears to mean choosing to have complete control over a natural process. As one midwife explains...

The C-section has sadly become part of our culture and an acceptable, if not desirable, way to give birth. Women find the control element of cesareans, particularly elective cesareans, appealing (in my opinion). In our society, generally, women are used to having control of their own bodies and their lives and to give up their body to labor and birth is very frightening to some women. Having 'a date' for their delivery day is another example of control—being able to plan it into their life/husband's vacation, etc., as opposed to waiting until the baby is ready.

It's interesting, though, that women who have both elective and emergency cesareans report feeling a *lack* of control when they comment on their own experience, as we shall soon see. In some sense then, the idea that a cesarean gives women control is just a fallacy... Although women might have superficial control in terms of timetables, when it comes to emotions they tend to feel out of their depth. And of course, although the birth itself may be scheduled, post-operative recovery for both mother and baby is highly unpredictable—far less so than with a vaginal birth.[60]

Apart from wanting to control the processes, why else do women actually choose to have a cesarean? One reason is that it appears to mean choosing to blot out pain and discomfort because anesthesia and painkillers have to be used. It is a fallacy, unfortunately, that no pain will be experienced, mainly because postpartum pain is usually a feature of a cesarean. If physical pain is not experienced (thanks to strong painkillers), often there is emotional pain, so one way or another cesarean moms seem to get a worse deal. Another common reason to have a cesarean is to avoid damaging the sexual organs. Unfortunately, this is also misguided because damage to a woman's pelvic floor usually occurs mainly during pregnancy. The best way to avoid damage to the vagina and perineum is actually to have a completely drug-free, physiological birth because less damage is likely to occur when the woman is aware of all sensation and in control while she's giving birth.[61]

From the point of view of caregivers and/or hospital managers, cesareans may mean more profits are made and they certainly mean they have control...

# The feeling

What does it actually feel like to have a cesarean?

## Birthframe 16: Nina Klose—How it felt, having a cesarean

First they put me on the strange, narrow, tipped operating table and increased the anesthetic in my epidural. I was so terrified my teeth were chattering. Two kind young anesthesiologists stayed with my husband and me at the head end of the drape so we wouldn't see me being wrenched open. The operating room had one of those strange fly-eye 7-faceted non-glare lamps overhead, lots of monitors checking my heart rate, blood pressure, blood oxygen, etc. They put an oxygen mask on me, to give the baby extra air, they said. When my lower body felt like nothing but lumps of dough, they got to work putting in a catheter and then slicing open the abdominal wall.

"What's happening now?" I asked, because it felt like I was being tugged in all directions. I wondered if the baby were coming out. "Oh, they're pulling apart some of the layers of your muscle wall," the anesthesiologist said brightly. "It heals better than when they cut it." I wished I hadn't asked.

Not much later we heard a small cough and a phlegmy squall. "There he is! He's breathing!" I said, and burst into tears. We didn't get to see him right away, because the pediatricians had to check him for any problems related to the meconium and the long labor. We heard him screaming for his life for several minutes. Then they brought him to us wrapped in a blanket. My husband held him and showed him to me. All I could see was a bit of a tiny face in a big white blanket. I didn't know what to do. (If this ever happens to me again, I will make sure that the drape is lowered so that I can see the baby lifted out, and have him brought to me right away to be comforted.)

"Can I nurse him?" I asked, since I knew that was what you were supposed to do. They said maybe not yet, since I was still swathed in operating sheets to my chest.

"Shall I give him his Vitamin K?" asked the midwife. I said no, I didn't want that to be the first thing he ever tasted.

"How much does he weigh?" I asked. Nobody had weighed him yet.

"Would you like me to weigh him for you?" the midwife offered.

I said okay, since I couldn't think of what else to do. So she took him away and I could hear him shrieking in terror as he was made to shiver for the sake of his parents' idle curiosity. I berated myself for being so thoughtless as to allow him to be taken away again so soon. Finally, he was returned to us, still terrified. My husband was led away with the baby to the delivery room. The anesthesiologist stayed with me while the doctors finished putting me back together. They lifted me from side to side to wash away the blood and iodine disinfectant that had sloshed onto the table, then they lifted me back onto my wheeled bed, and rolled me back along the corridor to the delivery suite.

Having the cesarean was like getting burglarized. Strange burglars, who ransack the house and leave behind the world's most priceless gift, my baby son. But a burglary nonetheless. I was robbed of an experience I'd anticipated my whole life.

Sometimes life forces us to face our worst fears. Perhaps it was necessary for me to go through the cesarean first so that I would be able to appreciate the gift of a normal birth at home. I can't take it for granted. Now over three years later I can say it was a valuable experience. After all, now I've been in an operating room and experienced major surgery. Gee whiz. Kind of interesting.

I've experienced all those hospital interventions and I know what it's like. If it's happened to you, I can relate. It is possible to do it another way the second time! From the day after my son's birth, I comforted myself by thinking about what I was going to do differently if I got a chance to try again. The day after I discovered I was pregnant a second time, I started looking for an independent midwife. I really did my homework.

I still regret the terror and pain that my son must have felt being yanked out into bright light by strangers. He is happy and healthy and no one could say, "Here is a child who had a difficult birth." But these things are not visible to the naked eye. Wouldn't it have been so much better for him if I had been able to welcome him to the world in a gentler way? Is that why he cried inconsolably for 36 hours after his birth? They say some babies 'just are' more unsettled than others. How can I help wondering if the cesarean was the cause? Maybe I was the cause of his distress, because I was so upset. Maybe if I hadn't wanted a natural birth so much, I wouldn't have caused him such distress. But I can't be different from who I am. My husband and I were both overjoyed about the baby, nothing could change that. It was a more abstract feeling than after the second, natural birth. After the second one, I felt a physical joy. My whole being was happy. The first time, we were filled with love and wonder for our beautiful son. But my body felt sad.

## Birthframe 17: Anonymous—Emotional and physical effects

There's an emotional trauma connected with going through it—thinking you'll deliver vaginally and then being told that you must have a C-section and minutes later having your baby handed to you. Pretty mind-blowing.

On the psychological front there's a nagging feeling of failure for not having delivered vaginally, as well as a sense of having 'missed out' on the 'real' experience of having a baby. Both sensations, however, pale in comparison to the general glory of having a baby (any which way you do) and raising a child.

On the recovery side, it took me a good while to recover. I found it difficult to stand up straight and walk for at least two weeks. I also had considerable pain in the incision area—at first, sharp and severe from the cut, and then more of an aching like I'd been seriously punched. The scar healed just fine, however. I will note, in case it's of interest, that I seem to have discoloration of the skin now. I feared it was cancerous, but my US doctor, who did not perform the C-section, thinks it is a result of the cesarean. What else? Perhaps it was age related, but the whole delivery process took the wind out of me for many months. I didn't really recover for an entire year, but that probably would have been the case, C-section or no.

## Birthframe 18: Anonymous—Two cesareans compared

Basically, I was an elective cesarean with my son because I was absolutely terrified of childbirth / have an issue with control (not helped by an unexpected pregnancy), and also was strongly advised to do so by three female friends who are doctors. I think they advised me to have a cesarean because we were all quite young at the time—mid/late 20s—and they'd done their OB-GYN placements and had been horrified by some of the things they'd seen: women screaming in pain, episiotomies, postpartum incontinence, etc. I expect they saw a skewed sample—the easy births don't require such intervention from doctors. Also doctors are generally calmer about surgery—they've seen so much—and it's not such a big thing for them. I wonder how they feel about it now... None of them have had babies yet.

Post my cesarean with my first baby, I was in pain for a very long time. Even three months after the birth my scar was still nagging and painful by the end of the day or if I went out for a long walk. I remember the horror I felt when the prescription drugs ran out. After five weeks, a string came out of my scar. The soluble sutures had in fact not dissolved, and my body had had to reject them by sending the length of suture up through the scar and to the surface. I hadn't been warned of this possibility. I was so squeamish and horrified by the whole thing that somehow I managed to ignore this, and didn't even visit the doctor. I think it could have been one of the reasons for the continuing pain. My scar didn't really heal properly and became keloid.

And then it was my daughter's turn. When I became pregnant with her I suddenly realized the downside of an elective cesarean—it makes it much more likely that you'll need one the next time, only this time round you have a heavy child who you're not allowed to pick up. I really debated whether to go VBAC but was told that I would be allowed only an hour in second stage labor, and effectively as a first timer my chances of an emergency section were considerable. So I booked a maternity nurse and went for it again.

Second time round was incredibly different. After 10 days I felt better than I had after three months with my first—I was going up and down stairs, I was off the prescription drugs. My scar healed fine. It made me realize how rough the first time had been, now I had something to compare it with.

Retrospectively, I think I'd have done better to go for natural childbirth the first time round: that six weeks after the second child is born when you can't pick up your jealous first child, or put them into the bathtub, or change them, is really hard. I didn't even think of it when I was making the original decision.

However, it's not the end of the world. My pelvic floor and perineum are intact and my son doesn't seem too traumatized. The actual experiences of the surgery were very weird, though: I had an awake one both times. It was okay the first time but the second time it was like a dreadful nightmare, this unending horror, pinned to a bed with tubes everywhere and an oxygen mask on, and all these faces looming in and out of my vision. I don't know if you're including a psychological look at the post-op recovery; it might be interesting to examine the mental trauma that may arise from the experience. I hope this is helpful.

" After the cesarean I was in a state of shock. Weeping uncontrollably for weeks and months. I became agoraphobic (normally I have to be outside), and couldn't travel in a car because it was far too scary. I was more vulnerable than a cream cake at a Bulimic Congress and the gory nightmares I was having were beyond the fright that most people are required to endure. Worst of all, I thought this was permanent. The good side is that everyone who has this Dark Night of the Soul learns from their experiences and gets stronger, and that's just what happened to me.

## Birthframe 19: Anonymous—Positive cesarean experience

In case any client of yours happens to be one of the women who needs a cesarean for medical reasons, here's a positive birth story (from *LLL GB News*). It's the story of a woman who was initially in favor of a vaginal birth, who changed her mind after a difficult first labor, which resulted in an emergency cesarean. She describes how she managed to 'optimize' outcomes for an elective cesarean (for her second child), by taking an intelligent, caring and proactive approach.

I feel happy for those mothers who have experienced natural births but concerned that mothers who have experienced other births might feel that their experiences may be seen as somehow less honourable, less heroic. I've had two cesarean deliveries: each time I felt that this birth was an incredibly heroic act. I think all births must be.

The first was performed as an emergency after an incredibly traumatic labor. When my beautiful baby, Tom, emerged he weighed in at almost 12 pounds! After having such a traumatic time, I chose to have my second baby by cesarean. I asked other mothers to share with me their experiences of planned cesareans (through a La Leche League newsletter).

I asked other mothers to share with me their experience

I wanted advice in general and also particularly wanted to know how well breastfeeding had gone for other cesarean mothers, without the hormonal kick-start of going through labor. I was touched and impressed to receive many letters in reply. We are all so busy bringing up young families, with very little time that we can spend writing letters, and yet here were all these La Leche League mothers taking the time and trouble to write long and detailed letters to a complete stranger! It was such a treat to read these handwritten, deeply personal stories and great to find that almost all these mothers reported a positive start to breastfeeding.

The day for my elective cesarean arrived and my husband and I arrived at the hospital, as planned, at 7:30AM—thanks to one tip I got from a letter to sleep at home the night before. Our hospital's policy is normally to admit a mother the night before for tests and to meet with an anesthesiologist, then they simply keep the woman in hospital overnight. The hospital was happy to comply with my request so Tom, our 3-year-old, had his granny with him to take him to kindergarten and give him lunch afterwards.

I'm writing this account 17 months later and find I can scarcely remember the hours before Louis' birth. I think I was very excited and also very anxious. When birth begins with labor, one gets swept along in the unfolding experience. Without labor, there is lots of room to reflect on what a truly extraordinary and somehow unbelievable miracle is about to happen—wonderful, but scary. A lovely midwife welcomed me at the door when I arrived back at the hospital, showed me where to put my things and took me down to the operating suite. She joked and chatted with me as the anesthesiologist did his work and informed me of what was happening all the way through. She knew I wanted to be well-informed as I'd shown her my birth plan on my visit the night before.

Once prepared, I lay on the operating table, numb from the waist down... well not quite numb, actually, but unable to feel pain. It's a strange sensation, hard to describe. My husband was beside me, dressed up in his little operating room hat and clothes.

It was a moment of intense feeling. I was aware that someone was about to cut into my body—okay, I couldn't feel it, but somewhere inside us all must be an instinctive sense that we should not let ourselves be cut. I was aware that of course I was about to meet my baby... Would it be a girl or a boy? Would it look like Tom? Would it be healthy?

I could feel the baby moving about inside me—it had been awake and vigorous all through the preparations. I felt rather pleased that it wasn't being roused from slumber! I also was aware that it would be quite a few weeks before I would begin to feel really recovered from the surgery—it's hard to welcome the prospect of weeks of discomfort and limited mobility.

And I was aware that here I was, about to begin an entirely new phase of my life as mother of two. I also realized that I was lying in exactly the same spot where I had been when Tom was born, exactly three years and two weeks before. Except that this time I was conscious, it wasn't an emergency and the circumstances surrounding the birth were different.

I was aware I was about to begin a new phase of my life

Suddenly, here he was! Our beautiful baby boy, perfect and chubby, and screaming. The midwife wrapped him up immediately and put him on my chest for me to hold while I was being sutured. Some mothers are able to nurse at this point but I was quite happy not to, feeling a bit strange and particularly immobile.

After a little while, we went through to the recovery room where, as I'd requested, I was able to watch as Louis was washed, checked and weighed (9lb).

Then came the breastfeeding... It's not the easiest thing to nurse a baby when lying flat on your back with IVs and tubes in the backs of your hands and no movement in your lower body, especially when it's a newborn baby who has never nursed before. It was agony! I kept taking him off to see if I could improve the positioning but nothing particularly helped. (Breastfeeding continued to often be quite painful for a couple of weeks or so. I have a private theory that some mothers like myself are particularly sensitive to the pressure that the baby's suck exerts on the breast and we interpret it as pain until we get very used to it.)

Louis didn't want to feed for the next 15 hours, although I kept offering him the breast. I was getting anxious. Would the staff insist he should have a bottle of formula? Would my milk take five days to 'come in' the way it had with my first baby? A midwife approached with a bottle of water and I squawked a protest. But she said she only wanted to drip a few drops on my nipple to encourage Louis to latch on. I'd never heard of this before (or since!) but, miraculously, it worked. The milk 'came in' early on Day 3 (and, of course, Louis had the valuable colostrum when he nursed in the first two days). Maybe there was no delay because I did not start off exhausted after a long labor. Louis grew steadily and has been such a delightful baby. I'm proud of us both.

## Are cesareans really necessary?

Dystocia (also called 'failure to progress') and fetal distress are reasons given for a great number of cesareans.[62] It is an unfortunate truth that in some cases women's bodies really don't function healthily and in these cases cesareans may be the only solution. After all even in these days of high technology and information still approximately 8 in every 1000 babies are born dead... and there is still a fairly high maternal mortality rate, which varies from country to country. According to the World Heath Organization, a woman's chance of dying at some point in her life as a result of pregnancy or childbirth can range from 1 in 7 (in Niger), 1 in 8 (in Afghanistan), 1 in 17,400 (in Sweden), 1 in 26,600 (in Italy) to 1 in 47,600 (in Ireland!)[63]. Interestingly—or worryingly—neither the US nor UK statistics are particularly good (1 in 4,800 and 8,200 respectively). Would the rates improve if more women had optimally optimal births—i.e. physiological ones, taking place in ideal circumstances? Given the inevitable problems which occur when unnecessary interventions are used (e.g. in order to arrange pharmaceutical pain relief) I have no doubt they would, simply because all drugs used for analgesia or anesthesia have side-effects, the most important of which may be that the birth process itself is compromised. (Of course, these side-effects are not denied by anesthesiologists. On the contrary, they are constantly being researched.)[64] Nevertheless, cesareans would almost certainly still be necessary in a tiny minority of cases, even with the best possible prenatal preparation and a massive change in culture, and in these cases they constitute an optimal outcome for those particular mothers and babies.[65]

When considering unnecessary cesareans, it is important to reflect on what precisely causes them. Disturbance in labor and birth must surely be one key factor, given the link between hormones and emotions and the effect on emotions of any kind of disturbance. This means that while caregivers are right to monitor the progress of a woman's labor and her baby's safety, it is essential that they do so without actually causing a labor to become extended or a baby distressed. (We need to remember the association between EFM and the increase in the cesarean rate.) In particular we need to remember that most of the practices and interventions which are common on labor wards all over the world may well disrupt women's labors and even cause fetal distress.[66] Lack of privacy, interruptions, alienation through the use of machines and impersonal practices, deprivation of food and drink, continuous or constant monitoring... all these practices disturb and therefore probably lengthen a woman's labor.

An atmosphere of fear and tension, fetal scalp monitors and inappropriate physical positions—such as when the mother lies on her back—are other elements which are highly likely to cause fetal distress.[67] It seems that the real, psychological and hormonal needs of a laboring woman, not to mention the fetus, are being disregarded in the name of safety. And it all seems to lead to what is diagnosed as dystocia. If you find all this difficult to accept, I would ask you to at least suspend judgment until you have read more birthframes.

One of the emailed conversations I had with Michel while writing this book relates to this problem. I had been wondering how often cesareans might need to be performed, in an ideal world...

**Michel, if every woman were to begin labor spontaneously and were to labor undisturbed and unobserved, without pain relief or interventions, what do you imagine the rates of mortality and morbidity, and cesarean might be?**

Imagine that the prerequisite to be an obstetrician or a midwife would be to be a mother who has a personal experience of vaginal unmedicated birth. I guess that the cesarean rate would drop below 10% with the same perinatal mortality rates and lower rates of transfer to pediatrics.

**Is nature really that inefficient, though? Surely even a 1% cesarean rate would indicate extreme inefficiency in nature...**

... or would be a reminder of the laws of natural selection. The huge development of the part of the brain called the neocortex (the brain of the intellect) is a handicap where childbirth is concerned. As long as the neocortex remains active, it tends to inhibit the activity of the primitive brain structures that are supposed to work hard during labor (releasing all the necessary hormones). That is the main reason why a difficult birth is an aspect of human nature. There are other reasons why humans beings are condemned to have difficult births. One is that the fetal head is so large that it has to spiral down in a complex way inside the maternal pelvis. If we add the deep-rooted cultural misunderstanding of birth physiology, one can explain why the rates of cesarean sections are far above 1%.

**Can you explain what you mean by 'the deep-rooted cultural misunderstanding of birth physiology'?**

I mean a widespread, deep-rooted failure to understand the basic needs of women in labor.

We shall return to a discussion of these needs in the rest of this book... For now, it perhaps enough to conclude that a cesarean does not seem to be the best way of giving birth unless it really is medically indicated. On the other hand, when cesareans are performed in order to safeguard the well-being of either the laboring woman or the baby, they are to be welcomed.

It is, after all, thanks to cesarean surgery (as well as improved nutrition and sanitation and advances in the fields of antibiotics and anesthesia, in particular) that childbirth is no longer the potentially life-threatening rite of passage for women that it once was.[68]

With this surgery as a safety guard, we can now confidently leave the physiological processes to take place and have an unmedicated vaginal birth, knowing that if the mom-to-be or her baby are in danger, there is fast and effective surgery to rescue them. This represents a new vision of obstetrics, as Michel explains:

My idea is to introduce a sort of futuristic strategy of childbirth which is partly what I have already practiced. It is based on having a good understanding of physiology and of the basic needs of human beings in labor—for privacy, a sense of security, etc. If the midwife behaves in a motherly way and disturbs the woman as little as possible, either it works or a cesarean is necessary. This strategy makes it possible to avoid the two main alternatives as much as possible: drugs (which all have side-effects) and difficult interventions by the vaginal route (difficult forceps and so on). If we avoid these alternatives, there are really only two possibilities: a completely undisturbed physiological labor or a cesarean section.

Women just need to find a caregiver who will support them in this choice... Could that be you?

*Obviously, building a good relationship at prenatal appointments is crucial for success*

## NOTES & REFERENCES

1   Music which a woman particularly likes (which is presumably not whale music?) may well be helpful. Some research has demonstrated that music helps to reduce anxiety and/or helps women cope with labor and stress, or give them a greater feeling of being in control. See:

- Spintge R. Some neuro-endocrinological effects of so-called anxiolytic music. *International Journal of Neurology*, 1989, 19/20:186-196

- Browning C. Using music during childbirth. *Birth*, 2000, 27(4):272-276

- Browning C. Music therapy in childbirth: research in practice. *Music Therapy Perceptions*, 2001, 19(2):74-81

2   We will explore this point in detail later in the book when we're discussing why physiological birth is best—and full references will be provided then.

3   As I mentioned in Note 4 of the Introduction, many women report their experience of pain in surprising ways. In addition, data on postpartum pain are rarely collected and analyzed seriously, but I suspect any such studies would reveal enormous differences between women who labor physiologically and women who have medicated and managed births of one kind or another.

Looking at only pain in labor, it is worth noting that some people have also been focusing on the potentially orgasmic nature of birth—and this is the focus of the DVD *Orgasmic Birth*. (See www.orgasmicbirth.com.) In the following article, Ina May Gaskin also points out that clitoral stimulation produces endorphin-like effects. (She reports that 20% of laboring women reported orgasm-like experiences.)

- Gaskin, 1. The frequency of reported orgasms in labor and birth in a population of unmedicated women. Abstract No 310, 26th ICM Congress, 2002

Many people involved in birth conclude that we should re-examine our attitudes towards sexuality during labor as a result. Of course, this needs to be done during pregnancy, as part of prenatal preparation. How women should be asked to re-examine their attitudes (in practical terms) and how they should reconsider their expectations of sensations in labor is another question altogether, of course. Having said all that, whatever women's attitudes are when they first present for a prenatal appointment I think it's vital that we don't swing from one extreme to another, i.e. while explaining to women that there might be positive sensations associated with birth I don't think it's helpful to deny that for some, if not most women it is excruciatingly painful. Overall, though, I would hope that the message any midwife gives to a mom-to-be—when discussing potentially orgasmic or painful experiences of labor—is that physiological birth is certainly healthiest for both mother and baby and also best in psychological terms. Is this the message you personally would want to convey? How would you explain your answer?

In any case, it is clear from various studies that anesthesia and analgesia options now available are far from being totally effective. See the following:

- Agaram R, Douglas MJ, McTaggart RA, Gunka V. Inadequate pain relief with labor epidurals: a multivariate analysis of associated factors. *International Journal of Obstetric Anesthesia*, 2009, Vol 18,10-14,2009

- Le Coq G, Ducot B, Benhamou D. Risk factors of inadequate pain relief during epidural analgesia for labor and delivery. *Canadian Journal of Anaesthesia*, 1998, Aug;45(8):719-23

In the first study (Agaram, *et al*) the researchers initially concluded (from women's reports) that 16.9% of women who'd been in the cohort (of 260 subjects) experienced inadequate pain relief. Problems were associated with insertion of the epidural at more than 7cm dilation, women's past experience of opioid tolerance, previous failed epidural and insertion of the epidural by a trainee anesthesiologist. After adjustments, the researchers concluded that epidurals were ineffective in only 9.3% of cases. The second study (Le Coq, *et al*), which *observed* 456 women (instead of interviewing them), found that epidurals provided inadequate pain relief in labor in 5.3% of cases, and during 'delivery' (birth) in 19.7% of cases. Reasons for inadequacy (in order of importance) included inadequate first doses (so that two top-ups were needed), posterior position, pain when the epidural was sited, epidural being in place for more than six hours, and the epidural being in place for *less* than one hour (meaning it was less effective for 'delivery').

Another related angle is to do with later memories. In one study (which clearly needs to be repeated), even when women reported good pain relief as a result of an epidural, they generally reported lower levels of satisfaction with the birth overall a year after the birth. Another study (by Cooper, *et al*, is due out soon on this too. See:

- Morgan BM, Bulpitt CJ, Clifton P, Lewis PJ. Analgesia and satisfaction in childbirth (the Queen Charlotte's 1000 mother survey) *Lancet*, 1982, 2 (Oct 9) 808-810

- Cooper, *et al* Satisfaction, control and pain relief: short and long term assessments in a randomized controlled trial of low-dose and traditional epidurals and a non-epidural comparison group. *International Journal of Obstetric Anesthesia*, 2009. [Full reference not available at time of going to press.]

4    See: Odent M. The fetus ejection reflex. *Birth* 1987; 14:104-5.

5    For more on this subject read the following:

- Is there sex after childbirth? by Juliet Rix (Thorsons 1995)

- Michel Odent. The perineal preoccupation. In: *The Caesarean* (p96-101). Free Association Books. London 2004

- *Perineal Care: An international issue* by Henderson C and Bick D, (eds.) (Quay Books 2006)

Logic tells us that an upright position is likely to help a woman avoid tearing since the coccyx can then move back and therefore increase the size of the vaginal opening. Research itself has so far been inconclusive on this point, though, and has even contradicted this idea, noting that a) a side-lying position may be slightly better than upright positions, b) that full squats did result in more perineal trauma, and—overall—that c) all-fours, standing, kneeling and semi-recumbent positions were all similar as regards ending up with an intact perineum. However, it must be noted that research carried out will have looked mostly at women who were using some kind of drug-based pain relief—because it is used so frequently. No studies have as yet been conducted exclusively on women who have birthed without any drugs in their system, who are also undisturbed but supported by a sensitive midwife. For the research carried out so far see:

- Shorten A, Donsante J, Shorten B. Birth position, accoucheur and perineal outcomes: informing women about choices for vaginal birth. *Birth,* 2002, 29 (1):18-27
- Eason E, Labrecque M, Wells G, Feldman P. Preventing perineal trauma during childbirth: a systematic review. *Obstetrics & Gynecology,* 2000, Mar; 95(3):464-71
- Soong B, Barnes M. Maternal position at midwife-attended birth and perineal trauma: is there an association? *Birth,* 2005, 3:164-169

A full commentary on episiotomy is provided later.

6    See:
- Odent M. New reasons and new ways to study birth physiology. *International Journal of Gynecology and Obstetrics* 2001; 75:S39-S45 and Goland, R. S. et al.

Biologically Active Corticotrophin-releasing Hormone in Maternal and Fetal Plasma during Pregnancy. *American Journal of Obstetrics and Gynecology* 159 (1984): 884-890.

The following studies strongly suggest that the physiological processes are dramatically compromised when a woman has pharmaceutical pain relief. It seems there is an inhibition of catecholamine release (Jones, 1985) as well as an inhibition of the oxytocin peak, which typically occurs during a physiological labor (Goodfellow, 1983).

- Bacigalupo G, Riese S, *et al.* Quantitative relationships between pain intensities during labor and beta-endorphin and cortisol concentrations in plasma. Decline of the hormone concentrations in the early postpartum period. *Journal of Perinatal Medicine,* 1990, 18(4):289-96.
- Goodfellow CF, *et al,* Oxytocin Deficiency at Delivery with Epidural Analgesia, *British Journal of Obstetrics & Gynaecology,* 1983, 90:214-219
- Jones CR, McCullouch J, *et al.* Plasma catecholamines and modes of delivery: the relation between catecholamine levels and in-vitro platelet aggregation and adrenoreceptor radioligand binding characteristics. *British Journal of Obstetrics & Gynaecology,* 1985, Jun; 92(6):593-9

7    If you're interested in considering how a baby's ability to love might develop also see: Niles Newton. The Influence of the Let-Down Reflex in Breast Feeding on the Mother-Child Relationship. *Marriage and Family Living* 1958; 20: 18-20.

8    In a retrospective study conducted by Jordan, *et al* (Jordan, *et al,* 2009), which looked at 48,366 healthy women birthing singleton babies at term (i.e. women having healthy births), it was found that at 48 hours after the birth, rates of breast-feeding definitely seemed to be affected by epidurals, opioid analgesia (Demerol, diamorphine, etc) and ergomentrine (used in the third stage of labor). The researchers point out that "failure to breastfeed increases morbidity and mortality in both mothers and children in developed and developing countries", so the impact of any possible effects of drugs used unnecessarily could be enormous. In this study, beyond sociological factors, which have long been known to affect breastfeeding rates, lower breast-feeding rates were associated with induction

with pessaries (prostaglandins), epidurals and opioid analgesia, and ergometrine used for the third stage of labor. (Oddly, they found that first-time mothers who'd had gas and air were more likely to nurse. Could this be because these mothers were determined to avoid drugs in labor as much as possible, so as to have as 'natural' a birth as possible, and to nurse successfully too? Any determination to avoid everything except gas and air could be a particularly British attitude, which is misguided, in my view, for other reasons, as I shall explain later. The view that it is considered 'nothing' is reflected in the off-hand statement made by many women postpartum: "Oh, I only had gas and air.") Anyway, the study by Jordan, *et al* does provide some evidence that drug use in labor and birth has an impact on breastfeeding rates at 48 hours postpartum, which obviously will affect longer-term rates too, although it must be said that this evidence is not accepted by all anesthesiologists as *prospective* randomized studies are seen as more reliable. After all, women usually request epidurals *because* of difficulties, so it is not necessarily epidurals *per se* which cause later problems. Cause-effect are difficult to establish.

Other (prospective) studies reported fairly clear problems with narcotics used in labor (Beilin, *et al,* 2005; Camann, *et al,* 2007; Torvaldsen, *et al,* 2006). In the study by Beilin, *et al* researchers concluded: "Among women who breastfed previously, those who were randomly assigned to receive high-dose labor epidural fentanyl were more likely to have stopped breastfeeding 6 weeks postpartum than women who were randomly assigned to receive less fentanyl or no fentanyl." (Fentanyl was added to the drug bupivacain, in the epidural cocktail as bupivacain causes paralysis in the lower part of the body; adding fentanyl reduces this effect. Clearly, though, it's a problem if too much is used.) The study by Torvaldsen, *et al* concluded: "Women in this cohort who had epidurals were less likely to fully breastfeed their infant in the few days after birth and more likely to stop breast-feeding in the first 24 weeks"... although the researchers felt they were unable to say whether there was a causal link between epidural anesthesia and difficulties. This was despite the fact that "Intrapartum analgesia and type of birth were associated with partial breastfeeding and breastfeeding difficulties in the first postpartum week" and the fact that women who had epidurals were more likely to stop breastfeeding than women who used non-pharmacologic methods of pain relief. Camann's editorial (below) provides a good overview of this topic. See:

- Jordan S, Emery, S, Watkins A, Evans JD, Storey M, Morgan G. Associations of drugs routinely given in labour with breastfeeding at 48 hours: analysis of the Cardiff Births Survey. *BJOG: International Journal of Obstetrics & Gynaecology,* 2009, online publication on 1 Sept

- Beilin Y, Bodian C, Weiser J, *et al*. Effect of labor epidural analgesia with and without fentanyl on infant breast-feeding: a prospective, randomized, double-blind study. *Anesthesiology,* 2005, Dec;103(6):1211-7

- Camann W. Labor analgesia and breast feeding: avoid parenteral narcotics and provide lactation support. *International Journal of Obstetric Anesthesia,* 2007, Jul; 16(3):199-201

- Torvaldsen S, Roberts CL, Simpson JM, *et al*. Intrapartum epidural analgesia and breastfeeding: a prospective cohort study. *International Breastfeeding Journal,* 2006,Dec11;1:24

9    If you would like to explore this issue yourself straight away, see the Primal Health Research database at www.primalhealthresearch.com. To consider the possible link between the use of analgesics in labor and later drug use in the grown baby see: Jacobson B, Nyberg K. Opiate addiction in adult offspring through possible imprinting after obstetric treatment. *BMJ* 1990; 301: 1067-70 and Nyberg K, Buka SL, Lipsitt LP. Perinatal medication as a potential risk factor for adult drug abuse in a North American cohort. *Epidemiology* 2000; 11(6): 715-16.

10   Actually, research into the causative associations between drug use in labor and things such as alertness postpartum and breastfeeding is in its infancy, but results are already suggesting that drugs do have side– or after-effects; these are constantly discussed in the literature on anesthesiology. Possible (or probable) side– or after-effects are widely accepted as including nausea and vomiting, feelings of confusion, lowering of the blood pressure, sedation, urinary retention, slower emptying of stomach contents and itching (pruritis)—and the pain relief is not always effective. Of course, in all cases, it's not easy to establish what causes what.

Years ago, some practitioners were apparently concerned about the observable depressive effects on newborns of analgesia used in labor. I deduce this because researchers (Bonta, *et al,* 1979) discovered that another drug (naloxone) provided an effective 'antidote' and restored what apparently seemed to be an acceptable level of alertness in newborns. However, is naloxone (or any equivalents) really a solution to reduced alertness or is the mother-baby dyad losing out when natural alertness is not present? (Would you prefer natural sexual arousal when you meet your life partner, or passion produced by sedatives, counteracted by Viagra? Of course, we need to remember that these early interactions can never be repeated, and also that they might have a significant effect on later interactions too.) Mothers, incidentally, are sometimes assessed by anesthesiologists for alertness using a four-point scale (1. awake/alert, 2. drowsy, but readily responsive, 3. drowsy and requires shaking to rouse, 4. unconscious). But how awake and alert do people expect women to be just after they've had a baby? Physiological labor usually results in extreme alertness, which is, of course, beneficial for the bonding process.

In a more recent study (Volikas, *et al,* 2005), which looked at potential side-effects of patient-controlled opioid analgesia (remifentanil) postpartum, 22 out of 50 women were reported to have experienced some drowsiness. (44% seems rather a high percentage...) The researchers reported that at the dose used in the study, remifentanil had 'an acceptable level of maternal side-effects and minimal effect on the neonate'. Personally, I question whether any level of drowsiness is acceptable during this one-time encounter between new mother and baby. And I wonder how it could be established that there was only a 'minimal' effect on the neonate if there was no control group, i.e. if researchers did not compare these 50 neonates with 50 others, who were born entirely physiologically. What seems a normal level of alertness in a neonate might change if researchers were to document the extreme alertness many people have anecdotally reported when babies have been born without any drugs in their systems. There is an enormous difference, I would suggest, between a dull-eyed look and a vibrant gaze, in terms of bonding and simple joy in new motherhood. And when a newborn looks at his or her mother it's helpful, perhaps, if the mother isn't one of the 22 out of 50 women who were reported in this study 'to have experienced some drowsiness'. Hill (2008) later strongly recommended remifentanil, although Van de Velde disagreed in a follow-up article. Although remifentanil is more effective than Demerol and diamorphine (which perhaps explains why it is popular in some places), what price are women paying for their reduced alertness in terms of effective early bonding?

Another study (Wittels, *et al*, 1997) compared the alertness (amongst other things) of newborns exposed to either epidural morphine or intravenous patient-controlled analgesia. Of course, because the focus was on newborns of mothers who'd had a cesarean it was impossible to compare the alertness of babies born with drugs in their systems and that of babies who'd been born with absolutely no drugs in their system, so only 'relative' alertness could be tracked. Yet another study, back in 1981 (Rosenblatt, *et al*) looked at the influence of maternal analgesia (epidural bupivacaine) on the newborn. Significant effects were found: "Immediately after delivery, infants with greater exposure to bupivacaine in utero were more likely to be cyanotic [blue-skinned] and unresponsive to their surroundings. Visual skills and alertness decreased significantly with increases in the cord blood concentration of bupivacaine, particularly on the first day of life but also throughout the next six weeks. Adverse effects of bupivacaine levels on the infant's motor organization, his ability to control his own state of consciousness and his physiological response to stress were also observed." A recent study by Henrichs, *et al* (2009) considered whether alertness could be affected by a factor such as fetal size in mid- or late pregnancy. (The conclusion was that it could.) In a study such as this, I would imagine there could be numerous confounding factors, the principal one being the use of anesthesia or analgesia (or not) during labor. Personally, I would only trust the results of this study if all fetuses measured in utero had been born without any drugs in their systems. After all, while the motivation of these researchers appears to have been a desire to investigate behavioral problems in newborns (e.g. infant irritability), they do not appear to have taken into account the fact that one of the primary characteristics of narcotics-addicted neonates is that they are 'substantially more irritable' (Strauss, *et al*, 1975).

Given the vital importance of good bonding in the sensitive one-hour period following birth (from the point of view of later mothering behavior), I very much hope that other researchers will look further into the issues of alertness and breastfeeding success (or lack thereof), particularly in relation to drug-use in labor. See:

- Bonta BW, Gagliardi JV, Williams V, Warshaw JB. Nalaxone reversal of mild neurobehavioral depression in normal newborn infants after routine obstetric analgesia. *Journal of Pediatrics,* 1979. Jan;94(1):102-5

- Volikas I, Butwick A, Wilkinson C, Pleming A, Nicholson G. Maternal and neonatal side-effects of remifentanil patient-controlled analgesia in labour. *British Journal of Anaesthesia*, 2005, Oct;95(4):504-9. Epub 2005 Aug 19

- Wittels B, Glosten B, Faure EA, Moawad AH, Ismail M, Hibbard J, Senal JA, Cox SM, Blackman SC, Karl L, Thisted RA. Postcesarean analgesia with both epidural morphine and intravenous patient-controlled analgesia: neurobehavioral outcomes among nursing neonates. *Anesthesia & Analgesia,* 1997. Sep;8 (3):600-6

- Rosenblatt DB, Belsey EM, Lieberman BA, Redshaw M, Caldwell J, Notarianni L,

Smith RL, Beard RW. The influence of maternal analgesia on neonatal behaviour: II. Epidural bupivacaine. *British Journal of Obstetric Gynaecology,* 1981. Apr;88(4): 407-13

- Henrichs J, Schenk JJ, Schmidt HG, Arends LR, Steegers EA, Hofman A, Jaddoe VW, Verhulst FC, Tiemeier H. Fetal size in mid- and late pregnancy is related to infant alertness: the generation R study. *Developmental Psychobiology,* 2009, Mar; 51(2):119-30

- Strauss ME, Lessen-Firestone JK, Starr RH Jr, Ostrea EM Jr. Behavior of narcotics-addicted newborns. *Child Development,* 1975. Dec;46(4):887-93

- Hill D. Remifentanil patient-controlled analgesia should be routinely available for use in labour. *International Journal of Obstetric Anesthesia,* 2008, 17(4),336-339

- Van de Velde M. Controversy. Remifentanil patient-controlled analgesia should be routinely available for use in labour. *International Journal of Obstetric Anesthesia,* 2008 Oct;17(4):339-42. Epub 2008 Jul 9

11    Some midwives and researchers have rightly expressed a concern over maternal dissatisfaction after labors in which the mother has felt 'abandoned' or 'neglected'. Other midwives also report that women sometimes feel a deep sense of shock after experiencing a physiological labor for the first time. However, the hormones and endorphins produced usually conspire to transform women's attitudes, even in cases where they have not planned to have a drug-free labor and many women become almost evangelical about physiological birth, after they've experienced it for themselves. It certainly is an experience which is far beyond any person's normal day-to-day experience and usually brings with it many positive benefits. Since a minority of women do not experience these even when they abstain from drug use during labor, research could usefully be conducted to investigate possible best ways of preparing women for unmedicated labors and facilitating the physiological processes so that both *support* and lack and disturbance are key features.

For an exploration of related issues see: *Evolution's End: Reclaiming the Potential of Our Intelligence* (San Francisco: Harper 1995): 178-179; Odent, M. Orgasmic states, Ecstatic states and Mystical emotions. In: *The Scientification of Love* (pp 75-79). Free Association Books. London 1999.

For an article exploring ways of facilitating the natural processes, see:

- Odent M. New reasons and new ways to study birth physiology. *International Journal of Gynecology & Obstetrics* 2001, 75:S39-S45

Michel Odent expands further on these ideas in his book *Birth & Breastfeeding* (Clairview Books, 2007). In case you feel wary of facilitating the physiological processes of labor and birth *without pain relief* you may be reassured to note that quite a few contributors to this book mention their surprise at the *lack* of pain in their labors or the surprising nature of the sensations. Others—including myself— experienced intense pain but found they were able to travel through it, thanks to the strange hormonal processes which were taking place, which inevitably have an effect on the mind as well as the body. There will be more in later chapters on why and how it's possible to support women through either painful or painfree births. It will soon become clear to you why a drug-free labor is preferable to a medicated, managed one in terms of experience, as well as safety, and why it's so easy to disturb the natural processes if any drugs or interventions are used.

12 If you're interested in filling in gaps here and elsewhere, note that in general randomized controlled trials are impossible in many cases in the perinatal period. Women would refuse to participate in the process of randomization because they usually have strong preferences at this time. This is why there is a lack of valuable hard data on some issues.

Also note that in *Preparing for a Healthy Birth* I do recommend a few well-tried and 'sensible' interventions:

- The use of folic acid preconceptually and in the first trimester so as to prevent spinal malformations

- The use of vitamin supplements (e.g. Sanatogen's *Pronatal*) from the second trimester onwards and Omega-3 supplements in the third trimester. This is because many modern women have inadequate diets. Taking a multivitamin designed for pregnancy does not seem to be dangerous, as long as normal dosages are followed. In the following study, which looked at the use of vitamin supplementation in the *first* trimester, researchers concluded "In the final adjusted model, any use of vitamins during pregnancy was associated with decreased odds of miscarriage ... in comparison with no exposure." I am not recommending vitamin-use until the second trimester simply because there is so much very delicate fetal development going on in the early weeks, which it could be very easy to disturb. It's probably best to encourage pregnant women to focus on eating healthily instead–and of course this needs to continue throughout pregnancy. See: Hasan R, Olshan AF, Herring AH, Savitz DA, Siega-Riz AM, Hartmann KE. Self-reported vitamin supplementation in early pregnancy and risk of miscarriage. *American Journal of Epidemiology,* 2009, Jun 1;169(11):1312-8. Epub 2009 Apr 16

- Drinking 1-3 cups of raspberry leaf tea per day during the third trimester—a tried and tested uterotonic

- The prophylactic use of homeopathic Arnica 30 in the last month of pregnancy, which is at worst harmless, but may help prevent or heal bruising

For a justification of the advice for pregnant women to take Omega-3 supplements, see:

- Odent M, McMillan L, Kimmel T. Prenatal care and sea fish. European *Journal of Obstetrics & Gynecology & Reproductive Biology,* 1996, 68:49-51

- Odent M, Colson S, De Reu P. Consumption of seafood and preterm delivery— Encouraging pregnant women to eat sea fish did not show effect. *British Medical Journal,* 2002, 324:1279

And the article in which he explains briefly why fats are so important during pregnancy at www.wombecology.com/nutritionpregnancy.html

13 For a review and comparison of obstetric approaches (either 'managed' or natural) read any or all of the books listed below. After doing so, it is very difficult to reach the conclusion that intervention enhances the natural processes, unless it's really needed for medical reasons.

- The book written and researched by former WHO director Marsden Wagner: *Born in the USA: How a broken maternity system must be fixed to put women and children first.* University of California Press, 2006

- The book written by Jennifer Block, former journalist and co-editor of the revised classic *Our Bodies Ourselves: Pushed.* Da Capo Lifelong, 2008
- The book co-authored by seven researchers (Enkin, *et al*) *Guide to Effective Care in Pregnancy and Childbirth.* Oxford University Press, 2000
- Enkin's book, written with another eminent researcher Jadad: *Randomised Control Trials: Questions, Answers and Musings.* Blackwell, 2007
- The book edited by professor of midwifery, Soo Downe: *Normal Childbirth: Evidence and Debate.* Churchill Livingstone, 2004
- Walsh, D. *Evidence-based Care for Normal Labour and Birth.* Routledge, 2007

14　I realize you might be thinking, "Ah, yes, but every pregnancy and birth are unique." Well, they are, but there's enough of a pattern for us to be able to look at typical scenarios. Soo Downe and McCourt (2004) have come up with the concept of 'unique normality'. In other words, they recognize that there is a *range of normality*, which means that women don't have to fit neatly onto charts or into 'mental boxes' so as to be considered normal. See:

- Downe S, McCourt C. From being to becoming: reconstructing childbirth knowledges. In Downe S (ed) *Normal Childbirth: Evidence and Debate.* Churchill Livingstone, 2004

15　Nolan (2005) concluded that prenatal education packages do not consistently reduce intervention rates, which in turn suggests that they do not help. However, carefully designed (somewhat non-typical) prenatal classes did result in an increase in the use of upright positions and/or a decrease in the number of epidurals. See:

- Nolan M. Childbirth and parenting education: what the research says and why we may ignore it. In Nolan M and Foster J (eds) *Birth and Parenting Skills: New Directions in Antenatal Education.* Churchill Livingstone, 2005
- Nolan M and Foster J (eds) *Birth and Parenting Skills: New Directions in Antenatal Education.* Churchill Livingstone, 2005
- Foster J. Innovative practice in birth education. In M. Nolan and J. Foster (eds) *Birth and Parenting Skills: New Directions in Antenatal Education.* Elsevier Science, 2005
- Walsh D, Harris M, Shuttlewood S. Changing midwifery birthing practice through audit. *British Journal of Midwifery,* 1999, 7(7):432-345

16　See: Gaskin I. The frequency of reported orgasms in labor and birth in a population of unmedicated women. Abstract No 310, 26th ICM Congress, 2002. The documentary film *Orgasmic Births* also looks at this issue. See: www.orgasmicbirth.com.

17　Michel S, Rake A, Treiber K. MR obstetric pelvimetry: effects of birthing position on pelvic bony dimensions. *American Journal of Roentgenology,* 2002, 179:1063-1067

18　Gould (2000) commented on a woman's apparent need to move around during a physiological labor. Many other researchers who have conducted anthropological studies of indigenous peoples have noted that upright positions are generally favored for birth. See:

- Gould D. Normal labour: a concept analysis. *Journal of Advanced Nursing,* 2000, 31(2):418-427

- Gupta J, Hofmeyr G. Position for women during second stage of labour. Cochrane Review in: *The Cochrane Library,* Issue 4. Chichester: John Wiley & Sons Ltd, 2006
- Jarcho J. *Postures and Practices During Labour Among Primitive Peoples.* Paul Hoeber, 1934
- Kitzinger S. *Rediscovering Birth.* Little, Brown & Company, 2000
- Coppen R. *Birthing Positions: Do Midwives Know Best?* Quay Books, 2005
- Balaskas J. *New Active Birth: A Concise Guide to Natural Childbirth.* Unwin, 1995
- Lavin J, McGregor J. Native American childbirth on the western plains. *International Journal of Feto-Maternal Medicine,* 1992, 5(3):125-133

The following studies also noted various positive effects of moving around in labor and using upright positions, including a reduction in the need for pharmacologic analgesia, an increased sense of control and increased satisfaction with birth:

- Simkin P, O'Hara M. Non-pharmacological relief of pain during labour: systematic review of five methods. *American Journal of Obstetrics & Gynaecology,* 2002, 186:S131-159
- Spiby H, Slade P, Escott D, Henderson B, Fraser R. Selected coping strategies in labour: an investigation of women's experiences. *Birth,* 2003, 30:189-194
- Albers L, Sedler K, Bedrick E, Teaf D, Peralta P. Midwifery care measures in the second stage of labour and reduction of genital tract trauma at birth: a randomised controlled trial. *Journal of Midwifery & Women's Health,* 2005, 50:365-372

Very specifically, research has shown that upright positions for birth mean a shorter second stage, a much smaller risk of a need for episiotomy, forceps (or ventouse), less pain, easier 'pushing' and fewer fetal heart abnormalities. All in all, being upright seems a very good idea! See:

- De Jonge A, Teunissen T, Lagro Janssen A. Supine position compared to other positions during the second stage of labour: a meta-analytic review. *Journal of Psychosomatic Obstetrics & Gynaecology,* 2004, 25:35-45
- Gupta J, Hofmeyr G. Position for women during second stage of labour (Cochrane Review). In: *The Cochrane Library,* Issue 4. John Wiley & Sons Ltd, 2006
- Johnstone F, Aboelmagd M, Harouny A. Maternal position in the second stage of labour and fetal acid base status. *British Journal of Obstetrics & Gynaecology,* 1987, 94(8):753-757
- Chalk A. Pushing in the second stage of labour: Part 1. *British Journal of Midwifery,* 2004, 12(8):502-508

The only outcome which was found to be better when lying down was blood loss, although it is possible this was because of differences in ease of estimation. (It did not affect outcomes.) This is perhaps because the research which reached this conclusion a) included women who were using anesthesia and/or analgesia, which might bring about different results, and b) included women who had a managed third stage (which would result in completely different patterns of blood loss, of course).

19 At the following link and in the article below Michel explains the perfect conditions for a fetus ejection reflex at www.wombecology.com/fetusejection.html:

- Odent M. The fetus ejection reflex. Birth, 1987, 14:104-5

20 Control and management at this point often occur because of the mistaken notion (by caregivers) that a time limit needs to be put on the second stage of labor. In fact, large reviews of research has revealed that there is no connection between a long labor and a poor Apgar score, or admissions to a neonatal unit (Saunders, et al, 1992; Menticoglou, et al, 1995; Janni, et al, 2002; Myles and Santolaya, 2003—see below for full references). Some of the women in these studies had second stages that lasted more than five hours!—but these long second stages still had no effect on birth outcomes. Only in cases where long stages were linked to special first stage factors or intervention, or when the baby stayed too long on the pelvic floor, was there a link with maternal infection or bleeding, or a deterioration in fetal heart patterns (Nordstrom, et al, 2001.) Moreover, research has revealed all kinds of problems with what is known as 'commanded pushing', as follows: 1. It results in the woman holding her breath for prolonged periods, which affects the placenta, and therefore the unborn baby. (Caldeyro-Barcia, 1979). 2. Less oxygen gets to the baby's brain (Aldrich, et al, 1995). 3. Babies whose mothers had 'commanded pushing' for more than an hour had a lower pH at birth—which is bad news (Thompson, 1993). 4. Commanded pushing results in longer second stages (Thompson, 1993; Parnell, et al, 1993). 5. Women tend to become exhausted (Knauth and Haloburdo, 1986; Roberts, 2002). 6. There are more forceps and ventouse births (Fraser, et al, 2000; Hansen, et al, 2002). 7. More episiotomies are performed and more tears occur (Sampselle and Hines, 1999). 8. The pelvic floor is damaged more often, which results in urinary stress incontinence (Handa, et al, 1996). Most surprisingly, perhaps, one hospital audit discovered that although only 8% of midwives attending the births of low-risk women encouraged the women to push as they wished (Walsh, et al, 1999), the practice of commanded pushing only served to shorten the second stage of labor by 14 minutes (Bloom, et al, 2006), which—as I mentioned before—was not helpful in terms of improving outcomes. After reviewing all the research in 2006, Bosomworth and Bettany-Saltikov concluded that pushing while holding one's breath should also be discontinued because of its documented effects on the fetal heartbeat and the perineum. This echoes the guidelines provided by Enkin, et al (2000), which—based on a review of the available evidence—advises against routine directed pushing, pushing by sustained bearing down and breath-holding. See:

- Saunders N, Paterson C, Wadsworth J. Neonatal and maternal morbidity in relation to the length of the second stage of labour. *British Journal of Obstetrics & Gynaecology*, 1992, 99(5):381-385
- Menticoglou S, Manning F, Harman C. Perinatal outcome in relation to second stage duration. *American Journal of Obstetrics & Gynaecology*, 1995, 173(3):90612
- Janni W, Schiessl B, Peschers U. The prognostic impact of a prolonged second stage of labour on maternal and fetal outcome. *Acta Obstetrica et Gynaecologica Scandinavica*, 2002, 81:214-221
- Myles T, Santolaya J. Maternal and neonatal outcomes in patients with a prolonged second stage of labour. *Obstetrics & Gynaecology*, 2003, 102:52-58

- Nordstrom L, Achanna S, Naka K, Arulkumaran S. Fetal and maternal lactate increase during active second stage of labour. *British Journal of Obstetrics & Gynaecology,* 2001, 108:263-268
- Caldeyro-Barcia R, Giussi G, Storch E. The influence of maternal bearing down efforts and their effects on fetal heart rate, oxygenation and acid base balance. *Journal of Perinatal Medicine,* 1979, 9: 63-67
- Aldrich C, D'Antona D, Spencer J. The effects of maternal pushing on fetal cerebral oxygenation and blood volume during the second stage of labour. *British Journal of Obstetrics & Gynaecology,* 1995, 102(6):448-453
- Thompson A. Pushing techniques in the second stage of labour. *Journal of Advanced Nursing,* 1993, 18:171-177
- Parnell C, Langhoff-Roos J, Iverson R. Pushing method in the expulsive phase of labour. A randomised trial. *Acta Obstetrica et Gynecologica Scandinavica,* 1993, 72(1):31-35
- Knauth D, Haloburdo E. Effects of pushing techniques in birthing chair on length of second stage of labour. *Nursing Research,* 1986, 35:49-5 1
- Roberts J. The 'push' for evidence: management of the second stage. *Journal of Midwifery & Women's Health,* 2002, 47(1):2-15
- Fraser W, Marcoux S, Krauss I, Douglas J. Multi-centre, randomised controlled trial of delayed pushing for nulliparous women in the second stage of labour with continuous epidural analgesia. *American Journal of Obstetrics & Gynaecology,* 2000, 182:1165-1172
- Hansen S, Clark S, Foster J. Active pushing versus passive fetal descent in the second stage of labour: a randomised controlled trial. *Obstetrics & Gynaecology,* 2002, 99:29-34
- Sampselle C, Hines S. Spontaneous pushing during labour: Relationship to perineal outcomes. *Journal of Nurse Midwifery,* 1999, 44(1):36-39
- Sampselle M, Miller J, Luecha Y, Fischer K, Rosten L. Provider support of spontaneous pushing during the second stage of labour. *Journal of Obstetrics, Gynaecology and Neonatal Nursing,* 2005, 34:695-702
- Soong, B. and Barnes, M. Maternal position at midwife-attended birth and perineal Soong, B. and Barnes, M. Maternal position at midwife-attended birth and perineal trauma: is there an association? *Birth,* 2005, 3:164-169
- Handa V, Harris T, Ostergard D. Protecting the pelvic floor: obstetric management to prevent incontinence and pelvic organ collapse. *Obstetrics & Gynaecology,* 1996, 88:470-478
- Walsh D, Harris M, Shuttlewood S. Changing midwifery birthing practice through audit. *British Journal of Midwifery,* 1999, 7(7):432-345
- Bloom S, Casey B, Schaffer J, McIntire D, Leveno K. A randomised trial of coached versus uncoached maternal pushing during the second stage of labour. *American Journal of Obstetrics & Gynaecology,* 2006, 194:10-13
- Bosomworth A, Bettany-Saltikov J. Just take a deep breath. *MIDIRS,* 2006, 16(2):157-165

• Enkin M, Kierse M, Neilson J, Crowther C, Duley L, Hodnett E, Hofmeyr J. *A Guide to Effective Care in Pregnancy and Childbirth.* Oxford University Press, 2000

21  Michel explains in detail why it's important that no-one talks to the mom-to-be while she's in labor at www.wombecology.com/physiological.html

22  In this book I am wholeheartedly advocating that women have a physiological third stage and describing that as optimal for all kinds of reasons. Firstly, it's part of the healthy process.  As we've seen countless times before, intevening in a healthy process can actually *cause* pathology. If a birth has gone well up till this point, I see no reason why we can't assume the healthy hormonal cascade will continue right through to the birth of the placenta. Secondly, it needs to be noted that the use of a synthetic hormone (ergometrine in the US) in order to speed up birth of the placenta comes with a price: side effects. In the case of ergometrine, side effects can be raised blood pressure and vomiting, dizziness and headaches. (Personally, I do not want to be feeling nauseous or dizzy when I first meet my newborn babies. As I know from experience, the alternative is so much better.) Thirdly, the alternative—a completely physiological third stage—actually features a peak of oxytocin production in the woman greater than any other she is likely to experience at other points in her life. (Surely, this exists for a reason, probably related to the need for bonding?) Fourthly, active management of the third stage of labor involves an enormous amount of disturbance from a caregiver... This is very unfortunate because this is precisely the time (never to be repeated) when a mother first meets her baby, when they first interact and, crucially, when the baby has his or her first feed of breastmilk. It is a magical time if there is no disturbance and the healthy processes are allowed to proceed smoothly, without unnecessary disturbance. Having said all that, synthetic oxytocin (e.g. pitocin, or ergometrine) needs to be available with a caregiver in case there is a postpartum hemorrhage (PPH), for any reason, because obviously synthetic hormones can stop the hemorrhage extremely quickly and effectively. Because the very slight risk of hemorrhage exists, the caregiver also needs to be discretely present until after the placenta has been born.

In case you're wondering about the *safety* of a physiological third stage, after examining the research I personally am convinced it is safe, provided pitocin or ergometrine is available, in case they are needed, and provided—of course—an expert caregiver (i.e. a midwife or obstetrician) is in attendance between the second and third stages of labor. The research is difficult to present, mainly because so many points in studies are contentious. For example, when 500ml is taken as being the ceiling for PPH, results 'show' that more women having a physiological third stage have a 'hemorrhage'. However, what is a hemorrhage? If a woman loses up to 1,000ml of blood, but blood loss does not continue, why should that be called a postpartum hemorrhage? And since it doesn't have a bad effect on a new mother, why do we need to worry about it? (It's only really a cause for concern in cases where new mothers are very anemic, which is often the case—because of inadequate diet—in developing countries. Women who are severely anemic are much more seriously affected by even small levels of blood loss.) Going back to talking about healthy women, researchers have hypothesized that perhaps a greater level of blood loss during the third stage (i.e. 1,000ml, rather than 500ml) may be a feature of a healthy physiological stage and they see possible reasons for this (Harris, 2001; Wickham, 1999; Buckley, 2009). Perhaps because of these considerations, some researchers (e.g. Burchell 1980) have advocated that the ceiling should be increased to 1,000ml (i.e. the point at which blood loss is considered 'hemorrhage'). If the level is increased to 1,000ml, two studies have shown (Rogers, *et al*, 1998 and Herschderfer, 1999) that the difference in outcomes between 'managed' and 'physiological' third stages is not significant. In any case, when considering all this, we need to bear in mind that blood loss is estimated *visually* and the accuracy of estimates is very variable (as confirmed by Glover, 2003; Read and Anderton, 1997; and Bose, *et al*, 2006). (Consider how easy—or difficult!—it is to judge how much blood is soaked into a bedsheet or blanket and how an estimate of this might differ from an estimate of blood pooled on a waterproof sheet or floating round a birthing pool.) In addition, we need to remember that in almost all the studies conducted most women who had a 'physiological' third stage had had all kinds of interventions (and drugs) in the first and second stages of labor, so the healthy processes had no doubt been severely disturbed already. (We could only really draw accurate conclusions if only fully physiological births were included in the 'physiological third stage' cohorts of studies.)

Finally, the whole issue is compounded by the possibility that *disturbance* to the mother of any kind between the second and third stages might well actually *cause* a hemorrhage, because the hormonal cascade would be disturbed. Since disturbances could include even one word comments to the new mother, there must be numerous cases of disturbance in the studies.

In conclusion, then, although some studies conclude that active management is safer (e.g. Prendiville, *et al*, 1988 and 2006), I question many of the issues surrounding set-up of studies and am concerned mostly with their conclusion that active management was *worse* than a physiological third stage in terms of nausea, vomiting and hypertension... Another study, which compared the effectiveness of different mixtures (McDonald, *et al*, 2006) found that syntometrine (the cocktail used in the UK, which is a combination of ergometrine and synthetic oxytocin, i.e. pitocin) caused more nausea, vomiting and raised blood pressure—even though it was found to be most effective in terms of reducing PPH. Of course, my reason is that I think it's rather a tragedy if a new mother is experiencing these symptoms when she first meets her newborn baby and I wouldn't like to guess what the babies think about it. Given how desperately young children look forward to special events, I can well imagine that a fetus also anticipates his or her first meeting with his or her new mother with great interest! See:

- Harris T. Changing the focus for the third stage of labour. *British Journal of Midwifery,* 2001, 9(1):7-12
- Wickham S. Further thoughts on the third stage. *The Practising Midwife,* 1999, 2 (10):14-15
- Buckley SJ. *Gentle Birth, Gentle Mothering: A Doctor's Guide to Natural Childbirth and Gentle Early Parenting Choices.* Celestial Arts, 2009
- Burchell R. Postpartum haemorrhage. In E. Quilligan (ed.) *Current Therapy in Obstetrics & Gynaecology.* Philadelphia: W.B. Saunders, 1980
- Rogers J, Wood J, McCandlish R, Ayers S, Truesdale A, Elbourne D. Active versus

expectant management of third stage of labour: the Hinchingbrooke randomised controlled trial. *Lancet,* 1998, 351:693-699

- Herschderfer K. Results of RCT expectant versus active management within setting of Dutch midwives' independent practices (home births). Presented at National Study Day on Third Stage Issues, Manchester, 1999
- Glover P. Blood loss at delivery: how accurate is your estimation. *Australian Journal of Midwifery,* 2003, 16:21-24
- Read M, Anderton J. Radioisotope dilution technique for measurement of blood loss associated with lower segment caesarean section. *British Journal of Obstetrics & Gynaecology,* 1997, 84:859-861
- Bose P, Regan F, Paterson-Brown S. Improving the accuracy of estimated blood loss at obstetric haemorrhage using clinical reconstructions. *BJOG: An International Journal of Obstetrics & Gynaecology,* 2006, 113:919-924
- Prendiville W, Harding J, Elbourne D, Stirrat G. The Bristol third stage trial: active vs physio-logical management of third stage of labour. *British Medical Journal,* 1988, 297:1295-1300
- Prendiville W, Elbourne D, McDonald S. Active versus expectant management in the third stage of labour (Cochrane Review). In: *The Cochrane Library,* Issue 3. Chichester: John Wiley & Sons Ltd, 2006
- McDonald S, Abbott J, Higgins S. Prophylactic ergometrine-oxytocin versus oxytocin for the third stage of labour (Cochrane Review). In: *The Cochrane Library,* Issue 3. Chichester: John Wiley & Sons Ltd, 2006

23   Two research projects established that late or no cord cutting is optimal—one concluding that the baby receives some 43% of the blood left in maternal/placental circulation (Gunther 1957; Wardrop, *et al*, 1995.) Other studies (Kinmond, *et al*, 1993; Piscane, 1996) questioned the assumption that it is either necessary or helpful to cut the cord at all as part of the third stage procedures. See:

- Gunther M. The transfer of blood between the baby and the placenta in the minutes after birth. *Lancet,* 1957; I:1277-1280
- Kinmond S, *et al.* Umbilical Cord Clamping and Preterm Infants: a randomized trial. *British Medical Journal,* 1993, (6871): 306:172-175
- Pisacane A. Neonatal prevention of iron deficiency, *British Medical Journal,* 1996, 312:136-7
- Wardrop CA, Holland BM. The roles and vital importance of placental blood to the newborn infant. *Journal of Perinatal Medicine,* 1995, 23(1-2):139-43

Delayed cord cutting by up to two minutes was found to be beneficial for premature babies because there was less intraventricular hemorrhage and they then needed less blood transfusion (Rabe, *et al*, 2006). McDonald and Abbott (2006) found that delaying clamping for full-term newborns meant the babies got 30% more blood and up to 60% more red blood cells—all of which is excellent news, of course. It means that babies start off life with optimal hematocrit and hemoglobin levels (Prendiville and Elbourne, 1989), better blood circulation to vital organs, better heart-lung adaptation and a better chance of being able to breastfeed for longer (Mercer, 2001). See:

- Rabe H, Reynolds G, Diaz-Rossello J. Early versus delayed umbilical cord clamping in preterm infants. *The Cochrane Database of Systematic Reviews*, 2006, Issue 3
- McDonald S, Abbott J. Effect of timing of umbilical cord clamping of term infants on maternal and neonatal outcomes. (Protocol) *Cochrane Database of Systematic Reviews*, 2006, Issue 3
- Prendiville WJ, Elbourne DR. Care during the third stage of labour. In Chalmers I, Enkin M and Kierse M (eds) *Effective Care in Pregnancy and Childbirth*. Oxford University Press, 1989
- Mercer J. Current best evidence: a review of the literature on umbilical cord clamping. *Journal of Midwifery & Women's Health*, 2001, 46(6):402-414

24    The term 'cascade of interventions' was first conceptualized and explained by Sally Inch in 1989. See: Inch S. *Birthrights: Parents' Guide to Modern Childbirth*. Green Print, 1989. In the 'Why' section I explain in detail what the cascade of hormones involves.

25    Many studies, such as the following, conclude that women resort to pain relief less often when they are outside large hospitals (e.g. at home or in a birth center) and also when they have continuity of care (i.e. one midwife right the way through their labor and birth):

- Olsen O. Meta-analysis of the safety of home birth. *Birth*, 1997, 24(1):4-13
- Walsh D, Downe S. Outcomes of free-standing, midwifery-led birth centres: a structured review of the evidence. *Birth*, 2004, 31(3):222-229
- Hodnett E, Downe S, Edwards N, Walsh D. Home-like versus conventional birth settings (Cochrane Review). In: *The Cochrane Library*, Issue 2. Chichester: John Wiley & Sons Ltd, 2006
- Waldenstrom U, Turnbull D. A systematic review comparing continuity of midwifery care with standard maternity services. *British Journal of Obstetrics & Gynaecology*, 1998, 105:1160-1170
- Wraight A, Ball J, Seccombe I, Stock J. *Mapping Team Midwifery: A Report to the Department of Health*. Brighton: Institute of Manpower Studies, University of Sussex, 1993
- Harvey S, Jarrell J, Brant R, Stainton C, Rach D. A randomised controlled trial of nurse/midwifery care. *Birth*, 1996, 23:128-135
- Page L, McCourt C, Beake S, Hewison J. Clinical interventions and outcomes of one-to-one midwifery practice. *Journal of Public Health Medicine*, 1999, 21 (3): 243-248

26    This is suggested by the fact that the fetus produces oxytocin, vasopressin and catecholamines during labor. If the fetus produces these substances, surely it must also experience their effects? See:

- Chard, T, *et al.* Release of Oxytocin and Vasopressin by the Human Fetus during Labour. *Nature*, 1971, 234:352-354
- Hagnevik K, Faxelius G, *et al.* Catecholamine surge and metabolic adaptation in the newborn after vaginal delivery and caesarean section. *Acta Paediatrica Scandinavica*, 1984, Sep;73(5):602-9

- Colson S. Womb to World: a metabolic perspective. *Midwifery Today*, 2002, Spring; 61:12-17
- Irestedt L, Lagercrantz H, *et al*. Causes and consequences of maternal and fetal sympathoadrenal activation during parturition. *Acta Obstetrica et Gynecologica Scandinavica*, Suppl, 1984, 118:111-5
- Lowe N, Reiss R. Parturition and Fetal Adaptation. *Journal of Obstetric, Gynecologic & Neonatal Nursing*, 1996, 25:339-349
- Leake RD, Weitzman RE, Fisher DA. Oxytocin concentrations during the neonatal period. *Biology of the Neonate*, 1981, 39(3-4):127-31.
- Eliot RJ, Lam R, *et al*. Plasma catecholamine concentrations in infants at birth and during the first 48 hours of life. *Journal of Pediatrics*, 1980, Feb; 96(2):311-5
- Eliot RJ, Lam R, *et al*. Plasma catecholamine concentrations in infants at birth and during the first 48 hours of life. *Journal of Pediatrics*, 1980, Feb; 96(2):311-5

27   Bystrova, *et al*,s study (2009) confirmed the earlier, very famous research of Kennel and Klaus (reviewed by the authors themselves in 1988), which indicated that babies and mothers should remain together in the early moments and minutes after birth. Bystrova, *et al* concluded: "Skin-to-skin contact, for 25 to 120 minutes after birth, early suckling, or both, positively influenced mother-infant interaction 1 year later when compared with routines involving separation of mother and infant." One of the original researchers, Marshall Klaus, also offered a commentary on this recently, as you will see from the reference below.

- Bystrova K, Ivanova V, Edhborg M, Matthiesen AS, Ransjö-Arvidson AB, Mukhamedrakhimov R, Uvnäs-Moberg K, Widström AM. Early contact versus separation: effects on mother-infant interaction one year later. *Birth*, 2009, Jun; 36(2):97-109
- Kennell JH, Klaus MH. The perinatal paradigm: is it time for a change? *Clinical Perinatology*, 1988, Dec;15(4):801-13
- Klaus MH. Commentary: An early, short, and useful sensitive period in the human infant. *Birth*, 2009, Jun;36(2):110-2

If either your client is reluctant to nurse or you don't feel confident about providing support for breastfeeding, you might find it useful to read breastfeeding expert Suzanne Colson's comments on breastfeeding attitudes and approaches at www.midwiferytoday.com/articles/breastfeednem.asp

28   Although research into EFM began quite early on, the first *review* of research was carried out in 2005 by Thacker, *et al,* and this prompted a wider re-examination of EFM, which had become routine in so many hospitals up till then. Alfirevic, *et al*'s review of research in 2006 (which covered 37,000 women in total!) concluded that EFM had few benefits and many drawbacks. The rate of cesareans and forceps/ventouse births rose significantly and it brought no improvement in Apgar scores when 'EFM women' were compared to those who'd just had periodic checks with a fetal monitor (e.g. a fetoscope or Doppler). (Another study by Madaan and Trivedi in 2006 found that women trying for a vaginal birth after a previous cesarean were more likely to end up with another cesarean if they had EFM, and there was no improvement in outcomes when they had EFM.) In the

review by Alfeirevic, *et al* one benefit was associated with EFM—the number of neonatal seizures dropped—but there was no significant difference in any long-term effects, such as cerebral palsy. (Interestingly, one earlier study—by Luthy, *et al* in 1987—found *higher* rates of cerebral palsy when EFM was used for pre-term labors.) Overall, given the many documented disadvantages of EFM, the conclusion remains that the procedure is only advisable in certain high risk cases—and even then it is recognized that EFM may not help prevent perinatal deaths or cerebral palsy, so some high risk women opt out too. See:

- Flynn AM, *et al*. A randomized controlled trial of non-stress antepartum cardiotocography. *British Journal of Obstetrics & Gynaecology*, 1982, 89:427-33
- Kidd LC, *et al*. Non-stress antenatal cardiotocography—a prospective randomized clinical trial. *British Journal of Obstetrics & Gynaecology*, 1985, 92:1156-59
- Impey L, Reynolds, M, *et al*. Admission cardiotocography : a randomized controlled trial. *Lancet*, 2003, 361:465-70
- Alfirevic Z, Devane D, Gyte G. Continuous cardiotocography (CTG) as a form of electronic fetal monitoring (EFM) for fetal assessment during labour. *Cochrane Database of Systematic Reviews,* Issue 3, 2006
- Haverkamp AD, *et al*. A controlled trial of the differential effects of intrapartum monitoring. *American Journal of Obstetrics & Gynecology*, 1976, 126:470-76
- Kelso IM, *et al*. An assessment of continuous fetal heart rate monitoring in labor. *American Journal of Obstetrics & Gynecology*, 1978, 131:526-32
- McDonald D, Chalmers I, *et al*. The Dublin randomised controlled trial of intrapartum fetal heart rate monitoring. *American Journal of Obstetrics & Gynecology*, 1985, 152:524-39
- Prentice A, Lind T. Fetal heart rate monitoring during labor—too frequent intervention, too little benefit. *Lancet*, 1987, 2:1375-77
- Wood C. A controlled trial of fetal heart rate monitoring in low-risk obstetric population. *American Journal of Obstetrics & Gynecology*, 1981, 141:527-34
- Thacker S, Stroup D, Peterson H. Continuous electronic fetal heart monitoring during labour. (Cochrane Review) In: *The Cochrane Library,* Issue 1. Oxford: Update Software, 2005

29  Graham (1997) charts the origins and evolution of the episiotomy; first it was performed as a result of notions about a woman's 'imperfection', then it continued because of a surgical mind-set and finally it just became institutionalized in Western birthing practices. Its use was (wrongly) justified because of fears of cerebral palsy (episiotomy, it was thought, would mean the fetal head was 'knocking' against the perineum for a shorter time), because it was felt it shortened the second stage of labor, and finally—and most persistently—because it was thought to prevent tears into the anal sphincter. Starting with Sleep, *et al* (1984), research has revealed that its use is inappropriate in most cases, mainly because it results in *more* (not less) tearing and suturing. (This was the conclusion of Carroli and Belizan's Cochrane review in 2006.) Also, it is now considered bad practice for most women because research has also shown that an episiotomy reduces strength in the pelvic floor (Sartore, *et al,*2004), leads to pain in this area (Bick, *et al*, 2002) and even painful sexual intercourse afterwards (Bick, *et al*, 2002). Some studies (Dipiazza, *et al*, 2006; Williams, 2003; Richter, *et al*, 2002)

have concluded that episiotomies even *facilitate* bad tears into the anal sphincter. It's also worth noting that some academic papers have condemned the practice of routine episiotomy after a woman has previously had a tear (Peleg and Zlatnik, 1999; Dandolu, *et al*, 2005). According to Dannecker, *et al* (2004), episiotomies should only be performed when the fetus is clearly distressed and needs to be born quickly (e.g. in the case of a breech birth which is not progressing smoothly). See:

- Graham I. *Episiotomy: Challenging Obstetric Interventions.* Blackwell, 1997
- Sleep J, Grant A, Garcia J, Elbourne D, Spencer J, Chalmers L. West Berkshire perineal management trial. *British Medical Journal*, 1984, 289:587-590
- Carroli G, Belizan J. Episiotomy for vaginal birth (Cochrane Review). In: *The Cochrane Library*, Issue 3. John Wiley & Sons Ltd, 2006
- Sartore A, De Seta F, Maso G. The effects of mediolateral episiotomy on pelvic floor function after vaginal delivery. *Obstetrics & Gynaecology*, 2004, 103:669-673
- Bick D, MacArthur C, Knowles H, Winter H. *Postnatal Care: Evidence and Guidelines for Management.* Churchill Livingstone, 2002
- DiPiazza D, Richter H, Chapman V, Cliver S, Neely C, Chen C, Burgio K. Risk factors for anal sphincter tear in multiparas. *Obstetrics & Gynaecology*, 2006, 107(6):1233-1236
- Williams A. Third-degree perineal tears: risk factors and outcome after primary repair. *Journal of Obstetrics & Gynaecology*, 2003, 23(6):611-614
- Richter H, Brumfield C, Cliver S, Burgio K. Risk factors associated with anal sphincter tear: a comparison of primiparous patients, vaginal births after caesarean deliveries and patients with previous vaginal delivery. *American Journal of Obstetrics & Gynaecology*, 2002, 187:1194-1198
- Peleg D, Zlatnik M. Risk of repetition of a severe perineal laceration. *Obstetrics & Gynaecology*, 1999, 93(6):1021-1024
- Dandolu V, Gaughan J, Chatwani A. Risk of recurrence of anal sphincter lacerations. *Obstetrics & Gynaecology*, 2005, 105:831-835
- Dannecker C, Hillemanns P, Strauss A. Episiotomy and perineal tears presumed to be imminent: randomised controlled trial. *Acta Obstetrica et Gynaecologica Scandinavica*, 2004, 83:364-368

30  Control and management at this point often occur because of the mistaken notion (by caregivers) that a time limit needs to be put on the second stage of labor. In fact, large reviews of research has revealed that there is no connection between a long labor and a poor Apgar score, or admissions to a neonatal unit (Saunders, *et al*, 1992; Menticoglou, *et al*, 1995; Janni, *et al*, 2002; Myles and Santolaya, 2003—see overleaf for full references). Some of the women in these studies had had second stages that had lasted more than five hours!—but these long second stages still had no effect on birth outcomes. Only in cases where long stages were linked to special first stage factors or intervention, or when the baby stayed too long on the pelvic floor, was there a link with maternal infection or bleeding, or a deterioration in fetal heart patterns (Nordstrom, *et al*, 2001.) Moreover, research has revealed all kinds of problems with what is known as 'commanded pushing', as follows: 1. It results in the woman holding her breath for prolonged periods,

which affects the placenta, and therefore the unborn baby. (Caldeyro-Barcia, 1979). 2. Less oxygen gets to the baby's brain (Aldrich, et al, 1995). 3. Babies whose mothers had 'commanded pushing' for more than an hour had a lower pH at birth—which is bad news (Thompson, 1993). 4. Commanded pushing results in longer second stages (Thompson, 1993; Parnell, et al, 1993). 5. Women tend to become exhausted (Knauth and Haloburdo, 1986; Roberts, 2002). 6. There are more forceps and ventouse births (Fraser, et al, 2000; Hansen, et al, 2002). 7. More episiotomies are performed and more tears occur (Sampselle and Hines, 1999). 8. The pelvic floor is damaged more often, which results in urinary stress incontinence (Handa, et al, 1996). Most surprisingly, perhaps, one hospital audit discovered that although only 8% of midwives attending the births of low-risk women encouraged the women to push as they wished (Walsh, et al, 1999), the practice of commanded pushing only served to shorten the second stage of labor by 14 minutes (Bloom, et al, 2006), which—as I mentioned before—was not helpful in terms of improving outcomes. After reviewing all the research in 2006, Bosomworth and Bettany-Saltikov concluded that pushing while holding one's breath should also be discontinued because of its documented effects on the fetal heartbeat and the perineum. This echoes the guidelines provided by Enkin, et al (2000), which—based on a review of the available evidence—advises against routine directed pushing, pushing by sustained bearing down and breath-holding. See:

- Saunders N, Paterson C, Wadsworth J. Neonatal and maternal morbidity in relation to the length of the second stage of labour. *British Journal of Obstetrics & Gynaecology,* 1992, 99(5):381-385
- Menticoglou S, Manning F, Harman C. Perinatal outcome in relation to second stage duration. *American Journal of Obstetrics & Gynaecology,* 1995, 173(3):906-
- Janni W, Schiessl B, Peschers U. The prognostic impact of a prolonged second stage of labour on maternal and fetal outcome. *Acta Obstetrica et Gynaecologica Scandinavica,* 2002, 81:214-221
- Myles T, Santolaya J. Maternal and neonatal outcomes in patients with a prolonged second stage of labour. *Obstetrics & Gynaecology,* 2003, 102:52-58
- Nordstrom L, Achanna S, Naka K, Arulkumaran S. Fetal and maternal lactate increase during active second stage of labour. *British Journal of Obstetrics & Gynaecology,* 2001, 108:263-268
- Caldeyro-Barcia R, Giussi G, Storch E. The influence of maternal bearing down efforts and their effects on fetal heart rate, oxygenation and acid base balance. *Journal of Perinatal Medicine,* 1979, 9:63-67
- Aldrich C, D'Antona D, Spencer J. The effects of maternal pushing on fetal cerebral oxygenation and blood volume during the second stage of labour. *British Journal of Obstetrics & Gynaecology,* 1995, 102(6):448-453
- Thompson A. Pushing techniques in the second stage of labour. *Journal of Advanced Nursing,* 1993, 18:171-177
- Parnell C, Langhoff-Roos J, Iverson R. Pushing method in the expulsive phase of labour. A randomised trial. *Acta Obstetrica et Gynecologica Scandinavica,* 1993, 72(1):31-35
- Knauth D, Haloburdo E. Effects of pushing techniques in birthing chair on

length of second stage of labour. *Nursing Research,* 1986, 35:49-5 1

- Roberts J. The 'push' for evidence: management of the second stage. *Journal of Midwifery & Women's Health,* 2002, 47(1):2-15
- Fraser W, Marcoux S, Krauss I, Douglas J. Multi-centre, randomised controlled trial of delayed pushing for nulliparous women in the second stage of labour with continuous epidural analgesia. *American Journal of Obstetrics & Gynaecology,* 2000, 182:1165-1172
- Hansen S, Clark S, Foster J. Active pushing versus passive fetal descent in the second stage of labour: a randomised controlled trial. *Obstetrics & Gynaecology,* 2002, 99:29-34
- Sampselle C, Hines S. Spontaneous pushing during labour: Relationship to perineal outcomes. *Journal of Nurse Midwifery,* 1999, 44(1):36-39
- Handa V, Harris T, Ostergard D. Protecting the pelvic floor: obstetric management to prevent incontinence and pelvic organ collapse. *Obstetrics & Gynaecology,* 1996, 88:470-478
- Walsh D, Harris M, Shuttlewood S. Changing midwifery birthing practice through audit. *British Journal of Midwifery,* 1999, 7(7):432-345
- Bloom S, Casey B, Schaffer J, McIntire D, Leveno K. A randomised trial of coached versus uncoached maternal pushing during the second stage of labour. *American Journal of Obstetrics & Gynaecology,* 2006, 194:10-13
- Bosomworth A, Bettany-Saltikov J. Just take a deep breath. *MIDIRS,* 2006, 16(2):157-165
- Enkin M, Kierse M, Neilson J, Crowther C, Duley L, Hodnett E, Hofmeyr J. *A Guide to Effective Care in Pregnancy and Childbirth.* Oxford University Press, 2000

31   See Note 22 in this section.

32   Of course, here, Michel was not following protocols typical of a big maternity hospital in the 1980s or 1990s, where constant monitoring was considered essential. Increasingly nowadays, midwives and obstetricians are recognizing that 'watchful waiting' is an excellent alternative, which produces better results. This is no doubt because of evidence which is gradually showing that monitoring can be a form of intervention. For example, De Jonge and Lagro-Janssen (2004) linked vaginal examinations in the second stage to the supine position for birth (because it's difficult for women to get up afterwards at this stage)—so they concluded by suggesting no exams be conducted unless there is cause for concern. Many years before, Carolyn Flint (1986) had recommended that vaginal exams should be conducted in whatever position the woman happened to be in at the time, and had described how this is possible. See:

- De Jonge A, Teunissen T, Lagro Janssen A. Supine position compared to other positions during the second stage of labour: a meta-analytic review. *Journal of Psychosomatic Obstetrics & Gynaecology,* 2004, 25:35-45
- De Jonge A, Lagro Janssen A. Birthing positions: a qualitative study into the views of women about various birthing positions. *Journal of Psychosomatic Obstetrics & Gynaecology,* 2004, 25:47-55
- Flint C. *Sensitive Midwifery.* Heinemann, 1986

33  If Michel Odent's approach seems negligent to you, you might like to check out the following studies, all of which focus on 'less' being 'more' when it comes to midwifery care:

- Leap N. The less we do, the more we give. In M. Kirkham (ed.) *The Midwife-Mother Relationship.* Macmillan, 2000 (pp 1-18)

- Kennedy H. A model of exemplary midwifery practice: results of a Delphi study, including commentary by Ernst K. *Journal of Midwifery & Women's Health,* 2000, 45(1):4-19

- Fahy K. Being a midwife or doing midwifery. *Australian Midwives College Journal,* 1998, 11(2):11-16

In a presentation at a conference in the UK in 2004 (the annual conference of the British National Childbirth Trust), the late researcher and midwife Trisha Anderson apparently also joked that good midwifery involves 'drinking tea intelligently'. (I'm sure coffee would be fine too!) In other words, in order to best facilitate the natural, healthy, physiological processes of birth a midwife needs to do nothing, while engaging in 'watchful waiting', just to check everything's okay as labor progresses—simply because a good midwife (or obstetrician) knows what to look out for. This contrasts with the obstetric model of care, in which caregivers often have what some researchers (e,g, Grol and Grimshow, 2003) have termed 'a compulsion to act'. See:

- Grol R, Grimshaw J. From best evidence to best practice: effective implementation of change in patient's care. *Lancet,* 2003, 362: 1225-1230

Finally, in the following book, researchers conclude (after conducting an extensive review of midwifery and obstetric practice that 'what really counts, cannot be counted':

- Chalmers I, Kierse M, Neilson J. *A Guide to Effective Care in Pregnancy and Childbirth.* Oxford University Press, 1989

34  After a detailed discussion of the evidence surrounding the use of EFM (and other forms of clinical observation) Denis Walsh has pointed out that 'surveillance' creates an atmosphere of anxiety in the labor room. He reminds us that in 1996 Downs commented that 'gains in technology seem to be resulting in the loss of human essence.' The particular problem with an atmosphere of anxiety is that it prevents the hormones of birth from being produced. (There will be more about this soon.) This explains why Michel wanted to leave me undisturbed. See:

- Walsh D. *Evidence-based Care for Normal Labour and Birth.* Routledge, 2007

- Downs F. Technical innovations: legal implications for nursing. *American Nurses' Association Clinical Sessions,* 1966, 232-237

- Odent M. *Birth and Breastfeeding.* Clairview Books 2003

35  Sheila Kitzinger has written about the need to 'guard the birth space' and Michel Odent has also written about this topic in detail. The topic of birth territory is also discussed by various writers (Kathleen Fahy, Carolyn Hastie, Jennifer A Parratt, Maralyn Foureur, Carolyn Hastie, Bianca Lepori, Lesley Page, Michel Odent, Pat Brodie and Nicky Leap in *Birth Territory and Midwifery Guardianship.* See:

- Fahy, K, Foureur, M, Hastie, C (eds). *Birth Territory and Midwifery Guardianship.* Elsevier, 2008

- Kitzinger S. *Rediscovering Birth.* Little, Brown & Company, 2000

- Odent M. *Birth and Breastfeeding.* Clairview Books, 2003

36  According to Denis Walsh, in 2007 Nottingham City Hospital, which has a large obstetric unit, recommended—and, presumably, continues to recommend—a minimum vaginal examination interval of 12 hours for first-time mothers. (Denis Walsh mentions this in his book *Evidence-based Care for Labor and Birth*, Routledge, 2007.) The following systematic literature review also failed to identify any research basis for vaginal examinations to assess the progress of labor:

- Devane D. Sexuality and midwifery. *British Journal of Midwifery*, 1996, 4(8):413-20

The following researchers also concluded that vaginal examinations were not necessary for facilitating physiological birth:

- Chalmers I, Kierse M, Neilson J. *A Guide to Effective Care in Pregnancy and Childbirth*. Oxford University Press, 1989

The practice of routinely conducting vaginal examinations was shown, in the following study, to be a ritual which legitimized intrusion into the very private birth place during the second stage of labor:

- Bergstrom L, Roberts J, Skillman L, Seidel J. "You'll feel me touching you, sweetie": Vaginal examinations during the second stage of labour. *Birth*, 1992, 19(1):10-18

Also see:

- Stewart M. "I'm just going to wash you down." Sanitizing the vaginal examination. *Journal of Advanced Nursing*, 2005, 51(6):587-594
- Warren C. Invaders of privacy. *Midwifery Matters*, 1999, 81:8-9

37  Michel explains in detail why it's important that no-one talks to the mom-to-be while she's in labor at www.wombecology.com/physiological.html

38  The importance of continuity of care has been confirmed by various studies. See:

- Haggerty J, Reid R, Freeman G, Starfield B, Adair C, McKendry R. Continuity of care: a multidisciplinary review. *British Medical Journal*, 2005, 327:1219-1221
- Flint C. *Midwifery Teams and Caseloads*. Butterworth-Heinemann, 1993
- Flynn A, Hollins K, Lynch P. Ambulation in labour. *British Medical Journal*, 1978, 2(6137):591-593
- Green J, Curtis P, Price H, Renfrew M. *Continuing to Care: The Organization of Midwifery Services in the UK A Structured Review of the Evidence*. Books for Midwives Press, 1998
- Walsh D. An ethnographic study of women's experience of partnership caseload midwifery practice: the professional as friend. *Midwifery*, 1999, 15(3):165-176
- Page L, McCourt C, Beake S, Hewison J. Clinical interventions and outcomes of one-to-one midwifery practice. *Journal of Public Health Medicine*, 1999, 21(3): 243-248
- North Staffordshire Changing Childbirth Research Team. A randomised study of midwifery caseload care and traditional 'shared-care'. *Midwifery*, 2000, 16(4):295-302

39  Michel discusses the impact of apparently small interventions, which change a woman's internal chemistry, in the chapter 'The scientification of love' in his book *The Farmer and the Obstetrician* (Free Association Books, 2002). He discusses

the problem of stimulating the neocortex (through talking) in the chapter 'Breaking a vicious circle' in *The Caesarean,* Free Association Books, 2004.

40   Many studies, such as the following, conclude that women resort to pain relief less often when they are outside large hospitals (e.g. at home or in a birth center) and also when they have continuity of care (i.e. one midwife right the way through their labor and birth):

- Olsen O. Meta-analysis of the safety of home birth. *Birth,* 1997, 24(1):4-13

- Walsh D, Downe S. Outcomes of free-standing, midwifery-led birth centres: a structured review of the evidence. *Birth,* 2004, 31(3):222-229

- Hodnett E, Downe S, Edwards N, Walsh D. Home-like versus conventional birth settings (Cochrane Review). In: *The Cochrane Library,* Issue 2. Chichester: John Wiley & Sons Ltd, 2006

- Waldenstrom U, Turnbull D. A systematic review comparing continuity of midwifery care with standard maternity services. *British Journal of Obstetrics & Gynaecology,* 1998, 105:1160-1170

- Wraight A, Ball J, Seccombe I, Stock J. *Mapping Team Midwifery: A Report to the Department of Health.* Brighton: Institute of Manpower Studies, University of Sussex, 1993

- Harvey S, Jarrell J, Brant R, Stainton C, Rach D. A randomised controlled trial of nurse/midwifery care. *Birth,* 1996, 23:128-135

- Page L, McCourt C, Beake S, Hewison J. Clinical interventions and outcomes of one-to-one midwifery practice. *Journal of Public Health Medicine,* 1999, 21 (3):243-248

41   Michel Odent has written about this phenomenon in the chapter called 'Antenatal scare' in his book *The Caesarean.* Free Association Books, 2004.

42   Focusing on ways of facilitating best outcomes actually seems to be more important than risk assessment. In Michel's opinion, the same principles apply to low and high risk laboring women: the amount of disturbance a woman experiences and the extent to which she feels safe are most important. In other words, leaving the laboring woman undisturbed might be more important than close monitoring because it will give the woman and her unborn baby the best possible conditions for successfully orchestrating the cocktail of hormones necessary for a safe birth. Again, Michel has written about this in the chapter called 'Breaking a vicious circle' in *The Caesarean.* Free Association Books, 2004.

43   As explained earlier, the only effect of EFM identified by research is to increase the cesarean rate; it does nothing to improve either fetal or maternal morbidity or mortality. See Note 28 in this section.

To consider the use of EFM in high risk pregnancies specifically, see the references below. Michel's view is that the conditions needed to optimize outcomes for both low and high-risk labors are the same—privacy, non-disturbance, and the feeling of being safe because a competent midwife (or obstetrician) is discretely in attendance. The references:

- Brown VA, *et al.* The value of antenatal cardiotocography in the management of high-risk pregnancy: a randomised controlled trial. *British Journal of Obstetrics & Gynaecology,* 1982, 89:716-22

- Haverkamp AD, *et al*. The evaluation of continuous fetal heart rate monitoring in high risk pregnancy. *American Journal of Obstetrics & Gynecology*, 1976, 125: 310-20
- Leveno KJ, *et al*. A prospective comparison of selective and universal electronic fetal monitoring in 34,995 pregnancies. *New England Journal of Medicine*, 1986, 315:615-19
- Lumley JC, Wood C, *et al*. A randomized trial of weekly cardiotocography in high-risk obstetric patients. *British Journal of Obstetrics & Gynaecology*, 1983, 90:1018-26
- Sky K, *et al*. Effects of electronic fetal heart rate monitoring, as compared with periodic auscultation, on the neurological development of premature infants. *New England Journal of Medicine*, 1990, (March 1):588-93
- Luthy D, Shy K, van Belle G, Larson E, *et al*. A randomised controlled trial of electronic fetal monitoring in preterm labour. *Obstetrics & Gynaecology*, 1987, 69:687-695
- Madaan M, Trivedi S. Intrapartum electronic fetal monitoring vs. intermittent auscultation in post caesarean pregnancies. *International Journal of Obstetrics & Gynaecology*, 2006, 94:123-125

In the case of EFM to detect cerebral palsy a false positive rate of 99.8% has been observed! See:

- Nelson K, Dambrosia J, Ting T, Grether J. Uncertain value of electronic fetal monitoring in predicting cerebral palsy. *New England Journal of Medicine*, 1996, 334:659-660

As noted elsewhere, attitudes towards monitoring really are changing amongst caregivers... For example, one large British obstetric unit has recommended a minimum vaginal examination interval of 12 hours for first-time mothers. See:

- Thorton J. Natural labour guidelines. Nottingham City Hospital. Personal Communication with Denis Walsh, 2006. Reported in *Evidence-based Care for Labour and Birth,* Routledge, 2007

44  Specific research is mentioned throughout this section, when relevant. For a more complete overview of research relating to cesareans, particularly as this operation compares with the physiological model of birth, see Michel Odent's book *The Caesarean* (Free Association Books 2004).

45  For more on this (including references to research), see the chapters called 'What mothers say' and 'The perineal preoccupation' in Michel's book *The Caesarean* (Free Association Books 2004). Also see Harrison's paper, as follows, which explains why the unborn baby needs to be exposed to the stress of labor if he or she is to be ready for life outside the womb:

- Harrison J. Fetal perspectives on labour. *British Journal of Midwifery*, 1999, 7 (10): 643-647

46  To explore the truth of this contention, visit the Primal Health Research Database at www.birthworks.org/primalhealth. Use the keyword 'cesarean' or 'caesarean' to locate the many articles which suggest correlations. One study in 2008 (Cardwell, *et al*) found that children born by cesarean were 20% more likely to get childhood-onset type 1 diabetes. Another study (Thavagnanam, *et al*, 2008) also found that

cesarean-born children were 20% more likely to be asthmatic. Another study (Koplin, *et al,* 2008) found that being born by cesarean increased a baby's chances of developing sensitization to food allergens, which might make them more prone to allergies. See:

- Cardwell CR, Stene LC, Joner G, Cinek O, Svensson J, Goldacre MJ, Parslow RC, Pozzilli P, Brigis G, Stoyanov D, Urbonaite B, Sipetić S, Schober E, Ionescu-Tirgoviste C, Devoti G, de Beaufort CE, Buschard K, Patterson CC. Caesarean section is associated with an increased risk of childhood-onset type 1 diabetes mellitus: a meta-analysis of observational studies. *Diabetologia,* 2008, May;51 (5): 726-35. Epub 2008 Feb 22

- Thavagnanam S, Fleming J, Bromley A, Shields MD, Cardwell CR. A meta-analysis of the association between Caesarean section and childhood asthma. *Clinical & Experimental Allergy,* 2008, Apr;38(4):629-33

- Koplin J, Allen K, Gurrin L, Osborne N, Tang ML, Dharmage S. Is caesarean delivery associated with sensitization to food allergens and IgE-mediated food allergy: a systematic review. *Pediatric Allergy & Immunology,* 2008 Dec; 19 (8):682-7

47   Newton gave the hormone this name in:

- Newton N. The Influence of the Let-Down Reflex in Breast Feeding on the Mother-Child Relationship. *Marriage & Family Living,* 1958, 20:18-20

In 1988 Marchini demonstrated that cesareans involve significantly lower levels of oxytocin. For more details on this see:

- Marchini G, Lagercrantz H, Winberg J, Uvnas-Moberg K. Fetal and maternal plasma levels of gastrin, somatostatin and oxytocin after vaginal delivery and elective cesarean section. *Early Human Development,* 1988, Nov;18(1):73-9

48   When considering the main factors causing urinary incontinence later on in life (which seems to relate to damage to the pelvic floor), researchers established that obesity was the second most important cause, with only heredity coming before that. See:

- Hannestad Y, Rortveit G, Dalveit A, Hunskaar S. Are smoking and other lifestyle factors associated with female urinary incontinence? The Norwegian EPINCONT study. *British Journal of Obstetrics & Gynaecology,* 2003, 110:247-254

It is true that ultrasound examinations have identified micro-damage to the pelvic floor after vaginal births (Dietz and Schierlitz 2005) but—as Denis Walsh points out—there is an enormous difference between mechanical and neural damage, mainly because neural damage (which is what we're talking about here) involves absolutely no symptoms, so in fact it doesn't affect women. As another research study concluded (Cardozo and Gleeson 1997), the best possible outcome is for a woman to have an intact perineum—i.e. for no tearing *or cutting* to take place—because this is associated with the least pain postpartum, the least amount of incontinence (usually none) and the earliest resumption of normal sexual intercourse after birth. It's also interesting (and perhaps a little perplexing) to note that women who've had an emergency cesarean have virtually the same rates of urinary incontinence postpartum as women who've given birth vaginally—and that even women who've had elective cesareans complain of incontinence postpartum (Chaliha, *et al,* 2002, Lal, *et al,* 2003, Rortveit, *et al,* 2003.) Research conducted by Gordon and Logue 1985, Jolleys 1988, Nygaard 1997 and Viktrup 1992 found no differences in muscle strength or urinary continence between women birthing vaginally and by cesarean. According to these research studies, the women who suffer most after giving birth are those who have forceps deliveries. See:

- Dietz H, Schierlitz L. Pelvic floor trauma in childbirth—myth or *reality? Australian & New Zealand Journal of Obstetrics & Gynaecology,* 2005, 45(1):3-11
- Walsh D. *Evidence-based Care for Normal Labour and Birth.* Routledge, 2007
- Cardozo L, Gleeson C. Pregnancy, childbirth and continence. *British Journal of Midwifery,* 1997, 5(5):277-281
- Chaliha C, Khullar V, Stanton S. Urinary symptoms in pregnancy: are they useful for diagnosis? *British Journal of Obstetrics & Gynaecology,* 2002, 109:1181-1183
- Lal M, Mann C, Callender R, Radley S. Does caesarean delivery prevent anal incontinence? *Obstetrics & Gynaecology,* 2003, 101:305-312
- Rortveit A, Kjersti D, Yugvild S, Hannestad S, Hunskaar S. Urinary incontinence after vaginal delivery or caesarean section. *New England Journal of Medicine,* 2003, 348: 900-907
- Gordon H, Logue M. Perineal muscle function after childbirth. *Lancet* 1985;2:123-5
- Jolleys JV. Reported prevalence of urinary incontinence in women in a general practice. *British Medical Journal,* 1988, 296:1300-2
- Nygaard IE, Rao SSC, Dawson JD. Anal incontinence after anal sphincter disruption: a 30-year retrospective cohort study. *Obstetrics & Gynecology,* 1997, 89(6):896-901
- Viktrup L, *et al.* The symptom of stress incontinence caused by pregnancy or delivery in primiparas. *Obstetrics & Gynecology,* 1992, 79(6):945-9

In conclusion, research has shown that the urinary incontinence often results in the following cases: having a forceps birth (Arya, *et al,* 2001), pushing actively for a long time (Kirkman 2000), having lots of babies, having babies who weigh more than 3,700g, and being over 30 at the time of giving birth (Mason, *et al,* 1999). (In case a woman finds herself in any of these categories, she might at least be reassured to note that none of these factors are associated with increased incidence of tearing.) Research has shown that urinary stress incontinence is best avoided by doing pelvic floor exercises prenatally (Morkved, *et al,* 2003, Sampselle, 2000) and a woman pushing her baby out herself, according to her own timetable (Kirkman, 2000); postpartum pelvic floor exercises have also been found to be helpful (Glazener, *et al,* 2001). See:

- Arya L, Jackson N, Myers D, Verma A. Risk of new onset urinary incontinence after forceps and vacuum delivery in primiparous women. *American Journal of Obstetrics & Gynaecology,* 2001, 185:1318-1324
- Kirkman S. The midwife and pelvic floor dysfunction. *The Practising Midwife,* 2000, 3(8):20-22
- Mason L, Glenn S, Walton I, Appleton C. The prevalence of stress incontinence during pregnancy and following birth. *Midwifery,* 1999, 15(2):120-127
- Mason L, Glenn S, Walton I, Appleton C. The experience of stress incontinence after birth. *Birth,* 1999, 26(3):164-171

- Morkved S, Bo K, Schei B, Salvesen K. Pelvic floor muscle training during pregnancy to prevent urinary incontinence: a single-blind randomized controlled trial. *Obstetrics & Gynaecology*, 2003, 101:313-319
- Sampselle C. Behavioural intervention for urinary incontinence in women: evidence for practice. *Journal of Midwifery & Women's Health*, 2000, 45(2):94-103
- Glazener C, Herbison G, Wilson P. Conservative management of persistent postnatal urinary and faecal incontinence: randomised controlled trial. *British Medical Journal*, 2001, 323:593

Looking at incontinence later on in life, researchers have established the following causes (in order of importance, starting with the most important) (Hannestad, *et al*, 2003, Rortveit, *et al*, 2003): heredity, obesity, smoking, HRT, the number of children a woman's had and how she's had them (noting that forceps is the worst way, from the point of view of incontinence). See:

- Hannestad Y, Rortveit G, Dalveit A, Hunskaar S. Are smoking and other lifestyle factors associated with female urinary incontinence? The Norwegian EPINCONT study. *British Journal of Obstetrics & Gynaecology*, 2003, 110:247-254
- Rortveit G, Daltveit A, Hannestad Y, Hunskaar S. (2003) Vaginal delivery parameters and urinary incontinence: the Norwegian EPINCONT study. *American Journal of Obstetrics & Gynecology*, 2003 Nov; 189(5):1268-74

Finally, in case all this isn't enough to put you off your dinner, painful sex (dyspareunia) and perineal pain after childbirth have been associated with forceps and ventouse births, episiotomy and anal spincter tears (Buhling, *et al*, 2005). See:

- Buhling K, Schmidt S, Robinson J, Klapp C, Siebert G, Dudenhausen J. Rate of dyspareunia after delivery in primiparae according to mode of delivery.
- *European Journal of Obstetrics & Gynecology*, 2005, 124:42-46

*Dyspareunia and perineal pain have been associated with forceps and ventouse births, episiotomy and anal tears*

49    There's something about a physiological birth which seems to empower women. This could be because a physiological birth is an *active* birth... or it could be because of the very different experience a woman has (in terms of hormones, endorphins, etc) when she labors undrugged. Then again, it could be something to do with control—and issue which seems to keep coming up when emotions around birth are investigated.

50    Amu O, Rajendran S and B, Ibrahim I. *British Medical Journal*, 1998, 317:462-465 (15 August)

51    See the following study, as well as Michel's comments on this (with references to research) in the chapter 'Once a caesarean always a caesarean' in *The Caesarean* (Free Association Books 2004):

- Van Ham M, van Dongen P, Mulder J. Maternal consequences of caesarean section. A retrospective study of intra-operative and postoperative maternal complications of caesarean section during a 10 year period. *European Journal of Obstetrics, Gynaecology & Reproductive Biology*, 1997, 74(1):1-6

52 Women who've had a cesarean are also more likely to have a cesarean at the end of their pregnancy, experience post-traumatic stress syndrome and have reduced breastfeeding rates. See:

- Hemminki E, Merilainen J. Long-term Effects of Caesarean Sections: Ectopic Pregnancies and Placental Problems. *American Journal of Obstetrics & Gynecology*, 1996, 174, No 5:1569-1574
- Hemminki E. Impact of caesarean section on future pregnancy—a review of cohort studies. *Paediatric & Perinatal Epidemiology*, 1996, 10(4):366-379
- Ananth C, Smulian J, Vintzeleos A. The association of placenta praevia with history of caesarean delivery and abortion: a meta-analysis. *American Journal Obstetrics & Gynaecology*, 1997, 177(5):1071-1078
- Ryding E, Wijma K, Wijma B. Experiences of emergency caesarean section: a phenomenological study of 53 women. *Birth*, 1998, 25(4): 246-251
- Sackett D. Evidence based medicine: what it is and what it isn't. *British Medical Journal*, 1996, 312:71-72
- Di Matteo M, Morton S, Lepper H, Damush T, Carney M, Pearson M, Kahn K. Caesarean childbirth and psychosocial outcomes: a meta-analysis. *Health Psychology*, 1996, 15(4):303-314

Also see the chapter 'Safer and safer' in *The Caesarean* (Free Association Books 2004).

53 In addition, because of the decreased catecholamine surge, babies born by cesarean are at increased risk of respiratory compromise (Faxelius 1983) as well as low blood sugar (Hagnevik 1984). See:

- Faxelius G, Hagnevik K, et al. Catecholamine surge and lung function after delivery. *Archives of Disease in Childhood*, 1983, Apr; 58(4):262-6
- Hagnevik K, Faxelius G, et al. Catecholamine surge and metabolic adaptation in the newborn after vaginal delivery and caesarean section. *Acta Paediatrica Scandinavica*, 1984, Sep; 73(5):602-9

54 Enkin 2000 concludes that the risk increases by four times. See:

- Enkin MW, Keirse MJNC, Neilson J, Crowther C, Duley L, Hodnett E, Hofmeyr J. *A Guide to Effective Care in Pregnancy & Childbirth.* 3rd edition. Oxford: Oxford University Press, 2000

55 If you still doubt this, see the following articles:

- Liu S, Liston RM, Joseph KS, et al. Maternal mortality and severe morbidity associated with low-risk planned cesarean delivery versus planned vaginal delivery at term. *Canadian Medical Association Journal*, 2007, 176(4):455-60
- Hannah ME, Hannah WJ, et al. Planned caesarean section versus planned vaginal birth for breech presentation at term: a randomised multicentre trial. *Lancet*, 2000, 356:1375-83
- Krebs L, Langhoff-Roos J. Elective cesarean delivery for term breech. *Obstetrics & Gynecology*, 2003, 101(4):690-6
- Cheng M, Hannah ME. Breech delivery at term : a critical review of the literature. *Obstetrics & Gynecology*, 1993, 82:605-18

• Lumley J. Any room left for disagreement about assisting breech births at term? *Lancet*, 2000, 356:1368-9

56  For the following account, I am indebted to *The Caesarean* (Free Assoc Books 2004), *Obstetrics by Ten Teachers* (Hodder Arnold 2000), *Myles Textbook for Midwives* (Churchill Livingstone 2000) and *Abdominal surgical incisions for caesarean section* (Mathai M, Hofmeyr GJ. Abdominal surgical incisions for caesarean section. *Cochrane Database of Systematic Reviews* 2007, Issue 1. Art. No.: CD004453. DOI:10.1002/14651858. CD004453. pub2). Some details are vague because of the variation in protocols in different hospitals or different countries. See:

• Mathai M, Hofmeyr GJ. Abdominal surgical incisions for caesarean section (Protocol for a Cochrane Review). In: *The Cochrane Library*, Issue 2. Oxford: Update Software, 2004

57  These have also been recommended by Joel-Cohen and Wallin.

58  This is especially important because one study (Nissen, *et al*, 1996) found that after cesareans the first suckling typically took place 240 minutes after the birth, which contrasted dramatically with the 75 minutes in the case of women who hadn't had cesareans. See:

• Nissen E, Uvnas-Moberg K, Svensson K, Stock S, Widstrom AM, Winberg J. Different patterns of oxytocin, prolactin but not cortisol release during breastfeeding in women delivered by caesarean section or by the vaginal route. *Early Human Development*, 1996, 45:103-18

If hormones are inevitably disrupted, it's important that a woman who needs an elective cesarean compensates by making a special request in advance to ensure that her baby gets put to her breast as soon as possible. Other studies (Salariya 1978, De Chateau 1977) found that early and frequent suckilng positively influences milk production and the duration of breastfeeding. The full references are as follows:

• Salariya EM, *et al*, Duration of Breastfeeding after Early Initiation and Frequent Feeding, *Lancet*, 2, No 8100, 1978, 1141-1143

• De Chateau P, Wiberg B. Long-term Effect on Mother-Infant Behaviour of Extra Contact during the First Hour Postpartum. *Acta Paediatrica Scandinavica*, 1977, 66:145-151

59  In this chapter there is no focus on possible financial reasons for the rise in statistics, but some writers have suggested finance might well be a factor. After all cesarean sections cost significantly more than a vaginal delivery and involve longer hospital stays (which tend to be around three times longer), which also need to be paid for. Only a 1% increase in the cesarean rate in a particular hospital will dramatically increase its income and, of course, individual obstetricians may benefit enormously financially over a period of time. In places where women have a choice between 'state-funded' or 'private' hospitals, it is not difficult to see how increasing the rate of cesareans might be financially appealing within the private sector. Of course this is often where fashions and trends are set for the rest of the population because what is 'expensive' is often seen as being desirable, even if it is also 'expensive' in terms of health and a newborn baby's welfare. Rather tellingly, in places where there are high rates of poverty (e.g. in

Brazil or China), there is often a dramatic difference in c-section rates between government and private hospitals, with private rates rising as high as 90%.

60    This issue is also discussed in *The Caesarean,* Free Association Books, 2004. Also, one study found that women who had a cesarean were more likely to experience a decline in mood and self-esteem postpartum (Fisher 1997). See:

- Fisher J, *et al,* Adverse Psychological Impact of Operative Obstetric Interventions: A Prospective Longitudinal Study, *Australia & New Zealand Journal of Psychiatry,* 31, 1997, 728-738

Elective cesareans involve lower levels of endorphins

61    Michel Odent has written about this topic in the chapter called 'The perineal preoccupation' in *The Caesarean* (Free Association Books, 2004). Other differences in the processes have been identified by other researchers. Elective cesareans involve lower levels of endorphins (Facchinetti, 1990), catecholamines (Jones, 1982), and prolactin (Heasman, 1997). The full references are as follows:

- Facchinetti F, Garuti G, Petraglia F, Mercantini F, Genazzani AR. Changes in beta-endorphin in fetal membranes and placenta in normal and pathological pregnancies. *Acta Obstetrica et Gynecologica Scandinavica,* 1990, 69(7-8): 603-7

- Jones CM 3rd, Greiss FC Jr. The effect of labor on maternal and fetal circulating cate-cholamines. *American Journal of Obstetrics & Gynecology,* 1982, Sep 15; 144(2):149-53

- Heasman L, Spencer JA, Symonds ME. Plasma prolactin concentrations after caesarean section or vaginal delivery. *Archives of Disease in Childhood, Fetal & Neonatal Edition,* 1997, Nov;77(3):F237-8

62    The strict protocols based on observations carried out as far back as 1954 (resulting in the 'Friedman curve') are questioned by more modern research. Many studies have concluded that first-time mothers tend to have longer labors than women who've already had at least one baby. The following study concluded that current diagnostic criteria for protracted or arrested labor may be too stringent:

- Zhang J, Troendle J, Yancey M. Reassessing the labour *curve. American Journal of Obstetrics & Gynaecology,* 2002, 187:824-828

This and other papers, such as the following one, suggest that there is more variation in women (laboring healthily) than was thought to be the case in the past:

- Gurewitsch E, Diament P, Fong J, *et al.* The labour curve of the grand multipara: Does progress of labour continue to improve with additional childbearing? *American Journal of Obstetrics & Gynaecology,* 2002, 186:1331-1338

Also see:

- Gross M, Haunschild T, Stoexen T, Methner V, Guenter H. Women's recognition of the spontaneous onset of labour. *Birth,* 2003, 30(4):2b7-271

- Gross M, Hecker H, Matterne A, Guenter H, Kierse M. Does the way that women experience the onset of labour influence the duration of labour? *British Journal of Obstetrics & Gynaecology,* 2006, 113:289-294

One research study pointed out that the last thing a busy labor ward needs is 'nigglers' (women whose labors aren't progressing 'fast enough') because they are seen to 'clog up the place'. See:

- Hunt S, Symonds A. *The Social Meaning of Midwifery.* Basingstoke: Macmillan, 1995

The following paper reports on latent periods (when nothing happened) during home births, which were (correctly) not interpreted by midwives as pathological:

- Davis B, Johnson K, Gaskin L. The MANA Curve—describing plateaus in labour using the MANA database. Abstract No 30, 26th Triennial Congress, ICM, Vienna, 2002

In the following paper, the conclusion is that midwives and other caregivers need to focus on 'unique normality', which varies from woman to woman:

- Downe S, McCourt C. From being to becoming: reconstructing childbirth knowledges. In Downe S (ed) *Normal Childbirth: Evidence and Debate.* Churchill Livingstone, 2004

Women were equally satisfied with longer labors

Another study demonstrated that women were equally satisfied with longer labors:

- Lavender T, Alfirevic Z, Walkinshaw S. Effects of different partogram action lines on birth outcomes: a randomised controlled trial. *Obstetrics & Gynaecology,* 2006, 108(2):295-302

One midwife, the late Trisha Anderson, at a conference in the UK in 2004, also proposed types of dystocias, which are clearly to be avoided—lack of continuity of care 'dystocia', inexperienced doctors at the start of their rotation 'dystocia', absence of expertise 'dystocia', disagreements between caregivers 'dystocia', inadequate handovers 'dystocia' and intimidation 'dystocia'.

63 See full statistics at the following URL:

 http://whqlibdoc.who.int/publications/2007/9789241596213_eng.pdf

64 If you doubt this idea at all now, explore www.birthworks.org/primalhealth—a website which catalogues academic articles according to topic (autism, anorexia, suicide, etc).

65 Nolan (2005) concluded that prenatal education packages do not consistently reduce intervention rates, which in turn suggests that they do not help. However, carefully designed (somewhat non-typical) prenatal classes did result in an increase in the use of upright positions and/or a decrease in the number of epidurals. See:

- Nolan M. Childbirth and parenting education: what the research says and why we may ignore it. In Nolan M and Foster J (eds) *Birth and Parenting Skills: New Directions in Antenatal Education.* Churchill Livingstone, 2005

- Nolan M and Foster J (eds) *Birth and Parenting Skills: New Directions in Antenatal Education.* Churchill Livingstone, 2005

- Foster J. Innovative practice in birth education. In M. Nolan and J. Foster (eds) *Birth and Parenting Skills: New Directions in Antenatal Education.* Elsevier Science, 2005

- Walsh D, Harris M, Shuttlewood S. Changing midwifery birthing practice through audit. *British Journal of Midwifery,* 1999, 7(7):432-345

66    For an explanation of this, see the following article:

- Odent M. The second stage as a disruption of the fetus ejection reflex. *Midwifery Today, International Midwife,* 2000, Autumn; (55): 12

It's useful to consider *all* forms of monitoring of labor, including vaginal examination to track cervical dilation

It's actually useful to consider *all* forms of common monitoring of labor, including vaginal examination to track cervical dilation. According to Denis Walsh, in 2007 Nottingham City Hospital, which has a large obstetric unit, recommended—and, presumably, continues to recommend—a minimum vaginal examination interval of 12 hours for first-time mothers. (Denis Walsh mentions this in his book *Evidence-based Care for Labour and Birth,* Routledge, 2007.) Of course, this recommendation reflects changing attitudes towards monitoring generally, which includes EFM. The following systematic literature review also failed to identify any research basis for vaginal examinations to assess the progress of labor:

- Devane D. Sexuality and midwifery. *British Journal of Midwifery,* 1996, 4(8):413-20

The following researchers also concluded that vaginal examinations were not necessary for facilitating physiological birth:

- Chalmers I, Kierse M, Neilson J. *A Guide to Effective Care in Pregnancy and Childbirth.* Oxford University Press, 1989

The practice of routinely conducting vaginal examinations was shown, in the following study, to be a ritual which legitimized intrusion into the very private birth place during the second stage of labor:

- Bergstrom L, Roberts J, Skillman L, Seidel J. "You'll feel me touching you, sweetie": Vaginal examinations during the second stage of labour. *Birth,* 1992, 19(1):10-18

Also see:

- Stewart M. "I'm just going to wash you down." Sanitizing the vaginal examination. *Journal of Advanced Nursing,* 2005, 51(6):587-594
- Warren C. Invaders of privacy. *Midwifery Matters,* 1999, 81:8-9

Michel Odent discusses the impact of apparently small interventions, which change a woman's internal chemistry, in the chapter 'The scientification of love' in his book *The Farmer and the Obstetrician* (Free Association Books, 2002). He discusses the problem of stimulating the neocortex (through talking) in the chapter 'Breaking a vicious circle' in *The Caesarean,* Free Association Books, 2004.

He discusses the problem of stimulating the neocortex...

67    See Michel's comments on this in the chapter called 'At the dawn of the post-electronic age' in the book *Birth and Breastfeeding* (Clairview, 2003).

68    Michel comments on this in the chapter 'Safer and safer' in *The Caesarean* (Free Association Books 2004).

*It's helpful for all people involved in birth to understand the reasons behind decisions, whatever they might be in the case of a particular woman's labor and birth*

# Optimal Birth: Why

## Why is maximally physiological birth best?

Before we get very intellectual and analytical about this question, let's go straight to the heart of the matter.

### Birthframe 20: Jan Tritten—Motivation to become a midwife

Like Jan Tritten, founder of American-based website and magazine *Midwifery Today*, from your own experience you may know why it's best to aim for physiological birth, avoiding what has been called the industrialization of birth. Here, Jan writes about what made her want to change what she calls 'the world of birth'.[1]

I am from the United States, a country where midwives have all but disappeared. Doctors did something about 'the midwife problem' by nearly eliminating them. However, women's experiences with birth have never been doctors' forte. That is the domain of midwives. It is how we can distinguish ourselves as the rightful helpers of women in birth. It is out of personal experience that women have begun to reclaim midwifery. For me, it was the pain from my first birth and the joy of my second one that put a fire in my belly that will not go away.

> For me, it was the pain from my first birth and the joy of my second one that put a fire in my belly that will not go away

Recently, a friend told me that I suffer from post-traumatic stress disorder. I spent years going over and over the experience, wishing it had been different. The most difficult to deal with were the flashbacks and the guilt.

I was having a lovely labor. Mostly, I was left alone with my sister and husband. I enjoyed the process of bringing forth my baby. Yes, labor can be complete joy. My three were. My anthropology background helped me know women could do this because billions of women before me had. In those days, at least for me, there was little interruption from staff. However, about half an hour before my daughter was born, they came in and insisted on a saddle block. "But I am having a natural birth," I murmured. "This is a natural birth," they countered. And with one needle in the back, my birth was ruined, leaving me to carry the guilt and pain still today, 30 years later.

My little girl came out blue, without my feeling her birth, to women screaming at me: "Push, or we'll have to take it out" (with forceps). They pinked her up with oxygen and gave me a quick glance at her, then put me in a big recovery room with all the other victims. My baby went to the nursery.

There I was, utterly alone, languishing without my baby. Although I would breastfeed her for three years, I struggled for two days getting started. I think I was blessed that no one tried to teach me how to do it.

My second birth was a powerful contrast. My doctor talked me into having a homebirth, having seen the light himself.

My second birth was a powerful contrast. My doctor talked me into having a homebirth, having seen the light himself. I enjoyed the pregnancy immensely. And not only was the birth a beautiful, easy and joyous birth, it was completely without intervention. I was joyous and riding high on the wave of that incredible experience for three years straight.

During my pregnancy, my enlightened doctor had taught me so much at each prenatal visit. He recommended several progressive books, which had surfaced by the mid '70s. And after the birth I kept reading my childbirth books. My husband said: "But you already had the baby." I said: "I think there is something more here." Within three months, I was practicing midwifery with The Birth Co-op. Marion Temple McClean, our mentor and now regular *Midwifery Today* writer, was teaching us at this time.

Birth experiences really were the catalyst for the revival of midwifery in the US. Many midwives have this motivation, though their birth stories are different. In my case, experience motivated me, after being a homebirth midwife, to start *Midwifery Today* to try to change birth practices around the world. Every cell in my body screams for the beautiful gift of undisturbed birth that a woman's being cries for, even if her mind has become confused with pregnancy myths and birth legends.

The contrast between my births has propelled me to spend my life working for women and their unborn babies, even when they don't understand the power and beauty they so easily give away.

This article was adapted from an article which appeared in the British professional magazine *The Practising Midwife* in January 2003. By the way, *Midwifery Today*—a quarterly journal and a weekly e-mailed newsletter—is a wonderful source of information and inspiration for both midwives and women interested in childbirth. To arrange free subscription to the newsletter see www.midwiferytoday.com/enews /

Coming back to the question of this chapter, let's now get a bit more specific. There's some overlap, but we can in fact tease out eight reasons...

# Reason No. 1:
## The physiological processes are complex and efficient

It's very easy to underestimate the complexity and efficiency of the physiological processes when they are not disturbed. I've already been through each stage of pregnancy, labor and birth but I haven't yet described the hormonal landscape while all these processes are taking place. It's quite incredible...[2]

It's very easy to underestimate the complexity of the physiological processes. The hormonal landscape is incredible...

## The hormonal cocktail of pregnancy

The first hormone noticed in pregnancy tests is, of course, human chorionic gonadotropin (hCG). Progesterone, which is initially produced by the ovaries (stimulated to do so by hCG) and later by the placenta, will already have prepared the womb for pregnancy by stimulating the uterine lining to thicken (so that the fertilized egg can implant itself).

Progesterone also has the important function of stopping the uterine muscles from contracting, thus allowing the fetus to grow undisturbed, and of stopping lactation from beginning until after the birth. It also protects the placenta by fighting off unwanted cells and strengthens the mucous plug, thereby preventing infection.

Fetal growth and maternal breast development are supported by another hormone called human placental lactogen (HPL) and this hormone also has the important job of ensuring that the mother doesn't absorb too much glucose. Estrogen, a group of hormones, is also vital and is thought to trigger the maturation of fetal lungs, kidneys, liver and adrenal glands, the production of prolactin, which prepares the woman for breastfeeding, and to regulate bone density in the fetus—amongst other things.

Other hormones—calcitonin, thyroxine, insulin, relaxin, cortisol and erythropoietin also play key roles. Oxytocin is vital in pregnancy too, although it is 'drowned out' by progesterone until the woman goes into labor. However, as well as stimulating the mammary glands to produce milk, it enhances nutrient absorption and helps the woman conserve her energy by making her sleepy![3]

## The hormonally-triggered switchover to 'instinctive'

Since the hormones of birth are produced mainly in the 'middle' brain (sometimes called the 'mammalian' brain or limbic system), an important switchover needs to take place sometime in the early stages of labor. Instead of being controlled by the neocortex (which makes us rational beings), as is normally the case, the woman must suddenly surrender control to this more primitive part of the brain. Hormones which she spontaneously produces will allow her to effect this important switchover.

Interestingly, these hormones (which allow a woman to experience labor as positive and even enjoyable) also trigger appropriate mothering behavior after the baby is born, so they have two important purposes. What's more, beta-endorphins, which are produced during pregnancy, begin to be produced at much higher levels, along with CRH (another hormone produced in stressful situations) at this stage. Rising levels of beta-endorphins now and during labor, which peak at birth and subside one to three days after the birth, help the mother to cope with pain and perhaps even experience it as positive. And they also help her to get into an appropriate instinctive state of mind...[4]

## The hormones of labor and birth

Despite all we that we know about birth, it is still not known precisely what triggers the onset of labor, but both estrogen and progesterone are thought to play a part, perhaps because of their changing ratios.[5] Of course, estrogen is the main hormone which prepares the uterus for the contractions of labor[6] and both hormones also serve the important function of activating the woman's natural opiate-producing mechanisms which operate in the brain and the spinal cord.[7]

Produced in the hypothalamus and released into the bloodstream in a pulsatile manner, oxytocin is associated with positive emotions and sensations. Often called 'the hormone of love',[8] it is produced during orgasm, birth itself, breastfeeding and also in social situations such as eating with other people.[9] (Interestingly, perhaps, the baby produces this hormone as well as the birthing woman and even the placenta, so he or she is swimming in oxytocin-flavored amniotic fluid![10] Because of this strange fetal hormonal production at the onset of labor some researchers have suggested that the baby may be responsible for triggering the onset of labor.)[11] Wherever it comes from (i.e. baby, mother or placenta), while oxytocin is thought to be the main initiator of rhythmic contractions, some researchers have hypothesized that prostaglandins assume this same role later on in labor.[12] Interestingly, oxytytocin has been shown to relieve pain in pregnant rats and mice.[13]

Oxytocin is responsible not only for the contractions which open up the laboring woman's cervix, but also for those which push the baby down and out of her body[14] and for the contractions which make her uterus contract back to its pre-pregnancy size postpartum. Intriguingly, though, something interesting occurs just before the moment of birth in the case of a woman who has been completely undisturbed while in labor. Catecholamines—the fight-or-flight hormones adrenaline and noradrenaline, along with dopamine—are produced alongside oxytocin.[15] (These hormones are usually antagonistic, but they aren't when they are produced spontaneously and smoothly in labor.) Instead of stimulating the body to fight or flee (as they usually do), just seconds before the birth these hormones activate what some researchers (particularly Michel Odent) have called the 'fetus ejection reflex'.

From the laboring woman's point of view, there is a rush of energy which makes her want to stand up and perhaps also grab hold of something. (Any birth attendants, including a much-loved midwife, might be abused or shouted at around this time! Frequently, the woman expresses intense fear, anger or resentment.) And then she flicks her hips forward and gives birth! The intense, highly efficient contractions stimulated by the high levels of oxytocin combine with these catecholamines to produce the perfect cocktail for birth. As for the baby, the sudden surge of catecholamines (especially noradrenaline) has a beneficial effect because it protects him or her from the effects of oxygen deprivation, helps metabolistic processes and temperature regulation and improves respiration—which will come in handy outside the womb, in the big wide world.[16]

## Hormonal production shortly after the birth

Postpartum uterine contractions are made possible because the newborn baby helps the mother to continue producing oxytocin. His or her stroking, licking or sucking actions, alongside skin-to-skin and eye contact, trigger a loving response in the mother, which results in continued hormonal production.[17] Most importantly, perhaps, all this pleasant 'bonding' which facilitates efficient uterine contraction has a safety function because it prevents a woman from suffering from a postpartum hemorrhage. (On this point, though, it's important to remember the role of catecholamines. While they were useful at the point of birth, their usefulness disappears straight afterwards. If levels do not drop—which is likely to occur if the new mother feels cold—oxytocin production will be inhibited, which will obviously increase the risk of PPH. When they do drop, although it's positive in terms of safety, the mother typically feels shaky or cold.

Research has found that oxytocin levels in the mother peak when the placenta is born and then subside over the next hour.[18] Levels of oxytocin in the newborn baby peak around 30 minutes after the birth, but are still high for at least four days after the birth.[19] This generally means that both mother and newborn baby enjoy the pleasant effects of oxytocin in the first hour after the birth and afterwards.

High catecholamine levels at the moment of birth also cause extreme alertness in the baby.[20] Levels typically remain high for around 12 hours,[21] then subside so a much-needed rest is possible.

Noradrenalin, another hormone which is produced spontaneously at this time, is thought to promote mothering behavior because mice who were bred to be deficient in this didn't care for their young unless they had an artificial injection of the hormone. (It's interesting to note that smell probably also plays a key role in early mothering behavior.[22] One research study discovered that monkeys who were given cesareans rejected their babies unless the newborns were swabbed with secretions from the mother monkey's vagina.)[23]

Prolactin, the main hormone stimulating breastfeeding, also causes mothering behavior,[24] as well as an increase in appetite (in the mother), suppression of fertility and changes in sleep cycles (which suddenly include more REM sleep), amongst other things. Of course, these changes support the new mother's abrupt shift in lifestyle, which is bound to include a need for a focus on the newborn baby (even in sleep), to the detriment of the mother's own needs.

Finally, it's interesting to note that the beta-endorphins which began rising at the onset of labor[25] continue to play an important role in the mother-infant breastfeeding relationship.[26] With their ability to induce feelings of pleasure and dependency, they usefully help to cement the mother-baby bond. Since they peak approximately 20 minutes after a woman starts breastfeeding at any one time, they continue to enhance the mother-infant relationship long after the birth took place.[27]

## Reason No. 2:
## Intervention is an uncontrolled experiment

Experimentation has become the standard approach over the last few centuries, despite the likelihood that any intervention is likely to disturb the physiological processes. Experimental procedures have become routine before research has confirmed they are safe or even helpful. How has this situation developed?

### From the birth of time...

Hundreds (and thousands) of years ago interventions were initiated when women or babies were in difficulty. Perhaps it seemed likely the mother was about to die in labor; sometimes the fetus was already known to be dead; sometimes it seemed to be having difficulty getting itself born. Anthropological data (collected by Sheila Kitzinger, Jacqueline Vincent Priya and others) make it clear that other non-life-saving interventions were also common and very culture-specific. They came about as a result of certain beliefs, which were often not based on any facts. These beliefs affected the ways in which birth was facilitated or unintentionally hindered. For example, women may have been isolated or, alternatively, surrounded by people, perhaps given a shock or kept away from all stress and there was often the belief that the pre-milk 'colostrum' was bad for the baby. (Of course, research has now shown us how wonderfully beneficial it is.)[29] Finally, interventions became necessary as a result of the use of pain relief, which usually created problems which then needed to be dealt with.

### 1800-1900

Intervention was clearly advocated by certain doctors as soon as they started taking over from midwives as the usual birth attendants. One midwife's diary suggests that doctors were already using laudanum for pain relief or anesthesia. A range of folklore methods had been employed over the centuries, but specific efforts to develop pain relief for laboring women were now being made. It is unclear when opiates began to be used, but they were probably already in use in the early 1800s. Ether (diethyl ether, to be precise) seems to have been first used in January 1847 in Edinburgh, Scotland, and in April or May of the same year in America. Soon after, English Queen Victoria famously allowed chloroform to be used for the birth of her eighth child in 1853. By the late 1800s, routine intervention had become commonplace in some American hospitals. For example, at one hospital in Philadelphia it was routine to administer quinine (because it induces uterine contractions), as well as drugs for constipation, sleeplessness and headaches; at the onset of labor women were given an enema and bath, the amniotic sac was ruptured, and at birth forceps were applied; the third stage was helped along by the use of ergot (ostensibly so as to minimize blood loss) with the attendant pressing down on the woman's abdomen.

## The 1900s and 1910s

By the early 1900s in both the USA and Britain, midwives were losing their power, partly because they were becoming associated with uneducated, illiterate women. (First hand experience was seen as being inferior to the institutionalized study of obstetrics, which meant that educated, wealthier women tended to avoid midwives.) In 1902 in the UK the Midwives Act effectively institutionalized a subservient role of midwives to physicians. Up until the 1920s, maternal mortality was still a major problem since 1 in 250 women died in childbirth. According to *Obstetrics by Ten Teachers* (Hodder Arnold 2000), this was because the modern pharmacologic treatments for postpartum hemorrhage and sepsis (which killed most women) were unavailable. Research was being conducted in Germany in the early 1910s into new forms of pain relief. The result—a mixture of morphine and scopolamine, which induced a condition known as 'twilight sleep', followed by the use of ether or chloroform—was so exciting to one group of American women that they actively promoted it.

# Pause to reflect...

## ...about the history of midwifery and obstetrics

1   How do you think it might have been to be a midwife 100 years ago? What about 1,000 years ago? Have you read any novels which guess at what midwifery might have been like in the past? *Clan of the Cave Bear* (Hodder 2006) includes a birth which was supposed to have taken place amongst Neanderthals (by a human woman!) and *The Red Tent* (Pan 2002) guesses at what life might have been like for women a couple of thousand years ago. What do you think of these fictional guesses?

2   How do you think doctors and male obstetricians gradually asserted more and more power over the years when the balance of power was changing? Have you read the novel which tries to guess at this process—*The Birth House* (Harper Perennial 2006)? If so, what did you think of it?

3   If you'd been a midwife in the early 1900s, how do you think you would have behaved? Do you think your answer would have been different if you'd had children yourself already? What if you'd had a natural birth?

4   How do you think midwives and other caregivers over the centuries and decades have managed to bring about change? What did they probably sacrifice or gain as a result, do you think? Could you be like them?

For material relevant to these questions, see Birthframes 15, 20, 26, 32, 39, 65, 66, 67, 68, 79, 89, 95, 98 and 99.

## The 1920s and 1930s

In 1920, an American professor of obstetrics, Joseph DeLee—a popular speaker, textbook writer and the inventor or modifier of many obstetric tools—gave a seminal speech to fellow obstetricians, recommending that forceps and episiotomy be routine for every birth, that women should be sedated, that ether should be administered as soon as the fetus entered the birth canal and that ergot or a similar drug should be used to speed up delivery of the placenta. These practices, known as 'prophylactic obstetrics', soon became routine in most US hospitals. From 1928 onwards active steps were taken in the UK to reduce risks through better education, although prenatal care itself was a fairly haphazard process until the 1940s, undertaken only by family doctors. In 1937 the antibiotics sulphonamides were discovered and penicillin came soon after. These discoveries quickly reduced the rates of puerperal sepsis.

## The 1940s

Before WWII probably more than 90% of women in the USA gave birth at home under the care of a midwife, and only occasionally with a doctor in attendance as well. The increased experimentation with different forms of pain relief from the 1940s onwards meant that hospitalized childbirth became preferable from both a choice and a safety point of view—since more was likely to go wrong when drugs and other interventions were involved. At that time, morphine was widely used, often alongside other drugs. In the US, it was usually used while women were in 'twilight sleep'. In France the morphine was part of a mixture called 'spasmalgine'. In the UK, ether and chloroform were still widely used. It was in this decade that the so-called 'natural childbirth movement' developed in the USA and with it the concept of the husband as 'labor coach'. It was also the time when blood transfusions became safe and when it was discovered that ergometrine could treat and prevent postpartum hemorrhage.

## The 1950s

Induction and augmentation of labor now became more of a focus so intramuscular injections of synthetic oxytocin (pitocin) became popular, used alongside different painkillers. Since the effects of intramuscular synthetic oxytocin were uncontrollable (because nothing can be done after an injection), IVs soon became common. With an IV in place, if the effect of the synthetic oxytocin already administered was too strong and therefore dangerous (i.e. causing too intense and continuous uterine contractions) it was possible to slow down or stop the IV altogether. It was in fact during this decade that IVs became safe because the original rubber tubes—which often caused intense allergic reactions—were replaced by plastic tubes. It was also the time when the vacuum extractor (ventouse) was invented in Sweden. This device appeared to present a good alternative to forceps, which had been in use since the sixteenth century because it seemed to offer less trauma to both mother and baby. Ultrasound scanning was also first adapted for obstetric use in the late 1950s. By the late 1970s and 80s it had become available in most hospitals. Techniques and treatments which we now take for granted (relating to anesthesiology, antibiotics and blood transfusion) were also further developed.

## The 1960s

In the 1960s IVs of synthetic oxytocin continued to be used widely as a means of inducing and augmenting labor. The main painkiller used was Demerol (a synthetic morphine, known as pethidine in the UK and Dolosal in France). Cesarean surgery, which had until this time been unpopular, suddenly became more appealing to women, thanks to the new use of the 'bikini line' incision (instead of the vertical incision which had been used until then), which was originally proposed by Muro-Kerr in the 1920s. This technique, which was developed in the 1950s, now became widely used, not only because of its increased popularity with women but also because surgically trained obstetricians were now available.

## The 1970s

The fast development of electronic fetal monitoring (first invented in 1957) radically changed the atmosphere of delivery rooms. Cesarean rates started to increase dramatically, partly because of this use of electronic fetal monitoring (EFM).[30] IVs of synthetic oxytocin and Demerol were still used. The growing public confidence in the technology and pain relief options offered by hospitals resulted in a rapid increase in the number of births taking place in hospital. In the UK between 1960 and 1979 the use of pain relief increased from 25% to 99%. (Nevertheless, as a result of increasing awareness of risks associated with pain management, there has since been a drop in the number of women using analgesics or anesthesia.)

## The 1980s and 1990s

In the 1980s the conventional technique of epidural anesthesia became available in large departments of obstetrics. So as to facilitate the administration of epidurals, which many women saw as extremely appealing, large, well-equipped maternity units were built in the 1990s. Size mattered because it was only economics of scale that made it possible to provide anesthesiologists and pediatricians 24 hours a day. Routines and protocols to support the large numbers of interventions now carried out had become commonplace. The injection of an oxytocic agent for the delivery of the placenta had also become routine and women wanting a physiological third stage were often seen as being unusual. Episiotomy was also still routine in many hospitals, although research has now conclusively discredited it.[31]

Only economics of scale made certain services possible

## How much intervention takes place now?

In the 21st century there seems to be the implicit assumption among many people that intervention only takes place to save lives. This is far from true since the use of pain relief still constitutes one of the main forms of intervention in modern childbirth and, as we've already mentioned, pain relief usually necessitates other forms of intervention. Here's a brief summary of practices which are still common in some parts of the world, looking at what takes place before any one pregnancy and birth, during it and afterwards. It's necessary to take this broad view because interventions taken at any stage can affect what occurs later on.[32] I've taken an international perspective because we do, of course, live in a global village, with plenty of movement from country to country and a mix of nationalities or heritages within any one.

### Before the birth

Many women experience intervention before pregnancy or during pregnancy. This may affect the health of their baby and how successful they can be in giving birth.

Before a woman becomes pregnant numerous old interventions might influence her ability to give birth and her baby's ability to get itself born easily, safely and smoothly. Episiotomies performed in previous labors which have not healed well, old forceps trauma and abortion injuries (especially those inflicted by unqualified surgeons) can all affect a woman's muscular control and comfort when subsequently giving birth. Circumcision (also called 'incision' and 'infibulation') is another practice which constitutes a pre-pregnancy intervention which can affect a woman's subsequent experiences. The World Health Organization (WHO) estimates that three million girls in Africa are at risk annually. In the notes accompanying the novel *Possessing the Secret of Joy* (Walker 1993), it is reported that an estimated 90 to 100 million women and girls living today in African, Far Eastern and Middle Eastern countries have been genitally mutilated. Recent articles in the media have reported on the growing practice of 'female circumcision' in the United States and Europe among immigrants from countries where it is part of the culture. This means a significant number of women are embarking on pregnancy with a compromised physical starting point. WHO (in 2008) estimated the figure at 100 to 140 million—i.e. these are women experiencing pregnancy and birth with the consequences of female genital mutilation. Many others start pregnancy with episiotomy and uterine scars from previous cesareans.

During pregnancy all kinds of screening and diagnostic tests are offered, as well as routine prenatal checks, which I regard as interventions because they must in some way affect either the woman or the baby. They include blood tests, repeated use of ultrasound with a Doppler machine, maternal serum screening (the alpha-fetoprotein test, the triple screen test or the quad-screen test), ultrasounds, chorionic villus sampling, nuchal translucency testing and amniocentesis.[33]

Psychological interventions might also have an influence
either before or during any one particular pregnancy

Psychological interventions might also have an influence either before or during any one particular pregnancy... After all, almost all women are seeing an enormous number of fear-inducing images relating to labor and birth on TV or in magazines and newspapers before they even become pregnant. Perhaps the negative images prevail because they are considered by planners to constitute more exciting viewing than the positive, natural alternatives, which many people wrongly associate with alternative lifestyles or primitive cultures. The prevailing fear of childbirth and tendency to leave responsibility to 'the professionals' mean that many women feel disempowered and incapable of giving birth successfully. A negative mind-set is created well before women even begin to actively explore real-life facts and figures, or read about other women's experiences with a view to forming their own opinions. In other words, modern myths about pregnancy and childbirth often replace the reality in people's minds and create a psychological environment of fear and foreboding. (This is even true of midwives, unfortunately.) Prenatal appointments filled with fear-inducing 'routine checks' reinforce this negative mind-set in the mom-to-be.

## Interventions during labor and birth

In the US as many as 60-70% of women have some sort of anesthesia during birth (according to research conducted in 2005).[35] Intervention of other kinds is also common. 'Listening to Mothers' (Maternity Center Association 2002),[36] a recent US survey apparently carried out with great care, found more than a 99% intervention rate. The cesarean rate is generally reported as being around 25% in the USA but is possibly around 40% in Latin America, and 90% in some private hospitals in major cities, such as Rio de Janeiro.

In Britain, of 3,000 women who responded to the *Mother and Baby* magazine survey in 2002, 94% used some form of pain relief involving drugs.[34] 24% of these respondents were induced. The cesarean rate varies enormously from 6 or 10% (amongst independent midwives) to 25% or more in some hospitals. According to the *Mother and Baby* survey, forceps are used in about 1 in 10 births. From my own experience I know that the use of oxytocics (i.e. syntometrine) in the third stage of labor (to speed up delivery of the placenta) is usually routine.

The situation is similar in Australia. Sarah Buckley, a family doctor and researcher, sourcing her information from a 2001 report from the NCHS [Australian National Center for Health Statistics] reports that in the USA 19.8% of women have their labor induced and 17.9% have an augmentation with synthetic oxytocin (pitocin). She says that according to a 2001 report from the AIHW [Australian Institute of Health and Welfare] 25% of women in Australia have their labor induced and another 20% have an augmentation.

In Israel, according to Wendy Blumfield (author of *Life After Birth,* Element 1992), an estimated 30% of first-time mothers and 40% of multiparas opt to give birth without pain relief. (Of course, this means that 60-70% of women do use pain relief.) The induction rates are comparable. Episiotomies were routine for first births up to three years ago. Now they are done according to need. (In other countries, episiotomy is often still routine, but not everywhere.) In Israel, forceps are little used, but together with ventouse, are used in about 4% of births. This figure is increasing because of the increasing use of epidural anesthesia, which gives women less control over the pushing stage of labor. Cesareans are performed in about 16% of births and are on the rise.

In most other countries—in particular in the developing world—health professionals are highly interventionist and pain relief is widely used. I was surprised to discover this to be true in both Sri Lanka and Oman when I was pregnant with my first and third babies. It seemed odd that so much intervention and so many drugs should be used in cultures which in many ways still seemed in tune with their cultural traditions. The Western model from a few decades ago is being copied, despite growing evidence that it is far from perfect. And I discovered that professionals in these countries often assert the 'need' to use these interventions even when research has clearly shown them to be harmful. Protocols, once they are in place, are difficult to get changed.

Here are specific practices which I see as interventions...

- **Induction or acceleration of labor by any means** Both 'natural' and artificial means are sometimes used to induce or speed up labor. Friends, relatives and even caregivers might coax the mom-to-be to ingest various substances (e.g. spicy food or cod liver oil), or they might encourage excessive exercise or laughter. Professionals try to intervene and trigger a woman's labor by administering prostaglandins (by pessary) or pitocin (intravenously);[37] alternatively, they might break a woman's water (i.e. perform an amniotomy) or 'strip her membranes'—with or without her consent, even in countries where consent is legally required. When a laboring woman has an IV inserted in her arm, it is possible that pitocin may be added to the original glucose mix to accelerate contractions; again, this is sometimes done without the woman's consent.[38]
- **Prepping** Hospitals and birthing centers always used to 'prepare' women for labor in some way, but now—hopefully—all or most of these routines have disappeared from most institutions. So-called preparation can include asking the woman to put on a hospital gown, asking her to lie down or stay on a bed, shaving off her pubic hair, administering an enema, and setting up an IV ('just in case' it's needed later) or slipping in a heparin lock for the same reason. All these practices constitute interventions because they can disturb the woman (by inhibiting her movements and perhaps by making her fearful) and they might consequently disturb the complex processes going on within her.

- **Electronic fetal monitoring (EFM)** This is often part of the routine admission procedures so may be considered a type of 'prepping'. Whether it takes place for 20 minutes when the woman first arrives at hospital or at any other stage of labor or birth, EFM is certainly an intervention which immobilizes the laboring woman either completely or partially. EFM may also encourage her to lie down (rather than lean forward, for example) and it's also likely to create a feeling of anxiety.[39]
- **Dietary controls** Women are often told not to eat or drink during labor in case a general anesthetic is required later. Stopping a woman from following her natural impulses interferes with her natural ability to regulate her intake of food and drink. Appropriate levels of hydration and the absence of ketones (which start being produced when a woman has not eaten for a while) are both important if labor is to progress well. In any case, research has not shown fasting to be helpful even when a general anesthetic is used.[40]
- **Directed activity** Women are often told to take a bath because it will 'help them relax'. (In fact, it's more likely to slow down contractions, which is unhelpful.) Women may also be encouraged to walk up and down hospital corridors or take a walk around the hospital. Any time a woman is told to do something in this way, an intervention is taking place because the suggestion is likely to distract the woman from tuning in to her instinctive knowledge of how to labor and give birth. The only non-intrusive kind of 'direction' anyone could give a woman would be, "Do whatever you feel like doing." This would simply serve to remind the woman that she can be confident in her own ability to choose appropriate activities and physical positions.

66 The nursing staff wanted me to stay on the trolley, to keep an eye on the monitors, but I put the equipment on a little table and got my mother to push them along while I walked up and down 'to the bathroom'.

- **Physical constraint** Even when no EFM is used, women are often told to lie down or stop walking around, or their movements are constricted indirectly through the use of some other medical procedure. Even stopping a woman from making noises is a form of intervention because it is a way of stopping her from releasing tension in a way which might help the natural processes to progress. Obviously, insisting that a woman give birth with her legs in lithotomy stirrups is also a form of physical constraint and an intervention, because it makes it impossible for her to use her natural ability to find a physical position which is helpful for both herself and her baby. Left to her own devices, a woman is likely to want to work with gravity, rather than against it, so will usually spontaneously choose a standing, kneeling or squatting position. Many women who have given birth with their feet in stirrups have complained that it was difficult 'pushing uphill'.

- **Any communication verbal or nonverbal** People often fail to see the role of comments or eye contact in disturbing a woman who is in labor or actually giving birth. Because any comment or communication (including a glance or eye contact) can disturb the woman or distract her, it constitutes an intervention. Giving a woman instructions on how to push her baby out into the world is an obvious form of communicative intervention—but by no means the only one. Merely mentioning within earshot of a laboring woman that a negative or frightening procedure might be necessary is a form of intervention, again because it is a way of disturbing the woman's state of mind as she moves through labor and birth. Negative or frightening possibilities which are often mentioned include breaking a woman's water sac, using drugs for pain relief and the so-called 'probable' need for an episiotomy, forceps or a cesarean.
- **The use of analgesia or anesthesia** The use of any drugs to alleviate pain constitutes an intervention.
- **Techniques for relieving pain** The use of any technique might disturb the normal course of a labor and birth, so must be considered an intervention. This includes breathing exercises, TENS machines, acupuncture, shiatsu, aromatherapy and any other complementary therapy.
- **The use of a fetal scalp electrode** This is often done in order to make EFM more reliable—which is considered necessary because EFM has been widely reported in research studies to be ineffective.
- **Episiotomy** Obviously, cutting a woman's sexual parts while she is giving birth is an intervention.
- **The use of forceps or ventouse** These are two other common and very obvious cases of intervention, which may be used out of impatience or as a result of a disregard for the rhythms of labor and the variation within the ranges as to what is 'normal'.

## After the birth

Only a few decades ago it was routine practice in hospitals around the world to put babies in a separate nursery from their new mothers, and it still is in some places. Nowadays, a lot of babies spend time in Intensive Care Units (ICUs). If the babies are premature, this intervention may well be life-saving and some full-term babies who are experiencing real difficulties may also need this kind of intensive care. .Others who are transfered to ICUs may experience considerable distress unnecessarily. How many babies are affected by this kind of intervention? All premature babies and most multiples are, but in *Immaculate Deception II* (Celestial Arts 1996), Suzanne Arms implies that many singleton babies may also be affected. She writes: "Various studies show that between 15% and 25% of full-term, healthy babies born to healthy mothers are also spending days or weeks in intensive care units."

Unfortunately, she doesn't refer us to any specific studies but we can at least guess that the percentage is still high in some hospitals around the world. Full-term healthy babies born to healthy mothers may be taken off to the ICU if there are any breathing difficulties at birth or if the mother's (and therefore the baby's) temperature is raised, because this could indicate the presence of an infection. The raised temperature is often the result of the mother using epidural anesthesia and indicates no problem in the baby but because of the possibility of an infection, babies with a temperature at birth are given what is called a 'septic workup'. Since this involves repeated blood tests and one or more spinal taps, it could be painful and unpleasant for the baby. As with interventions occurring during pregnancy and labor, there is perhaps the tendency amongst health care professionals to over-test in order to be sure that no problems need dealing with. The long-term effects of separating mothers and babies are less of a focus and even humanistic supplementary treatment, such as kangaroo care is not used as widely as we might hope it to be.

Whether or not a baby has started life in an ICU, one other significant intervention which many babies are experiencing is bottle-feeding. Programmed to consider breastmilk their birthright, numerous babies are forced to consume formula made from cows' milk for the first few months of their lives. Despite the fact that almost all women would theoretically be able to breastfeed if they tried, many decide to use formula. The negative experience of friends, relatives and health professionals (who have often not nursed their own babies themselves, or who gave up early), as well as the 'scientific' look of formula, make many feel they are succumbing to the inevitable. Aggressive promotional advertising by the major formula milk companies was widespread until very recently. Even without it, interventions which take place during labor and birth have an impact on the hormones a new mother does or doesn't produce as she is giving birth and therefore affect whether or not she will spontaneously nurse her baby immediately after the birth. Fortunately, worldwide awareness of the benefits of breastfeeding is triggering another kind of more positive intervention: various schemes and promotional campaigns are being set up worldwide to encourage mothers to nurse, even if they appear to have no desire to do so. This kind of intervention is necessary for mothers who have had a medicated birth because the natural processes have been so badly disturbed that they are unlikely to spontaneously put their newborn babies to their breast.[41]

Finally, writers and researchers such as myself are trying to intervene to create a new awareness of childbirth in the 21st century. Some are angry mothers who have had bad experiences themselves; some are midwives or obstetricians who are disenchanted with practices and outcomes they have witnessed; some are researchers who have uncovered disturbing new data which put old practices into a new perspective.

## How much intervention is needed?

How much intervention really is necessary from a health and safety point of view? How much of it is used because protocols are in place, because it is expected or because it impresses the client? How many lives are saved—or endangered—by all this intervention?

Of course, this is the $64 million dollar question. Nobody knows the answer. We have drifted so far away from the normal, undisturbed physiological processes of pregnancy, labor and birth that it's impossible to have a clear answer. We have never really allowed the natural processes to take place on a large scale within the safety net of our modern expertise and technology. Unfortunately, for psychological reasons as well as humanitarian ones, it would be impossible to conduct a randomized controlled trial.

There are indeed a small number of women who need the help of doctors when they conceive, gestate a baby or give birth. These are the women who—years ago—would have remained childless, who would have had dangerous pregnancies, who would have died in childbirth, or who would have given birth to dead babies. But the great majority of women would probably be better off without so much intervention... don't you think? It seems to me that too many procedures are over-used because of fear, misjudgment or inappropriate protocols. While I was researching this book, I often asked Michel Odent how much intervention had really been necessary, in his opinion, in the case of particular births. We were usually in agreement as to the appropriateness (or otherwise) of intervention. However, all his comments are tempered by his overall view (which also coincides with my own) which he expressed in his book, *Birth and Breastfeeding* (Clairview Books 2007):

66 Should the method of the mammals be inefficient in any particular instance, there must always be teams capable of doing epidurals to compensate for the lack of endorphins; to use IVs to compensate for a deficiency of hormones from the posterior pituitary; and to perform cesarean sections to rescue babies in distress.

That brings us back to the concept of optimal birth... In other words, perhaps it's time to change the approach typically taken by caregivers in modern birthing facilities.[42] Normality, not pathology, must become the new focus. Non-invasive tests utilized to identify the tiny minority of women and babies who really do need intervention should be just that—non-invasive. They should not impinge on a mom-to-be's sense of well-being or undermine her confidence in her ability to give birth. Empowering the great majority of women and supporting them unobtrusively as they labor through a completely physiological birth should be the new aim. Only this shift in focus will enable women and babies to experience safe, life-enhancing births, which—many more people will come to realize—actually result in much less pain overall, not more...

Of course, at times intervention is necessary for the safety of either mothers or babies. At other times, one intervention—to induce labor or use so-called 'pain relief', for example—leads to all kinds of other interventions, which would not otherwise have been necessary. Here, we see some life-saving intervention, carried out for the sake of a baby who was born prematurely, who would not have survived without it. See Birthframe 95 for the full story.

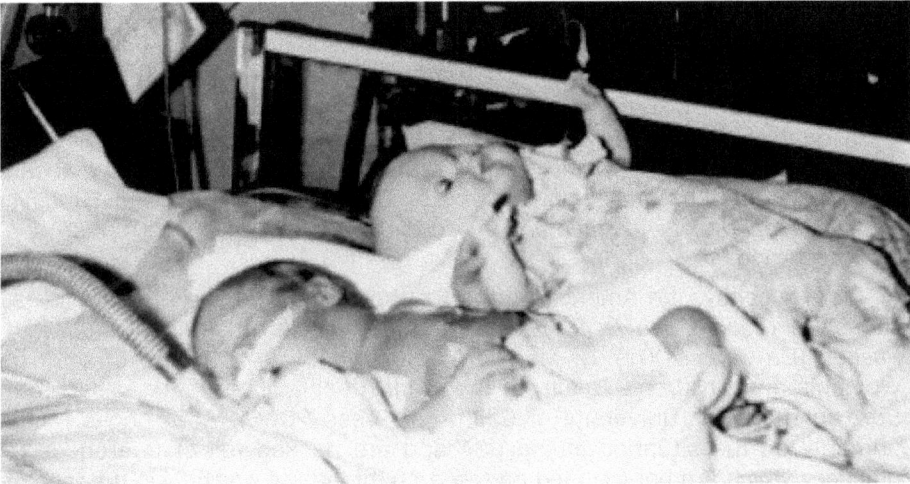

*At 8 weeks old preterm Kaia weighed just 2½lb. Here she's lying next to a small doll.*

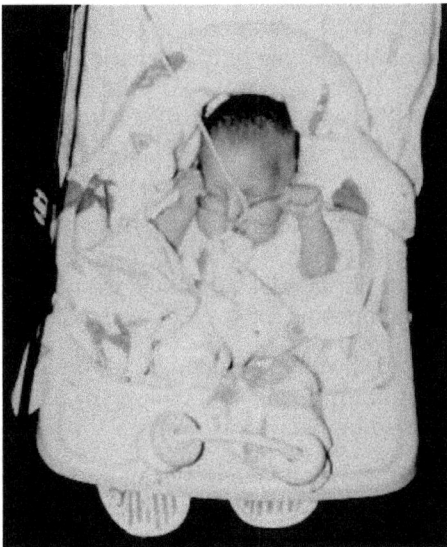

*Kaia's first time in a swing at 3 months*

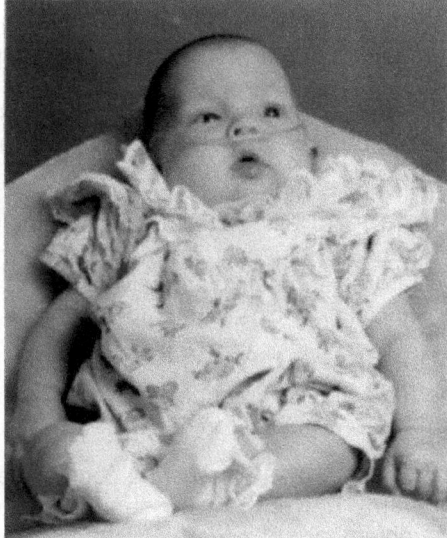

*At 4 months old this little girl still needed intervention to help with her breathing.*

## Reason No. 3:
## Research suggests it's best not to disturb the processes

As you probably know, we do already have a fair amount of research data which indicates that the physiological processes are best left to proceed undisturbed. Many practices associated with modern maternity care which imply 'management' and a deviation from spontaneous, healthy behavior are not recommended by research, whether you see them as 'interventions' or not.

As you probably know, a Canadian team of researchers, Enkin et al, did a great deal of work to survey research carried out into the effectiveness of midwifery and obstetric practices a few years ago. Their book, which has been periodically updated and supplemented by CD-ROMs, is *A Guide to Effective Care in Pregnancy and Childbirth* (Oxford University Press 2000). Also see www.cochrane.org.

More recently, the American editor of the revised edition of *Our Bodies, Ourselves* (Touchstone Books 2005), Jennifer Block, researched and wrote *Pushed* (Da Capo 2007). Dr Marsden Wagner, who was formerly Director of Women's and Children's Health at the World Health Organization, also wrote *Born in the USA* (University of California Press 2006). Both books focus primarily on the situation in the USA and are packed full of references to practices which are not founded on research, or indeed which fly in the face of recommendations based on research.

Here are a few commonly used interventions which research has shown to be unhelpful or harmful, which cannot be part of an 'optimal birth' scenario:

**In pregnancy...** assigning a mom-to-be to several caregivers; using ultrasound routinely to assess fetal growth;[43] screening for gestational diabetes

**In labor...** using EFM as a screening test on admission; confining a woman to a bed; stopping women from eating and drinking; not using one-to-one care or midwifery-led care; routinely inserting an IV or heparin lock 'just in case'; using artificial oxytocics (pitocin) to augment labor; imposing arbitrary time limits

**During the birth...** insisting on the lithotomy position; directing the woman's efforts with commanded pushing; imposing arbitrary time limits; routinely using episiotomy; using forceps or ventouse to speed up 'delivery'

**Postpartum...** restricting contact between the mother and her baby; taking healthy newborns to a separate room or nursery; routinely giving babies supplements of water or formula when mothers are nursing; restricting visits from younger children to hospital wards

It's easy to see how these practices can adversely affect the safety and success of physiological birth.

What is certain is that many interventions used are far from life-saving. The table below is an adaptation of Marsden Wagner's findings, included in his book *Born in the USA* (Uni of California Press 2006). It's striking how much practice can differ from research recommendations.

| The theory... | In practice in the USA today | The research says: |
|---|---|---|
| Women should have one caregiver for the whole of their labor and birth (rather than a series of caregivers on different shifts, with different duties) | Less than 10% of the time they do | Women should always have continuity of care |
| Midwives should be the routine caregivers, not obstetricians, because birth is a normal physiological process, which is usually healthy | They are in only 5% of cases | This should happen in 80% of cases, i.e. usually |
| Women should not have food or drink in labor | 86% don't | It's fine to do so |
| EFM should be carried out routinely on all women | 93% have it | It doesn't help |
| An IV should be set up routinely for all women | 86% have it | They shouldn't |
| Women should stay in bed for their labor | 69% do | None should |
| Women should lie back in bed and give birth with their legs in stirrups (in the lithotomy position) | Almost all women do | None should. They should be upright |
| An episiotomy should be carried out routinely so as to widen the outlet for the baby being born | 35% of the time it is | Less than 20% of women need one |
| Women should have labor induced with drugs | 44% do | Only 10% should |
| Labor should be accelerated with drugs too | 53% of labors are | Only 10% of labors need to be |
| Women need ventouse or forceps to get babies out of their bodies | 13% appear to need these | Less than 10% should, in fact |
| Cesareans should be carried out because nobody wants a vaginal birth anymore, do they? | 27% of births are sections | Only 10-15% of births should be |
| The new mother should hold her baby during the routine examination of her newborn | She almost never does | Well, she should. It's more humane. |

These data about practices were sourced from the national survey of obstetric practices in the USA, organized in 2002 and entitled 'Listening to Mothers'. (The survey results, published by the Maternity Center Association, is available at www.maternitywise.org.) The data from research comes from the Cochrane Library (www.cochrane.org). How much have things changed recently and how well do American birthing facilities provide evidence-based care?

What are the implications of any discrepancies in care between evidence-based recommendations and outdated protocols?

## Pregnancy

When a woman has no continuity of care in pregnancy, she is unlikely to develop a relaxed, confident relationship with her caregiver(s).[44] This will mean she is in danger of succumbing to what Michel Odent has called the 'nocebo effect', i.e. the situation when a woman becomes fearful as a result of caregivers' comments or concerns.[45] When various caregivers (or even the same one repeatedly) ask the mom-to-be to have ultrasounds in order to look for abnormalities or check for adequate growth, for example, the implicit message to the mom-to-be is that the caregivers are not actually really *expecting* things to progress normally.[46] Screening for gestational diabetes is problematic simply because the condition is considered non-existent by some researchers, or non-treatable by others. (Other screening tests, e.g. for anemia, provoke similar levels of controversy, as we shall see later on.)

Women are in danger of succumbing to 'the nocebo effect'

## Labor

As we've already mentioned, although EFM can ostensibly be justified on the grounds that it provides some useful information about a woman (and her fetus) on admission, in research studies it has only ever consistently been associated with a rise in the cesarean rate.[47] It has never been shown to improve maternal or infant morbidity or mortality rates. Printouts have, however, often been used in court cases. They also successfully put a woman in a position where she is subservient to technology and the decisions of her caregivers, all of which sets the scene for limiting the laboring woman's sense of freedom. Inhibiting a woman's movements and actions by confining her to a bed, stopping her from eating and drinking and inserting an IV or heparin lock also function to make women feel like a patient (who must be ill), as opposed to an empowered woman about to give birth. In any case, research has shown that these procedures serve no useful purpose and in fact that they inhibit the smooth progression of labor. This makes sense when we consider the close relationship between emotions and hormones and the importance of a 'correct' hormonal cocktail if the birthing processes are to proceed smoothly.[48] Of course, imposing arbitrary time limits simply exacerbates any psychological damage done by other procedures (by making the woman feel tense and observed) and artificial oxytocics give the woman higher levels of pain, which is likely to propel her down a cascade of interventions, after initially succumbing to some kind of pain relief.[49]

Imposing arbitrary time limits simply exacerbates any
psychological damage done by other procedures

**Birth and beyond...**

When a woman isn't allowed to move around and give birth in her own time, in a standing position, she will certainly not experience a fetus ejection reflex.[50] Episiotomy has been shown to cause more tears, not fewer, or to make them worse when they do occur and forceps may have lasting effects on babies.[51] The importance of bonding after any kind of birth has been well-publicized so it is tragic if it is inhibited in any way through 'routine' procedures...

# Questions for reflection...

1   Have births you yourself have witnessed or attended been evidence-based?
2   How much evidence and what kind of evidence is needed, in your opinion before practice should be changed? Are there any exceptions?
3   Do you have any personal experience of birth involving any disturbance?

## Disturbance as a result of attempts at pain relief

Pain medication constitutes the most significant intervention which can be predicted in many births, and is used when women do not think in advance about optimality for themselves and their babies. Research clearly suggests that this kind of intervention, whatever form it might take, causes problems, reduces the safety of birth, and also worsens the experience of birth for both mother and newborn, both physically and psychologically. Not only are the physiological processes of oxytocin and endorphin production disrupted within the body (meaning, perhaps, that they do not even take place), side-effects or after-effects of pain relief are experienced in all cases, with these causing significant disturbance in too many cases. Most importantly, when a woman has pain medication she loses her sense of fulfillment, as well as her alertness and mobility in the immediate postpartum period, so her birth is disturbed in this way too.

### Other disturbance which gets too little press

My own research into individual women's experiences has made me aware of other kinds of disturbances, which are not even (as yet) taken seriously by many caregivers, at whichever level of the hierarchy. Even when disturbance need not exist, caregivers sometimes interfere with the birthing processes by unwittingly upsetting or distracting the laboring woman, or by preventing her from entering a state of consciousness which will help her to experience optimal outcomes, without being aware of what they're doing. Given the extremely high complexity of the physiological processes—which it must be remembered only take place as described in a completely undisturbed birth— a disturbance can easily occur, as well as feelings of disempowerment.[28]

On the next few pages are two accounts which illustrate these ideas...

## Birthframe 21: Liliana Lammers—Learning from first time

In her first labor, Liliana—the doula we met earlier in this book— experienced the effects of having a lot of people watching while she was at the pushing stage. (One doctor and 10 students were watching!) This made her decide to avoid hospitals for any future births. With her first birth, the atmosphere of the hospital, the insensitivity of the people, the monitoring—all constituting an atmosphere of feeling very 'observed'—clearly stopped her labor from progressing. The psychological aspect of giving birth is very clear from her three subsequent births. Here, we see how positive birth experiences can be when there is no disturbance from outside.

When expecting my first child, I registered in a London hospital, as the most natural thing to do... I was born in hospital... and my mother too! It was 1983, I arrived smiling and I was told, after an internal, that I was 8cm. One hour later I found myself in a labor ward, surrounded by 10 students, a male doctor and a midwife. A monstrous fetal monitoring machine was attached to me and sounded very loud. Bright lights, voices everywhere, people coming in and out, my husband at my right... I was soon in a state of shock. Labor almost came to an end. Epidural, episiotomy and a very light and unnecessary forceps followed. "Never again! Never again," I kept repeating to myself, while holding my baby daughter.

> "Never again! Never again," I kept repeating to myself, while holding my baby daughter.

In 1987 I was expecting for the second time. I kept well away from doctors, hospitals and all their machinery. I was feeling wonderful—I knew my baby was well. I wanted to stay at home this time. A friend of mine mentioned Michel Odent... I read a bit about him... "Yes, he will understand me." He did. I had the most beautiful, undisturbed birth one can possibly imagine. Another daughter.

And two years later my son was born. Big boy, 9lb 8oz. Born between lunch and dessert! Literally. I was so lucky to be assisted by Michel Odent again.

I was in labor, but very hungry, to everyone's surprise. I had a big lunch and then, when the cake I had baked that morning was approaching the kitchen table, I stood up suddenly and with an unusually loud voice said, "No dessert. The baby's coming!"—and turning to my husband: "Do the dishes." After this, I rushed into a dark room I had prepared, went on my knees, on the floor, burying my head in cushions.

One and a half hours later I was lying down on the floor, my baby on my chest, both warmly wrapped, the cord not yet cut... and eating the cake! (This child is now 17 and has always had an incredible appetite.)

My fourth child was born at dawn, in a friend's sitting room, at a time when I could only hear the birds singing... and Michel Odent gently sleeping at the other end of the room. It was so magical, sweet, beautiful. My baby girl found the breast in minutes!

## Birthframe 22: Deborah Jackson—Unexpected disturbance

As well as interruptions, or having too many people around, a simple thing like eye contact or talking seems to disturb many births. The following account seems to illustrate the importance of not disturbing a woman in labor. These are the birth accounts of Deborah Jackson, the well-known journalist and writer of books on childcare. (She is the author of *Three in a Bed* (Bloomsbury 2003), *Letting Go as Children Grow* (Bloomsbury 2003), *Mother and Child* (Duncan Baird 2001) and *Baby Wisdom* (Hodder & Stoughton 2002).

Deborah's accounts suggest the importance of creating an environment where the healthy processes can proceed undisturbed and unobserved. After all, it seems to me that her first labor became established so effectively precisely because people initially took so little interest in her. And while I realize that talking to a woman in labor is not (yet!) illegal, I can certainly relate to Deborah's irritation at her midwife's chatter while she was in labor.

A friend of mine always says, "Happy birthday, Mom," to me on my children's birthdays. "It may be the day they were born, but you are the one who remembers it best," she says. My children think this is funny. A birthday isn't something you remember, it's the chance to be at the center of the known universe.

One of the rituals attached to my children's birthdays is that I recount their birth stories. When they were little, they used to cuddle up close, wide-eyed and excited, thrilled to imagine themselves emerging, cute and lovable, into my arms. Now that Frances is 15, Alice is 12 and Joe 8, they tease me instead. "I know, I know," they laugh, "you've told me that before." Nevertheless, the edited highlights are what they get, and the edited highlights are what will come back to them when they have their own children one day.

**Frances:** Frances's story became the introduction to my first book. She traveled through the night to be with me, as did her father, who was working away from home at the time. What made this birth special was that I was surrounded and supported entirely by women.

I'd had a day of regular but bearable contractions. As I was three weeks early, no one was taking much notice of me. But at 10:00PM, my friend Frederique, who had two darling children of her own, turned up at the door with an overnight bag in her hand. "I think you could be going into labor," she said, "I've come to stay with you."

Lucky she did, because by 11:00PM I could stand it no longer and begged to go to the hospital. It was going to be a long night. I thought I was about to give birth and phoned Paul, my husband, to ask him to catch the next train. "Am I in labor?" I asked the midwife as I stood on the phone, doubled over in pain. "If you know the answer to that, you're a better man than I am," she replied.

"Am I in labor?" I asked. "If you know the answer to that..."

I'd written out a birth plan and this, in the end, helped to shape the birth I wanted. Without it, I think I would have been unable to articulate anything but my knee-jerk reactions. I did a lot of stomping around, grunting and dancing and at some time in the night my mother arrived. Fred was going on holiday the next day and had to get some sleep. Mom had driven two hours to be with me and become my emotional support.

There were some interventions: they broke the water and I had a 'half-dose' of Demerol, before falling into a deep sleep, slumped over a bean bag. When I asked them to remind me what an epidural does, my mother soothed me, saying: "It's no more than a body can bear." When I stood up, my baby was born, tearing past me as if she was in a rush to be getting somewhere. The midwife walked into the room and caught her, like a soccer ball, in her hands. Within seconds, she was in my arms.

I looked up to see five women in the room: my mother, a female doctor, two midwives and a trainee. In this large teaching hospital, they had all gathered to see the 'physiological third stage' that I had requested on my birth plan. Births were routinely induced by drugs at that time and people wanted to know what happened if you left the placenta to its own devices.

What happened was a long wait, during which I crouched naked, holding Frances to my breast. Perhaps 25 minutes later, I felt a mighty contraction and the placenta came out easily.

Half an hour later, I was alone with my baby, a slice of toast and a hot drink. Frances nursed with all the instinct she was born with—luckily no one tried to intervene to show us how. Paul arrived on the morning train to find his daughter enjoying a big breakfast. And the trainee midwife came up later to thank me—she said it was the best birth she had witnessed in all her months of training.

**Alice:** Like most second births, this is a shorter story. Again, it was three weeks before my due date, but by now, we were living in a different part of the country. We had guests in the house: my mother- and brother-in-law. Braxton Hicks contractions had been cramping my style for a week or more, waking me up at night and keeping me on edge. In the afternoon, I drove Paul's grandmother to the station and had to stop for contractions. A couple of hours later, I was getting into a great rhythm and we phoned to alert the midwife.

Having given birth once, the fear had gone out of it. I had decided to have this baby at home. It wasn't so much the birth I was thinking about, but the aftermath— I particularly wanted to cuddle up with my new baby in my own bed, without any interruptions. At home, there are no sleep police patroling the ward and telling you to put your baby back into the crib. And after all, if there should be a crisis, we lived opposite the maternity hospital.

Everyone else went out for a meal and my midwife arrived—a friendly face I had met at a Michel Odent talk. She suggested I get into the bath and I recall the frisson of pleasure as I sat in my attic bathroom, knowing my baby was coming and looking out into the electric storm that raged that night. Then everything happened very quickly.

I got out of the bathtub and started to repeat the affirmations and visualizations I had been taught by another friend, a trainee hypnotherapist. I really could feel the contractions working in my favor, helping me to open up and my baby to be born. Paul arrived, but he seemed a thousand miles away. I was already completely immersed in waves of sensation. As I sat on the floor with my legs stretched out, watching them shake, I heard a distant voice say, "She's in transition." I thought, "You have to be joking, it takes much longer than this." They tried to help me up to the bed and as I moved forwards, Alice shot out with a yell. Then all was quiet.

Alice was so tiny that, stretched out on her back, she fitted perfectly along the crook of my arm to my fingers. Despite a fast, natural birth, she was a sleepy baby and I had to keep nudging her to remind her to nurse. That night, all four of us slept in our low, large family bed and it felt right.

**Joseph:** Everyone thought Joe's birth would be even faster than Alice's, but it wasn't like that. Four years had passed and now we were living in yet another part of the country. My doctor was in favor of home birth and all along I planned to have my baby in my bedroom. A friend asked if she could be there to see the birth and I half said 'Yes'. Two weeks before my due date and I already felt as though I was overdue. I was sure it was a girl.

That night, I was serving spaghetti with tomato sauce. I carried my plate to the table and, in my state of pregnant imbalance, managed to drop the lot on the carpet. I got down on my hands and knees to mop it up and immediately knew the contractions were starting. "I think the new baby will be born tonight," I told Frances and Alice. I asked Paul to take charge of the girls. I wanted to go into my own cocoon for this birth. When the midwife arrived, she was thrilled to hear that Alice was born in two hours. "That means I shall be able to get home to see my favorite program," she said. But labor went on for longer than we expected.

Unfortunately, the midwife also kept talking...

I tried getting into a warm bath, but this time the magic didn't work. I kept moving and visualizing, but unfortunately, the midwife also kept talking and I didn't have the guts to ask her to keep quiet. I found it impossible to 'let go' while she chatted to me and I felt obliged to smile politely and respond between contractions. Maybe four hours passed and I still felt far too much in control, but I knew it was time for the baby to be born. I decided not to invite my friend to observe after all. Somehow, I felt too observed already. I crouched by the side of the bed and tried to bear down. But it felt as though the baby, far from wanting to come out, was being sucked back inside me. Perhaps she's changed her mind, I thought. After a while, I could hear the panic in the midwife's voice. I was at the top of three flights of stairs, we lived a good 20 minutes' drive from the hospital and contractions seemed to have stopped. I felt very frightened. So I pushed with all the strength I could muster. No contraction, but I pushed my baby out. And out the baby came, screaming with annoyance at having been disturbed. This baby was bigger than the girls, but still tiny compared to 4-year-old and 7-year-old sisters. And the biggest surprise was yet to come: it was a boy.

## Reason No. 4:
## Pain relief seems to cause more problems than it solves

As already mentioned, pain relief causes significant disturbance and the inadequacy of pain relief currently available is not widely disputed.[52] The search for perfect pain relief continues... As yet, no drugs have been found which are both totally effective in labor and birth and which also have no side-effects. In many cases, side-effects or, worse, longer term after-effects may be substantial and in cases where they aren't, we must recognize that we may simply not yet know what they are. In fact, as Michel Odent has tried to demonstrate, both through his book *Primal Health* (Clairview Books 2002) and his Primal Health Research Database (www.primalhealthresearch.com), long-term effects may occur simply as a result of certain experiences in what he calls the 'primal period', which runs from conception until the baby's first birthday. The research he has found so far already suggests the effects of pain relief on longer-term experience may be significant. (Clearly, more research needs to be done.)

Given the implication that all kinds of pharmacologic attempts at pain relief (involving either analgesia or anesthesia) may bring more problems than they solve, the question arises as to why on earth women continue to request pain relief of all kinds...[53] We need to consider this question in the light of the wider struggle women have had in the last century to improve their experience of life in general. But an irony seems to be tied up with the issue of pain relief... While women themselves first saw its use as a basic right and even a form of liberation, pain relief itself made women more compliant and less assertive at a time when key figures in British and American patriarchal society must have found them difficult to control. (Do you know about the aggressive tactics of the suffragettes in the UK and even the less violent, but persistent approach of the suffragists and other campaigners in the UK, not only for the right to vote, but also for educational opportunities, the right to work and equal pay?) And when we consider why women still so consistently request pharmacologic pain relief we also need to remember that its primary objective is only to relieve pain; too often the effect on the fetus or newborn have been ignored or dismissed.

## Questions for reflection...

1  To what extent do you think women are fully informed about possible or actual side– and after-effects when they are offered pain relief?
2  To what extent are side– and after-effects of pain relief fully known?
3  How much do women typically consider the effect of drugs on their baby, in your opinion?
4  Why do women have such a strong conviction they will need pain relief?
5  Why do women who avoid drugs in labor often seem fulfilled postpartum?
6  How do you feel about pain relief yourself (either as a midwife or as a laboring woman? Do you routinely offer it? Do you avoid its use? Why?

## Pain relief offered

So as to be clear about what we're talking about, before looking at specific problems associated with it, perhaps first of all we need to define what we mean by 'pain relief'. Here's an overview of what's used...

- First there are sedatives and analgesics. A sedative (such as temazepam) is sometimes used to encourage a woman to sleep or rest. Analgesics (such as Tylenol and co-dydramol) act as mild to moderate painkillers. Inhalation analgesia (such as Entonox, i.e. 'gas and air') is used in the UK.
- Next we have narcotics—Demerol, Nisentil, Dolophine, diamorphine (which is actually 100% heroin), pentazocine (Fortral) and meptazinol (Meptid). These are even stronger drugs, which function as painkillers. Apparently, though, women widely complain that they don't mask the pain—they say they just make it more bearable. In cases where drugs theoretically have no pain 'ceiling' (i.e. they are all-powerful)—such as heroin—women lose alertness and awareness entirely. These narcotic drugs also weaken the newborn baby's suck, making breastfeeding difficult or impossible.
- Then there are tranquilizers... They include Valium, Vistaril and Penergan and are sometimes used in conjunction with narcotics (e.g. Demerol with Phenergen). They're given to women who are considered to be particularly tense.
- Barbiturates are rarely used nowadays because of the depressant effect they can have on the baby. Seconal and Nembutol are sometimes used if it's felt that the mother needs to be sedated.
- Amnestics are drugs which remove the memory of pain. The main example, scopolamine hydrobromide (known as hyoscine hydrobromide in the UK), was widely used in the 1950s and is still occasionally used today.
- Local anesthetics block sensation entirely in certain organs. However, they're not considered totally effective, so they're rarely used. The pudendal and perineal block are examples.
- Regional anesthetics include epidurals, spinals, saddle blocks and paracervicals. When they're successful, they numb sensation completely in a specific area. However, they are not always successful, they carry risks and are not considered useful or desirable at all stages of labor.
- Next, there's general anesthesia, which results in complete unconsciousness. It's used for emergency cesareans because it allows the medical team to work slightly more quickly than they can when only regional anesthesia is used. It's not used at other times because of the increased risks associated with general anesthesia over local varieties and also because it's considered desirable for mothers to be alert at the time of their babies' births—for the sake of bonding, of course, apart from anything else.
- Finally, there are various attempts to relieve labor pain which do not involve any drugs. These include TENS, complementary therapies (such as aromatheraphy, acupuncture, homeopathy, herbalism), and massage. Most, if not all, of these non-pharmacologic approaches also seem problematic in various ways as we shall see.[54]

Like obstetric interventions and protocols generally, most drugs used in labor have not been around for very long, so it's important to remember the limitations of what we know. Drugs only began to be used widely for childbirth during the second half of the 19th century after the English Queen Victoria famously allowed chloroform to be used for the birth of her eighth child in 1853. As we saw earlier, in the 20th century obstetricians continued to experiment with drugs against a backdrop of debates about religious, philosophical, social and political questions, as well as medical ones. The fact that drugs made women more compliant and less assertive in an age when women were still fighting  for basic rights needs to be taken into account. (Ironically, women themselves often mistakenly saw them as a form of liberation.) Also, we need to remember that drugs were primarily selected for their pain-relieving properties: the baby's welfare and the mother's wider experience were not seen as being a primary concern and breastfeeding has only recently been recognized for its health-giving benefits, so the effect on this was also not considered. The most popular pain-relief nowadays—the epidural—has only been widely available since the 1990s (after being developed in the 80s). No form of drug-based pain relief, including the epidural, has been thoroughly tested for its long-term effects on the baby. But side-effects for this and other forms of pain relief are already well-known and widely recognized, as we'll see in the next few pages.

## Epidurals

Epidurals are particularly worth considering because many women mistakenly seem to believe they promise no pain. They have been called the 'Cadillac' of pain relief.

The reality can be quite the opposite. First, taking a purely straightforward, successful epidural as our reference point, having an epidural involves accepting complete management of labor, which involves a loss of control and dignity... Repeatedly, in my research lack of 'control' in labor and birth seemed to be associated with low satisfaction levels with birth postpartum. There are other disadvantages too for either mother or baby...

- Electronic monitoring is necessary and this is associated with an increase in cesarean rates, possibly because caregivers misinterpret or overreact to readings on the monitors.
- Epidurals induce fetal heart decelerations which can be worrying enough to necessitate the use of fetal scalp blood sampling.[55]
- The woman's blood pressure needs to be constantly monitored because the epidural will cause it to fall to possibly dangerous levels. This fall in blood pressure can result in serious fetal distress because less oxygen is circulating in the woman's body.
- Intravenous fluids need to be administered to counteract the fall in blood pressure. These fluids can cause excessive swelling of the woman's feet, legs and breasts. Over-engorged breasts can make it difficult for the newborn to latch on.

- A catheter needs to be used to drain off the woman's urine, because the epidural numbs the woman's bladder too. The catheter must remain in place until the epidural has worn off. (Of course, catheterization also carries the risk of infection and may result in incontinence for as long as three months after the epidural has been used. A year after the epidural, incontinence is reported to be more common than for mothers who did not use one.)
- Since women's contractions often become weaker after the use of an epidural—probably because the woman feels disturbed, observed and 'managed', caregivers may feel it is necessary to use a synthetic oxytocic. Pitocin produces unnaturally strong and long contractions which can also cause fetal distress by depriving the baby of oxygen.

On a very basic level, all this 'management' means it's impossible for a woman to 'go to another planet' and experience a fetus ejection reflex, which makes the labor and birth faster, smoother and safer.

The newborn baby may have a very different first experience of life—since his or her heart function, breathing, metabolism and ability to adjust to heat or cold may all be affected.

According to some more research which Sarah Buckley has summarized, drugs administered by epidural go straight through to the unborn baby at equal and sometimes greater concentration levels than those experienced by the mother.

Some drugs go to the baby's brain and almost all drugs take longer to eliminate from the baby's immature system than from the mother's system, after the cord has been cut. One researcher found the drug bupivicaine and its breakdown products in the circulation of babies for three days after the birth. Drugs used in epidurals also cause more acidemia (high levels of acid in the blood) in babies than a general anesthetic—which is apparently a sign that epidurals compromise fetal blood and oxygen supply

In addition, there are numerous other problems with epidurals...

- Epidurals interfere with the woman's normal ability to produce endorphins, which are the body's natural painkillers. This is because of the pharmacologic products they contain.
- Some women report that their epidural only partially takes effect, e.g. only on one side, or leaves an area of pain. This means they merely experience the pain of labor and birth in a different way. One study (reported at a Midwifery Today conference in 1996) revealed that 15% of women who had an epidural got no pain relief at all and another study (published in 1995) reported a wide variation in the amount of pain relief obtained. It is suggested that this variation is not because of lack of expertise on the part of anesthesiologists but because of the different metabolic rates of the various women.

- Epidurals are usually administered after women have experienced contractions for a length of time and even then they only take effect for an hour or so.[56]

Epidurals are usually administered after contractions have been experienced for a length of time

- Most professionals also prefer epidurals to wear off before the actual birth. This all means that relief is only temporary or partial. If epidurals are given too early, i.e. before contractions become well established, they can have the effect of slowing down contractions. (One research team also reported a higher cesarean rate amongst women who had epidurals earlier on in their labor.) Of course, pain which suddenly returns after epidural relief is bound to be more shocking and difficult to cope with than pain which has gradually built up, along with normal endorphin levels.
- An epidural which takes effect for longer often results in a forceps delivery with an episiotomy because the numbing of sensation that takes place in the abdominal area also means the woman is unaware of any pushing sensations. This means she may not be able to push her baby out or that she will, at the very least, require instructions on when to push her baby out (known as 'commanded pushing'). The drugs in the epidural affect the baby's ability to rotate during the second stage of labor and also prevent the woman from spontaneously adopting the best physical position for pushing.
- Of course, women almost always have a cesarean when their posterior babies do not turn before the birth because the unusual position usually makes the labor and birth more painful. Even when the baby is in a good position for the birth, the birth takes longer with an epidural in place.[57]
- Some women have problems urinating after the birth, if they've had an epidural.
- If the mother develops an infection because of the catheter, the IV, the constant vaginal exams or the possible use of forceps, the baby will also be affected and would need to transfer to Intensive Care immediately after the birth. This would mean additional intervention and stress for both mother and baby, not to mention the father. Whether or not an infection develops, according to one study (Fusi *et al*, 1989) as many as 1 in 4 women who have epidurals develop a fever after four hours and almost half do after eight hours. This is because of the body's inability to regulate its temperature when half the body is numb.[58]
- After having an epidural some women report having migraine-type headaches which last for days or even weeks after the birth. This kind of headache occurs when the needle goes in too far and there's a leakage into the cerebrospinal fluid.
- A small number of women are paralyzed permanently.[59]

- According to another research team, the incidence of third degree tears is three times greater after an epidural. These deep tears can later result in fecal incontinence and chronic pain during sex. Fecal incontinence means having no control over doing poops or passing wind, so they are a cause of great embarrassment and discomfort. The other problem associated with these deep tears is that poop goes into the woman's vagina, which can cause infection as well as unpleasantness.[60]
- In the case of sheep which have been used for experiments, 7 out of 8 ewes which are given epidurals fail to show normal mothering behavior toward their lambs for at least 30 minutes.[61] Research has even confirmed that the same problem may exist in human mothers... In one study mothers who'd had epidurals described their babies as more difficult to care for one month after the birth.[62] This has led some researchers to conclude that human mothers who have an epidural may have to fall back on their conditioned maternal behavior, since it may not come instinctively.[63] Some recent studies of the effects of epidurals on breastfeeding have concluded that epidural babies have a weaker sucking reflex and capacity to suck.[64]

> In one study mothers who'd had epidurals
> described their babies as more difficult to care for
> one month after the birth.

6 6 I was admitted at 4:45PM and by 7PM I was getting very painful contractions, an epidural was mentioned and after a few more contractions I decided to have it. What wondrous relief except that you are literally numb from the waist down which makes any form of movement amusing. Things then slowed down as expected, so up went the fluid IV and the hormone IV to get things going again.

6 6 The contractions then came stronger and closer together, but, due to the effects of the epidural, all I could feel was pressure on my bowel when they were at their very strongest. I had to be told when to push by the other people in the room.

> I had to be told when to push by other people in the room

6 6 After the epidural I could only have ice chips, and I didn't have any because I was too busy screaming and trying to push. I was desperately thirsty, but now was not allowed to have any water because I was being 'prepped'. A few minutes later I was taken into surgery.

6 6 I had an epidural the first time, but didn't the second because there wasn't enough time to get it organized. I was really glad, afterwards, that I hadn't had it. I seemed to recover so much faster after the birth, not having had pain relief while I was in labor.

> I seemed to recover much faster without an epidural

If the idea of a woman avoiding an epidural still shocks you, consider the following email I received from Beth Dubois, who we met earlier on... (You can read more about her experience of birth soon.)

**From:** Beth Dubois

Hi Sylvie,

Your book sounds like it will be very encouraging for women. It makes me think of an interesting dream I had some months after the birth. A mom-to-be was telling a group of mothers how she was planning to have an epidural, which she said had been great for her during her first child's birth. When I heard her (despite the incredible birthing experience I'd had), I felt doubt about whether there was really any benefit to experiencing the pain of labor and birth. That night I had a dream in which Valerie, my midwife, was my fitness coach. She was coaching me on some particular physical fitness challenge, like an obstacle course. I had previously not been able to do it at all. Suddenly, now that I had given birth, I was able to fly through the fitness course nearly effortlessly. I was amazed at how what had once seemed nearly impossible was now easy. What I got from this dream was that the experience of giving birth naturally had changed me at a very profound level and was totally worthwhile.

## Opioid analgesics and sedatives

These are used in many hospitals around the country, so their effects need to be taken very seriously, even if many people have come to regard Demerol, diamorphine and other opiates as being harmless forms of pain relief.

Why are they a problem? From the laboring woman's point of view, it's worth noting that women have often reported feeling drunk and out of control after having doses of opioid analgesics, such as Fentanyl (which is 100 times more potent than morphine) or diamorphine (i.e. heroin). Even these drugs do not fully mask the pain of labor or birth. They can also cause problems for the woman, such as nausea, vomiting, sedation, itching, hypotension or respiratory depression and can prolong labor.[65] For the neonate, the effect of opioid analgesics and sedatives can be dramatic. Since they go straight through to the fetus and are usually still present in the baby's system at birth, they can cause early breathing problems, or convulsions in some cases, they can weaken the baby's 'suck' (so have a dramatic effect on breastfeeding)[66] and they can affect the fetal heart rate and change neonatal behavior. Research has also established a link between the presence of opiates in the baby's body at birth and later drug addiction in adulthood.[67]

Here's a bit more information about opioid analgesics from a mother, the New Zealand researcher and doctor, Sarah Buckley, who now lives in Australia...

Demerol (pethidine, meperidine) is the usual opiate administered in Australian labor wards; in 1998 here in Queensland Australia, around a third of laboring women used this drug. In the US, other narcotics such as Nalbuphine (Nubain), butorphanol (Stadol), alphaprodine (Nisentil), hydromorphone (Dilaudid) and fentanyl citrate (Sublimaze) have been traditional mainstays of labor analgesia, but have now been largely replaced by epidural analgesia, which may also contain opiates. [In the US Demerol is often used.] All opiates used in labor can cause side effects such as maternal nausea, vomiting, sedation, itching, low blood pressure, difficulty breathing; for the baby they include fetal heart rate abnormalities, breathing problems, difficulty with early breastfeeding and altered neonatal neurobehavior.[68]

As with oxytocin [pitocin], use of these drugs will reduce a woman's own opioid hormone production[69] which may be helpful if excessive levels are inhibiting labor. However the use of Demerol has been shown to slow labor, in a dose-response way;[70] in one randomized trial, morphine (but not naloxone) administered in labor directly reduced oxytocin release.[71]

And again, we must ask: What may be the effects for mother and baby of laboring and birthing without peak levels of these hormones of pleasure and transcendence?

What may be the effects of laboring and birthing without peak levels of these hormones of pleasure and transcendence?

Some researchers have nominated our endogenous opiates as the reward system for reproductive acts—that is, the endorphin fix keeps us making babies, having babies and breastfeeding.[72]

Anecdotally, I notice that women who reap pleasure from these activities—e.g. home birthers and La Leche League mothers—tend to have larger families. On a global scale, also, countries that have embraced the obstetric model of care, which prizes drugs and interventions above birthing pleasure and empowerment, have experienced steeply declining birth rates in recent years.

More serious are the implications of Swedish research into the use of opiates at birth, published in 1990[73] and recently replicated with a prospective US population.[74]

In the first study, researchers looked at the birth records of 200 opiate addicts born between 1945 and 1966, and compared them to their non-addicted siblings. Offspring whose mothers used analgesia in labor (opiates, barbiturates and nitrous oxide gas [i.e. Entonox]) were more likely to become addicted to drugs (opiates, amphetamines) as adults, especially when multiple doses were administered. For example, when a mother had received three doses of opiates in her labor, her child was 4.7 times more likely to become addicted to opiate drugs in adulthood.

Animal studies suggest a mechanism for this. It seems that drugs administered chronically in late pregnancy can affect brain structure and function (e.g. they can cause chemical and hormonal imbalance), which may not be obvious until young adulthood.[75]

Whether such effects apply to human babies who are exposed for shorter periods around the time of birth is not known but, as one researcher warns: "During this prenatal period of neuronal [brain cell] multiplication, migration and inter-connection, the brain is most vulnerable to irreversible damage."[76]

## Entonox (gas and air)

As you may know, this form of analgesia is very often used in labor wards in the UK and also in other countries, such as Sweden and India. You will read about it in this book in some of the birthframes and if you happen to be in another country you may well find it's available there too. It's portrayed in many old films as being a 'routine' part of birth, e.g. in Ingmar Bergman's film *Secrets of Women* and it is certainly used routinely in Britain. Out of 3,000 women who returned questionnaires in a British *Mother and Baby* magazine survey in 2002, as many as 79% of women in the UK reported having used it. Women often say, "Oh, I just had gas and air" when talking about their labors—as if it wasn't anything much. This seems strange, given the way a friend described her experience of using it. She said it was wonderful, like being in a helicopter, circling above her body as each contraction washed over her. Other women find its effect weak and almost unnoticeable. Irrespective of how women report its effects, the fact is that gas and air is a powerful form of pain relief. One manufacturer of Entonox states that it is a very effective analgesic agent, which is rapidly eliminated from the body *once inhalation stops* (the italics are mine). This 50/50 mixture of oxygen and nitrous oxide targets the opiate receptors of the brain, so it must work in a similar way to morphine and synthetic morphines such as diamorphine, even though these are usually given in higher doses all at once, rather than in short bursts over a period of time.

Why gas and air is perceived as being harmless, with no side effects, is a mystery, given that it is usually used repeatedly for half an hour or so for every one of very frequent contractions. The fact that the main ingredient of gas and air has been associated with increased incidence of drug addiction in adolescents who've experienced it as a fetus during their birth is either little known or inexplicably disregarded—even though, admittedly, this research has only focused on the use of nitrous oxide in isolation.[77]

No research has yet investigated the long-term effects of combined oxygen and nitrous oxide, but it seems likely that oxygen is the more 'innocent' ingredient.[78] Even if the mixture widely used in the UK and elsewhere really is 'harmless' in itself, which seems unlikely, I believe there are other good reasons to avoid it. Personally, I'd even ensure it's not in the house if I were having a home birth—for the same reason as I keep chocolate cookies well out of my way! Here's a summary of reasons why...

- It is mostly used at a critical time of labor, i.e. in the half hour or so before the baby is born. Although it only remains in the bloodstream for a very short time, if it's used for every contraction over half an hour or so its effect will not be so fleeting... and it's bound to have an effect on the baby.
- Using the mouthpiece or 'mask' to self-administer gas and air, the laboring woman will inevitably be distracted from the sensations she's experiencing. That's the whole point, after all, isn't it? In a distracted state—which is often more panicky than calm—she will be unable to tune into what is happening in her body. This means she is less likely to be able to coordinate her pushing effectively, which will result in a less smooth journey for the baby and a higher likelihood of injury for the mother. That means a much higher likelihood of tearing and sutures.[79]

- It is very unlikely a laboring woman who is using gas and air will be able to tune into her instinctive knowledge of how to give birth. This means she will not experience an authentic fetus ejection reflex—the natural reflex which enables us to birth our babies quickly, efficiently and, above all, safely.
- Any side effects the mother-to-be experiences while using gas and air will detract from and probably replace the state of mind and sensations she would otherwise have been experiencing. The mother's state of mind and perception of her bodily state is extremely important at this stage because it is just minutes or seconds before she will meet her new baby for the first time.
- If there is a 'break in transmission' in administering the gas and air (either because of the woman's uncoordinated efforts or because of some other factor—e.g. the gas and air running out or being taken away), the laboring woman is likely to be thrown into a panic. This is not the state of mind she should be in just before the birth of her baby. She should be tuning into her bodily sensations in a positive, accepting way—realizing that they are a signal that her baby is just about to be born.

I decided to ask Michel for his view...

**Why is Entonox used so often in British hospitals but not elsewhere? In a way it seems surprising, especially since nitrous oxide on its own is poisonous.**

There were probably some accidents with nitrous oxide, so it is possible that many people in countries other than the UK were not receptive when a safe formula was offered. Before the age of evidence-based medicine, the practice of midwifery and obstetrics was highly influenced by the opinions of a small number of authoritative professors or authors of books. This explains the differences between countries.

**Is it a safe drug, do you think?**

We have no data on that. Studies into the use of nitrous oxide suggest that this on its own is definitely not helpful from the baby's point of view.[80]

We know that all opiates suppress the sucking reflex

We do know that all synthetic morphines—all opiates, in fact—suppress the sucking reflex and since the target of nitrous oxide is the opioid brain receptors, there may be some similar effect in the case of Entonox. Nitrous oxide is not an opiate itself, but it does seem to interfere with the natural production of endorphins, which are natural opiates, probably because it targets the brain's opiate receptors.

So in a sense 'gas and air' has something to do with opiates—it's not a separate topic. It may be harmful in the same way as other opiates which are not produced directly by the woman's body.[81]

**However, do you think it might be a useful last resort for women who are having difficult labors—e.g. if their baby is in a posterior position?**

My feeling is that it is a substitute for privacy. In other words as long as the word privacy is not understood, gas and air will be useful. Once a woman asked us if we have gas and air. Liliana replied with a joke: "I am the gas and Michel is the air!"[82]

## Birthframe 23: Caroline Turner—Regrets about Entonox

If you're one of the people who believes that Entonox is innocuous, please consider the following birthframe. In this account the writer appears to regret only one thing: using gas and air. Ironically, perhaps this woman would have been able to tolerate the pain more and not felt the need for this artificial pain relief so compellingly, if her husband had not coached her through each contraction. In any case, this account does illustrate very well that even when all does not start out well, it is certainly possible to optimize outcomes by opting for physiological approaches and avoiding intervention and pain relief wherever possible.

In January 1998 we lost a baby, George, due to me having an incompetent cervix. It was a distressing time, to say the least, as I had to give birth to a 'stillborn' baby at 22 weeks. I then had three earlier miscarriages, which compounded our distress and disappointment. After all this, I fell pregnant with Douglas, who was born at 41 weeks, in July 1999. I had attended prenatal yoga as the whole prospect of carrying a baby at this point was incredibly stressful and I didn't think my nerves would be up to the job. Yoga helped me by relaxing me and I learned a lot about complementary therapies and alternative medicine in the discussion group that followed our yoga practice.

When I fell pregnant with Douglas I was advised that a simple procedure called a cervical suture was recommended. I agreed to this and this was put in at 13 weeks of pregnancy.

Doug's birth was lovely but long. I felt the first twinges on the Saturday evening, went to hospital on Sunday at 8AM to be told I was 4cm dilated. I had my water broken at 1PM and then went into transition, demanded meptid and pushed for 2½ hours. Douglas was born at 4:30PM. I was exhausted.

I then experienced three miscarriages before falling pregnant again. I was determined to be fit from the start and continued with my regular yoga practice. I went for my pre-surgery (for my cervical suture) check, which consists of an ultrasound at 13 weeks, and we found out that I was expecting identical twins—monochorionic, diamniotic twins. This posed no problems for the suture, although it was not removed until Week 38 with Douglas and would not be removed any earlier. If, however, I went into labor before this time I would have to rush to hospital as it would cause obvious complications. I had the suture removed at 37 weeks as the babies were growing well and, should labor be triggered by the removal of the suture, all should be well.

I actually went into labor naturally two days short of my due date. My water broke in bed at 2AM and we got to the hospital at 3AM. I was only 3cm dilated, which means no dilation at all, as after giving birth you may never go back completely anyway. The presenting baby was cephalic and the second breech. The hospital was comfortable with delivering a breech baby vaginally as long as it was second. Georgia was born at 8:39AM and Frances was born at 8:47AM. They weighed 7lb 9oz and 7lb 7½oz respectively.

Why had I wanted to avoid a cesarean for my twins?

♥ I wanted to keep my recovery time as short as possible due to having Douglas waiting at home.

♥ I live a little way out of town and need my car to get Douglas to many of his clubs. I could not afford not to drive. Friends had offered to take him places but I wanted to carry on as best I could as normally as possible—and that meant I took him everywhere. People could not offer me a ride as we would not all fit into a regular car, as those available already have toddlers themselves.

♥ I wanted to prove that I could do it naturally.

♥ I wanted the birth of the babies to be as fantastic as Douglas' birth.

Why did I want to avoid interventions for this birth?

♥ I was more experienced in yoga this time and knew that I could breathe through the pain. I knew what I was doing this time round and was more confident in my role as mother and wanted to assert my authority on the birth. I really regretted having my water broken with Douglas as up to this point I had used no pain relief at all. The contractions then came too quickly and I could not compose myself in between them as I had successfully been doing before my water had been broken. So this time I breathed into the pain—long deep breaths. My husband 'coached' me through each contraction, counting up to 45 each time because we had discovered that this was how long each contraction lasted. I used very little gas and air—the little I used, I used as the pain seemed to overwhelm me as one contraction blurred into another. Indeed I dilated from 2cm to 7cm in about $1\frac{1}{2}$ hours.

If there were a 'next time' what would I do differently?

♥ THERE WILL BE NOT BE A NEXT TIME!!!!!!!! However, on a purely hypothetical level, here's my answer: I would not use gas and air. I did the pushing without it, even the breech delivery. It is possible.

♥ Georgia was delivered with only the minor complication of meconium being present. Frances however gave us a start as the cord was prolapsed. There was a hint of movement towards the operating room, at which point I thought, "No way am I having a section now!" and started to push without waiting for any contractions. Contractions soon kicked in and out shot Frances with some assistance from the doctor, who had to be present for a breech delivery .

♥ I did hemorrhage that afternoon but my placenta was quite roughly pulled from me (I later found out) which may or may not have been the cause of it. I then had to have a blood transfusion, which picked me up marvellously. With all my births I had no sutures as I only suffered minor tears and am very grateful for this. I took arnica to help me recover and I was all back in the right place within 48 hours of giving birth this time!!

I think this is all that may be of interest to you. If you have any questions then let me know... I would love to be named in your book and am very keen to help advocate natural birth. If I can be of any more assistance then let me know.

## Key reasons for inadequacy...

To summarize, here are the key reasons why common forms of pain relief seem inadequate and even unhelpful in terms of facilitating optimal outcomes.

### Epidurals
• These necessitate EFM because the anesthetics used cause fetal heart decelerations; this may in turn increase the likelihood of a cesarean.[83]
• They necessitate continuous monitoring of the woman's blood pressure as it may fall to dangerous levels. So as to counteract problems with lowering of blood pressure, intravenous fluids have to be used, which can cause excessive swelling of the feet, legs and breasts. Swelling of the breasts can cause infant feeding problems.
• The catheter needed to drain off urine can easily introduce infection into the woman's body.
• Epidurals tend to increase the length of a woman's labor.[84]
• Epidurals, it seems, may have a subtle impact on human mothering.
• Women who have epidurals are likely to have subnormal alertness levels.[85]
• They may be followed by 6-week-long headaches, temporary or even permanent paralysis.[86]

### Demerol, diamorphine and other opioid analgesia
• Women have frequently reported feeling drunk and out of control with Demerol and even then it doesn't effectively mask pain in labor and birth.
• Like any other opiate used in labor, Demerol and diamorphine can cause problems for the woman, such as nausea, vomiting, sedation, itching, hypotension or respiratory depression.
• Opioid analgesia can have dramatic effects on the baby at birth. Since it goes straight through to the baby and is usually still present in the baby's system at birth, it can cause early breathing problems, it can weaken the baby's 'suck' (so has a dramatic effect on breastfeeding), it can affect the fetal heart rate and it can change neonatal behavior.
• Research has established a link between the presence of opiates in the baby's body at birth and later drug addiction in adulthood.[87]

### Entonox
• Women are disturbed by the process of self-administering gas and air, so cannot possibly 'go to another planet'.
• Using gas and air probably interferes with a woman's ability to produce natural endorphins.
• Experiments into the use of nitrous oxide (50% of gas and air, the other 50% being oxygen), showed it substantially increased the risk of drug dependency in adulthood.

Overall, all forms of drug-based pain relief interferes with the physiological processes and any long-term effects on both mother and baby are not known.

## What about complementary approaches?

Aromatherapy, herbs, homeopathy, TENS machines, shiatsu and acupuncture are often associated with 'natural' birth. They're generally assumed to be helpful but harmless. But do they disturb a healthy birth?

The assumption seems to be that complementary therapies are necessary for someone who wants to give birth without the use of drug-related pain relief. A book about natural childbirth is usually a book which gives guidance on the use of complementary therapies. So why—if we want a healthy birth—should complementary therapies be avoided?

The answer is simple: basically, something can't be useful, without also being powerful. Treatments of whatever kind given during the gestation and birth of a baby actually change what is happening. This is why I think it's extremely risky using them during pregnancy, labor and childbirth. Herbs, acupuncture, TENS machines, shiatsu, aromatherapy and all the other so-called 'natural' treatments may well disturb the physiological processes which would otherwise take place.

There's a problem here, though. What's 'complementary' and what's simply part of everyday life? After all, we're constantly ingesting food and drink and smelling smells... what if these are part of a herbalist's apothecary or an aromatherapist's box? This is a difficult question but, as a rule of thumb, take note of anything which is offered for medicinal or therapeutic use in any way—peppermint tea, a lavender comforter, a shiatsu pressure point—and discourage your clients from using it unless you have good reason to do otherwise. This means no herbal or 'fruit' teas during pregnancy or labor, however seemingly innocuous. It also means no ordinary coffee or tea—or at least very little—and no seemingly harmless luxury treatments. This is simply because they're all too potentially powerful and disruptive.

*What's 'complementary' and what's simply part of everyday life?*

Having said all this, let me now say that a handful of exceptions seem beneficial—in particular, two which have survived the test of time and two others which are the result of recent research. One is the use of raspberry leaf tea (a uterine tonic) in the third trimester and possibly also during labor. Another is the use of arnica—a homeopathic tablet (Arnica 30) taken daily to prevent bruising in the last month of pregnancy. The other two are the use of supplementary folic acid in the first trimester of pregnancy to prevent spinal tube defects, and the use of Omega 3 fish oil supplements in the third trimester to help fetal brain development.

Everything else is best avoided, it seems. When women's bodies are left alone, in most cases they will redress any imbalances automatically, provided the woman doesn't have any major health problems and provided she does everything she can to facilitate the normal, healthy processes.

Let's look more closely at a few complementary therapies...

## Aromatherapy

Have you ever used some bath salts or bubble bath and felt very sleepy afterwards? Some of the aromas used in commercially produced products may have this effect because they are derived from herbs or flora which are used medicinally for the same purpose. Also, there is the problem that organizing aromatherapy for a birth usually means inviting another person into the room. What's more, inevitably, it's likely to either distract the woman or make her feel she 'needs' something other than her own resources and/or make her feel like a patient who needs to have things 'done' for her. She's better off without it.[88]

## Herbs

Did you know that some innocent-seeming herb teas (such as peppermint) are produced from herbs from the same family as ones used to induce miscarriage (such as pennyroyal)? If you're not already aware of the power and usefulness of herbs in daily life, you might like to try a few out... if you're not pregnant.

Did you know that some herbs are used to induce miscarriage?

## TENS

This is generally marketed as a non-invasive method of pain relief—and this, no doubt, explains its popularity, as well as the fact that it's both relatively cheap and readily available. However, it is still very possible that TENS disturbs the healthy physiological processes of labor. And I think women are still better off without TENS machines.[89] Before we consider the reasons why, here's another extract from one of my conversations with Michel...

### What do you think of TENS?

TENS is a very minor intervention, perhaps not more than a placebo. In 1975, I described in *La Nouvelle Presse Médicale* the technique of what I call 'lumbar reflexotherapy', which is at the root of TENS.[90]

I already gave the physiological interpretations that are given today to explain how TENS works (if it works). With my technique you just need sterile water, while to use TENS you need to buy a device (so it's more commercial). The effects of lumbar reflexotherapy are spectacular, compared with the effects of TENS.

### So do you recommend TENS to pregnant women? Surely, recommending a machine such as TENS implies a lack of faith in women's instinctive ability to give birth unaided and couldn't its use distract women and stop them from 'going to another planet', as you call it?

I never took the initiative to recommend TENS, for the reasons that you mention. But when a woman has bought the device and intensely wants to use it, I don't dissuade her. This is true for other non-pharmacologic [complementary] treatments too. I am reluctant if the treatment implies the introduction of a therapist, i.e. another person in the birthing place.

I am reluctant if the treatment implies the introduction of a therapist, i.e. another person in the birthing place

Here are my own thoughts on TENS:

- As I've already suggested, depending on something 'outside' is likely to undermine a woman's confidence in her own ability to cope. I think a woman's belief that she can give birth is almost as important as her intrinsic ability to do so.
- A woman using a TENS machine is likely to be focusing on technical and practical details, rather than on what is going on within her—so it's less likely she'll 'go to another planet'.
- Inevitably, the machine will restrict her movements to some extent and this will make it less likely that she'll tune into her intuitive knowledge of which positions are best for her baby.
- Research has revealed that women perceive that the use of TENS machines increases, not decreases, the incidence of intense pain later on in labor.[91]

No doubt this is because using a machine which is helpful for contractions early on in labor makes later contractions (for which there is suddenly no relief) seem stronger. If using a TENS machine really does make later pain in labor seem more intense, it's likely to be a first step towards other forms of pain relief. (Perhaps this is why it's widely claimed by both women and midwives that TENS is not powerful enough for use through the whole of labor.) Doing without TENS will mean there's a gradual, more manageable build-up of intensity throughout labor.

There may be other as yet undetected effects on a person and on the birthing process. Just because no research project has as yet uncovered risks does not mean there aren't any.

## Acupuncture

Michel comments on the use of acupuncture in his book *Birth Reborn* (Souvenir Press 1994). In his experience, traditional Chinese acupuncture is seldom used for labor and birth in China, although it is believed to be useful for turning breech babies during pregnancy. He believes the Chinese' avoidance of using acupuncture during labor and birth is probably due to an acknowledgment on the part of acupuncturists of the extreme complexity of the physiological processes, which they realize are best left untouched. I must say, I have actually been told the opposite by an acupuncturist, i.e. that acupuncture is used by 90% of laboring women in China. Which is true? If acupuncture is being used more nowadays, is there a connection between this and the extremely high level of medical intervention which takes place in births in China?

While I was researching this book I found a couple of birth stories which mentioned the use of acupuncture. The women who sent these had a very positive attitude towards this complementary therapy... despite the fact that they also ended up having a very difficult labor and birth. In one case, the author of the birth account was a potential contributor. When she heard my interpretation of her birth story, she decided she wanted to withdraw her story.

She was worried the detail of the account would identify her acupuncturist, and she didn't want this to happen because she said acupuncture had helped her family in so many other ways. As far as she was concerned, she would never have got through such an awful labor if it hadn't been for the acupuncture treatments. In my view, it was the acupuncture which had disrupted her labor and made it so long and difficult. A chicken and egg situation, it seems...

Perhaps it's not always dangerous using acupuncture in certain cases, as Birthframe 39 demonstrates, but leaving the healthy processes to take place undisturbed does seem to be the best approach most of the time. Acupuncture, like other types of complementary medicine, is simply too powerful not to be considered a serious intervention.[92]

> I went to a friend's therapeutic masseur, who suggested I rub the acupuncture points on my ankles to induce birth and it did! That night, at 4:00AM my water broke and the contractions started. I'd planned to have a home birth with a government midwife through my health insurance provider and knowing they were probably off duty, I waited until 9:00AM to phone them. But they refused to come out as I'd had a show of fresh blood when the water broke. They said the fetus could be in danger because of internal bleeding around the placenta and I must come to the hospital.

## Massage

Massage can sometimes be an extension of the spontaneous touching which can occur between a laboring woman and her husband, or it could be touching which has been specifically requested by the woman (either from her husband or from a midwife or other birth attendant). If it's not really this at all, but a contrived pressuring of specific areas, 'massage' becomes 'shiatsu'—i.e. dramatic results are achieved by putting pressure on various pressure points— so it can become another form of disturbance. Nevertheless, if 'ordinary' massage is  carried out over many hours it can certainly be very useful and might in fact rescue a laboring woman who is otherwise having difficulty coping, either generally or because of posterior lie.[93]

# Where does it all leave us?

It's particularly important to remember how little we know about pain relief whenever we offer it, or provide it, and to remember the risk of disturbing the smooth and normally efficient processes of labor. Caregivers should perhaps focus primarily on facilitating these processes so that they really do happen in optimal ways (on all levels) because—as many of the birthframes in this book illustrate—the fulfillment experienced after a drug-free, undisturbed birth can far exceed the expectations of a woman who is initially apprehensive. It is striking that this is particularly the case when women have previously experienced pain relief... Only then do they realize the incredible contrast between a 'managed' labor (which needed to be managed and consequently disturbed as a result of the use of pain relief) and a labor and birth which are entirely spontaneous and physiological. I wonder what babies would say too?

## Birthframe 24: Sue Pakes—Relatively painfree births

In any case, on the topic of pain and pain relief, perhaps we sometimes forget the discovery of the doctor Grantly Dick-Read that some women actually don't experience birth as painful.[94] I got talking to the woman who contributed this account while I was at a playgroup, heavily pregnant with my third child...

I always enjoyed the conversations at playgroups when they invariably came round to the subject of childbirth, a bit like discussing exam results when you know you've done well! It seems that I have been one of the lucky ones and, yes, it does happen.

As the due date of my first child loomed, my mother did admit that she had not had too much trouble giving birth to my three siblings and myself, though you don't know how much gets forgotten over time. In those days it was customary to have the first child in hospital and any further births at home, so in the back of my mind I felt confident that if we had more than one child, I too could enjoy a home delivery. Also, I felt that at long last my wide hips may be coming into their own!

The delivery of your first baby is certainly a surprise. Nothing can prepare you for what is going to happen. I was well planned for the event: I had given up work six weeks before, had rested at home, spent sessions at the local swimming pool building up my stamina, requested a water birth at the hospital and had even spent time recording my favorite music ready for the hours I expected to be in labor.

Nothing can prepare you... but I was well-prepared for the event

When the time arrived, not even knowing that I was in labor but wondering why I had been sick and why was I suffering from what I thought to be bad period pains, we drove to the maternity hospital. On arrival at around 6:00AM, I was examined by a midwife. To her surprise, she found that I was sufficiently dilated to move to a delivery suite. Once we were there and I had climbed onto the bed, she went in search of pain relief. Within minutes, however, my water had gone and I was ready to push! As my husband called for someone 'quickly', I couldn't believe that it was all happening so fast. There wasn't sufficient time for any Demerol, so I was offered gas and air. When the moment came to push I pushed as hard as I could and, after a short amount of pain, our son was born at 7:10AM weighing in at 8lb 8oz—and I didn't even need any sutures.

So when it came to the due date of our second child, I wondered if I would have a similar easy delivery. Again, things seemed to get going early in the morning (after a good night's sleep) and I awoke again to uncomfortable period pains. I was adamant that I was going to have a shower before the midwife arrived to assess me, and even managed to wash my hair! In the meantime, slightly apprehensively, my husband called the midwife at 6:30AM. When she arrived around 7:20AM and asked me to lie down on the bed, I knew that as soon as I lay down I wouldn't be going anywhere. Again, after one massive push our daughter was born at 7.42am, weighing in at 9lb 8oz. Throughout the time, I had been totally calm and when I saw how relaxed the midwife was, I felt I too had little to worry about. My body took over.

So when it came to the due date of our third child, I felt we ought to be a little more organized at home, and get the midwife to drop off the home delivery pack. My midwife was more than happy with a home delivery, but my doctor certainly did not agree with it. (He believed that babies should be born in hospital.) So I told him that I would endeavor to get to hospital but was fairly sure that I would not make it in time and did not want to be one of those stories in women's magazines—"MOTHER GIVES BIRTH ON HER FRONT PORCH"!

So with everything in place, the due date arrived. My first child had been two days late, the second four days late, so how long did I have to wait for my third? As we got to three days late and me willing things to get going, I went to bed. During the night things started but, as the midwife advised, the baby was lying the wrong way round and I needed him to turn, so off she went. I returned to bed, lay on my side and left it to the baby to get himself into position and after one large contraction, he managed to and was ready to make his entrance. As we called the midwife again at around 8:15AM the next day, which was a Tuesday morning, we failed to realize that perhaps the morning traffic would influence her arrival. This time we were on our own, and the thought did cross my mind as to what I was doing. I was worried about being on my own and even began to think that perhaps I should have listened to the doctor and gone to the hospital. I felt let down by the midwife because instead of being a perfect delivery I was worried that something might go wrong. However, again, my body (and some say Mother Nature) took over. At 8:55AM the midwife arrived just as our son was born. His amniotic sac was still intact, which apparently means that this delivery was slow and controlled.

I had been so lucky. All three of my children's births had been easy and relatively painless. With the second and third I had not had any pain relief. Also, the fact that the deliveries had been quick was a real benefit for my recovery and also my babies', who had not suffered any trauma and were all calm and relaxed babies. It used to cross my mind occasionally whether I ought to offer to be a surrogate mother as really I had enjoyed easy pregnancies and very easy births.

## Birthframe 25: Gemma Shepherd—Psyching up to the pain

Other women try to psyche themselves up for birth during their pregnancy.[95]

When my first baby Kizzy was born in hospital I felt very empowered by the birth. I had used no drugs of any sort and had no interventions, yet I couldn't imagine wanting to repeat it. "How does anyone ever have more than one baby?" I asked my mom the next day. However, a year or so later a friend at a La Leche League meeting about birth recommended *Spiritual Midwifery* by Ina May Gaskin. The book is full of stories about peaceful, joyful home births. It was a complete revelation to me that birth could be totally painfree and enjoyable. I thought a lot about changing my view of pain and began to see contractions as 'rushes of energy'. I learned to relax my body and to think of pain as 'an interesting sensation that I needed to concentrate on'.

In the book it is often said that if the mouth is 'soft' (i.e. relaxed), then the pelvic floor area will be too. I practiced this by blowing out through my mouth with very relaxed lips when I went to the bathroom. I also tried not to tense up when I experienced any sort of pain. For instance, if I stubbed my toe I would relax myself and the pain lessened much more quickly than normally. I was able to combine Ina May Gaskin's theories about pain with Michel Odent's writings about a woman's instinctive understanding of what positions to adopt during labor and birth. (After all, most—if not all?—of the women in *Spiritual Midwifery* had given birth in bed, on their backs or semi-reclined.)

I had a sort of practice run at birth when I had a miscarriage when Kizzy was 20 months old. After two weeks of spotting blood, a mini-labor began when I was 12 weeks' pregnant. I knew the baby had died because I had stopped feeling pregnant around the time the bleeding had started. (I had also had an ultrasound where they 'informed' me there was no heartbeat, as if I didn't know what was going on in my body.) I relaxed through the miscarriage and it was painless. I could sense that the feelings would have been painful, much like strong period cramps, if I had been tense. I felt elated when it was over, as if I had really achieved something by not feeling any pain. I wasn't ready for another baby yet and felt that the miscarriage had been for the best. (I had another ultrasound afterwards where they 'let me know' that my uterus was now empty!)

My next pregnancy was planned and the baby was due when Kizzy was $3\frac{1}{2}$. By this time we had moved away from the big city, where the maternity care was very impersonal, to a small, fairly isolated town. The local hospital has a midwife-run maternity unit and the midwives are friendly, relaxed and have plenty of time to chat. Even so, I knew I would have a home birth this time. I planned a water birth and rented a pool as (even though I was much more confident I could cope with the contractions) I still felt that a pool would provide some pain relief if I needed it and the calm, enveloping warmth of the water was appealing. My team of four midwives seemed keen on the water birth and had special training sessions at another hospital to prepare them. None had delivered a baby in water before and I think they had little experience of home birth. By the end of my pregnancy I felt fully prepared for the birthing to come. Dora was born at home on a beautiful sunny morning in June 2000. The labor was fairly quick. My water broke at midnight and I tried to sleep, but was too excited. I had a long bath instead and meditated and visualized my womb opening up to let my baby out. I managed some sleep, but by 4 o'clock the contractions were coming regularly. I felt very peaceful and safe as I pottered around the silent, sleeping house, picking up and arranging things for the midwives. When a contraction came, I would lean on a piece of furniture or the kitchen counter and sway gently, breathing slowly, out through my mouth. Each contraction was like an exciting journey—enjoyable in itself, but also promising even better to come. I felt the love of friends and family like a warm glow, especially when I was looking at various photos around the house. Even though I was alone physically, I knew there were many people with me in spirit. I often smiled and laughed during contractions and when one was over I waited eagerly for the next.

Jim, my husband, called the midwife at about 7 o'clock, when I felt things were progressing, then began to fill the pool. The midwife arrived and checked me just before 8:00AM—I was 3cm dilated. This surprised me as I thought I was further along, but as things turned out maybe I was.

After a few more contractions things began to intensify. I remember moaning 'Woaah' as the feelings suddenly became much stronger. There was still no pain, though I had to work much harder to stay relaxed enough. (I realized after that this was probably transition.) I decided to get into the pool now, thinking I might not be able to cope if things got any stronger. The water felt wonderful, like a big, warm hug. I had two more contractions then, out of the blue, I had an overwhelming urge to push with the next. The midwife, who was waiting for the other midwife to arrive, shouted "No!" when I told her. I understood then that she was scared to be on her own. I didn't let her fear affect me, but when I pushed during that contraction some diarrhea came out. The midwife insisted I leave the pool as she thought it was too contaminated to give birth in.

Although I had expected to want to squat to give birth, my body knew that it needed to stand and my knees straightened up automatically. I leaned on the side of the pool. During the first push I had felt the baby move down rapidly. With the next, the head crowned and I reached down to touch Dora's soft hair. I had to slow myself down at this stage to be sure I didn't push too hard. I stopped pushing mid-contraction and breathed slowly. I was aware of the beauty of the sunny, blue sky outside and the wonderment of the moment. I felt a strong connection with all of nature. Then the final contraction, and the head was born. I looked down and clearly recall how surreal it seemed to see her head between my legs. I reminded myself to go slowly and within a few seconds her whole body slid gently out.

I held her and she cried with her eyes tight shut for several minutes—it was probably very bright for her. Dora was born at 9:18AM, less than an hour and a half since I was examined at 3cm dilated. Kizzy and Jim came into the room, amazed to see a baby. They had been making breakfast in the kitchen when Kizzy said she could hear a baby crying. Jim wasn't sure at first as he thought there was still some way to go!

Dora nursed for the first time and the placenta was delivered a few minutes later. Unfortunately, the midwife pulled on the cord as the placenta was coming out. This was totally unnecessary, felt quite uncomfortable to me and could have created problems.

Dora's birth was serene and peaceful, full of joy and love, something to be cherished. I was on such a high to have achieved my dream of a totally painfree birth and to have my beautiful baby girl in my arms. The only thing to mar the experience was the midwives' interference, however minor. Luckily, I was strong and aware of my body to the extent that I was able to ignore the negative aspects of the midwifery 'care' and still feel I had had the best experience of my life. I was eager to give birth again!

I was eager to give birth again!

## Birthframe 26: Anonymous—Vaginal vs cesarean birth

Here's a more generalized comment, comparing the pain of a cesarean with that of a birth with no drugs whatsoever...

A friend recently asked me what it was like giving birth at home. She sounded as though she was afraid of the pain, and skeptical about why she should choose to subject herself to it. I found I was at a loss to explain why I thought it was so worthwhile. I got as far as 'It was important to me to know what birth felt like.' I could pretty safely say it was the best experience of my life. Sure it hurt, but it was more than worth it. But I couldn't think of a reason why *everyone* would want to.

Here's one: quick recovery. After my son's birth by cesarean section, I wasn't allowed to get out of bed—I couldn't even pee for myself, I was hooked up to a catheter. The day my daughter was born, I laughed, cried, slept with her snoring on my chest, and went out for a walk that afternoon. Family and friends came over for a party that evening. The baby and I were up celebrating with everyone else. The next day my diaphragm felt bruised as though I'd run a marathon from the exertion of pushing. A day after that, I felt nearly normal.

Here's another: it's very satisfying. It makes you feel very proud to push a baby into the world all by yourself. It's a deeply carnal satisfaction.

# Pause to reflect...
## ...about fast and/or painfree births

1   Why do you think the term 'precipitous birth' is used in a negative sense to refer to a birth which is actually very smooth and fast? Why is this not considered preferable? Would you consider it preferable in some cases?

2   Should women routinely be prepared for the eventuality that the midwife might not arrive before they give birth? Should they be prepared for potentially being alone at a key stage (even in hospital), when all midwives and other caregivers are busy with other duties?

3   Do you believe it's possible for women to really prepare for a painfree birth? What kinds of preparation would be helpful?

4   Do you think that birth can be pleasurable and even orgasmic for certain women.... Or even for *all* women?

5   In what ways can caregivers 'mar the experience' of birth for women?

For other relevant material see Birthframes 3, 21, 24, 38, 39, 48, 56, 65, 67, 69, 71, 90, 93, 94, 97 and 99.

## Reason No. 5:
## Babies are disadvantaged, perhaps far into the future

### The baby's perspective

The baby's perspective is often forgotten when childbirth is under discussion. Years ago, the baby was often regarded as being an insensitive, unfeeling part of the equation of childbirth. Discussion has always taken place as to when a fetus or baby becomes a truly sentient being, with sensations, thoughts and emotions. This debate still continues but research is increasingly providing evidence that our estimate of a person's development of consciousness should be placed earlier, rather than later, perhaps even at a few weeks of gestation... if not before. A simple glance through the week-by-week guide to pregnancy at the back of this book will confirm that fetal development suggests experiences which in many respects are similar to those of a live baby or even adult person.

Whatever your own view, from the 1920s on some European psychologists and clinicians were researching the effect of birth experiences on growth and development. Otto Rank's contention in 1923 that adult psychological problems might stem from birth separation anxiety, implying the primacy of the *mother*-infant relationship (over that of the child's relationship with his or her father) was seen (by Freud in particular) as an extreme and unacceptable claim. Today, of course, it is accepted without question, especially as a result the work by Marshall Klaus *et al* (first published in 1976) on bonding.

Frederick Leboyer, a French obstetrician born in 1918, certainly felt that a baby was completely sentient at birth and that he or she could be affected by birth experiences. He successfully raised many people's awareness of what he saw as the baby's perspective through his book *Birth Without Violence* (Inner Traditions Bear and Company 2002), which was first published in 1974. To do this, he used a poetic approach, not one based on hard data, and his book spawned a gentle birthing approach which involved subdued lighting and gentle touch, as well as a bath directly after the birth in lukewarm water.[96]

David Chamberlain, a psychotherapist based in North America, also came to believe that birth experiences are crucially important to babies. He came to this conclusion after hypnotizing adults and attempting to gain access to their birth experiences, which in many cases he found surprising and important in terms of their later lives. In his book *Babies Remember Birth* (Ballantine Books 1990) he details some of the hypnotherapy sessions he conducted, which seemed to suggest that babies are even sensitive to comments people make at the time of their birth. His research through psychotherapy also led him to conclude that babies became frightened when things go wrong at birth, for example when they have difficulty breathing, when they are suddenly subjected to bright lights or when they are taken away from their mothers. Along with Thomas Verny, he co-founded the Pre- and Perinatal Psychology Association. (See www.birthpsychology.com for more information on research and events.)

The idea that babies can be scarred by words and actions at their birth might seem extreme to you, but people in other cultures are more accepting of this idea. Nina Klose, who's done research in Russia comments...

❝ There's a feeling in natural childbirth circles in Russia that we have a responsibility to our children to try to give them the gift of a natural birth because of the lifelong physical and emotional after-effects of a traumatic birth.

If you're skeptical, imagine for one moment what it must be like to be born under pressure, with artificially-induced contractions (which are stronger and which often compromise the baby's oxygen supply), in an unfamiliar drugged state, with an electrode pierced through your head or suctioned to it. Consider what it must feel like to meet an unresponsive, dopey or merely passive mother, after hearing her lively voice for weeks, and indeed months in the womb. Imagine the reality of having difficulty establishing breathing because of a hurriedly severed umbilical cord or an excess of drugs in your system. Imagine what it must feel like to have unnecessary difficulties with your circulation or with basic temperature regulation. And how must it feel to have something thrust up your nose, or to have something put into your eyes at the moment of birth? Imagine almost immediately being laid on your back so as to be weighed on cold, hard scales... How would you feel as a newborn if you were taken away from your mother so as to be washed and examined? What might you feel when put to your mother's breast only to find that you can't suck because of drugs in your system affecting your instinctive sucking reflex?[97] Do you think you might feel any panic, confusion, sadness, pain or distress? Might you not feel a strange longing for a better birth, which is somehow more sensitive and beautiful?

Some midwives and obstetricians dismiss the idea that the baby's state of mind at birth is important, or their actions or words show a refusal to take it really seriously. This seems strange because interventions or drugs are surely highly likely to affect a baby's feelings and perception of the world.[98] If contractions were artificially accelerated, they were probably more stressful for the baby so he or she is likely to be more tense or tired at birth. If forceps were used, a headache at birth is a probability. If drugs were used for pain relief, the baby is likely to feel drowsy, woozy, sleepy or nauseous and anesthesia of any kind may well desensitize a baby for up to a month after the birth. Not a very nice way to start life really. And what could the effects be on early bonding?[99]

Imagine, for a moment, the alternative: a fresh, alert, happy sort of feeling, with a usefully strong suck; heightened sensitivity to your mother's soft, responsive arms, in a room full of kindness, with dimmed lights and subdued sounds. Imagine, if you were the baby, your first experience with your mother, looking up into her eyes, your first attempt to suck at her breast—it's likely to be fulfilling if no drugs are affecting your innate ability to suck.

Consider the relief at having completed a journey which might perhaps have been a little frightening and physically tiring... Consider this very different transition from life in the womb to life outside it.

When drugs are not used, mothers are clearly attuned to their babies' movements, and right after birth a kind of entrainment may be established to set up behavioral patterns and language. And yet women are still medicated during the birth process, leaving them incapable of responding to the cues of their newborns, while babies—separated from their mothers—are left crying for the expected touch of parental skin. Why are so many caregivers not facilitating what should be a very natural path to effective parental care by encouraging the avoidance of drugs and unnecessary interventions? Is it because of a lack of convincing research data?

## Hard data

There is actually quite a bit of hard data to confirm that events before, during and after birth might have long-term effects.

For example, researchers found that obstetric complications (which often result after unnecessary interventions) were associated with a higher incidence of childhood asthma.[100] Children who experienced certain procedures at birth, i.e. cesarean section, vacuum extraction, the use of forceps, or who were literally pulled out by a care provider manually, were found to be particularly at risk. Rates of asthma amongst children birthed normally were significantly lower.[101]

Many other studies have shown links between a baby's birth experiences and his or her late experiences or health in life. Michel Odent reports:

Recently, there has been an accumulation of hard data confirming the lifelong consequences of the prenatal environment and also of the way we are born. In other words, the branch of epidemiology I have called 'primal health research' (since 1986) has developed dramatically. It developed at such a pace that in 1997 we found it necessary to establish and to continuously update a Primal Health Research Database so that everybody can refer to this data. [See www.wombecology.com or www.birthworks.org/primalhealth]

Today the database contains hundreds of abstracts of articles published in authoritative scientific and medical journals. All of them show correlations between what happened during the 'primal period' (from conception until the first birthday) and what happens later on in life in terms of health and behavior. It is not easy to detect such articles because they do not fit into the current classifications. This is the main reason for trying to bring them together.

From an overview of the data we see immediately that, in all fields of medicine, there have been studies detecting correlations between an adult disease and what happened when the mother was pregnant. It is even possible to conclude that our health is to a great extent shaped in the womb. There are many studies confirming that the emotional states of pregnant women may have lifelong effects on their children. This might lead us to conclude that the first duty of health professionals should be to deal tactfully with the emotional state of pregnant women. This is not easy in the framework of industrialized childbirth, which implies a certain style of prenatal care, constantly focusing on potential problems. Health professionals need to avoid the common mistake of doing more harm than good by interfering with the imagination and belief system of the person they are taking care of.

Through the data provided by primal health research, we are can try to forecast what sort of disaster might be induced by the industrialization of childbirth [i.e. excessive intervention]. The data indicates there might be more violent young criminals, more teenage suicides, more drug-addicted adults, more anorexic girls, more autistic children... All these conditions have never been as frequent as they are today. We don't know why this is the case. But, for all of them, the Primal Health Research Database reveals studies detecting risk factors in the period surrounding birth.

Here is some more detail on research in the Primal Health Research database:

- Anorexia nervosa has been found to be more common in girls who have had difficult births, i.e. births where there was clearly a great deal of intervention. Girls were found to have been more at risk for anorexia nervosa after having a cephalhematoma (which is a marker of a highly traumatic birth) or a vaginal instrumental delivery, i.e. forceps or vacuum. Interestingly, the fact of having had a cesarean birth did not predispose girls to anorexia.
- Adults who have committed suicide using asphyxiation were usually babies who had breathing difficulties at birth. (This kind of difficulty is more likely when babies have drugs in their system at birth.)[102]
- Adults who have committed suicide by violent mechanical means usually had some kind of mechanical birth trauma caused, for example, by the use of forceps.[103]
- Autism has been associated with various aspects of the period surrounding birth: induction of labor, 'deep forceps' delivery, birth under anesthesia and resuscitation at birth.[104] Babies born in certain hospitals where induction of labor and the use of a mixture of sedatives, anesthesia agents and analgesics during labor were routine were particularly at risk.[105]
- A marked increase in the incidence of schizophrenia was shown in adults whose birth showed an excess of complications of both pregnancy and delivery. Although pre-eclampsia (which the mother can perhaps do nothing about) was associated with schizophrenia, other complications would certainly have resulted from inappropriate or excessive interventions during labor and birth.[106]
- One study showed a clear link between left-handedness (which is often associated with behavioral disorders) and complications either during pregnancy or birth.[107] Again, complications at birth might well have been caused or exacerbated by inappropriate or excessive intervention.
- As we have already mentioned, babies born under the influence of nitrous oxide (a component of Entonox, i.e. gas and air) have been found to be more susceptible to amphetamine addiction later in life. The use of opiates or barbiturates (or both), alongside nitrous oxide, has been associated with opiate or other drug addiction (e.g. amphetamines) in offspring later in life. Researchers think that the clear link established could be the result of 'imprinting' during labor. I wonder if there's any link between the widespread use of ecstasy and cocaine in nightclubs and the way the same young people were born? Our contemporary adult drink/drug culture two decades after births became more medicated on a widespread basis should prompt us to reflect on what we're potentially doing to babies at birth.[108]

As if this weren't all enough, there's one other enormous area of possible repercussions to consider...

## What about a baby's capacity to love?

It's possible that a person's capacity to love may develop around the time of birth. Michel Odent has argued that this is the case in his book *The Scientification of Love* (Free Association Books 1999). All his hypotheses are backed up by research data, drawn from a wide range of sources. He is of the opinion that even a baby's lifelong capacity to love might be affected by an interventionist birth involving drugs and alienating procedures which disrupt the natural production of hormones in both mother and baby. It's easy to understand how. In a fully undisturbed labor, a woman produces various hormones which may 'imprint' on the baby's mind as he or she is being born. These hormones—oxytocin and prolactin, in particular—produce loving, mothering behavior in the mother, and they may well also prepare the newborn for a new loving relationship and for relationships in the future. Makes sense somehow.

> Michel Odent has argued that a person's capacity
> to love may develop around the time of birth.

If these ideas seem extreme, at least allow them to stay at the back of your mind for a while. I remember my own reaction when my obstetrician in Sri Lanka wrote some comments in a letter to me after the birth of my first daughter. "For a long time," he wrote, "and this is no exaggeration, I have wondered if violence at birth contributes to making a violent society." I must admit, it seemed an almost ridiculous notion to me at the time. But the more I think about it and the more I study the research data, the more I feel inclined to think a link may not be so outrageous an idea after all.

If you are about to interrupt me to say that babies born in the developing world don't always grow up to be gentle and peace-loving, do note that these countries may be the places where *most* intervention and disturbance is going on. I only learned this through living in Sri Lanka and Oman and through extensive travel elsewhere. In Sri Lanka, in a very rural place called Ella—a tiny mountain hamlet seven hours by train from Kandy—I was astonished to hear from the landlady of our guest house that all women in Ella travel to their local hospital to give birth. (Apparently, it was only eight miles away.) When I was pregnant in both Sri Lanka and Oman I find it an enormous challenge to avoid interventions prenatally (as well as intrapartum) because these were forced on women if they weren't accepted routinely. This is partly why I started researching birth so carefully, because I needed ammunition to back up my refusals! Gradually, I realized that in highly undeveloped places technology and 'Western' ideas are sometimes almost worshipped as gods. It seems our idyllic notions of developing countries eschewing all technology are very misplaced. Too many human beings around the world are being born in unloving births. Perhaps it is not surprising if, as a result, these babies are unable to love.

Photo © Jill Furmanovsky

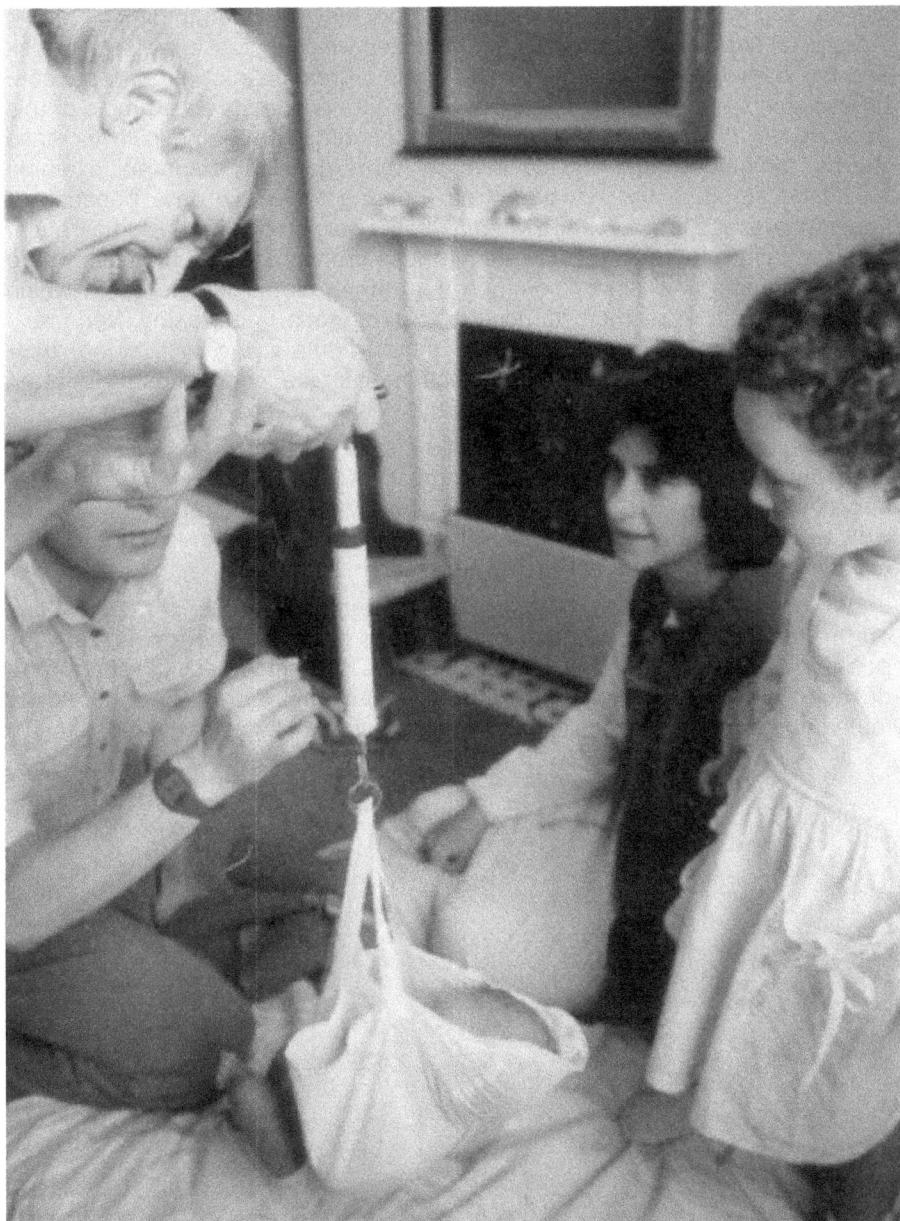

*How can you make everything more personal and caring?*
*Here Michel Odent is weighing Liliana's baby after the birth*

How can you make everything more personal and caring?

In some developing countries hygiene may be poor due to lack of resources, climate and lack of training, and these factors may also lead to difficult births. Traditional superstitions also sustain practices which increase the level of risk and the subsequent necessity for intervention. Creating a lot of noise to scare off evil spirits after the birth, before the placenta is born, is an example of this because any disturbance to the mother at this time can dramatically increase the risk of hemorrhage and consequent maternal death. Another example would be stopping the newborn from nursing for a few days because of superstitions associated with the early 'milk', colostrum; obviously this might result in failure to breastfeed at all. All these problems occur in places where female circumcision is very common (e.g. Africa or the Middle East), which makes birth an even more difficult and dangerous undertaking. Very early marriage (from the age of 12) also results in girls giving birth before their bodies are physically ready, which results in injury (i.e. perineal damage, often including fistula) and disadvantages for the babies.

It would be good if researchers could continue to study any links between birth and later behavior or health more systematically over the next few decades and beyond. Links which have already been established suggest it's well worth discouraging women from using drugs or interventions during childbirth unless they are absolutely necessary.

Whatever our views on the possible long-term side-effects of interventions during labor and birth, we can only be sure that we have minimized any possible long-term effects if we allow babies the smoothest possible introduction to this world, involving an absolute minimum of drugs or unnecessary interventions, and no insensitive comments around the time of the birth.

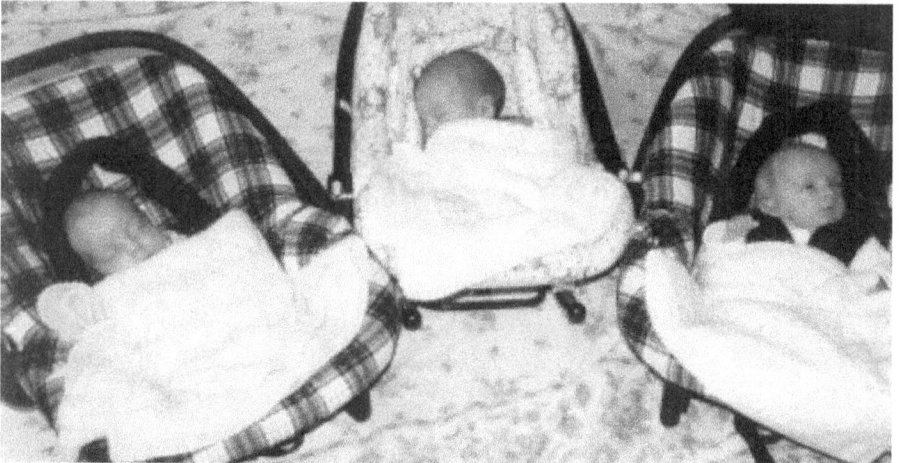

*The mother who gave birth to these triplets (by cesarean) had such a bad time in the hospital that afterwards she said: "I wouldn't go back 'to visit' if you paid me."*

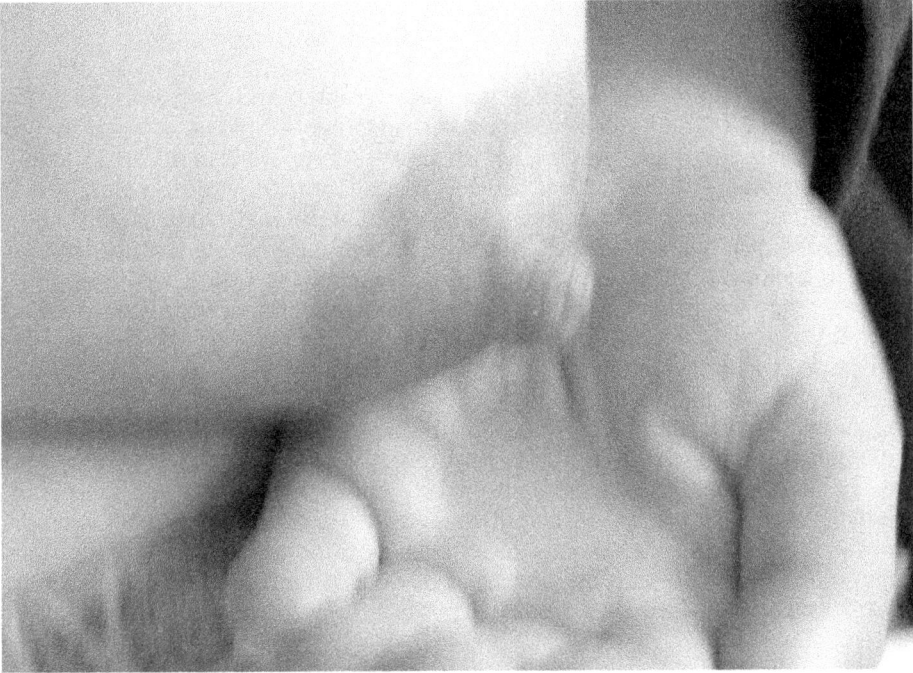

*Milk's come in... But is baby able to suck?...
or does he have drugs in his system which make it impossible?*

## Reason No. 6:
## Women end up disempowered, upset and depressed

### Negative emotions postpartum...

Visiting many baby groups in different places has confirmed my early impression that many mothers experience early motherhood as negative and disempowering. Women typically sit around, barely talking to each other or only exchanging superficial information and comments, in an apparent effort to display an ability to cope, despite behavior which suggests difficulties. Many women have told me they are too overwhelmed to go to groups. Of course, a lack of openness could be attributed to tiredness or a kind of identity crisis after passing into the uncharted territory of motherhood. Certain information and statistics, however, confirm the notion there might be deeper reasons.

Certain information and statistics confirm the notion there might be deeper reasons for women feeling upset and depressed

A survey commissioned by the British *Mother & Baby magazine* in 2002, which collated the responses from 3,000 women, found that 86% of respondents reported they were in pain after giving birth for an average of 24 days; almost a third of these said they were in 'considerable pain'. According to MIND, the leading UK mental health charity (see www.mind.org.uk) new mothers *usually* get the 'baby blues' for two to four days after the birth. MIND reports that doctors usually put this down to hormonal changes or to 'the experience of being in hospital'. Negative feelings after giving birth—which are said to include feeling emotional, difficulties sleeping and loss of appetite, anxiety, sadness, as well as feelings of guilt and inadequacy—are apparently so common, they are considered normal. I would guess the situation is very similar in the US... Have modern 'developed' societies come to terms with the lack of empowerment which results when women don't experience the natural endorphins produced in a physiological birth? When women don't experience the usual postpartum hormones—particularly oxytocin (which is also the hormone of orgasms) and prolactin (which induces relaxation and mothering feelings)—are women plunged into unnatural depression? After all, women who've had drug-free physiological births have reported minimal postpartum pain (both emotional and physical), but obviously more research needs to be done to confirm that outcomes certainly are better postpartum without interventions.

The long-term effects of a failure to experience the natural endorphins and hormones of birth and breastfeeding are difficult to estimate, but one cannot help but wonder why so many millions of anti-depressants are prescribed in the USA each year, many of them to adult women of childbearing age. Apart from the many women who go to their doctors complaining of general depression, perhaps as many as 1 in 10 women are diagnosed with full-blown postpartum depression. This apparently leaves them feeling depressed, irritable, tired, sleepless, not hungry, unable to enjoy anything (including sex), unable to cope, guilty and anxious.

How many women who don't ever get to their family doctors also experience feelings of disempowerment, sadness and depression? How many women go into hospital to give birth feeling optimistic and possibly even excited, but later come out feeling angry, disillusioned and even betrayed? In that survey conducted by *Mother & Baby magazine* over 81% of respondents admitted being frightened during their labor and 53% said that childbirth had been 'far more shocking' than they'd expected. 57% felt their prenatal classes had not revealed the truth about the experience of giving birth in the modern world. That was in 2002. Has the situation changed since then? Are these women representative of women in the USA too?[109] What *is* the truth, which those women found so shocking? Judging from many birth stories I've read (some of them published, others sent to me while I was researching birth), I conclude that a great number of women feel regrets about some of the interventions they have while they are in labor or giving birth. The simple, but uncomfortable, truth seems to be that many interventions are felt to have been unnecessary, perhaps done for the convenience or control issues of caregivers, not safety.

Photo © Colin Smith

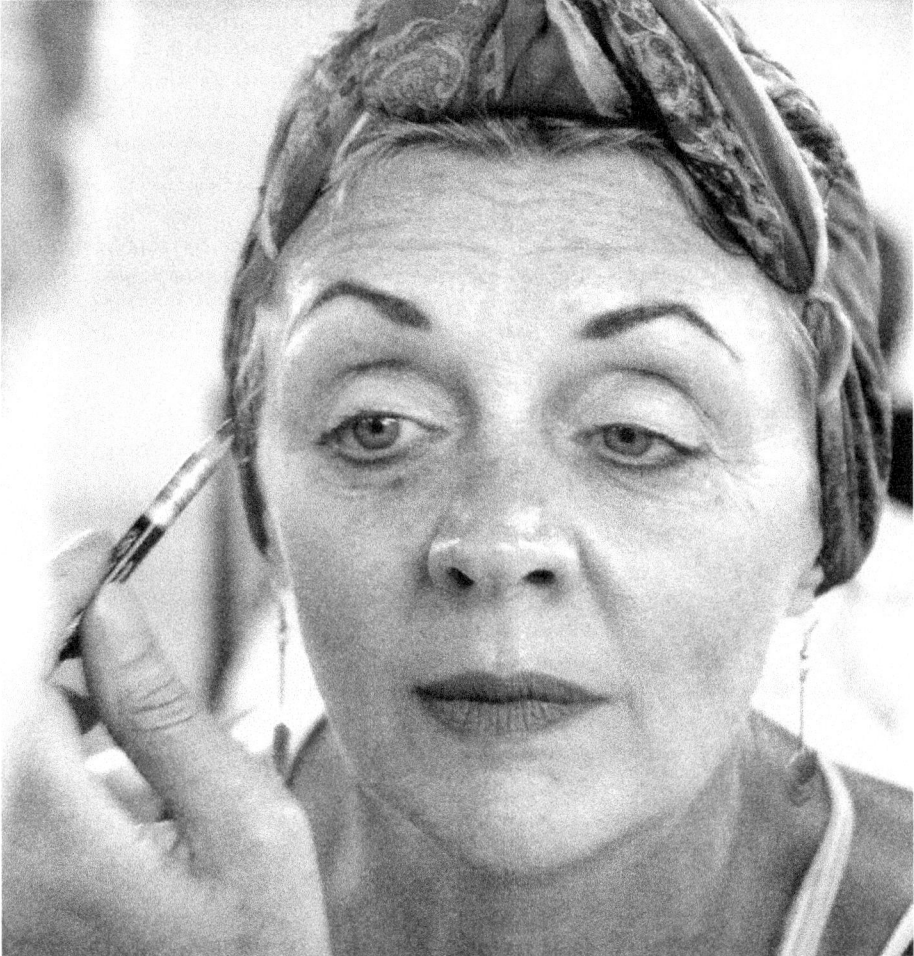

*How long do women continue to experience discomfort and just put a brave face on things after they've given birth? And why do so many older women seemed resigned to their problems 'down there'? Are they all really an inevitable part of childbearing?*

What makes me most angry is the sense of complacency amongst women. It's as if they have given up fighting for their rights even though so many people realize there are problems with so many aspects of birth. (Yes, enormous improvements have been made in some places over the last few decades, but don't you agree we've still got a long way to go?) Repeatedly, I am taken aback when I meet people. Only last week I was being polite to someone I was next to while standing in line and eventually she 'mentioned' the impact motherhood had had on her life: "I can't work myself," she said. "You see, I've got mental health problems. It all started after I had my second baby." She was in her late 50s. I'd said nothing at all about the work I'm doing to research birth.

## Birthframe 27: Anonymous—Extreme anger postpartum

Here's some commentary I found on the Internet. After corresponding with the author, for very obvious reasons we decided it would be better if she remained anonymous. Unfortunately, her cynicism proved to be well-founded because the care she was given was later also evaluated negatively by other caregivers.

I am coming to terms with my anger and bitterness. They eat at me and paralyze my body, while those I am angry and bitter at go blissfully about their work. I believe that the people I entrusted with my care believed that they were doing their best, that they were doing their job. I also believe that in the circumstances under which it was performed my cesarean was necessary. I also certainly believe that those circumstances were, for the most part, caused by my caretakers and were completely preventable. I was caught up in a system that has it all WRONG.

> I believe that those circumstances were, for the most part, caused by my caretakers and were completely preventable

When I realize how much fear I carried with me throughout the pregnancy, how many things I did because of fear, and how little REAL information I got, it makes me heartsick. I had a cascade of interventions leading, finally, to the cesarean. When my baby was taken out, he showed absolutely no signs of the postmaturity that I was threatened with along with insinuations that I was an irresponsible ignoramus for simply wanting to let him come when he was ready. I was so tense and/or terrified for the last three weeks of my pregnancy, it's no wonder that my body would not allow the baby to come. My body knew that babies should not be born where it is not safe. No mammal will give birth in a dangerous place.

I am free to seek expert opinions, but no one else has our interests at heart like I do. I am now a doubter, never taking any statement at face value and now I trust myself more. I may never learn to trust myself completely, as I have been taught from my earliest moments, like most of us, that the authorities know best. At least I know now that I am worthy of trust.

> I am now a doubter, never taking any statement at face value

I have learned that just because a doctor or organization of doctors says something or holds a belief it does not mean that it has a root in hard science. I have also learned that when something has been discounted as 'never scientifically proven', it may simply mean that no one has yet bothered to do, or been able to get funding for, the research. I have learned that Science doesn't have all the answers. I have learned that most statistics can be cleverly twisted to anyone's end. I have learned that people are lied to by people who have been lied to themselves, ad infinitum, until they really believe the lie to be true. I have given up the naive belief that a woman practitioner or midwife, because she is a woman, will be a sympathetic and knowledgeable advocate.

## Birthframe 28: Christina Mansi—Different types of cascades

An unsatisfactory first birth experience often leads a mother to research and campaign for a better birth for a second birth. This perhaps explains the trend for requests for 'natural' birth after an interventionist first experience. And in many cases, women do find that a different midwife (or obstetrician) will take a surprisingly different approach...

I had my first child, Samantha, in January 1985 but what influenced my expectations and hopes was a TV program I'd watched about four years before. The program had Miriam Stoppard as its hostess and the subject was Michel Odent and his work in Pithiviers, France as a natural childbirth pioneer. It inspired me to read books about the subject (home births, natural birth—by Sheila Kitzinger and Michel Odent)—and my enthusiasm grew. When I first became pregnant I sought out a doctor who was sympathetic and experienced, who would attend me at home.

The labor started on Saturday morning. The first sign was a little gush of amniotic fluid—but no pain. The doctor examined me, and assessed that I was not dilated at all. He seemed a bit put out that this was happening on a Saturday—he had to visit his sick wife in the hospital. I tried the usual tricks to bring the labor on—taking castor oil and scrubbing floors.

The doctor said at 3:30PM that we had better go to the hospital as some hours had elapsed since the water had leaked; he said that there was a risk of infection and that the baby could become blind if that happened. We went to the hospital very disappointed and frightened. The resident doctor at the hospital read the letter of introduction from my 'home birth doctor' (who had to go to visit his wife). He looked and acted very disdainfully towards me—that I'd hoped for a home birth and now here I was needing professional help after all. I was wired up to a machine to measure contractions... there were none. I'd heard about the method called stripping the cervix, whereby the doctor uses his fingers to manipulate the cervix into loosening, and maybe rupturing the membranes. I wanted to try this first before resorting to drugs. It was done very painfully and some contractions started.

Over the hours, I was put on a pitocin IV—the contractions were not strong enough and the baby had to be delivered within 12 hours, according to the hospital's rule, because of the risk of infection. I had the water sac broken so that a monitor could be attached to the baby's head. This was very upsetting and painful. I felt raped. So much water came out—I realized that before it was only a leak and was not as dangerous for infection as the full water emptying. Then I was persuaded to have an epidural because they wanted to increase the pitocin to speed things up. The baby's pulse was now dropping so low they were worried, so they phoned their top doctor who advised a cesarean. The epidural was wearing off now and I could feel the baby low down in the canal, almost ready to come out. They still went ahead with a general anesthetic and I woke up to see my husband showing my lovely baby wrapped up tight in a shawl. She was born at 12:30AM on Sunday, nine hours after I went to the hospital.

I was so happy to have a lovely baby that I soon forgot about my ordeal. On Samantha's first birthday I relived the humiliating, disappointing and painful time of her birth. I duly wrote a letter to the hospital. I complained about my treatment. I did receive a reply and I was invited to meet the doctors at the hospital for a talk/discussion. I did not go in the end as I felt that I would be out of my depth at the meeting—and it would be me against them. I had already been through a lot of grief with Samantha's health since she was 8 months old. A doctor at the hospital had told me she had a heart murmur. She was also taking a pediatric steroid in tablet form to try to combat the low blood count they had discovered she had because of a rare blood disease called Diamond Blackfan anemia. I had enough battles to win ahead of me. I discovered that a suggested allergy to wheat could be her problem. We discovered that in fact her blood count went up when she didn't eat wheat or other wheat gluten products. It took me three years to finally wean her off steroids. She's now a healthy young lady, though.

My second daughter, Kathryn, was born in February 1988. I made contact with Michel Odent, who was living in London then. I wanted to try again for a home birth and he agreed to help and support me.

The first sign of labor was a gush of amniotic fluid, as before. This was at 8:30PM. The real labor pains came at 10:30PM. Kathryn was born on my bed at 2:30PM after four hours of labor. It was painful and I was glad to be on my own in the dark of my bedroom. My husband and Michel Odent were there but kept out of the way and didn't interfere.

At the last moment Michel Odent came over and helped, while my husband supported me under the arms as I gave birth in a supported squat position. She weighed 7lb 2oz. I held Kathryn immediately and put her to my breast. About half an hour later the afterbirth was expelled naturally.

I was in shock for the first half hour after Kathryn's birth and was shivering—but that is apparently normal—so we turned on the electric fan heater. Kathryn did not cry when she was born. She lay on my thighs—skin-to-skin—for a long time, contentedly looking around in the half light. I nursed Kathryn exclusively (no solids) until she was 6-8 months old. I didn't have my first period until she was 18 months old and carried on feeding her until two months before my third child was born, when she was 2 years 4 months.

My son Anthony was born in August 1990. I had put on more weight with this pregnancy than before. He was overdue by a week or so. We had visitors—relatives—staying in our home at the time. The night that the guests went to spend one night away my labor started at 10:30PM. It was a long and hard labor—six hours. The second stage only really started when I forced myself to get up into a squatting position. Michel Odent, who had been outside the room listening to the progress of my labor, came in as I was starting to push. He helped to deliver Anthony's shoulders. He later said that his shoulders were broad—he was a bigger baby than the others: 8lb 4oz. I recovered quickly again and nursed him until my second son (my fourth child) Jonathan was born in October '92.

Jonathan was four weeks overdue. My labor started at 8:30PM and he was born at 10:30PM. I wanted to try a water birth so I was in my extra large bathtub with lots of lavender oil when the midwife arrived at 9:30PM. (Michel Odent was not in the country at the time.) She didn't think that I was in real labor as I seemed so calm in the water. When I got out of the bathtub to call her from my bedroom she was surprised. Jonathan was born on the floor of the bathroom, my husband again helping to support me in a squat position. I had no tears, he didn't look overdue and weighed 7lb 14oz. I nursed him till he was over 3. I felt very emotional when I tried to stop. He was my last baby. He's now 9 years old!

Although it's possible to write off the ease of Christina's second, third and fourth births (putting it down to 'luck'), it's equally plausible to suggest that the behavior of her caregivers may have had an influence on the types of labors and births she experienced. The behavior of professionals is a critical element in any scenario because midwives and obstetricians (not to mention doctors and any other caregivers) can obviously do a great deal to influence how a woman feels about herself and, critically, whether or not she feels like a patient or a powerful woman, giving birth. Clearly, the first time Christina thought she was going into labor she very quickly felt disempowered and fearful. The disempowerment which many women feel during their labors often occurs gradually and unexpectedly, as happened with Christina.

There are many cases where the approach taken to caring for a woman is open to discussion or even dispute. Although each of Christina's first two labors began in exactly the same way, the care provided was entirely different. Christina herself certainly felt that the intervention in her first labor was inappropriate and unhelpful, even though it was no doubt within the protocols of the particular institutions concerned at the time of the birth.

Here's a list of interventions which some professionals might consider to have been unhelpful and inappropriate, based on experience and research:[110]

1  **The vaginal exam** Bearing in mind that the main risk was of an infection developing, this may have been inadvisable because it increased that risk. Christina could have checked herself for infection by taking her own temperature every hour.

2  **Transfering Christina to hospital** Again, a hospital is a place which is full of germs, so it could be argued that she would be safer from the risk of infection at home, where her body was at least used to the germs around. In addition, given the way most people associate hospitals with illness and emergencies, quite apart from the fact that it was an unfamiliar place, it's likely that transfering Christina to hospital made her fearful and tense. Since this would have prompted her body to produce adrenaline (which normally inhibits the production of oxytocin, the hormone needed for contractions to continue) it's not surprising that her labor stopped on admission to hospital. The doctor's negative behavior towards Christina—

she said 'He looked and acted very disdainfully towards me'—would have contributed towards Christina's anxiety.

3   **EFM** Wiring a woman up to a machine which produces a computer readout—and usually thereby immobilizing her too—is a very effective way of making a normal person immediately feel like a 'patient'. Since routine use of EFM on admission to hospital has only ever been associated with a rise in the cesarean rate in all research studies (with no improvement in outcomes), there are more arguments against its use than for it.

4   **Stripping the cervix** This would again have increased Christina's risk of infection. As an artificial method for inducing (or augmenting) labor contractions this intervention also carries with it all the risks and disadvantages of induction, i.e. it often starts off what many have called 'the cascade of interventions'. This is simply because it is unlikely to be completely effective, so further interventions (to augment labor further) are highly likely to be necessary.

5   **Pitocin** Putting Christina on an IV must have confirmed her feeling of being a 'patient' because it implied that she was in need of artificial treatment. In other words, whether or not it was necessary, putting Christina on an IV must have communicated to her nonverbally that her bodily functions were ineffective. This would have accentuated any fear she felt, along with her feeling of disempowerment—and, as we know, adrenaline (produced in response to fear) stops oxytocin production. In addition, the IV will have increased her risk of developing an infection.

6   **Imposing a time limit** Protocols vary widely, being based on different views. In any case, since the risk was related to an infection developing, checking for infection specifically—rather than imposing an arbitrary limit—may have given Christina the extra time she needed to be able to give birth vaginally.

7   **Continuous monitoring throughout labor and birth** Of course, this—as opposed to intermittent, discrete monitoring—became necessary after the initial interventions because they *increased* the level of risk. The use of artificial oxytocics (pitocin) to her augment labor made monitoring particularly necessary because of the unpredictability of effects on any particular woman.

8   **The cesarean** If a different approach had been taken—either because of a different professional attitude or because of the 'patient' actually *refusing* treatment—the cesarean might not have been necessary. Giving a woman a C-section for the birth of her first baby may increase her risk of infertility (so may mean she can have no more children);[111] it also inevitably makes any subsequent pregnancy, labor and birth higher risk because if a vaginal birth is countenanced it would necessarily be a VBAC, rather than a straightforward second birth. Given Christina's comments about her feelings on her first daughter's birthday, it's clear that performing a cesarean on a woman the first time she goes into labor may also change the psychological landscape if she gets pregnant again afterwards.

Overall, 'watchful waiting' might have led to a much more empowering birth...

*This is part of a mural at Pithiviers hospital near Paris. What kind of visual and social environment helps to promote the right kind of intervention or watchful waiting? What kinds of systems and protocols best support the physiological processes of birth?*

## Do you want an empowered woman?

A woman who felt empowered by the processes going on within her might have refused these interventions. Or if she had been well-informed about birth, she may have hoped for the empowerment which often occurs postpartum after an unmedicated, undisturbed vaginal birth. But is this really what you want as a caregiver?[112]

Firstly, it's important to say that having empowered women around, giving birth, isn't such a bad thing. If you know that the women you are caring for are having a 'real' experience of childbirth which is likely to result in less pain for them and their babies overall, surely you must be happy? If you know that the women's choices, even if a little inconvenient, are in line with research and are likely to ensure safer outcomes for both the women and their babies, surely again you can't feel bad about those choices?[113]

Secondly, it must be said that the role of a caregiver should really be to support a woman in her choices. When I asked an experienced midwife if she felt it was irresponsible for a woman to refuse treatment she replied:

It's not exactly irresponsible as I do not believe women will put their babies at risk. I think they know their bodies and we have to work with them and support them if they go against medical or official guidance and not threaten them or bully them into doing what they feel is not right for them or their unborn child. Most women are sensible and some make decisions that the medical and some of the midwifery profession don't agree with but it is the midwives' role to support and give care whatever the choices are that the woman makes. However, it is important that the midwife explains any potential risk in a reassuring, accurate and professional way.[114]

As a researcher who has been impressed by the complexity of the natural processes, I personally feel that refusing interventions in cases where there is a choice is in no way irresponsible. In fact, it is the opposite because it is respecting something which is enormously complex and easily disturbed—i.e. the cascade of processes, which occurs through pregnancy, labor and birth.

As an author who has read material from literally hundreds of women, and interviewed many more, I feel sure that refusing pain relief or treatments to apparently ease pain is also wise because it means that women are likely to experience less pain overall, if postpartum pain is also taken into account. Physical and emotional damage caused by episiotomies, forceps bruising and perineal damage, cesareans, catheterization and disappearing babies (who are whisked away 'for observation' or treatment), can last for years. When breastfeeding proves to be difficult or even impossible because of drug use or disturbance of hormones during the intrapartum period I can't see that the pain relief used provides any overall advantage. (The woman may have saved five hours' worth of pain in order to create six weeks' worth.) We need to remember the advantages for women of full alertness, privacy and dignity, as well as a naturally balanced hormonal scenario, as they meet their new babies.

As a mother who has asserted herself when faced with similar situations, I have experienced the empowerment that results from entirely optimal, i.e. safe, physiological births, taking place with supportive caregivers and a loving husband. So saying 'no' to interventions seems necessary at times.

# Birthframe 29: Sylvie Donna—Refusing interventions

You see, I had a problem similar to Christina's...

At 38 weeks in my second pregnancy I had a leak of something and I also had the grippe. Naturally, I phoned my doctor because I wanted to be checked out. He said it was probably amniotic fluid and that I should go to the hospital immediately. I asked him what would be done there. Firstly, he said, a few cells would be scraped from around my cervix to confirm whether or not the leak had been amniotic fluid. After that, I would probably be induced because I was, after all, at 'term'.

Knowing that a vaginal exam would increase my risk of infection (which is the main danger after a leak of amniotic fluid) I said I would rather not go in. The doctor repeated his instruction to me five times, and five times I refused! (A letter followed in the mail the next day suggesting he wanted to strike me off his list but a well-considered reply from me managed to persuade him against this.)

That afternoon and evening I searched through my pregnancy books. I also called an independent midwife as well as Michel himself, who happened to be at home. (Most of the time Michel is traveling round the world lecturing at conferences or setting up research projects. I was lucky that my due date occurred in his Christmas vacation.) Both the midwife and Michel said the leak could well have been urine—I did, after all have a bad cough so it could have been 'stress incontinence'—or it could have been a leak of hindwater, which would 'heal' up. I was told one option was to keep taking my temperature to check it didn't rise dangerously and indicate an infection. Both Michel and the midwife agreed my risk of infection would increase if I went into the hospital and had a vaginal exam.

Two and a half weeks later, after recovering from the grippe and after attending my Master's graduation ceremony—looking extremely large!—I had a very straightforward two-hour labor, as I explained in Birthframe 3. Nina-Jay, my second daughter, was—and still is!—beautiful and very healthy.

When intervention is inappropriate, it's as if midwives (or obstetricians) are intervening out of fear and a desire to control. Or is it just that they're intervening because they want to get women through their births—this 'awful experience'—as fast as possible, with as little (visible) pain as possible? Whatever the reasons for *inappropriate* intervention, unhelpful interference in birthing processes (which are very delicate) often results in more problems and more risk. This is particularly the case when artificial hormones or interventions are used for induction and augmentation of labor, and when drugs are used for pain relief (either analgesia or anesthesia) because further interventions later become necessary, and of course, more pain may well be experienced postpartum as a result. In any case, these interventions disturb what's going on already...

More pain may be experienced postpartum after interventions

## Reason No. 7:
## Hidden problems are truly horrendous

Most people like to think that the USA is a modern place, which is free and open. Most people feel proud of health services available, even though there's increasingly been talk recently about the possible need for improvements. People who are healthy, with good health insurance, probably secretly believe that doctors are dealing with everyone else's problems in the background.

Although I think standards are reasonably high in the USA (and also Latin America), compared to elsewhere in the world, particularly the developing world, I'm afraid I'm not of the opinion that all women's problems are busily being solved. To remind yourself of the unresolved problems you only need to visit certain coffee shops during the day in the middle of the week or go to certain drop-in centres, or even churches, where the dispossessed women of society hang out, but often rarely open up their hearts. To get a glimpse of the many women who don't even get that far (perhaps because they're too depressed to go out) you just need to look out of your window as you're driving round virtually any area in 'Middle America'. (Okay, I admit that in the wealthier areas women are experts at self-disguise, so they're less easy to spot.) But, seriously... haven't you also noticed those grim-looking, shapeless women pushing baby buggies up streets, waiting with empty eyes outside stores, or looking around shops when they have no money to buy anything? Don't you have any 'old friends' who you feel have 'let themselves go'? Don't you ever see a woman postpartum who hints to you that all might not be well?

One of the biggest taboos is incontinence. Although it's possible to see ads about products for this in many magazines, it's rare for women to talk about it openly. And when one woman does have an unresolved problem, she really is ostracized from society, in the USA just as she would be in Africa.

*In Kiberia (the largest slum in Nairobi in Kenya) there are over 700,000 people.*
*We assume many women have problems which are taboo in their society.*
*How many 'mommies' have similar, unreported problems in the USA? Why are so many incontinence products bought privately by mail order? [Photo: Jenny Matthews/CARE]*

## Perineal damage, urinary, fecal and flatus incontinence

When I had my first baby and started ordering products by mail order I was very struck by the easy availability of cushions of various types to support a damaged perineum. Then, after quite a few playgroups, over coffee or on long walks home up hills and along busy streets, quite a few women were very open with me about what they'd experienced postpartum as a result of perineal damage. For days and sometimes weeks—and in some cases much, much longer—these women experienced real discomfort doing virtually everything, but particularly when they sat down.[115]

One friend confided to me that her daughter had been torn very badly after a very interventionist birth and she described (in somewhat too vivid detail!) how she's found out about the tear and the poor repair that had been done on it afterwards. The daughter (a modern, outgoing woman, with no personality problems) had put up with enormous discomfort for months, not only when she sat down but also when her husband tried to make love with her. Finally, not feeling able to face any kind of medical professional (who, she felt, had caused the problem in the first place), she somehow found some courage and said: "Mom, I just can't stand it any longer. Could you have a look at it down there?" Apparently, the damage was very obvious and the mother helped the daughter get help. (Of course, it was always readily available, but it's not the easiest problem to bring up with a family doctor a year or so after giving birth.) The daughter had the tear resutured and some scar tissue was also removed. At last, a few months later, the daughter was able to get back to normal life.

Considering other aspects of postpartum problems, I personally felt I was above respecting taboos about incontinence (or any other taboo, for that matter). I'm such an open, straightforward person, after all! Then I became aware of one woman who never seemed to be talking to anyone else. For a few weeks I watched her from a distance and confirmed my suspicion of her solitude. When the opportunity arose to chat to her, I took it enthusiastically but was immediately overwhelmed by a stench which left me speechless. And although, since then, when I'm in that area, I do sometimes see that woman, I must admit I find it difficult to get talking to her. I'm as bad as the rest. Fecal incontinence really isn't any fun. And this woman seems to realize this because she never spontaneously goes up to anyone, including me... It's very sad.

And what about flatus incontinence? Not being able to control farts is not a very sociable problem to have and, after initially being a joke, must become a real source of humiliation for so many women. How many women report their problems to their family doctor, I wonder? Not many, I would guess...[116]

## Sexual dysfunction

All these problems often lead to sexual dysfunction, as I've hinted before. Again, we have such taboos about sex—even though we pretend to be open— that too many problems must go unnoticed. But, of course, the problems certainly wouldn't be 'unnoticed' by the women who are suffering from them.[117]

# Pause to reflect...

## ...about perineal damage and sexual problems

1 Have you ever heard friends talking about perineal damage or incontinence socially?

2 When women report problems of any kind relating to perineal damage (for example in postpartum appointments or at prenatal appointments in subsequent pregnancies) how difficult do you think it is for them to bring up the subject?

3 When a woman appears to be mentioning something 'small' to do with perineal damage, how do you respond to her, to help her to open up?

4 Have you ever had any embarrassing problems yourself? (Ah! Caught you out there! How easy or difficult it is to talk about them?)

5 As far as sex goes, how open do you think people really are with both friends and strangers?

6 How open do you think couples are, between themselves, when it comes to talking about sex?

7 Do you think mothers and daughters are usually open with each other?

8 In a situation where a woman is in a difficult marriage, where she is either estranged from her mother or the mother has died, and/or where the woman has recently moved house (so has little face-to-face contact with her usual confidantes), how can you best speak during any kind of appointment or meeting so as to be as supportive as possible? What would not be helpful?

9 How widespread do you think the problem of perineal trauma is amongst women in various age groups, taking into account their possible experience of birth, their health status (e.g. in terms of obesity, which has been associated with higher rates of perineal injury) and their likely openness (or not) about their true experience?

10 What can you do as an effective caregiver?

Both photos below © Colin Smith

*Awaiting a new phase of life—fatherhood—with a new set of challenges?*

*How might men be affected by women's problems?*

## Ineffective anesthesia

Anesthesia is an area of medicine which we generally consider to be effective and efficient.[118] When I was sent an account of *ineffective anesthesia,* although I was honestly open to anything, I must admit I did find it rather unbelievable. Surely this couldn't still be a possibility in the 21st century? In any case, could the after-effects really be that dramatic? Nevertheless, I later had a long phone chat with Robin Russell, the editor of the *International Journal of Anaesthesia,* who was fantastically helpful, and the subject somehow came up. To my astonishment he wasn't at all dismissive but agreed it did occur at times.

If it's true that anesthesia really is ineffective at times (could I have misunderstood what I heard?), we really should take the next account very seriously indeed, particularly since the cesarean rate has risen so dramatically in many countries around the world. It's an account of a cesarean where the anesthetic failed to work properly. According to Carol Weihrer, the woman who set up the Anesthesia Awareness Campaign (www.anesthesiaawareness.com), this scenario was not at all rare in 2004, and I assume the situation hasn't improved dramatically over the last few years. Carol says that C-sections and open heart surgery are the operations most at risk of unwanted awareness during surgery. According to data she has collected, anesthesia awareness occurs about 100 times a day in the US alone.

What's all this got to do with optimizing outcomes?

♥ The fact that this problem (amongst others) exists is another reason to promote normal, physiological birth, wherever possible.

♥ When an operation is necessary (for a medically-indicated cesarean), Carol is convinced the risk of awareness can be reduced or virtually eliminated when brain activity monitors (e.g. BIS monitors) are used. Are these monitors routinely used where you work? If so, why not?

## Questions for reflection...

1   Do cost factors provide a sufficient response to any complaints received?

2   Are there in fact ever complaints in the area where you work?

3   Have you ever heard of any reports of ineffective anesthesia, either officially or anecdotally?

4   If ever reports occur, what do you imagine might be the consequences of unwanted awareness during surgery?

Do you have any other thoughts or questions?

## Birthframe 30: June Worcester—Ineffective anesthesia

The next contributor's account might help you to answer the last question in the 'Questions for reflection'... Although from a statistical point of view this woman may seem easy to dismiss, considering her as a real individual, who perhaps represents others, her voice really does need to be heard.

I will tell you of my experience. I really feel it was just yesterday instead of 18 years ago. The daughter whose birth I shall describe will hopefully someday give us grandchildren. I don't want to make her—or anyone else—scared of the birth process, whether it be vaginal or by C-section. After all, her experience, and other people's experiences will probably be very different. (That's why I haven't told her face-to-face about my own experiences of giving birth.) But since my story might help somebody, even though it's very difficult to tell, I'm going to have a go at explaining what happened. I pray this account may help somebody somewhere.

When I went into labor with my first child, I was hoping to have a normal, vaginal birth. However, after 36 hours of contractions, it was discovered that there was something wrong with my pelvis. Where 'normal' women have a pelvis that will give during the birthing process to accommodate the head and shoulders, mine was totally fused. This meant that a vaginal delivery would be impossible for this or any other birth. So I am at the opposite end of the spectrum to some of the births you are describing in your book. This third cesarean section, which I'm about to describe, put me the very end of the spectrum.

So what happened? At 6:30AM I checked into the hospital and was prepped for surgery. I was wheeled into the operating room and was asked if I would mind if student nurses were present. I didn't, so they came in. Next the anesthesiologist approached and asked several questions, routine ones about allergies etc. He then explained the 'process' of how I would be given drugs. It was going to be via IV, of course. I knew the procedure. They strapped my arms and wrists down on each side, applied the electrode pads on my chest and the anesthesiologist said, "Okay, June. We're going to put you to sleep now. You'll need to count down backwards for me." I did and my eyes closed and the nightmare began. The next question I heard was from my doctor: "Are we ready?" I am inside my head thinking, "Hey! I shouldn't be hearing you. I'm not asleep yet!" I remember trying to talk but couldn't. That was the start of what I call hell. My doctor continued and I assume he was talking to the student nurses. He said, "Scalpel. We're going to start by..." There were medical terms which I had no clue about.

And THAT was when I felt the cutting of my belly. I do mean *felt*. They pulled that layer back and cut the next layer and finally the uterus. The pain was unreal. I panicked at this point my mind was racing to every inch of my body to command it to move, to twitch anything to let them know that something was wrong. I heard the doctor ask for retractors. I felt the skin being pulled back. It felt as if I would rip.

And that was when I felt the cutting of my belly. I do mean *felt*.

What happened next was an intense pressure pushing on my stomach, just pushing and pushing. Then finally I heard the doctor say, "Oh, damn!" and I immediately panicked, thinking, "Did he just cut the baby?" Then, "It's a girl." And one of the student nurses said, "That is what she said she wanted." Then I felt what was like my OB pulling my whole darn insides out. Oh my God did that hurt. He then asked, "How's the baby? Nurse?" I was still wondering if my baby had been hurt. Then I heard my daughter cry for the first time. I remember thinking, "Thank you, Lord, for letting my baby live!" That first cry was the most beautiful sound I'd ever heard. If I could have cried at any point it would have been with joy over hearing that little baby. I was also relieved to hear the nurse saying, "She's doing just fine."

Those feelings were so short-lived, though, because of the other more intense sensations I was experiencing. He continued to do something that made me feel like my insides were being pulled out of me. Then I heard him tell the nurses that I was also having a tubiligation. A deep burning pain bolted through my body. I thought I was going to die. I wanted to die. I can only figure that he was cutting my tube on the right side. Then I felt the suturing of the tube into the tissue. I was still desperately trying to move, talk, communicate, anything because I knew that the left tube had to be cut too. But to no avail. Nobody heard, saw or knew I was being tortured.

When the other side was finished, he called for instrument count, then close. I also heard him ask for a 'gauge'. He must have used sutures inside because I felt the entering of a needle, the pull and tie of those. He then called, "Stapler!" Of course, I knew that he was almost done. I heard and felt each click of the stapler, and felt each pain of the staples as they were put in.

I was horrified throughout the whole ordeal. I couldn't talk. Early in the surgery when he first came in and asked, "Are we ready?" I had tried to talk and had realized I had a tube in my throat. I also knew it was there when I heard the pump breathing for me. I thought, "If I could hold my breath, or change my breathing—something that would indicate I have a problem." But no. I felt like I'd be sick when they pushed on my stomach but realized that the tube would not let me be, that it wouldn't even let me utter cries or screams.

I wish, Sylvie, I could find words that would be strong enough, descriptive enough and hellish enough to paint a picture in someone's mind who has not been through this so that they can see through words what I felt in pain throughout that whole ordeal. You asked if or how it has affected my life... I am being seen for depression and anxiety. I am getting treatment for this but so far it hasn't worked. I have changed in personality since this delivery. I don't sleep well, have nightmares, can't sleep in a bed, and am deathly afraid of having another surgery. The list goes on. I am living a nightmare. To this day, 18 years later, I have flashbacks of the day I had this surgery and of myself in pain. I have to try and shake those thoughts out of my mind. They only last for a short time but they creep back in.

When I spoke to the doctor, nurses and of course to the anesthesiologist they made me feel I must be crazy. I am not crazy and there are so many others, I have discovered, just like me, who are struggling with this kind of experience.

Photo © Colin Smith

*What other reasons cause so many other mothers to be disempowered or depressed?*

## Reason No. 8:
## Optimal birth is more beautiful...

### The dignity and experience of optimal birth

There is a quiet dignity to optimal birth—birth without any unnecessary drugs or interventions—which managed births simply do not appear to have. While the woman is in labor although she may appear to be wild and out of control—difficult *to* control, in fact!—she is certainly not a pathetic creature. Instead, as long as she feels secure and supported (by good, but undisturbing midwifery care—'watchful waiting'), there is a quiet, purposeful strength about her.

> There is a quiet dignity to optimal birth—
> birth without any unnecessary drugs and interventions—
> which managed births simply do not appear to have

### Birthframe 31: Fiona Lucy Stoppard—Absorbing other world

Here are some comments from a mother who experienced an undistributed birth, Fiona Lucy Stoppard, followed by her account of her actual experience of birth. As you will see, her words beautifully capture the memory of 'going to another planet'... the process which allows women to tune into their instinctive knowledge of how to give birth.

The world is a wonderful place and there are some beautiful things. Great people. Nature. Art. Poetry and music. Writing. Friendships and love. But, basically, what I feel is that at the core of it all—the one real thing, the crux of it all—is pregnancy, labor and birth... and also breastfeeding and bonding. It's like the sun coming out. It really illuminates everything. I'm not saying that if a woman chooses not to have children, she can't have a wonderful life and do a lot—I don't mean that. But it's just that it's like everything else is pushed away, all these wonderful things. It's like the core of life. When women give birth without drugs, they're really connecting then. It's so subtle, what's going on, and drugs just spoil it. It's like some wonderful beautiful flower that's incredibly delicate and beautiful. It just can't be a production line. A production line's for making bread, for very practical things. This is just the absolute... it all needs to be treated with the greatest of care.

> What I feel is that at the core of life—the one real thing,
> the crux of it all—is pregnancy, labor and birth... and also
> breastfeeding and bonding. It's like the sun coming out.
> It really illuminates everything.

I know there are all sorts of different things happening during pregnancy and labor. It would be wonderful if midwives could just look at a mom-to-be and know exactly what was going on and know just how to deal with it. But so often they say, "Oh, something's not right, we'd better induce you"—and maybe that might not have been the right thing at all. I know there are times when it is true the case is "Quick! Hospital! Get the baby out." Great, that's wonderful—you've got a beautiful baby, if that had to happen. But if the woman is healthy and it's a healthy pregnancy then you have to let the baby have the best possible experience. And it's not just the experience at the time, because it affects the future of the mother and the child.

## The actual birth:

My water broke very, very early in the morning and then she wasn't born until after midnight. The labor was all that day—all morning, afternoon and evening—and she was born after midnight. So it was less than 24 hours. It was very slow for a long time. I phoned Michel and he said "Oh it's fine. It'll be alright. I'll come along later." I don't think I was having contractions then—only very slight ones. He came to see me, then he went home again. He was able to go backwards and forwards a little bit, although there was one point when he knew it was getting close and he said he'd stay.

I don't know what I can remember of the labor—it was just very hard work. A lot of resting. Very hard work. Completely closed off from the rest of the world. I remember asking David once to get me a mango and he went out and bought me a mango which I didn't finish. It was actually too much to eat a mango because I had to concentrate. I went really deep, deep within. Quite spontaneously.

And I remember that the next morning when I woke up I thought, "Oh, I'm here, back in the world again." I looked out of the window and people were doing normal, everyday things and I felt like saying, "Don't they realize what's just happened here?!"

The next morning I thought, "Oh, I'm here, back in the world." People were doing normal, everyday things. I felt like saying, "Don't they realize what's just happened here?!"

## Birthframe 32: Pauline Farrance and partner—A resolution

Here is an account to give us a little intuitive insight into the reasons for avoiding the alternatives to normal, physiological birth if at all possible. This account came in very late on while I was writing this book, with this note:

Dear Sylvie,

At last I have managed to find the report and your address at the same time! I am so sorry to have taken so long—I am really ashamed of myself! I do hope this will be of some use to you. I hadn't read it for some time and it really brought tears to my eyes. Rosie's birth really was a magical experience and I so much wish I could have repeated it.

*Pauline*

I sat down to write about the wonder of Rosie's birth at home, but felt it all started over two years ago with Sean's birth in the hospital, which affected both myself and my family so much that we were determined our next baby's introduction to our world would be a happier and easier event. So, firstly, I am writing about the birth of Sean Henry on October 17, 1986.

*We were determined our next baby's introduction to our world would be a happier and easier event*

I attended our excellent prenatal classes which helped me to look forward with excitement to my baby's arrival, instead of with the fear of childbirth which I'd always held. I had hoped for a natural and gentle birth but due to slightly raised blood pressure following a family crisis, at 36 weeks I found myself on the local hospital baby extraction line. Although I instinctively and intellectually felt that all the intervention was unnecessary, and indeed dangerous, and that the baby was the best judge of when he would be 'better out than in' (the prenatal ward catchphrase), I did not have the confidence to totally resist the daily pressure from the doctors during my two weeks on the ward. Being told I was killing my baby did not help the situation—or my blood pressure!—although some of the midwives were very kind and supportive and were obviously concerned at the number of inductions being performed. However, I did have the knowledge and courage to decline their frequent offers to break my water (at 1cm dilation) and 'pop' me down to the operating room for a nice, convenient, planned cesarean.

Unfortunately, due to the constant harassment and worry, my blood pressure rose even higher and I finally agreed to prostaglandin pessaries and had seven inserted over seven days. Sean still showed no sign of budging, which simply meant that he was not ready for the outside world. (All my other tests on urine and placenta, etc were perfectly normal.) I was then a Failed Induction and subjected to the full panoply of modern obstetrics: water broken, numerous IVs (including pitocin), gas and air, and meptid. I didn't really know what was going on during 15 hours of labor, except that everything stopped at one point—not surprising with that cocktail of drugs which made me quite unable to stray from the bed.

The resultant birth was a 'normal delivery' with a bullying, insensitive midwife who took Sean away immediately (despite my written and oral requests to the contrary) and returned him to me clean and swaddled, so that I couldn't even get to see or touch his fingers. I didn't have a chance to put him to the breast for four hours, despite frequent requests for help, and Sean and I had great difficulty with feeding for ages afterwards. He was sleepy for a few days due to the drugs he received through me, and I was in such a state trying to fend off the bottles from the night staff, it is amazing we ever got the hang of breastfeeding! Needless to say, it took me nearly a year to get over the trauma of those three weeks, so when Mike and I found we were expecting another baby, we realized that we could not risk the same things happening again, especially as I would have to be on top form to cope with a baby and our extremely boisterous 2-year-old.

The reactions of the medical profession to my request for a home birth could take me through several paragraphs. Suffice to say, it made everyone very nervous. Tactics used to dissuade me ranged from coercion to disbelief ("There is no such thing as a home birth in this country") and from the usual threats ("What if something should go wrong?") to the sudden discovery of a problem with my antibodies which meant the baby should be checked by a pediatrician at birth!... none of which were valid.

## My request for a home birth made everyone nervous

I pursued every avenue imaginable and pestered all sorts of people in my determination to ensure that this time my baby and I would not start off our relationship emotionally battered. Time was running out when I read a newspaper article about Michel Odent. It appeared that he had attended some home births, so I attempted to contact him via his publisher. A few days later I answered the phone to a gentle French accent—I couldn't believe my luck. Much to our amazement, Michel was willing to attend our baby's birth if possible (he refused to use the word 'deliver'). From then on we just hoped that baby would have the good sense to be born out of rush hour times as we knew that, with Michel, we would be able to truly get to know our new baby during that very sensitive period immediately after the birth and she would be treated with gentleness and respect.

I wake up on October 9 feeling that this would be a good day to have a baby—the sun is shining and I feel well and reasonably energetic for a change. Baby due today, though, so I'm sure nothing will happen. Cathy and daughter Emily (aged 2) are coming to have lunch and spend the afternoon with us.

Around mid morning I feel very mild period-type pains but I had them Friday night so I take no notice. About 1:30PM I have to stop for a moment through the quickening—I couldn't really describe it as a pain—but manage to prepare a simple lunch quite easily. We sit down to lunch about 2PM and I finally tell Cathy that I am having 'pains' but think it is probably a false alarm. About an hour later Mike comes in from painting the gates outside to have lunch and I tell him about the 'pains'. He says he'll finish the gates but I tell him I'd rather he didn't risk welcoming our baby into the world covered in black gloss. It then registers that something might really be happening so he looks a bit panicky and rushes about in confusion.

4:00PM: I still can't be convinced I'm definitely in labor as I'm not really in pain and I don't want to call Michel unless I'm sure, so we start writing down times. Contractions seem to be about three minutes apart. Mike gets electric heaters and plastic sheets (the only special equipment needed) and looks agitated. I keep laughing because I can't believe this is the real thing. Everything is so normal and Sean is rushing about terrorizing Emily, as usual.

4:40PM: Mike decides it's time to call Michel—he's getting worried about being 'in charge', I think. Luck is with us and he is at home and says he will be right over. Cathy sits reading a book with Sean and Emily and I join in, but have to stop suddenly to lean on the back of the chair. Emily asks, "What's Auntie Pauline doing?" "Having a baby, darling". We both can't stop laughing, which is a bit tricky during a contraction. My own sensitive little angel hasn't noticed a thing!

**5:30PM:** Cathy prepares a meal for the children as Mike is not capable in present state of agitation. I go upstairs to change but keep getting delayed by a contraction. It's cool and peaceful upstairs and contractions seem stronger.

**5:40PM:** Mike is looking out of the window and announces Michel's arrival with great relief. I greet him with a smile then wish I hadn't as he tells me I don't look like a woman in labor. We show him our list of times of contractions but he isn't interested. (Michel told us that he can tell if labor is normal and how it is progressing simply by listening to the noises an uninhibited woman makes, and I remember clearly how I could not stop myself from making different noises as labor got stronger.)

**5:50PM:** I go upstairs with Michel and he checks baby's heart and my blood pressure as I tell him it has been raised... in fact to the same level as when I was admitted with Sean. It's now the lowest it's been for two weeks.

**6:00PM:** Children are eating with Cathy. Michel joins them to chat to Cathy and asks about their bedtime. (He had previously told us that it is best not to have anyone else around for the birth as the number of people present is directly proportional to the length of labor, and the longer the labor the more likely complications occur.) Mike orders pizzas for himself and Michel due to vivid recollection of no food or drink for 12 hours before Sean's birth, and we feel it will be a long night.

**6:45PM:** Cathy and Emily leave. Mike takes Sean up to bed—skips the bath today! Pizzas arrive.

**7:00PM:** Sean asleep in record time. Contractions getting stronger now. I try out the newly acquired bean bag but it radiates heat back at me so I kneel in front of the couch with my head on my arms, leaning on the seat, swaying my hips through contractions. I had intended to wear my 'Singapore happy coat' for the birth but don't feel like moving now! It is incredible how the contractions have taken off now it is quiet and there are no distractions. Mike has organized our birthing music—Irish harp and Clannad (a favourite band), which is very soothing.

**7:15PM:** Contractions stronger so I start moaning through them. Mike and Michel chomping away on pizzas at the other end of the room. Mike comes over to see how I am and reminds me to 'breathe', which I had forgotten! He gets me a cushion to lean on which is lovely and cool. I ask him to hurry and finish his pizza as I'm in pain!

**7:25PM:** Making a lot of noise now as contractions really hurt, and I'm worried about coping for hours like this. I hear Mike make a move to get up but Michel whispers that I am best left 'in my own world'. Any outside interference will delay things.

**7:30PM:** Mike now joins me. He is very calm and reassuring and I hold him through contractions.

**7:45PM:** I feel I want to go to the bathroom. I remember this feeling before Sean was born but don't allow myself to possibly imagine baby is coming as I couldn't bear the disappointment if it's a false alarm. Mike helps me upstairs between contractions. I sit on the toilet and feel very constipated and confused. Water breaks. Maybe baby is coming... I put my hand down thinking I might feel the head and find blood on my hand.

Michel comes upstairs and tells Mike to get the electric heaters on at the bathroom door—we have a very small bathroom!—so we know this must be IT. Mike helps me off the toilet and they pull my clothes off and Michel tells him to hold me from behind in the supported squat. I feel my uterus push and a burning sensation and give a little cry as (at 7:55PM) baby's head is born. Then my uterus pushes again. Mike and I are still standing in suspended animation as Michel tells us to look down at our baby daughter lying in his hands! We can't believe she has arrived so quickly and easily. Michel hands her to me and I cradle our beautiful daughter in my arms, her lovely virgin skin touching mine. The wonder and joy of that moment, to be holding my newborn babe so close and feeling her perfect little body next to mine, I can never describe. I kept saying to Mike, "We've got a little girl!" and his face was alight with happiness.

I sat in the bathroom amid the goo and mess for about 30 minutes while Michel and Mike sorted things out. I was blissfully unaware of what was going on, having eyes only for my baby. I remember Michel tying the cord after about 10 minutes and Mike cut it and found it quite tough! Then Mike took Rose Eleanor while I went to lie on the bed.

About 9PM Michel felt my stomach and said the placenta had separated. He massaged my stomach and the placenta came away with a contraction. Mike was trying to clear up the bathroom and was pleased the birth hadn't been on a carpet! Michel left about 10PM, and we sat in bed marvelling at our new baby daughter. She was very alert and never cried, although she gave the occasional whimper. We had the lights very low for her all the time. She didn't want to nurse immediately after the birth but kept nuzzling my breast and looking around. Mike opened a bottle of champagne but I didn't like to drink more than a glass so I contented myself with two packets of mints! We sat and caressed our little miracle and talked over and over about her birth, we were both so excited. About 2AM I thought we should get some sleep and maybe Rose Eleanor was tired, so we all snuggled up in our bed. It was incredible that Sean had slept through all this but at 5AM he woke and came into our bed. His little face was a picture when he heard Rosie stirring and he laughed and said, "Baby sister!"—he was so happy to see her.

## And now over to the father...

Whenever there was a choice I have always opted for nature's way in areas such as food, environment and general health. However, this had never extended to having a baby and I preferred to abdicate the responsibility to the medical profession, who appeared more than willing to take over.

I attended the local classes and that's when doubts began to creep in, especially as I learned of all the medical hardware, paraphernalia and drugs that were deemed essential nowadays for giving birth. I began to realize that the medical profession was perhaps, after all, just as horrified as I at having the responsibility for the birth.

I began to realize that the medical profession was perhaps just as horrified as I at having the responsibility for the birth

However, we went ahead with the hospital birth of our firstborn as it was decided by a hospital screening process that Pauline's blood pressure was outside the norm, for which she was immediately admitted. Sean was eventually extracted in a very workmanlike manner, some two and a half weeks of protestations, seven pessaries, two gas tanks, two doses of chemicals, ten yards of graph paper, one cup of tea and 15 hours of narcotic trance later. It would be an understatement to say it was a very traumatic time for all of us and I was disappointed at not experiencing the joy one has heard about as the father in attendance at the birth of his first child. It was more a sense of overwhelming relief that the whole episode was over with and all three parties (physically) healthy and alive.

16 months later we found we were expecting our second baby and this time we were determined from the outset that this baby would be born at home so that we could have more control and responsibility. Our doctor reluctantly agreed to cover the birth but opposition grew from all quarters of the medical profession and eventually no doctor would agree to cover the birth at home. This again I would attribute to their horror at having the responsibility for the birth. However, in desperation and clutching at any possible straw, Pauline managed to meet Michel Odent, who agreed to be present at the birth, if he was available.

The day of the birth thankfully arrived on a Sunday. Although we had read every possible piece of literature on the subject, I was filled with dread and my heart began to palpitate. I thought I might possibly have to deliver the baby myself if Michel was held up in the traffic... I began to race around looking for textbooks, plastic sheets and the bucket. (The requirement for the bucket had somehow lodged in my mind from the prenatal classes.)

I was greatly relieved when a French-registered Renault pulled up outside, and the calming presence of Michel entered the house. I had by this time got my act together and put on some soothing music. Later, we turned the lights down very low and Michel told me to let Pauline get into her own world, and we chatted casually as we ate the enormous home delivery pizzas while Pauline got on with the business of having a baby, leaning on the couch with her head in a cushion. For some reason I was convinced it would be a long night and I was building up my reserves of energy in anticipation!

Minutes after finishing the pizzas we were all in our tiny bathroom with me holding Pauline in the supported squat and Michel holding a beautiful baby girl in his hands. She gave what seemed to be an obligatory whimper to let us know she was okay and settled blissfully into her mother's arms and there they stayed for about half an hour. I was totally dumbstruck by the ease and beauty of the birth and just gazed in disbelief. After coming down to Earth, I began clearing up operations and gave thanks for lino tiles instead of carpet in the bathroom, but wished I had not eaten the pizza!

Michel's presence was quite unobtrusive throughout his visit, but it was his confidence and experience in allowing nature to take its course without any unnecessary interference which made the birth such a trouble-free and memorable event, and gave us so much joy and confidence during the weeks that followed.

## And afterwards...

We have obviously talked endlessly about the birth and were so happy and grateful that everything had gone so well, but we felt so sad for the vast majority of mothers (and fathers) who are denied the joy of such a precious and unique experience in our so-called civilized society today. It is widely accepted that about 90% of women are able to give birth normally, without complications, so why don't we keep the costly expertise and machinery for those who really need it? Of course, many men and women are happier in the hospital environment, and we would never try and dissuade them from this as the labor will always be prolonged by fear and anxiety. But the present day fear of complications during childbirth is largely unfounded and we should look at the event in perspective as a part of family life.

We were so excited by Rosie's birth that we were on a high for at least a week and had loads of energy to carry us through the sleepless nights, etc. Added benefits of the birth at home include much less disruption and confusion for older brothers and sisters, and it meant so much to my parents to be able to hold their second grandchild so soon after she was born—a privilege not allowed after Sean's birth. In addition, the baby will not succumb to any of the infections which abound in hospital.

Quite simply, three days after Rosie's birth we were feeling a little down because we had enjoyed the birth so much. We wanted the clock to turn back to Sunday so that we could do it all again!

*Why do women sometimes prefer to stay in their own home, rather than move into a hospital labor ward or delivery room to give birth? Is it an unwise choice or a wise one?*

# Pause to reflect...

## ...about feelings around the time of a birth

1  When people—both men and women—are dissatisfied with births they experience in a hospital environment, what is it precisely, do you think, that usually makes them so dissatisfied?

2  After I gave birth in a private hospital in Sri Lanka I was aware that staff were being extremely flexible and accommodating on my account. Help was provided (e.g. when I was unsure about how to change diapers or bathe my new baby) but apart from that I was left completely to my own devices (with my baby rooming in) and no routines were insisted upon. At times I would order food and not get round to eating it. Although staff came to collect my tray, no one ever got annoyed that I hadn't yet eaten food (mostly because I'd been busy breastfeeding) and no one ever put pressure on me to hurry up. I was allowed any number of visitors and there was no restriction on what they brought with them (within reason, I suppose!) and how long they stayed. My husband was allowed and encouraged to stay with me as much as he wanted to and even to stay overnight with me in the same private room. Can you think of any reasons why it's helpful for this kind of flexibility to exist when a woman's had a baby? Can you think of any practical problems that need to be overcome in order to facilitate this kind of in-hospital care? Is it possible to be equally accommodating and supportive to women when they first arrive and when they're in labor? What would need to change in order for it to be possible?

3  Why do some women campaign with such determination to have a home birth? Why are some professionals not supportive of the idea?

4  Have you ever attended a home birth yourself? How was it?

5  Have you heard people talking about home birth? Were the comments positive or negative? Why do you think they spoke as they did?

6  How can the beauty of birth be maintained, in whatever place, home or hospital or elsewhere, alongside safety?

For material relevant to these questions, see Birthframes 2, 3, 4, 7, 11, 12, 15, 21, 22, 24, 25, 27, 28, 31, 32, 37, 38, 39, 48, 59, 65, 66, 67, 69, 70, 71, 72, 73, 78, 81, 82, 84, 85, 90, 91, 92, 93, 94, 95, 96, 97, 98 and 99.

Photo © Paul Mann

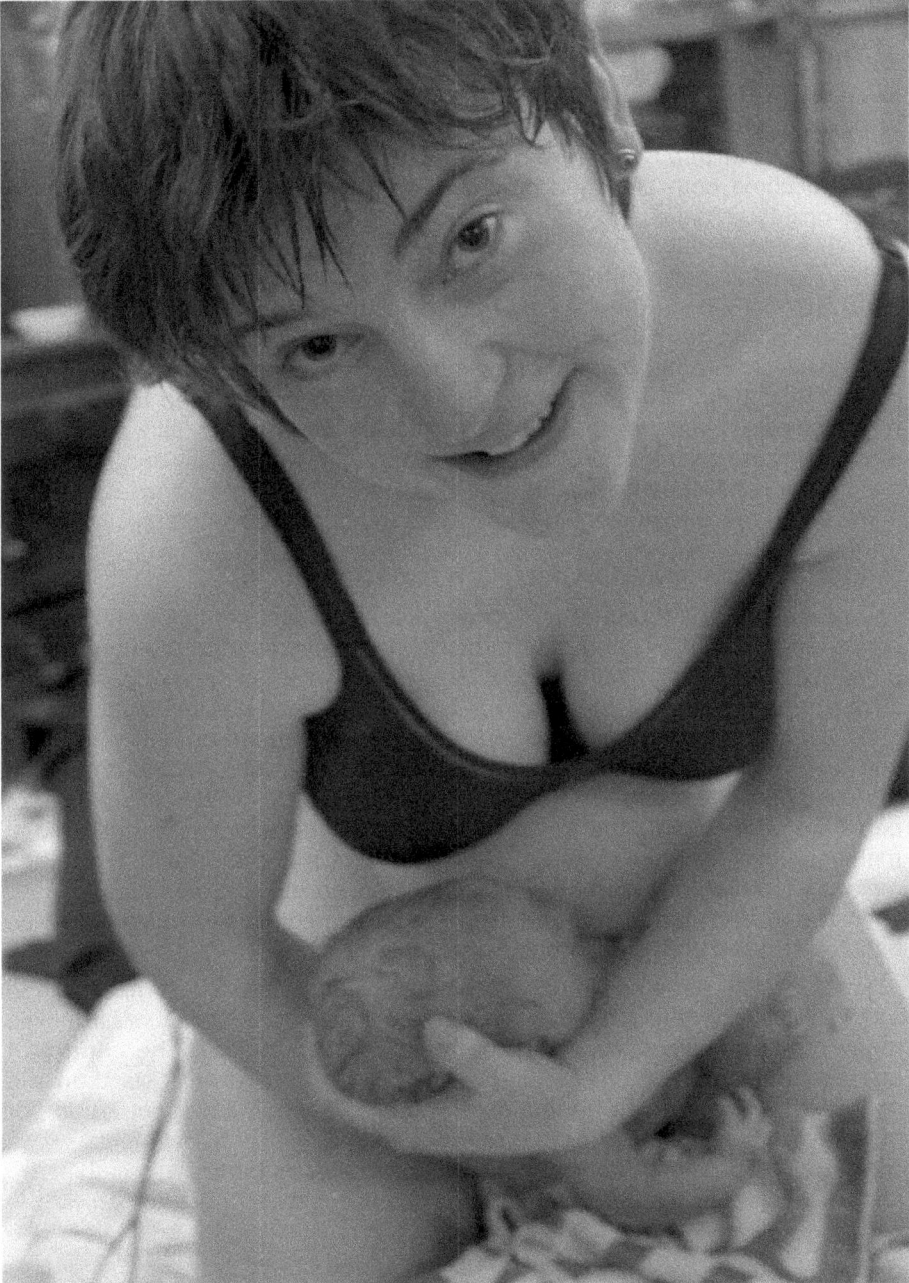

How do you think women feel after giving birth and why are they so fragile at this time?
This woman (Joanne Whistler) had a wonderful birth on a boat... Would your answer be
different if the photo were of someone who hadn't? What would she look like exactly?

## A summary of reasons why physiological is best:

### 1 The physiological processes are complex and efficient

As we saw earlier in this section, the physiological processes are not only complex and efficient, they are also fascinating in that they lead a laboring and birthing woman through an experience—usually involving pain—which enables her to become both alert and empowered postpartum. If she labors without drugs or interventions and without any disturbance, a woman will usually drift off into a completely 'other' state of consciousness, which allows her to cope with her pain and also emerge from it fulfilled and ecstatic. The hormones and endorphins produced naturally by her body immediately after the birth allow her to bond with her baby and initiate breastfeeding. This in turn kick-starts her very positively into her role as a new mother. The neonate is also advantaged by a physiological birth in that he or she is ideally predisposed to cope with physiological changes necessary in the 'outside world' which allow a smooth and safe transition into a new life. Most importantly, without any drugs in his or her system, the neonate begins life in a state of extreme alertness and is immediately ready to nurse and relate to his or her mother and father.

### 2 Intervention is an uncontrolled experiment

The experimental attitude of obstetricians over the last century or so has no doubt led to important and useful discoveries. Nevertheless, we cannot fool ourselves that the development was in any way systematic or compassionate, since the real needs of the laboring woman and her fetus or neonate were often ignored or horribly disregarded. The interventions we are left with today and which linger on in hospitals in the USA and elsewhere in the developed world (and in developing countries too, which aspire to a 'Western' model of birth) are often merely the result of a series of haphazard experiments and marketing exercises, or they are the outcome of the enthusiasm of a gifted male obstetrician speaker, who obviously had no direct experience of birth himself and who had little interest in researching conditions which would be best for either women or their babies.

The question as to how much intervention is really needed is impossible to address since we have moved so far from the physiological model of birth in so many institutionalized contexts. Management of birth and the control of unpredictable and worrying processes are seen as higher goals than the optimization of outcomes (in statistical, physical and psychological terms) for both mothers and babies. However, an exploration of improvements which might well be possible through supporting physiological births is something we should consider as a means of optimizing conditions for birthing women.

Management of birth and the control of unpredictable and worrying processes are seen as higher goals than the optimization of statistical, physical and psychological outcomes

### 3 Research suggests it's best not to disturb the processes

Data which challenges many current practices is now widely available. The drive towards evidence-based practice is a very positive step in the USA but we should note that an evidence-based approach would guide us towards promoting a physiological model of birth, without many of the practices which have become standard in most hospitals. New models of care and working practices which support the needs of laboring women clearly need to be developed in all states so that evidence-based practice can become a reality.

### 4 Pain relief seems to cause more problems than it solves

Although women generally have a high expectation of successful pain management (by medics) while they are in labor and giving birth, the reality of what is possible probably falls far short of expectations in the majority of cases.

### 5 Babies are disadvantaged, perhaps far into the future

This is a poorly researched area, but one which we would do well to explore further because of the possible implications for all levels of society.

### 6 Women end up disempowered, upset, depressed

We unfortunately tend to disregard the negative psychological outcomes of birth, perhaps because of our general fear of personal consequences and litigation, which can be crippling in professional terms. However, the problems of negative outcomes remains, and urgently needs to be given more attention.

### 7 Hidden problems are truly horrendous

For very obvious reasons, it's likely that many problems remain unreported. Problems which are reported sound truly horrible.

### 8 Optimal birth is more beautiful...

Midwives and other caregivers who've witnessed undisturbed physiological births are usually totally convinced of their optimality. In case you're still unconvinced, here is an extract from an email I received:

> I recently talked with a man who has been an anesthesiologist for over 30 years at a medical center in the USA. He has been at many, many births administering anesthesia and has never seen a woman give birth without medication. When I briefly described my experience I could tell he just had no context for it. I think it's so totally different to see a woman birthing in an environment where she feels comfortable, has whatever support she needs, and is not medicated. He probably cannot even imagine the power that moves in the uninterfered-with birthing woman.

Can you imagine what it's like to give birth completely undrugged and undisturbed, but supported? Have you ever experienced it? Why do people who have campaign passionately for a change toward what they see as optimality?

# Pause to reflect...
## ...about optimality and how to justify it and explore it

1   Of all the reasons you've read about in this section of the book, which reasons do you accept and which do you disagree with?

2   Which of the reasons presented urgently need to be researched more, in your opinion?

3   Which of the reasons do you already discuss with pregnant women? Which of the reasons would you consider giving, if you wanted to dissuade a mom-to-be from having pain relief for medical reasons in a particular case? Which of the reasons (if any) would you feel uncomfortable about discussing with clients?

4   Which of the reasons are least discussed in the medical literature? Which are least discussed in the media? Which involve fallacies or misconceptions on the part of either moms-to-be or journalists? Which reasons would you be happy to talk about on the radio or TV?

5   When considering each of the reasons presented in this section did any make you feel intensely uncomfortable? If so, why?

6   Would you want to add any reasons for promoting physiological birth as a path towards optimality?

7   Would you want to prepare a list of other reasons to promote a different kind of 'optimality'? If so, what kind of 'optimality' would that be and what would your reasons be?

8   How often have you considered optimality when actually attending births, or experiencing childbirth yourself (i.e. when giving birth)?

9   To what extent do you think it is important for medical professionals to consider the 'why' of what they do, generally speaking? If you believe it's important, how can colleagues' or line managers' reasons be investigated and challenged in normal working environments? What problems are you likely to encounter? What are the possible solutions?

10  How much do you think pregnant women should become involved in discussions about 'why' things are done? How different are their views and those of their unborn babies likely to be on optimality in general?

## NOTES & REFERENCES

1   For more about other women's attempts to sustain the art and science of midwifery, see the following articles:

  - Boyer EL. Midwifery in America: A Profession Reaffirmed, Journal of Nurse-Midwifery, 1990, Vol 35,No 4,214-19

  - Declercq E. The Trials of Hanna Porn: The Campaign to Abolish Midwifery in Massachusetts, American Journal of Public Health, 1994, Vol 84,No 6,1022-8

  - Myers S, *et al.* Unlicensed Midwifery Practice in Washington state, *American Journal of Public Health*, 1990, Vol 80,726-8

  - Schrader C. The Memoirs (1693-1740) of the Frisian Midwife Catharina Schrader. 1987. Translated and annotated by Hilary Marland, with introductory essays by M.J. van Lieburg and G. J. Kloosterman. Rodopi, Amsterdam (available from Mrs. L. J. Brooke, 4595 Club Dr, NE, Atlanta, GA 30319

  - Tyson H, Outcomes of 1001 Midwife-Attended Home Births in Toronto, 1983-1988, *Birth*, 1991, Vol 18, No 1,14-9

  - Van Alten DM, Treffers PE, E. Midwifery in the Netherlands. The Wormerveer Study; Selection Mode of Delivery, Perinatal Mortality and Infant Morbidity, *British Journal of Obstetrics & Gynaecology*, 1989, Vol 96,656-62

2   I realize you might be thinking, "Ah, yes, but every pregnancy and birth are unique." Well, they are, but there's enough of a pattern for us to be able to look at typical scenarios. Soo Downe and McCourt (2004) have come up with the concept of 'unique normality'. In other words, they recognize that there is a *range of normality*, which means that women don't have to fit neatly onto charts or into 'mental boxes' so as to be considered normal. See:

  - Downe S, McCourt C. From being to becoming: reconstructing childbirth knowledges. In Downe S (ed) *Normal Childbirth: Evidence and Debate*. Churchill Livingstone, 2004

  By the way, for the research behind this detailed account I am grateful to Sarah J Buckley.

3   Uvnas-Moberg K, quoted in a report of the Australian Lactation Consultant's Conference, Gold Coast, Australia 1998, published in *Australian Doctor*, 1998, July 8:38

4   Michel describes this process of transformation in the following article:

  - Odent M. New reasons and new ways to study birth physiology. *International Journal of Gynecology & Obstetrics*, 2001, 75:S39-S45

5   Weiss G. Endocrinology of Parturition. *Journal of Clinical Endocrinology & Metabolism*, 2000, 85(12):4421-5

6   See the following articles:

  - Jackson M, Dudley D. Endocrine assays to predict preterm delivery. *Clinics in Perinatology*, 1998, 4:837-857

  - Petrocelli T, Lye S. Regulation of transcripts encoding the myometrial gap junction protein, connexin-43, by estrogen and progesterone. *Endocrinology*, 1993, 133: 284-90

7     Russell JA, et al, Brain Preparations for Maternity-Adaptive Changes in Behavioral and Neuroendocrine Systems during Pregnancy and Lactation. *Progressive Brain Research*, 2001, 133:1-38

8     By Michel Odent and Niles Newton.

9     Verbalis JG, et al, Oxytocin Secretion in Response to Cholecystokinin and Food: Differentiation of Nausea from Satiety, *Science*, 1986, 232:1417-1419

10    Fuchs A, Fuchs F. Endocrinology of human parturition: a review. *British Journal of Obstetrics & Gynaecology*, 1984, 91:948-67

11    Chard T, et al, Release of Oxytocin and Vasopressin by the Human Fetus during Labour. *Nature*, 1971, 234:352-354

12    Fuchs A, Fuchs F. Endocrinology of human parturition: a review. *British Journal of Obstetrics & Gynaecology*, 1984, 91:948-67

13    Lundenberg T, Uvnas-Moberg K, et al. Anti-nociceptive effects of oxytocin in rats and mice. *Neuroscience Letters,* 1994, Mar 28; 170(1):153-7

14    Dawood MY, et al, Oxytocin in Human Pregnancy and Parturition, *Obstetrics & Gynecology*, 1978, 51:138-143

15    See the following articles:

      • Costa A, De Filippis V, et al. Adrenocorticotrophic hormone and catecholamines in maternal, umbilical and neonatal plasma in relation to vaginal delivery. *Journal of Endocrinological Investigation*, 1988, 11:703-9

      • Eliot RJ, Lam R, et al. Plasma catecholamine concentrations in infants at birth and during the first 48 hours of life. *Journal of Pediatrics*, 1980, Feb; 96(2):311-5

      • Lagercrantz H, Bistoletti H. Catecholamine Release in the Newborn Infant at Birth. *Pediatric Research,* 1977, 11,No 8: 889-893

16    See the following articles:

      • Hagnevik K, Faxelius G, et al. Catecholamine surge and metabolic adaptation in the newborn after vaginal delivery and caesarean section. *Acta Paediatrica Scandinavica*, 1984, Sep;73(5):602-9

      • Colson S. Womb to World: a metabolic perspective. *Midwifery Today*, 2002, Spring; 61:12-17

      • Irestedt L, Lagercrantz H, et al. Causes and consequences of maternal and fetal sympathoadrenal activation during parturition. *Acta Obstetrica et Gynecologica Scandinavica,* Suppl, 1984, 118:111-5

      • Lowe N, Reiss R. Parturition and Fetal Adaptation. *Journal of Obstetric, Gynecologic & Neonatal Nursing*, 1996, 25:339-349

17    Matthiesen AS, Ransjo-Arvidson AB, et al. Postpartum maternal oxytocin release by newborns: effects of infant hand massage and sucking. *Birth*, 2001, Mar;28 (1):13-9

A recent review of research has confirmed that skin-to-skin contact at birth has all kinds of benefits (Anderson, et al, 2006): longer breastfeeding and less crying from the baby. Finigan and Davies (2005) also concluded that a better emotional connection was established with skin-to-skin contact. See:

- Anderson GC, Moore E, Hepworth J, Bergman N. Early skin-to–skin contact for mothers and their healthy newborn infants. *The Cochrane Database Systematic Reviews*, 2006, Issue 3

- Finigan V, Davies S. "I just wanted to love him forever"—women's lived experience of skin-to-skin contact with their baby immediately after birth. *Evidence Based Midwifery*, 2005, 2(2):59-65

18  Nissen E, Widstrom AM, *et al*. Elevation of oxytocin levels early post-partum in women. *Acta Obstetrica et Gynecologica Scandinavica*, 1995, 74(7);530-3

19  Leake RD, Weitzman RE, Fisher DA. Oxytocin concentrations during the neonatal period. *Biology of the Neonate*, 1981, 39(3-4):127-31

20  Eliot RJ, Lam R, *et al*. Plasma catecholamine concentrations in infants at birth and during the first 48 hours of life. *Journal of Pediatrics*, 1980, Feb; 96(2):311-5

21  Eliot RJ, Lam R, *et al*. Plasma catecholamine concentrations in infants at birth and during the first 48 hours of life. *Journal of Pediatrics*, 1980, Feb; 96(2):311-5

22  See the following articles:

- Russell JA, *et al*, Brain Preparations for Maternity-Adaptive Changes in Behavioral and Neuroendocrine Systems during Pregnancy and Lactation, *Progressive Brain Research*, 2001, 133:1-38

- Axel R. The molecular logic of smell. *Scientific American*, 1995, 130-7

23  Lundblad EG, Hodgen GD. Induction of maternal-infant bonding in rhesus and cynomolgus monkeys after cesarean delivery. *Laboratory Animal Science*, 1980, Oct;30(5):91

24  Grattan DR. The actions of prolactin in the brain during pregnancy and lactation. *Progress in Brain Research*, 2001, 135:153-171.

25  Brinsmead M, *et al*, Peripartum Concentrations of Beta Endorphin and Cortisol and Maternal Mood States, *Australian & New Zealand Journal of Obstetrics & Gynaecology*, 1985. 25:194-197

26  See the following articles:

- Franceschini R, *et al*, Plasma Beta-endorphin Concentrations during Suckling in Lactating Women, *British Journal of Obstetrics & Gynaecology*, 1989, 96,No 6: 711-713

- Zanardo V, *et al*, Beta Endorphin Concentrations in Human Milk, Journal *of Pediatric Gastroenterology & Nutrition*, 2001, 33,No 2:160-164

27  Bacigalupo's research (1990) suggests that levels subside slowly, reaching normal levels one to three days after the birth. See:

- Bacigalupo G, Riese S, *et al*. Quantitative relationships between pain intensities during labor and beta-endorphin and cortisol concentrations in plasma. Decline of the hormone concentrations in the early postpartum period. *Journal of Perinatal Medicine*, 1990, 18(4):289-96

28  A Cochrane review conducted in 2006 discussed the possible complications of epidurals, which makes clear their potential to disturb birth. See:

- Anim-Somuah M, Smyth R, Howell C. Epidural vs non-epidural or no analgesia in labour. *The Cochrane Database of Systematic Reviews*, 2006, Issue 2

Now that the opioid fentanyl has been incorporated into epidurals (alongside bupivacain), I assume the woman is likely to be in an even less alert state because Marucci, et al (2003) conclude: "Intrathecal bupivacaine-fentanyl dose produces a larger alertness decrease than single bupivacaine, because the anesthetic block density increases." See:

- Marucci M, Diele C, Bruno F, Fiore T. Subarachnoid anaesthesia in caesarean delivery: effects on alertness. *Minerva Anestesiologica*, 2003 Nov; 69(11):809-19, 819-24

29    Stress itself is something to be taken seriously indeed, not only during pregnancy but even before conception. Studies by Khashan, et al (2008 and 2009) found that maternal exposure to severe life events, particularly in the six months before pregnancy, may increase the risk of preterm and very preterm birth; and that mothers exposed to severe life events before conception or during pregnancy have babies with a significantly lower birthweight. (Note, their conclusion is not that this *may* happen, but that it does.) Another study by a slightly different team of researchers (Khashan, et al, 2008) found that babies of pregnant women exposed to stress during the first trimester were at greater risk of having schizophrenia later on.

- Khashan AS, McNamee R, Abel KM, Mortensen PB, Kenny LC, Pedersen MG, Webb RT, Baker PN. Rates of preterm birth following antenatal maternal exposure to severe life events: a population-based cohort study. *Human Reproduction*, 2009, Feb;24(2):429-37. Epub 2008 Dec 3
- Khashan AS, McNamee R, Abel KM, Pedersen MG, Webb RT, Kenny LC, Mortensen PB, Baker PN. Reduced infant birthweight consequent upon maternal exposure to severe life events. *Psychosomatic Medicine*, 2008, Jul;70(6):688-94
- Khashan AS, Abel KM, McNamee R, Pedersen MG, Webb RT, Baker PN, Kenny LC, Mortensen PB. Higher risk of offspring schizophrenia following antenatal maternal exposure to severe adverse life events. *Archives of General Psychiatry*. 2008 Feb;65(2):146-52
- Short SJ, Lubach GR, Karasin AI, Olsen CW, Styner M, Knickmeyer RC, Gilmore JH, Coe CL. Maternal influenz infection during pregancy impacts postnatal brain development in the rhesus monkey. *Biological Psychiatry*. 2010 Jan 13. (Epublication ahead of print.)
- Selten JP, Frissen A, Lensvelt-Mulders G, Morgan VA. Schizophrenia and 1957 pandemic of influenza: meta-analysis. *Schizophrenia Bulletin*. 2009 Dec 3. (Epublication ahead of print.)
- Brown AS, Vinogradov S, Kremen WS, Poole JH, Deicken RF, Penner JD, McKeague IW, Kochetkova A, Kern D, Schaefer CA. Prenatal exposure to maternal infection and executive dysfunction in adult schizophrenia. *American Journal of Psychiatry*. 2009 Jun;166(6):683-90. Epub 2009 Apr 15
- Boksa P. Maternal infection during pregnancy and schizophrenia. *Journal of Psychiatry & Neuroscience*. 2008 May;33(3):183-5

30    The study by Haverkamp, et al (1976) found that cesarean rates increased by 160%. The use of forceps or ventouse also increased by up to 30% (MacDonald, et al, 1985), which in turn increases the rate of urinary stress incontinence (Arya, et

*al*, 2001), anal sphincter tears (MacArthur, *et al*, 2005) and increased difficulty—or pain—when having sex (Bick, *et al*, 2002). See:

- Haverkamp A, Thompson H, McFee J. The evaluation of continuous fetal heart rate monitoring in high-risk pregnancy. *American Journal of Obstetrics & Gynaecology*, 1976, 125(3):310-320

- MacDonald D, Grant A, Sheridan-Pereira M. The Dublin randomised control trial of intrapartum fetal heart rate monitoring. *American Journal of Obstetrics & Gynaecology*, 1985, 152(5):524-539

- Arya L, Jackson N, Myers D, Verma A. Risk of new onset urinary incontinence after forceps and vacuum delivery in primiparous women. *American Journal of Obstetrics & Gynaecology*, 2001, 185:1318-1324

- MacArthur C, Glazener C, Lancashire R. Faecal incontinence and mode of first and subsequent delivery: a six year longitudinal study. *BYOG: An International Journal of Obstetrics & Gynaecology*, 2005, 112:1075-1082

- Bick D, MacArthur C, Knowles H, Winter H. *Postnatal Care: Evidence and Guidelines for Management.* Churchill Livingstone, 2002

- Byrne M, Agerbo E, Bennedsen B, Eaton WW, Mortensen PB. Obstetric conditions and risk of first admission with schizophrenia: a Danish national register based study. *Schizophrenia Research.* 2007 Dec;97(1-3):51-9. Epub 2007 Aug 31

31   Graham (1997) charts the origins and evolution of the episiotomy; first it was performed as a result of notions about a woman's 'imperfection', then it continued because of a surgical mind-set and finally it just became institutionalized in Western birthing practices. Its use was (wrongly) justified because of fears of cerebral palsy (episiotomy, it was thought, would mean the fetal head was 'knocking' against the perineum for a shorter time), because it was felt it shortened the second stage of labor, and finally—and most persistently—because it was thought to prevent tears into the anal sphincter. Starting with Sleep, *et al* (1984), research has revealed that its use is inappropriate in most cases, mainly because it results in *more* (not less) tearing and suturing. (This was the conclusion of Carroli and Belizan's Cochrane review in 2006.) Also, it is now considered bad practice for most women because research has also shown that an episiotomy reduces strength in the pelvic floor (Sartore, *et al*,2004), leads to pain in this area (Bick, *et al*, 2002) and even painful sexual intercourse afterwards (Bick, *et al*, 2002). Some studies (Dipiazza, *et al*, 2006; Williams, 2003; Richter, *et al*, 2002) have concluded that episiotomies even *facilitate* bad tears into the anal sphincter. It's also worth noting that some academic papers have condemned the practice of routine episiotomy after a woman has previously had a tear (Peleg and Zlatnik, 1999; Dandolu, *et al*, 2005). According to Dannecker, *et al* (2004), episiotomies should only be performed when the fetus is clearly distressed and needs to be born quickly (e.g. in the case of a breech birth which is not progressing smoothly). See:

- Graham I. *Episiotomy: Challenging Obstetric Interventions.* Blackwell, 1997

- Sleep J, Grant A, Garcia J, Elbourne D, Spencer J, Chalmers L. West Berkshire perineal management trial. *British Medical Journal*, 1984, 289:587-590

- Carroli G, Belizan J. Episiotomy for vaginal birth (Cochrane Review). In: *The*

*Cochrane Library,* Issue 3. John Wiley & Sons Ltd, 2006

- Sartore A, De Seta F, Maso G. The effects of mediolateral episiotomy on pelvic floor function after vaginal delivery. *Obstetrics & Gynaecology,* 2004, 103:669-673

- Bick D, MacArthur C, Knowles H, Winter H. *Postnatal Care: Evidence and Guidelines for Management.* Churchill Livingstone, 2002

- DiPiazza D, Richter H, Chapman V, Cliver S, Neely C, Chen C, Burgio K. Risk factors for anal sphincter tear in multiparas. *Obstetrics & Gynaecology,* 2006, 107(6):1233-1236

- Williams A. Third-degree perineal tears: risk factors and outcome after primary repair. *Journal of Obstetrics & Gynaecology,* 2003, 23(6):611-614

- Richter H, Brumfield C, Cliver S, Burgio K. Risk factors associated with anal sphincter tear: a comparison of primiparous patients, vaginal births after caesarean deliveries and patients with previous vaginal delivery. *American Journal of Obstetrics & Gynaecology,* 2002, 187:1194-1198

- Peleg D, Zlatnik M. Risk of repetition of a severe perineal laceration. *Obstetrics & Gynaecology,* 1999, 93(6):1021-1024

- Dandolu V, Gaughan J, Chatwani A. Risk of recurrence of anal sphincter lacerations. *Obstetrics & Gynaecology,* 2005, 105:831-835

- Dannecker C, Hillemanns P, Strauss A. Episiotomy and perineal tears presumed to be imminent: randomised controlled trial. *Acta Obstetrica et Gynaecologica Scandinavica,* 2004, 83:364-368

32   Michel Odent discusses the impact of apparently small interventions, which change a woman's internal chemistry, in the chapter 'The scientification of love' in his book *The Farmer and the Obstetrician* (Free Association Books, 2002). He discusses the problem of stimulating the neocortex (through talking) in the chapter 'Breaking a vicious circle' in *The Caesarean,* Free Association Books, 2004.

33   As you may already know, many caregivers and researchers still insist that ultrasound is non-invasive. The situation does not seem clear-cut, though... A review of 16 studies which had all been published since 1990 was conducted in 2008 (Chaimay, *et al*). Surprisingly, I think, the researchers concluded that ultrasound has no adverse effect on child development outcomes (when given during pregnancy) despite noting that "all studies demonstrated that ultrasound examinations during pregnancy increased the risk of undesirable developmental outcomes". It seems the researchers' reasons for dismissing the importance of these possible undesirable outcomes were a) effects were considered 'minimal' and b) the studies were criticized on methodological grounds. (The studies which were reviewed by these researchers were 13 randomized controlled trials, one cohort study, and two case-control studies—but they were all deficient in some respect, apparently.) Oddly, perhaps, even though the abstract for this review includes the statement: "Presently, it is not clear whether [ultrasound] has a negative effect on the health and development of children" the conclusion is that ultrasound is safe, after all. Surely more research is needed, particularly since ultrasound has become so common? See:

- Chaimay B, Woradet S. Does prenatal ultrasound exposure influence the development of children? Asia Pacific Journal of Public Health, 2008, Oct; 20 Suppl: 31-8

For other specific research studies on ultrasound see the following:

- Ewigman BG, Crane JP, *et al*. Effect of prenatal ultrasound screening on perinatal outcome. *New England Journal of Medicine*, 1993, 329:821-7
- Bucher HC, Schmidt JG. Does routine ultrasound scanning improve outcome in pregnancy? Meta-analysis of various outcome measures. *British Medical Journal*, 1993, 307:13-7
- Larsen T, Larson JF, *et al*. Detection of small-for-gestational-age fetuses by ultrasound screening in a high risk population: a randomized controlled study. *British Journal of Obstetrics & Gynaecology*, 1992, 99:469-74
- Secher NJ, Kern Hansen P, *et al*. A randomized study of fetal abdominal diameter and fetal weight estimation for detection of light-for-gestation infants in low-risk pregnancy. *British Journal of Obstetrics & Gynaecology*, 1987, 94:105-9
- Johnstone FD, Prescott RJ, *et al*. Clinical and ultrasound prediction of macrosomia in diabetic pregnancy. *British Journal of Obstetrics & Gynaecology*, 1996, 103:747-54

34 Many studies, such as the following, conclude that women resort to pain relief less often when they are outside large hospitals (e.g. at home or in a birth center) and also when they have continuity of care (i.e. one midwife right the way through their labor and birth):

- Olsen O. Meta-analysis of the safety of home birth. *Birth*, 1997, 24(1):4-13
- Walsh D, Downe S. Outcomes of free-standing, midwifery-led birth centres: a structured review of the evidence. *Birth*, 2004, 31(3):222-229
- Hodnett E, Downe S, Edwards N, Walsh D. Home-like versus conventional birth settings (Cochrane Review). In: *The Cochrane Library*, Issue 2. Chichester: John Wiley & Sons Ltd, 2006
- Waldenstrom U, Turnbull D. A systematic review comparing continuity of midwifery care with standard maternity services. *British Journal of Obstetrics & Gynaecology*, 1998, 105:1160-1170
- Wraight A, Ball J, Seccombe I, Stock J. *Mapping Team Midwifery: A Report to the Department of Health*. Brighton: Institute of Manpower Studies, University of Sussex, 1993
- Harvey S, Jarrell J, Brant R, Stainton C, Rach D. A randomised controlled trial of nurse/midwifery care. *Birth*, 1996, 23:128-135
- Page L, McCourt C, Beake S, Hewison J. Clinical interventions and outcomes of one-to-one midwifery practice. *Journal of Public Health Medicine*, 1999, 21 (3): 243-248

35 This was an informal study by Perez P and Snedeker C, who are the authors of *Special Women: The Role of the Professional Labor Assistant*. Cutting Edge Press, 2000.

36 Maternity Center Association. *Listening to mothers: report of first national US survey of women's childbearing experiences*. New York: MCA, 2002 (www.maternitywise.org/listeningtomothers/index.html).

37 Many people have written about the dangers of administering pitocin. To provide just two reasons, Freidman 1978 found that artificial oxytocin (pitocin) causes the resting tone of the uterus to increase and this can result in abnormal fetal heart rate patterns, fetal distress and even uterine rupture (Stubbs 2000). What's more, research has found that artificial oxytocin has minimal effects on cervical dilation,

compared to labor without drugs (Bidgood 1987) and other research showed that women who'd had artificial oxytocin did not experience an increase in beta endorphin levels in labor (Genazzani 1985).

- Freidman EA, Sachtleben MR. Effect of oxytocin and oral prostaglandin E2 on uterine contractility and fetal heart rate patterns. *American Journal of Obstetrics & Gynecology*, 1978, Feb 15; 130(4):403-7

- Stubbs TM. Oxytocin for labor induction. *Clinical Obstetrics & Gynecology*, 2000, Sept; 43(3):489-94

- Bidgood KA, Steer PJ. A randomized control study of oxytocin augmentation of labour. 2. Uterine activity. *British Journal of Obstetrics & Gynaecology*, 1987, Jun; 94(6):518-22

- Genazzani AR, Petraglia F, *et al*. Lack of beta-endorphin plasma level rise in oxytocin-induced labor. *Gynecologic & Obstetric Investigation*, 1985, 19(3):130-4

38  I hope you will agree that inductions do not always—or often—mean that labor suddenly starts efficiently and continues without interruption. This perhaps explains why some studies have revealed that as many as 57% of low-risk first-time mothers have syntocinon (artificial oxytocin through a drip) to speed up their labors. Walsh (2007) says this suggests 'a collapse in physiological ability to labor spontaneously'. See:

- Mead M. Midwives' perspectives in 11 UK maternity units. In S. Downe (ed.) *Normal Childbirth: Evidence and Debate*. Churchill Livingstone, 2004

- Walsh D. *Evidence-based Care for Normal Labour and Birth: A guide for midwives*. Routledge, 2007

39  The only effect of EFM identified by research is to increase the cesarean rate; it does nothing to improve either fetal or maternal morbidity or mortality. See Notes 28 and 43 in the 'What' section.

40  There really is no scientific basis for stopping women from eating and drinking in labor, as research and one systematic review has shown (Goer, *et al*, 2007). In this review we read: "The likelihood of a fed woman undergoing an unplanned cesarean under general anesthesia dying of pulmonary aspiration calculates to 8 per 10 million or 1 in 1,250,000 ... A study of 13,400 emergency surgeries under general anesthesia reported no deaths from aspiration in patients in reasonably good health (Warner, 1993)."

- Goer H, Sagady Leslie M, Romano A (The Coalition for Improving Maternity Services). Evidence Basis for the Ten Steps of Mother-Friendly Care. Step 6: Does Not Routinely Employ Practices, Procedures Unsupported by Scientific Evidence: The Coalition for Improving Maternity Services. *Journal of Perinatal Education*, 2007, Winter;16(Suppl 1):36S

- Warner MA, Warner ME, Weber JG. Clinical significance of pulmonary aspiration during the perioperative period. *Anesthesiology,* 1993, Jan;78(1):56-62

41  Michel Odent discusses this further in the chapter called 'Colostrum and civilization' in his book *Birth and Breastfeeding*, Clairview Books, 2003.

42  Grol and Grimshaw (2003) have written about doctors' impulse to act. Other writers have written about the way in which midwives tend to be dominated by

obstetricians, when they work closely with them (Donnison 1998, Coombs and Ersser 2004). See:

- Grol R, Grimshaw J. From best evidence to best practice: effective implementation of change in patient's care. *Lancet*, 2003, 362:1225-1230
- Donnison J. *Midwives and Medical Men: A History of the Struggle for the Control of Childbirth.* Historical Publications, 1988
- Coombs M, Ersser S. Medoca; hegemony in decision-making—a barrier to interdisciplinary working in intensive care. *Journal of Advanced Nursing*, 2004, 46(3):245-252

43  The following study challenged the view that ultrasound is actually useful for accurately establishing fetal weight. See:

- Rogers MS, Chung TK, Chang AM. Ultrasound fetal weight estimation: precision or guess work? *Australian & New Zealand Journal of Obstetrics & Gynaecology*. 1993 May;33(2):142-4

44  Continuity of care is essential. A study by Haggerty, *et al* (2005) emphasized the importance of different types of continuity—informational, management and relational. See:

- Haggerty J, Reid R, Freeman G, Starfield B, Adair C, McKendry R. Continuity of care: a multidisciplinary review. *British Medical Journal*, 2005, 327:1219-1221

The first of the following studies emphasized the importance of working in small team of midwives (numbering at the most six), while the other studies concluded that continuity was most important during labor, birth and postpartum (and perhaps not such a concern prenatally):

- Flint C. *Midwifery Teams and Caseloads.* Butterworth-Heinemann, 1993
- Flynn A, Hollins K, Lynch P. Ambulation in labour. *British Medical Journal*, 1978, 2(6137):591-593
- Green J, Curtis P, Price H, Renfrew M. *Continuing to Care: The Organization of Midwifery Services in the UK A Structured Review of the Evidence.* Books for Midwives Press, 1998
- Walsh D. An ethnographic study of women's experience of partnership caseload midwifery practice: the professional as friend. *Midwifery*, 1999, 15(3):165-176
- Page L, McCourt C, Beake S, Hewison J. Clinical interventions and outcomes of one-to-one midwifery practice. *Journal of Public Health Medicine*, 1999, 21(3): 243-248
- North Staffordshire Changing Childbirth Research Team. A randomised study of midwifery caseload care and traditional 'shared-care'. *Midwifery*, 2000, 16(4):295-302

45  See the following articles:

- Odent M. The Nocebo effect in prenatal care. *Primal Heath Research Newsletter*, 1994, 2(2)
- Odent M. Back to the Nocebo effect. *Primal Heath Research Newsletter*, 1995, 5(4)
- Odent M. Antenatal scare. *Primal Heath Research Newsletter*, 2000, 7(4)

46  See Note 33 of this section.

47 The only effect of EFM identified by research is to increase the cesarean rate; it does nothing to improve either fetal or maternal morbidity or mortality. See Notes 28 and 43 in the 'What' section.

48 Research relating to each practice has already been discussed earlier in these notes. The main way of optimizing outcomes seems to be to ensure that the physiological flow of hormones truly becomes a cascade of hormones because just one interruption can transmute it into the cascade of interventions which has become so familiar in maternity hospitals. Since emotions can be affected by both speech and nonverbal behavior, the need for caregivers to ensure that each laboring and birthing woman has an undisturbed and protected atmosphere is paramount. In this respect, I am always astonished that women are often transferred from one room to another during labor, often at a crucial stage, when the woman most needs to be left alone, in order for the physiological processes to proceed smoothly.

49 The idea that a labor should begin and then carry on continuously until the baby has been born is not one which has always existed. As well as many anecdotal accounts I have heard about laid-back attitudes towards stop-start labors only 50 years ago. Ina May Gaskin also discovered a Portuguese word *pasmo*, used in a 19th century textbook on midwifery. Basically, *pasmo* meant that labor stopped and everyone went back to their daily lives until it started up again. See:

- Gaskin IM. Going backwards: the concept of 'pasmo'. *The Practising Alidwija*, 2003, 6(8):34-36

50 Various researchers have demonstrated the benefits of mobility during labor and birth. The following researchers concluded that movement meant that women needed less pain relief and generally ended up feeling more satisfied with their labors:

- Bloom S, McIntyre D, Beimer M. Lack of effect of walking on labour and delivery. *New England Journal of Medicine,* 1998, 339(2):76-79
- MacLennan A, Crowther C, Derham R. Does the option to ambulate during spontaneous labour confer any advantage or disadvantage? *Journal of Maternal and Fetal Medicine,* 1994, 3(1):43-48
- Hemminki E, Saarikoski S. Ambulation and delayed amniotomy in the first stage of labour. *European Journal of Obstetrics, Gynaecology & Reproductive Medicine,* 1983, 15:129-139

The following studies noted that mobility resulted in more effective contractions, shorter labor, less augmentation, fewer operative deliveries and less fetal distress:

- Flynn A, Hollins K, Lynch P. Ambulation in labour. *British Medical Journal,* 1978, 2(6137):591-593
- Read J, Miller F, Paul R. Randomised trial of ambulation versus oxytocin for labour enhancement: a preliminary report. *American Journal of Obstetrics & Gynaecology,* 1981, 139:669-672
- Albers L, Anderson D, Cragin L. The relationship of ambulation in labour to operative delivery. *Journal of Nurse Midwifery,* 1997, 42(1):4-8
- Caldeyro-Barcia R et al (1979) showed that movement was beneficial for the baby:
- Caldeyro-Barcia R. Influence of maternal bearing down efforts during second

stage on fetal well-being. *Birth & Family Journal*, (1979), 6(i):7-15

- Caldeyro-Barcia R, Giussi G, Storch E. The influence of maternal bearing down efforts and their effects on fetal heart rate, oxygenation and acid base balance. *Journal of Perinatal Medicine*, 1979, 9: 63-67

51 See Note 31 of this section.

52 One example of a problem widely acknowledged is the impact on drug use in labor (either for pain relief or during the third stage) and reduced breastfeeding rates. It's possible there may be other after-effects too, which still need to be researched. Various other connections between drug use and side- or after-effects are also widely known and are detailed below under separate types of pain relief. See:

- Jordan S, Emery, S, Watkins A, Evans JD, Storey M, Morgan G. Associations of drugs routinely given in labour with breastfeeding at 48 hours: analysis of the Cardiff Births Survey. *BJOG: International Journal of Obstetrics & Gynaecology*, 2009, online publication on 1 Sept

53 See Note 34 in this section.

54 The main exception seems to be massage. Detailed guidance on massage is provided at www.midwiferytoday.com/articles/midwifestouch.asp and also in *Beautiful Birth* by Suzanne Yates (Carroll & Brown, 2008). One research study found that women who received massage during labor (abdominal effleurage, sacral pressure and shoulder/back kneading) experienced significantly less pain and anxiety. See:

- Chang M, Wang S, Chen C. Effects of massage on pain and anxiety during labour: a randomised controlled trial in Taiwan. *Journal of Advanced Nursing*, 2002, 38 (1): 68-73

Another study found that with massage, women experienced decreased depressed mood, anxiety and pain, less agitated activity, shorter labors, shorter hospital stays and less postpartum depression. I would certainly recommend massage personally... Massage is one excellent reason to encourage your clients to have a doula or other birth assistant with them while they're in labor. This non-medical birth attendant should be willing to perform continuous massage for as little or as much time as the pregnant woman requests! See:

- Field T, Hernandez-Reif M. Labour pain is reduced by massage therapy. *Journal of Psychosomatic Obstetrics & Gynaecology*, 1997, 18:286-291

55 In fact, a review of research conducted in 2005 (by Thacker, *et al*) concluded by recommending the use of fetal blood sampling. However, a new review conducted a year later (by Alfirevic, *et al*) reached the opposite conclusion and did not endorse the practice. These researchers concluded that it does not help to reduce cesearean rates. See:

- Thacker S, Stroup D, Peterson H. Continuous electronic fetal heart monitoring during labour. (Cochrane Review) In: *The Cochrane Library*, Issue 1. Oxford: Update Software, 2005

- Alfirevic Z, Devane D, Gyte G. Continuous cardiotocography (CTG) as a form of electronic fetal monitoring (EFM) for fetal assessment during labour. *Cochrane Database of Systematic Reviews*, 2006, Issue 3

56 The following researcher reported on an increase in the cesarean rate when an epidural is sited in early labur:

- Klein M. In the literature: epidural analgesia: does it or doesn't it? *Birth,* 2006, 33(1):74-76

57 In one study (Thorp et al, 1989) babies lying in the posterior position early on in labor were less likely to turn into a more favorable position for the birth when the mother had an epidural. (Without an epidural, only 4% of babies were still in the posterior position when they were ready to be born; with an epidural the percentage was 19%.) See:

- Thorp J, *et al.* The effect of continuous epidural anesthesia on caesarean section for dystocia in nulliparous women. *American Journal of Obstetrics & Gynecologists,* 1989, 161(3), September:670-674
- Lieberman E, Davidson K, *et al.* Changes in fetal position during labor and their Lieberman E, Davidson K, *et al.* Changes in fetal position during labor and their association with epidural analgesia. *Obstetrics & Gynecology,* 2005, 105:974-82

Without an epdiural sited and with the woman moving around, it seems a baby is much more likely to turn to an anterior position. Researchers (Gardberg and Tuppurainen 1994) found that when labor began, 10-15% of women's babies were in the posterior position, but by the end of labor (before the actual birth), this percentage went down to just 6%. Basing their comments on anecdotal evidence only, Sutton and Scott (1996) suggested the number of babies presenting in posterior position is increasing. One systematic review which considered the use of the knee-chest position during pregnancy could not draw any hard and fast conclusions. However, Sutton and Scott's conclusions and recommendations do tie in well with what we already know about the mechanical elements of labor (which are outlined in a 2004 *MIDIRS* article on the subject). The various sources for these documents are as follows:

- Gardberg M, Tuppurainen M. Anterior placental location predisposes for occipito posterior presentation near term. *Acta Obstetrica et Gynecologica Scandinavica,* 1994, 73:151-152
- Sutton J, Scott P. *Understanding and Teaching Optimal Fetal Positioning.* Tauranga, New Zealand: Birth Concepts, 1996 (available from www.amazon.co.uk)
- Hofmeyr G, Kulier R. Hands/knees posture in late pregnancy or labour for fetal malposition (lateral or posterior) (Cochrane Review). In: *The Cochrane Library,* Issue 3. John Wiley & Sons Ltd, 2006
- *MIDIRS* and the NHS Centre for Reviews and Dissemination. *Positions in Labour and Delivery.* Informed Choice for Professionals leaflet, 2004

The following researchers found that adopting the knee-chest position for 30 minutes during labor significantly reduced persistent back pain in posterior labors:

- Stremler R, Hodnett E, Petryshen P. Randomised controlled trial of hands-knees positioning for occipitoposterior position in labour. *Birth,* 2005, 32(4):243-251

58 Fusi L, *et al.* Maternal pyrexia associated with the use of epidural anesthesia in labour. *Lancet,* 1989, June 3:1250-1252

Also see:

- Apantaku O, Mulik V. Maternal intra-partum fever. J Obstet Gynaecol, 2007, Jan; 27(1):12-5

59　Sudlow C, Warlow C. Epidural blood patching for preventing and treating post-dural puncture headache. *Cochrane Database Systematic Review*, 2002, (2):CD001791

60　See the following articles:

- Rortveit A, Kjersti D, Yugvild S, Hannestad S, Hunskaar S. Urinary incontinence after vaginal delivery or caesarean section. *New England Journal of Medicine*, 2003, 348:900-907
- Ramin SM, *et al*. Randomized trial of epidural versus intravenous analgesia during labor. *Obstetrics & Gynecology*, 1995, 86:783-789

61　Anesthesia Disturbs Maternal Behavior in Primiparous and Multiparous Parturient Ewes, *Physiology & Behavior*, 1987, 40:463-472

62　See:

- Sepkoski CB, Lester G, *et al*. The effects of maternal epidural anesthesia on neonatal behavior during the first month. *Developmental Medicine & Child Neurology*, 1992; 34:1072-80
- Murray AD, *et al*. Effects of epidural anesthesia on newborns and their mothers. *Child Development*, 1981, 52:71-82

63　Walker M. Do labor medications affect breastfeeding? *Journal of Human Lactation*, 1997, Jun; 13(2):131-7

64　Riordan J, *et al*, Effect of Labor Pain Relief Medication on Neonatal Suckling and Breastfeeding Duration, *Journal of Human Lactation*, 2000, 16,No 1:7-12

- Ransjo-Arvidson AB, *et al*. Maternal Analgesia during Labor Disturbs Newborn Behavior: Effects on Breastfeeding, Temperature, and Crying, *Birth*, 2001, 28,No 1:20-21

Michel makes some brief comments about epidurals at www.wombecology.com/epidural.html. For more on epidurals see:

- Beilin Y, Bodian C, Weiser J, *et al*. Effect of labor epidural analgesia with and without fentanyl on infant breast-feeding: a prospective, randomized, double-blind study. *Anesthesiology*, 2005, Dec;103(6):1211-7
- Henderson J, Dickenson J, Evans S. Impact of intrapartum analgesia on breastfeeding duration. *Australian & New Zealand Journal of Obstetrics & Gynaecology*, 2003, 43(5):372
- Ransjo-Arvidson A, Matthiesen A, Lilja G. Maternal analgesia during labour disturbs newborn behaviour: effects on breastfeeding, temperature and crying. *Birth*, 2001, 23(3):136-143
- Torvaldsen S, Roberts CL, Simpson JM, *et al*. Intrapartum epidural analgesia and breastfeeding: a prospective cohort study. *International Breastfeeding Journal*, 2006, Dec11;1:24

Other studies conclude that tachycardia is more likely, as well as hypoglycemia. See:

- Lieberman E, O'Donoghue C. Unintended effects of epidural analgesia during labour. *American Journal of Obstetrics & Gynaecology,* 2002, 186:S31-68

65   See: Mander R. *Pain in Childbirth and its Control.* Blackwell Science, 1998.

66   This is because sucking intensity, duration and frequency has been shown to have a dramatic effect on prolactin levels, which are necessary for lactation and breastfeeding. See:

   • Grattan DR. The actions of prolactin in the brain during pregnancy and lactation. *Progress in Brain Research,* 2001, 135:153-171

   • Ransjo-Arvidson A, Matthiesen A, Lilja G. Maternal analgesia during labour disturbs newborn behaviour: effects on breastfeeding, temperature and crying. *Birth,* 2001, 23(3):136-143

   • Jordan S, Emery S, Bradshaw C. The impact of intrapartum analgesia on infant feeding. *British Journal of Obstetrics & Gynaecology,* 2005, 112(7):927-930

67   A Cochrane review conducted in 2006 into the use of opioids administered intra-muscularly concluded there was insufficient evidence to evaluate the comparative efficacy and safety of different opioids. See:

   • Elbourne D, Wiseman R. Types of intra-muscular opioids for maternal pain relief in labour (Cochrane Review). In: *The Cochrane Library,* Issue 2. Chichester: John Wiley & Sons Ltd, 2006

68   American College of Obstetricians & Gynecologists, Obstetric Analgesia and Anesthesia, 1996, Technical Bulletin No 225 (July).

69   Thomas TA, *et al,* Influence of Medication, Pain and Progress in Labour on Plasma Beta-endorphin like Immunoreactivity, *British Journal of Anaesthesia,* 1982, 54:401-408

70   Thomson AM. A Re-evaluation of the Effect of Pethidine on the Length of Labour, *Journal of Advanced Nursing,* 1994, 19,No 3:448-456

71   Lindow SW, Van der Spuy ZM, *et al.* The effect of morphine and naloxone administration on plasma oxytocin concentrations in the first stage of labour. *Clinical Endocrinology,* 1992, 37(4):349-53

72   Kimball CD, Do Endorphin Residues of Beta Lipotrophin in Hormones Reinforce Reproductive Functions? *American Journal of Obstetrics & Gynecology,* 1979, 134, No 2:127-132

73   Jacobsen B, *et al,* Opiate Addiction in Adult Offspring through Possible Imprinting Jacobsen B, *et al,* Opiate Addiction in Adult Offspring through Possible Imprinting after Obstetric Treatment, *British Medical Journal,* 1990, 301:1067-1070

74   Nyberg K, Buka SL, Lipsitt LP. Perinatal medication as a potential risk factor for adult drug abuse in a North American cohort. *Epidemiology,* 2000, 11(6): 715-16

75   See the following studies:

   • Myerson BJ. Influence of Early B-endorphin Treatment on the Behavior and Reaction to B-endorphin in the Adult Male Rat, *Psychoneuroendocrinology,* 1985, 10:135-147

   • Kellogg CK, *et al,* Sexually Dimorphic Influence of Prenatal Exposure to Diazepam on Behavioral Responses to Environmental Challenge and on Gamma

Aminobutyric Acid (GABA)-Stimulated Chloride Uptake in the Brain, *Journal of Pharmacology and Experimental Therapeutics*, 1991, 256,No 1:259-265

- Liversay GT, *et al*, Prenatal Exposure to Phenobarbital and Quantifiable Alterations in the Electroencephalogram of Adult Rat Offspring, *American Journal of Obstetrics & Gynecology*, 1992, 167,No 6:1611-1615

ı Mirmiran M, DF. Swaab, Effects of Perinatal Medication on the Developing Brain. In *Fetal Behaviour*, Nijhuis JG (ed). Oxford University Press, 1992

76   Mirmiran M, Swaab DF. Effects of Perinatal Medication on the Developing Brain. In Fetal *Behaviour*, Nijhuis JG (ed). Oxford University Press, 1992

77   See the following articles:

- Jacobson B, Nyberg K. Opiate addiction in adult offspring through possible imprinting after obstetric treatment. *British Medical Journal*, 1990, 301:1067-70

- Nyberg K, Buka SL, Lipsitt LP. Perinatal medication as a potential risk factor for adult drug abuse in a North American cohort. *Epidemiology*, 2000, 11(6):715-16

78   Robertson (1997) is one researcher who is concerned that women are inhaling a drug with unknown side-effects. See:

- Robertson A. *The Midwife Companion*. Ace Graphics, 1997

After conducting a systematic review of Entonox, Rosen (2002) concluded with a muted endorsement. However, although women generally speak positively of 'gas and air' when they've used it (usually because they're proud not to have needed anything stronger), they are not commenting from the perspective of a woman who has experienced the alternative— a completely alert, empowered birth, involving only naturally-induced states of mind. Many, who are dismissive of the effectiveness of Entonox as a form of pain relief say its primary role was as a distraction. If this is all it is, it's a shame these women gave up the possibility (for themselves and their babies) of experiencing birth with natural endorphins and—most importantly—the surge of oxytocin (the hormone of love), which is a feature following physiological births, i.e. births without any drugs. See:

- Rosen M. Nitrous oxide for relief of labour pain: a systematic review. *American Journal of Obstetrics & Gynaecology*, 2002, 186:S 110-126

79   Avoiding sutures seems to be a priority from the point of optimizing women's experience postpartum because women report that they are extremely painful (Salmon 1999, Sanders *et al*, 2002). But how easily can they be avoided? In one study in which women filled out questionnaires 12 months after giving birth, women who had had no suturing for second-degree tears reported *no problems* (Clement and Reed 1999). Another study (Lundquist *et al*, 2000) demonstrated that leaving second-degree tears unsutured had a good effect on breastfeeding—presumably because women could sit more comfortably. Yet another study on suturing for larger second-degree tears (e.g. Fleming *et al*, 2003) had more mixed results, leading to the conclusion that perhaps only tears less than 2cm x 2cm, which are not bleeding—and which definitely don't affect the anus—should be left unsutured. However, a more recent study, which looked at outcomes after a year, (Langley *et al*, 2006) found no significant differences between sutured and unsutured women. See:

- Head, M. Dropping stitches. *Nursing Times,* 1993, 89(33):64-65
- Salmon, D. A feminist analysis of women's experiences of perineal trauma in the immediate post-delivery period. *Midwifery,* 1999, 15(4):247-256
- Sanders, J, Campbell, R. and Peters, T. Effectiveness of pain relief during perineal suturing. *British Journal of Obstetrics & Gynaecology,* 2002, 109:1066-1068
- Clement, S. and Reed, B. To stitch or not to stitch. *The Practising Midwife,* 1999, 2(4):20-28
- Lundquist, M, Olsson, A, Nissen, E. and Norman, M. Is it necessary to suture all lacerations after a vaginal delivery? *Birth,* 2000, 27(2):79-85
- Fleming, V, Hagen, S. and Niven, C. Does perineal suturing make a difference: the SUNS trial. *British Journal of Obstetrics & Gynaecology,* 2003, 110:684-689
- Langley, V, Thoburn, A, Shaw, S. and Barton, A. Second degree tears: to suture or not? A randomised controlled trial. *British Journal of Midwifery,* 2006, 14(9): 550-554

80  We do know that all synthetic morphines—all opiates, in fact—suppress the sucking reflex and since the target of nitrous oxide is the opioid brain receptor there may be cause for concern in this respect. (This was noted in a study by Righard and Alade, for example, who concluded: "It is suggested that contact between mother and infant should be uninterrupted during the first hour after birth or until the first breast-feed has been accomplished, and that use of drugs such as pethidine should be restricted.") A study on other effects of opiates, which may be relevant, is that by Jacobsen, *et al,* as follows:

- Jacobson B, Nyberg K. Opiate addiction in adult offspring through possible imprinting after obstetric treatment, *British Medical Journal,* 1990, 301:1067-70
- Righard L, Alade MO. Effect of delivery room routines on success of first breast-feed. *Lancet,* 1990, Nov 3; 336(8723):1105-7

81  For clarification, see:

- Emmanouil DE, Quock RM. Advances in understanding the actions of nitrous oxide. *Anesthesia Progress,* 2007, Spring;54(1):9-18, Review

82  Researchers (Gardberg and Tuppurainen 1994) found that when labor began, 10-15% of women's babies were in the posterior position, but by the end of labor (before the actual birth), this percentage went down to just 6%. Basing their comments on anecdotal evidence only, Sutton and Scott (1996) suggested the number of babies presenting in posterior position is increasing. One systematic review which considered the use of the knee-chest position during pregnancy could not draw any hard and fast conclusions. However, Sutton and Scott's conclusions and recommendations do tie in well with what we already know about the mechanical elements of labor (which are outlined in a 2004 *MIDIRS* article on the subject). The various sources for these documents are as follows:

- Gardberg M, Tuppurainen M. Anterior placental location predisposes for occipito posterior presentation near term. *Acta Obstetrica et Gynecologica Scandinavica,* 1994, 73:151-152
- Sutton J, Scott P. *Understanding and Teaching Optimal Fetal Positioning.*

Tauranga, New Zealand: Birth Concepts, 1996 (available from www.amazon.co.uk)

- Hofmeyr G, Kulier R. Hands/knees posture in late pregnancy or labour for fetal malposition (lateral or posterior) (Cochrane Review). In: *The Cochrane Library*, Issue 3. John Wiley & Sons Ltd, 2006
- *MIDIRS* and the NHS Centre for Reviews and Dissemination. *Positions in Labour and Delivery.* Informed Choice for Professionals leaflet, 2004

The following researchers found that adopting the knee-chest position for 30 minutes during labor significantly reduced persistent back pain in posterior labors:

- Stremler R, Hodnett E, Petryshen P. Randomised controlled trial of hands-knees positioning for occipitoposterior position in labour. *Birth,* 2005, 32(4):243-251

83    The only effect of EFM identified by research is to increase the cesarean rate; it does nothing to improve either fetal or maternal morbidity or mortality. See Notes 28 and 43 in the 'What' section.

84    Liang, *et al* (2007) found that women who had epidurals had a significantly longer first and second stage of labor (which increased the likelihood of forceps or a cesarean), but did not conclude that the epidural *per se* caused the urinary incontinence. In a study by Behrens, *et al* (1993), average labor times increased from 4.7 to 7.8 hours. See:

- Liang CC. Wong SY, Chang YL, Tsay PK, Chang SD, Lo LM. *Chang Gung Medical Journal,* 2007, 30(2):161-167
- McRae-Bergeron CE, *et al*, The Effect of Epidural Analgesia on the Second Stage of Labour, *Journal of the American Association of Anesthetic Nurses*, 1998, 66, No 2 :177-182
- Behrens O, *et al*, Effects of Lumbar Epidural Analgesia on Prostaglandin F2 Alpha Release and Oxytocin Secretion during Labour. *Prostaglandins*, 1993, 45, No 3: 285-296

An epidural which takes effect for longer often results in a forceps delivery. For confirmation of this and a discussion of related issues, see:

- Hughes D, Simmons S, Brown J, Cyna A. Combined spinal-epidural versus epidural analgesia in labour. (Cochrane Review). In: *The Cochrane Library*, Issue 3. John Wiley & Sons, 2006
- Anim-Somuah M, Smyth R, Howell C. Epidural versus non-epidural or no analgesia in labour. *The Cochrane Database of Systematic Reviews*, 2006, Issue 2
- Annandale, E. Dimensions of patient control in a free-standing birth centre. *Social Science & Medicine*, 1987, 25(11):1235-1248

McRae-Bergeron (1998) documented longer second stages of labor and extra need for forceps after epidurals

- McRae-Bergeron CE, *et al*, The Effect of Epidural Analgesia on the Second Stage of Labour, *Journal of the American Association of Anesthetic Nurses*, 1998, 66, No 2:177-182
- Torvaldsen S, Roberts C, Bell J, Raynes-Greenow C. Discontinuation of epidural analgesia late in labour for reducing the adverse delivery outcomes associated with epidural analgesia. *The Cochrane Database of Systematic Reviews*, 2006, Issue 3

85 Now that the opioid fentanyl has been incorporated into epidurals (alongside bupivacain), I assume the woman is likely to be in an even less alert state because Marucci, *et al* (2003) conclude: "Intrathecal bupivacaine-fentanyl dose produces a larger alertness decrease than single bupivacaine, because the anesthetic block density increases." See:

- Marucci M, Diele C, Bruno F, Fiore T. Subarachnoid anaesthesia in caesarean delivery: effects on alertness. *Minerva Anestesiologica,* 2003 Nov; 69(11):809-19, 819-24

86 Denis Walsh also points out the other inevitable effects of having an epidural. These, he says, include the woman becoming a passive patient, restrictions on mobility and tethering to a bed, zero second stage physiology, meaning that 'commanded pushing' becomes necessary, and an increase in the amount of time spent in technical measurement and record-keeping—as well as the disadvantages already listed in the text. This all results in what he calls 'the profound medicalisation of labor and birth'. See:

- Walsh D. *Evidence-based Care for Normal Labour and Birth.* Routledge, 2007

Other studies suggest there may be other effects, although data is as yet inconclusive. Even one company which produces bupivacaine (Astra Zenec) acknowledges in its literature that local anesthetics rapidly cross the placenta and can cause varying degrees of maternal, fetal and neonatal toxicity. See the following articles:

- Fernando R. and Bonello E. Placental and Maternal Plasma Concentrations of Fentanyl and Bupivacaine after Ambulatory Combined Spinal Epidural (CSE) Analgesia during Labour. *International Journal of Obstetric Anesthesia,* 1995, 4:178-179

- Brinsmead M. Fetal and Neonatal Effects of Drugs Administered in Labour, *Medical Journal of Australia,* 1987, 146:481-486

- Mueller MD, Bruhwiler H, *et al*. Higher rate of fetal acidemia after regional anesthesia for elective cesarean delivery. *Obstetrics & Gynecology,* 1997, Jul; 90(1): 131-4

A Cochrane review conducted in 2006 (by Elbourne, *et al*) into the use of the same opioids administered intra-muscularly (rather than via an epidural) concluded there was insufficient evidence to evaluate the comparative efficacy and safety of different opioids. See:

- Elbourne D, Wiseman R. Types of intra-muscular opioids for maternal pain relief in labour. In: *The Cochrane Library,* Issue 2. Chichester: John Wiley & Sons Ltd, 2006

87 Other studies have suggested other possible effects of opiates. A Cochrane review conducted in 2006 (by Elbourne, *et al*) into the use of opioids administered intra-muscularly (rather than via an epidural) concluded there was insufficient evidence to evaluate the comparative efficacy and safety of different opioids. See:

- Brinsmead M. Fetal and Neonatal Effects of Drugs Administered in Labour, *Medical Journal of Australia,* 1987, 146:481-486

- Elbourne D, Wiseman R. Types of intra-muscular opioids for maternal pain relief in labour. In: *The Cochrane Library,* Issue 2. Chichester: John Wiley & Sons Ltd, 2006

Photo © Jill Furmanovsky

*Should we perhaps help to free women of the tyranny of fear of labor and birth and consequently, the tyranny of the side– and/or after-effects of pain relief used in labor?*

Photo © Virgilio Ponce

*After all, what does the average woman (or man) know about drugs used in labor?*
*By the way, if you consider this text and the images to be propaganda, think for a*
*moment how pharmaceutical companies promote their drugs to medical personnel...*

88 This is the advice I'm providing, based on the idea that's it's important not to distract the laboring woman and have her depend on outside means of support (which might prevent her from getting into the right state of mind to give birth). Nevertheless, it must be said that research has found aromatherapy to be useful. One large study (Burns, et al, 2000) found that women used fewer opioids as a result (Demerol, diamorphine, etc) and another researcher (Mousely, 2005) found similar results and also noted that caregivers were enthusiastic about the use of aromatherapy. See:

- Burns E, Blamey C, Ersser S. The use of aromatherapy in intrapartum midwifery practice: an observational study. *Complementary Therapies in Nursing & Midwifery,* 2000, 6:33-34

- Mousely S. Audit of an aromatherapy service in a maternity unit. *Complementary Therapies in Clinical Practice,* 2005, 11:205-210

89 Research has indeed backed up this assertion. The following review of research concluded that studies of TENS provided no compelling evidence for TENS having any analgesic effect:

- Carroll D, Tramer M, McQuay H, Nye B, Moore A. Transcutaneous electrical nerve stimulation in labour pain: a systematic review. *British Journal of Obstetrics & Gynaecology,* 1997, 104:169-175

90 See the following articles:

- Odent M. La réflexothérapie lombaire. Efficacité dans le traitement de la colique néphrétique et en analgésie obstétricale. *La Nouvelle Presse Médicale,* 1975, 4 (3):188

For more on this, see:

- Bahasadri S, Ahmadi-Abhari S, Dehghani-Nik M, Habibi GR. Subcutaneous sterile water injection for labour pain: A randomised controlled trial. *Australian & New Zealand Journal of Obstetrics & Gynaecology,* 2006, 46:102-6

- Lytzen T, Cederberg L, Moller-Nielsen J. Relief of low back pain in labor by using intracutaneous nerve stimulation (INS) with sterile water papules. *Acta Obstetrica et Gynecologica Scandinavica,* 1989, 68:341-3

- Martensson L, Ader L, Wallin G. Sterile water papules against labor pain. A simple, safe, effective method. *Lakartidningen,* 1995, 92:2395-6

- Matensson L, Wallin G. Labour pain treated with cutaneous injections of sterile water: a randomised controlled trial. *BJOG: International Journal of Obstetrics & Gynaecology,* 1999, 106:633-7

91 This is referred to by Enkin, et al on page 252 of their review of research. See:

- Enkin M. Keirse MJNC, Renfrew M, Neilson J. *Guide to Effective Care in Pregnancy and Childbirth.* Oxford University Press 1995

92 This is a difficult topic because, admittedly, research studies into the use of acupuncture during labor have concluded that women receiving acupuncture use less pain relief and have less augmentation. I wonder if this is potentially because women who arrange acupuncture in advance of their labors set out to have a less medicated birth? As with other topics, it would be very difficult to conduct a randomized controlled trial to test out something like acupuncture because in the

USA women can exercise choice and they would probably be very unwilling to be 'randomly' assigned to an 'acupuncture' group for pain relief, particularly if they had no expectation of it being effective. For more information, see:

- Ramnero A, Hanson U, Kihlgren M. Acupuncture treatment during labour—a randomised controlled trial. *British Journal of Obstetrics & Gynaecology*, 2002, 109:637-644
- Neisheim B, Kinge R, Berg R. Acupuncture during labour can reduce the use of Merperidine: a controlled study. *Clinical Journal of Pain*, 2003, 19(3):187-191
- Skilnand E, Fossen D, Heiberg E. Acupuncture in the manage-ment of pain in labour. *Acta Obstetrica et Gynaecologica Scandinavica*, 2002, 81(10):943-948
- Ternov N, Buchhave P, Svensson G, Akeson J. Acupuncture during childbirth reduces use of conventional analgesia without major adverse effects: a retrospective study. *American Journal of Acupuncture*, 1998, 26(4):233-239
- Martoudis S, Christofides K. Electro-acupuncture for pain relief in labour. *Acupuncture in Medicine*, 1990, 8(2):51

If you're asked to arrange acupuncture, you could perhaps refer to the following book:

- Yelland S. *Acupuncture in Midwifery*. Blackwell, 2004

Research into the use of acupressure (shiatsu) or reflexology also drew positive conclusions. See:

- Kyeong Lee M, Bok Chang S, Kang D. Effects of SP6 acupressure on labour pain and length of delivery time in women during labour. *Journal of Alternative & Complementary Medicine*, 2004, 10(6):959-965
- Waters B, Raisler J. Ice water for the reduction of labour pain. *Journal of Midwifery & Women's Health*, 2003, 48:317-321
- Yates S. *Shiatsu for Midwives*. Books for Midwives Press, 2003
- Liisberg G. Easier births using reflexology. *Tidsskrift for Jordemodre*, 1989
- Motha G, McGrath G. The effects of reflexology on labour outcomes. *Journal of the Association of Reflexologists*, 1993, June,2-4

93 Tips on massage (for caregivers) are provided at www.midwiferytoday.com/articles/midwifestouch.asp and also in *Beautiful Birth* by Suzanne Yates (Carroll & Brown, 2008). This is one complementary therapy (if you would like to call it that), which can be used without disturbing a physiological labor. Of course, it is best if the 'masseur' or 'masseuse' is sensitive and familiar to the laboring woman and is in the birthing environment *anyway*. (This could be a husband, good friend or midwife.) It also goes without saying, I hope, that anyone doing massage needs to be silent. One research study found that women who received massage during labor (abdominal effleurage, sacral pressure and shoulder/back kneading) experienced significantly less pain and anxiety. Perhaps this is because massage is one way of making women feel *supported,* but not disturbed. See:

- Chang M, Wang S, Chen C. Effects of massage on pain and anxiety during labour: a randomised controlled trial in Taiwan. *Journal of Advanced Nursing*, 2002, 38 (1): 68-73

Another study found that with massage, women experienced decreased depressed mood, anxiety and pain, less agitated activity, shorter labors, shorter hospital stays and less postpartum depression. I would certainly recommend massage personally... It's one excellent reason for a doula or other birth assistant, to be present, who is willing to perform continuous massage for as little or as much time as the laboring woman requests. See:

- Field T, Hernandez-Reif M. Labour pain is reduced by massage therapy. *Journal of Psychosomatic Obstetrics & Gynaecology,* 1997, 18:286-291

On the subject of posterior lie, researchers (Gardberg and Tuppurainen 1994) found that when labor began, 10-15% of women's babies were in the posterior position, but by the end of labor (before the actual birth), this percentage went down to just 6%. Basing their comments on anecdotal evidence only, Sutton and Scott (1996) suggested that the number of babies presenting in posterior position is increasing. One systematic review which considered the use of the knee-chest position during pregnancy could not draw any hard and fast conclusions. However, Sutton and Scott's conclusions and recommendations do tie in well with what we already know about the mechanical elements of labor (which are outlined in a 2004 *MIDIRS* article on the subject—see below). The various sources for these documents are as follows:

- Gardberg M, Tuppurainen M. Anterior placental location predisposes for occipito posterior presentation near term. *Acta Obstetrica et Gynecologica Scandinavica,* 1994, 73:151-152
- Sutton J, Scott P. *Understanding and Teaching Optimal Fetal Positioning.* Tauranga, New Zealand: Birth Concepts, 1996 (available from www.amazon.co.uk)
- Hofmeyr G, Kulier R. Hands/knees posture in late pregnancy or labour for fetal malposition (lateral or posterior) (Cochrane Review). In: *The Cochrane Library,* Issue 3. John Wiley & Sons Ltd, 2006
- *MIDIRS* and the NHS Centre for Reviews and Dissemination. *Positions in Labour and Delivery.* Informed Choice for Professionals leaflet, 2004

The following researchers found that adopting the knee-chest position for 30 minutes during labor significantly reduced persistent back pain in posterior labors:

- Stremler R, Hodnett E, Petryshen P. Randomised controlled trial of hands-knees positioning for occipitoposterior position in labour. *Birth,* 2005, 32(4):243-251

94  In case you don't know, the obstetrician Grantly Dick-Read (1890-1959) was perhaps the first medical person to suggest the possibility that labor needn't be painful, since he had observed as much (quite by chance) one day in a woman who hadn't expected labor and birth to be painful. He was also one of the first obstetricians to voice the opinion that obstetric practices often caused more problems for women in birth than they solved. Another obstetrician similarly focused on optimizing women's experience, rather than spoiling it (through inappropriate and unnecessary interventions) is Ricardo Herbert Jones who is based in Brazil, in an urban area where he says the cesarean rate is 90%. Although obstetricians are sometimes seen as difficult to deal with by some midwives (no doubt rightly so), it is reassuring that some very good progress has resulted from the work of certain obstetricians. Grantly Dick-Read inspired the foundation of the British National Childbirth Trust (the NCT) and was its first president (in 1956). See *Childbirth Without Fear* by Grantly Dick-Read (Pinter & Martin, 2004).

95 Another factor to take into account when reading this account is the fact that the author, Gemma, had multiple sclerosis (MS). Michel Odent has noticed (only anecdotally) that women with MS tend to give birth easily. He had a particular motivation to research this topic since he attended the labor and birth of his former partner, Judy Graham (the mother of his son, Pascal)–when she too had MS. Another issue which this account might prompt us to consider is whether or not it's a good idea for pregnant women to 'prepare' for birth at all. Michel himself seems to consider the idea absurd but I very much doubt that he would if he himself had experienced pregnancies without support from caregivers who 'believed' in physiological birth (i.e. that women are at all capable of giving birth without 'help' and 'management'). If he had actually been pregnant himself and if he had 'given himself up' to a hospital which had been very interventionist or 'controlling', I think he would be evangelical about the need for preparation! As it is, he and others like him are very aware of the ease with which some teenagers (who appear to have no preparation for birth) give birth. Other people, for example other female writers who have spent long years writing books to help women 'prepare', who have perhaps faced what even Michel himself calls 'the industrialization of childbirth', have a very clear notion of the need to prepare. After all, modern woman is subjected to so much *negative preparation* simply because of exposure to the media in all its forms, especially movies made by people who have never read a book like this, and because of exposure to other women who have very bad memories of their own births. In addition they may well read pregnancy books which promote a fearful or 'passive patient' attitude. No doubt you will have your own very clear opinion on the need to prepare or not... My own view is that 'methods' for preparation are misguided. However, I do believe that preparation in the form of reading for information and inspiration can be enormously helpful. That was my main impetus for writing a book for pregnant women. After all, if women go into labor the first (or indeed the fifth) time without information about pain and pain relief, and without a knowledge of what physiological birth can be like, how on earth are they going to be able to make informed choices? There is certainly not enough time for caregivers to provide information to women while they are in labor and the emphasis on checking for problems during prenatal appointments is likely to mean that the only 'preparation' which occurs is to build up a woman's fear and awareness of what might go wrong. I think we need something far more positive and constructive. Perhaps a book–such as *Preparing for a Healthy Birth* (Fresh Heart 2010), which pregnant women can read in calm, private moments—and supplementary prenatal communication and activities organized by caregivers and teachers—are the form that preparation needs to take.

96 In 1978 Salter conducted a small study (with only 12 subjects) confirming that newborns who'd had the Leboyer treatment fared better in terms of alertness than those who hadn't; in the same year Oliver and Oliver conducted another small study (with 37 subjects) which also found that Leboyer babies seemed more relaxed (since they less often had tense hands, etc.) and spent more time with their eyes open immediately after their birth. See:

- Salter A. Birth without violence: a medical controversy. *Nursing Research,* 1978. Mar-Apr; 27(2):84-8

- Oliver CM, Oliver GM. Gentle birth: its safety and its effect on neonatal behaviour. *Journal of Obstetric, Gynecologic and Neonatal Nursing,* 1978. Sep-Oct: 7(5):35-40

97  Hale T. The Effects on Breastfeeding Women of Anaesthetic Medications Used during Labour, paper presented at Passage to Motherhood Conference, Brisbane, Australia, 1998

98  Belfrage P, *et al,* Lumbar Epidural Analgesia with Bupivacaine in Labor, *American Journal of Obstetrics & Gynecology,* 1975, 123:839-844

A study conducted back in 1977 (Caldwell, *et a*) found bupivacaine in babies 18 hours after the birth, compared to just one and a half hours in the case of the mother. See:

- Caldwell J, Moffatt JR, Smith RL, Lieberman BA, Beard RW, Snedden W, Wilson BW. Determination of bupivacaine in human fetal and neonatal blood samples by quantitative single ion monitoring. *Biomedical Mass Spectrometry,* 1977 Oct;4(5):322-5

99  For research which focuses on the effects on the neonate of various types of delivery, see the following:

- Lundell BP, Hagnevik K, Faxelius G, Irestedt L, Lagercrantz K. Neonatal left ventricular performance after vaginal delivery and cesarean section under general or epidural anesthesia. *American Journal of Perinatology,* 1984, 1 (2):152-7

- Hagnevik K, Faxelius G, Irestedt L, Lagercrantz K, Lundell BP, Persson B. Catecholamine surge and metabolic adaptation in the newborn after vaginal delivery. *Acta Paediatrica Scandinavica,* 1984, 73(5):602-9

- Christensson K, Siles C, *et al.* Lower body temperatures in infants delivered by caesarean section than in vaginally delivered infants. *Acta Paediatrica Scandinavica,* 1993, 82(2):128-31

- Thilaganathan B, Meher-Homji N, Nicolaides KH. Labor: an immunologically beneficial process for the neonate. *American Journal of Obstetrics & Gynecology,* 1991, 171(5):1271-2

- Molloy EJ, O'Neill AJ, Grantham JJ, Sheridan-Pereira M, Fitzpatrick JM, Webb DW, Watson RW. Labor Promotes Neonatal Neutrophil Survival and Lipopolysaccharide Responsiveness. *Pediatric Research,* 2004, May 5

- Gronlund MM, Nuutila J, Pelto L, Lilius EM, Isolauri E, Salminen S, Kero P, Lehtonen OP. Mode of delivery directs the phagocyte functions of infants for the first 6 months of life. *Clinical & Expermental Immunology,* 1999, 116(3):521-6

- Gasparani A, Maccario R, *et al.* Neonatal B lymphocyte subpopulations and method of delivery. *Biology of the Neonate,* 1992, 61(3):137-41

- Fujimura A, Morimoto S, *et al.* The influence of delivery mode on biological inactive renin level in umbilical cord blood. *American Journal of Hypertension,* 1990, 3(1): 23-6

- Gemelli M, Mami C, *et al.* Effects of the mode of delivery on ANP and renin-aldosterone system in the fetus and the neonate. *European Journal of Obstetrics & Gynecology & Reproductive Biology,* 1992, 43(3):181-4

- Steverson DK, Bucalo LR, *et al.* Increased immunoreactivity erythropoietin in cord plasma and bilirubin production in normal term infants after labor.

*Obstetrics & Gynecology*, 1986, 67(1):69-73

- Lubetzky R, Ben-Shachar S, *et al*. Mode of delivery and neonatal hematocrit. *American Journal of Perinatology*, 2000, 17(3):163-5

- Asien AO, Towobola AO, *et al*. Umbilical cord venous progesterone at term delivery in relation to mode of delivery. *International Journal of Gynaecology & Obstetrics*, 1994, 47(1):27-31

- Lao TT, Panesar NS. Neonatal thyrotropin and mode of delivery. *British Journal of Obstetrics & Gynaecology*, 1989, 96(10):1224-7

- Mongelli M, Kwan Y, *et al*. Effect of labour and delivery on plasma hepatic enzymes in the newborn. *Journal of Obstetric & Gynecological Research*, 2000, 26(1):61-3

- Miclat NN, Hodgkinson R, Marx GT. Neonatal gastric pH. *Anesthesia & Analgesia*, 1972, 57(1):98-101

100 Xu B, Pekkanen J, Jarvelin MR. Obstretric complications and asthma in childhood. *Journal of Asthma*, 2000, 37(7):589-94

> The amount of beta-endorphin in the colostral milk of mothers who'd given birth vaginally was significantly higher than colostrum levels of mothers who had a section

101 It's perhaps also worth noting that a study by Zanardo, *et al* (2001) found the amount of beta-endorphin in the colstral milk of mothers who'd given birth vaginally was significantly higher than colostrum levels of mothers who had a cesarean section. Having beta-endorphins in milk is thought to be important as these opiate-like substances make the newborn baby 'addicted' to its mother's milk. See:

- Zanardo V, Nicolussi S, Giacomin C, Faggian D, Favaro F, Plebani M. Labor pain effects on colostral milk beta endorphin concentrations of lactating mothers. *Biology of the Neonate* 2001, 79(2):79-86)

- DiMatteo MR, Morton S, *et al*. Cesarean Childbirth and Psychosocial Outcomes—A Meta-Analysis. *Health Psychology*, 1996, 15(4):303-14

102 Jacobson B, Eklund G, *et al*. Perinatal origin of adult self destructive behaviour. *Acta Psychiatriatrica Scandinavica*, 1987, 76:364-371

After attending approx 16,000 births—in Paris and London—Michel has reported almost never observing breathing difficulties in naturally-born newborns. Getting breathing started effectively and ensuring that any difficulties are overcome is a major preoccupation of caregivers when women have drugs of any kind during labor and/or birth because obviously death or brain damage can quickly follow a lack of breathing after the baby's been born.

103 Jacobson B, Eklund G, *et al*. Perinatal origin of adult self destructive behaviour. *Acta Psychiatrica Scandinavica*, 1987, 76:364-371

104 Tinbergen N and EA. *Autistic Children: New hope for a cure*. Allen & Unwin, 1983. In correspondence with Michel Odent before his death, Tinbergen wrote that he was considering how to test his hypothesis that there are links between pregnancy and birth practices or situations and autism. He was particularly interested in

some children's inability to establish eye contact, which he hypothesized they may not have learned from their mothers immediately after their birth. It would be nice if someone else could follow up on these ideas.

105 Hattori R, *et al*. Autistic and developmental disorders after general anaesthetic delivery. *Lancet*, 1991, 337:1357-1358 (letter).

106 Kendell RE, Juszczak E, Cole SK. Obstetric complications and schizophrenia: A case control study based on standardised obstetric records. *British Journal of Psychiatry*, 1996, 168:556-61

Also see:

- Khashan AS, McNamee R, Abel KM, Mortensen PB, Kenny LC, Pedersen MG, Webb RT, Baker PN. Rates of preterm birth following antenatal maternal exposure to severe life events: a population-based cohort study. *Human Reproduction,* 2009, Feb;24(2):429-37. Epub 2008 Dec 3

- Khashan AS, McNamee R, Abel KM, Pedersen MG, Webb RT, Kenny LC, Mortensen PB, Baker PN. Reduced infant birthweight consequent upon maternal exposure to severe life events. *Psychosomatic Medicine*, 2008, Jul;70(6):688-94

- Khashan AS, Abel KM, McNamee R, Pedersen MG, Webb RT, Baker PN, Kenny LC, Mortensen PB. Higher risk of offspring schizophrenia following antenatal maternal exposure to severe adverse life events. *Archives of General Psychiatry.* 2008 Feb;65(2):146-52

107 Bakan P, Dibb G, Reed P. Handedness and birth stress. Neuropsychology, 1973, 11: 363-366.

108 Consider the following research studies:

- Jacobson B, Eklund G, *et al* . Perinatal origin of adult self destructive behaviour. *Acta Psychiatrica Scandinavica*, 1987, 76:364-371

- Jacobson B, Nyberg K. Opiate addiction in adult offspring through possible imprinting after obstetric treatment, *British Medical Journal*, 1990, 301:1067-70

- Nyberg K, Allebeck P, Eklund G, Jacobson B. Socio-economic versus obstetric risk factors for drug addiction in offspring. *British Journal of Addiction*, 1992, 87:1669-1676

- Nyberg K, Allebeck P, Eklund G, Jacobson B. Obstetric medication versus residential area as perinatal risk factors for subsequent adult drug addiction in offspring. *Paediatric & Perinatal Epidemiology*, 1993, 7:23-32

- Nyberg K. *Studies of perinatal events as potential risk factors for adult drug abuse*. Thesis: Dept of Clin Alcohol & Drug Addiction Research, Karolinska Institute, Stockholm, Sweden, 1993

- Nyberg K, Buka SL, Lipsitt LP. Perinatal medication as a potential risk factor for adult drug abuse in a North American cohort. *Epidemiology*, 2000, 11(6):715-16

109 Nolan (2005) concluded that prenatal education packages do not consistently reduce intervention rates, which in turn suggests that they do not help. However, carefully designed (somewhat non-typical) prenatal classes did result in an increase in the use of upright positions and/or a decrease in the number of epidurals. See:

- Nolan M. Childbirth and parenting education: what the research says and why we may ignore it. In Nolan M and Foster J (eds) *Birth and Parenting Skills: New Directions in Antenatal Education.* Churchill Livingstone, 2005
- Nolan M and Foster J (eds) *Birth and Parenting Skills: New Directions in Antenatal Education.* Churchill Livingstone, 2005
- Foster J. Innovative practice in birth education. In M. Nolan and J. Foster (eds) *Birth and Parenting Skills: New Directions in Antenatal Education.* Elsevier Science, 2005
- Walsh D, Harris M, Shuttlewood S. Changing midwifery birthing practice through audit. *British Journal of Midwifery,* 1999, 7(7):432-345

110   Specifically for data relating to the PROM issue, see:
- Ladfors L, Mattsson LA, Eriksson M, Fall O. A randomised trial of two expectant managements of prelabour rupture of the membranes at 34 to 42 weeks. *British Journal of Obstetrics & Gynaecology,* 1996, Aug; 103(8):755-62
- Other issues are explored throughout the book, but note in particular the following studies, which suggest that the risk of intervention would have risen and the increase in 'management' would probably have occurred considerably simply because of the transfer to hospital:
- Machin D, Scamell M. The experience of labour: using ethnography to explore the irresistible nature of the bio-medical metaphor during labour. *Midwifery,* 1997, 13: 78-84
- Kirkham M. The culture of midwifery in the National Health Service in England. *Journal of Advanced Nursing,* 1999, 30:732-739
- Ball L, Curtis P, Kirkham M. *Why Do Midwives Leave?* Royal College of Midwives, 2002
- Stapleton H, Kirkham M, Thomas G, Curtis P. Midwives in the middle: balance and vulnerability. *British Journal of Midwifery,* 2002, 10(10):607-611
- For research on the necessity (or otherwise) of EFM in high risk pregnancies, see:
- Brown VA, *et al.* The value of antenatal cardiotocography in the management of high-risk pregnancy: a randomised controlled trial. *British Journal of Obstetrics & Gynaecology,* 1982, 89:716-22
- Haverkamp AD, *et al.* The evaluation of continuous fetal heart rate monitoring in high risk pregnancy. *American Journal of Obstetrics & Gynecology,* 1976, 125: 310-20
- Leveno KJ, *et al.* A prospective comparison of selective and universal electronic fetal monitoring in 34,995 pregnancies. *New England Journal of Medicine,* 1986, 315:615-19
- Lumley JC, Wood C, *et al.* A randomized trial of weekly cardiotocography in high-risk obstetric patients. *British Journal of Obstetrics & Gynaecology,* 1983, 90:1018-26
- Sky K, *et al.* Effects of electronic fetal heart rate monitoring, as compared with periodic auscultation, on the neurological development of premature infants. *New England Journal of Medicine,* 1990, (March 1):588-93
- Luthy D, Shy K, van Belle G, Larson E, *et al.* A randomised controlled trial of

electronic fetal monitoring in preterm labour. *Obstetrics & Gynaecology*, 1987, 69:687-695

- Madaan M, Trivedi S. Intrapartum electronic fetal monitoring vs. intermittent auscultation in post caesarean pregnancies. *International Journal of Obstetrics & Gynaecology*, 2006, 94:123-125

111 See the following research, which presents mixed results:

- Saraswat L, Porter M, Bhattacharya S, Bhattacharya S. Caesarean section and tubal infertility: is there an association? *Reproductive Biomedicine Online*. 2008 Aug;17(2):259-64
- Wolf ME, Daling JR, Voigt LF. Prior cesarean delivery in women with secondary tubal infertility. *American Journal of Public Health*. 1990 Nov;80(11):1382-3
- LaSala AP, Berkeley AS. Primary cesarean section and subsequent fertility. *American Journal of Obstetric Gynecology*. 1987 Aug;157(2):379-83.

112 Research has actually confirmed the inability of caregivers to appraise research (Veeramah 2004) and—perhaps needlessly!—research has also revealed that lack of time is a factor (Hundley 2000). Another research study (Richens 2002) mentioned the problem that an individual caregiver may feel he or she does not have the necessary authority or autonomy to change his or her practice, and another mentioned more generalized institutional constraints (Scott, *et al*, 2003). Nevertheless, as Denis Walsh has exemplified (in the case of one patient who insisted on having a lotus birth, despite extreme pressure not to), when women assert themselves they can actually succeed in demonstrating to caregivers that fears are ungrounded and that there are possible improvements to current practice. See:

- Veeramah V. Utilisation of research findings by graduate nurses and midwives. *Journal of Advanced Nursing*, 2004, 47(2):183-191
- Hundley V. Raising research awareness among midwives and nurses: does it work? *Journal of Advanced Nursing*, 2000, 31(1):78-86
- Richens Y. Are midwives using research evidence in practice? *British Journal of Midwifery*, 2002, 10(1):11-16
- Scott T, Mannion R, Marshall M, Davies H. Does organisational culture influence health care performance? A review of the evidence. *Journal of Health Service Research and Policy*, 2003, 8(2):105-117
- Walsh D. *Evidence-based Care for Normal Labour and Birth*. Routledge, 2007

113 It is actually difficult to prove *directly* the improved safety of physiological birth. There is limited research data for this since randomized controlled trials are impossible in the perinatal period because women will not agree to participate in research at this sensitive time. However, it is easy to see the superiority and 'healthiness' of physiological birth over other approaches involving unnecessary interventions if a) we realize how complex the normal, healthy processes are (and consequently understand their fragility), b) we understand the consequences of unnecessary interventions (including drugs), which involve upsetting the delicate balance of the natural processes, and c) we remember that women having an undisturbed physiological third stage experience a peak of oxytocin higher than any other in their lives... since this must have a huge impact on mothering

behavior.

114 Even if you feel uncertain, you need to communicate what you know to moms-to-be, as accurately and as honestly as possible. While doing this, you don't even need to feel bad about any feelings of uncertainty you may have. After all, there are very few areas of life where we know everything and as a human being you cannot possibly be expected to have every possible bit of information. Presenting ourselves with honesty, openness and integrity usually prompts a very positive response and if you also 'hand the process' of research over to your clients you will be doing something very constructive in terms of developing a mutually respectful relationship.

115 Epidurals, which are increasingly popular, are associated with fecal and so-called 'flatus' incontinence (i.e. having no control over passing wind or doing poops). This is because they result in longer labors and higher rates of forceps births, although—as with other issues around intervention—while problems increase, it's not easy to identify categorically what causes what, i.e. whether epidurals specifically are the cause, because they're often requested when many other problems are present. This makes research results difficult to interpret accurately.

- McRae-Bergeron CE, et al, The Effect of Epidural Analgesia on the Second Stage of Labour, *Journal of the American Association of Anesthetic Nurses*, 1998, 66, No 2:177-182

- Torvaldsen S, Roberts C, Bell J, Raynes-Greenow C. Discontinuation of epidural analgesia late in labour for reducing the adverse delivery outcomes associated with epidural analgesia. *The Cochrane Database of Systematic Reviews*, 2006, Issue 3

116 Research has shown that the urinary incontinence often results in the following cases: having a forceps birth (Arya, et al, 2001), pushing actively for a long time (Kirkman 2000), having lots of babies, having babies who weigh more than 3,700g, and being over 30 at the time of giving birth (Mason, et al, 1999). Research has also shown that urinary stress incontinence is best avoided by getting the woman to do pelvic floor exercises antepartum (Morkved, et al, 2003, Sampselle, 2000) and by 'allowing' her to push her baby out herself, according to her own timetable (Kirkman, 2000); postpartum pelvic floor exercises have also been found to be helpful (Glazener, et al, 2001).

- Arya L, Jackson N, Myers D, Verma A. Risk of new onset urinary incontinence after forceps and vacuum delivery in primiparous women. *American Journal of Obstetrics & Gynaecology*, 2001, 185:1318-1324

- Kirkman S. The midwife and pelvic floor dysfunction. *The Practising Midwife*, 2000, 3(8):20-22

- Mason L, Glenn S, Walton I, Appleton C. The prevalence of stress incontinence during pregnancy and following birth. *Midwifery*, 1999, 15(2):120-127

- Mason L, Glenn S, Walton I, Appleton C. The experience of stress incontinence after birth. *Birth*, 1999, 26(3):164-171

- Morkved S, Bo K, Schei B, Salvesen K. Pelvic floor muscle training during pregnancy to prevent urinary incontinence: a single-blind randomized controlled trial. *Obstetrics & Gynaecology*, 2003, 101:313-319.

- Sampselle C. Behavioural intervention for urinary incontinence in women: evidence for practice. *Journal of Midwifery & Women's Health,* 2000, 45(2):94-103

- Glazener C, Herbison G, Wilson P. Conservative management of persistent postnatal urinary and faecal incontinence: randomised controlled trial. *British Medical Journal,* 2001, 323:593

Looking at incontinence later on in life, researchers have established the following causes (in order of importance, starting with the most important) (Hannestad, *et al,* 2003, Rortveit, *et al,* 2003): heredity, obesity, smoking, HRT, the number of children a woman's had and how she's had them (noting that forceps is the worst way, from the point of view of incontinence). See:

- Hannestad Y, Rortveit G, Dalveit A, Hunskaar S. Are smoking and other lifestyle factors associated with female urinary incontinence? The Norwegian EPINCONT study. *British Journal of Obstetrics & Gynaecology,* 2003, 110:247-254

- Rortveit G, Daltveit A, Hannestad Y, Hunskaar S. (2003) Vaginal delivery parameters and urinary incontinence: the Norwegian EPINCONT study. *American Journal of Obstetrics & Gynecology,* 2003 Nov; 189(5):1268-74

Considering protection of the perineum from damage during childbirth, research has actually confirmed prenatal perineal massage as being useful for protecting the perineum from damage by three randomized controlled trials:

- Shipman M, Boniface D, Tefft M, McCloghry F. Antenatal perineal massage and subsequent perineal outcomes: a randomised controlled trial. *British Journal of Obstetrics & Gynaecology,* 1997, 104:787-791

- Labrecque M, Eason E, Marcoux S. Randomised controlled trial of prevention of perineal trauma by perineal massage during pregnancy. *American Journal of Obstetrics & Gynaecology,* 1999, 180:593-600

- Davidson K, Jacoby S, Scott Brown M. Prenatal perineal massage: preventing lacerations during delivery. *Journal of Obstetric, Gynaecological & Neonatal Nursing,* 2000, 29(5):474-479

117 Painful sex (dyspareunia) and perineal pain after childbirth have been associated with forceps and ventouse births, episiotomy and anal spincter tears (Buhling, *et al,* 2005). See:

- Buhling K, Schmidt S, Robinson J, Klapp C, Siebert G, Dudenhausen J. Rate of dyspareunia after delivery in primiparae according to mode of delivery. *European Journal of Obstetrics & Gynecology,* 2005, 124:42-46

118 This may well be the popular concensus, but it is not necessarily what is generally felt amongst professionals, who often have to deal with problems when they arise. A Cochrane review conducted in 2006 identified only one study which confirmed that epidurals were more effective than other forms of pain relief—but this does perhaps reflect a generally-held view that epidurals are effective. However, the rest of the review discussed the possible complications of epidurals, which makes clear their potential to disturb birth. See:

- Anim-Somuah M, Smyth R, Howell C. Epidural versus non-epidural or no analgesia in labour. *The Cochrane Database of Systematic Reviews,* 2006, Issue 2

- Annandale E. Dimensions of patient control in a free-standing birth centre. *Social Science & Medicine,* 1987, 25(11):1235-1248

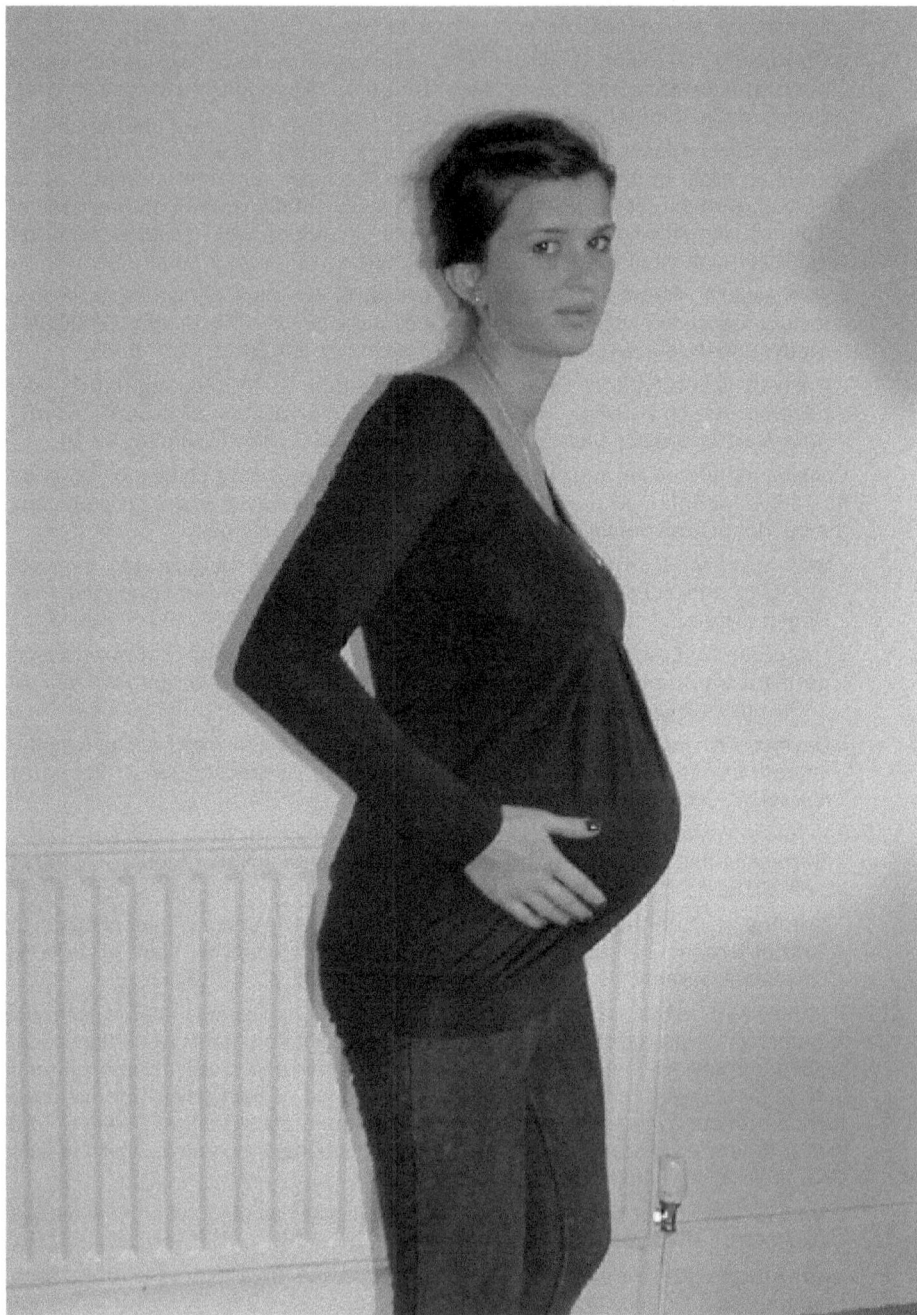

*How can you help?*

# Optimal Birth: HOW

## How can optimal birth be facilitated?

In this section I present some suggestions, birthframes, notes and opinions (not all my own), based on both research and experience and informed by current protocols. I also ask you to consider a few issues (by presenting questions); the answers to many of these could make an enormous difference to any one individual woman's experience of childbirth. I hope that by focusing on the questions you will get a clearer idea of your own views as they stand now and be in a better position to evaluate what you read in the rest of this section of the book. When you think through each issue, please also remember what any life coach will tell you: problems can also be opportunities.

## Pause to reflect...

### ...about training

1   What should midwifery and obstetric training ideally include?
2   Do you think anything was missing from your training?
3   To what extent did your training focus on pathology?
4   How much time was spent talking about pain relief and concomitant safety issues?
5   To what extent did you focus on the study of normality?
6   How much did you study ways of facilitating birth?
7   How many drug-free births were you required to attend?
8   How many undisturbed physiological births did you see?

### ...about experience

1   Is your experience of being a caregiver generally a good one or could you improve your professional lifestyle?
2   Has any experience ever influenced your practice?
3   Has any experience in your career ever saddened you?
4   Has any experience in your career ever inspired you?
5   To what extent does previous experience inspire or limit the decisions you make?

6 How do you try to communicate your experience to more newly-qualified staff?

7 Are you open-minded about having new experiences?

8 Do you ever try and tap into the experience of colleagues or other professionals you meet at conferences, etc.?

9 Do you ever try to learn from non-professionals?

## ...about research

1 To what extent do you keep up with the latest research?

2 Where do you go to access research on any given topic?

3 What are the quickest ways of accessing research?

4 What are the most effective ways?

5 How can you evaluate the validity of research studies?

6 How much do you trust the conclusions of researchers?

7 How would you go about conducting your own research?

8 In which cases has research changed or influenced clinical practice?

## ...about the law

1 Are there any references to the law in any protocols or guidelines you are required to follow?

2 Does a fear of litigation influence the way you practice or intend to practice?

3 Do you know anyone who's had to deal with litigation?

4 How much are you influenced by insurance?

5 What would you do if a client threatened to sue you?

6 Have you ever refused to support a woman, explaining to her that you're afraid she might sue you afterwards?

7 Do you ever explain or discuss legal issues with clients?

8 Are there any laws you think should be changed in your state?

9 How do you think clients could prepare the way for non-intervention and non-disturbance, in legal terms?

10 Are there any laws which you would be prepared to break either in the US or abroad a) while practicing or b) when pregnant?

11 How do you feel about the ACOG and ACNM guidelines?

## ...about power

1 Do you feel that power has shifted at all between midwives, obstetricians, family doctors and other caregivers during your own career?

2 Have there ever been cases where you have been frustrated by any professional hierarchies?

3 How much power do you feel you have within your own working context?

4 To what extent do you think a mom-to-be should determine the course of her own care?

5 How do you feel about a doula influencing decisions which a laboring woman may be making about her care?

6 Is there an adequate dialog between midwives, obstetricians and any other relevant caregivers?

7 To what extent are you informed or controlled by ACOG or ACNM guidelines, or by another professional body?

8 To what extent is there a dialog between your line managers (if you have any)—amongst themselves and with you?

9 To what extent do specialist caregivers communicate with local family doctors?

10 How much do you exchange ideas and information with moms-to-be?

## ...about protocols

1 How do you think protocols are established?

2 Are protocols always in line with research findings?

3 How often do protocols change?

4 How long have specific protocols been in place at your particular place of work?

5 Who influences changes made to protocols?

6 How much power do you personally feel you have?

7 How much power can a group of midwives or other caregivers exert?

8 Do protocols serve the needs of moms-to-be?

9 Do protocols ever serve the financial needs of an institution, or any other interests?

10 Do protocols ever focus primarily on caregivers' needs?

11 Do protocols ever help you to exert authority over a laboring woman?

12 Which protocols do you like?

13 Which protocols would you really like to change?

14 Are there any protocols which need to be changed for the sake of safety or to ensure better outcomes?

15 To what extent are protocols observed?

16 When is it possible to get permission not to follow protocol and what's the procedure for doing this?

## ...about personalities

1   Have you ever felt that anybody's personality has made following protocols either more difficult or easier?

2   Do you think clashes between personalities within a workplace are ever more likely to prompt problems which result in litigation?

3   Do you think certain personality types might be more suited to certain types of work or facilitating certain types of outcome?

4   To what extent do you think an individual can become aware of his or her personality strengths, weaknesses and idiosyncrasies?

5   Do you think systems and protocols can effectively limit any damage?

6   Some managers have said that working on weaknesses is futile because it leads to mediocrity. To what extent do you think a person can usefully work on their strengths and to what extent do you think it's time-wasting and unproductive to focus on weaknesses?

## ...about optimal care at all stages of the process

1   How should midwives and other caregivers ideally behave?

2   How much monitoring do you need to do, in your opinion, in order to ensure safest outcomes, and what type of monitoring should this be?

3   In your opinion, what are the keys to promoting optimality?

4   How often have you witnessed optimal births so far, in your view?

5   How often have optimal births occurred while you personally have been the main caregiver?

6   Can optimality ever be compromised by an assistant caregiver?

7   In what kind of circumstances have you so far experienced optimality?

8   What do you think optimality means from the point of view of a) a primp (a first-time mom), b) a woman approaching her second labor or c) a woman approaching her third or subsequent labor?

9   How is it possible to communicate to pregnant clients your own view of optimality, based on your experience and what you've found out about research? How much do you need to take on other key people's views?

Photo © Colin Smith

*Do you think any of your ideas might meet with anybody's disapproval?*

Whatever your answers to each question, it's interesting to consider a few comments from women on caregivers. Do they make you change any answers?

66 My doctor was aware from the start that I again planned a home birth. The obstetrician at the hospital told me that I could not have prenatal shared care at the hospital if I was planning to have the baby at home and pressurized me into agreeing to come to the hospital for the birth. The midwife afterwards told me not to worry— that shared prenatal care with delivery at home if all was well, or at the hospital if not, was fine.

66 We moved when I was 5 weeks pregnant, and the first thing I did was find a midwife who would deliver my baby at home. At first I approached the community hospital, and spoke to my doctor, who I shan't name, because although he read me the riot act and said officially a home birth would be endangering my life and that of the baby, he, unofficially, was glad I wanted a home birth, and said he would support me. I was well pleased. Next, I went to see a midwife who is also pro-home birth. But as soon as she heard of my previous cesarean she declined her services. I didn't want to press my case because I wanted to make sure my midwife was 100% on my side. I understand that some caregivers advertise that they attend home labor, but my faith in the system had been totally shattered, because I found this was just not true. And I didn't want to be hauled from home to hospital in labor under any spurious pretext. So I hired an independent midwife who really supported me.

66 I booked with community midwives who are generally supportive of home birth but when my history of a PPH [postpartum hemorrhage] was revealed (by me, voluntarily) I was told that a home birth would be out of the question. I was made to feel that my body was at fault for pouring out a life threatening amount of blood after giving birth—the truth, however, was that mismanagement had caused the heavy blood loss. I soon booked with independent midwives who had the confidence in my body which I had and I gave birth to a second daughter at home in under three hours.

66 At each check-up at the hospital it became apparent that none of the midwives had any experience of delivering twins and after further research and general reading it was clear that a first-time mother having a drug-free natural delivery was quite rare. I was told that my babies were cephalic [head down, i.e. not breech] so a cesarean wasn't absolutely necessary but was told to 'keep my options open'. Even my prenatal class teacher (who was wonderful) said a natural delivery was, of course, possible but she wasn't sure whether it was realistic. The only positive response was from the obstetrician himself who, in an evening talk at the hospital, made it 'loud and clear' that he believes it is every woman's right to have an obstetrician present in a birth of this kind. However, he also said it is more likely there won't be one available, so whoever is there is likely to have had little, if any, experience of delivering twins; therefore he said it would be safer to have a cesarean. He was the most amazing man I have ever met. After the talk, when I told him how important it was to me to at least try for a natural delivery, he wrote on my notes to inform him when I was in labor and he would come in. He gave me his vacation dates, which I somehow managed to avoid, and he stuck to his word and came in his 'free time' and delivered my daughters. I know I couldn't have done it without him!

❝ Elaine visited my house throughout the pregnancy for tests, and listened to baby's heartbeat through her midwife's trumpet [fetoscope]. We got to know her, and she us, and it was a gentle and caring relationship. By the time I was in labor, my trust in Elaine was complete.

We got to know our midwife and she us, and it was a gentle
and caring relationship. By the time I was in labor,
my trust in her was complete

❝ I felt very confident in my ability as a woman to give birth and this decision also let me feel in control of who was to be present during the labor and birth. I was not prepared to let into my home any midwives that I felt didn't respect my rights as a mom-to-be whilst giving birth. This for me was very significant emotionally as I believe that a midwife who does not revere childbirth and show consideration for the laboring woman has no place in being a midwife. As long as there is no danger to the mother or child, the midwife should be led by the laboring woman. If this does not happen it can seriously undermine a woman's confidence in her abilities within herself to give birth, which I think is evident with my first birthing experience.

❝ The birth of my first child felt very institutionalized and governed by hospital policy. The midwife was definitely in control of the entire situation with me as her charge. My second birth was a truly wonderful experience that was influenced by an extremely intuitive midwife who encouraged me to do what felt to be right for me. These experiences have made me highly aware of the dynamics between the birthing mother and the midwife, whose main purpose is to provide support for the mother and her husband in labor, and who needs to understand both the physical processes and the emotional needs of the mother. I feel that this relationship is one of the most important factors during labor and birth.

I feel that this relationship is one of the most
important factors during labor and birth

❝ Basically, the midwives made themselves invisible. I certainly hardly heard them discussing anything. I do remember hearing them ask Simon to bring clothes and diapers—I couldn't believe I was giving birth already!

❝ Whenever I asked the midwife for guidance about what I was supposed to be doing or how far she thought I was dilated she just told me I was doing really well, coping remarkably well and asked me to feel what my body was telling me to do and to follow that.

❝ Michel was great. He was very calm.

❝ Julie [the midwife] hadn't even got her gloves dirty. She had done nothing fantastically.

# The prenatal period

Again, let's go back to basics and consider some key questions.[1]

## What's the purpose of prenatal care?

After consulting various textbooks and professionals, I managed to determine that, generally speaking, you and other healthcare practitioners are aiming to:

- Confirm a woman's pregnancy and estimate when she is likely to give birth
- Assess any possible risks, based on her previous medical history, age or current circumstances
- Encourage the woman to take folic acid supplements in the first trimester
- Talk about diet, i.e. encourage the mom-to-be to eat healthily and avoid certain foods
- Confirm the woman's blood group in case a blood transfusion is needed later, and establish whether or not she is Rhesus negative, in which case you will probably recommend a few additional procedures
- Offer the mom-to-be various checks to look for abnormalities and subsequently, if any are found or suspected, offer an abortion
- Offer the woman ultrasounds to monitor her baby's size and positioning in the womb, to check for multiples and to check the position of the placenta and any fibroids
- Screen for rubella immunity, sexually transmitted diseases, diabetes and gestational diabetes, Strep B, as well as genetic disorders (if there is a family history)
- Track the woman's blood pressure and urine in case hypertension develops or protein appears in the urine, so as to detect pre-eclampsia and ward off eclampsia
- Track the baby's positioning and the quantity of amniotic fluid, as well as fetal growth, through ongoing manual 'palpation' or the use of repeated ultrasounds

## What are the potential problems?

There are a few, it seems...

- Moms-to-be are very emotional, suggestible beings
- Pregnancy, labor and birth does not proceed in a mechanical way, but is affected by psychological factors—because hormones are sensitive to emotions
- One intervention tends to lead to another
- Women often come to check-ups wanting to ask numerous questions and leave feeling that they've had no opportunity to ask them

- Moms-to-be often feel undermined by caregivers' use of technical terms (either intentionally or unwittingly)... or even belittled by caregivers' use of 'babyish' terms, when they already know the proper terms
- There are often disagreements between women and midwives about approaches to care
- There can be serious personality mismatches
- Women often 'shop around' for suitable caregivers, so midwives sometimes feel defensive
- While some women want to avoid any intervention, others come hoping for every possible form of intervention so that they can avoid as much pain as possible and 'control' what seems an unpredictable and scary process

## What are the potential opportunities?

This could be the source of some professional satisfaction:

- You have the opportunity to get to know many different types of women
- You may be able to take a 'whole person' approach to any problems by paying attention in a different way, or trying new approaches to tests
- Some women may push you to extend your normal practice by making requests which are new to you
- You can experiment with different ways of communicating with women
- Potentially, you have the power to exert a great deal of influence over any individual woman and the choices she makes as she progresses through her pregnancy, labor and birth
- You can prepare women who are afraid and make them end up feeling empowered and profoundly satisfied
- You can act as a mentor or guide as women pass through this key transition in their lives

## What are the potential risks?

Of course, risks aren't always negative, but an intrinsic part of life (as in crossing the road). But do any of these potential risks ever limit what you do? (How many traffic accidents would it take to stop you crossing roads?)

- You may lose a client because of your approach
- You may get sued and lose your job altogether
- You may become too popular! (That might cause jealousy.)
- You may have problems with colleagues or superiors
- You may open yourself up to criticism—can you face it?
- You may get promoted and gain power to inspire others
- You may learn new things from different clients
- You may start viewing things differently if you experiment with different approaches...

# Pause to reflect...

## ...about the need to prepare the woman for birth

1   To what extent do you think pregnant women need to prepare themselves for birth?

2   What kinds of prenatal preparation can be harmful?

3   What kinds of prenatal preparation take part *anyway* as part of a woman's immersion in her own culture? (Think here about the role of the media in 'preparing' women's expectations for birth and the role of pregnancy magazines.)

4   To what extent do prenatal classes vary in their approach? Do hospital-run classes differ significantly from classes organized by organizations such as BirthWorks or Birthing from Within, for example?

5   To what extent are pregnant women able to contribute to the 'syllabus' of prenatal classes and how many opportunities do they get to address their own, very personal needs?

6   Why is it that teenage pregnancies often culminate in a birth which appears to be very instinctive, while births which take place later on in a woman's life often tend to 'go wrong' in some way, with the woman apparently losing her apparent 'instinctive' ability to give birth?

7   Do pregnant women in their 20s tend to prepare more or less than older women? What differences are there in their preparation?

8   How helpful are books which present a 'method' for preparing for birth? To what extent do they cause you problems as a caregiver? Thinking about all the books you know, what kind of prenatal preparation course does each one present?

9   To what extent can positive (and negative) birth stories help a mom-to-be prepare for birth?

10  To what extent should a woman's prenatal preparation be 'emotional' (e.g. involve drawing pictures, exploring emotions) and to what extent is it best if it's 'rational' (e.g. involve learning about the physiological processes and forms of pain relief available)?

How much do pregnant women need to prepare for labor?

*How much do you know about what women do to prepare for their own labors and births? How much do they tell you? How much do you want to help them prepare?*

## About the companion book, for pregnant women...

The book which was written for pregnant women who are hoping to optimize their experience of birth—*Preparing for a Healthy Birth* (Fresh Heart 2010)—is not a 'preparation course' in the usual sense. In a way it's a kind of 'anti-preparation course' in that it primarily suggests that women prepare for their upcoming birth in terms of their knowledge and practicalities. There is no method, there are no breathing exercises, physical exercises or psychological exercises, although all of these are discussed at various points. (None of them are considered necessary for optimizing the woman's birth experience.) As the following contents pages suggest, the emphasis is simply on presenting information (both factual and personal, in the form of birth stories) so that women can make informed choices about their own upcoming labor and birth.

### Introduction: A personal journey...

This explains the importance of becoming informed so that informed choice is possible and begins to explain ideas relating to birth.

### Step 1: Understand 'healthy'

This chapter simply describes the physiological processes of pregnancy, labor and birth, so that they aren't a surprise when they start happening.

### Step 2: Consider your assumptions

Here, common assumptions are challenged about what constitutes high risk. The rise of the cesarean rate is considered, so that women become better motivated to research their personal situation and consider their options.

### Step 3: Do not disturb

This chapter describes in detail the hormonal processes of pregnancy, labor and birth and explains what the possible effects of intervention and analgesia or anesthesia might be on the birthing experience.

### Step 4: Help your baby

Here women are given a week-by-week guide to pregnancy so as to help them 'bond' with their unborn baby as he or she is growing on a week-by-week basis.

### Step 5: Care about care

This chapter considers the role of prenatal care and outlines key issues (e.g. prenatal testing and due dates), which it is helpful for women to have thought through before they come to you. Again, if they're well informed, their choices will also be made with full knowledge of what they're letting themselves in for.

## Step 6: Think ahead

Here the practicalities of birth are considered in much more detail and some fairly common 'difficult' situations are considered. The idea here is to ensure that women who experience difficult situations (such as premature labor or stillbirth) are not completely unprepared, simply because outcomes are optimized if women have some knowledge about these situations in advance.

## Step 7: Choose who

Hopefully, you are the ideal caregiver for each of your clients! This chapter encourages women to make sure they are in tune with their caregivers (whether they are midwives or obstetricians). It also helps them think through who they might like to have with them at the time of the birth (e.g. doula, husband, friend, family...).

## Step 8: Choose where

Of course, the main focus of this chapter is the home vs hospital debate. However, the 'dry land' vs 'water birth' debate is also considered, again so that women can make better-informed choices.

## Step 9: Help your body

This chapter considers some of the obvious physical aspects of pregnancy relating to diet and exercise and also typical symptoms or physical problems.

## Step 10: Help your mind

This is an unusual chapter in that it considers a wide range of psychological problems which may arise antepartum and provides advice about how a woman might deal with these. The chapter also provides a review of key points covered elsewhere in the book so that women approaching their due date can reassure themselves and approach their upcoming labor and birth with confidence.

## Postscript: Your new life

This very brief end-section looks at some of the keys to optimize the early postpartum period.

Throughout the book there are birthframes—as in this book, although many birthframes are different—and many more comments from other women. The idea is that women will become more confident about their own labor and birth by finding out about how other women coped in a range of situations. A glossary, index and useful contacts pages at the end of the book are intended to help pregnant women in your care find relevant information and support quickly, whatever situation they might find themselves in. The idea is to make your life easier and help to open up the channels of communication.

# The importance of communication

Do you doubt the need to communicate well each and every time you encounter a mom-to-be? If so—or even if you're already convinced of the need to take care—consider the cases in this section, which I assure you are all entirely true.

### Birthframe 33: Anonymous—Undermining prenatal care

Here's an account from a contributor who was not happy with her midwife's approach during her first appointment. (In this case, the midwife didn't take up various opportunities, but instead created new problems.) By the way, this appointment took place in the USA but it could have been anywhere...

I had worked for a publishing company that specialized in some parenting and childcare books, and had published several on active birth, breastfeeding, taking part in one's own care, etc, so I felt happily enlightened. I went to my first prenatal check-up thinking I would know what questions to ask, that I would be an informed patient. I was relieved to see that the head of the 'normal OB' department was a woman, a Certified Nurse Midwife. That sounded proper and competent. Hoping to ask lots of questions, I actually could not get a word in edgewise.

The usual first questions: age, marital status; blood pressure taken, weighed, measured, the pelvic exam. She listened to my heart, then with a shocked look, she said, "Did you know you have a heart murmur?!" "Well, no," I said. "I have had regular medical care my whole life and no one has ever found any heart irregularities." She looked at me in disbelief. I was wondering what the hell she was talking about and was afraid that it might hurt the baby. Later on, another practitioner told me that it is very common for women to develop a mild heart murmur in pregnancy, and it isn't anything to worry about.

Then we had the following conversation...

CNM:     Was this a planned pregnancy?
Me:      (ready to laugh and joke about our spontaneity)
         Well, ha, not exactly, but...
CNM:     What birth control were you using?
Me:      Uh, condoms.
CNM:     Did the condom break?
Me:      Uh, no... We were not using one.
CNM:     Well, why not?! Surely you knew what could happen!

I had no response. I was shocked, I think. I just looked at her. I was wondering what about me had given her the impression that I was so young and ignorant, or unworthy of respect. I had come straight from work, was dressed and groomed respectably, she knew I was married. I had no idea.

> I was wondering what about me had given her the impression
> that I was so young and ignorant, or unworthy of respect

She told me that she had something that could make the pregnancy "more real for people like you". I thought: People Like Me? Exactly what category was that? Middle Class Working Married People in Their Mid-Twenties Who Dare to Be Happy When They are Pregnant? I had a feeling instead I had been put in the Irresponsible Idiots category, which was distinctly shameful. Suddenly, I was 6 years old, caught drawing on the walls.

She wheeled a machine on a little cart into the room and told me to lie down. I was uneasy and didn't like the situation at all anymore, but I was also not going to stand up for myself. I had to be a good patient. I was not a troublemaker. The machine was a brand new portable ultrasound, which she had been "wanting to try out". So I was instructed to look at the little screen, to see the baby, to make it 'real' for me.

She didn't know anything about me, hadn't asked me anything about myself or my feelings, yet she seemed certain that I was in denial and this experience wasn't sinking in for me. Looking at the nondescript image on a small greyish-green screen, I thought that my changing body and swelling, tender uterus, my nausea and cravings, were certainly more real than anything she had shown me.

I then got a lecture about nutrition, even though she didn't seem even vaguely interested in my personal diet or knowledge of the topic, and was sent on my way with an order to make my next appointment at the front desk on my way out. The whole thing took about 20 minutes and this was my 'long, personal, initial interview'.

I was on the verge of tears all the way back to work. I was suddenly not so excited about this pregnancy and felt ashamed of myself both for letting her use ultrasound on me and my baby for no good reason (even though I didn't know at the time that there were any risks associated with it) and for being so 'irresponsible' as to get pregnant without 'trying'. I tried to remember why I had been so cheerful on my way in.

*How can you captivate your clients, while also informing and inspiring them?*

## Birthframe 34: 'Jackie'—Nocebo effect with one comment

While I was researching this book, I came across various cases where a disturbance as apparently minor as talking appeared to have disturbed the natural cascade of hormones throughout pregnancy, labor and birth. It seems to me that the main protocol caregivers should follow should be: "Do not disturb the magic and wonder of birth!" If this seems extreme, please at least consider again the connection between emotions and hormonal production. If we make pregnant or laboring women fearful, what's the outcome likely to be in terms of their ability to produce oxytocin, the hormone of love, which cannot usually be produced while adrenaline is present? Consider the following cases:

### Background:

Age 33, 5'4". Primipara, no health problems

### Situation:

She's attending a 'routine' ultrasound at 20 weeks. She returns from the ultrasound to speak to the obstetrician...

Obstetrician: What's your shoe size?

Jackie:         Er... 8½ . Why?

Obstetrician: You're going to have a section.

> "You're going to have a section."

### Outcome:

Jackie doesn't respond. She feels distressed and confused, initially because she doesn't know what a 'section' is—although it sounds bad (associations with being 'sectioned', etc. meaning being psychologically ill)... and later, when she's found out what it means, because she's dreading the idea of having a cesarean section, and feeling afraid.

Jackie never sees that obstetrician again. She cries every day for the rest of her pregnancy. "I was in a real state. Ask my sister." She goes two weeks overdue. She then goes into labor spontaneously. After laboring unproductively for over 24 hours she (predictably perhaps?) has an emergency in-labor cesarean. She *believed* the obstetrician. She blames her bad outcome on the prediction. "That's why I didn't have any more children," she says. "It was such an awful experience." (Note: I chanced upon this case study because I was in a shop with this woman, who kept on admiring my own three children, but repeatedly saying: "I've only had one baby and I definitely won't be having any more". Initially, I felt sure there must be some other 'ordinary' reason—finance, husband's view, etc—but she assured me it was because of this obstetrician's comment and the fact that she *therefore* had a cesarean—in her opinion.)

## Birthframe 35: Elise Hansen—Caregiver inspiring confidence

In the following interview, Elise Hansen—a midwife, working in Oregon—explains how and why she had a healthy birth. Could it be that the caregiver's words served to reassure the mom-to-be and help facilitate a positive outcome? Perhaps, given the fact that adrenaline (the hormone associated with fear) is antagonistic towards oxytocin—i.e. it stops it from being produced!—it's not such an extreme idea. In this case, no adrenaline production was stimulated!

**Elise, I understand you gave birth to your baby vaginally, even though she was breech. Why? Why not simply have a C-section?**

I gave birth to my daughter in 1977. At the time I was not a midwife. The only thing I knew about breech birth was that it could be difficult, but I didn't know anything more specific than that. I was living in France at the time and it never even OCCURRED to me to have a C-section. I had had two easy and fast vaginal births before that. (When I returned to the States, I then found out that she would have been an automatic C-section here.) I had searched out a local doctor who specialized in 'Leboyer births' and who attended at a very small clinic. Once we knew that my daughter was breech, I asked this doctor about the difficulty associated with a breech birth, but he responded that there really wasn't any difficulty at all, just that it was 'a little more complicated'. His matter-of-fact, no-fear response just reinforced my trust in my ability to birth easily.

**When did you find out your baby was breech? What kind of breech position was your baby in?**

I think she was breech from the very beginning. I don't remember her ever being vertex. She was a footling, but I'm not sure we knew that part before the birth. If the doctor knew, he didn't mention it.

**Did you try to turn her at all?**

No, the issue never came up. If there was no difficulty associated with a breech birth, why bother changing it?

**Did you have any trouble finding someone to attend the birth?**

No, as I mentioned before, I had chosen the caregiver because he was a 'Leboyer' doctor; the breech position seemed incidental to him (and, consequently, to me).

**And how did the birth go?**

GREAT! My husband woke me up in the middle of the night wondering if I was okay because I guess I was moving around in the bed. I woke up but didn't notice anything unusual. Got up to pee and my water broke. Went and took a shower, then we woke up my 4-year-old daughter and got into the car. We had to stop by the post office to use the phone (we had no phone in our house) to call a friend to meet us at the clinic to take care of the 4-year-old. By the time my husband got back into the car, I could no longer sit down, so I grabbed onto the cross bars on the inside of the roof of our car (an old 2CV with a peel-back roof) and just hung from them for the 10-minute ride.

By the time we arrived at the clinic, I could feel my daughter's foot at the entrance of my vagina. However, I had to walk up two flights of stairs to get to the birthing unit. During the whole walk up with a foot dangling out under my dress, I had images in my head of my baby bouncing down the stairs behind me! It was kind of funny, really, but that was the only time I was a little nervous, wondering if I would be able to catch her. As soon as I hopped up onto the exam table, the resident midwife lifted my dress and shouted in a very surprised voice (in French, of course), "But, Madame, you didn't tell me this baby was breech!" I replied, "But, Madame, you didn't ask!" By then, she just had time to grab a towel and hang onto my daughter as she came out in one quick contraction. About 45 minutes had passed since my husband had shaken me awake.

### Would you do the same again?

You bet! (Though, knowing what I do now about breech birth in general, I would probably have a little more fear.) But I would have a midwife AND a home birth! And probably STILL do it outside this country.

### Do you have any advice for women whose baby is presenting breech?

Just that they continue to have faith and trust that their bodies know how to give birth. There is nothing abnormal about a breech birth. Those babies just choose to come into this world walking. Find a caregiver who believes in healthy birth and who honors every woman's (and baby's) choices and who has experience, but not fear.

## Birthframe 36: 'Sarah'—Mother not understanding high risk

I met the woman who is the focus of the following case study just after I'd done a training course as a doula. (Childcare issues and even financial difficulties have always prevented me from actually working as a doula, or indeed training as a midwife—so you're stuck with me as a writer and researcher, I'm afraid!) Anyway, having been asked to visit this woman as a doula (while I was in another part of the country) I found myself in a very difficult situation. This woman clearly hadn't received some crucial messages about her risk situation and as a doula I was not supposed to either have any authority or give her any advice. Nevertheless, I could see her life and that of her baby were in danger...

### Background:

Primip. 38 weeks pregnant. Straightforward pregnancy.

### Situation:

At 38 weeks, she experiences sudden, continuous, severe abdominal pain and continuous bleeding for the first time. Phones hospital midwives for advice.

Experiences severe abdominal pain and continuous bleeding for the first time at 38 weeks. Phones hospital for advice.

**Woman's report:**

"I told them I wanted to come in. They told me it was just early labor—that I should just have a bath and relax. They said it'd be a long while yet."

**Outcome:**

Sarah does as instructed. A lot of blood in the bath. Severe abdominal pain continues. Husband is so stressed out—"This can't be right!"—that he just calls 911 and an ambulance takes Sarah to the Emergency Department. No fetal heartbeat on arrival. Sarah has an emergency cesarean and the baby is stillborn. (It was a case of placental abruption.) She is pregnant again a year later and is given the option of: a) a cesarean at 36 weeks (with no information about risks for premature babies), or b) an induction at 38 weeks —ignoring the considerably increased risk of uterine rupture after the use of synthetic oxytocics?

Most importantly, and this is where my sense of panic came in, Sarah does not realize that a repeat occurrence of bleeding might endanger her own life and that she must get to *any* hospital not just the one she's registered at ASAP. I explain it to her. I'm not sure if she believes me because I'm not her midwife. She continues to argue that she's not registered at other hospitals, so if she's out shopping in the local town (which I know), she insists she wouldn't be allowed to go there. I return home worried and pessimistic about her upcoming experience. I'm afraid I don't know what happened in the end because I don't live nearby and didn't feel I could make a nuisance of myself any more, after Sarah's initial response to my comments. I would have to trust that this time Sarah would get the message about the risk to herself, as well as her baby, at one of her prenatal check-ups. Bizarrely (given the history), Sarah's husband (who was present when I met Sarah) is also in denial that Sarah's life is in danger if she should start bleeding again.

I hope these glimpses into other people's worlds help you to realize how important communication may be prenatally and intrapartum. (We'll consider some more cases soon.) I believe that carelessness in communication can have a dramatic impact on outcomes, either physical or psychological.

## Birthframe 37: Pat T from the UK—Nobody listening

In case all this seems a load of nonsense, please consider the next account... Here, communication problems are the main focus. The cesarean that was eventually performed may have been much less traumatic if this woman's caregivers—not to mention her husband—had been more supportive of her wishes. At the very least, good preparation, teamed with good communication, including full information and honesty, whether for a physiological birth or for the possibility of an in-labor cesarean, is always likely to ease the emotional pain women experience when things don't work out as they hope.

I've spent my whole life listening to other people. First as a dutiful daughter and attentive student. Later as a researcher, journalist and psychotherapist. Later still, as the wife of a man whose communication skills left something to be desired, and the mother of an articulate and intellectually precocious toddler. Listening is a skill. You can get it down to a fine art if you practice. You can learn to listen to words as well as beyond words. You can even 'listen' to bodies and facial expressions. Listening. It's not the simple, passive activity of the powerless which it is made out to be.

What's all this got to do with having babies? More than I might have realized five years ago. I've told my story before—in the fluid, chatty atmosphere of groups of women and in the stark, irretrievable black and white of print. The birth of my son and the events surrounding it are burnt into my brain. They have made me who I am and yet, as I grow and change, these events, or rather the way I perceive them, shifts. Events and emotions which seemed so prominent in my early experience of motherhood move in and out of focus. Same events, different perspective.

So how do I tell it this time? What feels important? It dawned on me early one morning when my son had awakened me for the second time to adjust his covers. Hot with adrenaline, brittly awake and wondering why I ever wanted a child in the first place, my mind began to spin. Among the most important aspects of pregnancy for me had been negotiating the gap between listening and speaking up, and weathering the soul-destroying impact of not being heard.

> Among the most important aspects of pregnancy for me had been negotiating the gap between listening and speaking up, and weathering the soul-destroying impact of not being heard.

As a good listener, pregnancy was the single most startling experience of my life. It was when I first discovered I had something to say. I read all the books (another way of listening) by Sheila Kitzinger and Janet Balaskas and was grateful for their encouragement to 'speak up'. The first person I spoke to was my husband.

**"I'm pregnant."**

Actually, I said it through tears. At the time I didn't understand why I was crying, but I do now. The implication of those two little words, the effect it would have on my life, was simply enormous. Jake's response was silence, then ambivalence bordering on disinterest. It lasted for months. He told none of his friends, nor did he tell his family. I told the people I knew and loved and met their individual responses, ranging from 'How wonderful' to, curiously, 'Are you going to keep it?' as best I could. I'd heard of mothers not acknowledging their pregnancies until some prenatal test or other confirmed the baby was 'perfect', but even after he saw proof of his perfect creation on the ultrasound screen, Jake remained silent, and seemed resentful and defended against the mention of babies or baby books. When I was nearly five months pregnant, I practically had to threaten to leave him before he would telephone his family to tell them the news.

**"I'd like to give birth at home."**

There was the now familiar silence from my husband. My doctor responded though: "Sorry, we don't do those. Here's a list of hospitals in the area for you to consider." Did she hear me? I walked away, clutching the list, utterly dejected. I hadn't just requested a home birth, I'd told this woman something about myself. Seven words this time, but again the message was enormous. Seven words which also said: I care about my health and my baby's health; I will not simply be directed into the system. (I'm not even sure if I trust the 'system'.) I want to give birth in familiar surroundings with few interventions. I am an individual... Didn't this deserve some respect?

The community midwife (whose brief is, after all, to support women in their choices) didn't hear me either. She dropped in on me one day and said she didn't like doing home births (not encouraging news), especially with first-time mothers. "It's a matter of obstetrics, dear. You see?" What on earth did that mean? Was the word 'obstetrics' meant to frighten me? Actually, it did, but not for the reasons she probably hoped. Obstetrics is a medical speciality—the care of women with complicated pregnancies. I didn't fall into that category and it certainly was scary to imagine what kind of care I might receive at the hands of a man (probably) whose job it is to treat unhealthy women. I was healthy. My baby was healthy. Too bad the midwife wasn't. Today, I can chuckle at her parting shot. She apologized for coming to my home alone and said, astonishingly: "I usually bring another midwife with me. That way, if there is a dispute over what's been said, I have someone to back me up. Moms-to-be don't always *listen* very well."

Actually, I think moms-to-be listen too well, and usually to the wrong people. In order of importance I recommend moms-to-be should listen to the following people: 1) themselves and 2) their babies. Everything else is just noise. Tales told by idiots. Sound and fury signifying nothing, or, if you prefer, as one modern author calls it, 'Birthcrap'.

Somewhere around three months I stopped listening and began learning how to speak. It was a painful process for a lifelong listener. 'I-want-a-home-birth' took up a lot of my time, although the process was helped along by transfering to a helpful doctor for the duration of my pregnancy.

My mother's response was silence, uncharacteristic for her, but she quickly reverted to type: "Are you doing this because you can't afford proper care?" She's an American (while I'm 'British'), so her bottom line is always money. That the American medical system regards midwifery and home birth as some kind of second-rate compromise is both sad and shocking.

Then there was the classic parental question: "Are you going to get married?" I told her that focusing on parenthood seemed a daunting enough job for the moment. I just wanted to put my energy into that, and not into plans for a wedding. She began to make the plans for me (or herself?) anyway—without my knowledge or consent. At 32 years of age, this was a 'favor' I could do without.

"When will you be getting married?"

"I don't want to get married."

"Summery weddings are always nice."

"I don't want to get married."

"But I've made plans."

**"I DO NOT WANT TO GET MARRIED!"**

I tried to shrug it off, but the truth is my unexpected pregnancy, along with learning to speak up for myself, caused a rift between myself and my mother which has never really healed. She has never acknowledged her status as grandmother (a role too frightening for her to contemplate) and never seen her grandson, and we are all the poorer for it. As for getting married, well, let's just say I should have listened to myself a bit more carefully.

> I should have listened to myself a bit more carefully...

Out went the happy fantasy of sharing this time with my mother. We had less in common than at any other time in our lives. She hated pregnancy and birth and resented the demands of motherhood. I found it, and still do, exhilarating, challenging and surprisingly fulfilling. She feels more secure when things remain stationary. I try to stay open to the cyclical turning of Life's wheel, even when it's taking me in a direction I'm unsure of. Happily, there were enough 'mother substitutes' in the world of pregnancy, childbirth and mothering to sustain me. They appeared in the obvious forms of Kitzinger and Balaskas, as well as in the women at local maternity organizations and, curiously, even in some of the men I met: Michel Odent, Yehudi Gordon and Norman Stannard. Okay. So maybe I was privileged. More importantly, years of journalistic training had taught me how to seek out the information and support I needed—but, like listening, it's a skill anybody can acquire.

My healthy baby and I ambled along happily enough through all this. Whatever was going on outside me, inside I was content and enjoying myself. I confess I loved being pregnant. The night I first felt Alexander move, I was co-facilitating a women's dream group and I was so knocked-out that I had to withdraw quietly from the proceedings. I enjoyed eating what I liked, when I liked. As women we're taught from an early age to deny and control our bodies, but I found the inevitability of pregnancy very reassuring. Life was growing inside me and there was nothing I could do to stop it. This was a creative act that would reach its own conclusion without any conscious act of will on my part.

There were also breaks in my increasingly stormy relationship with my partner. I can recall lazy afternoons when Jake would read *Just So Stories* to my ever-increasing baby belly. I genuinely enjoyed our trip up to the Cerne Abbas Giant (an ancient fertility symbol carved into a hillside—you can see his penis for miles!) when I was eight months pregnant. Jake was in his element and laughed as he regaled passers by. "It works!" he shouted to their curious smiles. In these brief moments I enjoyed a much-needed taste of 'happy families'.

Hardened feminists should cover their eyes at this point because, social connotations aside, I also loved being barefoot. In fact, being barefoot and pregnant was quite an important part of my process. Both feet firmly on the ground. Finding my feet. After years of wearing high heels, they were quite a revelation. Moving slowly. Finally coming to terms with the fact that my natural rhythm, physically, mentally and emotionally, is quite slow was also a difficult, but valuable experience.

As it turned out, I needed both feet firmly on the ground because at 32 weeks my baby turned into the breech position. Although the community midwives said there was nothing to worry about, they quietly set about sabotaging my plans for birth at home and transfered me, without my knowledge, to care under an obstetrician. I only found out about it when I showed up at the clinic for a 34-week check-up and was greeted with: "What are you doing here?! You're not supposed to be here!" Anyone would think I was carrying an infectious disease rather than a baby.

As the breech persisted, I had a harder time making people hear me.

**"I want a trial of labor."**

"Of course you can have a trial of labor—in the hospital."

**"No. At home."**

One day, some of the midwives at the local clinic hustled me into a side room. To me, they were suddenly like menacing witches. "You'll never be allowed to have a home birth now," cautioned one. "You're lucky to have been allowed a home birth at all," and then adding insult to injury—"at your age," snided another. "I'm afraid you'll have to do what the doctor says," remarked the third. I left the 'meeting' in tears. I few years later, I was surprised to see this occasion recorded in my case notes as 'Patricia has been reassured...' Reassured?! Were they serious or just plain stupid? My call to the hospital a few days later to check the policy on breech deliveries was recorded in a similarly snotty and grudging manner: '[Midwives] apparently told her *(so client says)* that she has to have an elective cesarean section.' I recommend every mother get hold of her case notes and read them. They can be rollicking good fiction. Sadly, they also show how inept most healthcare practitioners are at listening.

> "Frankly, if you don't consent to have an elective cesarean next week, I will withdraw my care. Then where will you be?" Better off, I thought, but said nothing and left in tears.

At 37 weeks my obstetrician said, "Frankly, if you don't consent to have an elective cesarean next week, I will withdraw my care. Then where will you be?" "Better off," I thought, but said nothing and once again left in tears. Let's be clear. I didn't want his care in the first place. It was foisted on me by midwives so unskilled in almost every department that it was a toss-up as to who would provide the most inappropriate care, them or the obstetrician. Also, how truly 'elective' would an elective cesarean be, given the amount of emotional blackmail involved? True to his word, the obstetrician withdrew his care.

"I want a trial of labor. At home."

I said it so many times, occasionally very loudly, that eventually someone heard. First, it was my husband, whose involvement thus far in my plans for a home birth had been to speak to a woman he knew, who had a friend, whose sister's cousin's wife had a home birth and her baby needed to be transfered to hospital care (or something equally convoluted). It was never revealed why, but he used this as irrefutable proof that our baby would be born dead or damaged. (Years later, when we divorced, he would still be claiming that I tried to kill our baby by wanting to have him at home.) However, one day, unexpectedly, touched by my increasing distress, he phoned the NCT [the British National Childbirth Trust] for advice, and they gave him the number of an independent midwifery practice with experience of breech deliveries at home. It was the kindest thing he ever did for me and I'll always be grateful. Having engaged these two powerful women, I found that things began to fall into place.

They read my birth plan—the first ones to do so, as far as I knew. They talked to my doctor, the hospital and my husband on my behalf. By 38 weeks, I felt a great weight had been lifted from my shoulders. They listened to me as I recounted the difficulties I'd had being heard and carried the burden for me. They were there on the other end of the phone when, 10 days before the birth, my father died and I wasn't sure what to do. Should I go ahead and grieve his loss now, or file it away and get on with the equally demanding business of birth? Which of these legitimate impulses do I follow?

In fact, my father's death crystallized certain things for me. Pregnancy, and now the impending birth of my son, had hurled me into the unknown. First there was this speaking-up stuff. Then there was the matter of learning to set boundaries and prioritize. All of a sudden, it wasn't so simple and I didn't have much of a role model to refer to. My own mother always believed that as long as she was happy and fulfilled—'doing her own thing' it was called when I was growing up—I would also be happy, but the feminism of the day played a cruel trick on us both. Neglecting the needs and experiences of mothers—90% of women, after all—has left a yawning chasm in our psyches. If this is what women are built to do and we deny the process, where does that leave us? My mother's philosophy also failed to acknowledge that a mother needs to be happy in the same room as her child—or at least the same metaphorical space—or it doesn't work. Mom was happy doing her own thing, but she was never at home.

For myself, I noticed very early on how many things I had to negotiate now, just to make a single choice, and how often my needs seemed to conflict with my baby's. By this time it was clear my son had 'chosen', by whatever unfathomable, primal process, to be a breech baby. Maybe he misheard all those Beach Boys records I listened to as a girl in California (no, no, darling boy, they were singing about *beach* babies, not *breech* babies). Whatever it was, his choice was in direct conflict with my need to have a straightforward, non-interventionist birth at home. How do we resolve this one? "Our children don't always do what we want them to," said the very wise doctor who, three times, tried and failed to turn him. Amen.

Our children don't always do what we want them to do

Maybe it was a gift to learn this so early. I noticed this conflict elsewhere too. For instance, prenatally: some of the procedures (CVS, amniocentesis and possibly ultrasound) which are used to detect abnormality and thus reassure mothers (and, I suspect, doctors) that their babies are perfect, can end up damaging babies. Another example: the more effective the pharmacological method of pain relief, the greater the risk of harmful side-effects for the baby. Some mothers can successfully shut their minds to the conflicts but we listeners aren't built that way.

Labor was another surprise. I knew it was near because I'd had Braxton Hicks-like contractions and a niggling lower backache for a week. I'd lost my interest in sex and spicy food—both potent labor inducers—a couple of weeks before, but one day, impatient to move things along, I found myself masturbating on and off all day. It must have done the trick. That night, unable to sleep (again) I lay in bed trying to figure out whether I should go to the bathroom (again) or whether I could hold on and stay horizontal (my then favorite position) for just a little longer. I decided to wait. Ooops! Suddenly there was a pop and a dramatic gush of water.

**"Jake, wake up! I need a towel."**

He didn't hear me the first couple of times and I don't think he quite comprehended what was happening. He stumbled to the bathroom, muttering about being in the middle of a dream ("Eternally," I thought to myself) and returned with the towel.

This was it. My contractions established themselves soon afterwards and were enough to take my breath away by the 12-hour mark. The TENS machine broke down. "Never mind. It was useless anyway." To my husband this was irrelevant. We'd paid for the damn thing, after all. So, at 10 o'clock at night, he engaged himself in one of those pointless, heroic struggles to get the company which supplied it to send us another one in a taxi. (They did as well!) "It doesn't matter," I said. But my midwife, who had guessed the score, winked at me. "It'll keep him busy." "Hallelujah," I thought.

Later I wanted a bath. I needed to be in the water and feel some heat on my back. I called for my husband for help.

**"Could you squeeze the water over the small of my back?"**

He began to rub the sponge on my back. It felt excruciating, like broken glass on my sensitive skin.

"No. Just hold it above my back and squeeze the water out."

He continued to rub. "Like that?"

**"No. Don't touch me."**

"What do you mean?" I can't describe how much I *did not* want to be having this conversation, so typical of our communication difficulties, with him at that moment.

So typical of our communication difficulties...

That's where the account ends... but, as I mentioned earlier, the real ending was an emergency cesarean. Could things have been better in any way?

## Birthframe 38: Karen Low—Not being respected or heard

The next account is particularly interesting because of the dissonance between the mother-to-be and the healthcare professionals, particularly since the mother's focus was on adopting a research-based approach and following official guidelines. The problems of bringing protocols in line with research and guidelines are obvious to anyone working within a large maternity hospital, but you can perhaps nevertheless relate to this woman's frustration that the care she was being offered did not appear to be in line with any official recommendations, particularly since her own past experience of labor was also completely disregarded. The fact that this mom-to-be met with such great resistance to her idea of having a homebirth, despite a good medical history, is also worthy of discussion. Why are many midwives so focused on negatives, instead of on building the confidence of women who want a physiological labor and birth for the sake of their unborn babies? Finally, in the light of this account, it's interesting to consider the issues of monitoring—how and why it should take place, the availability of pain relief (or not) and the possible value of training mothers-to-be in basic emergency procedures.

After two successful home births, one water and one dry land I had planned another home water birth for No. 3. I found things had changed, but not necessarily for the better, during the 12 years since I'd had my last home birth and 19 years since the first one! I hadn't particularly seen eye to eye with the head midwife at my local hospital, as she had initially refused to "book me in" for a home birth until I was 36 weeks! A letter to the head of midwives had sorted that out, but she continued to be generally negative and hell bent on pointing out all the possible reasons why I wouldn't end up having this baby at home. This wasn't helped by the fact that apart from a very early ultrasound I had declined any others as well as the triple blood test. Although I had no reason to think my membranes would rupture prematurely, I do remember being warned during one of my prenatal appointments that this would result in fairly immediate induction. I didn't think it was worth arguing this one as I felt it was an unlikely possibility.

I had had another healthy pregnancy and although I was now at the grand old age of 41 I felt well and confident. I'd been practicing positive thinking pain control methods using 'hypnobirthing' and planned another low-key labor at home with all the family present, as well as a couple of close women friends to support me and keep our daughters company, if need be.

On Thursday, January 10 at 3:00AM my water broke with a pop and a gush. I got up and had a hot drink, expecting labor to be imminent—but nothing further, apart from continuing heavy water leakage for a couple of days, after which it subsided to a trickle. A midwife phoned in the evening to cancel my imminent appointment, which was lucky! She called again the following day to make another appointment for the same day, but husband Al managed to put her off until Monday. I felt confident that things would move along soon and knew that I would face pressure to be induced if I said my water had broken. Friday was uneventful, but I was feeling more concerned that nothing was happening and was starting to worry.

The next day (Saturday), I had a show and the head was now feeling very low, making walking uncomfortable at times. I was walking up lots of hills trying to get things moving and eating lots of pineapple! We looked up the official guidelines regarding spontaneous rupture of membranes and found they recommend offering intervention after 96 hours as opposed to our local 18-hour hospital policy for induction. We decided not to contact the community midwives as we were aware that pressure would start to have an induction, but took my temperature every four hours as a raised temperature is a possible indicator of infection. I felt that the possibility of infection was less at home than in hospital and the risks of induction were greater than those of 'watchful waiting' as long as I felt well and happy and was continuing to feel lots of movement.

I woke up at 2AM on Monday, January 14, at last feeling a first mild contraction and lots more wriggling! After a couple of hours I woke Al and announced that pains were about 25 minutes apart and of varying strength. Al made us some hot drinks and we watched the end of a movie that we'd started watching the night before. (I can't remember what it was about!) At 6AM we called one of my best friends to announce the good news and told her we'd call again when things had hotted up a bit. We showered and began to fill the birth pool in our bedroom. At around 9AM I woke my younger daughter who was very excited. She gave me a lovely pedicure and amused me with some board games for awhile. Contractions continued but didn't get much closer or much stronger and were very manageable. Around midday we thought we'd better let the midwives know and I called the number on my notes, only to find that the number was wrong (good job it wasn't an emergency, I thought). Eventually we spoke to a member of staff who advised that a midwife would call us back.

Half an hour later my least favorite midwife phoned. She asked about the speed and intensity of the contractions and if I had lost any water. I lied and said that water had broken on Saturday morning but that I had monitored my temperature and said we believed the membranes had somehow sealed up again. I was told: "This changes everything!" Apparently, I was no longer 'normal' because the water had broken over 14-18 hours ago (little did she know). She said I should have been induced the day before (Sunday). I was also told off for not letting them know as soon as my water had broken. She then said: "You will not like what I have to say... You would have been induced at the hospital yesterday if you had reported water breaking." I politely told her that I would have refused her 'offer' but felt livid to be spoken to in that way. I was then informed that I was not in labor and that my cervix was just 'irritable' because contractions were sporadic. I'd already told her this was my modus operandi during my labors and that there were other signs of labor too—but she wasn't interested. She suggested we make a note of times of contractions and strength, which we then started to do. She then said that she would need to speak to her supervisor and would call me back.

I was so angry my contractions stopped completely for awhile. Al tried to call the Head of Midwifery but she was constantly engaged (probably speaking to the midwife). Eventually, he got through to someone else (the Head of Midwifery's assistant) and he explained to her the problem we'd had with the midwife. The Head of Midwifery had apparently not received our letter posted the previous week, informing her that this particular midwife was not welcome at our birth. She was very apologetic and helpful and stated that my tormentor was in fact on her way to our house but that she would stop her. (She wouldn't have got through the door anyway!) She also said that our named and prefered midwife (Linda) would arrive a little later on.

Al then received a call from our prefered midwife, but she was a little frosty and had clearly been briefed. She stated she was on another job and would get to us at some stage. Al didn't tell me that she had a little bit of attitude as he didn't want the contractions to stop again and for me to get upset.

At approx. 15:45 Linda arrived. Husband supplied her with a list of timings and information on the duration of contractions, which were decreasing in the time apart and increasing in intensity. The last four contractions had been five to six minutes apart and each one had lasted approximately 90 seconds to two minutes. Al and I both suggested they believe I was really in labor, with Al guessing at the actual birth time to be around 18:30 based upon his knowledge and experience of our two previous births. Linda wrote up her notes but didn't appear to believe the birth was imminent, stating that she would be very surprised if Al was correct. I said I would appreciate some gas and air and she assured me that it would arrive with the midwife who would soon attend me. Linda said she was going off duty around 17:00 and would be leaving us at 16:30, but reassured us she would make arrangements for another midwife to attend and would also call back within the hour to inform us of when she would arrive. She also said that if we had any issues we should call the hospital. When she left I decided I would definitely benefit from the pain numbing comfort of the birthing pool and both Al and I climbed in—he with a beer and me with a nutritious fruit cocktail! It wasn't long, however, before any relief I originally felt was superceded by increasingly painful contractions and I really was looking forward to some gas and air to help me cope.

At 17:15 one of my girlfriend birth attendants called the hospital to tell them that I wanted a midwife as soon as possible, as we all believed the labor had now progressed to an advanced stage. Al also called the hospital when there was no news half an hour later to ask when the midwife was due and also to ensure they would have Entonox with them when they arrived. He was told that a midwife was on her way to us and would be with us within 40 minutes, so she was expected before 6:00PM. Linda called back asking how things were going. Al told her we'd contacted the hospital as per her instruction and requested Entonox. Linda said she would return almost immediately with Entonox— but unfortunately this was the last contact we had with Linda. At 6:10PM Al contacted the hospital again, requesting information and stating that the labor was now well advanced—and asking: "Where is the midwife?" He was then told that two midwives had been dispatched: the midwife we already knew about, and another who had to go to the hospital to pick up the Entonox. Expected Time of Arrival for the first was 30 minutes later and the second was due after 40 minutes. Husband was furious but also anxious not to convey this to me (as I was listening). This explains why he didn't press for the reasons for the delay.

At 6:45PM my 18-year-old daughter, Cara, heatedly called the hospital and told them the birth was very imminent and where were the midwives?! The response was that they were on their way. I realized at this point that we were going to be going it alone, but felt quietly confident and in safe hands. I felt like I wanted to get out of the pool and onto the toilet after feeling the overwhelming urge to push. After sitting for awhile on the toilet I changed to an all-fours position in front of it. My female birth assistants were invited in to help and were soon followed by both our daughters, Cara and Christy.

At 6:55PM, after several huge pushes, Hubby said he could just see a head crowning, then gradually eyes, a nose and a mouth. He was very encouraging and told me exactly what was happening second by second. After another contraction a shoulder emerged, quickly followed by the rest of body. I was absolutely exhausted and looked down between my legs to see a clear full set of male genitalia and not much else! My friend held him while she was sitting on the toilet behind me and I tried to get my breath back! Our son was quiet and slightly grey, although obviously quite adorable. I realized with hindsight that this gave my 'assistants' a bit of a shock as they weren't sure he was breathing, or what to do if he wasn't. Al and my friend attended to preparations for cutting the cord, which was completed by my younger daughter, Christy. Between them they successfully clamped the maternal end but were unable to secure the baby end, so left it secured with scissors. I decided to sit back in the pool as I felt unable to sit comfortably.... I felt as if I had been turned inside out, down below! I nursed our son sitting there for the first time. He seemed a more normal color now and was very alert and interested in everything.

At 7:30 we received a telephone call from the midwife, who was looking for the house. Hubby went outside with a flashlight and waved her in, closely followed by the other midwife. They had a quick look to make sure everything was fine and clamped the other end of the cord, which just needed a bit more force applied! Christy assisted in weighing the new arrival and he was a healthy 7lb 4oz. Both midwives apologized for the delay but said they hadn't received a call until 6:05PM. Ironically, both had been available prior to receiving the calls. Finally, at 8:30PM the third stage completed naturally on dry land and Christy had the workings explained to her, and was also able to handle the placenta. I even managed to take some photos of it! Both midwives were very friendly and supportive and commented on how well we'd all done, and of course how gorgeous Josh was!

Both friends who were present at the birth suffered a bit of post-traumatic shock following the birth and it was only later that I fully appreciated the extremely stressful position this put them in, as well as my wonderful husband, of course. Christy also admitted that she had never felt so frightened and was amazed that she had continued to be capable of standing as she had been shaking so much!

Thankfully, we were both okay, but it could have very easily gone badly wrong if the circumstances of my first birth had repeated themselves and I had hemorrhaged. Although we would have been well prepared for this practically (we had syntometrine) we didn't have anyone to administer it. My husband never tires of telling people how he finally got to fulfill his lifelong dream of uttering the immortal line 'Clamp, please!' before cutting the cord. I don't regret deciding for our third (and last) birth to have the baby at home. I trust that after our complaint and the subsequent investigation, the series of errors resulting in this situation will be impossible for any more home births.

*Just after the birth, before the midwives arrived, in the pool with daughter Christy...*
*The scissors can just about be seen, still attached to the umbilical cord.*

*A very recent photo of the three Low children, aged almost 2, 20 and 13.*
*(The woman on the far left was Karen's birthing partner when Josh was born.)*

## Good or bad listening

Listening is an enormous topic and whole courses have been designed to train people to listen better. Based on my own experience and training, I would suggest the following:

1 On an ongoing basis, be aware of how much you are listening in proportion to the amount you are talking.

2 Consider whether your responses to what moms-to-be say might really be satisfactory and helpful to the women themselves.

3 Encourage women to really talk openly and honestly by using what the Japanese call *aizuchi*. These are words (such as 'Yes', 'I see', 'Uhu') and body language (such as nods, eye contact and warm facial expressions) which make it clear to listeners that you want them to continue speaking.

4 Reflect back what you think you've heard before you respond to anything which seems potentially significant. You can do this by simply paraphrasing what you hear, either in simpler more direct words or—in cases where the topic may be sensitive—in a very vague way which invites further comment.

5 Ask open questions as well as closed questions. As you may know, a closed question is a question with one, specific answer (e.g. "What's your name?"), while an open question could prompt many answers or a whole monolog (e.g. "How do you feel about that?").

6 Make sure you always ask permission to 'do' anything to women which invades their personal space in any way. For example: "Can I ask you to lie down up here to be examined? I'd like to use an ultrasound Doppler to check the baby's heartbeat, if that's okay—or would you prefer me to use a fetoscope, which doesn't use ultrasound?")

7 Never talk about a woman to a colleague in front of her, as if she isn't there.

8 Check that women understand technical terms when you use them, if they come up or—if you're worried it might seem condescending at times—be open to your clients' reactions. (Yes, I'm calling them 'clients' not 'patients', and I hope you are too. If they're patients how on earth are they going to believe they're doing something healthy in being pregnant and giving birth? We are patients when we're ill.)

9 Always conclude any phase of a check-up with an open invitation for women to speak more. For example, you could ask: "Is there anything else you'd like to talk about?" Note that 'talk' is better than 'ask' because the latter implies you're the expert and the mom-to-be must defer to you at all times. While this may well be true, use of the word itself might stop women from asserting their own wishes for the birth. This is important because as well as monitoring safety, a midwife or other caregiver is supposed to be an advocate...

Always conclude with an open invitation to your client to speak

*Always remember the effect your words might have, positively or negatively*

## Words, words, words...

As you may know, my background is in language teaching—i.e. teaching English to professionals, business people or academics of other nationalities. Doing this prompted me to consider the ways in which we use words and how they might possibly affect outcomes. (I'm interested in all kinds of situations, not just birth. I was particularly fascinated by the power of words I had when I was in managerial positions—to either get people to do things or make them very angry!) Teaching a gynecologist one-on-one one year, was also particularly interesting and prompted a lot of thought, particularly about communication. As an enthusiastic language student (of French, German and Japanese, which I'm fluent in—if you catch me on a good day!—and a reluctant student of other languages, which I was forced to study a little just to survive more happily) I have also become aware of certain seemingly significant differences between certain languages. For all of these reasons I am going to make some comments about the language used for birth which—if nothing else—I hope will prompt you to think through your own use of language. Given that women are in such a fragile state emotionally, if not also potentially physically (since they are embarking on something life-threatening, i.e. childbirth) how you speak at any time in the whole process of childbirth is extremely important, in my opinion. So I'm going to focus here on words and expressions used antepartum, intrapartum and postpartum.

### Be aware how you speak to different women and adapt wisely

Be particularly aware of how your moms-to-be react to words you use about their anatomy. Try to copy the words they use themselves, unless they seem really derogatory (like 'cunt') because speaking as they do is likely to help them relax and feel at ease. Be aware when you do this that all words related to genitals tend to be associated with shame in one sense or another... So you may not always be listening to comfortable, confident women!

While doing this, also be careful not to 'talk down' or 'talk up' to women, depending on how you view them in terms of social class, education, etc. As you may realize, poorly educated women from disadvantaged backgrounds can be very smart and wise, while women at the other end of the social spectrum (wealthy and/or well-educated) can be remarkably naïve or silly at times! Since a caregiver is supposed to be an advocate for women, it is important to establish an atmosphere of egalitarian exchange (of information and thoughts), rather than a hierarchical, 'obedient' caregiver-patient relationship.

### Avoid language with negative imagery

Words and phrases which are particularly dangerous include the following:
- Birth canal: This suggests the baby is going to be attempting to exit through a concrete tunnel. Say instead, 'when the baby makes its way out through your body'—or something which evokes the idea of a possible process!
- Pain: Why assume the woman's going to experience pain? If you do make this assumption, she almost certainly will. Women are surprisingly obedient.
- Contractions: This word suggests the cervix is gradually getting smaller and smaller... Why not talk about 'openings' or expansions instead? Even give her some positive imagery (opening flowers, ripples on ponds, etc).
- Care: This word is astonishingly overused in the medical literature and suggests a woman's inherent inability to give birth without outside help. However strong our need to feel we're indispensible, the simple fact is that the vast majority of women, if left completely alone, would just give birth without any instruction or outside help—and both they and their babies would be safe. Monitoring is carried for the sake of the small minority of women who have problems. Let's not forget all those healthy bodies, which know not only how to grow a baby, but also how to get it born.
- Delivery: Again, I think it's preposterous that caregivers claim they are 'delivering' a baby (as if it were a FedEx package) when, in fact, a baby usually emerges all on its own—particularly when caregivers don't interfere.
- Letdown: The release of breastmilk from the breast is not a 'letdown' or disappointment at all... It's a delight for both mother and baby. Don't use this negative term ever, *please*... (How about 'milkflow reflex' instead?) Help women to celebrate, not feel disappointed by their efforts at mothering.

In the companion book for moms-to-be—*Preparing for a Healthy Birth* (Fresh Heart, 2010) language is used in a different way from in other books on pregnancy. Since the book is intended to provide reassurance and inspiration for moms-to-be, as well as lots of useful information on which to base decisions, I made an effort to only include positive imagery through words.[2]

## Create new terms

You too can join me in this drive for positivity, which could well help to optimize outcomes, by developing new ways of speaking to counteract negative mind-sets—whether your own, your colleagues' or your clients'! Why not, after all? Can you not see the advantage of replacing our usual arsenal of upsetting words and expressions, with positive ones, which create a more positive mind-set? After all, somebody created the more negative approach in the first place! And perhaps needless to say, you can make changes in both the way you speak and the way you write up notes, reports and academic articles...

For example, in this book, the term 'optimal birth' is a reminder that research has repeatedly confirmed that normal, physiological birth is best for both mother and baby. The upbeat term 'fizzy-logical' (which I use elsewhere), which I thought up to refer to 'physiological birth', is also an attempt to make something which initially sounds very boring and inaccessible easy to understand. (It's quick to explain and easy to remember.) As you may have noticed—hopefully not with a sense of outrage!—I have even redefined the word 'intervention' to mean literally any action which disturbs the processes of birth, including a one-word comment. I coined the terms 'birthframe' and 'care guide' because I felt it was necessary to avoid the phrases 'birth story' and 'birth plan' because of the negative associations both of these have acquired. The word 'birthframe' is obviously an extension of the phrase 'birth story' but replacing the word 'story' (which suggests that something is fiction, and therefore not to be taken very seriously), makes an account seem something that should be honored and taken note of. To me, the term 'care guide' is preferable to the phrase 'birth plan' because while some words go together, some just don't work alongside each other. If you don't believe me, consider how we say 'damp socks' and 'moist cake', even though 'damp' and 'moist' essentially mean the same thing (containing a certain amount of water). Like 'damp cake', the phrase 'birth plan' can provoke negative reactions—as I found with many midwives during each one of my pregnancies. When they retorted "But you just can't plan birth" they very effectively undermined all my hopes, preparations and even research. Since a pregnant woman really can constructively provide guidelines on the type of care preferences she has, the phrase 'care guide' seems very appropriate—and it even uses the word 'care' to positive effect!

So why not be creative in the same way? The first step is to be aware of how you're speaking whenever you speak to anyone in a professional capacity. (Of course, you could also transfer this idea to your personal life, if you really want to transform your relationships!) If at any time you notice a negative reaction in the person you're speaking to, consider how you could improve how you're speaking. It's a long process, of course, but one which is well worth engaging with. This is because—in my experience—changes can transform outcomes.

Finally, a warning... Avoid doublespeak because it can be a pernicious form of inaccuracy which changes people's perceptions. For example, why not call diamorphine 'heroin' (a word which every woman knows) and artificial oxytocin (pitocin) 'artificial oxytocin'—not 'oxytocin'? After all, real oxytocin seems to be very different in its effects and the artificial version of the hormone is certainly not the 'hormone of love'.

## The chance to develop even better relationships

As well as taking care over communication, you could actually aim much higher and work to establish a truly excellent emotional climate through a different approach entirely. (Hear me out, please... it might be fun!)

Since emotions and hormones are inextricably linked, your relationship with clients is perhaps the single most important factor in determining safe outcomes. Your clients will eventually need to be producing oxytocin in your presence (so as to stimulate or sustain contractions), rather than adrenaline, which would be produced in a stressful or frightening environment. How is this possible if their experiences of you are not terribly positive? (By the way, in case you're worried about triggering a mom-to-be's labor by getting her to produce oxytocin during her prenatal check-ups, let me hasten to add that you're aiming here to get her producing the amount someone might produce when having dinner with a friend—not the amount which is usual during an earth-shattering orgasm!)[3]

How is it possible to make a woman feel comfortable when a fair amount of your time during each prenatal check-up will be spent performing checks (auscultation, palpation, urine checks, etc), which carry the implicit message that there is a risk that something may well have gone wrong since the woman's last check-up? (The word 'check-up' also needs to be changed!)

### Some guidelines for constructive, collaborative communication...

Based on my experience of teaching adults in companies for around 20 years, here are a few tips...

- Ask women about situations or symptoms, don't *tell* them. In other words, never say, "I expect you're feeling..."
- Ask plenty of open questions, e.g. "How are you feeling?" "How have things been?" "Do you have any preferences?"
- Talk about positives, not just negatives, remembering that talking about risks or potential problems can cause fear.
- Never suggest negative outcomes. For example, don't say "If *you* get, *you* need..." but use distancing language instead. For example, you may need to say, "When *a woman* gets pre-eclampsia, *she* needs to be monitored closely in case..."
- Explain any technical terms you use, without being patronizing. Avoid using 'babyish' language, but consider slang with some clients, as I've suggested before, perhaps even to make them laugh at times.
- Never put a woman down for asking 'silly' questions. Positively encourage her, keeping a very serious expression on your face, remembering the steepness of the learning curve for a first-time mom.
- Always give a woman an opportunity to add another question...

Always give a woman an opportunity to add another question...

- Try and draw out any other concerns or questions, remembering that hesitation may mean the client disagrees with you!
- Explain everything you want to do before you do it, and as you're doing it too, giving the woman a chance to refuse.
- Respect a woman's right to choose her own care and encourage her to plan for a positive birthing experience.
- If at all possible, help women to leave each prenatal check-up on a positive note.

Here are some other, more general ideas, some of which might seem sensible to you, some weird. Some you may have already tried. Why not try out some more and think of others too?

- **Maternity fashion shows** You could have afternoon sessions (while women wait for their prenatal check-ups). Each fashion show could feature maternity wear. Both local and national companies would probably quite easily be persuaded to sponsor the shows.
- **Singalongs** Michel Odent used to have singing sessions with moms-to-be and their husbands every week. Why not try something similar? (It could even be a karaoke session involving songs from movies which your clients request. There could be a sign-up sheet for suggested songs or movies to sing along to or you could provide a selection.)
- **Partner picnics** These would be 'pot-luck' meals (everyone brings a dish of some kind—savory or sweet, with no forward planning) to take place at lunch—or dinner-time.
- **Foodie fetus lunches** Nutritious food, prepared according to a sign-up list, could be eaten before a teaching session.
- **Morning munch-time** I've suggested this 'term' rather than 'coffee mornings' for the simple reason that I believe pregnant women should avoid drinking too much coffee. However, a non-caffeine coffee morning would facilitate informal chatting between staff and moms-to-be.
- **Beautiful belly boasts** These would be workshops in which women could express and discuss any feelings they have about their changing self-image. Photos of trendy pregnant celebrities could be used as discussion starters.
- **Fear focus groups** No prizes for guessing that these could be groups in which women discuss and explore their fears.
- **World pregnancy workshops** These could be sessions in which women find out (from you) about birthing practices in other countries. Of course, the books *Ourselves as Mothers* (Doubleday 1992), *Ever Since Eve* (Oxford University Press 1984) and *Birth Traditions and Modern Pregnancy Care* (Element Books 1992) would provide plenty of material.
- **Local pregnancy groups** As the name suggests, these could be groups that look into the past—and future!—of a particular locality. Pregnant women could interview local mothers, talk to older people, meet up with a few local midwives and generally become more acquainted with their community.

### Birthframe 39: Jennifer Jacoby—The value of relationships

Finally, here's an account which shows the differences between various types of relationships between caregivers and clients. The mother who contributed this account says, "I feel that the mother-midwife relationship is one of the most important factors during labor and birth." From her account we can see why not only this relationship, but also others, can play a key part in optimality.

## Questions for reflection as you read...

1  Of all the caregivers you read about here, which are you most like?
2  How effective is each of the caregivers?
3  What is the mom-to-be's response to each type of communication?
4  What other approaches could caregivers have taken?
5  Do the caregivers in this situation seem to be working together?
6  How could the situation have been improved?
7  To what extent should communication in birth be down to 'luck' and to what extent should it be a question of training and protocol?

The birth of my first child felt very institutionalized and governed by hospital policy. The midwife was definitely in control of the entire situation with me as her charge. My second birth was a truly wonderful experience that was influenced by an extremely intuitive midwife who encouraged me to do what I felt to be right for me. These experiences have made me highly aware of the dynamics between the birthing mother and the midwife, whose main purpose is to provide support for the mother and her husband in labor, and who needs to understand both the physical processes and the emotional needs of the mother. I feel that this relationship is one of the most important factors during labor and birth.

> The midwife needs to provide support for the mother and her husband in labor and she also needs to understand both the physical processes and the emotional needs of the mother

In my second pregnancy, when I was fairly overdue, I decided to try acupuncture on Wednesday evening to get my labor started. That night I was woken at about 2:00AM with fairly mild contractions and they had completely fizzled out by 6:00AM. My husband took Thursday off work as we were hoping that something might happen. I had woken him up at 4:00AM to put the TENS machine on me, which I found to be more annoying than helpful. I had been convinced throughout my pregnancy that my labor would be very short and worried that my husband wouldn't make it home in time for the birth as we live in a long way out of London and he works at a hospital over the other side of the city. Of course, nothing happened that day and we went to bed on Thursday evening thoroughly disappointed.

The hospital staff phoned me all week, harassing me
because I was refusing to go and see an obstetrician

I had had irate hospital staff phoning me all week and harassing me because I was refusing to go and see an obstetrician to talk about induction. We had planned a home birth and had hired a birthing pool and I was really looking forward to it. I had started to feel somewhat despondent and felt that my whole labor and birthing experience was about to be taken over by hospital staff. I reluctantly agreed to go to the hospital on Friday morning to be monitored and see an obstetrician, as I didn't think anyone would try and induce me then and there and wouldn't be rushed into any decisions as I would have the weekend to think any choices over.

I was woken up at 2:00AM on Friday morning with mild contractions and slept in between them until 4:00AM when I felt I had to get up and move around. I spent the early hours of the morning walking around the garden and picking up in the house. I woke my husband at 6:00AM and we played Scrabble for an hour and then phoned the midwife at 7:00AM to say we wouldn't be going to the hospital that morning as my contractions had started up again. She phoned back half an hour later to say she would come and visit us on her way into work. By the time the midwife arrived (at 9:30AM) my contractions had stopped. She stayed with us for about 40 minutes and just before she left I had a fairly mild contraction. She asked me to go to the hospital to be monitored and we agreed and set off at about 10:30 with my first daughter, Milly (aged 2).

We arrived at the hospital and I was monitored in the Day Unit. Although my contractions had started up again (at seven-minute intervals) they were relatively mild and not getting closer together. I thought they were probably Braxton Hicks contractions as I was beginning to think this baby was never going to be born.

The obstetrician came to see me at 1:00PM and booked me in for an induction the following Friday because he only does inductions on a Friday. As he was talking he looked at me and asked if I was okay. I replied that I was fine, just having a contraction. He wanted to know why no one had told him I was in labor and I told him I wasn't but I was just having very mild contractions. He wanted to examine me, which he did. I knew something was up by the look on his face and then he said, "If you want a home birth you had better get in the car and hurry because you are 5cm dilated and your water bag is bulging." It seemed as though a cheer went up with all the midwives in the Day Unit and everyone, including us, was excited, not to mention surprised. One of the midwives phoned my midwife to get her to meet us at home and I dressed, got our things together and hurried back to the car.

Before you start to think that I have an incredible pain threshold you should be aware that the contractions I had with the birth of my first daughter were started artificially with pitocin and were so incredibly painful (as any unlucky person who has had it can tell you). These were the only comparison of contractions I had.

It seemed as if a cheer went up with all the midwives

We got home at 1:30PM and my husband put Milly to bed (very quickly) and then set about filling the birthing pool we had rented. I got the baby's clothes, etc ready, phoned my mother-in-law to come around and look after Milly and then phoned my Mom in Australia and a couple of friends to let them know my pregnancy was finally coming to an end.

My midwife arrived at 2:15PM and just after that my contractions got serious. The pool was ready at 3:00PM and I got into it. Throughout the whole experience I was amazed with the role the midwife took, as it was so different to my last birthing experience in hospital. She didn't examine me internally and only monitored the baby about three or four times. I kept asking what I should be doing and every time she replied: "What do you feel like doing?"

Rosa was born under the water at 4:11PM weighing in at 9lb 4oz (4.2kg) and measuring 56cm. I had a small tear along my episiotomy scar that didn't require sutures. By 6:00PM all the family had arrived (truck loads of them) and we were all downstairs singing 'Happy Birthday' to Rosa and eating birthday cake.

Rosa is now 11 weeks old and is the most peaceful, relaxed baby I have ever known. She has always slept well (probably due to her incredible size) but now a bad night's sleep for Rosa is sleeping from 9:00PM to 6:30AM.

The midwife who delivered Rosa works out of our local city hospital and is proof that a terrific midwife makes all the difference. I can't even begin to describe how fantastic she was! We owe our perfect birth experience to her expertise.

*Jennifer Jacoby with baby Rosa, just after the birth*

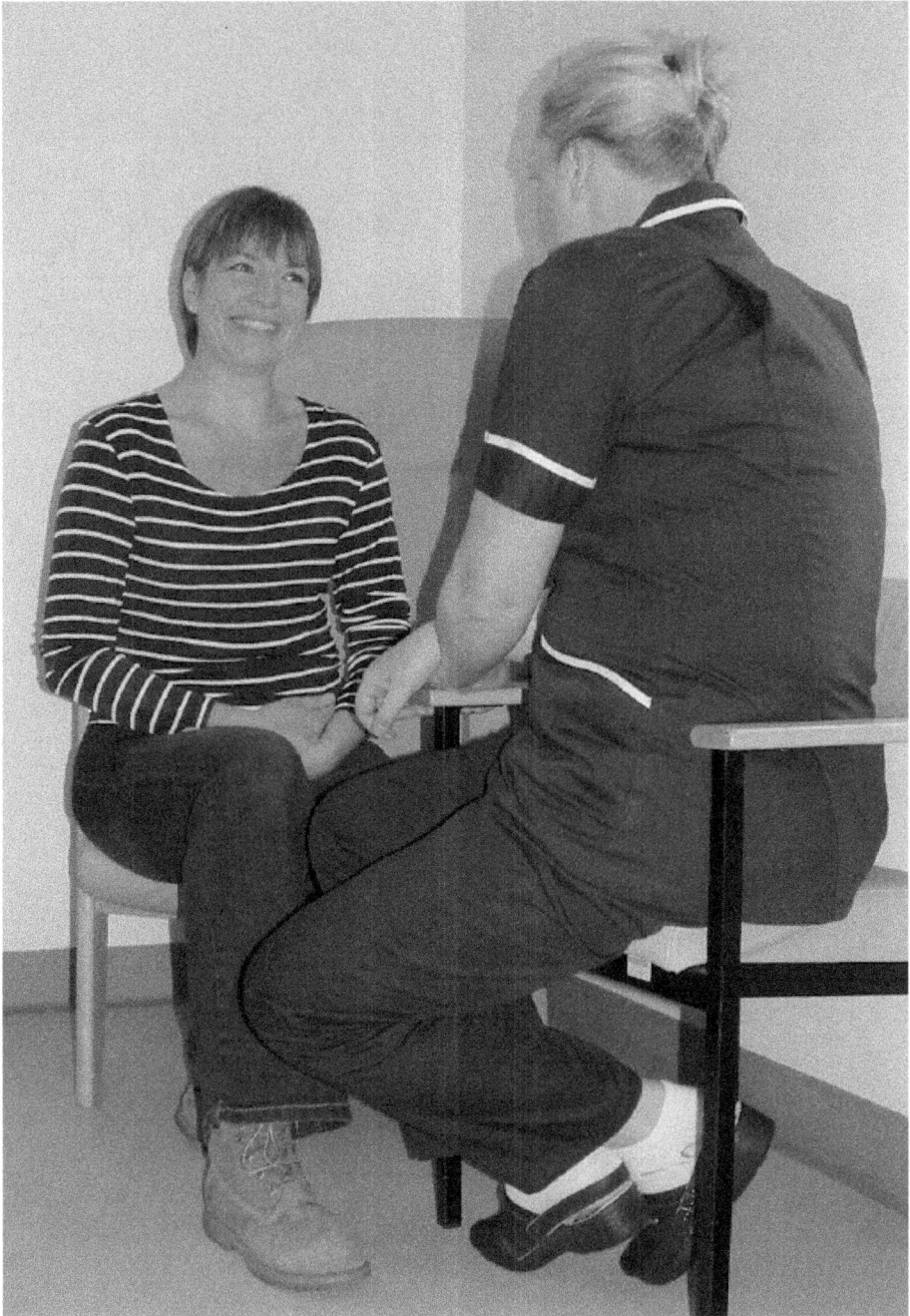

*What best facilitates good rapport? What impedes it?*

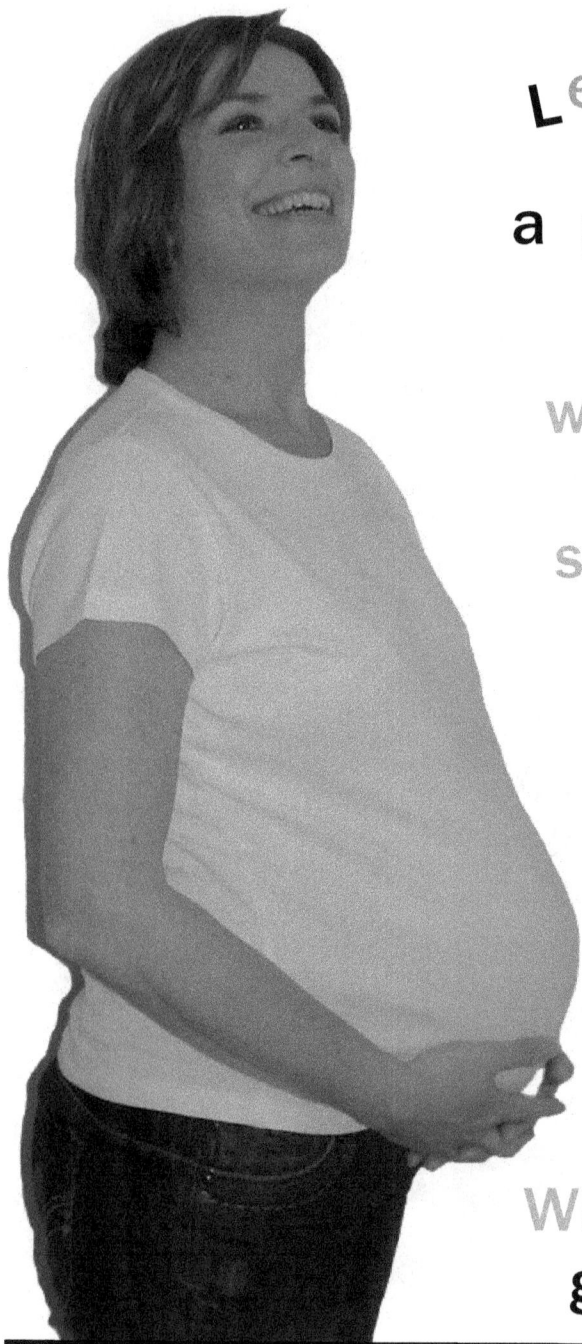

L<sup>e</sup>t's h<sub>a</sub>v<sup>e</sup>
a bit <sub>o</sub>f f<sub>u</sub>n

while we also
sort out the
serious stuff!

What have we
got to gain?!

## And finally...

Let's not forget the importance of giving women a really warm welcome whenever we meet them, and particularly when we see them in labor. Consider the following comment:

> With my first birth, I just felt there were too many people around... I had about five different midwives, one after the other. Also, the approach from the one midwife who started off with me was all wrong. When I arrived at the hospital after my water had broken, she seemed put out that I'd come in to be checked. She didn't give me a good welcome—just made me feel I was in the way. When another midwife put me on an electronic fetal monitor, she said she wasn't sure how it worked. It was new equipment in a new hospital. I ended up having lots of pain relief. First I had gas and air, then a couple of shots of diamorphine and then—since they didn't think I was dilating enough—an epidural, along with a pitocin IV. After several vaginal exams, the midwives and doctor who were in the room decided I needed an emergency episiotomy and forceps as the cord was around my daughter's neck. The forceps damaged her left eye.
>
> When I went into labor with my second daughter I had a relaxed, friendly welcome from the midwife. She stayed with me all the way through my labor, right till the birth (which was only 2½ hours altogether). Everything seemed better and the birth itself was a fantastic experience for both me and my husband.

# Pause to reflect...

## ...about information

1 How much information do you think you should give each woman about her own condition and situation?
2 How can you communicate information about risks, without inducing fear?
3 Are there any cases where it's justifiable to keep information secret?
4 What if women ask you questions which make you feel uncomfortable?
5 What will you say if a mom-to-be says you've made a mistake?

## ...about respect

1 Will you trust a woman's self-reports of her own labor?
2 Are there any cases where you would attempt to override a woman's decisions?
3 What if a woman seems even better-informed about research than you are yourself... How will you cope with that? How can you value the woman's input while also contributing your own knowledge and experience and also checking up on details?

## Risk assessment

### The importance of an accurate due date

Given the sense of anticipation most women have about their due date (even though over 95% of births don't take place on it) and given the pressure many women are under once they apparently become 'overdue', obviously it's vital to take this date seriously. When I was pregnant myself I privately felt upset that this calculation was undertaken with only a piece of unimpressive card and that no particular attention was ever paid to the fact that my usual menstrual cycle was 33 days. Despite widespread mentions in the literature that this would mean a strong likelihood of my babies being born five days after the date the charts indicate, this later date was never fixed as the official due date. Isn't this strange?

> Given the sense of anticipation most women have about their due date and given the pressure many women are under once they become 'overdue', it's vital to take this date seriously

When I was collecting material for this and earlier books I received an alarming number of comments and accounts about problems to do with due dates which were in dispute. So as to avoid confusion or unnecessary anxiety later on in a woman's pregnancy, when you give a client her 'due date'...

- Double-check it's possible for her to have conceived on the date her due date suggests. If her husband was away on business that weekend and only around the weekend after, this might need discussing.
- Double-check the length of her cycle, if she knows what it is, adding on or subtracting days, as necessary.
- Explain that many women experience light bleeding at the time when they would normally have their period in the first and second months of pregnancy.
- Tell your client that very few women actually give birth on their due date.
- Explain that anything from 37 weeks to 41 weeks is considered 'term' (or 42 weeks in some practices).
- Tell her she may later want to read up on prematurity, postmaturity and induction, in case it ever becomes relevant.

## Questions for reflection...

1 Why do you think many women become so determined not to be induced?
2 How many cases of real postmaturity have you seen so far?
3 What do you consider to be the most effective ways of checking for safety?
4 What role can anxiety and over-checking play in the labor-birth process?

| Month | 1 | 2 | 3 | 4 | 5 | 6 | 7 | 8 | 9 | 10 | 11 | 12 | 13 | 14 | 15 | 16 | 17 | 18 | 19 | 20 | 21 | 22 | 23 | 24 | 25 | 26 | 27 | 28 | 29 | 30 | 31 |
|---|---|---|---|---|---|---|---|---|---|---|---|---|---|---|---|---|---|---|---|---|---|---|---|---|---|---|---|---|---|---|---|
| Jan | 1 | 2 | 3 | 4 | 5 | 6 | 7 | 8 | 9 | 10 | 11 | 12 | 13 | 14 | 15 | 16 | 17 | 18 | 19 | 20 | 21 | 22 | 23 | 24 | 25 | 26 | 27 | 28 | 29 | 30 | 31 |
| Oct | 8 | 9 | 10 | 11 | 12 | 13 | 14 | 15 | 16 | 17 | 18 | 19 | 20 | 21 | 22 | 23 | 24 | 25 | 26 | 27 | 28 | 29 | 30 | 31 | 1 | 2 | 3 | 4 | 5 | 6 | 7 |
| Feb | 1 | 2 | 3 | 4 | 5 | 6 | 7 | 8 | 9 | 10 | 11 | 12 | 13 | 14 | 15 | 16 | 17 | 18 | 19 | 20 | 21 | 22 | 23 | 24 | 25 | 26 | 27 | 28 | | | |
| Nov | 8 | 9 | 10 | 11 | 12 | 13 | 14 | 15 | 16 | 17 | 18 | 19 | 20 | 21 | 22 | 23 | 24 | 25 | 26 | 27 | 28 | 29 | 30 | 1 | 2 | 3 | 4 | 5 | | | |
| Mar | 1 | 2 | 3 | 4 | 5 | 6 | 7 | 8 | 9 | 10 | 11 | 12 | 13 | 14 | 15 | 16 | 17 | 18 | 19 | 20 | 21 | 22 | 23 | 24 | 25 | 26 | 27 | 28 | 29 | 30 | 31 |
| Dec | 6 | 7 | 8 | 9 | 10 | 11 | 12 | 13 | 14 | 15 | 16 | 17 | 18 | 19 | 20 | 21 | 22 | 23 | 24 | 25 | 26 | 27 | 28 | 29 | 30 | 31 | 1 | 2 | 3 | 4 | 5 |
| April | 1 | 2 | 3 | 4 | 5 | 6 | 7 | 8 | 9 | 10 | 11 | 12 | 13 | 14 | 15 | 16 | 17 | 18 | 19 | 20 | 21 | 22 | 23 | 24 | 25 | 26 | 27 | 28 | 29 | 30 | |
| Jan | 6 | 7 | 8 | 9 | 10 | 11 | 12 | 13 | 14 | 15 | 16 | 17 | 18 | 19 | 20 | 21 | 22 | 23 | 24 | 25 | 26 | 27 | 28 | 29 | 30 | 31 | 1 | 2 | 3 | 4 | |
| May | 1 | 2 | 3 | 4 | 5 | 6 | 7 | 8 | 9 | 10 | 11 | 12 | 13 | 14 | 15 | 16 | 17 | 18 | 19 | 20 | 21 | 22 | 23 | 24 | 25 | 26 | 27 | 28 | 29 | 30 | 31 |
| Feb | 5 | 6 | 7 | 8 | 9 | 10 | 11 | 12 | 13 | 14 | 15 | 16 | 17 | 18 | 19 | 20 | 21 | 22 | 23 | 24 | 25 | 26 | 27 | 28 | 1 | 2 | 3 | 4 | 5 | 6 | 7 |
| June | 1 | 2 | 3 | 4 | 5 | 6 | 7 | 8 | 9 | 10 | 11 | 12 | 13 | 14 | 15 | 16 | 17 | 18 | 19 | 20 | 21 | 22 | 23 | 24 | 25 | 26 | 27 | 28 | 29 | 30 | |
| Mar | 8 | 9 | 10 | 11 | 12 | 13 | 14 | 15 | 16 | 17 | 18 | 19 | 20 | 21 | 22 | 23 | 24 | 25 | 26 | 27 | 28 | 29 | 30 | 31 | 1 | 2 | 3 | 4 | 5 | 6 | |
| July | 1 | 2 | 3 | 4 | 5 | 6 | 7 | 8 | 9 | 10 | 11 | 12 | 13 | 14 | 15 | 16 | 17 | 18 | 19 | 20 | 21 | 22 | 23 | 24 | 25 | 26 | 27 | 28 | 29 | 30 | 31 |
| April | 7 | 8 | 9 | 10 | 11 | 12 | 13 | 14 | 15 | 16 | 17 | 18 | 19 | 20 | 21 | 22 | 23 | 24 | 25 | 26 | 27 | 28 | 29 | 30 | 1 | 2 | 3 | 4 | 5 | 6 | 7 |
| Aug | 1 | 2 | 3 | 4 | 5 | 6 | 7 | 8 | 9 | 10 | 11 | 12 | 13 | 14 | 15 | 16 | 17 | 18 | 19 | 20 | 21 | 22 | 23 | 24 | 25 | 26 | 27 | 28 | 29 | 30 | 31 |
| May | 8 | 9 | 10 | 11 | 12 | 13 | 14 | 15 | 16 | 17 | 18 | 19 | 20 | 21 | 22 | 23 | 24 | 25 | 26 | 27 | 28 | 29 | 30 | 31 | 1 | 2 | 3 | 4 | 5 | 6 | 7 |
| Sept | 1 | 2 | 3 | 4 | 5 | 6 | 7 | 8 | 9 | 10 | 11 | 12 | 13 | 14 | 15 | 16 | 17 | 18 | 19 | 20 | 21 | 22 | 23 | 24 | 25 | 26 | 27 | 28 | 29 | 30 | |
| June | 8 | 9 | 10 | 11 | 12 | 13 | 14 | 15 | 16 | 17 | 18 | 19 | 20 | 21 | 22 | 23 | 24 | 25 | 26 | 27 | 28 | 29 | 30 | 1 | 2 | 3 | 4 | 5 | 6 | 7 | |
| Oct | 1 | 2 | 3 | 4 | 5 | 6 | 7 | 8 | 9 | 10 | 11 | 12 | 13 | 14 | 15 | 16 | 17 | 18 | 19 | 20 | 21 | 22 | 23 | 24 | 25 | 26 | 27 | 28 | 29 | 30 | 31 |
| July | 8 | 9 | 10 | 11 | 12 | 13 | 14 | 15 | 16 | 17 | 18 | 19 | 20 | 21 | 22 | 23 | 24 | 25 | 26 | 27 | 28 | 29 | 30 | 31 | 1 | 2 | 3 | 4 | 5 | 6 | 7 |
| Nov | 1 | 2 | 3 | 4 | 5 | 6 | 7 | 8 | 9 | 10 | 11 | 12 | 13 | 14 | 15 | 16 | 17 | 18 | 19 | 20 | 21 | 22 | 23 | 24 | 25 | 26 | 27 | 28 | 29 | 30 | |
| Aug | 8 | 9 | 10 | 11 | 12 | 13 | 14 | 15 | 16 | 17 | 18 | 19 | 20 | 21 | 22 | 23 | 24 | 25 | 26 | 27 | 28 | 29 | 30 | 31 | 1 | 2 | 3 | 4 | 5 | 6 | |
| Dec | 1 | 2 | 3 | 4 | 5 | 6 | 7 | 8 | 9 | 10 | 11 | 12 | 13 | 14 | 15 | 16 | 17 | 18 | 19 | 20 | 21 | 22 | 23 | 24 | 25 | 26 | 27 | 28 | 29 | 30 | 31 |
| Sept | 7 | 8 | 9 | 10 | 11 | 12 | 13 | 14 | 15 | 16 | 17 | 18 | 19 | 20 | 21 | 22 | 23 | 24 | 25 | 26 | 27 | 28 | 29 | 30 | 1 | 2 | 3 | 4 | 5 | 6 | 7 |

*In case you would like something to replace an old and unimpressive cardboard wheel, here's a reference chart, based—of course—on women who have a 28-day cycle. In addition, you might like to consult www.redbabybook.com to get an adjusted EDD.*

## The risks of risk assessment

From the very first appointment onwards, you may feel negative about a woman's chances of having a physiological birth because of risks you perceive. Like many women nowadays, given changing trends, your new client may be over 35 (i.e. 'elderly'), or she may come to you with a tale of woe about her first experience of birth. If your first reaction is to immediately conclude she is incapable of having a physiological birth, why not instead take a few moments to think how her previous birth might have been *disturbed* by her previous caregivers (perhaps even at the woman's own request)? You can even explore this openly with your new client because—after all—you do need to educate her to some extent if she is to have a better experience this time. Or do you find yourself doubting that optimality really is possible for most women you see?

> You may well feel negative about a woman's chances
> because of the risks you perceive, rightly or wrongly

Actually, I think optimal birth is possible in many more situations than many people expect. It is actually a *safer* option than a birth which involves the use of drugs for induction, augmentation or pain 'relief', or which includes interventions which are initially not primarily carried out for safety reasons. In many caregivers' opinion, the same principles apply to low and high risk laboring women: the amount of disturbance a woman experiences and the extent to which she feels safe are most important. In other words, leaving the laboring woman undisturbed might be more important than close monitoring because it will give the woman and her unborn baby the best possible conditions for successfully orchestrating the cocktail of hormones necessary for a safe birth.

In any case, it's very unfortunate when a woman is labelled 'high risk' early on in her pregnancy... From that moment onwards she will inevitably worry about the birth, which will change her entire experience of pregnancy. Instead of being a time of joy, discovery and expectation, it will become a time of dread, anxiety and tension. Since psychological reactions so often have a physical impact (taking into account the close link between emotions and hormonal production), this is a real tragedy, especially if the mom-to-be was wrongly labelled 'high risk' in the first place.

Perhaps many caregivers have very low expectations of women's safety prospects because they are accustomed to scenarios which involve the use of drugs (usually for hoped-for pain relief or induction/augmentation). Many have not attended births of 'high risk' women who choose to avoid drugs and interventions, if at all possible. As many professionals have found, if the physiological processes are not just left undisturbed, but are *facilitated* (with the midwife or other caregiver creating an atmosphere around the laboring woman of confidence, security and privacy) beautiful outcomes are often the result. Perhaps these births need to be seen to believed... so why not see them?

On the next few pages, you can read a few birthframes about women who would normally be labelled 'high risk', many of whom went on to have extremely successful—and joyful—physiological births. These births were truly optimal because as well as being safe, they were beautiful. In other cases, I ask you to think through certain elements of risk assessment by considering information which is not always included.

These births were truly optimal because
as well as being safe, they were beautiful

When you read the cases of births which represent optimal high risk births (in my opinion) please consider whether these cases might be replicated many times if more 'high risk' women were 'allowed' to give birth physiologically. With the infrastructure in the USA (roads, telephones, healthcare facilities), as well as expertise and resources, I believe we will achieve much better birthing outcomes overall, if we respect the possibilities of the undisturbed physiological processes, while being ready to intervene if absolutely necessary. In other words, if caregivers and pregnant clients all choose to do so, we can have the best of both worlds—beautiful physiological birth (which I have called 'fizzy-logical' because of its ecstatic and evidence-based elements), with hi-tech emergency intervention any time it's really needed... but *only* then.

# A couple of elderly primigravidae...

Women who have their first baby late are becoming increasingly common...

### Birthframe 40: Anonymous—Optimizing the physical side

This 'elderly' primipara did everything she could think of to optimize her pregnancy. Sometimes, I think, we are apt to forget just how motivated older mothers can be to make their experience as good as it possibly can be. Have the 'older' moms-to-be you've met been similarly motivated? If not, could you encourage them to be? As you will see, this woman's labor was fast, easy and safe—and, in conventional obstetric terms, it was non-interventionist. This contributor (who prefers to remain anonymous) took up her midwives' suggestion to use TENS and gas and air even though I feel, having read a longer version of this account, that her body—and mind, for that matter—would have been plenty strong enough to travel through this labor without any pain relief whatsoever. In any case, she mentioned that the TENS pads actually fell off at some point, long before she realized this had happened! What is particularly interesting is the physical preparation that this woman undertook. It no doubt facilitated her easy, fast and safe labor and postpartum recovery, especially since it was clearly underpinned by a very positive mental attitude.

Her physical preparation facilitated her easy, fast and safe labor

Once I found out I was pregnant I wanted to make sure that I did everything possible to help me with the labor. I was 36 but it was becoming the trend to have babies later in life so I didn't really think about my age. Whilst trying for a baby I took folic acid, as recommended. I did have quite bad sickness for the first three months or so but once that was over I kept as fit as possible. I had always regularly attended a gym and generally used to keep fit and active so couldn't see any reason to change that. I used to enjoy weight training but obviously changed to much lighter weights and started swimming a lot more. A friend of mine had swum regularly throughout her pregnancy and had quite an easy labor so I decided to follow her example. I went to the gym three times a week throughout my pregnancy and swam two times a week. I swam up to 40 lengths at a time, obviously getting slower as my pregnancy advanced! This was the breaststroke, incidentally, which Sylvie tells me would have helped the baby get into a good position for the birth.

> A friend of mine had swum regularly in her pregnancy and had quite an easy labor, so I decided to follow her example

I ate as healthily as possible and bought a blender to make fruit smoothies to up my intake of fruit. Constipation is sometimes an unwelcome side-effect of pregnancy! I also read an article in a magazine about raspberry leaf tea, which is a herbal remedy and is supposed to help shorten labor. I had nothing to lose so decided to try this too, taking it in tablet form, which I prefered. Each tablet was 400mg and the label said to have one three to six times a day. It said the tablet could also be broken up and dissolved in hot water—but it tasted awful this way! It seems raspberry leaf is recommended for the last three months of pregnancy but not before. I also read that rubbing Vitamin E oil onto your perineum regularly can help stretch this area and lessen the chance of having to be cut whilst giving birth. Again, I thought, "Why not?" so I used to massage the oil in every night throughout the last six to eight weeks.

I worked full-time right up to seven days before giving birth and although I was tired toward the end I am glad I did because I believe it is important to keep active for as long as possible. The baby arrived three days before my due date. At 10 o'clock at night my water broke naturally. I was at home and I had no warning at all—I just got up from my chair to take a glass into the kitchen and that was it! I was relatively calm and was in no pain or discomfort whatsoever. I gathered my bags and got to the hospital for around 11 o'clock.

At about 1:00AM I started getting contractions—they were quite strong and started coming regularly. It seemed to happen very quickly and I was a bit shocked at how strong they were. At about 3:00AM the nurses on the delivery ward examined me and discovered I was around 7cm dilated. I was quite shocked and things started happening very, very quickly by then. The contractions just got stronger and stronger and by about 3:45AM I was fully dilated.

> I started getting contractions... It seemed to happen very quickly and I was a bit shocked at how strong they were.

I had decided I wanted to try and squat to give birth and not lie flat on my back but I found that I was not really strong enough

I had decided beforehand that I wanted to try and squat to give birth and not lie flat on my back but I found that I was not really strong enough because it is very hard on the thighs! The nurses raised the bed up so that I was practically sitting and I had both feet on the bed, which was a huge help. I gave birth at 4:10AM, exactly 1 hour 10 minutes after arriving in the delivery suite. I was totally alert and felt absolutely wonderful, apart from being a bit sore. I had no sutures at all, just a graze, which healed itself. The whole experience was amazing.

My body healed very quickly indeed and I had no problems whatsoever. I was back at the gym 10 days later and regained my figure very quickly. I am sure that my easy labor was a direct result of the preparation I did throughout my pregnancy.

My body healed very quickly and I had no problems whatsoever. I am sure that my easy labor was a direct result of the preparation I did throughout my pregnancy.

*Optimal labor is a short, but unusual trip, which takes us down hitherto unexplored routes, then back onto firm land again, without undue stress or trauma. I hope this photo (which I took in Venice, Italy) is not a metaphor for the way birth always needs to be... the man in charge, with women only inactive observers... What do you think yourself?*

## Birthframe 41: Sarah Cave—Avoiding drugs to help the baby

As you'll see, this next first-time mom somehow knew what was best for her labor and birth quite intuitively—and she was right, of course. What is interesting, I think, from the point of view of providing optimal care, is her reason for wanting to avoid the use of pain relief: she simply felt it would be best for the baby. Bearing this in mind as you read, consider the following:

- If a woman writes in her care guide or tells you verbally she doesn't want to be offered pain relief, how literally do you think you should stick to this?
- To what extent would your answer be honoring or dishonoring your client?

I gave birth the day after those wedding photos were taken. I'm 36 and Alice is my first baby—she's now 19 weeks old. I had a fabulous pregnancy. Loved all of it, even though I had slight morning sickness between Weeks 4 and 12. I was never actually sick—I just felt queasy and sensitive to smells. Amazingly, I also went off coffee and chocolate! Looking ahead to the birth, I decided I wanted to do without painkillers if I could because I felt it would be the best thing for the baby. And it was a fantastic birth. Not painless exactly, but really, really good.

My labor started when I walked in my front door, after I got back from that wedding! My water seemed to be going. I felt wet and thought, "Surely that's not my water breaking?" All night long I had contractions and I spent a lot of the time in bed, sort of asleep, but I kept on getting up for the contractions too. My husband wanted to keep me company and massage me, or whatever, but I just told him to go back to bed. "Just let me be. I just want to do this on my own," I told him.

We eventually went into the hospital at about 8:00 the next morning. When we got there, a midwife examined me. "I'll just go and get someone else to have a look at you," she said. "I'm not sure if my fingers are long enough." The second midwife said that, yes, I was definitely 10cm dilated. She asked me why we hadn't come in sooner. "I just wanted to be at home," I said, "just wanted to be quiet." They thought they'd better get me to the delivery room and suggested I get in a wheelchair. "Oh, I'll walk!" I said. Of course, I was naked by this time and one of the midwives tried to hold a towel or something round me, but I told them not to bother. I really couldn't care less who saw me by that stage! The midwife was fantastic—very calm. It was like having Mom at the birth. She looked after me and didn't let the doctors in. It was just me, my husband and the midwife. It was a really intimate and relaxed atmosphere. At one point, the midwife offered me gas and air, but I thought, "No, I'll just keep going and see if I can manage," remembering it would be the best thing for the baby if I could. And of course, I did. Somehow, the time just went past.

Alice is an 'angel baby'. She's feeding and sleeping well. She's very contented. The only blip was getting mastitis on four occasions from the second week after the birth until last week. I'd never heard of the condition until it happened. My biggest mistake was leaving the mastitis on the first occasion for over a week *before* going to the doctor's office—and it was too late by then. So my advice to other new moms would be to get help any time you feel pain in the breast. Apart from that, I'd say enjoy every day! Don't give up on breastfeeding. I didn't and it's going great now. Alice has nearly doubled her birth weight and she's not even 6 months old yet! Good stuff, breastmilk.

This is Sarah Cave, a 36-year-old primip chatting at a wedding a few hours before going into labor, almost as soon as she got back home

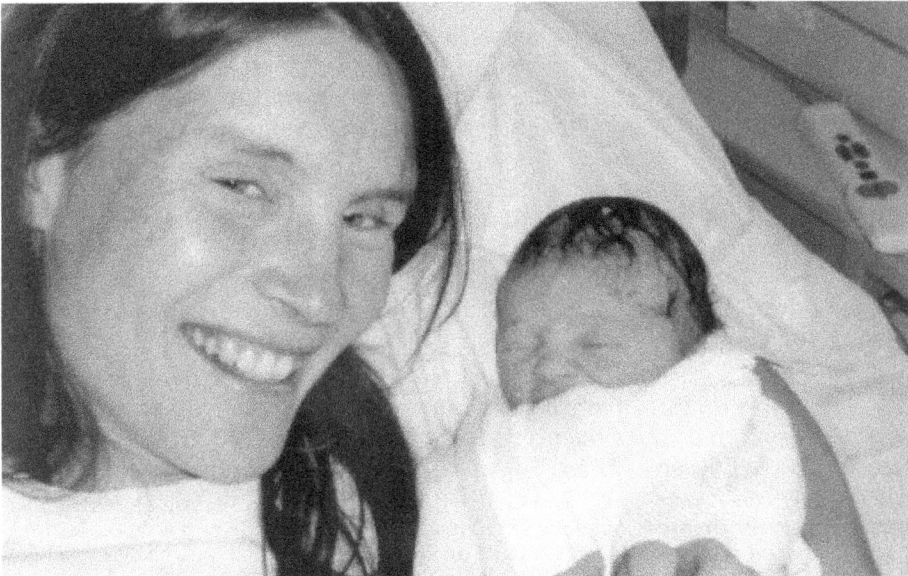

Here she is again, the next day, at her local hospital, showing off her new baby after a smooth, unmedicated birth, with no interventions

## Women whose birth went wrong last time

As I've already mentioned (earlier in this book), I was contacted by many women whose second (or third) experience of birth was much better than their first. And, as I also pointed out earlier, two elements seemed to be involved in the better outcomes afterwards:

♥ Learning from what had not gone smoothly before

♥ Improving conditions—or using a caregiver who improved them through his or her approach (especially in terms of non-disturbance)

When a woman requests a VBAC this is perhaps one very common and clear case where a woman is wanting to improve on a past experience. But do VBACs get optimal results?[4] What if a woman in your care asks for one? Since the cesarean rate has increased dramatically, it's important to consider these questions because more and more mothers have a cesarean for their first baby, so may consider a VBAC. Needless to say, with the rising fear of birth due to the falling birth rate (and consequent lack of familiarity with the processes) and negative portrayals of birth in the media, the question about optimality may be psychological as well as clinical. The fact that so many celebrity moms choose to book themselves into a private clinic for an elective cesarean and give the impression they can 'bounce back' with no apparent difficulties postpartum also does nothing to encourage women to go the vaginal route.

Of course, whether considering VBACs from a psychological or clinical point of view, reasons for the original cesarean may be very relevant. Nevertheless, we must remember when considering these that many commentators consider the rising cesarean rate to be due to cascades of intervention or failure to progress because of poor conditions during labor, not to a change in basic female physiology. Poor management of various stages of labor is also a reason suggested by many dissatisfied women who've had emergency cesareans, so this also needs to be seen as a possibility. Even if caregivers themselves (particularly midwives) are competent and sensitive, the systems they work within may prevent them from offering the high quality of care they might prefer. This is particularly true if their working systems make continuity of care or privacy and non-disturbance during labor and birth an impossibility.[5]

I asked perinatal researcher Hélène Vadeboncoeur about VBAC... As well as experiencing a successful VBAC herself, Hélène has researched relevant issues for a book on VBAC in French (*Une autre césarienne ou un accouchement naturel? S'informer pour mieux decider*, Carte Blanche 2008)—published by Fresh Heart in English in 2011. She made the following comments...

Giving birth vaginally after a cesarean (a VBAC) can represent an extremely important event in some women's lives. It also represents a less risky option for a woman than a repeat cesarean because of the risks associated with surgery, which relate to the use of anesthesia, the surgery itself and complications like infection, all of which have implications for mortality risk and postpartum recovery. After VBACs hospital stays are shorter, recuperation is faster, there is less postpartum pain and many women experience exhilarating feelings of empowerment and fulfilment. Seeing the joy of a woman at the birth of her first baby born vaginally after a previous cesarean can be very rewarding for midwives and other caregivers too.

A VBAC is possible for most women who've previously had a cesarean. It is even possible when it is assumed the woman is carrying a 'big' baby, when she has had two cesareans before, when the baby is lying breech and when a woman is expecting twins.

Some risk factors may mean the baby needs to be born in hospital, e.g. two or more previous cesareans, a uterine suture technique involving one seam and not two, and a birth interval (between the cesarean and the planned VBAC) of less than 18-24 months.

A VBAC is *not* recommended when the uterine scar is vertical in the body of the uterus because of the higher risk of uterine rupture in this case—although it is still a possibility when the vertical incision is very low. (Although a vertical incision is known as a 'classical incision' only a very small percentage of cesareans involve this kind of incision nowadays.) A cesarean carried out by low transverse uterine incision (i.e. a 'bikini line incision') is better from the risk point of view.

When conditions are good 75-95 % of women have a successful VBAC. Conditions are optimal when a woman goes into labor spontaneously, when a midwife is in attendance, when the environment generally is supportive and when no time limit is placed on either the labor or the birth (provided labor progresses well), and when emergency facilities are available nearby. Supportive care is particularly important because a woman who has previously had a cesarean often lacks confidence in her ability to give birth, particularly when the reason for her previous cesarean has a link with the labor itself (slowing down, stopping, malpositioning of the baby during labor, etc.)

Notes:

♥ It is particularly important to avoid inducing a woman who has previously had a cesarean because this increases the risk of uterine rupture. All prostaglandins to ripen the cervix—especially misoprostol (Cytotec)—are particularly inadvisable because they significantly increase the risk of uterine rupture. Inducing with synthetic oxytocin (pitocin) may also increase the risk, to a lesser degree.

♥ In cases where women plan to give birth at home or in a birthing center it is essential for an agreement to be made in advance with a nearby hospital where surgical facilities are available. This is because a repeat cesarean would need to be carried out very fast, if uterine rupture were suspected at any point during the labor. The most reliable sign of uterine rupture is prolonged anomalies in the baby's heart rhythm.

♥ The risk of uterine rupture does <u>not</u> increase during the second stage of labor. Overall, during a VBAC with spontaneous onset of labor between 2 and 6 women in every 1,000 (0.2 to 0.6% of women) experience uterine rupture. This compares with 2.7% of women who need an emergency cesarean during a labor of any kind. In cases where uterine rupture does occur, 5% of babies involved suffer from a lack of oxygen or die, but the other 95% are fine. The mothers themselves may experience a hemorrhage and need a hysterectomy if rupture occurs. In comparison, a woman has a risk of dying during of after a cesarean between 2 and 11 times higher than with a vaginal birth, a higher risk of various serious health complications, and the baby also has an increased risk of dying or of suffering from complications, especially serious respiratory problems.

♥ It is very important to pay attention to the moment in labor when the previous cesarean took place, because this can be a particular psychological hurdle for the woman in labor. At this point, the laboring woman may well need to verbalize how she feels or express any fears she may have.

♥ A midwife's presence, discretion and confidence in the woman's ability to give birth can be essential ingredients for success.

Hélène also provided two birthframes from her own research.

## Birthframe 42: Marie-Claude—VBAC better than a section

First, here's a woman, who is talking about the end of her VBAC labor...

I push with all my might, more determined that ever. I know I'll soon see my little baby. It's not necessary to use the vacuum cup. At 4:06, my baby is here. The doctor asks me to stretch my arms to pick him up. I take him and put him on my abdomen. He gives me a first look. I am living the most wonderful moments in my life! What a sensation to feel this small human being cuddled skin-to-skin against my stomach. I am so proud of myself for having succeeded.

A few hours later, I get up. I feel no pain. My baby is crying and I go and fetch him myself, for I'm able to do this. I settle down comfortably to nurse him. All goes superbly well! I have sutures and hemorrhoids but after only 24 hours, the pain has already receded. On the second day, I leave the hospital in really great shape... No comparison can be made between this birth and that of my first baby.

## Birthframe 43: Céline—Feeling strong after a VBAC

Now here's an account from a VBAC mother—a nurse—who gave birth vaginally after two previous cesareans.

I'm giving birth to a baby who is bigger than my two first babies and I pull his shoulders towards me as he emerges from my body, just as my doctor suggests. (This is what I wanted to do but in the heat of the moment it didn't occur to me.) We keep our new baby with us for two hours after the birth. This is my most beautiful memory of all.

After the birth, Daniel (my husband) is very moved and he starts crying. Not me. I feel strong and not at all weepy. I had the feeling that I should give birth alone. I wanted to be able to say, "It was me who gave birth." I didn't want anyone to touch me or egg me on during labor. I was proud of myself. It's true I had a lot of support, but it was me who gave birth. This was truly an accomplishment for me. I have the feeling that this changed something in my life, that this somehow liberated a blocked energy inside me. I can now move on to other things. I have the impression I am capable of achieving a whole load of things. I've seen a lot of my old doubts disappear!

## Questions for reflection...

1   Do you think these accounts represent VBAC women's typical experience?
2   Why do you think different people in a birth scene (woman, her husband, her midwife, her obstetrician, her doula...) might have very different perceptions of a particular birthing scene? Whose is most valid?
3   If you were the mom-to-be or new mother, how might you feel?

For other perspectives on VBAC, see the birthframes which follow in this chapter, as well as the section on cesareans at the end of the 'What' section.

Here are a few other perspectives on VBAC from some other contributors...

## Birthframe 44: Anonymous—Failed VBAC despite good care

With so many positive second-birth accounts and so much publicity about the feasibility of a vaginal birth after a cesarean (a VBAC), it is sometimes assumed that it must *always* be possible. However, as the next account illustrates, sometimes a planned VBAC turns out to be another cesarean. This can be extremely disappointing, as the following account illustrates. Here we meet up again with one of the anonymous contributors we've met before, earlier in this book. I know who the caregiver was in this case and I also know that she is not only an outstanding midwife, she is also perhaps the most committed person to VBAC it would be possible to find. Nevertheless...

For my second pregnancy and birth, I had wonderful, skilled, loving care from a home birth midwife. Labor started naturally, I dilated quickly, and the urge to push came after just a few hours.

But then I pushed... and pushed... and pushed... and pushed... The three different midwives reached in to try to turn the baby's head to facilitate delivery, to no avail.

After six hours of pushing, as the sun rose, my midwife said my scar just didn't look right, and I walked down the stairs with my midwife and husband, got in the car, and we drove to the hospital in rush hour traffic! (At the time I lived in a western suburb of the nearby city.) I was having immense and powerful pushing contractions, kneeling backward in the car with my midwife helping me.

I was admitted to the hospital, and we all agreed that after the now 7+ hours of pushing, that perhaps an epidural was in order so I could rest a bit. Everyone was very calm and professional. My midwife stayed with me the whole time. I and the baby were both doing fine according to the monitors. They catheterized me to empty my bladder, as I had been unable to urinate much during labor. That was a relief, actually.

In three or so hours, at 3PM, a doctor came in to check me. The baby had not moved down at all, to everyone's great surprise. I had continued to have pushing contractions. I was able to keep that up because I had been so well fed and watered during labor! I was informed that a cesarean would be necessary. I knew they were right, and even my midwife, who was an expert on VBAC births, agreed. I cried. She cried. My husband held me. The nurses were confused, but polite. The anesthesiologist was kind and professional, as was everyone else.

My midwife sat with me and described to me what was happening step by step. The doctor opened me, looked up and said, "Who did this?" I for some reason thought it was funny he asked, but his voice was dead serious as he repeated, "I mean it, WHO did this? Who did the previous C-section?" I told him, and he kind of snorted angrily. As my daughter's head was exposed and pulled out he said, "WOW, look at the size of that head!" I think he said it just to make me feel better. Her head just looked like a baby head to me! She was beautifully chubby and healthy, (three ounces heavier than my son at 8lb 14oz) and her Apgar score was perfect.

I held her after they fiddled with her for a few moments. Of course, I would have prefered to hold her immediately, but it was wonderful to see her. Then my husband held her as they sutured me.

My recovery was much less painful and traumatic than after the first C-section. There was one nurse who came in and lectured me as I emerged groggy from my anesthesia. She told me I was lucky I came in when I did and said how dare I try a home birth after a 'failed' first birth. She also said the uterus was so thin when they did the surgery I was lucky it hadn't ruptured. I laughed. Did she know I had been in the hospital for over eight hours before the surgery was performed? I didn't actually say anything, I was too groggy. She huffed out.

The doctor came in and told me that I had massive scar tissue from the previous surgery, which he had managed to 'clean up'. My bladder had been abnormally adhered to my uterus after the first C-section, which is why I could not urinate during labor. I was catheterized for five days so that the bladder could heal properly. The doctor was amazed that I hadn't had any symptoms from that problem. Neither doctor nor midwife could give any specific reason why I could not deliver vaginally. It was apparently possible that the scar tissue had caused an impediment. My bone structure seemed normal. I dilated normally. My contractions were very strong and coordinated, for hours and hours and hours! She was a baby on the large side of things, but certainly not abnormally large.

I put up with a few annoying staff comments such as "She'll never sleep in a crib if you keep holding her." But as a better-informed, second-time mother no procedures were performed or drugs given to my daughter that I did not want. I wish I hadn't had to have any drugs, but I'll take what good things I can.

I cannot describe that feeling of loss—really a physical ache in my vagina, as strange as that sounds. The great expectation of holding the baby there, but then not, again. I don't know what to make of it emotionally. I sometimes don't even want to think of it, the disappointment is still strangely fresh. At least this time, I wasn't pressured into anything, I didn't make decisions out of fear. I had the prenatal care and labor I wanted until the very end. My care was respectful and loving, and I was NOT a particularly 'good' patient! Thank heavens.

I am trying to accept that even when I do everything right, there are some variables that I don't have control over. There is some reason for everything.

**Some thoughts on optimality:** I found this account very sad to read and the emails I exchanged with this contributor confirmed that. However, I do think this birth was optimal since the woman did at least do everything to optimize this particular VBAC, including getting an excellent caregiver. Do you agree? If you were the caregiver at this birth would you have behaved in the same way? How can women who have such a bad second experience through no fault of their own (in terms of 'misguided' decisions) be made to feel better afterwards, i.e. more *comfortable* with themselves as they continue to live?

## Birthframe 45: Michel Odent—Facilitating a VBAC properly

When there are no medical impediments to a successful birth, non-disturbance seems to be a key factor in the success of a VBAC. Here Michel Odent provides an account of a birth which illustrates this principle...

Recently, I attended a birth that could be used to illustrate my assumption that the current rates of cesareans would drop dramatically if the basic needs of women in labor were better understood. The woman who gave birth belongs to a family that is familiar with cesarean birth. Her brother, her sister and herself were born by cesarean. When she gave birth for the first time, it was decided after a long trial of labor in hospital that the baby was too big for her and that she needed a cesarean: the baby was 9lb (4kg).

While she was expecting her second baby, she asked me if I would come to her home when the labor started, because she wanted to try to give birth vaginally. My answer was: "Yes, if I'm not in Costa Rica".

The labor started during the night preceding my flight to Costa Rica. I could stay in her home until she was in advanced labor, so that when she arrived at the hospital with Liliana (her doula), she was not far from a point of no return. She eventually gave birth with the help of ventouse to another 9lb baby.

When she was expecting her third baby, she asked me again if I might come to her home when she was in labor. My answer was: "Yes, if I'm not in Italy".

The labor started two days before my flight to Italy. There were ideal conditions of privacy. I was following the progress of labor from another room, through the sound. Liliana who, as a doula, behaves like a cat, was around, evaluating the progress of labor with her own criteria—postures, sound, breathing patterns, etc.

I did only one vaginal exam. At 12 noon, the father left in order to make some arrangements so that the children could stay in the house of friends. Soon after, there was a series of powerful contractions—a real 'fetus ejection reflex'. At 12:45 the ecstatic mother gave birth to an 11lb (5kg) girl... no drugs, no tears, no episiotomy.

## Birthframe 46: Nina Klose—A detailed account of a VBAC

The final account here provides details about a VBAC woman's perspective. Nina's first baby was born by emergency cesarean. Despite her apprehensions about pushing during her second labor, she found her body knew exactly what to do when it came to it...

My mother says that newborns look like little angels that haven't fully descended to Earth yet. Newborns do have a faraway look in their eyes. But if you've given birth, you'd hardly say they come from the sky. There is nothing ethereal about pushing a baby into the world. It is just the opposite, a wholly carnal and earthy experience. Pushing out my daughter felt like doing a very large and difficult poop out my vagina.

My mother says that newborns look like little angels. If you've given birth you'd hardly say that newborns came from the sky.

People keep telling me I had an easy labor. I'm sure it was, relatively speaking. Now that it's over, I'd be ready to do it again next week. But it didn't feel easy. It hurt like anything. The thing is, I was so thrilled that my body was doing what it needed to do, that I hardly minded the pain. With our first child I never went into established labor. During his birth, I had had fairly regular, painful contractions from early in the morning, but labor never became 'well established,' according to the midwives who examined me.

It's hard to pinpoint exactly what went wrong. Was it because I hadn't fully visualized what it meant for the fetus to descend through my pelvis and out into the world? That I couldn't see myself becoming a mother? Was it because I allowed the doctors to 're-date' my pregnancy based on the 12-week ultrasound, even though I *knew* when I had conceived? They pushed the official due date two weeks earlier. This made me more anxious about being induced, so I tried to initiate the labor with nipple stimulation early in the morning of the day my labor started. Maybe trying to induce was the starting point of the problems. Maybe he hadn't got into the right position, and wasn't quite ready to be born yet.

My water broke at 6:00AM. I phoned the birth center, but since nobody asked me, I never mentioned to the midwife that my water had broken. I figured, since they didn't ask, it couldn't be important. I was in what I thought was labor all day long, but we only went to the birth center in the evening. I discovered a light green blush of fresh meconium on my sanitary napkin. But I had just changed the pad, so when I got there— a long drive all the way across town—the midwives didn't believe me, and sent me home. I phoned again when we got back—I was sure it was meconium. We drove back, this time in the small hours. The birth center confirmed that it was meconium and sent me straight back across town again to the hospital. Hospital staff were unsympathetic to my desire for a normal, unmedicated birth.

I will never forget the calm before the storm, the two hours' respite they agreed to give us before starting an induction. My contractions had ceased entirely. I stood at the window of the hospital praying for the labor to resume. My womb was still resting as the sun rose. I realized I would soon be forced to acquiesce to the hospital's interventions. I requested an epidural before the induction... but no one told me that the chances of a C-section are over five times higher with an epidural! (Isn't there a legal requirement to inform patients about the risks they face??) The baby couldn't tolerate the heavy spasms of induced contractions. His heart-rate plummeted. They turned off the IV. His heart rate settled. But there were still no effective contractions, the water had been broken for over 24 hours and I had not dilated past a few centimeters. If I had been at home, I probably would have gone calmly to sleep, rested up, and then given birth. Or perhaps he was in a funny position. Who knows? The doctor who delivered him couldn't find any clear explanation for the meconium or why he hadn't descended.

For me, agreeing to the cesarean felt like a death sentence. I had so longed to give my son a good birth, and to experience that life-changing moment of pushing him into the world. Why did the cesarean have to happen? I still ask myself. If I had it to do over again, I believe the most important problem was no continuity of care. If I'd been attended by one midwife throughout the pregnancy and birth, I'm sure I wouldn't have felt pressured into trying to induce the birth. And everything might have been different.

## Technically, my second labor took only five hours... but I'd call it a 10-hour labor with a day's break in the middle

Technically, my second labor took only five hours early Thursday morning, from when the first strong contractions began after midnight, until the placenta was born around 5:00AM. But I'd call it a 10-hour labor with a day's break in the middle. I was awake most of Tuesday night with contractions every 10 minutes that were strong enough to get me out of bed, but not strong enough to have to shout and holler. I sat up on the couch and breathed hard until about 6:00AM when they started to ease off.

"I'm in labor!" I announced proudly to my midwife on Wednesday morning. "I told my husband he could stay home from work. Was that the right thing to do?" "You're not in labor, and he has to go to work," the midwife replied firmly. "You're probably going to have several nights like this before you have a baby." Crestfallen, my husband went off to work, and I went back to bed. I took several long naps during the day, while our nanny took our 2-year-old son to his playgroup and fed him lunch. My back ached, and I had occasional mild contractions during the day, but nothing more.

"It won't be tonight," I thought on Wednesday evening. But just in case, I put the birthing pool together. It was sort of like a blue canvas tent on a 5-foot-wide hexagonal frame, turned upside down. It filled to about 2 feet deep through a garden hose attached to a bathroom faucet.

I went to bed at 10:30PM. Around midnight I found myself sitting up in bed in the middle of a contraction. I flopped back to sleep again between a few more contractions. Then I felt a trickle of water as though I'd wet the bed as my water broke. I went to the bathroom for a pad, and the contractions suddenly became very strong. For a while I sat on the couch in our bedroom huffing and puffing. My husband kept on snoring. There didn't seem much point in waking him up—what could he do? It was still going to hurt. I went downstairs to try to distract myself.

## There didn't seem much point in waking him up...

I guess contractions feel different for different people. I felt a stabbing pain in the small of my back that came in waves. One deep breath in, blow it away, another, another, soon it'll pass. Phew. It's gone. Then a rest before the next one begins. The day after my daughter was born, I told my brother, "It's like running a race," because my diaphragm felt bruised afterwards from breathing hard. But that's not quite correct. During a road race, the pain is psychological as much as physical. You have to force yourself to keep running, because if you don't, you will simply stop running and quit the race. But in labor, there's no quitting.

## It's like running a race... but actually that's not quite correct

I could watch TV—would that help? I still hadn't looked at the instructional video that came with the birthing pool, so I put that on. Some stupid woman holding a baby was explaining in great detail how wonderful her water birth was. I switched her off and tried the end of *Singing in the Rain*, which we'd started watching a few days before. Couples in beautiful clothes waltzed to a brass band. Infuriating creatures. No, that wasn't what I needed.

## I stayed sitting and swayed back and forth

The contractions were coming every few minutes. If I was sitting when one arrived, I stayed sitting and swayed back and forth, taking deep breaths and breathing out with a loud 'Ooooh, ahhhhhh'. Sitting seemed to lessen the pain along the base of my belly. If I was standing, I paced to the end of the room and back hollering the words to the *Battle Hymn of the Republic*. After awhile, it didn't help to walk away from the pain. "This is no fun," I said to myself. "Remind me I said so next time I have the stupid idea that I want a baby. Nothing can possibly make this worthwhile!"

"Call me if you feel like you don't want to be by yourself any more," the midwife had said. I called her around 1:30AM. "I don't like this anymore." It was such a relief to do something besides pacing up and down the living room that I didn't even notice the next contraction. "If I can talk through a contraction, does that mean it's not that strong?" I asked.

"Well, usually that's right. But I can come out now if you want me to," she said, polite but unconvinced.

Oh dear. It's going to get much worse than this, I thought. "No, that's okay. Maybe I can survive awhile longer."

I survived exactly 15 more minutes before I called her back. This time I was careful to make lots of convincing puffing noises during a contraction.

"I'll come right now," she said.

I can't remember what I did during the next 45 minutes. Knowing that help was on the way made the time fly. When I judged that she was about to arrive, I woke my husband up. His sleepy sense of decorum dictated that he should be fully dressed for her arrival. I went downstairs again to wait. I still remember the elation and relief I felt at the sound of the front gate creaking, then a clunk as she deposited her midwifery equipment on the doorstep.

The midwife kissed me when I opened the door. Another contraction arrived. "Breathe it out. You're doing great," she said, pressing her hands hard against the small of my back. My back felt less like it was going to fall off with her hands pressing against it. My husband came downstairs.

"We haven't filled the pool yet," I said, "Do you think we should?"

"I'll fill it," my husband offered.

"The hose attachment is in the top drawer in the bathroom, along with the monkey wrench," I said. "It attaches to the shower." I had tried attaching the hose a few times myself to make sure it worked, but it hadn't occurred to me that someone else might need to know how to work it.

My husband was up on a stool in the bathtub when I made it upstairs. "The shower head won't come off," he said. "The monkey wrench doesn't fit." "It fit when I tried it," I snapped, then grabbed for my back again as another contraction arrived.

"Shall I have a go?" The midwife climbed up on the stool, but couldn't make it work, either.

"Who do I have here? Two idiots?" I grumbled. "I don't believe this. I'm going to have to do it myself." I must have been in transition.

But five minutes later, he and the midwife attached the hose. "You can turn it on now," the midwife called from our bedroom. We could hear the water swishing through the hose. A minute later I heard myself exhale hard at the end of a contraction. I was hugging my husband. "That sounds weird," I thought to myself. "Am I pushing? I don't feel like I am. Am I supposed to be doing this?"

"That sounded like a pushing noise!" said the midwife, running in from the other room.

"I think I have to pee." I tried to sit on the toilet. "No, I think I have to poop."

"It sounds like you're pushing," she repeated. "We'd better check how far dilated you are."

After a few more contractions I made it to our bed, where the midwife felt my cervix. "You're fully dilated!" Seeing I didn't look pleased, she added: "That's good news!" I was thinking, "Oh no, now I'm going to have to push the baby out. What if I can't do it?"

"Would you like to get in the pool?" The midwife suggested. It was only half full, but the water was rising quickly. The idea of moving or doing anything different sounded impossible, but I said I'd like to because I figured I was supposed to. I stepped out of my pajamas and into the pool. The instant I felt the warm water over my ankles I felt better. The warm water was instantly soothing, just as all the water birth brochures promised. It was even better than the firm pressure of someone's hands against my back. It was like being hugged all over.

> The instant I felt the warm water over my ankles
> I felt better. The warm water was instantly soothing.

The next contractions still hurt in the small of my back, but they were completely different. Before pushing begins, it is a matter of passively enduring the pain. But when pushing starts, you feel like you're doing something. It felt sort of like throwing up—or should I say, throwing down, because the involuntary hurling of the stomach muscles went down into my gut, not up into my throat. Pushing was scary, uncontrollable. I didn't know how long I could keep it up before my body ripped apart. "Just go with it," the midwife said. "You're doing great." She was kneeling next to the pool. I held onto the rim and crouched in the water. Before, looking straight into her eyes helped me remember that a contraction would end in another moment. Now I wasn't looking at her anymore. I just held onto the rim of the pool and puked my stomach into my pelvis, then closed my eyes and took deep breaths when a break came. I was making heavy, pig-like exhaling noises. Soon after I got into the pool, our son woke up next door. "Mama!" My husband went in to sing him back to sleep. "No…. MAMA!" he protested.

Through the open door I tried to sing him a song between contractions, but quickly realized that he was just as happy listening to his favorite Papa story about the rabbit and the hedghog.

> I tried to sing him a song between contractions…

I started to feel more pressure in my bowels. It sounds strange, but the most pleasurable feeling was the hot sting of the perineum tearing as the baby's head crowned. The feeling of climax. The next push came immediately, and the baby's body squelched out into the water. And then everything stopped. No more pain. Instant peace. The midwife reached down and lifted the baby to the surface. Its face was a purple maroon, with slippery black hair matted down on its head. The baby looked very big to have just come out of my body. It breathed, then cried.

To avoid being disappointed if we ended up with two of the same, we'd convinced ourselves that we were having another son. "Now I get to check that it really is a boy," I thought. "The part I missed last time." I never got to see my first child as he was pulled out of my belly behind the screen on the operating table. The doctors took him away, suctioned his lungs, wrapped him up and presented him like a Christmas present, with nothing but a red, terrified and screaming face sticking out. He might as well have come from a factory. It took months before I was convinced he was my baby.

I lifted the baby's bottom toward the surface. I saw what looked like a scrotum. The rest of the crotch was blocked by the umbilical cord running from the navel, down between the legs and into the water. "Yep, a boy," I decided.

"You have a little girl," the midwife said. I took another look. Indeed, that was no penis I was looking at. Because of all the hormones in the mother's body, the baby's genitals come out all swollen. A daughter! I couldn't believe it. I put her to my breast, and she stopped crying. The midwife tiptoed to the door of our son's room.

"And then Rabbit said to Hedghog..." my husband was saying.

"You have a little girl," she announced quietly.

My husband said later that he'd heard the baby crying, but couldn't really believe that it had been born already.

"Our client isn't going back to sleep," he reported to me. Our son appeared in the doorway. "Ooh!" he said, taking in the blue pool, the midwife, and his mother in the water. He padded up to the pool. "Baby!" he announced, looking at the slippery beet-red person in my arms.

"That's your sister," my husband said to him, his voice cracking. A few minutes later, sitting in the pool, I was on the phone to my mother and brother in Washington, DC.

My daughter and I sat in the water and nursed for nearly an hour until the cord stopped pulsing. The midwife clamped the cord next to the baby's navel, then cut through the tough, horny tube. A moment later the placenta slid out like a big jellyfish.

"Is it true that eating a piece of placenta helps the uterus contract?"

"Yes, though I only know two people who've actually tried it."

"I did last time, but it was probably pointless, because it was after I came back from the hospital. And it must have been filled with epidural drugs and other awful chemicals. I ate a bit raw."

"What did you do with the rest of it? Bury it in the garden?"

"Um, no. I fried it up with onions and ate the whole thing."

"Oh," she said politely. We examined the placenta to make sure no part had been left in the uterus. She pulled off a small, meaty piece. "This is probably enough to help your uterus contract." It was like eating warm, raw liver.

My husband and son came back from cooking up a pot of oatmeal. I climbed out of the pool carrying my newborn daughter. We all watched as the midwife weighed her in a sling: 8lb 6oz—nearly a pound over her brother's birth weight. Then the midwife sewed me up. My perineum had a fairly long but shallow tear from the bottom of the vagina toward the anus, which the midwife said looked like it wasn't sitting together tidily and might heal in an uncomfortable way if it wasn't sewn up. She injected a surface anesthetic, the only drug which appears in my birth notes, then put in four sutures. It was nearly 7:00AM when we finished our oatmeal. I can't remember now who was holding the baby. All I can remember is how famished I was. I was watching the clock. I couldn't wait until the nanny arrived, so I could show off my good work. My husband had paged her to come at 8:00AM, an hour earlier than usual. When I heard her key in the lock, I wanted to leap down the stairs with the baby in my arms. "But she'll tell me to go right back to bed," I thought. I suddenly realized I was exhausted.

I was on such a high that day, I even briefly went out. We invited a few friends over in the evening to celebrate with a glass of champagne. Afterwards, I slightly regretted disturbing our baby by inviting other people into her new home. I am ashamed to confess that in the photos of her birthday party, she is screaming her head off as she's passed around. By the time I descended from Cloud 9 enough to feel like resting, the baby had got over her initial post-birth calm and wanted to breastfeed frequently. I should have slept when she slept! (Mental note: Stay in bed next time and keep visitors away.)

During the day my daughter was born, I could still remember exactly what the sickening wave of a contraction felt like. By the next morning I wasn't sure anymore. Did it really hurt that much?

Some days I feel weepy and nostalgic. I'd like to be back in the blue plastic pool, pushing my daughter—my daughter!—into the world. She's three weeks old now. Some days I miss being pregnant. How strange from one moment to the next to change from a beautiful, rotund, mom-to-be into a tired, empty-bellied mother. Now that life has gone back to normal, I wish I could be pregnant and special again. I miss the feeling of somebody's feet kicking inside me. But then I reach out my hand to touch our daughter's soft, downy head and I remember how I pushed her out into the world. Life won't ever go back to 'normal'—our daughter is here for keeps.

## Postscript by Nina:

Our third child was born after a similar five hours of established labor, five and a half years later. The contractions in this labor felt much more painful than in the second one. Boy, did it hurt! (They do say third births are sometimes tougher. Or maybe I noticed the pain more. I was expecting it to be a piece of cake!) This time, I did take it easy and let people wait on me. Like her big sister, the new baby was calm and happy after the birth, and so were her older brother and sister.

I would gladly re-live the births of either of my daughters. There is nothing more intensely satisfying than laboring and birthing a baby. I wish I could do it again a dozen times more. It's raising them that's hard work!

*Here's Nina, feeding her second baby in the woods. As a financial editor, Nina has taken up some activities (such as breastfeeding) with a very different pace since giving birth. What other kinds of adjustments do women make?*

# Pause to reflect...

## ...about vaginal birth after cesarean (VBAC)

1   Given the accounts you've read, what do you think are the keys to facilitating VBACs—what are the essential components?

2   What should you avoid doing at all costs?

3   What is it desirable for a caregiver to do, in order to cover everything possible to facilitate a VBAC?

4   Do you think you need to focus on anything particularly with the client while she is pregnant, before she goes into labor?

5   Why is it, do you think, that reading VBAC birthframes and other women's accounts of physiological births is often so inspiring and reassuring to potential VBAC women?

6   If the woman has a panic while she's in labor, what can you do about it?

7   Why do you think so many VBAC mothers find their VBAC so immensely satisfying?

8   Do you think optimality means the use of pain relief in a VBAC or a physiological approach? If your answer is 'pain relief', what are the potential risks and drawbacks?

9   If you suspect that a woman who is requesting a VBAC was previously a 'victim' of a cascade of unnecessary intervention, how can you identify where things went wrong and why? Are there any other methods you could use?

10  Why do you think women request VBACs even after several cesareans?

For material relevant to these questions, see Birthframes 12, 13, 14, 15, 16, 17, 18, 19, 27, 28, 30, 32, 34, 35, 36, 37, 42, 43, 44, 45, 46, 48, 49, 50, 52, 53, 54, 55, 56, 57, 58, 59, 70, 75, 76, 77, 78, 79, 81, 82, 83, 84, 86, 87, 88, 95, 96, 97, 98 and 99.

Before we go any further considering risk assessment, let's take a break and move on to a related topic...

# Prenatal testing

You are probably required to offer your clients many tests. In some cases, either you or the hospital will even assume that a woman is going to have certain tests, without even consulting her on her wishes or preferences. In other cases, you may well put pressure on moms-to-be in your charge because you may personally feel that specific tests dramatically reduce risks for women. While this may well be the case, and while hospitals may be behaving extremely efficiently in assuming that all women will want tests, I would like to ask you here to reconsider these protocols...

## Reassurance vs. worry

While some tests may reassure clients—because a negative test result is always pleasant to receive—the wait and/or a positive test result may have an altogether different effect. Of course, this is the reason that chorionic villus sampling is sometimes offered in preference to amniocentesis (since it can be offered earlier and the wait is shorter).[6] But is the test reliable enough in terms of identifying problems or predicting outcomes and might the worry caused by the prospect of side-effects (e.g. the risk of miscarriage) outweigh the advantage of the reassurance the test seeks to give? The 'nocebo effect', i.e. where procedures carried out by caregivers cause fear and worry in clients, may be more significant than any additional information obtained.[7] Should pregnancy be a period of testing, waiting, worrying and relief (or discussion), or should we consider the advantages of a more hands-off approach instead? Even if tests continue to be used, how is it possible to safeguard the joy, anticipation and sense of discovery a pregnancy should optimally involve? What is your role as a caregiver? To what extent should you be a 'checker' and to what extent should you be a person who inspires and educates? How could you give women more choice in terms of the care-giving style they experience?

**Michel says...**

Think about what a screening test really is. When you go into the airport, you go via the metal detector. In some cases it rings but it doesn't mean you have a weapon in your pocket; it may just be a piece of wire. So if a test is positive, it simply means a risk which is more than 1 in 200, for example. That means there is an almost 99.5% chance that there's no problem.

**Here's a comment from a woman who ended up with an unnecessary section:**

❝ Almost every visit I was given some test for something that I was at 'very low risk' for, but that I should have the test 'just in case'. They would say it was my choice, but it was obvious that they really thought I should take the test, so I would do it. I was a good patient. I was afraid, both of being seen as irresponsible or a troublemaker, and of the serious diseases and situations I was supposedly being tested for. I would wait with anxiety for the test results.

# Implications of communication

The way in which you present a test is highly significant in both personal and legal terms. Not only are you responsible for effectively and clearly communicating information (so that true informed consent will be possible), you are also responsible for conveying more nebulous impressions, which will affect the mom-to-be's feelings and perceptions. And while you communicate information, you need to be very aware of the controversies surrounding different kinds of tests. (Just how invasive are they? And why do some women get so upset about the prospect of a test which seems so innocuous to other people?)[8] If you find yourself at the 'dismissive' end of the spectrum in terms of evaluating the invasiveness of tests, perhaps you should consider some observations from researchers who have considered the possible effect of amniocentesis on the fetus. A range of reactions were noted from unborn babies, including an accelerated heart rate, staying motionless for two minutes after insertion of the needle and even slower breathing movements for up to two days after the amniocentesis test. One fetus who was accidentally hit by a needle was filmed twisting away and then repeatedly striking out at the needle barrel.[9] Another retracted a limb and turned a somersault when accidentally hit by the needle and one even knocked the needle away with its tiny hand.[10] One scientist even found that the catecholamine levels of a fetus whose mother had amniocentesis rose after the test—indicating that the growing baby felt under stress.[11]

### A couple of comments from other women...

❝ As I was nearing 40 when pregnant with my fourth baby, we read about amniocentesis and decided to have it done. The information we read had informed us that the positioning of the needle was carefully monitored by ultrasound. We were both very surprised when staff identified what they considered to be a pool of amniotic fluid then switched the scanner off before inserting the needle. Our baby was very active at this stage and could easily have changed position during the time taken to prepare the needle and insert it—a risk we would not have taken had we been given ALL the information.

❝ The medics told me my bloods were wrong and I needed amniocentesis to check for Down's syndrome, but I declined, saying I'd rather have another ultrasound or check my blood for fetal blood cells to check its chromosomes. My husband didn't agree, but I decided that at this late stage (five months) I wouldn't have an abortion anyway, so what was the point in giving the fetus a headache or worse? Every trip to the clinic made me feel worried as they found 'inconsistencies' so I stopped going—I felt absolutely wonderful, the fetus was growing and moving, if I was out of their standard ranges then what of it? Although I understood how the system worked, it was extremely hard to withstand the pressure that the community places on you. You are so vulnerable when pregnant and it's hard to go against the flow.

A woman is so vulnerable when she's pregnant...

# The outcome of a 'positive' result

In a case where the result of a test is positive it is obviously only useful if it influences treatment. If a woman has decided in advance that she wouldn't have an abortion why offer her tests which check for Down's syndrome, etc? Do you spend enough time discussing outcomes of test results with women before they agree to have the test in the first place?

If a dating ultrasound is carried out to confirm the EDD, what use is this if the woman feels strongly about being induced? Of course, as in other cases, women need information so that they can make an informed choice.

### Here's an extract from one of Michel's articles:

In many countries about 10 prenatal visits is routine. Each visit offers an opportunity for a battery of tests. These traditional patterns of medical care are based on the belief that more prenatal visits mean better outcomes. They are not based on scientific data.

Studies made in the UK failed to find any association between late enrolment in prenatal care (after 28 weeks' gestation) and either adverse maternal or neonatal outcomes[12] or between the number of visits and the onset of eclampsia. This casts doubts on the efficacy of such protocols.[13]

I asked Michel for more specific comments...

### What advice would you give someone who's just discovered she's pregnant?

I rarely give advice. It's not in my nature to give advice. Hmmm. I suppose I'd say, if you know that you are pregnant and if you know when you have conceived your baby and you think that everything's okay, doctors can probably do nothing for you. Women need to realize that the role of medicine in pregnancy is very limited. Really, if a woman feels she's in good shape, the only thing doctors can do is to detect a gross abnormality and offer an abortion. That's all. And even then, there are false positives. What's important is for a mom-to-be to be happy, to eat well, to adapt her lifestyle to her pregnancy, to do whatever she likes to do. If a woman has a passion for her job, perhaps it's better for her to go on working... I think that's what we have to explain to women. They have to realize that doctors have very limited power.

What advice would you give a mom-to-be yourself? What's your view on prenatal testing? Would your view be different in your own pregnancies?

### A comment from a mother of three born 1967-72, and grandmother of seven, born between 1984 and 2006:

66 It seems there's a lot more worry involved in being pregnant for women nowadays. Pregnancy was an awful lot simpler in our day without all the prenatal testing. There was a lot less worry. Since there weren't any tests, we just had to trust that everything would be fine. You can always find people who will tell you horror stories about what can go wrong, but you just have to block them out and believe that the best will happen. Most of the time, it does.

## Our modern need to create the 'perfect' baby

I find it interesting how *different* women's attitudes are during pregnancy and postpartum. While there are no doubt numerous cases of babies who are rejected at birth and even given up for adoption, there are also many, many cases of parents who survive the birth of a baby who is not 'perfect'. Down's syndrome seems a particular case in point because children born with this condition sometimes have a *different* range of skill sets to 'normal' children or their development is slower. These children are also well-known for being affectionate and loving and many of them now attend regular schools.

Perhaps this raises some more philosophical issues too... What are two people really doing when they plan to have a baby? What outcomes do they dare to hope for? What are their motives? How much are they prepared to adapt their current lifestyle in order to accommodate the new baby (or babies!) into their lives and meet his or her (or their) needs?

Of course, these questions also raise much bigger questions, which are perhaps much more important from your point of view:

1 Should society be more careful about preparing people for pregnancy, birth and parenthood? How well is sex education fulfilling this function?
2 What expectations (and fears) of birth and parenthood is sex education currently creating? What practicalities are being routinely taught, if any?
3 To what extent does a caregiver need to prepare a mom-to-be and her husband for parenthood—and for all its potential joys and disappointments?

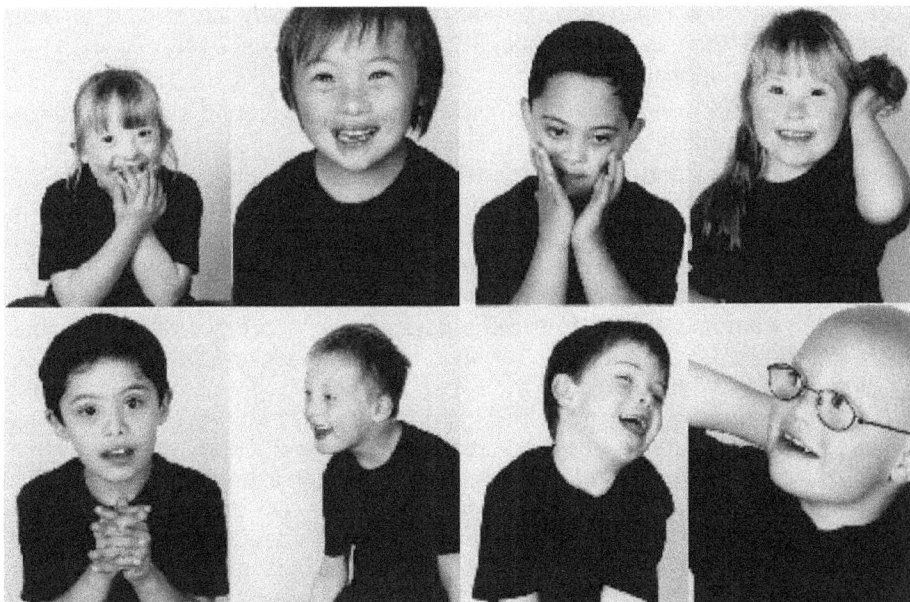

*Here are a few photos of children with Down's syndrome taken by Richard Bailey.*
Photos © Richard Bailey. See www.ds2005.com for an update on Richard's work.

**A comment from another contributor, who was an elderly primigravida:**

66 I was 36 and didn't want a lot of medical intervention so I didn't go to the doctor until at least 14 weeks were up, just in case the fetus dropped out beforehand. I used to be a nurse and had delivered babies in developing countries, and knew that the medical system in the UK is not very good at reducing stress in the mother. In fact I knew that lots of trips to the hospital, clinics or even prenatal yoga classes would make me more anxious. Pregnant women are forever being told what to do and not do, loads of prohibitions on what we must eat, drink, smoke, work at, etc and treated as if we are to blame for everything that might go wrong. I figured that if it was so hard to have a healthy baby there wouldn't be an over-population problem in the world, although most first-time mothers are a lot younger than I was.

## Monitoring procedures

Monitoring is another aspect of prenatal care which needs to be thought about because it can lead to additional tests, interventions and radically different birthing experiences. The way in which it's conducted can also dramatically affect any individual woman's 'emotional landscape', which in turn may have an impact on hormonal processes—because, as we've seen, these are intrinsically interlinked. (As we've already noted, it's no coincidence that we produce adrenaline when we're afraid and oxytocin when we're having a meal with a friend or our partner.)

It's even useful to consider the *definition* of 'monitoring'. Personally, I prefer a broad definition: it can be anything from a few questions and palpation to the use of ultrasound. Obviously, the idea is to track both maternal and fetal progress and intervene with additional checks when any concerns arise, all so as to ensure optimal outcomes.

Let's consider some typical forms of prenatal monitoring so as to gain a greater understanding of them:

- **Blood pressure and urine check** While the aim is to check and record the woman's blood pressure and urine, does anything else happen at the same time? Since women are told by their caregivers that these checks are carried out so as to check for symptoms of pre-eclampsia and the life-threatening condition eclampsia, which very occasionally follows it, many (if not all) women are likely to travel a psychological journey. I well remember my feeling of anxiety each and every time these checks were performed on me when I was pregnant and the sense of relief I experienced when I heard I'd again 'passed' the tests. The importance of distracting and reassuring women while carrying out these tests is clear. And if ever you discover a high blood pressure reading late on in a woman's pregnancy, it's worth remembering that this alone needn't necessarily constitute a problem. In fact, research shows that a woman's blood pressure is *supposed* to rise later on in pregnancy (back up to normal levels)—and that the rise is associated with good outcomes![14] Also, while remembering that few medical treatments are available to effectively reduce a woman's blood pressure, simply advising a woman to lie on her left side might help to reduce any pathologic rise.

- **Palpation** While the objective is to establish fetal size, positioning and the amount of amniotic fluid, again there can be a psychological effect on the woman. If an unusual position is discovered, although there is value in informing the woman, it's obviously also important to give her information about the likelihood that the fetal position may well change later on in her pregnancy. Also, as we noted in an earlier chapter, it's also helpful to give her advice on how she might be able to influence the position of her baby, through the day-to-day positions she adopts (see pp 469 and 512-3). While measuring a woman's 'baby belly' manually (with a tape measure), it's also important to remember that bellies grow outwards at the sides, as well as at the front! If ever you suspect the fetus may be SGA, consider whether or not it's really constructive offering an ultrasound in a case where the woman has refused ultrasounds before, or when she is unlikely to accept an induction.[15]
- **Auscultation** It's worth remembering that a) many women might prefer to have this carried out without the help of ultrasound (using a fetoscope) and b) it can be done with the woman sitting up, rather than lying on her back.

I decided to ask Michel Odent about auscultation...

Michel, why is auscultation a routine part of prenatal care? Obviously, it's to listen to the fetal heartbeat... but why? A dead fetus would be detected, but the woman would have found out a few days later anyway, since presumably she would go into labor if that were the case. Since auscultation is carried out only once each month/two weeks/week (depending on the stage of pregnancy), the chances of catching a fetus while in distress, but early enough to carry out an emergency cesarean, seems remote. So what's really the reason that auscultation is carried out? I can understand it better in labor—but even then, a minimum of disturbance to the laboring woman seems ideal—hence your preference for the Doppler, of course, which allows more discrete auscultation.

Sylvie, you are right. Prenatal auscultation is not a very useful ritual and cannot change outcomes. Here is a list of studies regarding EFM.[16] References 2, 6, 8 are about the non-stress test, which simply means prenatal auscultation. None of these studies could detect an effect on birth statistics. Reference 13 is about admission cardiotocography (a 20-minute trace when the woman in labor enters the maternity unit). Once more, no detectable effect on statistics.

## Questions for reflection...

1  What's your view on ultrasound, in full ultrasounds and in Doppler technology?
2  To what extent should you listen to women's concerns about side-effects?
3  How much information do you have yourself and how well do you keep up with the latest research?
4  How much information do you give women before using ultrasound?
5  What would your choices be for your own pregnancy?

## Ultrasound

Ultrasounds have quickly come to be seen as a non-invasive, routine form of monitoring even though many research studies have detected changes in outcomes when ultrasound is used. (For example, one study noted an increased incidence of left-handedness, which has been associated with behavioral disorders.) Although none of the changes seem to be specifically pathologic, surely the fact that different outcomes have been detected in some studies should make us hesitate to use this technology unless it is really likely to change treatment offered and accepted? Shouldn't our memory of other recent mistakes—e.g. X-raying—give us reason to be more hesitant to use artificial technologies when they aren't strictly necessary? And shouldn't we note that studies focusing on how ultrasound has improved outcomes have so far drawn a blank? Here's an extract from an article by Michel Odent:

Routine ultrasound scanning in pregnancy has become the symbol of modern prenatal care. It is also its most expensive component. A series of studies compared the effects on birth outcomes of routine ultrasound screening versus the selective use of ultrasounds. One of these randomized trials, published in the *New England Journal of Medicine*, involved 15,151 pregnant women.[17] The last sentence of the article is unequivocal: "Whatever the explanation proposed for its lack of effect, the findings of this study clearly indicate that ultrasound screening does not improve perinatal outcome in current US practice".

Around the same time, an article in the *British Medical Journal* assembled data from four other comparable randomized trials.[18] The authors concluded: "Routine ultrasound scanning does not improve the outcome of pregnancy in terms of an increased number of live births or of reduced perinatal morbidity. Routine ultrasound scanning may be effective and useful as a screening for malformation. Its use for this purpose, however, should be made explicit and take into account the risk of false positive diagnosis in addition to ethical issues".

It is possible that in the future a new generation of studies will cast doubts on the absolute safety of repeated exposure to ultrasound during fetal life. One of the effects of this might be to reduce dramatically the number of ultrasounds, particularly in the vulnerable phase of early pregnancy.

Even in a high risk population of pregnant women, ultrasounds are not as useful as is commonly believed. Evidence from randomized controlled trials suggests that sonographic identification of fetal growth retardation does not improve outcome despite increased medical surveillance.[19] In diabetic pregnancies it has been demonstrated that ultrasound measurements are not more accurate than clinical examination to identify high birthweight babies.[20]

Ultrasounds are not as useful as is commonly believed

### An anonymous comment, gleaned from the Internet:

66 I am appalled at the practice of renting hand Dopplers out to expectant mothers, some of whom are listening to their babies several times a day. Talk about exposure! Also, if a woman isn't really trained at using one correctly, I would assume that every time she can't find the heartbeat it causes undue distress.

**Here's another comment from a woman who contacted me:**

66 I have a currently pregnant friend who has had Weeks 8-20 of her pregnancy completely ruined by a false positive ultrasound which gave her all sorts of alarming 'news', none of which has now been found to be true.

## What is ultrasound and why is it used?

An imaging technique dating back to the Second World War, ultrasound involves the use of ultra-high frequency sound waves traveling at 10-20 million cycles per second.[21] (Note that this is dramatically faster than audible sound, which only travels at 10-20 *thousand* cycles per second.) Initially developed to detect enemy submarines and later used in the steel industry, ultrasound soon became the focus of research conducted by a surgeon in Glasgow, Scotland. In July 1955 this surgeon—Ian Donald—tried using it on abdominal tumors which he'd removed from his patients. Realizing that different tissues gave different patterns of 'echo', he soon realized this represented a new way of looking into the world of the growing fetus.[22] The new technology quickly spread into clinical obstetrics and commercial machines were already available in 1963.[23] By the late 1970s ultrasounds had become a fairly routine part of obstetric care.[24]

When a sonographer carries out an ultrasound the hand-held transducer emits ultra-high frequency soundwaves. The echo caused when the waves hit shapes inside the woman cause an image to build up, since hard surfaces (such as bone) cause a stronger echo than that produced by soft tissues. An ultrasound machine uses pulses of ultrasound lasting a mere fraction of a second each, and uses the gaps to interpret the echo that is produced. Other devices, such as fetal monitors, Dopplers and vaginal scanners use continuous waves, which mean the woman has longer exposure to the ultrasound.

You are probably well aware of the various uses for ultrasound in prenatal clinics nowadays. It seems, though, that ultrasound is only really useful for confirming whether or not a woman is expecting more than one baby, for determining whether or not a baby is breech, very close to 40 weeks, and for checking for placenta previa. (Of course, you can identify breech and multiple pregnancies through palpation too.) As you probably know, in 19 out of 20 cases placenta previa diagnosed by ultrasound is actually *misdiagnosed* because it's checked for too early: the placenta will effectively move up later in the pregnancy and will not cause problems at the birth. In any case, a study conducted in 1990 concluded that detection of placenta previa by scanning is not safer than detection in labor.[25] Finally, using ultrasounds to check fetal growth also seems unnecessary: research shows that dating by ultrasound may be inaccurate, especially for ultrasounds later in pregnancy.[26] It seems the best way of checking fetal growth may still be palpation.

The best way of checking fetal growth is still palpation

Carrying out ultrasounds for abnormalities, although widespread, also seems to be unreliable. Apparently, only between 17% and 85% of the 1 in 50 babies that have major abnormalities at birth are identified in advance.[27] A recent study from Brisbane showed that ultrasound at a major women's hospital missed around 40% of abnormalities,[28] and major causes of intellectual disability such as cerebral palsy and Down's syndrome are unlikely to be picked up on a routine ultrasound, as are heart and kidney abnormalities.[29] Of course, ultrasounds, like other kinds of testing, also produce false positives. A UK survey showed that, for 1 in 200 babies aborted for major abnormalities, the diagnosis on post-mortem was less severe than predicted by ultrasound and the termination was probably unjustified.

In this survey, 2.4% of the babies diagnosed with major malformations, but not aborted, had conditions that were significantly over or under-diagnosed.[30] There are also many cases of error with more minor abnormalities, which can cause anxiety and repeated ultrasounds, and there are some conditions which have been seen to spontaneously resolve.

As well as false positives, there are also uncertain cases where the ultrasound findings cannot be easily interpreted and the outcome for the baby is not known. In one study involving women at high risk, almost 10% of ultrasounds gave uncertain results.[31] This can create immense anxiety for the woman and her family, and the worry may not be allayed by the birth of a normal baby. In the same study, mothers with 'questionable' diagnoses still had this anxiety three months after the birth of their baby.[32]

In 1975 a study of scanning unborn babies using Doppler ultrasound was published in the *British Medical Journal*. The researchers didn't tell the mothers whether the ultrasound machine was switched on, but when it was the fetuses were found to move about much more. In the UK in 1993, the Association for Improvement in Midwifery Services' journal, the *AIMS Quarterly*, published a selection of comments from mothers about their babies' responses to being scanned.[33] They included:

66 The baby was moving around so much the technician could not take any measurements...

66 It had both hands up to its ears, fist fashion.

66 The gynecologist got very frustrated because he could not get a clear picture because she (the baby) would not sit still. At first she would move to a totally different part of my womb, then when she was bigger, turn around and around.

66 [The baby] was extremely active when we wanted a picture of her. Then she put her head as low as possible in my pelvis where the ultrasound seemed to have difficulty getting a clear picture.

And what of the known effects of ultrasound? I learned from Sarah Buckley, who has researched this area, that ultrasound waves are known to affect tissues in two main ways. Firstly, the sonar beam causes heating of the highlighted area by about 1 degree Celsius. (Apparently, this is presumed to be non-significant, based on whole-body heating in pregnancy, which seems to be safe up to 2.5 degrees Celsius above normal.)[34] Secondly, ultrasound apparently causes 'cavitation'. This is where the small pockets of gas (which exist within mammalian tissue) vibrate and then collapse. In this situation, according to the American Institute of Ultrasound Medicine Bioeffects Report (1988) "temperatures of many thousands of degrees Celsius in the gas create a wide range of chemical products, some of which are potentially toxic. These violent processes may be produced by micro-second pulses of the kind which are used in medical diagnosis..." The significance of cavitation effects in human tissue is unknown.[35]

A number of studies have indeed suggested cause for concern. Studies not involving humans have shown that cell abnormalities persist for several generations,[36] that the myelin that covers nerves is damaged[37] and that rates of cell division are reduced, while 'aptosis' (programmed cell death of the small intestine) happen two times as often.[38]

Studies on humans exposed to ultrasound have shown that possible adverse effects include premature ovulation,[39] preterm labor or miscarriage,[40] low birth weight,[41] poorer condition at birth,[42] perinatal death,[43] dyslexia,[44] delayed speech development[45] and less right-handedness.[46] (Non right-handedness is, in other circumstances, seen as a marker of damage to the developing brain.)[47] One Australian study showed that babies exposed to five or more Doppler ultrasounds were 30% more likely to develop intrauterine growth retardation—a condition that ultrasound is often used to detect.[48]

For Sarah Buckley, ultrasound represents yet another way in which the deep internal knowledge that a mother has of her body and her baby is made secondary to technological information that comes from an 'expert' using a machine. She says this is how the 'cult of the expert' is imprinted from the earliest weeks of life. She feels that by treating the baby as a separate being, ultrasound artificially splits mother from baby well before this is a physiological or psychic reality. For these reasons, Sarah urges all pregnant women to think deeply before they choose to have a routine ultrasound. It is not compulsory, even though ultrasounds are sometimes pre-booked for women on the assumption that all pregnant women will want them. Sarah says the risks, benefits and implications of scanning need to be considered for each mother and baby, according to their specific situation. She encourages pregnant women to seek out experienced and highly-skilled operators and suggests they request that the ultrasound be performed as quickly as possible. If an abnormality is found, she advises pregnant women to ask for counselling and a second opinion as soon as is practical, remembering that it's her baby, her body and her choice at stake.[49]

## Birthframe 47: Nina Klose—To get the ultrasound or not?

Here are some contemplations from one woman who was particularly well-informed during her pregnancies. To what extent can you relate to her experiences? In what ways do you think the ultrasounds she had enhanced or spoiled her pregnancies? Would you have made the same decisions if you had been pregnant yourself? Will you advise your own daughters (if you have any!) to have ultrasounds or would you be at all concerned about the effect they might have on your grandchildren? What advice or information do you think you should give women in your care, considering both ethics and safety factors?

I had plenty of ultrasounds during my first two pregnancies. I'm not sure now it was such a great idea. During each pregnancy, I had a false positive for a serious defect! Not only did these false positives cause endless extra anxiety, which can only have been bad for the fetus, they also probably contributed to an unnecessary cesarean section for my son.

During my first pregnancy, the 'triple test' for neural tube defects showed an 'abnormally high' AFP level, which is supposed to indicate a higher likelihood of spina bifida. I was rushed in for an extra ultrasound the next day, at 18 weeks. The ultrasound indicated no sign of any problem. "The best way to explain why the AFP level was high is that your baby is actually older than we thought," the doctors said. Even though I was certain I knew the date of conception, I agreed to let the doctors re-date the pregnancy based on the first (12-week) ultrasound, giving me a due date two weeks earlier. Later, I went 'overdue' past the new, earlier due date. I got nervous I would be induced. I tried to bring on labor. Things went awry. The rest is history: my water broke, the baby passed meconium, I ended up in hospital: induction, epidural, cesarean, the classic sequence.

The second time around, the 12-week ultrasound showed choroid plexus cysts, according to the doctors. Apparently, cpc's—which appear in the ultrasound as flower-like empty spaces in the developing brain—are a normal stage of fetal brain development. However, they're 'supposed' to disappear by 12 weeks. If they are still visible at 12 weeks, they can be 'associated' with Trisomy 18, a very rare, fatal genetic disorder. "We can't see the fetus too well, but it looks as though there could be something unusual in the brain. Do you mind if we use a vaginal scanner?" Next, I had the pleasure of a roomful of male doctors stuffing a dildo-shaped scanner up my vagina while they searched for evidence of my baby's ostensible abnormality. (My husband was abroad, so I went to the ultrasound alone, expecting it to be entirely routine.)

They decided 12 weeks was too early to tell conclusively if there was a problem. "Don't worry," they said, "you have only a very small chance of having a baby with T-18. Just go home, relax, and come back in three weeks for another ultrasound, at which point we'll be able to tell more clearly if there's any problem. Meanwhile, don't let this spoil your early pregnancy." Giving me no more information than this to go on—not even the spelling of 'choroid plexus cyst,' or any statistics on the occurrence of cpc's and their link to Trisomy-18, I was sent home to panic alone.

I researched cpc's and learned that the chance of any problem really was very low, at most only 1-3%. I trusted my intuition that everything was fine and calmed down. Not surprisingly, the next ultrasound indicated that everything was fine, and we soon forgot all about it. Why did we put ourselves through such an ordeal in the first place? It was all in the interests of our child's future health, or so we thought. But what did we gain by it?

Even though I felt doubtful about ultrasound by the time I got pregnant for the third time, when it came to it I just couldn't withstand the pressure! I felt that a third pregnancy was an even bigger responsibility than a first or second, because now I already had three other people (husband and two kids) involved. I wasn't 100% sure I would decide to carry a Down's baby to term. I wanted to be able to demonstrate to myself and my husband that I had been responsible in trying to have a healthy baby. I decided against amnio, which was routinely offered at the age of 38, in favor of the 'integrated test', which includes one ultrasound and two blood tests. I decided to do the ultrasound because it was less invasive than amnio. So I made compromises, accepting some tests and declining others. A funny approach? Perhaps.

A baby is so much a gift from God. No matter how many tests you do, there's only so much you can know in advance. There are plenty of other things that can go wrong either before, during or after birth that can't be measured or tested for. Does it really make sense to put the baby on trial? Isn't it better to have faith that things will go well, to love the baby and let it develop as it's meant to?

*Nina relaxing with her baby, after her third child's birth*

## Prenatal checks

There are two other commonly performed tests which seem to cause a great many pregnant women a great deal of worry! Treatment for a positive result seems to be fairly standard, but it is actually quite controversial...

- **Anemia check** As you know, it's a routine part of prenatal care to check the amount of red blood cells pigment (the hemoglobin concentration). This is because there is a widespread belief that this test can effectively detect anemia and iron deficiency. In fact, it cannot diagnose iron deficiency because the blood volume of pregnant women is *supposed* to increase dramatically. The hemoglobin concentration indicates primarily the degree of blood dilution, which is an effect of placental activity. In fact, a large British study, involving 153,602 pregnancies,[50] found that the highest average birth weight was achieved in the group of women who had a hemoglobin concentration between 8.5 and 9.5. These are the women who would normally be called 'anemic'! Michel Odent says that although health professionals tell women they are at risk of anemia unless their hemoglobin concentration is above 10.5, when the hemoglobin concentration *fails* to fall below 10.5 research shows there is an increased risk of low birth weight, preterm birth and pre-eclampsia![51] He says it is a regrettable consequence of routine evaluation of hemoglobin concentration that women are told they are anemic—when they aren't—and are given iron supplements. There is a tendency both to overlook the side-effects of iron (constipation, diarrhea, heartburn, etc) and also to forget that iron inhibits the absorption of such an important growth factor as zinc.[52] Research has shown that iron supplementation can also exacerbate lipid peroxidation (a process which leads to the development of free radicals, which is obviously bad news) and it can even increase the risk of pre-eclampsia.[53]

## Questions for reflection...

1 How often do you (or do you think you would) prescribe iron tablets?
2 What are the side effects of iron? Do you think it's important that iron inhibits the absorption of zinc (which is essential for growth)?
3 What do you think of the idea of encouraging dietary changes or a food-based solution, such as that offered by 'Floradix'?
4 Do you ever run any more tests, after the standard one, to double-check for anemia?
5 How often have you observed hemorrhage as a result of anemia?
6 Do you think there are any differences in outcome in different countries, i.e. developing countries vs developed countries because of nutritional status?

There is a tendency both to overlook the side-effects of iron

- **Gestational diabetes check** Another test which many practices give routinely is for so-called gestational diabetes. Michel Odent says this diagnosis is useless because it merely leads to simple recommendations that should be given to all pregnant women, i.e. they are told to avoid sugar (including soda, etc.), to choose complex carbohydrates (wholemeal pasta, bread, etc) and to make sure they get enough physical exercise. A huge Canadian study demonstrated that the only effect of routine glucose tolerance screening was to inform 2.7% of pregnant women that they have gestational diabetes; the diagnosis did not change birth outcomes.[54] This issue should be of particular concern when we consider how anxious women become when they are identified with some kind of 'illness' during their pregnancies. Although common sense suggests it must be more difficult to give birth to larger babies (which may result from women who fail the glucose tolerance test) there are plenty of accounts which suggest that size is not the issue since all babies' heads tend to have a diameter of no more than 10cm. Even if the risk of shoulder dystocia really is higher, would it not be better to take another approach, which doesn't involve unhelpfully labeling women? Is it not time that more nutritional studies were conducted to determine how nutrition during pregnancy might affect a growing fetus? Should we perhaps even evaluate the Russian practice of fasting one day a week and eating plenty of fruit and veg on other days?!

This all leads us to a very important question: what can a health professional do in order to influence outcomes? Over to Michel again...

From the point of view of the mom-to-be, the primary question should be: "What can the doctor do for me and my baby, since I already know I am pregnant and I can feel the baby growing?" The doctor should answer with humility: "Not a lot, apart from detecting a gross abnormality and offering an abortion."[55]

We should not conclude that there is no need at all for medical visits in pregnancy: we cannot make a comprehensive list of all the reasons why women might need the advice or the help of a qualified health professional before giving birth. It is the word 'routine' that should be discarded. It is easy to explain why current habits are a waste of time and money; it is also easy to explain why they are potentially dangerous. It is dangerous to misinterpret the results of a routine test and to tell a healthy mom-to-be that she is anemic and that she needs iron supplements. It is dangerous to present an isolated increased blood pressure measurement as bad news. It is dangerous to tell a mom-to-be that she has 'gestational diabetes'. In general, it is the very style of medicalized prenatal care, constantly focusing on potential problems, which causes problems by making pregnant women worry.

The fall of routine medicalized prenatal care should take place alongside a rediscovery of the basic needs of pregnant women. I well remember the atmosphere of happiness that accumulated at singing evenings in the maternity unit at the Pithiviers hospital, in France. These singing sessions probably had a more positive effect on the development of babies in the womb than a series of ultrasounds. Pregnant women need to socialize and share their experiences. It is easy to create occasions for this: swimming, yoga, prenatal exercise sessions... Let us dream of the potential of specialized restaurants for parents-to-be!

## Grey areas of risk assessment...

Coming back to risk assessment, the next two births would be considered high risk by some caregivers, but fairly low risk for others. What is your own view? Why is it, do you think, that some midwives feel confident in their breech birth skills, while others avoid breech births whenever they can? Why do you think some people were so easily persuaded by the results of the huge Hannah trial into the safety of vaginal breech birth (see page 597), while others dismiss this trial because of problems with the research design? Do you believe the results of this study might have been different if the attendants who 'allowed' the women to give birth vaginally were a) well-trained in breech birth, b) experienced in breech birth, and c) not afraid? Could the results have been still different if the women concerned had been laboring and birthing in entirely undisturbed circumstances, as the next account implies might be the case?

### Birthframe 48: Esther Culpin

Here, a midwife decided she would prefer to arrange a vaginal home birth, even though she knew her baby was in the breech position.

At a time when breech presentation is almost synonymous with cesarean section, I find it useful to write up the story of the easiest of my four deliveries. In this case, changing the birth environment absolutely transformed the way I gave birth. I set out to create a new birthing environment this time around as my previous births, although loosely defined as 'normal', were definitely not. The births of my three sons followed long and traumatic labors and I experienced excessive blood loss immediately afterwards. Overall, birth appeared to be extremely risky and to contemplate it all over again did not seem like a good idea. The solution to my difficult birth experiences could have been an elective cesarean section. Instead, I had the opportunity to look more closely at important issues that might have affected the way my first three births had turned out.

Giving birth at home was a really important factor for me because of the absolute freedom it gave me to do as I wished in labor. I needed to arrange for a midwife to be in attendance who would respect my need for a calm and undisturbed environment. Michel Odent was happy to assume this role and was noticeably unperturbed by my traumatic labor and delivery record!

When, after 30 weeks of pregnancy, my daughter was persistently a breech presentation, I made no change of plan. I was still happy, confident and looking forward to an easier birth in the privacy of my home. At this point technology could have taken over, but all the required information seemed to be available, literally through the midwife's hands. There was never a suggestion that I should undergo an external cephalic version (ECV) and, having experienced that procedure 14 years previously, I did not feel I wished to undergo it again. Although I was still aiming for the birth to be at home, I booked in at the maternity hospital just one mile away, in case admission to hospital and emergency treatment should be required. Local midwives, although acquainted with home birth, indicated their preference not to be involved with a breech birth if it was to be at home.

Local midwives indicated their preference not to be involved

I went into labor a week after my due date. This time around, as part of a strategy for giving birth easily, I aimed to keep myself rested. This was achieved by not doing too many things in a day so that I would be able to cope with the rigors of labor whenever it started. As the process got underway, I found that being at home had a direct effect on the way I coped and on the optimism I felt. (Remember, the baby was breech, I had never experienced an easy birth before, and I was at home!)

I had never experienced an easy birth before and I was at home

On the domestic front I was assisted by my husband. Again, considering my history of traumas associated with giving birth, he was superb. His responsibilities were wide-ranging, but his priority was to maintain a safe, dark and secure environment for me, so that I should experience no disturbances. A birthing pool in the living room was filled with warm water, in case it should be needed for pain relief. The children went out to breakfast with their grandparents. For this birth, because I badly wanted things to progress easily, I did not wish to have any distractions at all.

For me, labor in any circumstances remains hard, but given that this time I would be able to adopt any position, and there were no outsiders coming in and out of the room (as there could easily be in the hospital setting), I felt that I was on the way to giving birth quickly and easily.

In fact, for the first time in my experience, labor progressed extremely quickly and I found myself trying to slow things down so that I would not give birth before assistance arrived! What was noticeable at this point was the lack of instructions I was given: Michel gathered silently all the information he needed to assess the situation. I was obviously ready to deliver and, because I was not directed in any way, I decided to get into the pool!

At the next contraction and whilst standing upright, my baby's body was born. I was assisted out of the pool and supported from behind so I could maintain a standing position. Now there was a long pause while the cord, which was wrapped tightly three times around her neck, was unwound. Her head was deflected by inserting a finger into her mouth before her head was delivered. During these few critical minutes there was no discussion about whether I had a girl or a boy.

My daughter lay on the carpet, motionless at first. But, in that situation, I was her life support, just as I had been all along. The fact that I was personally and actively involved in those early moments was of prime importance to me, whatever the outcome would be. I trusted deeply that my daughter would live and I felt that my participation in every way would give her an optimal chance. But, whatever, we were together in this and that's how it should be, whatever the outcome.

The position that I adopted at this point, immediately after her birth, was also extremely advantageous. I was leaning over my daughter, who was lying on the ground. This was the optimal position in the early moments of her life: it aided the natural compression of the uterus and meant that the baby could be readily gathered up as soon as this became appropriate. There was no cutting of the cord, no touching of my abdomen and no administration of artificial hormones.

After a little while, I moved on to a nearby couch and instinctively lay on my side with the baby. The move from floor to couch was easily undertaken, the cord remaining slack in the process. I was now in that wonderful time following birth but had not experienced any preceding trauma.

Probably within the hour, the placenta separated and by that time the cord was lifeless and could be cut and tied. I really prefered the idea that the cord would not be cut in the early moments after birth since this would maximize all the benefits to the baby. Blood loss was minimal. I had experienced a totally physiological third stage.

This birth turned out to be easy and untraumatic

This birth turned out to be easy and untraumatic. Very simple measures were taken to change the factors surrounding birth and these appeared to make a huge difference: I was at home, I had freedom to move around as I wished, I was not watched by anybody (including my husband) and I had faith in the 'midwife' [Michel Odent]. From my perspective as the mother, the fact that my baby was in a breech position proved to be a secondary consideration.

Very simple measures were taken to change factors surrounding birth and these seemed to make a huge difference

I asked Michel about his approach. Here are his comments...

Esther's daughter was born several years before the publication in *The Lancet* in the year 2000 of the huge randomized multicenter trial that is considered a landmark in the history of breech births.[56] It is easy to summarize what we learned from this study: we learned that a breech birth in a conventional hospital and in the presence of an obstetrician is dangerous. The case of Esther's delivery does not belong to this framework. It occured outside the conventional hospital environment and, in the mind of Esther, I was probably an old friend of the family with an experience of home births and breech births, rather than an obstetrician.

What can we say today to women who want to avoid a cesarean section in spite of a breech presentation at term? Here are the rules that I gradually adopted after supervising about 300 breech births by the vaginal route:

- The best possible environment is usually a place with nobody else around other than an experienced, motherly and low profile midwife who is not scared by a breech birth.
- The first stage of labor is a trial. If it is straightforward, easy and fast, the vaginal route is possible. If the first stage is long and difficult, a cesarean section should be decided without any delay, before a point of no return is reached.
- Because the first stage is a trial, it is important not to make it artificially too easy, either with drugs, or even with water immersion.
- After the point of no return, privacy remains the key word.
- It is permissible to be more audacious with a frank breech than with a footling breech. A cord prolapse outside the hospital environment can be a disaster.

Here, as in other cases where risk is perceived as being high, the danger of creating a fearful, anxious environment around the laboring woman is very high and it's likely that this in itself may well increase the level of risk, rather than help caregivers reduce it in any way.

*Little Ms Culpin having fun on the slide at the age of 2*

By the way, as you probably know, external cephalic version is a controversial topic when it comes to optimality. Quite apart from safety issues, it rarely seems to be successful and—since it implicitly communicates to a mom-to-be that 'something's wrong' which 'needs' to be corrected—it can create unnecessary anxiety. Because of the risk that the umbilical cord could get twisted round the fetus' head during this procedure, it is vital for this procedure to be conducted with continuous monitoring, in a place where a cesarean could be carried out immediately, if necessary. Moxabustion, the Chinese procedure which pre-dates acupuncture, is also controversial and I heard about a couple of cases where a labor did not proceed smoothly after its use. The procedure involves burning a herb close to the skin.[57] Here, as in other cases, it may be best just to facilitate the physiological processes, rather than attempt to change them in some way. But when is 'facilitating' itself a type of change? Perhaps that's a question we have to hold in our hearts and heads for awhile...

## Birthframe 49: Liz Woolley—VBAC for second breech baby

Here's another account from a woman who decided that the way risks are usually assessed might not always take all the facts into account. It's interesting how she mentions thoughts and states of mind at various points in her account. Perhaps these are more powerful than we usually assume, whatever the potential physical constraints.

I have had two breech babies. The first, a boy, was born in 1999 by cesarean, the second, a girl, was born naturally in 2001.

I was a breech baby and so was my brother, so it shouldn't have been a surprise when it became clear that my second child had no intention of turning round, but I kept thinking she would. I tried external cephalic version (ECV) and moxabustion, as well as lots of undignified butt-in-the-air positions. But as nothing was working, my midwife and I began to plan for a breech birth.

With my first baby, I had encountered a lot of opposition to having my son vaginally at the hospital where I was registered. As he only turned breech a few days before my due date, I had little opportunity to find any alternatives, let alone research whether what I was being told about risks was accurate. After he was born by cesarean I felt very upset emotionally, as well as physically. Obviously, I was happy to have him, but I felt 'all wrong'. I had gone into labor and then had an emergency section and my body felt strange, literally as though it was still pent up, still had something left to do. I can't explain it in words. I felt like I had let my body down and let my baby down.

Second time around I had an independent midwife, Judith, with lots of experience of breech birth and had also switched hospitals and booked with an obstetrician who was prepared to support my efforts for a vaginal birth. Both Judith and the obstetrician felt that the studies on breech birth had not given enough weight to the impact different skills and experience of midwives/doctors had in determining outcomes. However we were all clear that whilst we were trying for a breech birth, at any sign of problems we would go straight for a cesarean. Given this, and the small risk of a problem with my previous section, we felt we had to opt for a hospital birth.

A couple of days before my due date, my water broke as I was cooking the evening meal. There were no contractions so we went ahead with our dinner with my husband's brother, who then took our son to his grandparents. I phoned Judith, who called round to test if it was definitely amniotic fluid. I was happy for her to leave while we awaited developments. At about 10:00PM I started having contractions. They were anything from 10 to two minutes apart and quite strong. In between, I tried to read to keep calm. We wrote down times. Past midnight I was sick a few times and my husband managed to doze off for a bit. As it approached 6:00AM I decided I wanted to leave for the hospital. Although it's only about 12 miles away, I wanted to beat the terrible rush-hour traffic jams.

There were no contractions so we went ahead with dinner

Judith met us at the door of the hospital. Walking brought contractions thick and fast; I had one in the three yards between car and door, one in the elevator and a couple in the corridor. We went into a large room with a long row of windows, giving a good view over the city—we were four floors up. I said to Judith that after a whole night of contractions the last thing I needed was to be told I was only 2cm dilated. She said, "Shall we not look then?" So we didn't.

For a long time, I crouched on the floor and held on to a metal chair leg with each contraction, staring intently at the pattern on the hard tiles. After a while, Judith said she wanted me to get things moving by walking about more. With her on one side and my husband on the other we processed up and down the long room, lifting our feet up as high as we could. As before, walking brought a rush of strong contractions.

> I crouched on the floor and held on to a chair leg with each contraction, staring intently at the pattern on the hard tiles

At around midday Judith examined me and pronounced me 6-7cm dilated—good job she hadn't checked earlier. We kept walking but I refused to leave the room to take to the corridors. We had been left completely alone by the hospital staff and I couldn't bear the idea of seeing anyone else. Every now and then I was sick. As the afternoon wore on, Judith started getting me holding the end of the bed, squatting and imagining the baby moving down. At this stage, I don't think I actually believed that this baby would really be born and be born through my efforts.

Another examination showed that I was almost fully dilated. Judith quietly told me that the next part would be very hard work, but that if I wanted an epidural I could still do that. Having done so much already, I felt determined not to do this and risk the interventions that could follow. I told her I was scared of what was to come. She said being scared would make it harder, so I decided not to be. It seems incredible that you can decide not to be scared, but that is what I seemed to do.

I was very hot and sweaty and Judith suggested changing clothes. Once I got my big T-shirt off, I refused to put another one on. It felt like it would be a distraction.

Judith told me to start pushing with contractions. I couldn't get the hang of it at all. I felt no urge to push, I felt like I was pretending. I tried to think about a beautiful baby coming down. After awhile I started to understand that I needed to push really hard and for a long time each time. I think I finally realized that it was really up to me to do this and that I would have to work harder than I believed possible.

Judith said she could tell what sex the baby was (a quirk of having a breech). Then, after a big push, I finally felt the baby moving through me. I was leaning over the back of the chair when another push brought the baby's bottom out. I shouted to Judith as I was scared the baby would drop out onto the floor. The obstetrician was now in the room. She, my midwife and husband all lifted me onto the bed. This felt awful as it felt like the baby was being pushed back inside. Judith said to me: "The more of this you can do yourself the better."

The next contraction I gave a big push and the legs came out; the next brought the arms and shoulders spinning out.

Then there was what felt like a very long silent moment. I could see the sunset through the big windows. I could see the midwife and obstetrician looking at me. I wondered if they were worried, if there was a time limit on this bit. I didn't wait for a contraction but decided to push with all my might, to get the baby's head out as quickly as I could. Then she was in my arms. A bit blue but soon pink, not crying, very calm, big eyes open in the dimly lit room as night began to fall.

I felt a bit sore for about half an hour and then felt fine. Physically and emotionally I felt great then and for about three days afterwards. Really great, super-happy, full of energy. I awoke each morning with what felt like a hangover. Judith said it was coming down from the endorphins. Partly, I felt great too because I was amazingly proud of myself. I had no pain relief, no interventions and no sutures and a breech baby! Particularly after having a section for breech position the first time around, it seemed almost incredible.

## Birthframe 50: Anonymous—Bleeding during pregnancy

The next anonymous account is very interesting because it's about a woman whose risk was actually very high, but who didn't realize this. Somehow this information was never communicated to her and she made decisions about her care accordingly. As you will see, outcomes were good, although the woman felt angry retrospectively about the way one of the midwives spoke to her, which she felt was disrespectful. How would you personally have behaved if you had been looking after this woman? Do you think it's helpful or unhelpful for a woman to be told, on admission, that things might take a very long time? Do you think it's good to ask women if they want pain relief in a routine way, as was done here? (When you're on a diet, do you like to be offered cake and chocolate cookies?) Do you think it's good for a woman to be left alone (with her mother or husband, or another inexperienced birth attendant) during her first labor? What are the advantages and potential pitfalls of doing this? What are the possible alternatives?

Zack's birth was as peaceful as a hospital birth can be. I wasn't eligible for a home birth because of bleeding during the pregnancy. I had been given steroid injections to mature the baby's lungs but fortunately the bleeding had stopped and I was home for most of the pregnancy. I was a few days shy of my due date when the real contractions started. I returned to my mother's house and had a meal and waited to be ready to go to the hospital. A few hours later we called and told them my pains were a few minutes apart but my water hadn't broken. We headed in and were met by a midwife who stuck me on the monitor and was very careful to explain that things could take a long time as this was my first baby, but that I would do just fine.

As soon as I got my own room I was given a mattress on the floor, a birth ball and some cushions and encouraged to move around as much as possible, breathe through my pains and ask if I wanted any pain relief. (I had written in my birth plan that I wanted to be able to move around). We were left to ourselves for as long as we wanted, I was provided with plenty of water to drink and food was available but I did not want any. By the time my husband arrived I was ready for some pain relief so the midwife suggested I get in to the big bathtub (not a birth pool). I did this, but quickly became uncomfortable. For some reason being in the water made me feel a bit claustrophobic and edgy. I chose to have gas and air, which was provided quickly by the midwife, who took that chance to listen to the baby's heart before she left us alone again.

The only time the midwife came in the room was when I wanted her for something. She would then ask if she could listen in to the heartbeat with the Doppler, but she only did that if I was okay with it. (Sometimes I didn't want her near my belly.) I was only examined vaginally when I felt like pushing. My water still hadn't broken at that point, so she asked me if I wanted her to break the sac or leave it alone. I decided to get the water sac broken because I was feeling very tired as it was nearly midnight. The bag of water was broken quickly and without causing me discomfort, so it wasn't as bad as I'd expected. Then we established that it was definitely time to push. (I did need telling with my first—I wasn't sure I was ready!) I'd been very confident that I would get through the labor and my midwife's attitude really helped. If I became discouraged at any point she was great at getting me to focus and concentrate on the end goal of having my baby! Once he was born she helped me latch him straight onto my breast, as I had requested in my birth plan, which she had clearly read because she did everything I had asked.

Once the placenta had been born I went for a quick wash while Simon and Mom dressed the baby. The midwife left us alone again and told us that she would write up her notes outside so we could have some peace and quiet.

The birth was really exactly as I wanted it. I was active and left alone by the midwife as much as possible. She kept her voice low and spoke to me directly rather than to my mom or husband. I don't think anything could have been better, given that it was not a home birth.

## Birthframe 51: Anonymous—Midwife not reading birth plan

Here's the same woman's next birth story. In this case, what do you think of the midwife's behavior? Do you personally always read women's birth plans? Do you ever have problems remembering details? Do you respect requests about pain relief, even if it's just a question of whether or not the woman wants to have it offered? How do you feel about watching a woman who appears to be in a lot of pain? How would feel if you were watching the same woman making love? Would you be able to differentiate easily between pain and pleasure? When you attend a woman in labor, what assumptions do you think you bring to the laboring and birthing process? Do you change your behavior to accommodate individual differences? Do you think ARM—routine or otherwise— is a good idea? Why might you try to avoid this? Why do you think the British Association of Radical Midwives specifically chose the initials 'ARM' to name their new, different society? (It was because they wanted their non-interventionist preference to be similarly well-known!) To what extent do you think ARM symbolizes unnecessary intervention in maternity units? Can you think of any other ways of speeding up labor which do not involve any drugs or artificial interventions, such as ARM? When a labor has indeed speeded up, as desired, and you're not ready, do you think it's justifiable and safe to ask a woman to stop pushing? What is likely to happen if the baby comes out before you're ready? What may happen if you insist the woman waits? To what extent do you believe in allowing birth to take place undisturbed? How do you think interruptions will change the hormonal cascade? What impact could this have on safety? What do you think of people who say it's important to keep the atmosphere of a birth sacred? What precautions can be taken to ensure that mother and neonate stay warm immediately after the birth? Quite a few questions to consider while we read this account!

Lena's birth was an induction at 37 weeks. I had some bleeding throughout the pregnancy and had been admitted to the hospital on several occasions, given steroids, etc again. I was admitted and released every day for a week and eventually the obstetrician suggested that it was time to induce the baby. I had an ultrasound to check her size and state of health, which all went well

As I was already at the hospital they started the induction late at night rather than waking me up early. I had a vaginal exam to see I was a suitable candidate for the gel. Everything was okay with the baby and the induction was started by the obstetrician and a midwife. I thought this was done in a very matter of fact manner, on the ward with the drapes drawn, rather than in a private room. The obstetrician gave me a vaginal exam and the midwife inserted the gel. Within a few minutes I could feel my contractions starting so I walked round to bring them on even more. One thing that hadn't really been explained to me was that once I started to labor I would need continuous monitoring. I was advised by the midwife to stay on the ward for as long as possible to avoid this. After six hours I was examined again to see if the gel had worked. I was feeling a lot of contractions but the exam showed I hadn't progressed much. A second gel was added, which caused an almost instant increase in the frequency and intensity of my contractions. I asked to go to the labor ward so I could have my own room... Laboring on the ward is not pleasant. All women who are induced should, in my opinion, be given a private room.

The midwife on the labor ward was okay, but didn't read my birth plan and kept asking me about Demerol. I had written in my plan that I didn't want to be asked about that type of drug and that I was only interested in gas and air. My water hadn't broken so she did that for me and left us alone after my mom asked her to leave!

She left us alone after my mom asked her to leave!

Once I was ready to push the baby came very fast—I had an overwhelming urge to push. The midwife didn't have the birth kit ready and kept asking me to stop pushing, I was shouting at her that I couldn't and the baby was born and caught by a student midwife who had asked to join us for the birth.

After such a quick birth my daughter was in shock and became cold so we stayed on the labor ward for a few hours to warm her up so she didn't have to go off to special care. I nursed her and she was also placed in the warmer.

This birth was not as I had wanted, of course. I think that induction can be handled well but in this case it was pretty chaotic and the midwife was totally unprepared for my baby to arrive quickly. She also upset me by not reading my plan or talking to me about what I wanted. I was glad my mom and Simon were there to help me stay calm and focused as much as possible.

## Birthframe 52: Anonymous—Monitoring, mockery... panic?

As you will see, for her third labor this woman initially decided she would not be induced, but she ended up with another induction. Again, we need to consider various issues as we read this account... What is the evidence base for induction and for allowing labor to begin spontaneously? To what extent does anxiety play a role in decisions regarding induction? Why do you think bed rest was recommended in this case? What is the evidence base for bed rest? Why are more and more women experiencing back labors these days? What steps can be taken to prevent these? What is the evidence base for carrying out vaginal exams?[58] Do you ever encourage women to have VEs or discourage them from having them? When do you think it's justifiable to use EFM? Why do you think some women try to avoid having more vaginal exams or having electronic monitoring? Do you think women who do so are being irresponsible or that they are taking an even more responsible approach towards their own labors and births? What do you think of the differing assessments of labor of the midwife and the laboring woman? Are laboring women always wrong, in your opinion? What effect do you think it might have on a woman emotionally if she is told that she is not in labor, when she is convinced that she is? What do you think of the husband's presence at this birth? Could or should the midwife have done anything to intervene and ask him to leave (or distract him), in your opinion? Do you think the obstetrician was right in pressurizing this woman to accept a pitocin IV? (Are there any ethical issues involved in calling pitocin 'oxytocin', in your opinion, given that pitocin is an artificial version of the hormone?) What do you think of the midwife's attitude to obtaining maternal consent for insertion of the fetal scalp monitor? Why do you think many women want to ignore midwives' instructions to push or not? Why did this woman want to get out of hospital as fast as she possibly could? What could have been done to improve the mother's experience of this birth?

Henry was also an induction at 37 weeks for the same reasons as Lena. I was already in the hospital—again I'd been given steroids—and I'd decided that this time I would just wait longer and see if I went into labor naturally. I was on bed rest and had been having what I would describe as back labor for a few days. I had been on the monitor a few times a day every day but the contractions I was having were not showing up on the monitor, so I was able to just keep quiet and avoid examinations, etc.

On the day he was born I had been having stronger pains and thought I had started to go into labor for sure. I was put on the monitor and, again, nothing showed up. The midwife agreed to my request for an examination because I was sure something was going on! She found I was 3cm dilated and said, "See! I told you you weren't in labor. You were 3cm when you came to the hospital a week ago!" I pointed out that I was not dilated at all when I'd been admitted and that she should go away and read my notes if she wanted to be sure.

Soon after that I was moved down to the ward and Mom arrived. Simon was working and came along later. He was actually a huge disturbance during this birth because he was very stressed about some things that were happening with his job and was not able to concentrate on helping me out. I didn't feel I should ask him to leave but with hindsight this probably would have been best because his stress didn't help me.

The obstetrician came to see me several times and was looking at the trace from the monitor. The contractions weren't very strong so they started me on a pitocin IV. I wasn't sure if I wanted to go ahead but the doctor convinced me it was the best plan. I think this is the point where I really lost control of this birth altogether. The obstetrician used the expression: "We need to get some dynamite behind this one!"

A midwife and student came to start the IV and the midwife couldn't find a vein easily. I don't like needles and this stressed me out even more. Eventually, the student, who was actually a more experienced nurse, was able to do the IV. The IV was started and almost immediately I began to lose control of my ability to cope with the pain. The midwife and student didn't leave the room like I wanted them to but I didn't really ask them because I was using a lot of the gas and air and couldn't communicate what I wanted properly.

I lost track of time for awhile and was dimly aware of other midwives and the obstetrician coming and going. All I could really concentrate on was the sound of the baby's heart on the monitor, which was turned up very loud because he was in distress.

The next thing I knew a midwife was clipping the monitor to the baby's scalp because I was moving around so much. She didn't ask me if she could do this and was actually laughing when I asked her what she was doing. She made a comment about me being 'away with the fairies anyhow'. Very soon after that I had the urge to push and just did it without telling anyone. They noticed I was pushing and wanted to examine me first because I had only been 8cm dilated a couple of minutes before, when the scalp monitor was attached. I refused and just pushed the baby out. I really had to concentrate to ignore everything going on around me and actually the pushing phase was very easy. He arrived in a much more controlled manner than his sister, which was a surprise, given the hectic atmosphere.

Once he was born I nursed him while the placenta was born and took him home after a few hours. I couldn't get out of the hospital fast enough! Although I realized that the monitoring etc was necessary for this and my other births, given our circumstances, the emotional side of things was not dealt with, especially during this labor and birth. It was really the comment about me being 'away with the fairies' and the general atmosphere that made me feel very disappointed with this labor.

## Birthframe 53: Anonymous—Feeling overly observed

Finally, in order to consider optimality from even more angles, let's consider the same woman's fourth and final birthing experience. Again, because of her medical situation, this woman needed to be closely monitored. While you're reading, consider the following questions... Why do you think women get frustrated with caregivers focusing only on monitors and electronic equipment and ignoring their own reports or palpation of their 'baby belly'? When is a contraction 'real' and when is it an illusion?! Why do you think many (if not most) women want to give birth in privacy, without strangers traipsing through their birth space, or being noisy nearby? How many attendants do you think ideal? Using modern technology (e.g. cell phones), what kind of systems could be set up to minimize disturbance to women in labor? Are there any advantages, do you think, in women getting to know pediatricians before a birth takes place? What can be done to help a woman get to a premature baby more quickly? What are the possibilities for providing Special Care 'rooming in' facilities for babies who are premature? How could kangaroo mother care be optimally used in places where you work? How could parents be better accommodated and supported postpartum, when they have a premature baby?

Jamie was born at 33 weeks. I had been admitted to the hospital 10 days before his birth with bleeding. I was given daily ultrasounds because every time I went on the monitor his heartbeat was dipping down very low. I spent most of every day lying on my left side. I was allowed to head home for an hour or two here and there to see my other children.

I had been admitted 10 days before his birth with bleeding

The obstetrician had said that he thought my bleeding was coming from my placenta and probably had been in all my pregnancies—but eventually decided it must have been caused by micro-abruptions. Previous doctors had not been so sure and had simply described it as 'unexplained'. The plan had been to leave me on bed rest until 37 weeks and then to induce labor. The only thing that was complicating this plan was the dips in my baby's heartbeat. I was told that if they started to last any longer I would be rushed off for a C-section. I had also been given steroid injections again to mature the baby's lungs.

I had been having some contractions for the whole time I had been in the hospital but nothing that really showed up on the monitors. I was pleasantly surprised to find that the doctors and midwives looking after me were prepared to actually feel my baby belly and look and see that the contractions were real!

I was pleasantly surprised to find that the doctors and midwives looking after me were prepared to actually feel my baby belly and look and see that the contractions were real!

At just 33 weeks I woke up in the morning to more pains. I stood up to go to the bathroom and I felt a gush of water. I had never had my water break spontaneously before so headed off to the bathroom to see if I had wet myself, I was so unsure! Once I was sure my water had broken I called my husband and told him to come in. There was nothing to be done to slow my labor because of the bleeding I was getting. The midwife and doctor were very calm and collected and examined me to see how fast things were going. I was already at 3-4cm so we moved down to the labor ward. This was the first time it was just my husband and me at the birth. My mom was at our house with the other children.

Once I was settled into the right room things went pretty fast. I was ready to push after an hour or so. Unfortunately, the pediatricians were very impatient to get into the room as I was still pushing. The door to my room was open and although the midwife drew a drape across I was very aware that there were several people standing around waiting for me to give birth!

I asked loudly for everyone to "GET OUT!" and the midwife went and told them to leave and close the door behind them and that she would tell them when the baby arrived.

I was so nervous when I was pushing because I was not sure what to expect from such an early baby. The doctors had been to see me and told me some of what we could possibly expect but really nothing can prepare you for going into labor early like that.

The midwife was great for keeping me focused, but the baby's heartbeat was dipping very low for long periods so she had to tell me to push as hard as I could to get him out fast. I had wanted to do it in my own time again but I don't think there was much choice. He was pretty small, so was easier to push out than my other babies.

He cried when he was born and was very active. He was put onto my chest for a minute or so. After that he was taken away to Special Care to be assessed. I decided to have the injection to speed my placenta up so I could get to him as soon as possible. I was desperate to go and see him but as soon as the placenta was born I passed some huge blood clots so I was advised to wait awhile. A nurse came with a photo for me to see and at that point I persuaded them to let me go and see him, using a wheelchair!

Jamie was in Special Care for 11 days. I held him after two days and nursed him after three. I was home two days after the birth because there was nowhere for me to stay. The Special Care Unit was not geared up for a parent to stay with the baby, which is a huge shame.

Above: *The four children who were born after this unexplained bleeding.*
Below: *The youngest and the oldest of the four children.*

## Other 'clear-cut' cases

Many caregivers consider certain pregnancies to be automatic cesareans because of their assessment of risk. However, what about psychological factors? Do women always have a better experience when they have a very 'medically managed' birth? And what if a mom-to-be actively chooses a less interventionist approach and can justify it to you? Of course, when considering these questions we need to remember that optimality involves looking not only at safety factors, but also at the mother's physical and emotional state after the birth. In each of the following cases, these issues arise in some way. Could any of these women be your own clients? How do you behave differently?

### Birthframe 54: Anonymous—Complete lack of bonding

When women don't think ahead the outcome is often far from ideal, as the next account shows. Sometimes, tragically, life gives us no second chances. What could you do, in a case where you are assigned a similar woman, in order to help her prepare more successfully for birth and the postpartum period?

No family history on either side of multiples... I suppose history has to start somewhere. I continued to work full-time in a leisure center on my feet all day. Felt well throughout. A little breathless on dog walks up hills. Sickness always in the late afternoons. I didn't feel I got very big but people tell me otherwise. I would have to roll off the couch, though!

Hospital appointments—ultrasounds took longer to ensure they didn't do the same baby more than once! I never bothered to learn any doctors' names as I did not see the same person two times. My named obstetrician was on 'long-term sick', I later discovered. Three heart monitors across my belly was always long-winded. One doctor said it was a waste of time because of interference between them... It didn't stop them doing it.

The local midwife eventually made contact. We had been overlooked. She talked about prenatal classes. I was aware the kids would probably come at around 33-36 weeks. But there seemed no sense of urgency from anyone. I never heard from or saw the midwife again. So a rapport was never made with any professional. We seemed to be continually 'overlooked' or forgotten about.

At 28 weeks (only 10 weeks after finding out I was expecting triplets) I went into premature labor and had an emergency C-section.

I had another week at work before maternity leave started. I had been at work all day. Came home in the evening and couldn't get comfortable. At 10PM I noticed a pink tinge going to the bathroom and had 'period pains'. Called the hospital for advice and they said, "Come in." The journey was very uncomfortable. John wanted to drive quickly but every bump in the road jarred. Arrived 11:30PM-ish. Heart monitors x3 and vaginal exam. "You are 4cm dilated. Do you know what an epidural is?" I was then shaved. Very uncomfortable.

It all happened very quickly with no suggestion of trying to stop labor. I was wheeled into the operating room with John. At 1:49AM, 1:50AM and 1:51AM Friday Triplet 1, 2 and 3 were taken out of me. As far as I'm concerned, I didn't give birth. We were not shown the babies. Not even a quick lift up over the screen to see. Not a "One boy, okay." Nothing. Didn't hear them cry. John saw them four or five hours later. My first look was Polaroid photos in my ward room. Taken down to Intensive Care to see them later Friday afternoon, about 12 hours after having them. They were pointed out to me. It felt like: "That one, that one and that one are yours." 2lb 3oz, 2lb 5oz, 3lb 1oz birth weight. All wires, netting around the heads, feed tubes down the noses. They could have pointed to any incubator in the room and said it was mine. I felt nothing. Which one do you look at first? What are you supposed to do? What are you expected to do? That was the start of three months' hospital care.

I was given a breast pump. Express, bottle, label, fridge. This continued in the hospital and with a double electric pump on loan at home as well. I came home Monday—two nights after having them. The nurses care for them 24 hours a day. In an attempt to involve you, they hold back. How am I supposed to know what they want or need when walking in for only a few hours? An idea of timing for you: April, wedding—May, pregnant—July, found out triplets—October 15, had kids—October 22, moved house—January, kids come home.

*I felt nothing. Which one do you look at first?*

Bill developed hydrocephalis. We were told he was going to have another lumbar puncture—we didn't know he had had the first one! He needed a series of six. This was to reduce pressure and stop the head diameter increasing. If it didn't work, a shunt would be required. He was okay. Nine months later we were told he had cerebral palsy.

As they improved and gained weight, and were able to breathe without help, they were moved from Intensive Care to High Care, then Special Care. We were going through the motions of caring for these kids—learning how to do everything. Still no feeling of bonding or attachment.

The day before bringing them home John and I spent the first night alone—in charge—in control—responsible for our children. Hard work, no sleep, but satisfying. Our departure from the hospital involved being let out one door and being asked to bring the cribs back. Not escorted or helped to the door. Three car seats. One three-door car. John in the back, jammed between two babies. Me driving with another alongside. No looking back. (I wouldn't go back 'to visit' if you paid me.) Life can start now. Note: Need another car before our backs give out.

We put a big whiteboard on the wall in our living room. This kept track of who had been fed and changed and when. No sleep, not eating proper meals, needing help but not wanting it all resulted in postpartum depression. I asked for help from a visiting nurse three or four months after the kids came home. I was given anti-depressants. For a couple of days I felt dizzy but began to feel more 'myself'. I took one course or packet, but didn't go back for more. I didn't like needing help. I should have kept with them—they really did make a big difference.

Three and a half years on. We still go through the motions of caring for our children. We have a bigger family than we wanted. We have never had a rush of love for them. We still look longingly at parents with singletons, seeing all the things we couldn't, can't and will never be able to do. I feel the whole premature birth and experience described has played a huge part in this lack of emotion.

Things I wish other women pregnant with triplets could do:

♥ Go and see a Special Care Baby Unit now—just in case. Ask about 'kangaroo care'. It was never mentioned and now I think that skin-to-skin contact with each child may have helped 'connect us' a little. [See Birthframe 95.]

♥ Rest and read as much as possible now. Pregnancy, birth, and ideas for later. You won't get time when they arrive.

♥ Arrange help, especially regular week in, week out (same time, same day). It's a lifeline. Knowing when someone is going to walk in is great.

♥ Go outside with them. Being stopped and asked questions is much better than being stuck indoors.

♥ Try and make time to have a bath or shower!

♥ Do something for yourself.

Come to terms with the fact you will never be able to sit and chat with other parents in the same way as parents of singletons. You are not rude—just preoccupied!

## Birthframe 55: Mave Denyer—Vaginal triplet birth—why not?

Here's an account of a vaginal triplet birth which took place in the UK in 1961. As you will see, this mom-to-be was prepared for her uncomplicated vaginal birth simply because of the confidence she had in her body as a result of having given birth two times before, to singletons. How many other triplet moms-to-be could have similarly straightforward births nowadays, do you think, with the correct psychological support prenatally? Are the risks of vaginal births with multiples always greater than those associated with a cesarean, given the postpartum problems a mother is likely to have after surgery? Should more women be given a choice? What issues would need to be discussed?

I found out I was expecting triplets a few weeks before I went into labor. I was a bit shocked and I didn't quite believe it. I was having visits from the community nurse and she put on my card: "Lots of limbs, go for an x-ray". I had to see the doctor before the x-ray. But that's how they discovered it. It wasn't as a result of infertility treatment and there was no family history of multiple births. There was no treatment of any sort. It was just absolutely ordinary. I was warned by the health officer not to count my chickens before they were hatched because she said I might find that there were two normal ones and one very tiny one.

*I was warned not to count my chickens...*

I already had two children. My son was then 3½ and my daughter was 20 months. I'd chosen to have them at home, as most people did 40 years ago. They'd both been very good experiences. I'd had a different community nurse for each one and both were excellent. I remember the first one saying, "For goodness sake, walk about and let gravity do its work!" So I did. This was when I'd started having pains at home, and I was already grunting a bit. That was with Gavin, the first one. Nicky, the second one, was much quicker. I don't remember how quick—it was just quicker and a little bit easier, and I was delighted to have a daughter.

As for the triplets' birth, obviously I would have liked to have had them at home too but when they discovered it would be triplets, they said it had to be in the hospital. There was no talk about having a cesarean. Nowadays, people seem to do cesareans at the drop of a hat. They are necessary at times, of course, but there was no need for one in my case.

Nowadays, people seem to do cesareans at the drop of a hat

It all started a week before Christmas, when I began having pangs, you know, contractions. They were born a bit early, as multiple births almost always are, but I can't remember how many weeks—about three, I think. Anyway, I went into hospital for about 24 hours or so and they said, "Nothing's happening. Go home and come back again after your Christmas dinner." So I did. I had my Christmas dinner, as instructed, at a friends' house. She very bravely invited my husband, myself, my two children and my mom, who was staying with us, to join them for Christmas dinner. This was all at a few hours' notice. And we had two Brussels sprouts each, bless her! It was lovely. She was so good. Anyhow, I had Christmas dinner and then went into the hospital that night, when the water broke. So it was a case of call the ambulance immediately. I'm not sure of the exact time but it was probably about 3 or 4 o'clock in the morning that I went in. And they were actually born just after 2 o'clock in the afternoon, about 12 hours later, on the next day.

While I was having contractions I lay there, fidgeting occasionally, listening to the sounds of people around me. There was a woman in the next room making terrible noises, which terrified me. Every now and then someone came in and looked at me and said, "Ring the bell if anything happens". Eventually, something must have happened, but I don't remember what the first thing was. Whether it was just the increase in pains, you know in contractions...

They gave me Demerol and gas and air. I don't remember whether I asked for it or whether they thought I looked as if I needed it, but I really appreciated it. There were loads of Italian people about at the time, 'mama mia-ing', you know. Well, not loads, several. Some people make a terrible noise. I was trying not to. And I didn't have anybody with me. There wouldn't have been room. We did think about the possibility of my husband being there but they said it wouldn't be a good idea because of all the doctors and other staff that would be in the delivery room. I was only too happy to let them decide. I didn't feel I could do much except relax and try not to worry. I thought, "If anything goes wrong, they'll have to cope. I won't know how to". But nothing did. It was fine. They were born within about 45 minutes of each other.

They were all born vaginally. I'm very happy that they were as normal as possible. No cesarean. No messing about. For both the babies' sake and for mine, too. There were no instruments or tearing or anything like that. No, I was lucky—when you hear of what some people have. But of course, they were tiny, so in that respect I suppose I was more likely to have less damage. When the first one was born I said, "What's going to happen now?" and they said, "Well, we don't let you have three labors because it's three children. We'll make sure you have them all now more or less, within a reasonable time." But there was a certain amount of—if you can visualize a magician putting his hand into a hat and bringing out a rabbit—it wasn't quite like that but it was similar! There was a lot of manhandling going on and I can distinctly remember one of the doctors saying to one of the nurses something like, "Can you burst that bubble?". I presume that the first one was in his own bubble, which had burst, and that the two identical ones were in their own bubble, which hadn't, and there was a lot of fishing around going on while I was lying on my back.

> I'm very happy that they were as normal as possible

There were several members of staff around. I remember one girl was terribly excited. She was a student nurse and she'd got to her 98th baby and in those days apparently they had to help out at 100 births to get their certificate, step-up, whatever, and of course with me, she was so thrilled—it took her over the 100. She was more thrilled than I was! Of course, I was very excited too. I felt as if I could push a bus over. I think you do when you've had a baby. I felt the same with Gavin and Nicky—I felt as if I could leap out of bed and do anything. But I didn't. There were no buses going by at the time. Yes, it was very exciting. It was just a case of hoping they'd be all right and they were fine. There was nothing apart from being sort of premature.

They were all very much the same weight, which was a good thing, just over 4lb. Can't remember which was which but one was 4lb 6oz, one was 4lb 5oz and one was 4lb 2oz. They very quickly evened themselves out. Jonathan, the non-identical one, was the eldest and he was the first one to come home from hospital. They came home one after the other with a gap in between, to make it easier for me really. They had them in incubators at the hospital. I can remember going to see them one day and they'd suddenly got tubes up their noses, which frightened me to death, because I'd seen them lying there asleep. Tiny little old men, actually. They really looked like old men. And nothing will make me retract that, they really do, when they're very premature and very tiny. But they were fine. I was quickly told that the tubes up their noses were for feeding. It was just easier and quicker and they were taking the food better. Somebody didn't have to spend hours and hours feeding them. They stayed in hospital until they weighed about 6lb.

I had to pump off breastmilk and a neighbor took that in and deposited it every day. I nursed them all for awhile, but I had to supplementary feed as well. Then one day, when they were roughly two months old the milk just dried up, it suddenly wasn't there anymore at all. It was probably because I was so tired. They'd had two months. I think that was very good for them. Mind you, it took all day, virtually, changing and bathing them in between.

Top: *Mave with her newborns*  Bottom: *The triplets having fun in the snow*

There used to be diapers hanging up everywhere, including on my neighbors' washing lines! I was incredibly busy in the early days but I had a lot of help from our visiting nurse. She would come in and bathe one. She used to say, "Don't stop what you're doing!" I was so lucky to have friends and family and neighbors. They were all a bit intrigued, but they all gave a hand. They didn't just look at them, they helped. And when the babies needed shots the doctor came and did them at home so I didn't have to cart them all the way round to the doctor's office. When they were old enough to sit up, they would sit one each end of a twin stroller, with one sitting sideways in the middle. Then, as they grew too big for that, my husband made a triplet stroller for them—because, of course, there were no triplet strollers in those days. I used to push them in that round to the shops. Lots of neighbors were very good at taking the others to school or kindergarten (or whatever they were at) at the time. So I was really lucky to know so many people who were so useful. And I had a home-help for awhile.

We had to pay part of her wages but some of it was subsidized by the state. I was also very lucky with Cow and Gate. They sent me free milk for them for a year— powdered milk. I think they read about me in the newspapers. I was very famous for about 20 minutes and then a lady who lived not too far away, whose name was Mrs Meacham, had quads a few days later, so I was nobody then! I've never forgotten her— I felt very upstaged!

## It was nice having three little ones around

It was nice having three little ones around, though I'm very glad they didn't come first. I think it would have put me off trying to have any more to have three in one go. It would have been a bit frightening I think, to cope with three at once but... no, it was great. I mean, they improved with keeping. But I didn't know any other mothers of triplets, although I did know a mother of twins, who gave me her stroller. Sometime during their very young childhood my doctor asked me to go and talk to another mother in the village who had twins and who wasn't coping. She was very fond of one and not of the other, which was difficult. I just went and talked to her and let her talk to me, but they eventually moved away. I heard that one of them died later, as a baby, some time after they'd gone from here.

What advice would I give to a mother of twins or triplets? This takes some thinking about! I'd say the perfect mother hasn't been invented yet so don't expect miracles of yourself. Take every hour as it comes—some things just *have* to be done *now*, others can and must wait.

We tried to be fair to the older brother and sister as well as the triplets—all five needed feeding, and as many cuddles as time would allow (which was never enough). As they became older and more mobile the 'No' word became frequent—usually to avert danger. But usually a 'No' to one toddler heading for an electrical outlet would warn the other two not to attempt to do the same.

Of course, there were times when the boys were noisy, naughty or augumentative, but thank goodness, not too often. We moms are sure to get mad at times; we're as bad as our children when it comes to being perfect, but making an effort to keep calm is always a good idea. And I was so lucky to have so many blessings to count.

*Back in 1961 triplet strollers weren't readily available as they are now, so Mave's husband had to build one! In the 21st century things have certainly improved in terms of convenience but have they improved in terms of a mother's overall experience?*

### Birthframe 56: Janet Hanton—Opting for vaginal triplet birth

Actually, when we talk about vaginal triplet births, we're not only talking about births which took place in the far distant past. The next account is of a vaginal birth which took place in London, in the UK. The psychological aspect is again very important, as you will see. Would you attend a vaginal triplet birth?

In August 1999 in a small ultrasound room in a London hospital we were told some life-changing news. At 35 years old I was pregnant with my third child. Actually, third and fourth children, as I had discovered at a previous ultrasound that it was twins.

Since then, we had been away on vacation for three weeks, and had gotten used to the idea that things were not quite going as we had expected. We returned that day, just off the plane and very jet-lagged, for a further ultrasound. My husband jokingly said, "Just don't tell us there's another one!"—at which we were stopped in our tracks by those unforgettable words, "Um... well, actually there is." I was pregnant with triplets. No IVF, no family history, we simply were that freak statistic—a spontaneous triplet pregnancy.

At 9:00PM on November 30, the evening of my daughter's fifth birthday, my water broke. I couldn't really believe it. I was only 28 weeks pregnant. Though tired during the pregnancy, things had gone well. I had been resting in bed for two hours each day, and had been feeling healthy and full of confidence. I had been told at the hospital: "As long as you get to 32 weeks, we're not too worried." I had been so sure that this pregnancy would go way beyond that date. I rationalized that I had already had two full-term pregnancies without any problems or complications. Clearly, my triplets would be born later rather than earlier. So confident was I that I used to pass over the 'premature baby' section in the books on multiple birth, sure that I wouldn't be needing that information.

We had taken on Caroline Flint to be our independent midwife as soon as I became pregnant with what we assumed to be our single third child. Caroline had been our midwife for our second child, and I had given birth to him at home. When planning to have a third child, I had thought that we would probably have another home birth. However, with the news that I was expecting triplets everything had to change.

I had thought we would probably have another home birth

Caroline tried, on our behalf, to find an obstetrician who was willing to consider a vaginal delivery of triplets. She did not have any success. Whilst we were not against the idea of having a cesarean if it were necessary for the babies' safety, we wanted my case considered on an individual basis. I had given birth two times before without any need for medical intervention and wanted someone to look at my particular case, consider all the options, and make a safe and sensible judgment about the mode of delivery. The view seemed to be that triplets should always be born by planned cesarean and the fact that two of our babies were 'monochorionic' was given as a further reason for not having a vaginal delivery. (Incidentally, we have since discovered that all three babies are identical.)

## One obstetrician was prepared to consider a vaginal delivery

One obstetrician, Donald Gibb, was prepared to consider a vaginal delivery, but his contract with my community hospital was to expire before my due date, and so he was unable to take my case on. We accepted that the babies would be born by cesarean.

When my water broke, I immediately phoned Caroline, who said that she would come to the house and check me over. 15 minutes later I started to have contractions and so we arranged to meet at the hospital. The contractions were very quickly becoming regular and urgent-seeming. I felt panic-stricken, every minute seeming an eternity as we waited for my mother-in-law to arrive to look after our older two children.

We arrived at the check-in desk of the labor ward...

"I'm having contractions. I'm having triplets—I'm only 28 weeks."

"Take a seat. We don't have any rooms at the moment."

"But I think it's an emergency—I'm meant to be having a cesarean."

"We have other emergencies to deal with. Take a seat in the admissions area."

The situation seemed surreal. I was certain that I was in established labor and that things were happening at speed, and yet we couldn't seem to persuade the hospital staff to take us seriously. We went down to the admissions area, which was out of sight of all staff. I felt quite despairing, pacing up and down, convinced that I would give birth to our babies right there.

Then, to our utter, utter relief, Caroline arrived, took one look at me and dove off down the corridor. She came rushing back, looking relieved. "Donald Gibb is here."

By a quirk of fate Donald Gibb was there on his second-to-last night at the hospital. Immediately, everything started to happen. We were rushed into a room as some poor woman was wheeled out into the corridor, and I have a hazy memory of Mr Gibb tearing the sheets off the bed. There were no intensive care cribs available at the hospital and the plan was to transfer me by ambulance to another hospital, where there would be places for the babies once they were born. However, on examination I was already 9cm dilated (this only one hour after my water had broken). It was then inevitable that the babies were to be delivered there and then.

I felt disbelief that this was happening to me. I was full of fear at what might happen to the babies, and indeed to me, and yet couldn't quite take the situation seriously. I found myself giggling inappropriately at Mr Gibb's rubber boots. The room filled up with people and equipment, and I felt as though I was appearing in some bad hospital TV drama. At the same time, I was terrified. Instinctively, I turned my back on the room and knelt on all fours on the bed. This was the same position in which I had delivered my other two children, and I couldn't imagine any other way. I remember Mr Gibb saying, "Do you want to deliver the first one like that?" and replying that I wanted to deliver them all like that. He said that he would see how things went.

Instinctively, I turned my back on the room and knelt on all

The resident pediatrician kept appearing by my head to report on progress in the search for intensive care cribs for the babies. It was mostly bad news. They couldn't find three intensive care cribs available in one hospital; the babies might have to go to different hospitals, one might have to go to another town, etc... I became very agitated about this and remember Mr Gibb telling me not to worry, that it was their job, not ours, to sort this out, that our job was to get the babies out safely.

> I became very agitated about this and remember Mr Gibb telling me not to worry, that it was their job, not ours, to sort this out, that our job was to get the babies out safely.

From that moment on, I felt a sense of calm. I felt that I had done everything I could. I was in expert hands and what now happened to the babies was out of my hands. I felt totally focused on giving birth. Although the room was full of people and I had a tube in my hand in case I needed a cesarean, I was able to block almost everything out. I had an ultrasound during the labor, but it was so unobtrusive that I hardly noticed it. I knelt on the bed with my back to everyone and was aware only of Bruce (my husband) in front of me, Mr Gibb's voice, and Pam Wild (the other midwife) rubbing my back and reassuring me. I felt such faith in the people looking after me, and in my own ability to give birth that in spite of everything I felt calm and relaxed. I leant on Bruce, used gas and air, and concentrated on counting my breaths, keeping them long and even. Exactly the same as I had done in giving birth to my other two children. I remember saying to Pam and Caroline, "I'm pretending to have a home birth here."

At 10:30PM Kate was born. She was so tiny that it was not like the second stage of labor with my other children. It was more a question of trying to hardly push at all, to make her delivery really gentle. I didn't see or touch her, as she was taken straight to the resusitaire to be surrounded by a pediatric team and ventilated. I just had a chance for a quick look at her as she was wheeled past me on her way to SCBU [the Special Care Baby Unit], a scrap of humanity amidst a mass of tubes. I wasn't really able to focus on her birth as my mind and body both knew there were two more ahead. Then everything stopped. There were no contractions, and I lay against Bruce and felt as though I was almost asleep while we all waited. I had no sense of time at all, though in fact it was 40 minutes before Sophie was born. After 30 minutes Mr Gibb ruptured the membranes because she had some bradycardia [an abnormally slow or unsteady heart rhythm], and she was born shortly afterwards. 10 minutes after that, Louisa was born. All were born head down, and I was able to deliver all three kneeling on the bed with my back to the room. As with Kate, Sophie and Louisa were taken away immediately by their pediatric teams to be ventilated and we caught a brief look at them as they were wheeled away to SCBU. They weighed 2lb 9oz, 2lb 10oz and 2lb 11oz.

I asked for syntometrine, and shortly after the placenta had been delivered Bruce and I found ourselves alone in the room wondering whether it had all been a dream. We had no babies with us, all the medical staff had gone. It was only two and half hours since my water had broken.

A little later Caroline and Pam took us down to SCBU to see the babies, where they were being held until transfer to another London hospital. We had been lucky as three intensive care cribs had been found in one unit. I was in a wheelchair and felt as though I couldn't breathe properly. I couldn't really take it all in. The birth had left me feeling on a high, as though I could do anything. I felt fantastic, exhausted and confused. I couldn't relate emotionally to the three tiny bright red bodies in their incubators, covered with tubes, and hooked up to all kinds of machines. These didn't seem to be my babies. I was given a Polaroid picture of each one, and taken back to the ward. I didn't want to think about what had happened, and asked for a sleeping tablet. That night, Kate and Sophie were transfered and Louisa was transfered the next morning.

The babies spent the next two weeks in the neonatal intensive care unit. It was a very difficult and frightening time. The almost hourly ups and downs of each baby's progress meant an experience which could be likened to being on three roller coasters at once. And during this time we could do nothing but watch and wait. We couldn't hold them for many days and could do very little to help them. I was sustained during this time by the memory of their birth. The way in which I had  given birth to them gave me a physical connection with them which I was not able to have during the first weeks of their lives. I also found it helpful that I was able to start expressing breastmilk for them. This was frozen until they were ready to begin nursing.

After two weeks the babies were transfered back to the hospital where they were born. They spent a further five weeks in hospital before being discharged home at 36 weeks' gestation. I continued to express milk for them throughout their time in hospital, and for some weeks once they came home. For 10 weeks they were exclusively fed with breastmilk, initially by tube and later from a bottle.

Kate, Sophie and Louisa are now happy, healthy 3½-year-olds.

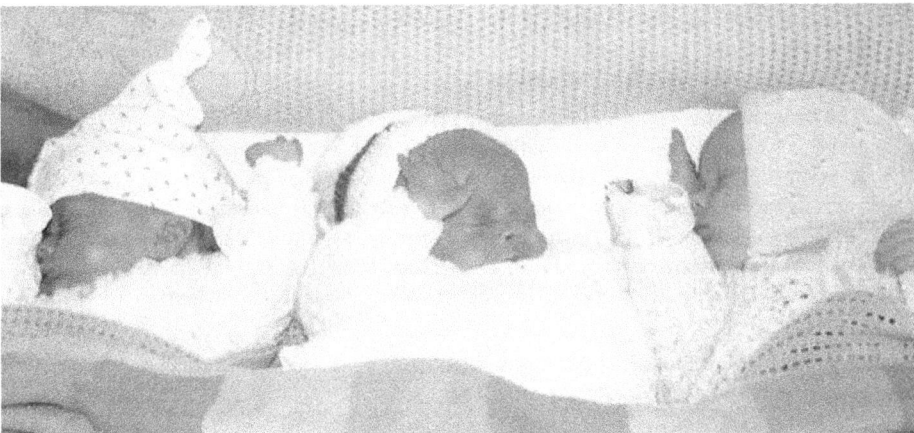

*Kate, Sophie and Louisa (left to right) at 7 weeks old, just after they had come home*

## Birthframe 57: Gaia Pollini—Dealing with negativity

Here's another account of a multiple birth. As you will see, the mom-to-be was consistently positive and proactive in dealing with psychological issues.

It was really important to me to have a healthy birth because I'm a rebirther. Perhaps I should explain a bit about rebirthing. It's a breathing technique which opens a person up emotionally and energetically, i.e. physically.

Going through the process of rebirthing allowed me to access deep feelings and old memories, including that of my own birth experience. Rebirthers have actually discovered that the circumstances of a person's birth are mirrored in his or her life patterns. (For example, induced people feel often rushed or pushed into things and people who've had emergency cesareans give things up half way through.) This is why rebirthers feel that the way babies are born is so important, and that a healthy birth is a good start to life—it simply results in less emotional baggage. People who are drawn to rebirthing are often aware of birth patterns repeating themselves in their adult life, or they are simply people who feel stuck and want to move on from old patterns and have heard that rebirthing works wonders in unblocking stuff, however old or deep it is (not just birth trauma). It is possible to heal almost anything with breathing and love. Rebirthing also makes you deeply aware of how fully aware a baby is before birth, as during sessions people often revisit prenatal states as well as birth and realize how much they were sensing and feeling and how much they knew about what was happening to their parents and around them.

I believe the positive attitude of my independent midwives also helped me because comments about 'twins always being early' can act like a negative mantra. In fact, our twins were born five days past their due date! I'm sure my midwives' positivity and faith were crucial. People are usually scared of being positive in case things go wrong. But there's really nothing to lose. If things go wrong, people can deal with it and then at least they haven't spent months worrying. It's best to stay positive and open and just have awareness of what's happening.

The only prenatal tests I had were two ultrasounds. I had the second one—another ordinary ultrasound (i.e. not a nuchal one)—because I was bleeding, as I was the first time. However, the ultrasound showed us everything was okay. A great deal of pressure was put on me to have more ultrasounds but I resisted this because my husband and I had decided we wouldn't have a termination anyway, even if it were recommended. In any case, the early ultrasound had at least made us aware that the babies were non-identical (with two placentas and two sacs), which is useful information for a twin pregnancy.

I had a fantastic pregnancy. I was extremely open, which meant I wasn't always feeling great! However, I was very much in tune with myself—even in terms of what I ate. Although I'd been vegetarian before I got pregnant, when—a couple of months into the pregnancy—I really started fancying meat, I ate it whenever I wanted to. The love of being pregnant can help you be yourself even more strongly than before.

I think babies are living beings the moment they are conceived

I think babies are living beings the moment they are conceived and what you experience of the world, they also experience. I visualized myself giving birth at 40 weeks because I really wanted them to have plenty of time to grow. I also talked to them a lot, using any positive words I could think of. I also let myself feel excited and curious about the magical process of giving birth. I expected to have a fantastic pregnancy, so I did!

Overall, I'd say the most important thing for me was to stay away from other people's fears. I did not want to be near anybody being negative—which is why I didn't go to the hospital or near some of my friends. We just said, "We know what we're doing. We don't need your opinion." I hardly had any negative feelings myself. As far as comments about my due date went, every time anybody said, "When's your due date?" and I would tell them and they'd reply, "Oh, they're going to be early because they're twins." I'd say, "No! Mine are not!" And they weren't.

My labor lasted about 15 hours, then I had this amazing urge to start pushing. At that point I just wanted to give birth. My heart was set on giving my two babies a beautiful birth, so I made sure I stayed positive—even though I knew one of my babies was breech. (The midwives had been able to tell this just by palpating my belly. Fortunately, they also knew how to deliver breech babies.) Anyway, it went very smoothly—although I was exhausted afterwards! There was no tearing. I started nursing immediately. It really couldn't have gone any better.

Then, when the twins were 8 months old, we bought a big camper van and drove to Italy. The idea was to look for our dream: a property and land where we could start a new life and business. We did find a beautiful house and land, and a year and a half later we moved to Italy permanently.

The twins are now 4 and I'm still nursing them! (They have wanted a lot less milk since they turned 3, so it's only occasionally now.) I became vegan a year after the twins were born, and they were vegan too until they were 3. Then we all started eating other things—I'm far too relaxed to be too restrictive. I just don't eat meat, and neither do my kids, only because by now we are not used to it anymore. We now eat cheese, eggs and fish—with pulses every day (which the kids adore)—as well as seeds and a few nuts, soya, quinoa, etc.

If any other mother of twins—or singletons, for that matter—wants to breastfeed, I'd say stick with it, because it really does get easier and easier. The first few months are the hardest because the babies can't hold themselves up at all and with two it is harder, but it becomes second nature quite soon, both for Mom and the babies. One day you won't even remember what was so difficult. I had big problems in the first few days—it was so painful and I was so incredibly tired. Do contact a breastfeeding counselor if you need to, even in the first week or two. Even if you have to pay it's well worth the money. A breastfeeding counselor will teach you positions and all the tricks, including nursing lying down—which is a real lifesaver! It is actually easier to nurse twins, than prepare bottle feeds for two. Remember, it does soon become easy.

After the initial problems, I loved it, and so did my kids. Within a week, all the pain was gone for me. I was just tired then! The advice and support I got at the beginning really helped make it a positive experience. And once it became positive and comfortable, there really was no reason to stop.

Before I sign off, I really want to add a bit about John, my husband. He was so, so important in the whole process. He was really positive and supportive and didn't get involved with the fear stuff. If we ever had different opinions, which was rare, he very openly talked about it and we always found solutions that felt right for both of us. He was respectful of my feelings and needs, without forgetting his own, which also needed to be worked through. We actually did the rebirthing sessions together, and a couple of times the sessions ended up with me and the rebirther holding and cuddling and supporting him. Everybody's birth issues come up around birth, not just the mom's! It was so beautiful and free and loving that he could have his process and space around giving birth. It also allowed me to have more support when I needed it, particularly at the birth itself. At that time, he was just so, so amazing. He was there all the way, not just in body, but really with his whole self. We were truly together then just as we are now, as parents. I'm sure our children really feel this. We're in this together, the four of us.

I truly believe that whatever happens to anyone is just the perfect thing for them. Even if I had had a terrible birth, in the end it would have been okay. But, man, am I glad it went the way it did! Life is beautiful. If you're going through all this pregnancy thing now, I'd just say TRUST YOURSELF and allow yourself to feel all you need to feel. Allow yourself to be whatever you are. If you can, surround yourself with people who listen to you and who will support you in what you do. And listen to yourself at least as much as you listen to others, and possibly more. If other people's opinions clash with how you feel, don't dismiss yourself. You're important and special.

# Pause to reflect...

## ...about information, decision-making and outcomes

1 When considering both singleton and multiple births, to what extent do you think it's helpful and safe to give women all possible information relating to their particular case? How does your view tie in with the views and guidelines of ACOG, ACNM and WHO (the World Health Organization)?

2 Do you disclose information and opinions about risk assessment differently depending on the type of woman you're dealing with? Should you?

3 Do you encourage women to go away and research topics for themselves?

4 To what extent are your own judgments about risk real judgments and to what extent are they defensive ones, influenced by a worry about litigation?

5 How greatly do you value the woman's quality of experience during the birth and postpartum? Do you think there are any problems you don't hear about?

6 Do you think births might be different for babies born in different ways?

*We need to consider carefully the possible effects on natural processes whenever we use new drugs or technology. Thalidomide, DES and X-rays are very recent memories, but what ongoing effects might other interventions be having, short– and long-term?*

On this topic, an email exchange I had with Michel Odent is very relevant...

**I attach a twin birth story which is full of interventions, along with detailed questions about each intervention. Was all this intervention really necessary?**
You cannot imagine the number of emails and phone calls I have about twins. Midwives practicing home birth before 'the industrialization of childbirth' were not scared by twin births. In general those who know about privacy as a basic need in labor are not scared by this sort of birth. It is the art of doing nothing. First you wait for the first baby. Then you wait for the second baby and finally you wait for the placenta. The point is to make sure that there is not too much excitation around after the birth of the first twin, so that the mother is not distracted and has nothing else to do than to look at her baby in a sacred atmosphere. The same after the birth of the second one, while waiting for the placenta. It is important to know that a twin delivery is often less violent, less intense and longer than a singleton delivery. Those who don't know about the importance of privacy are so scared about twin births that they create a cascade of interventions... if they have not chosen the easy way, that is to say a cesarean section. Today many practitioners are right to prefer a cesarean section. Giving birth without any privacy among scared people can be dangerous. This twin story you sent me is one among many others. You know my answers to all your questions.

## The importance of a balanced view

A large number of people (moms-to-be, their relatives and even professionals) assume it's necessary to take a very interventionist approach to prenatal care. This is because they feel that non-intervention (as happens in the developing world) results in high mortality and morbidity rates. The feeling of anxiety about outcomes is perhaps exacerbated by the fact that most people nowadays only have one or two children, so a lot of hopes are embodied in each pregnancy. But is this anxiety appropriate? Is it irresponsible to leave nature to take its course? Shouldn't we worry about things going wrong?

*Is it irresponsible to leave nature to take its course?*

Actually, your clients' lifestyles are quite dramatically different from those of women living in developing countries. Your clients have readily available, clean water; they have a good diet; they live in a climate which is less conducive to all kinds of disease and they have quick access to high quality healthcare if ever anything should go wrong, provided, of course, they have medical insurance. What's more, unlike many African women, American women are rarely circumcised, incised or infibulated—i.e. they don't have their genitals ritually damaged in any way—and their healthy diet in childhood means their body grew properly. (Even diets which Americans consider 'unhealthy' may be 'healthy' by African standards! At least most American children have proper nutrition.)

Bill Bryson, an American travel writer, reported on the conditions that many women experience in his book *African Diary* (Doubleday 2002), which was commissioned by the charity CARE International. Here's an extract:

To step into Kibera is to be lost at once in a random, seemingly endless warren of rank, narrow passages wandering between rows of frail, dirt-floored hovels made of tin and mud and twigs and holes. Each shanty, on average, is 10 feet by 10 and home to five or six people. Down the center of each lane runs a shallow trench filled with a trickle of water and things you don't want to see or step in. There are no services in Kibera—no running water, no garbage collection, virtually no electricity, not a single flush toilet. In one section of Kibera called Laini Saba until recently there were just ten pit latrines for 40,000 people. Especially at night when it is unsafe to venture out, many residents rely on what are known as 'flying toilets', which is to say they go into a plastic bag, then open their door and throw it as far as possible.

In the rainy season, the whole becomes a liquid ooze. In the dry season it has the charm and healthfulness of a garbage tip. In all seasons it smells of rot. It's a little like wandering through a privy. Whatever is the most awful place you have ever experienced, Kibera is worse.

Kibera is only one of about a hundred slums in Nairobi, and it is by no means the worst. Altogether more than half of Nairobi's three million people are packed into these immensely squalid zones, which together occupy only about 1.5% of the city's land. In wonder I asked David Sanderson what made Kibera superior. [David is CARE International's regional manager for southern and western Africa.] "There are a lot of factories around here," he said, "so there's work, though nearly all of it is casual. If you're lucky you might make a few dollars a day, enough to buy a little food and a jerry can of water and to put something aside for your rent."

"How much is rent?"

"Oh, not much. $10 or $12 a month. But the average annual income in Kenya is $280, so $120 or $140 in rent every year is a big slice of your income. And nearly everything else is expensive here, too, even water. The average person in a slum like Kibera pays five times what people in the developed world pay for the same volume of water piped to their homes."

"That's amazing," I said.

He nodded. "Every time you flush a toilet you use more water than the average person in the developing world has for all purposes in a day—cooking, cleaning drinking, everything. It's very tough. For a lot of people Kibera is essentially a life sentence. Unless you are exceptionally lucky with employment, it's very, very difficult to get ahead."

Every day around the world 180,000 people fetch up in or are born into cities like Nairobi, mostly into slums like Kibera. 90% of the world's population growth in the 21st century will be in cities.

*Bill Bryson*

Is it surprising, given this environment, that the developing world has such poor birth statistics? Incidentally, if you would like to support the work going on to improve living conditions for people in places like this, go to the website www.care.org

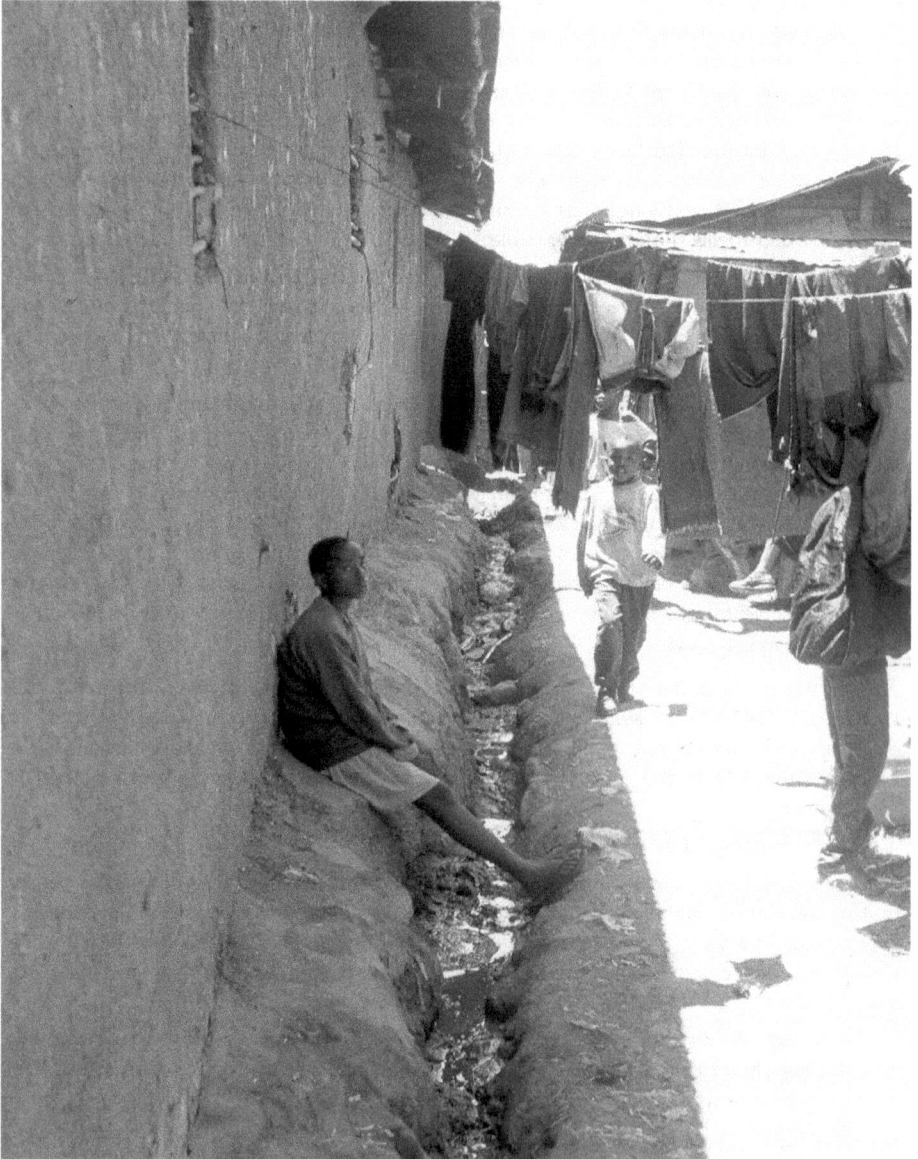

Photo © Jenny Matthews/CARE

*Over 700,000 people live in Kibera, Nairobi in Kenya, the largest slum in Africa.*
*Open drains and poor sanitation are an everyday problem, which has clear implications*
*for safety when women give birth. if you would like to support the work to improve living*
*conditions for people in places like this, go to www.care.org*

There are also other reasons for the 500,000 maternal deaths in developing countries—the figure reported by the UN Department of Public Information in 2005:

## EARLY MARRIAGE

In certain regions of the world, especially sub-Saharan Africa and the Middle East, adolescent marriages are very common. This is a particular problem because with insufficiently grown bodies, girls aged 15-20 who get pregnant are two times as likely to die in childbirth as those in their 20s. Girls under the age of 15 (who are also married off in many countries) are five times as likely to die in childbirth. Young girls who survive pregnancy and childbirth often suffer another problem because of their size—they are often left with a fistula (a rupture in the birth canal which occurs during prolonged, obstructed labor, which leaves the girl incontinent and ostracized from her community). Although 9 out of 10 fistulas can be successfully repaired surgically, this rarely occurs simply because of a lack of funds and trained surgeons. The UNFPA-led Campaign to End Fistula is working in 40 countries to rectify this situation.

## DISASTERS, SHORTAGES, POLITICAL UPHEAVAL & WAR

Maternity services may be dramatically affected by natural disasters, accidents, terrorism, shortages, politics and war.

## PREVENTABLE MEDICAL CAUSES

According to UNICEF, the UN Population Fund (UNFPA) and WHO, more than 80% of maternal deaths take place as a result of medical problems which could be solved through additional funding. These preventable causes include hemorrhage, sepsis, unsafe abortion, obstructed labor and hypertensive disease in pregnancy. Hemorrhage alone, which may well often be caused as a result of ritual disturbance occuring before the third stage of labor, accounts for 21% of the 500,000 deaths occuring each year. Although this and other problems can be treated in the US—e.g. by the use of oxytocics, blood transfusions and other skilled midwifery or obstetric care—they are rarely treated in poorly resourced places, simply because of a lack of skilled personnel (especially midwives), a lack of roads, clinics and blood banks, to mention only a few reasons.

## LACK OF EXPERTISE

Without skilled professionals teaching family planning, attending births and acting when problems occur during labor, any problems which do arise quickly result in deaths.

## A better starting point for birth...

Even in cases where your clients are not wealthy, they are probably in a much better situation than the women featured below. Poor living conditions in the States are usually much better than those experienced by *most* people in many developing countries, in cities or in the countryside. Most of your clients don't have to face the daily threat of infection and disease, and even though the US isn't perfect, there is generally more work available and better security. When things do go wrong during a birth, good communications and transport systems, as well as the easy availability of drugs, technology and expertise put Americans at a huge advantage, because it really is possible to get help.

*As you can see from the background, this Kenyan woman's living conditions are quite different from those in the USA*

*Uganda is battling HIV, AIDS, TB & Malaria. In 2009 there were 1.2 million AIDS orphans and 940,000 people live with HIV or AIDS.*

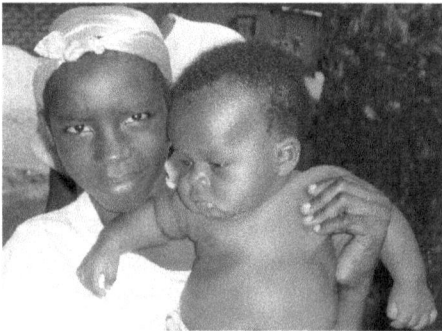

*Consider the experience of this woman from Angola. She may or may not be one of the lucky 30% of people who have access to government health care facilities, which were severely damaged during the Civil War up to 2002. Apart from difficulties getting treatment, she and her children will also be in danger of dying from malaria or other diseases.*

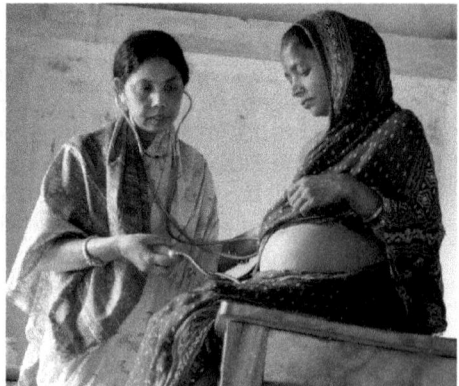

*Many women in India and other South Asian countries have arranged marriages (when they're as young as 12 years old) and might be having their 5th child by the age of 30. What's more, few of them have access to clinics because of transport problems or financial difficulties.*

# Pause to reflect...

## ...about mortality statistics and intervention

1   How much do you think maternal and neonatal mortality statistics from developing countries affect decisions you make about risk?

2   What do you think are the main reasons for mortality in poor-resource contexts? What kinds of women are affected and why?

3   Do you have the idea that birth in developing countries mostly takes place in extremely natural, supportive circumstances? If so, how can you explain the high level of hospital births in many poorer countries?

4   Why do you think hospitals in some developing countries tend to be even more interventionist than in wealthier countries? Why are many practices not at all evidence-based?

5   What kinds of political and personal issues are involved, do you think?

*Easy access to healthy food means that women in the US can be well-nourished during pregnancy. Even if women are poor, so much more food is available.*

## Providing psychological support

Also, consider the role of a caregiver when a woman has any kinds of psychological difficulties to work through. Note, of course, that this probably means 99% of the population! Not only are depressants frequently prescribed by doctors nowadays, women also frequently resort to alcohol abuse, recreational drug use and online gambling as a way of avoiding or dealing with their anxieties. In addition, more and more we hear about conditions such as 'tokophobia', which—as you may know—is a mortal fear of pregnancy and birth. Most women have some form of fear of the upcoming birth process (for whatever reason) and they often have all kinds of weird notions about birth to deal with, which may be based more on a soap opera screenwriter's view of birth than any real data. The question, quite simply, is this: How can you help?

Here are a few suggested approaches:

♥ Provide opportunities for your client to *talk.* This is a vastly underrated activity nowadays, but simple human contact with someone who can really help is a wonderful comfort for someone who is experiencing difficulties. I'm sure, on reading this, you probably immediately worry about time constraints... It's true, this aspect of care can be very time-consuming, but it's not necessarily time you need, but attention. When I was working as a manager I came to realize that even two minutes of focused attention on someone's needs (rather than on my own) reaped enormous benefits. And that's not mentioning the incredible power of simple alert listening and really *hearing.* Think how much time can be saved when misunderstandings are avoided! At least consider how you can improve the opportunities you provide for communication in terms of quality, if not quantity.

♥ Encourage your clients to explore their own feelings and experiences of pregnancy. Some women manage to make contact with their true thoughts and feelings by going on walks alone, others might have more success by doing something creative (such as painting, drawing or dance) and many others will simply benefit from sitting around listening to favorite music. The important thing is to encourage your clients to give themselves *time.*

♥ Consider suggesting a course in self-development, such as the ones provided in the books *Creating a Joyful Birth Experience* (Simon & Schuster 1994) or *Birthing from Within* (Souvenir Press 2007). Your clients may even be able to find a class based on this type of book in their area. In any such course, suggest that your clients leave out any activity which they intuitively feel may not be helpful or may disturb their baby. In *Birthing from Within*, for example, I would not recommend any of the exercises involving ice cubes! (They're used to simulate the experience of pain before the birth.) Doing the exercises might have a negative effect on both the mom-to-be and her baby, in my view. Similarly, I would beware of any breathing exercises—they're both unnecessary and unhelpful because the preparation done may override and distract the woman from her instinctive knowledge while she's actually in labor and giving birth.

- If you know your client is religious, and you feel this is appropriate in your professional working context, you could perhaps mention prayer or meditation. The Bible promises remarkable results, after all, and the idea that birth should be painful is argued against in some scholarly Christian works. The form of meditation called Vipassana, based on the original teachings of the Buddha, is also something you might like to suggest because it's very effective and teachers claim it is an entirely non-sectarian technique. [See: www.prayerguide.org.uk and www.dipa.dhamma.org].
- Encourage your clients to address any practical issues they're worrying about. Resolving these (e.g. place of birth, who's going to look after the children, etc) can have an enormously calming effect. The 10 preparatory steps outlined in the book *Preparing for a Healthy Birth* (Fresh Heart 2010) include support with this kind of decision-making (by providing relevant reading material about other women's experiences) and the psychological preparation provided through the other chapters should also help women to learn about and come to terms with issues surrounding birth.
- In some cases you might suggest that your client consult a counselor, therapist or psychologist. Hypnotherapy is becoming increasingly popular as a method for working through psychological fears about birth.[59]

## Birthframe 58: Nicolette Lawson—Hypnotherapy for birth

Here Dr Nicolette Lawson talks about her hypnotherapy practice and provides more details, so you can get a better idea what it's all about. She has a degree in mechanical engineering and used to work as a manufacturing systems engineer. She then moved on to become an environmental manager and gained an engineering doctorate in Environmental Technology. Now she sells baby carriers and works as a hypnotherapist and self-hypnosis teacher. I asked Nicolette a few more questions to try and find out why she had so dramatically changed her lifestyle. It seems the services she offers can make a significant difference to women who have worries about childbirth.

### What services do you offer your clients?

Using hypnosis I can help women to be healthy and happy during pregnancy, calm and relaxed throughout childbirth and confident in their abilities as a mother, once the baby is born. Occasionally, I also help pregnant women who may have specific phobias related to hospitals, blood or needles.

### Can you describe a typical client?

Typically, my pregnant clients come to me with only three or four weeks to go before their baby is due—although it would be even more beneficial if they started earlier. At this stage, they are really starting to worry about the actual birth.

Women are bombarded with bad news stories about childbirth

Women are constantly bombarded with all the bad news stories about childbirth and babies—tales of pain and intervention, tales of disrupted sleep and postpartum depression. It's no wonder women approach childbirth with trepidation and anxiety. Women with muscles tense from fear will undoubtedly feel pain, because they will be resisting the natural process that their body is going through. Women who can be calm and relaxed throughout childbirth, who can go with the flow, are naturally more likely to have a problem-free, healthy birth.

### Is there anybody you would refuse, if they came to you wanting hypnotherapy for pregnancy and childbirth?

There are no guarantees with hypnotherapy because everyone has a different belief system and different perceptions. However, anyone can be successfully hypnotized, if they really want to be. So I would be willing to work with anyone that wanted to give it a go—provided they were female!

### What first made you interested in hypnotherapy for birth?

I first learned self-hypnosis to help me finish my doctorate—which of course I did. Then, when I became pregnant with my second child, I decided to use self-hypnosis for both the pregnancy and birth. The birth of my first child was a long and complicated experience that I did not want to repeat.

### And did you find it helped with the birth of your next child?

Definitely! During the pregnancy I was much less tired and I had no acne—I had loads with my first pregnancy. Then the birth, which was at home, was calm and relaxed and my 10½lb baby was born with no pain relief—as I couldn't describe any of the process as painful. The midwives were quite impressed! My son was so calm and quiet when he was born—the birth had been so calm—and he was perfectly clean.

### Why do you think the birth of your first child was so much more difficult, without self-hypnosis?

I went into hospital 10 days before my due date, because my water broke. However, it turned out it was only my hindwater—no one had told me about that possibility—so they didn't start my contractions off. I wanted to go home, but the hospital said I had to stay—because of the danger of infection!—so I did. By the next day, there were still no contractions, so I was induced. This resulted in massive contractions and high blood pressure. The medical staff then insisted that I have an epidural, which was not what I wanted, to bring down my blood pressure and so protect me from the risk of pre-eclampsia. This resulted in me having to lie on my back with monitors strapped on me. The contractions slowed down and I labored all night, my 7½lb daughter finally being born at 6AM with the help of forceps, ventouse and an episiotomy.

I felt that each intervention by the hospital just led to the next in a long line of problems. The whole situation undermined my dream of having a healthy birth, shattered my confidence, and left me feeling totally out of control. With my second birth, I was determined to stay in control. Self-hypnosis helped me to have the confidence to do it, and to remain calm and relaxed throughout the process, which of course I did!

## How many sessions does a pregnant client need?

The majority of pregnant clients—assuming they have not had hypnotherapy before—require two sessions. The first session is to get the client used to going into hypnosis; it is a deep relaxation session that encourages general calmness, relaxation, confidence, competence and better sleep. The second session is specifically for the remainder of the pregnancy—being strong, healthy and sleeping well; the birth—remaining calm, relaxed and in control, going with the flow (with the contractions and the birth process as a whole); and after the birth—recovering easily, feeling confident, happy and competent, getting back to pre-pregnancy size, breastfeeding easily and coping well with everything and everyone. I record both these sessions and advise that the childbirth session is listened to every day up to the birth. This has the added benefit of ensuring that pregnant women take some well-earned rest and recovery each day. The first session can then be used whenever some calmness and relaxation is required after the birth.

## And what kind of results have you achieved?

When I first started, I visited all my clients after they had given birth to see how they got on. Now I simply don't have the time to follow them up, and most new moms don't have the time to let me know how they got on.

However, based on the feedback I collected when I first started and the ongoing calls I get from some of my clients, I can report that all of my clients thought the hypnosis did benefit them. All reported feeling more relaxed and calm. Many had healthy births, with minimal pain-relief—even a couple of women who had cesareans for their first baby went on to have a healthy, vaginal birth for their second child. Inevitably, some women had to have cesareans, for various reasons, but all that I spoke to reported feeling calm and relaxed, despite the surgery, and notably, they healed up and recovered better and quicker than expected. One woman, who hadn't expected to breastfeed because her mother and sister had had difficulties, went on to experience easy and enjoyable breastfeeding. The other common trend was in the babies, who tend to be very calm, happy and easy to look after. I really must collect some more information and write a book!

Even women who'd had cesareans for their first baby went on to have a healthy, vaginal birth for their second... I really must collect some info for a book!

### And do you have any regrets about giving up your life as an engineer?

Well, occasionally I regret the loss of money... I could have been earning a very handsome salary by now! But I don't regret giving up the long working hours and time away from home. Being a self-employed hypnotherapist means that I can work from home—hours that suit me and my family—and I have freedom and flexibility that's worth more than any salary. Added to that, I have incredible job satisfaction, knowing that I'm helping so many people to enjoy better lives and experience childbirth as it should be—natural and wonderful. I think being an engineer by background has actually helped me gain credibility with a lot of my clients. They reason that if an engineer—an intelligent, logical, rational, practical person—thinks hypnosis works, then there must be something in it!

### Finally, what advice would you give a woman who is not interested in hypnotherapy but who would like to have a positive birth experience?

The most important thing to remember is that childbirth is normal. Women's bodies were designed for it. It's not a medical procedure. Your body knows what it has to do. So, you need to trust your body and trust your instincts. Also...

- ♥ Don't believe all the negative things that other people tell you about pregnancy, childbirth and raising babies. Whatever they've experienced was their own experience—it doesn't mean yours will be the same.
- ♥ Learn all you can about the birthing process so that you know what's happening to you and you know what to expect.
- ♥ Stay calm and relaxed. Use whatever methods you want to help you—apart from drugs and alcohol!—and have a birthing partner who is calm and supportive too.
- ♥ Consider using hypnotherapy or learning self-hypnosis after all. It really is a very safe, natural and pleasant experience with so many beneficial side-effects: calmness, confidence, better sleep, more positive thinking and so on. How can you be negative about something unless you've tried it?

For more on Nicolette's work see www.m-power-me.co.uk

*A life-changing text result is not always good news... It takes some getting used to.*

Some women are terrified of giving up their lives to a baby

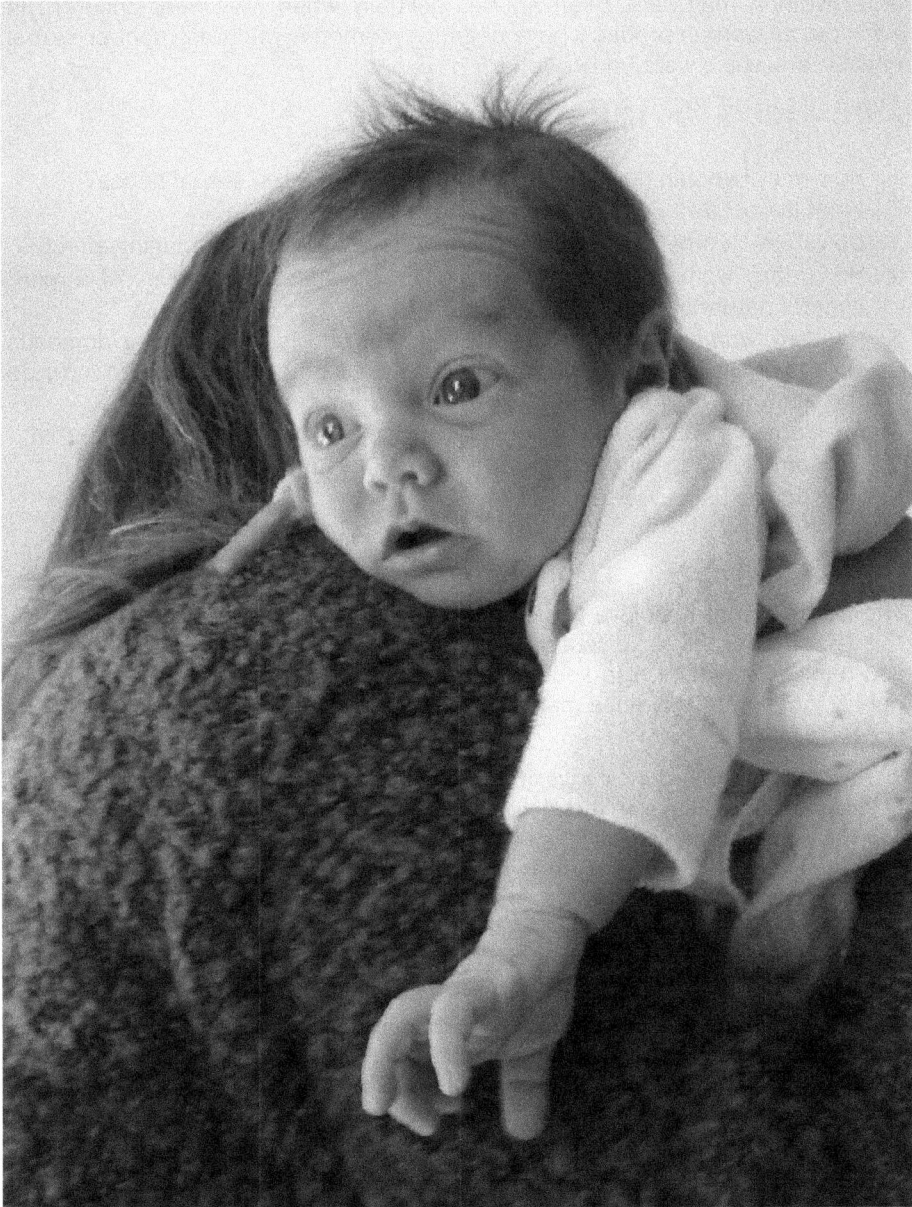

*This baby is a VBAC baby, born entirely healthily. Someone obviously confronted her fears!*

Sexual abuse is a particular problem. As you know, it is being increasingly reported in the media and some experts estimate that as many as 25% of all adult women may have been abused sexually when they were children. In addition, *all* women probably have negative memories of harassment or sexual encounters which went badly wrong.[60]

## Questions for reflection...

1   How many women do you know of who've experienced sexual abuse?
2   What is your own personal definition of sexual abuse?
3   Do you think other types of abuse might have a similarly damaging effect?
4   How many women do you imagine might be affected by abuse of a more general nature?
5   Has any woman ever confided to you that she is experiencing domestic violence? Given that it is apparently quite common during pregnancy, do you think you hear about all the women it happens to?
6   Are women's past failed relationships with men likely to affect their births?
7   Do women in your care usually open up on any of these subjects to you?
8   How do you usually respond if they do?

### Birthframe 59: Beth Dubois—Optimality after sexual abuse

The next account may give you some clues as to appropriate care and it also raises some issues for discussion. Would you have behaved differently? Can you think of anything else you could do which might help women who either openly, or *secretly* have bad memories or anxieties?

I have been enthralled by stories of empowering births since my college years, when I made the acquaintance of several home birth midwives. However, I was disappointed by the so-called natural childbirth I witnessed during my education as a nurse practitioner, when I was required to attend several hospital births. The nurses were zealously cheering, "Push! Push! I want to see this baby! Push!" No one was encouraging the woman to follow what her body and baby were doing naturally. It seemed the woman was expected to birth her baby for the nurses rather than to be present in her own experience.

When I became pregnant, I knew I wanted something different. I wanted to feel safe and comfortable so that I could open and allow my baby to be born. In addition, I am a survivor of childhood sexual abuse and I knew that the 'wrong' environment could trigger flashbacks and definitely impede the birthing process, possibly even creating further emotional trauma. On the other hand, I hoped that an empowering birth experience could promote healing at a very deep level. In a way my background as a survivor was a gift, as it led me on an intensive inner and outer quest for a healing pregnancy and birth. As I can see now, the process of undergoing the journey, in itself, was healing.

My husband and I spent endless hours reading and discussing aspects of prenatal care and birthing. We interviewed five home birth midwives and two nurse-midwife group practices. We changed caregivers three times, wavered for months when trying to decide between a home birth, birth center birth, and hospital birth. Finally, we decided we felt most comfortable with a planned home birth, with a nurse-midwife practice as our back-up, should a hospital birth become necessary. We took a great childbirth class based on the book *Birthing from Within* (Souvenir Press 2007), which prepared us for the birth in a very experiential way. In the class, we practiced pain-coping techniques (while holding ice cubes to simulate the pain of contractions!), discussed and faced fears, did birth artwork, and watched videos of women laboring and birthing in very non-interventionist settings.

I delighted in the physical closeness I had with my baby during my last few months of pregnancy. I loved to feel him move and to touch and massage his various parts. In fact, even when I was 40 weeks pregnant I felt very comfortable with my baby inside me; I didn't experience the feeling of 'wanting this baby out' or 'wanting to get my body back' that many women describe. And then 41 weeks came and went. As comfortable and wonderful as it was having him so close, I also longed to meet this new person and become a mother. Actually, there was also a sense of urgency because we were due to move to another state two weeks later! During this period of waiting (and trying many techniques to induce labor) I had plenty of time to worry about various things, mostly related to the birth and the move. At a certain point, my anxieties began to center on breastfeeding. As a nurse practitioner I was well versed in the benefits of breastfeeding and had a lot of 'book knowledge' on the subject; I was certain that I wanted to breastfeed. I was extremely concerned, however, that I would not be able to comfortably maintain the constant physical closeness and literally the suckling that breastfeeding would require. I was concerned that breastfeeding would trigger memories of my childhood abuse.

My midwives knew about the sexual abuse and we had discussed it in terms of the birth, but I had felt ashamed and afraid to bring up the topic of how my background might affect breastfeeding. As is common with abuse survivors, I carried the shame of the abuse. I felt I was the only one who had ever faced these issues and also, illogically, that admitting their existence might 'jinx' the breastfeeding. I was especially distraught because I believed from the core of my being that I would be failing my child and myself in a very profound way if I were not able to breastfeed.

Eventually, at 41 weeks and still counting, I knew I needed help! I could find no information about survivors of sexual abuse and breastfeeding in any of the literature. I took a courageous step and arranged for a meeting with a lactation consultant who came highly recommended. She listened to my fears and reflections and gave me a lot of reassurance. She said that I had a great chance of successfully breastfeeding. Women who have been sexually abused sometimes have trouble tolerating the physical closeness breastfeeding requires, as was my fear for myself, but different women behave differently when faced with new motherhood as a survivor of abuse. Some lose confidence in their bodies and worry that they will not be capable of producing enough milk. For others, the physical sensations involved with breastfeeding may remind them of the abuse they suffered, which was also something I was concerned about.

After speaking to the lactation consultant, I felt so much freer... my 'secret' was out. I then felt comfortable discussing the issues with my midwives—it turned out they were not at all surprised and had faced these same issues with other women before. I felt so relieved. Both my shame and fear decreased.

Our home birth midwife, Ann, said that many women with abuse backgrounds and/or who do not feel comfortable with nipple stimulation by their husband have no problem when they nurse a baby. She said that the hormones released by breastfeeding make women able to tolerate and even enjoy the close contact with their babies. Our nurse-midwife, Valerie, told me that it was the love that mattered in feeding the baby, whether from breast or bottle. She said I should try to nurse and see what happened. Of course I still desperately wanted to be able to breastfeed, but I felt less pressure after hearing her words.

At Valerie's suggestion, I took some time to write about my fears and realized that nursing a baby really would be different than the abusive situation I was in as a child. My baby, tiny and in need of food, warmth, and love, would be totally different than the adult who had abused me. And as a breastfeeding mother, I would be in a totally different position than I was as a child suffering the abuse. Instead of being a small child overpowered by someone huge and terrifying, I would be a grown woman responding very naturally to my baby's needs. These insights gave me hope that I would succeed at breastfeeding.

I also prayed that I would be able to breastfeed. To me, being able to successfully nurse seemed like a quantum leap. It seemed like it truly would be nothing short of a miracle. I didn't have time to do more therapy, more healing because my baby would be here any day... I needed grace! So I prayed that somehow, just somehow I would be able to successfully breastfeed.

As the days wore on I continued to be anxious that our baby was not yet born. Our move date was rapidly approaching and, to make things worse, I knew that at two weeks after our due date the policy of both our home birth midwife and nurse-midwife practice was for us to have our birth in the hospital. 'Nesting' was difficult, to say the least!

Suddenly at 13 days after our due date, I had a surprising inner shift and felt that the right thing for me to do was go to the hospital and try a very gentle labor induction—prostaglandin gel on my cervix (which is not the same as Cytotec, incidentally). My husband and I spoke to our midwives, who agreed with our plan. It was in fact Valerie who was the nurse-midwife on call. From our earlier conversations, I knew that she was very experienced in working with women with sexual abuse backgrounds. I also knew she recognized and appreciated the inner work I had done and that she welcomed the opportunity to attend our birth. I felt safe with the thought of being in Valerie's care. She said, "We'll make it just like a home birth. Come at 7:30 or 8:00PM and decorate your room. We can start the induction after that." My husband and I spent the day grieving that we weren't having the home birth we had hoped for and packing up everything we wanted to have with us.

Right: *During a non-worrying moment*

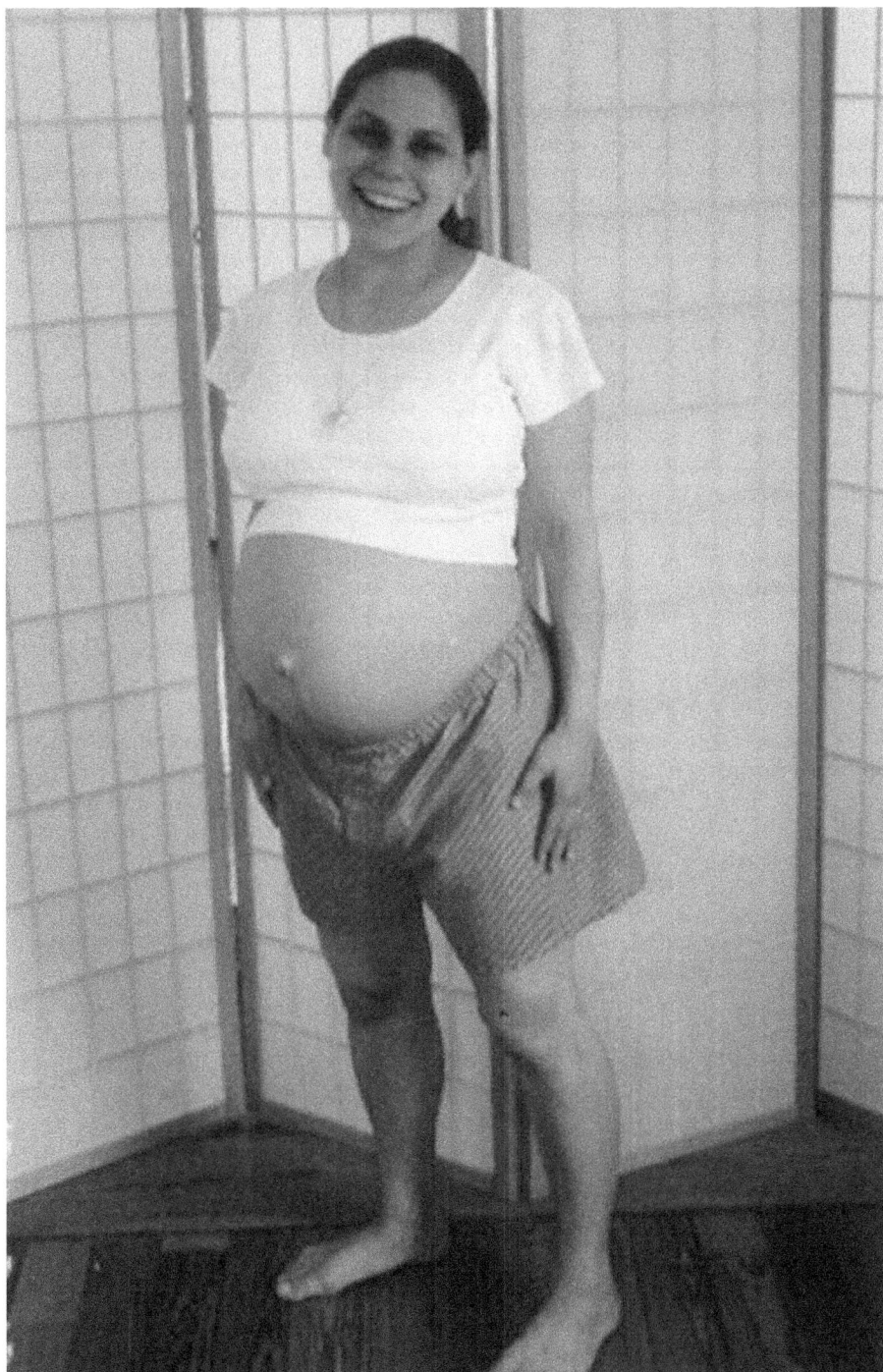

## We were welcomed warmly by the nursing staff

We were welcomed warmly by the nursing staff. Valerie had even 'reserved' a specific room for us which had a huge jacuzzi tub. Our home birth midwife, Ann, met us at the hospital which gave me a strong sense of protection. When we saw Valerie and I told her that we didn't want to be there, that we wanted a home birth, she said, "Okay. Then go home." But I was clear I did want to try to have my baby right then, so we stayed. However, knowing that no one was insisting on my being at the hospital kept me feeling in control (which was clearly Valerie's intention). We decorated our room a bit. Then Valerie inserted the prostaglandin gel and explained it was a 12-hour slow release. I would need to have fetal monitoring continuously for two hours (to make sure it wasn't releasing too quickly) and then we could go to sleep and see if anything happened in the morning. We never did go to sleep... To my great delight, within two hours I was having strong regular contractions. After the monitor was unhooked, Joel (my husband) and I walked up and down the halls, up and down the stairs, talking and laughing. I was so excited to finally be having regular contractions. I was even happy about the pain.

Wanting to just be alone with Joel, I asked him to ask Ann to leave, that we would call her when labor really got underway (as we were still under the impression that nothing much would happen until the morning). As soon as Ann left, I felt the intensity of the contractions increase. I got in the jacuzzi tub. Lying there in the dark, warm water and listening to the whirring of the jets, I felt like I myself was in the womb. As the contractions got stronger my focus became completely inward.

I remembered a slogan from *Birthing from Within*: 'Relax, breathe, do nothing extra'. I chanted 'Om' during the contractions and then put all my effort into releasing all thoughts and physical sensations in the space between the contractions. Joel chanted along with me and that really kept me going. He was incredible, consistently reading my cues perfectly. One word or one gesture and he knew what to do—rub my back in a certain place with a certain amount of pressure, bring me water... I could really feel that the two of us together were bringing our baby into the world. And quite literally no one else was around. We were in a sleepy hospital on a Friday night and actually Valerie and the nurses were in with another woman who was birthing her baby! At a certain point the pain was very strong and I had a clear sense the prostaglandin gel should be removed. Joel managed to find a nurse who was doubtful that I could be progressing so fast, but after consulting Valerie, did agree to remove it. My contractions continued to progress. At one point a nurse I'd never seen before strode into the room looking for a piece of equipment. She seemed totally unaware she was entering our space. Later she again strode in, this time probably to listen to our baby's heartbeat. Again, her energy did not feel 'in synch' to me. I felt very 'seated' in my power, like a lioness. I said emphatically to Joel, "Who is that? Tell her to leave." (I later learned that he then took her aside and explained that I was a survivor of childhood sexual abuse and that I needed to feel very safe in my space. And she did improve.)

Eventually I remember making my way to the bathroom. During my pregnancy I had the sense I would like to labor on the toilet because my pelvic floor muscles really could relax well there. This proved to be true. Through all the contractions my repetition of 'Om' continued, although the chanting eventually gave way to yelling! I would yell 'Om' as soon as the contraction began and continue all the way through it. The sounds that were coming out of me were incredible—they were deep, powerful sounds. I felt myself releasing on a very, very deep level what seemed like lifetimes of accumulated blockages. The process seemed to be freeing me from many of the wounds the abuse had left. I felt so much darkness that had previously taken up residence in my pelvic region being transmuted into light and being released through the sounds.

Unlike the abuse during which I 'left' my body, during the labor and birth, I was fully present. As this moaning and purging was happening, I knew that a miraculous process was occuring. I was simultaneously reaching into the depths of where the abuse memories were stored in my physical body and into a depth of healing power which now seemed to interpenetrate the very same location. Quite literally the site of my childhood abuse was now the canal that my baby was soon to travel through. I tangibly felt that as the darkness was released through the sounds, my pelvis was opening, freeing everything up for my baby to move through. I was also aware even then that the impact of this experience would be far-reaching in my life and that I would be able to offer my child a much healthier mother. I felt grateful for this miraculous process.

At a certain point, I noticed myself very naturally starting to push at the end of each contraction. I had read that grunty pushes at the end of a contraction can help the cervix dilate. I had the sense I was getting close. I told my husband to get Valerie. She arrived and sat on the floor, Indian style, in front of me—and the toilet! She asked me if the sounds I was making were scaring me. I said, "No". In fact, inside I felt very happy about the sounds as I knew what an incredible healing was taking place. I then mentally told the baby that everything was fine and not to be worried about the sounds, but my sense was that he or she was fine with it. From atop the toilet, I yelled and writhed and pushed my feet into Valerie's thighs during each contraction and then when the contraction was over I leaned forward and collapsed into her arms. This was an incredible part of the labor for me because I have never allowed myself to be mothered the way I allowed Valerie to mother me. I had felt engulfed by several women during my childhood and as an adult mostly avoided anything but loose hugs with other women. And here I was collapsing into Valerie's arms! This in itself was a whole other profound healing for me... I felt worthy of being mothered and I realized I wasn't defective in terms of allowing myself to be mothered when I was with a person I truly felt safe with.

Interestingly, at this point, my husband disappeared into the background for me. I specifically needed a woman to travel with me down this leg of the path, someone who had traveled this path before herself and accompanied other women down it. I later learned Joel had taken a long-needed bathroom break!

I think Valerie noticed those grunty pushes I was doing at the end of the contractions and she asked to check my cervical dilation for the first time since labor had started. She also probably noticed me saying, "I can't do it...," which is apparently a typical sign of being in transition. She checked and I was about 9cm dilated. I said, "Thank God!" Valerie requested I move over to the bed, asking Ann and Joel to 'make it like a seat' since I was so reluctant to leave the toilet.

I labored on all-fours for a while continuing to use movement and sound to cope with the pain of the contractions. I distinctly noticed my contractions become half as frequent. I understood I was in 'second stage' (fully dilated and pushing). My urge to push now took up much more of each contraction. It was so clear what my body was doing, and so easy for me to follow along. At one point I spontaneously reached my hand into my vagina and I felt this mushiness which I knew to be the baby's head; I was so excited!

At some point I noticed the nurse was concerned that the baby's heartbeat was low. I remembered Ann explaining to us during a prenatal visit that sometimes when the birth is imminent the heartbeat will drop from head compression and that it was not dangerous. I was not concerned and said aloud, "It's just head compression".

Apparently, the baby's heartbeat continued to be low (at times 60 beats per minute). Now Valerie said, "Bring the sound down lower out through your vagina" so I did. Then she said that we needed to have the baby born. By then I had flipped over onto my back. She asked me to pull up on my knees and do some forced pushes. Ann and Joel helped pull my legs up toward my chest. Of course, I didn't want to do this. I wanted to take my time and ease the baby out. I knew that both this position and forced pushing would be more likely to cause a tear and would decrease the oxygenation to the baby. Not wanting to do it and still wanting to do everything 'my way' I said, "I can't". Valerie and Ann said firmly, "Do this for your baby." I pushed about two times and his head was out. One more push and his body swooshed out. The room was dark. Valerie held him up, he let out a cry, she said, "The baby's fine," and immediately placed him on my belly. She said, "It was just head compression."

(It was only two years later, when consulting with another midwife, that I came to believe it to be a sensible practice for a midwife to 'want the baby born' when the heart rate has dropped for quite some time and is not resurging. My husband also told me that Ann said she would have done the same thing had our baby been born at home. I will never know for sure whether pushing him out before I felt ready was medically necessary. However, it was quite a profound moment for me. It was the moment that I did something I didn't want to do because the two midwives, who I knew wanted to avoid all unnecessary interventions, both made it clear that it was for my baby. I felt Mother Nature was giving me a nudge to step out of the role of the individual 'Beth' and into the role of motherhood. I felt the birth of my motherhood was in that moment.)

Our baby was so full and strong, not like the scrawny newborn I had expected. I rubbed his back and talked and talked and talked to him. It was so amazing to have skin-to-skin contact with him after having him inside me for so long. Since he was face down, I had no idea what he looked like or even if he was a boy or a girl. I just felt his body against mine and held him. Time stood still. After a while my husband suggested we cut the cord and see if we had a son or a daughter. (The placenta had already plopped right out after the baby was born). Then we discovered we had our Theodore and saw his beautiful face and beautiful body.

About 20 minutes after the birth, Ann helped me to nurse Theodore. He knew exactly what to do, which thrilled and amazed me and gave me confidence that breastfeeding would work out for us. Ann told us not to be surprised if he didn't feed as well the next few times, that in her experience babies often latch on very well right after the birth and then may need some help. I was so worried it somehow might not work out. I worried about whether Theodore's latch-on was correct. I leaked huge amounts of milk. I struggled with engorgement. I began to feel physically weak and emotionally overwhelmed five days after the birth. All of this while getting ready to move halfway across the country! Luckily, when my friends asked how I was doing I honestly said, "Not well." We asked for help. Friends began coming to our home day and night helping pack, bringing food, feeding me, and giving me emotional support. Their loving assistance touches me to this day.

By the time he was 12 days old, Theodore and I had got into a good rhythm with breastfeeding, my physical strength was gradually returning... and we were living in a new city! I struggled for several months adjusting to the changes in our lives— becoming a mother, leaving my previous employment, caring full-time for a new baby, living in a new city where I didn't know anyone, my husband working at a new job... And as I struggled with these changes, I would remember aspects of my birth experience and feel inspired and proud of my attainment. I also felt very proud and powerful that I was breastfeeding, that my body was nourishing my baby.

For the most part, I was able to tolerate the nearly constant physical closeness that my baby required. Occasionally, I felt overwhelmed at having him attached to my breast so often. During those times I sometimes imagined that the milk was flowing right out of my chest wall into him rather than coming out of my breast. I would concentrate on the sound and feeling of the milk gushing out of me and him gulping it down. This visualization helped me forget about my breast and the emotional discomfort I was having with his suckling. I imagined myself giving him a person-to-person, life-generating transfusion.

I was amazed and grateful to find that most of the time I relished the physical closeness with Theodore. He wanted either to be held or nursed day and night. Although it was exhausting to carry and nurse him constantly and I did not experience the kind of productivity I was used to in the working world, I was surprised at how often I felt a sense of fulfillment and accomplishment. I attribute this, in large part, to the mothering hormones released as a result of the frequent breastfeeding and nearly constant physical contact. I was surprised to find that my need to be with him was just as intense as his need to be with me.

I was surprised to find my need was as intense as his

By the time Theodore was 13 months old, I was able to appreciate how gradually, gently, and naturally the process of him becoming independent and separating occurs. For example, in the early months he nursed very, very often and was only content if we held him constantly. Then he began to crawl and sometimes prefered exploring to being held and even sometimes to nursing! By 10 months he could walk and eat a little bit of food and could nap without being held, but he still relied very much on breastfeeding, being held and sleeping with us during the night.

> By the time Theodore was 13 months old, I was able to appreciate how gradually, gently, and naturally the process of him becoming independent and separating occurs

Then at 13 months he would nurse less frequently because he was really busy doing all his work and explorations. At that age, he would eat some solids but would rely mostly on breastmilk for his main source of nutrition. He slept with us and nursed about 1-3 times during the night. Although at times he acted very independently such as when he, on his own initiative, went upstairs when I was downstairs or darted off when we were in a shop, I could tell he very much still needed me to be there. I think that his confidence in knowing I was nearby helped him to feel safe to explore.

The key for me with nursing a toddler is to make sure that I am comfortable emotionally and physically. As Theodore has grown, I have taken note of any aspects of our breastfeeding relationship that are uncomfortable for me and have sought to make adjustments as necessary. For example, when Theodore was under a year I enjoyed bathing with him and happily nursed him in the bath. When he was about 18 months I began to nurse him in the bath and found that now with his bigger body and both of us undressed I didn't feel comfortable. As soon as I became aware of my discomfort, I stopped nursing, got him involved in something else, and got myself dressed. In general, I noticed that when he nursed with the intention of drinking milk, usually the milk flowed easily and his suck was comfortable for me. But sometimes when he 'hung out' on the breast I found it irritating. At these times he was usually willing to accept a snack, story, or outing instead of nursing. He often seemed to appreciate a substitution for nursing because it actually better met his current need for attention, food, drink, or interesting activity.

When I wasn't keen on nursing him but could tell he had a real need, I would nurse him then, but with awareness. Rather than sinking into feeling like a victim, I would remember that I'm his mother, a grown woman, choosing to meet my toddler son's needs. Then, as I watched him nurse and saw the relaxation and reprieve it brought him, and the relaxation it brought me, my choice was affirmed. Sometimes, when I was nursing a bit reluctantly, it helped if I read while I did it. Reading helped me feel that I was giving something to myself as I nursed Theodore. At other times I found that focusing fully on Theodore helped me shift into more positive feelings.

> I was giving something to myself as I nursed

Things changed for me when he was about 30 months. One night during that time, he asked to nurse. I was sleeping fairly deeply at the time and, and as he nursed, I had the perception that someone was performing a sexual act on me. It turned out that Theodore's sucking on my breast was triggering a memory of my childhood abuse. Of course, Theodore was still his same little 2-year-old self and doing nothing unusual. But the size of his body (no longer a baby or a young toddler), his larger mouth and the difference in his suck, as well as it being during the night with me half-asleep all resembled the abuse too closely for my subconscious mind to differentiate. After that I gently and firmly discontinued night feeds, making it clear to him that although we wouldn't nurse at night, I'd still be there for him, responding to him. Whenever he requested the breast, I offered him a choice of water, a snuggle, or a snack. My husband offered to carry or rock him. We told him he could nurse again when it was light outside. Soon he often slept through the night.

As Theodore neared 3 years old, I began to resent his frequent requests to nurse throughout the day. Since, when he was younger, I had offered him the breast readily many, many times a day meeting his needs for food, drink, soothing, etc, he had come to think of nursing as the thing to do whenever a need of almost any kind arose. I think this kind of thing is something many mothers (not only those who are abuse survivors) go through with nursing children of his age. Theodore is quite verbal and I intuitively felt I was short-changing him if I didn't work with him to teach him other ways to meet his needs, verbalizing his frustration or need for food or drink, seeking help in completing a project (rather than tossing the project aside in frustration and reflexively yelling, "Milk!"), asking for attention from me (rather than trying to get a 'piece' of me by nursing).

Also, I felt I was short-changing myself since, at this point, I did not enjoy nursing very much. Now, even when I'm fully awake while nursing him, I often feel emotionally uncomfortable feeding such a big boy. My subconscious mind still 'reads' the breastfeeding as abuse and I usually want the feed to end as soon as possible. Two things have helped me during this stage. First, I have explained to him that he can nurse three specific times during the day: morning, after lunch, and after dinner. No longer do I have to respond to requests on a case-by-case basis, 20 times a day! I tell him clearly, "It's not time for milk," and offer other options instead. Second, I control the length of the breastfeeding session. Whenever I want it to end, I sing a specific song at a speed I determine. When the song is finished, Theodore stops feeding. He then often proudly says, "I stopped!"

Now, several months after his third birthday, Theodore continues to nurse three times a day but has, indeed, found other ways to meet his needs. For comfort he says, "Cuddle me!" When hurt, he calls out woefully, "I am hurt." He readily accepts snacks and drinks. I feel proud to see him growing up, yet continuing to express his needs so clearly and relationally.

I feel proud to see him continuing to express his needs

This process of being attuned to my own experience and making adjustments as needed has been immensely healing for me, a survivor of childhood sexual abuse. When I was abused as a child, my own needs were not considered; I felt incapable and unworthy of asserting my needs.

Now, in relation to Theodore (now that he's 3—this didn't apply when he was a baby), I am able to assert my own needs as being important. Not more important than his needs, but also important. I have found, again and again, that we can find a solution that works to meet both our needs. Through our negotiations, not only am I teaching Theodore about the 'give and take' of healthy relationships, but I am learning about it myself.

I now see that not only has breastfeeding been possible for me, a survivor of childhood sexual abuse, it has been immensely healing. My desire to have an empowering birth and fulfilling breastfeeding relationship has forced me to face emotional territory I would probably have otherwise avoided. One wound left by the abuse is an underlying sense of 'I can't do it. It's not even worth trying.' Birthing, breastfeeding, and now the process of weaning Theodore have helped to replace this with a very real sense of capability and trust in myself. I am now confident that I can and will stand up for the well-being of myself and my child whenever necessary. I have also gained a heightened sensitivity to both myself and my son, which continues to serve us in our relationship as we both continue to grow and change.

*Comfortably nursing*

The process of being attuned to my own experience and making adjustments as needed has been immensely healing for me

*Here's a recent photo of Beth with her two children. That arc in the sky is a rainbow...*

### Birthframe 60: Sylvie Donna—Motivation for writing

I've thought long and hard about whether or not I ought to include what follows, but after I ended up revealing some of what's below at a conference (because nobody else seemed able or willing to respond to a question about sexual abuse) I have decided to reveal more. It's something which is not easy for me because of the stigma our society attaches to certain things. We seem to operate so many double standards... First we have the 'majority of people' who are 'normal' and who don't have major psychological problems or major blots in their history. (Is this why we like to have perfect politicians? We get so upset when any of their flaws are revealed, even though we must surely be expecting them to also be human beings, like us.) Then we have the second category: people who are 'not like us', who are 'weird' in some way... Each person perhaps has his or her own idea of who is not acceptable, but we all have categories of people who we consider 'other'. (Simone de Beauvoir wrote very eloquently on this concept of 'otherness' in *The Second Sex,* first published in 1949 and reissued by Jonathan Cape in 2009.) Anyway, to satisfy your curiosity, perhaps, about who I am and what my motives are, and also to set the record straight, here are some more details and further commentary...

Someone behaved inappropriately towards me when I was a child and until I corresponded with Beth Dubois I'd never before had any real contact with another victim of sexual abuse. I did once attend a small, one-day workshop but looking back I realize how polite and non-communicative the other seven participants were. It's astonishing and sad how little was exchanged and communicated. We'd all taken the step to register for something which might help us, but then neither we nor the person who ran the session dared to actually talk about anything which really affected us. It was a remarkably sanitized and 'polite' gathering, and also remarkably ineffectual.

I'm convinced self-disclosure about abuse is very rare

Anyway, perhaps I had actually met many other 'victims' because many experts estimate that percentages of women with a history of childhood abuse are high, as I've already said. (Nevertheless, I'm not sure how it's possible to come up with any reliable figure, since self-disclosure is so rare.) Actually, when I first started corresponding with Beth I was surprised that she talked not about being a 'victim' but about being a 'survivor' of sexual abuse. What a difference a word makes! I'm also aware that the term 'sexual abuse' can conjure up extremes of behavior for so many people but I'm using it to describe any kind of behavior which relates to sexuality, which is abusive—i.e. not respectful of a person's feelings and need to be loved and honored. This is because I know from personal experience how negative the impact can be of any kind of abuse, whether involving the full sexual act, or not.

Thinking about women who've been sexually abused and about the rest of the female population I finally concluded that *all* women are probably victims of abuse of some kind—sexual, emotional or intellectual—because of the incredible inequalities that exist in our world between the two sexes. Which adult woman hasn't been undermined by a man, touched or talked to inappropriately or rejected personally or professionally—simply because she's female? Personally, I was molested repeatedly by a teenage boy, a babysitter, at the age of 5 and was later rejected by the young man I lost my virginity to, straight after 'penetration'. Both experiences were totally devastating for me, making me feel worthless, and despairing about the future. For many years I was completely unable to talk about either experience.

I finally concluded we're all probably victims of abuse...

At the time of the first set of bad experiences (when I was repeatedly caressed on the genitals by that babysitter, who I couldn't avoid even though I tried to in many ways) I suppose I intuited at the time how dramatically and inappropriately my personal space was being invaded and how 'special' genitals were. In the second case, when I agreed to have sex for the first time and was then not asked out again, I was so terribly devastated (I realize now) because penetration seemed the only thing left that was sacred and then even that element of my sexuality was treated without respect. Both these experiences (which I think I have now come to terms with) have made me realize just how awful inappropriate touch might be for some (or all?) women in childbirth. Perhaps my experiences also explain why I was so focused on making sure I was not touched at all inappropriately while I gave birth myself—even to the point of prefering no one else to catch my baby.

I eventually realized that my motivation for publishing
books about birth came as a result of my history

Anyway, after much reflection I have eventually realized that my motivation for publishing books about birth came as a result of my history. From long experience, I knew the psychological and behavioral devastation any kind of abuse could wreak and when I realized (through my reading) the potential for suffering around the time of childbirth I became very proactive about my own situation and *protective* of other women's futures. After I'd given birth (and experienced how special it could be) and repeatedly heard about other women's awful experiences it was as if I became a tigress, defending her cubs... Those cubs of mine are all the women in the world at risk of damage from future negative birthing experiences!

I have become a tigress defending her cubs—other women!

So now I have two answers to a question which once made me aghast with rage. "What good has come from being abused?" (This was in a self-help book which was supposed to promote a sexual healing journey.) After my anger eventually subsided and I later gave birth to three daughters, one day it dawned on me that my experience would make me more sensitive to my daughters' need for protection. Now I realize that my status as a survivor who's had time to heal over many, many years gives me the motivation and righteous indignation to do research to protect other women. I just do not want women to suffer when they give birth. It's hard enough as it is! No... I want to do whatever I can to help make women stronger and more empowered. After all, I realize that it's only by supporting the world's women that both men and women will discover joy and peace of mind and only when women experience birth consciously and positively that children will have mothers who are mentally and physically strong. If that means that men and women will need to negotiate new ways of being together in terms of day-to-day power and interpersonal relations, then so be it. At the moment we have a model of family and personal interaction based on profound past private suffering and that just isn't a good foundation for the future of humanity.

*Men and women will have to discover new ways of interacting which are based on mutual respect, sensitivity and love*

Men and women will have to discover new ways of interacting which are based on mutual respect, sensitivity and love. Perhaps this will all come about naturally if we really do rediscover the 'hormone of love' (oxytocin) and stop calling its imposter cousin (i.e. synthetic oxytocin) 'oxytocin', when it should really correctly be called 'pitocin'.

*We must stop calling synthetic oxytocin 'oxytocin'. It's an imposter and is nothing to do with the hormone of love!*

Research collated by Michel Odent in his book *The Scientification of Love* (Free Association Books 1999) has shown that experience of the real hormone at the time of birth may well make babies capable of loving and, consequently, spontaneously aware of how to interact more beautifully with other members of the same species—whether they happen to be male or female, strangers or life partners. Needless to say, this is an area for which more research is urgently needed.

*Experience of real oxytocin at birth may well make babies capable of loving and, consequently, aware of how to interact more beautifully with other members of humankind*

Above and below: *Sylvie Donna with her three daughters, here aged 7, 9 and 11. Is there a woman alive who doesn't have some kind of history which might influence her perceptions and expectations of birth? These issues need to be addressed.*

# Pause to reflect...

## ...about motivations prompting choices and action

1   How often do you think caregivers working within the field of midwifery and obstetrics are motivated by personal experience? When it occurs is it a good thing or a very bad thing?

2   What kinds of personal experience might motivate people to work in the field of birth? How is the care they provide for women likely to be different from that provided by people who either have no personal motivations and/or have no personal experience of pregnancy and birth? (Is it possible for any person to be completely objective?)

3   Do you think it would be valuable for caregivers' motivations to be explored in periodic workshops? Are people's motivations for continuing to work in the field of birth likely to remain the same, or are they more likely to change over time, as they attend more births and perhaps also experience giving birth themselves?

4   How might a male caregiver's perceptions and motivations differ from those of a female caregiver? Is the difference (if any) important?

5   What kinds of support or interaction might be possible with colleagues whose motivations differ from your own quite radically? Do you think it would be helpful for staff working together to regularly discuss their feelings and motivations for the approach they tend to take?

6   In your opinion or in your experience is it useful to explore a client's motivation for the choices she makes during pregnancy and labor? If her choices are motivated by fear or self-doubt do you think you either can or *should* help her to overcome these negative emotions?

7   What methods might you use to help a woman explore both her motivations and her options? As a woman's emotions develop through her pregnancy do you think it's worth exploring the woman's feelings again later on?

8   When a mom-to-be is motivated to work through a previous bad experience in preparation for her upcoming labor and birth, in what ways can you be helpful and supportive? When might you need to also seek outside help (e.g. from other professionals)?

Perhaps birth is always a challenging time, whatever the woman's experience, simply because of the baby she is carrying inside her. Is it possible that different babies behave differently during pregnancy (because of their emerging personalities) and that their births are similarly affected by their own unique qualities? Also, perhaps we cannot hope for perfection in life. Nevertheless, from your point of view as a caregiver, it is well worth considering what could promote optimality for women... don't you think? (Think of the job satisfaction!)

## Birthframe 61: Beth Dubois—Struggling with the midwife

The same woman whose first birth story we have recently considered, now tells us about her second birth experience in more detail. (There were some brief comments on her care in Birthframe 6, as you may remember.) This account should again help us to consider what optimality really means in practice.

Because of what Beth describes in this account as 'struggling with the midwife' (when the midwife, in trying to help, began to direct Beth) and due to the sheer physical and emotional intensity of the difficult part of this labor, for a long time the memory of the difficulty and intensity of the labor overshadowed the memory of her power in birthing the baby and the wonder of the baby's birth. Over time, though, Beth has come to view the experience more neutrally: she was having difficulty, she tried to get help from the midwife, that didn't work out, so she got free of the midwife's intervention, stood up, and that really did work—she pushed her baby out.

Understandably, midwives and women, including Beth, would not typically choose to experience difficulties in labor. However, Beth relates that the experience of undergoing extreme intensity and yet going through it—standing up and powerfully giving birth made a certain 'imprint' on her. She says she has drawn upon this experience to face other difficulties since then. Looking back, she considers this birth experience a great gift.

## Questions for reflection as you read...

1  To what extent do you think all women might resent feeling controled?
2  Are there any situations when you feel it's important to 'control' a birth?
3  Do you agree that different babies might create different labors?
4  Since the previous question errs into the area of spirituality (which has come up in quite a few accounts), do you believe that any spiritual issues affect birth and how it might take place?
5  What role do you think birth has in a woman's life?
6  To what extent is it inevitable for birth to be a struggle?
7  To what extent can a struggle be made into a transformative experience and what kind of care do you think might support any transformation?
8  How typical do you think Beth's frustrations and postpartum thoughts are?

On Tuesday night, two and a half weeks before my due date, Theodore, Joel, and I attended the 'Catch Your Baby' class our midwife, Sandi, gave in our home. She advised us to stay well rested and well fed so we'd be ready at any time. She brought us the birthing pool so we could have it ready for the birth. We decided to have it in our bedroom since that was the place in our home that seemed most relaxing and peaceful and the least over-stimulating. I'd also spent so much of the pregnancy in our bedroom... while I was sick in the first trimester, napping later, sleeping and, towards the end of the pregnancy, where I often lay awake during the night... so it seemed the natural choice for where to give birth. And this is where Zoe was conceived so it seemed natural to come full circle and birth her there!

On Friday I woke up and felt 'like a grouch'. We had planned for a friend, Angie, to give us a tour of the local medical center so we'd feel as comfortable as possible with the layout and know where to go if we needed to go there during the birth. We were to meet her at 9:00AM and Joel had a meeting on campus at 10:00AM. However, I had a 'breakdown'—I just felt totally overwhelmed about how to manage. It seemed like just too much for me to coordinate dropping Theodore (our older child) off at a friend's house to play and meet Joel, who planned to ride his bicycle and complete the tour in time for his 10:00AM meeting. I explained to Joel and he agreed to get to his meeting late. I just had a feeling I was experiencing some kind of pre-labor, even though I was a two and a half weeks away from my due date and Theodore had arrived two weeks after his due date! We brought Joel home to get his bike to ride to school. Theodore and I went grocery shopping, which seemed unusually overwhelming for me. As soon as we arrived, I ate lunch at the café! Then the shopping seemed very arduous, taking much more effort than usual. I kind of felt like I was 'losing it' the whole time, although I don't think this was evident to anyone else.

After we got home, I immediately carried the groceries in... It was as though if I delayed I'd be unable to do it. Once I was inside the house, putting away the groceries and the mess in the living room (books and toys strewn about) was simply too much for me. I phoned Joel, interrupting his meeting, and asked him to come home, saying I felt like something was 'up' with me. He insisted I call Sandi and let her know what was happening. I then ate almost an entire baguette!—carbo-loading, I later presumed. That evening around sunset I went on a nice, slow walk around the park, enjoying the beautiful spring weather and the opening of nature all around me. I thought, "What a natural and beautiful time to have a baby." It seemed as though the baby would burst forth from me just as was happening with the blossoms and buds all around. I enjoyed the sunset and felt a sense of happiness, peace and reverence.

When I got home, I called my friend Shelly, who we planned would watch Theodore during the birth, chatted with her and was reassured that she was available to watch Theodore that weekend.

On Friday night I had about three strong contractions in the night and felt like they were working on my cervix. I had felt ready and longing to go into labor for about a week. I just had the sense I was waiting for the baby to be ready and 'give the signal' to my body so my uterus would start to do the necessary contractions.

On Saturday morning I woke up and felt great—really different than I'd felt the day before. Physically and emotionally I felt very well. I had thought I'd go to my prenatal yoga class, but somehow that didn't seem quite right. I also felt a compulsion to return Theodore's library books, which I did. I kept noticing a vaginal discharge, occasionally blood-tinged, which I suspected was a slow leak of amniotic fluid. I went out for a walk in the park and felt really light and happy. I was having contractions and felt I was in early labor. I was really savoring the atmosphere: the sun, the lively Saturday park setting... People I walked by commented about my pregnancy much more than usual. One man smiled and waved to me like we were old friends and said he remembered me from my walk the night before. I had the feeling of being loved and embraced by the whole world—by people and nature. One woman with a 12-week-old baby struck up a conversation with me, told me all about her, and wished me well. That took my mind off my own contractions for awhile.

Remembering to stay well fed, I ate lunch. Joel had vacuumed the whole house that morning and everything was sparkling and in order, just how I love it to be. I was very caught up with my paperwork in our home office and had even completed in advance Theodore's homeschooling paperwork through to the end of March.

At a certain point I knew it was time for Theodore to go to Shelly's. We packed him some clothes, snacks, toothbrush, books and his comforter and pacifier. Theodore said it was a 'good idea' to move his bed into the office so we could put the birthing pool where his bed had been. That felt like the perfect place for it—right next to our own bed—with plenty of space still around it. I put the shower drape and extra sheets on our bed. Everything flowed so precisely and amazingly: Shelly picked up Theodore and we all agreed he'd sleep over at her house. A few minutes later Sandi, our midwife, arrived.

Initially, I could talk through the contractions but soon I no longer could. I noticed this when I was in the middle of telling Sandi a story, got to a contraction, and couldn't hold my train of thought. Sandi relaxedly went about setting up all her supplies around her house. We went about the house. I cleaned up a few last things, went through the mail, etc, stopping when I had a contraction. It felt fun and relaxed. It was SO incredible to be at home, just doing our usual things.

As the contractions got stronger, I got the red birthing ball and Joel and I relaxed in the living room. Then, whenever I had a contraction, I'd lean forward over the ball. Initially, I did deep breathing through the contractions. As they got more intense, I chanted 'Om'. Sandi came and went, sometimes quietly observing me and timing the contractions to get a sense of how the labor was unfolding. I had gushes of amniotic fluid with the contractions and finally just took off my underwear and stationed myself on a sanitary pad near the birthing ball.

At a certain point Sandi described how long and close together the contractions were and said she was going to start setting up the birthing pool. She also called the other midwife sometime around then to ensure she'd be close by. I really wanted to have the fewest people possible and so Sandi asked Libby, the other midwife, to come when the birth was closer.

It was lucky she started setting up the pool when she did because once it was ready, I was very ready to get in to have some pain relief from my intensifying contractions. When I got in, it felt great; I felt like I was at a fancy spa to have a hot tub all set up for me—in my own bedroom! Nevertheless, as time went by, the contractions intensified further and the water didn't remove the pain, although it must have eased it. I cannot imagine how women can labor without water. For awhile I continued to chant 'Om' with the contractions. With each contraction, I'd lean forward onto my hands and knees in the water—similar to the position I'd taken on the birthing ball. As the contractions further intensified, I made varying sounds to let the pain out. I had a lot of heartburn and often started having intense burping towards the end of a contraction.

The contractions became so painful I didn't think I could stand it. I called for Sandi and told her. In the space between the contractions, there was no pain. But it got to the point where when the contraction started to come on, I'd feel panicked, not wanting to have to undergo the pain that would follow! Sandi said I was 'getting close... so close'. She offered to check my dilation and said I was a 'stretchy 9cm'. There was apparently a lip of cervix on one side holding back the baby's head. I agreed for her to massage that part of the cervix while I pushed and then I was completely dilated. I could feel the baby's head in the birth canal. The contractions continued to be just as painful. I yelled through them... whatever sounds I made seemed to help and sometimes I panicked and cried and yelled out loud.

During this time I was also struggling with Sandi. She was telling me to push, to push out my bottom, to bring the sound down lower, etc... I really didn't like that. I found it unhelpful and it did not support me in tuning into what would work for me, into how I could best work with my body. I experienced it as though she was criticizing what I did with each contraction. It seemed that no matter what I did, there was her voice during a contraction telling me to do something specific, which I guess she thought was what I wasn't currently doing, but was what she thought I should be doing to get the baby born. The baby's heartbeat was always perfect. At one point, utterly frustrated and irritated with her misguided 'help', I said, "Do you think I'm not doing it right?" According to my husband, my tone conveyed, "Shut the hell up!" She faded into the background.

Joel leaned in then and said, "You're doing everything exactly right." Although the pain of the contractions really had become unbearable (so I did want the baby to be born as soon as possible to end the pain), it was still very upsetting for her to give directives during contractions. I found it confusing and infuriating because I would try to follow what she was saying but then not be able to follow my body, or vice versa. On the one hand, during the birth, I literally thought, "So that's why people get epidurals. Why didn't I get one?! I never want to do this again." And on the other hand, I had a sense that this could possibly be like when one is close to dying and that having this experience and coming through it might serve as a guiding light later at the time of death and possibly during other times of intense difficulty.

I had a sense this could be like when one is close to dying...

By the way, the reason Sandi got so involved, I think, was that at a certain point I had given my power over to her, wanting her to somehow make things better and take away my pain, which is what she was trying to do. Anyway, I did eventually re-empower myself and then found a new position which worked to push out the baby. I got UP, standing with knees bent, leaning forward and holding the tops of my thighs. I yelled and pushed and pushed. I felt the ring of fire two times. I was determined to push out the baby's head with that push. What I hadn't expected was to push out the whole baby with that push. I had my eyes shut as I intensely concentrated, pushed, and yelled.

Then my eyes flew open when I heard a baby crying. There was Sandi, holding my baby. Sandi handed her to me and helped me sit down in the pool. I couldn't believe I was holding this whole, beautiful baby, who cried immediately and cried as I held her. Her cry didn't sound distressed to me—it was just a cry saying something like "I'm here". She was covered in vernix and surprised me by looking nothing like Theodore (my first baby). She had hair and dark eyes and a nose I didn't recognize from my side of the family. (Joel, my husband, says it's a nose like his and his grandma's.) She was beautiful in her completeness and simplicity. She cried and I shook—something I had read can happen to moms after giving birth. Joel turned up the heat and the midwives added hot water to the pool. They put towels and blankets around me and the baby to help warm us. Unfortunately, my perineum hurt and I took Ibuprofen right away.

I noticed that even when I was shaking and the baby was crying I was still strong enough to hold her. I felt there was a subtle teaching that, as I am, I can mother her... and that the shakes and crying (literally and metaphorically) would pass.

After a while, Joel and I checked to see if she was a boy or girl. We saw girl parts and a moment later a bulging blue thing (the umbilical cord) which both of us thought for a second could be boy parts! But we looked again and it was definitely a girl and an umbilical cord. (Phew!)

Zoe was born at 6:02PM, at twilight. It was magical how the daylight faded gently into night at the time she was born. Libby brought in some beeswax candles and everything was done by candlelight until it was time to suture my tear.

When Sandi came to our home at around 2:00PM, I told Joel how I'd been for a walk the night before at sunset and had said, "Wouldn't it be wonderful if our baby was born tonight, at that same time?"... and she was!

Eventually, the cord stopped pulsing and I—grudgingly, since my perineum was sore—pushed the placenta out, with Sandi applying gentle traction. (I didn't like that she applied traction. Why didn't she just let me push it out, I wondered.) Then, when I was ready, I was helped out of the pool. Joel took his shirt off and held Zoe skin-to-skin.

Sandi did Zoe's newborn exam. It was very gentle and thorough. Then she did the suturing for me. Finally, she and the other midwife cleaned up, gave us instructions, and threw the laundry into the washing machine.

Then it was Zoe, Joel, and me together for our first night. Theodore came back the next morning at 8:00AM and had a big smile on his face when he saw Zoe.

*Here's Zoe, with her big brother Theodore, whose birth we read about earlier*

The main thing that strikes me is that most births are like life, in that they are mixed experiences, with joyful and challenging aspects. My birth with Zoe was quite difficult and though I did find Sandi's directives unhelpful and infuriating, I may very well have had the same daunted feeling after giving birth even if she had not done anything I didn't like; it was just way more arduous and intense than I had been prepared for. Or perhaps, no matter what she did, I would not have liked it. Even if she said the perfect thing about tuning into my body, perhaps I would have hated that and yelled at her. Or if she had not come in at all perhaps I would have then blamed her for abandoning me and thought that my suffering was compounded because of that! It may be that I went through a certain phase of trying out things that ended up not working and then, with my husband's encouragement, I just figured out what to do. If Sandi hadn't been there, I might have gone through a similar phase of trying things on my own that didn't work and then figuring out what to do. It's impossible to know how it would have played out. It's also hard to know what the more spiritual aspects are... I know this book doesn't go into this kind of thing but Zoe was not an easy baby to carry. I was very sick in bed (vomiting, extreme tiredness) for the first three months. I had to have my friend Shelly watch Theodore because I was too sick to watch him. I had about two and a half months of doing fine, though with a very, very active baby inside, before I started having preterm labor and had to take it easy again. So the intensity of that one part of the labor certainly went hand-in-hand with the challenging pregnancy. And Zoe continues to be precious and quite a whirlwind! So perhaps there is some connection between the baby and the mom and lessons they are bringing to each other which is reflected in the birth. Who knows?!

## Building rapport and open communication

As many of the last few birthframes have revealed, communication and rapport between caregivers and their clients is a key to optimizing birthing experiences. Nevertheless, many of the birthframes I was sent gave me the impression that various circumstances cause problems in these respects...

# Pause to reflect...

## ...about your relationships with pregnant women

1 Have you ever wondered how honest women's responses are to you, when you ask them questions? Do you ever suspect any dishonesty?

2 How do you respond to a woman when you feel that she is being secretive or not entirely honest with you?

3 Do you ever cross-check information?

4 Does it ever occur to you that misunderstandings may have arisen which need to be clarified? What steps can you take to avoid misunderstandings of any kind during prenatal care?

5 When you disagree with a client, do you openly disclose why? Do you actually really understand your own reasons for disagreeing?

6 When a client disagrees with you, do you encourage her to explain why? How can you stop this kind of conversation from becoming confrontational?

7 Whenever there is any disagreement, how can you resolve it so that all parties (including other people you have to work with) are happy?

8 When a woman makes a request which makes no sense to you, either prenatally (in a birth plan, or in conversation) or intrapartum (between load groans), what do you usually do? How can you increase your understanding of other people's worlds?

9 Although you have probably been told you should never disclose anything about yourself when you are dealing with a client, as a human being do you think it's ever appropriate to do so?

10 To what extent can your relationships with clients be loving and caring?

## Develop trust and rapport!

This is an easy thing for an author of a book or a line manager to say and a much more difficult thing to actually achieve while working within the constraints of a busy maternity unit... but do you think you're really managing to develop good relationships with your moms-to-be? If you don't manage to, you may well find that communication becomes something of a game and that information is not provided by the mom-to-be openly and willingly. Consider the following accounts, both of which reveal odd mother-midwife relationships. How could the midwives concerned have really found out about the views and 'background' activities of these moms-to-be and how could they have made themselves more trusted and better used? I'm sure you don't need me to tell you that the answers to these questions are not merely of passing interest... they could well mean the difference between life and death for some other mothers or babies because we know from research (if not from simple common sense) that births which are attended by midwives or other experts have the potential—in an ideal world—to be much safer than those which are unassisted. (Of course, I am assuming that good, sensitive care is provided, without unnecessary, potentially iatrogenic interventions.) As for pregnancies and the safety factor, it is surely best if moms-to-be tell their midwives (or other caregivers) about treatments they are undergoing which may well have clinical implications... But how can clients be made aware of what needs to be said and what doesn't need to be, and how can prenatal check-ups generally be made more a forum for a free and trusting exchange of information and feelings? Is it worth explaining the rationale behind the law? Needless to say, these are big questions, but we do need to find the answers and implement any changes necessary to our day-to-day systems so as to improve outcomes, not to mention caregiver satisfaction levels.

### Birthframe 62: Helen Arundell—Depending on homeopathy

In this first birthframe we see how one woman mixed conventional prenatal care with 'complementary therapies'. Here we step into the private world of one mother's twin pregnancy. At the beginning of this diary, the mom-to-be is 38 weeks + 1 day...

Mar 25  It's Mother's Day. Up really early. Feeling bunged up and snotty. Had a few Braxton Hicks. Feel quite down again. Want to be left alone. Took Sepia (constitutional) before bed.

Mar 26  Feeling really down and have been in tears on and off all morning. Feel really emotional. Have not felt like this in a long, long time. (Mom and Pop are moving next Wednesday.) Had a few Braxton Hicks. Still bunged up and absolutely shattered all day. No motivation to do anything at all.

Mar 27   Feeling much better in myself today. Had a few stronger Braxton Hicks. Tired and still bunged up.

Mar 28   Woke up at 1:00AM in a hot sweat! Really tired during the day. A few Braxton Hicks. Feel okay.

Mar 29   Slept well. Have taken Kali bic for sinuses—three a day for four days. Saw midwife. All fine. Head has moved up a bit from 2/5 to 3/5 engaged. Had quite a few niggles. Babies moving well. Slight low backache. Took Caulyphylum at 9:45PM and went straight to bed!

Mar 30   Felt a bit teary in the morning. Fine in the afternoon. No twinges but a few Braxton Hicks. Have had a good day. Sun has shone all day and it's been warm. Babies moving a lot. Still bunged up but feel okay.

Mar 31   Bad night's sleep. Really uncomfortable all night. Woke up early morning with a strong pain across my belly, with a Braxton Hicks which went away after a bit. Feel alright so far, I suppose!

Apr 1    Babies moving quite a lot. A few Braxton Hicks. Feel really fed up and grouchy. Slept for one and a half hours in the afternoon and did feel better afterwards. I now measure 48 inches round my waist!

Apr 2    Babies moving loads. A couple of twinges. Really tired. Feel okay but really fed up. Legs feel really heavy and dragging in the evening by about 7:30PM.

Apr 3    Felt down but okay. Said goodbye to Mom and Pop. Had a few Braxton Hicks.

Apr 4    Had some diarrhea at 7:30AM. Had a few twinges—one light popping sensation about 2:20PM. No water or leakage, though. Felt a bit flat in the morning but better in the afternoon.

Apr 5    Feel okay. Saw obstetrician. Happy to let me go a bit longer. Babies growing and moving well.

Apr 6    No twinges. No signs. Scott off work. It really helped, him being here. Took Caulyphylum at 2:15PM. Had a bath in the evening and had hot flushes for about an hour afterwards.

Apr 7    40 weeks!!!!!! Holly woke at 5:00AM. Been up since then. Metallic taste in mouth. Couple of twinges in the morning. Really tired and a bit teary.

Apr 8    Due date. Feel okay. A bit flat and low. Really tired, but restless too. Couldn't sleep in the daytime. Babies moving well. Can't get comfortable unless lying down.

Apr 9    Feel low and teary. First day of the Easter vacation. Not coping with the girls too well. Want to be left on my own. Don't want to do anything. No energy. Want to just lie down and go to sleep. Legs feel heavy. Right nostril bleeding slightly. Had a few twinges. Felt better later in the afternoon. Took a Sepia at 10:00PM and then went to bed immediately.

Apr 10   Feel fine. Took Caulyphylum 200 at 9:00PM. Woke at 12:45AM and took another Caulyphylum 200.

Apr 11   Woke up feeling fine. Was sick. Just bile. Metallic taste in my mouth. Had membranes stripped. 2-3cm dilated and cervix 75% effaced. READY TO GO!!!! Feel okay... apprehensive, excited, nervous, scared... want to be on my own, I think! WATCH THIS SPACE!!!!!!!!!!!!

Apr 12  My birthday! Had a great night's sleep. It has been a lovely sunny day. Felt really good in the morning. Had a bit more of a show. A few niggles, but nothing much. Went out for lunch. Really thought that today would be the day. Really fed up by the end, but had had a nice day. Crushed two Sepia 30 and mixed them with water. Two teaspoons every two hours until bedtime, stirring 20 times each time. Babies moving a fair bit. Took Pulsatilla 30 at 9:00PM then went to bed.

Apr 13  Still having a bit more of a show. Had sex, but it didn't help. Really fed up and teary. Took Caulyphylum 200 at 10:30AM. Bit more of a show. Pulsatilla 30 at 12:30PM. Caulyphylum 200 at 2:30PM. Pulsatilla 30 at 5:30PM. Caulyphylum 200 at 6:30PM. Pulsatilla 30 at 8:30PM. Caulyphylum 200 mixture at 9:30PM, then bed. Also went to the hospital and went on the monitor to check that the babies were okay. Both fine. Very similar heartbeats.

Apr 14  Slept really well and dreamed too. Feel much better today. Still having a show. More mucous than anything. Took Caulyphylum 200 mixture every two hours or so. Had a few Braxton Hicks. Not so many as before. Babies moving well, but not so much.

Apr 15  Fine and fed up.

Apr 16  Fine and fed up.

40 weeks + 8 days: "Fine and fed up"

Apr 17  Pulsatilla 200 at 9:30AM. Caulyphylum 200 at 12 noon. Pulsatilla 200 at 3:00PM. Caulyphylum 200 at 6:30PM. Pulsatilla 200 at 9:30PM, then bed. Really fed up. Booked into hospital tomorrow to have water broken to get things going.

**Postscript:** Actually, this story does have a very happy ending, so is it important that the woman used all these extra treatments? After being induced on the next day—now 10 days after her due date—Helen finally gave birth (vaginally) to a girl weighing 7lb 4oz and a boy weighing 7lb 10oz. They showed no signs of postmaturity and both had excellent Apgar scores at 1 minute. The mother was delighted to report that she needed no sutures since there was just 'a tiny nick' where the second baby's nose had caught her. Excellent outcomes, with an exhilarated mother... but could any other complementary treatments promoted in the popular press, for example, have caused her or her babies big problems? How well-informed are you about complementary therapies and the extent to which women use them during their pregnancies? To what extent does the use of 'additional' treatments suggest a lack of faith in the normal, physiological processes and to what extent might it be advisable to coax moms-to-be out of this distrust? Is it a good idea and if you think it is, how could it be done?

How well-informed are you about complementary therapies and the extent to which women use them?

## Birthframe 63: Anonymous—Secret plan to birth unassisted

Here's an account by a woman who wasn't entirely open with her midwife about her plans for the birth... In case you're wondering, I received this account from an anonymous contributor who obviously wanted to share her experience.

**June:** A friend phones to tell me the wonderful news of her third daughter's birth at home. I realize once off the phone that I might be pregnant and if I am I'll probably start to feel sick soon. Next day I feel that familiar nausea! Another baby on its way! We knew we wanted a third child but had been unable to make a clear decision about when—now we didn't need to. It felt so right, we were overjoyed.

**July:** I read an article about a book called *Unassisted Childbirth* (by Laura Shanley) and get very excited. It's something I've liked the idea of before, but now I can read a whole book about it that will lay any worries to rest. I know this is what I want, no midwives this time, just John and me there to greet our new baby. I book in with the midwives for a home birth. (I don't mention plans for an unassisted birth).

**September:** Laura's second birthday—she's still an avid breastfeeder. My hemoglobin levels are low and midwives recommend supplementing with iron. I start to take Floradix. I feel great though.

**January:** I am having a wonderful pregnancy, blooming as people say. My official due date is January 22, though I was having regular periods every 40 to 42 days before I became pregnant, so my real due date should be about 10 days later. After three months of taking Floradix my hemoglobin levels are still considered to be too low. I try iron tablets and throw up after the first one. I resort to eating a few raisins and dried apricots daily. I read that Michel Odent says hemoglobin levels should be fairly low this late on in pregnancy as it is a sign of healthy placental function. I am no longer worried about being put under any pressure not to have a homebirth if my next blood test results are low... They're not.

**38-week check:** Midwife thinks baby may be breech and recommends an ultrasound. At first I refuse. I've had no ultrasounds so far and I am reluctant because their safety is unproven. Another meeting with the midwife—she tells me everything that can go wrong in a breech birth, that the midwives are all scared of attending a breech home delivery and that the doctors don't do breech deliveries in the hospital, they section all breeches. I still feel I want to go ahead with a home delivery, but decide to have the ultrasound to put my mind at rest. The baby is cephalic, so I can finish the pregnancy in peace—the breech scare has really shaken me. If this is what happens when I have routine prenatal care, I'm glad there won't be any midwives at the birth.

**January 31:** My water breaks as I get into bed. I put a cloth diaper in my panties and go to sleep feeling excited and a little nervous.

**February 1:** I wake occasionally and get up to change my pad. I have some contractions but nothing regular or strong and I get quite a good sleep. At 4:40AM I am woken by a strong contraction. I get up and prepare my birthing space with candles, cushions and towels. I continue to have strong contractions and I wake John to be with me. I find his presence a reassurance, feeling bound by the love and strength of the decision to be totally responsible for the birth of our baby—we are in this together.

I find the contractions hard to deal with. I can't seem to get to a relaxed place in my mind or body and hold onto it. I try lots of positions, ways of moving, breathing. Nothing works consistently and I feel almost like I have a split personality. On the one hand I am almost panicky, very edgy and unsettled. I am physically restless between contractions as well as unable to relax well during them. I walk around, flapping my arms and thinking and saying negative things to myself: "I'm not dealing very well with this, why is it taking so long? I'm never doing this again," etc. But on the other hand I feel a sense of calm and composure and I keep reminding myself that all is going well and I can do it. I look at the clock but I don't take it in—time is totally unreal. The feelings overwhelm me, but never become unbearable or agonizing. There is definitely pain, but it's okay.

I feel I might be in transition so I kneel on the towels I have prepared. When the need to push comes I have no control and can't help myself. I am making a lot of noise now, quite low, long moans. I am still trying to calm myself, stay relaxed, but it's like I've gone over the hill on a roller coaster and nothing can hold me back. At times I am convinced there is no way the head will fit through me, it feels enormous inside. But I still maintain an inner certainty and faith in my ability to give birth. I feel the head move down and it crowns. Between contractions I touch the 'huge', soft head between my legs. John momentarily panics, thinking the baby is just going to drop out onto the floor. I assure him it's not going anywhere until the next contraction. Then here it comes, the head is out and another pause then the body slides out. Our son is born into his parents' waiting, loving (ungloved) hands. He is making small choking sounds and I notice the cord is wrapped tightly round his neck. I remove it and he cries out, coughing and spluttering at the same time. He is white and blue and red and very slippery and so beautiful. He nuzzles at my breast and fingers but doesn't want to nurse yet. Time is still playing tricks on me. I ask John what time he was born, feeling like it must be at least 10 minutes ago and he says, "Nine minutes past seven"—about a minute ago! John takes some video of us and then we cover him with a warm towel and John phones the midwife.

Molly (nearly 6) wakes and comes into the room. She is awestruck, excited, yet quiet and still. The midwife arrives. She asks if I'm feeling shocked and I am confused. Then I remember she thinks it all happened so fast it took us by surprise. I play along. The labor was only two and a half hours long, even if it felt like forever at times. Laura (nearly 2½) comes in. I show her the baby and she is amazed. She whispers over and over, "A baby!"

She thinks it all happened so fast it took us by surprise

John cuts the now still cord and I hand him the baby. I stand and the cord reaches my ankles! I deliver the placenta in the bathtub after the midwife holds the baby to my breast for his first feed. It is 45 minutes since the birth. The baby is weighed: 9lb 1oz. More than two pounds heavier than either of the girls.

Baby and I get into bed and Laura and he share their first tandem feed. Five days later we name him Oliver Matthew, but decide to call him Matt.

# Freebirth

This is being talked about in the media more and more, so perhaps it is important to consider it here. While we are not talking about unassisted birth when we are considering optimality (because birth needs to be attended by a qualified professional if outcomes are to be optimized), since freebirth can be confused with 'undisturbed birth' and since it is often a reaction to an extreme, unnecessarily interventionist model of birth, we do need to take it into account and think about how we can enhance any freebirth we are either directly or indirectly involved with, as in the last birthframe.

Perhaps surprisingly, given that they are an official regulatory body, the British Nursing & Midwifery Council (NMC) make the following statement about freebirth: "A midwife has no right to be at a baby's birth and if a woman chooses not to contact or engage a midwife that is her right to do so." They also stipulate that a midwife needs to do the following. (Some points also apply to family doctors, of course.)

- Discuss any benefits, risks or concerns with the woman prenatally, if possible and document these discussions.
- Inform your line manager and a Supervisor of Midwives if you are concerned about any aspects of safety, or the woman's psychological well-being or mental capacity.
- Respect the woman's choice if she does decide to give birth unassisted, irrespective of discussions beforehand.
- Adhere to the NMC code of conduct (as usual) and the NMC miwives' rules and standards (in the case where the caregiver is a midwife).
- Don't be critical of either the woman's choice or even of the outcome.
- Support the woman and her family whenever you are called out.

Beyond this, you probably also know that a non-qualified person (such as a husband, friend or doula) is not allowed to offer a woman any medical or midwifery care in the UK, so this is made clear to anybody who asks about freebirth, or seems to be considering it. A non-qualified attendant is nevertheless free to attend the birth in the sense of being present, although this may not extend to assistance unless a midwife (or other professional) cannot be present in time, for whatever reason. What is the law in your state?

The NMC also stipulates that caregivers provide care whenever they are summoned to a birth, requesting additional care if necessary. They remind caregivers of their responsibility to inform the relevant bodies, as appropriate, depending on the situation on arrival and they specify the information caregivers are required to give the mother. Do you agree with this advice? What are the ethics and morals involved of behaving differently? Of course, wherever you are, whatever the law, as a caregiver you may find a very *unprepared* woman having a freebirth or, on the other hand, the mother may be extremely well informed and prepared...

Anyway, quite apart from the logistics, what is a freebirth actually like and why does it occur? Perhaps the reasons for the following freebirths may surprise you, wherever you live, whatever the legal situation in your state.

*This is Laura Shanley with her grown-up son, Willie, who was born in a freebirth. Laura, who has birthed all of her four children 'unassisted' (at least by the medical profession) suggests that every woman (modern or primitive) has the inborn ability to give birth, provided any ingrained fear, guilt or shame can be overcome. She helped overcome her own negative feelings through what she calls 'belief suggestions'; in other words, she repeatedly 'thought' statements which reflected what she wanted to believe—either on an ongoing basis or at key moments. Her husband completely supports this approach; he also believes that natural, undisturbed, home-based birthing is best because he says traumatic childbirth diminishes the mother-child relationship, starts a person off on a rocky start and is ultimately to blame for many of society's ills.*

## Birthframe 64: Laura Shanley—Unassisted birth by choice

Here's Laura's second birth story, in which she births a footling breech baby without any medical assistance, with excellent outcomes. Whatever our personal position on the issue of freebirth, it is truly remarkable to consider how straightforward a birth can be when a female body and her baby are left to labor entirely without disturbance and when both mother and baby experience birth without fear. (Laura says that this baby, who is now grown up, is still totally fearless!) The other interesting aspect of this account is the psychological approach of the mother before and during labor...

A year and a half after giving birth to my first son John, I realized I was pregnant with Willie. Once again, I had a very healthy pregnancy. I never experienced morning sickness or had any of the other so-called 'symptoms' of pregnancy (which I believe are often fear-induced). I decided I would give birth on my hands and knees because that had worked so well with John, but in a dream I was shown otherwise. In the dream, I was watching a woman giving birth standing up. She was straddling a little plastic baby bathtub and catching the baby herself. As I watched her, I heard another woman very gently say to me, "Tell her to remember not to do too much." I understood what the woman was saying and the peaceful feeling of the dream stayed with me for the remainder of my pregnancy. I want to say at this point that I don't follow every dream I have, but this dream was different. It seemed to be coming from the deepest part of my being. And so, I decided to follow it faithfully—I would catch my baby myself as I stood over my little bathtub, and I would move out of the way and essentially do nothing to interfere with the process.

On the morning of August 17, I began to feel contractions. David and I made love and I remember feeling an orgasm followed immediately by a contraction. The rhythmic contracting of my uterus during the orgasm felt almost identical to that of the contraction. They seemed to have the same pattern, although I must admit the orgasm felt better! A few minutes later, I was walking across the room when my water broke. I took out my little bathtub and stood over it as I had been shown in the dream. At that point I couldn't feel the contractions but I knew I was having them because when I put my hand inside my vagina I could feel my pelvic muscles rhythmically contracting around it. A few minutes later a foot appeared between my legs. I wasn't expecting a breech birth, although a friend of mine told me during my pregnancy that he had dreamed he saw the baby inside me standing right-side-up.

The contracting of my uterus during the orgasm felt almost identical to that of the contraction... but the orgasm felt better!

David and I said 'belief suggestions' that everything would be all right and then we patiently waited for Willie to be born. Little by little his foot got lower and soon his other foot popped out. When I felt the time was right I gave one push and gently pulled him out by the feet. David and John had been in the other room but walked in just as Willie emerged. David yelled excitedly, "You did it!" and Willie immediately began to nurse. Incidentally, 12 years later I read that Michel Odent says that for a breech delivery a woman should always be in a 'standing squat' or 'upright' position and an attendant should do absolutely nothing to interfere, if at all possible. This was the message of the dream.

Soon after the birth I dreamed that Willie was speaking to me. He told me that part of the reason his birth had been so fast and easy—he was born two hours after the first contraction—was that he hadn't been afraid either. Babies are always picking up on our beliefs, both before and after birth, so a fearless mother means a fearless baby.

# Pause to reflect...

## ...about freebirth and control issues

1   Do you think women who choose not to seek the help of professionals always have the same kinds of reasons and are the same kinds of people? If not, what other reasons might a woman have for going it alone?

2   To what extent do you think women who choose to give birth unassisted challenge any feeling you might have that you—as a qualified caregiver—have a right to be in control of a birth?

3   Why do you think 'control' is an issue in many experiences of birth?

4   To what extent can a woman take control of her pregnancy and birth herself, and to what extent does she have to surrender control? In a normal birthing scenario, what would be the optimal balance in terms of control?

5   Why do some women get extremely angry when they feel that their midwife has taken over their birth, even if only in the way in which she talks?

6   Why is it so important for some women that they are not treated like a patient when they are admitted to hospital for a birth which they hope will be normal and unmedicated?

7   Do you think issues of control are significant with respect to cesareans? In terms of the woman's feelings and sense of control what is the difference between 1) an in-labor cesarean, which is required for medical reasons, 2) an in-labor cesarean which is the result of an unfortunate cascade of interventions, and 3) an elective cesarean?

8   If you were going to give birth very soon, how much control would you like to retain and how much would you be happy to hand over to your caregivers?

For material relevant to these questions, see Birthframes 1, 2, 3, 4, 6, 11, 12, 14, 15, 16, 20, 21, 25, 27, 28, 29, 32, 33, 34, 43, 47, 48, 49, 52, 56, 57, 59, 61, 63, 65, 66, 67, 71, 75, 76, 78, 82, 86, 87, 89, 90, 94, 95, 97 and 98.

## Birthframe 65: Justine Rowan—Illegal home birth in the UAE

The following birth, which was ultimately successful in terms of a positive outcome for mother and baby, took place somewhat against the odds because the woman did not have the support of healthcare professionals when she needed it most. The woman's courage and her faith in her own body's ability to give birth are both impressive, although the support she received from her husband and her sister clearly helped her enormously in sustaining these. But do you in fact perceive it as courage? Or is it just rebellion? In this situation, in the UAE, the mom-to-be faced a dilemma: either she would have to give herself up to an extremely interventionist system, where the idea of client choice is not really a familiar one, or she would have to break the law and give birth alone. This dilemma existed simply because homebirth is not legal in this and in many other countries. In these places, how can a woman stay safe if she wants to optimize her birth by avoiding unnecessary drugs and intervention?

We live in the Middle East so our decision to have a home birth was complicated by a lack of options and the need for secrecy, since home birth is illegal here. We found a very senior and experienced doctor who agreed to attend our homebirth and so decided to proceed this way.

At 4AM on a Friday, I had a 'show' and a few hours after, my water seemed to break. Not a huge deluge, but enough, with more bursts over the next few hours. I went out for breakfast at a coffee shop with my husband, Stephen, and my sister, Sara, and her family. Then we came home, relaxed and during the late afternoon walked along the beach briskly for an hour. No contractions, but walking made the baby's head feel very low. I had strong feelings of excitement, and nervous anticipation all day. Whenever I thought about what was coming I had a surge in the pit of my stomach, pure adrenaline, but for the most part I was quite calm and positive.

At about 8PM I called the doctor who had agreed to attend the birth at home, to tell him things were starting, but slowly. He said he had been trying to contact me, to say that because I was now a week late, he thought a homebirth was too risky and I would have to go into the hospital. I was shocked, but quickly thought there was no point arguing with him. I just thanked him, finished the conversation and said goodbye.

Panicked discussions with Stephen. I called my sister, my other labor support person and general mentor throughout pregnancy. We discussed options. See how things go at home and then go into the hospital when labor was going strong? I felt nervous, but Stephen and I decided to see how things went at home. If we felt things weren't going well, then the hospital wasn't far. We went to bed at 10PM.

By 10:30 I was up having contractions. I lay on the couch to avoid disturbing Stephen who would need energy later! With each contraction, every 20 minutes or so, I would jump up and walk up and down the room breathing deeply, then lie back down on the couch and doze until the next one. This went on all night. At 6-ish I went to wake Stephen to tell him how exhausted I was.

I seemed to have a break for a bit then and lay in bed. Contractions seemed to fade for a while and then continued all day every 5-15 minutes. The day passed in a haze. I didn't feel like eating but I kept drinking and tried to keep busy. I had a bath, lay down for a bit. Stephen annoyed me by ironing 10 work shirts, when actually I wanted all his attention! By mid-afternoon contractions were every five minutes or so and a bit stronger. For each contraction I would say, "It's coming!", Stephen would rush over and I would stand and lean into him, arms around his neck and 'aahh' on the outbreath, through the contraction. I tried leaning on furniture and other positions but only leaning on Stephen helped. It really soothed me and helped me with the increasing contractions. I still felt quite calm and positive, like I was coping well, and we had a really good rhythm going.

By 6PM, after 24 hours already, the contractions were getting painful. Stephen was writing the time of each one and the duration. They were about three to five minutes apart now. Time passed quickly, in that I had already lost a sense of time; an hour would just go. We stayed in the bedroom mostly now, I'd 'aahh' through the contraction and then comment, "That one was stronger," "That was the strongest yet," and so on. Stephen said little, and kept his notes. By 9-ish I had a bath to see if it would help as the contractions were PAINFUL, but with decent breaks between. The bath was not good. Away from my human standing support, they seemed much harder to cope with and were getting more painful. I told Stephen to call my sister. First feeling of panic. How am I going to deal with this pain? It's bad. Sara seemed to appear instantly. Brave face from me, I got out of the bathtub and onto the bed, on all fours leaning into a pile of cushions to 'ahhh!!!!' through the contractions. We checked the heartbeat, which was fine. Stephen was sent to the backend, to rub my lower back through contractions and apply hot washcloths. This really seemed to soothe the pain. The pain became horrendous. Deep breath, down into the pillows to scream for all I was worth, up for air, quick rest, deep breath and down again. Sara disappeared, I realized, at one point. She had gone for a snooze on the couch. I called her back, to sit in front of me and witness my pain! I panicked when Stephen was going to leave the room to get ice chips. I needed him right there, big warm hand on my back at all times. I asked Sara several times if this was normal. She just kept saying, "Yes, you're doing fine. It's going well." She said to Stephen, after a bad one, "That will be working. They are strong now", which really encouraged me.

Contractions were now almost back-to-back. I struggled to get enough air in between screams. I actually found the screaming a good coping mechanism, and quite cathartic, although it must have been awful to listen to. I have never ever screamed like it before. It was an amazing release. Still, I started saying, "I can't do this for much longer! Am I near the end?" etc. Although, in fact, I was aware I was saying this to get reassurance and I knew I just had to keep going. I noticed that a really bad contraction tended to be followed by a slightly shorter less intense one and my scream would die to a groan... On the less intense ones I finally tried to relax my body and concentrate on opening up. Stephen suggested I try to relax my shoulders and legs and not tense up so much. Easier said than done, but finally right at the end I seemed to be able to do it one or two times.

Before the contractions were back-to-back, in between them I grabbed handfuls of iced apple juice shavings, which were great. I was very thirsty, but my jaw was grinding constantly like a cow chewing cud—a reflex I had read about somewhere! The ice was perfect. All this stage on the bed felt like a couple of hours, I suppose, but was actually much longer.

Then the contractions stopped. I got up, went to the bathroom, and felt a pressure—like I needed to do a poop. I walked around a bit at Sara's suggestion I think, and then sat on the toilet. I felt exhausted but quite calm inside. I was fully 'there' with what I was doing, but it was like being in a womb. In the darkened quiet room, my vision only seemed to extend a few feet around me into this warm dim secure place. I heard Sara and Stephen's reassuring voices, but I didn't see them or their faces. It was the perfect environment, nothing intruded on my concentration on my body, what it was doing and my response to it.

Eventually, Sara, in a worried voice, said that it was probably the baby bearing down that I could feel and to get back on the bed, which I did, back on all fours. This break felt like about 10 minutes. I couldn't believe it when they told me afterwards that it was well over an hour.

On the bed I couldn't get comfortable. I wanted something quite tall to lean on as I was upright on my knees, to help push, as I had a little bit of a pushing urge. We tried Stephen supporting me at the front as I sat on my knees, but it wasn't right. I didn't want to stand or squat, as I had suddenly decided that I wanted the baby to come out behind me, because I was too scared to see it coming out of me. In the end, we added more cushions, but I was still nearly on all fours. The pushing urge was weak as I was so tired. Sara told me not to force pushing, but I did. I wanted progress. The head finally crowned but kept slipping back. I could hear in their voices that Sara and Stephen were getting worried and finally Sara told me to do a big push. I did but as the baby's head crowned and I got the ring of fire, I could feel it was too tight and I said "I'm going to tear!" Sara said that I wouldn't, and to just keep going. I did, and with a nasty pain in my pubic bones and the back of the perineum, Hana's head came out. Deep breath and with the next push she was out. It was 5:10PM. I had pushed her out in about an hour, with almost no pushing urge. I stayed on all fours and just kept repeating, "Is she alright? Is she alright?" I was too scared to turn around and look. I just wanted my baby to be safe. Sara said, "It's a girl!" and they told me to turn around. I was just in shock. She was a girl. She was okay. She was a wet, slimy little thing with a swollen face and these bright eyes. She had a panicked expression on her face, I felt. Sara gave her to me in a cloth, after I tried to sit back against some cushions. I couldn't get comfortable and couldn't seem to get Hana high enough to my breast and close enough to look at her properly.She was stuck at my waist. I just looked at her, feeling relieved and thankful. No feelings of elation or joy even. It had been a trauma and the relief was just so strong. Stephen then cuddled in, pointing out that I hadn't looked at him or spoken to him. I was in such a state of shock and exhaustion, I hadn't even looked at him since our child had emerged! We cuddled up and sat there for an hour, talking about the birth, looking at her and talking to her.

Hana was very quiet and alert. She didn't want to feed. The placenta still hadn't come out so eventually I got into the bathtub, where Sara clamped the cord and Stephen cut it with prepared boiled scissors. Still no placenta. I think that I was so shocked that my cervix just clamped up. I had torn slightly, so we decided to go to the hospital, for a couple of sutures and for help with the placenta, if it still hadn't emerged.

I think that I was so shocked that my cervix just clamped up

I felt so weak I could hardly walk. I was at the hospital until about 4:00PM that day. Not a good experience and one which made me soooo glad I had given birth at home.

Soon after our arrival at the hospital, Sara told me to take Hana home. The nurse had weighed and measured her and I didn't want them to take her to the nursery or do any other procedures on her. The staff went crazy when they realized the baby was gone. It upset all their admission rules. I wasn't worried about being parted from her, as I strongly felt I had to get her out of the hospital to protect her.

I had a painful and not very sensitive manual removal of the placenta, which was very low down, in fact. Then I waited for over an hour to be sutured, in a room where I could hear someone at the end of labor in the next room. As the doctor sutured, I had a student watching and various nurses coming in and out having a quick look too. A cleaning lady even came in at one point. I cried out, quite quietly, when the doctor gave me an internal before suturing to check all the placenta was out, and I was told to be quiet by a nurse. I'm sure this horrible internal was unnecessary—they could have checked the placenta. Why didn't I object? Too exhausted. After the suturing I was told I had to be put in a bed on the ward. I insisted I was leaving. They said, "Not possible". The baby would have to be there and they would need our marriage certificate and passport copies for me to be discharged. Stephen went off to get these. Hana was brought back. She had been asleep since leaving our house and had carried on sleeping with Sara at her house. I cuddled her and tried to feed her but still she wasn't interested. After some argument, Stephen finally got a doctor to say we could leave. I had got out of bed and got dressed surrounded by nurses telling me I couldn't leave. We had to go to the admin department to get a birth certificate and were sent from pillar to post by a nasty official for two hours.

It seemed a very long time later that we were finally all cuddled up in bed. The day had been an unpleasant ordeal, an unnecessary stress that had exhausted me even further. Hana finally fed at 4:00AM, nearly 24 hours after she was born.

The one word I would choose to describe the birth is SAVAGE. I felt like an animal at the mercy of my physical imperatives. And yet it was, I know, a good birth, and with the benefit of distance and deep pleasure in my child, I would say it was great and, of course, I would do it all the same way again. Hana's now 2 and I'm already pregnant with my second child... This next time I hope I'll be more relaxed and able to enjoy it more.

The one word I would choose to describe the birth is SAVAGE

## Birthframe 66: Katya Korochantseva—No midwives around

Here's another story of unassisted birth, this time from Russia.

I first heard about homebirth from my friend Yulya. I was 19 at the time, and expecting my first baby. My mother told me I'd be better off at the hospital, but after I watched a video recording of a homebirth I knew it was what I wanted to do. At the time, I was in my third year in the Geology Department of Moscow State University, and in the evenings I enroled in an 'express' childbirth preparation class, two times a week for a month, since the baby was due soon. I did my best to prepare for the labor—ate a healthy diet, did plenty of exercises and stretches. During the last two to three weeks before the birth—it was July—I stayed at our summer house outside Moscow. I took long walks, swam two times a day, ate very little, and spent a lot of time talking to the baby. I was all by myself at our summer house, so I had a lot of time to devote to myself and the baby. One day before I went into labor, I came home to Moscow.

My water broke at 5 in the morning. We were living with my parents at the time. I really didn't want to birth there, but that's the way things were. I waited until my father left for work at 7.

Then we started calling midwives. We had a whole list of people to call. I was hoping that my instructor from the birth course would come. I called everyone on the list, but it was summer, and most of them were out of town. Finally, I reached one midwife, but I didn't like the way she sounded on the phone. We were sure there was no rush. Another midwife I liked better was expected to get home around lunchtime.

My husband was a great help. He was very confident that we could do everything ourselves. My mother told me I should go to the hospital, but we told her everything would be fine, that there was a hospital right next door, and that we'd go there if necessary. She went into the other room to start getting the baby clothes ready. My husband gave me massages, and when the contractions got very strong I got into the bathtub. We had gotten a 'delivery first-aid kit' ready beforehand. It included herbs to stop bleeding and to stimulate contractions, a rubber bulb to do an enema, an antiseptic for the umbilical cord and any sutures, thread for tying off the umbilical cord, etc. At my birth course they recommended putting a Band-Aid over the end of the cord so it wouldn't poke the baby, since the end of it is very sharp when it dries.

I started pushing, and our son was born around 9AM. When my husband went to tell my mother, "Come see your grandson," she was so nervous and absorbed in what she was doing that at first she didn't even hear him, and answered my husband only the third time he called her. She couldn't believe her daughter could have a baby like that at home and have everything be fine.

My husband pulled him out of the water right away. He suctioned the baby's nose and mouth with his own mouth, as we were taught in the birth course, and then we wrapped him in his daddy's shirt, like our folk wisdom says you should. He weighed 3.6kg.

The placenta came out easily, it was perfect—it looked like the closed bud of a flower. My midwife told me it looked just like a lotus. We kept it in a bowl with some water in it until it dried out. It was summer, and so it dried out very quickly. By around midnight it was all dried out, and we cut it off. It didn't need to be tied, since it was dry already.

We called the midwife back and told her we'd already had our baby. She was amazed that everything had gone so quickly, and said I must have slept through early labor. He was born very easily. In the evening of the same day, our 'spiritual obstetrician'—the same one who taught some of the birth preparation courses—came to check both me and the baby, and did some baby yoga with my son. By the next day, I was ready to have another one! My son and I went for our first walk together the next day. I nursed for 18 months. I had a small tear, probably because of the position I had lying in the bathtub. But it healed fine.

The second pregnancy wasn't quite as good. I think I must have caught a chill in my kidneys between my first and second pregnancies, because I didn't feel as good. I did all the exercises, took cold showers, everything they recommended in my course. I didn't take any antibiotics during my pregnancy, just used homeopathy. The baby was born right on its due date. I think she was the first baby born in Moscow on January 1, 2001. She was born at 4:00AM. The newspapers listed the first birth in hospital on January 1 as being at 5:00AM.

The labor began around 2:00AM. New Year's I hadn't been feeling too good. My belly felt hard, and we thought labor might be starting. Apparently, the strains of the new Russian national anthem, which we listened to on TV at the stroke of midnight, inspired the baby to make an appearance on New Year's during the night. My husband went to sleep saying, "It'll probably take a long time. Wake me up when you need me." I sat with my belly between my knees and massaged my belly. I sat during most of the contractions. When things got intense, I woke my husband and told him, "Wash the bathtub." He started running the water and I got in as soon as he was done, but there wasn't much water in it yet. The baby came out very fast. She was quite small, only 3.2kg. There was only about 10cm of water in the tub. I was kneeling with my legs to either side. It was a very comfortable position for me.

The baby came right out on the first push. I didn't even feel like I was pushing.

After she was born, we brought her into the bedroom. Our son woke up, climbed up on the bed and asked, "What's this? Is this my New Year's present from Father Frost?" "Yes, Father Frost brought you a baby sister!" we told him.

The same spiritual obstetrician came to check the baby. He said, "It's great that you had your baby at home, but you should have had a midwife attending you, especially the first time." He checked over the baby, did some baby yoga with her, and rather reluctantly conceded the birth had gone very well.

My husband and I are both very interested in naturopathy and folk medicine. We felt very confident that everything would go well. And it did.

We felt very confident that everything would go well. And it did.

## Birthframe 67: Katya S from Moscow—Uncaring caregivers

The importance of good midwifery becomes clear when we read the next account. Uncaring treatment and an unwillingness to adapt to this woman's personal circumstances made the woman conclude her only option was to give birth unassisted. Although this birthframe does not by any means describe ideal home births, the woman herself seems happy that she chose to have her second and third babies at home. Also, it's worth noting that the births took place without incident despite the fact that they were both unassisted.

I first heard about homebirth a long time ago, about seven to eight years ago, before I was pregnant with my first child. At the time, there were quite a few informal 'birth clubs' organized on an amateur basis—childbirth preparation classes for people who wanted to have a natural birth at home. They were often organized by people with a medical background, but it wasn't commercial in any sense. They weren't making a living off it, just doing it on their own, because it was important to them. Some of my friends went to one of these clubs, then had their babies at home, and told me and other friends about it. I kind of watched from the side, but didn't get involved. I was afraid of having my first baby at home.

Then I experienced birth at a maternity hospital, where no one cared at all. The nurses, midwives and doctors acted as if I and the other patients were unpleasant interruptions in their very important business of drinking tea and watching television. This was at a maternity hospital that was known for being fairly progressive. I had one room for the entire labor and delivery. It had two beds in it, one for laboring on and one for the delivery. After the birth, mother and baby were taken together to another room to recover in. The normal stay was five days.

During labor, I was mostly on my own in the room. The nurses would look in now and then. The doctor on the ward would check on me once in awhile. Other than that, I was pretty much on my own. I could have had my husband or another family member there if I had really fought for it. A friend had her baby at the same hospital a year before, and had her mother with her. But it was the winter, in the middle of a grippe epidemic, so they were pretty much discouraging any 'extra' people from being there.

The nurses and doctors did what they were 'supposed' to do, and no more. They gave me routine shots, I don't even know what they gave me. I asked them not to, but was told, "Everyone gets this. It is necessary." They wouldn't tell me what it was, but said it was to make the labor 'more effective,' so I assume it was some kind of drug to stimulate contractions.

The birth itself was okay. Everything was normal. I was given an episiotomy, just because that was the protocol at the time. According to the doctors, it was better to cut the perineum than to let it tear—they believed it healed better that way. I was sewn up under general anesthesia. Later on, my doctor asked me, "Who sewed you up? They did a terrible job!" "Beats me!" I told him. I have no idea who did it. It was a difficult birth in the sense that I must have pinched a nerve during labor. My legs ached for days afterwards, and the blood flow to my legs wasn't good for a long time. When you give birth on your own, you can move around between contractions.

It's difficult to know what's going on the first time you're giving birth. You don't know what to expect. You're in an unfamiliar place, alone, with strangers coming in and out. It all has a big effect. Each person reacts differently. I wanted to be left alone.

My first child was born with two red marks on her head, one under her hairline and one at the base of her skull, I believe it was from the stimulants they gave me. All the children born in hospital during that era had similar spots; my daughters born at home didn't have them. I think it's because the stimulants they gave us caused the contractions to be unnaturally strong, which left a mark on my daughter's head. You can still see the marks if you look really hard, though they are nearly gone now. (My daughter was born in 1995).

The second time around was completely different. Contractions began in the evening. I woke up a few times during the night. Then I got up in the morning and the baby was born 40 minutes later. I had had an agreement with a midwife. But by the time she arrived, the baby had already been born—about 20 minutes beforehand. The placenta was born very soon after.

My husband was there in the apartment, but not in the room with me. I wanted him nearby, but not in the room with me. He helped me after the birth.

You might say that we were very adventurous to have our baby at home, all by ourselves. But really, with a second labor it's completely different from the first time. You know what it's all about. I had had a healthy pregnancy, with no risk factors.

The third time was also very different. The pregnancy was unplanned, and I felt very differently about it. It was a joint decision to keep the pregnancy, but I was in a different place emotionally. We were sure it would be another girl. What was the use of yet another girl? Both of us had a fairly fatalistic attitude toward the whole thing.

We had been living in a very small apartment. At the end of the pregnancy we were right in the middle of moving into a bigger place. It was on the eighth floor. The elevator was broken. It was the middle of winter. My husband was working very hard, traveling for work quite often. He was trying to earn enough to support such a big family. One day I woke up and everything hurt, I felt achy. My husband returned from his business trip. A week later, he was out visiting friends in the evening. I called him up and said, "Come home!" because I could tell I was going into labor. The older ones were sleeping. We were still in the process of moving. The place was a mess, with boxes everywhere. My contractions began about 10PM, and our daughter was born at 2AM.

If the second labor was easy, the third was tough. I didn't plan to call anyone to help us. There wasn't any point. It was a painful labor, but what was the alternative? A couple of weeks before, I had tried to get the local doctor to come to check on my daughter, who had a very bad bout of grippe. "Who's going to help me carry my equipment upstairs?" he asked, when I told him we were on the eighth floor with a broken elevator. I could just see myself, eight months pregnant, trying to help him lug a huge metal box full equipment up eight flights of stairs. I told him not to bother.

So when I went into labor, I knew there was no point in calling a doctor. Another alternative was to go to the hospital. But I just couldn't face trying to walk down those eight flights of stairs, getting into a taxi, traveling to some hospital. Forget it. I wasn't going anywhere. All I wanted was for it to be over. It was more painful the third time around. I couldn't find a comfortable position. With the second, she just slid out all by herself. With the third one, I really had to push. I was kneeling when she was born, because it was the least painful position. That time I felt I needed my husband's help to get her out. He was there, and he caught the baby. I told him, "I can't do this by myself anymore. You've got to help me."

## Interviewer's/translator's comment:

From Katya's description of this birth, I think the baby was posterior, from the description of it being a more difficult and painful labor, and difficult to find a comfortable position. Neither Katya nor her husband can remember exactly how the baby came out. I am also struck by Katya's description of being ambivalent about having another baby, being in the process of moving and money worries, and how all these feelings could have contributed to the labor itself being more difficult.

## Birthframe 68: Sonia Winter—Promoting empowerment

From the UAE, where home birth is illegal, as mentioned before, I received the following comment. Sonia is a prenatal teacher in Dubai and is promoting positive experiences of hospital birth (alongside the work of many midwives).

I have been there and done that with fighting the system out here in order for women to achieve natural and/or normal births. Still, I persevere by continuing to teach women how to achieve a half-decent birth...

Last week I had a telephone call from a woman who had given birth the previous day. She had been attending my classes for eight weeks and had just had a totally natural birth in one of the private (highly interventionist) hospitals out here. She was on Cloud 9, ecstatic, thrilled, out of breath with excitement. Her phone call gave me goose bumps and left me happy all day. It also gave me the impetus to carry on. Just one normal or natural labor every once in a while, is all I ask. I know that sometimes labor deviates from the norm, but this woman totally believed in what I had shown her and trusted her own instincts.

Sometimes this process can backfire. Another woman I know had the labor of her dreams (following one of the worst labor stories I have heard for the birth of her first child). She was so empowered by her experience in being able to labor the way she wanted, that now she and her husband are having great marital problems! She is no longer the subservient wife she once was.

It was the birth of my second son at home, following the birth of my first in hospital (in the UK), that pointed me in the direction of prenatal teaching. My second birth was a natural birth which left me feeling extremely empowered. It changed the direction my life was heading in. I simply had to pass on this time honored almost forgotten 'skill' to modern-day women.

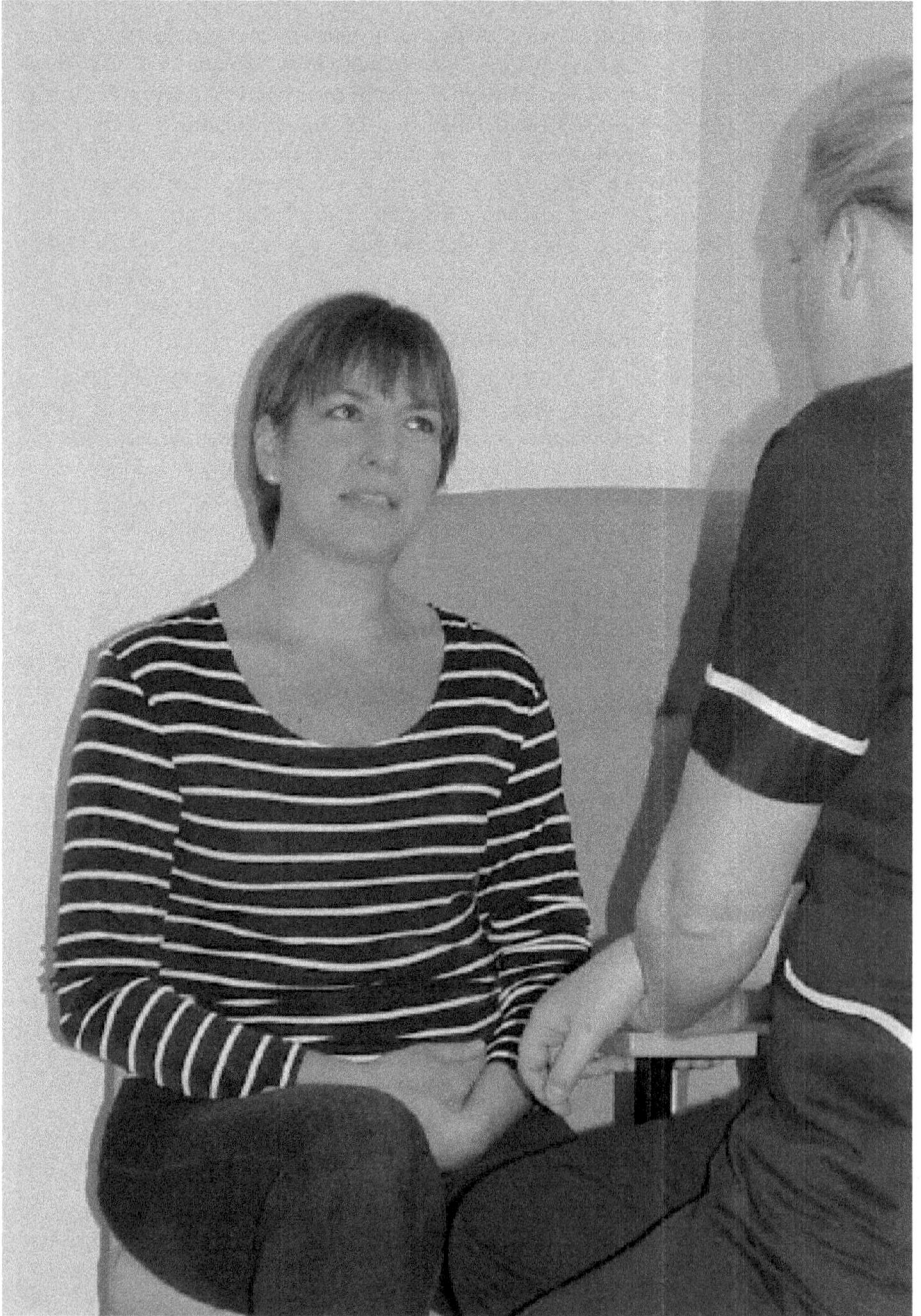

*Reaching agreement showing mutual respect is extremely important*

# The importance and role of birth plans

Perhaps one way of dissuading women to go the 'freebirth' route is to show them the advantages of having an attended birth by really focusing on meeting their requests. This can perhaps be achieved through the use of a birth plan.

## Reaching agreement

Anybody who's ever had a partner or been married will understand the importance of clear communication, and anyone who's ever tried to run a business will appreciate the importance of getting things in writing. In cases where it's easy to assume understanding or agreement, sadly we find it's often not there at all. Rapport is all very well, but sometimes we find we also need words, and they also need to be clearly set out in black and white on paper.

Communication, discussion and agreement is particularly important when it comes to preparing for birth because preferences can vary so widely. In this book I'm assuming you're supporting women who are choosing to give birth physiologically, but how does each individual woman understand that? How do you? There is, of course, wide scope for variation, so details need to be discussed.

In cases where there is any disagreement, some kind of compromise needs to be reached. If this is impossible, it will of course be necessary for your client to find another caregiver. However, just as *she* has the right to choose her caregiver, so you also have the right to limit what you do... or extend it beyond what some women would spontaneously request (e.g. by providing more discrete support and *less* disturbance). For this reason, there's no reason to be upset when a compromise cannot be reached. In any case, most of the time it should be possible to reach agreement.

Doing this preparatory planning will, of course, help your client to feel secure when she goes into labor. She will know that her caregiver is fully behind her wishes, ready to support her or help her if there is any kind of problem. She will also know—if she has opted for a physiological labor and birth—that her caregiver will do everything possible to facilitate her birth so as to optimize outcomes in all respects.

## Writing things down

Writing things down is useful so as to clarify issues and also so as to create a document which can be refered to by other people, if necessary. Another caregiver (e.g. an obstetrician) might need to look through a woman's birth plan, as well as her husband or a doula. Of course, having something written down means that the woman's preferences can be conveyed with a minimum of fuss or explanation and there is no danger of any key details being forgotten.

Having something written down minimizes misunderstandings

When your client does this, I would encourage you to advise her to call the document not a 'birth plan' but a 'care guide', as I've explained already. After all, the words 'birth plan' often seem to trigger negativity in caregivers and the words perhaps unhelpfully give women the idea that birth really can be *planned,* right down to the last detail. (Later on I discuss how we should all be positive about unexpected eventualities and not consider everything a disaster if it has not gone to plan...) To summarize, a care guide can do the following:

- ♥ Facilitate constructive discussion as to care during labor and birth
- ♥ Remind you of any one client's precise wishes when she goes into labor
- ♥ Help you to communicate your client's wishes to your colleagues, if you can't attend any particular woman while she's in labor and giving birth
- ♥ Serve to confirm agreements reached about the kind of care you will provide during the woman's labor and birth
- ♥ Reassure the mom-to-be that she has been listened to and *heard*, and her wishes accepted by all relevant parties

## Tips for success

Like any other document, a care guide can be useful or a waste of paper, depending on how it's prepared and used. Here are a few ideas for optimizing the effectiveness of any one care guide:

1 Make sure it's typed so it can be easily and quickly read.

2 Encourage your client to consider how to express her preferences most effectively. (Various approaches are exemplified in the companion volume to this book for pregnant women *Preparing for a Healthy Birth.*)

3 If you find anything vague or confusing, help your client to rewrite it so it doesn't cause anybody else confusion.

4 If you disagree with any points in the care guide, be very honest with yourself about the reasons for your disagreement. Discussing feelings and problems openly is likely to result in much stronger agreements and better relationships.

5 Express any concerns about risk cautiously, remembering the very real (and immediate) risk of inculcating fear and worry which wasn't there before, which wouldn't be helpful.

6 Sign the care guide when it's confirmed so as to indicate clearly to your client that you agree with it and accept it.

7 Ask your client to make several copies of her care guide.

8 Encourage your client to give a copy of the care guide to anyone who might be present at the birth, e.g. her mother, her husband, a doula, a friend, or her obstetrician.

9 Encourage your client to keep a copy of the care guide on her person at all times (e.g. in her purse) in case she suddenly goes into labor.

10 If your client is having a home birth, suggest she stick a copy of her care guide on a door near the front door inside her home so that it will immediately be seen by any caregivers arriving to attend her in labor.

**A few comments from mothers:**

" It wasn't really the risk of hemorrhaging, he finally admitted, that was preventing him from agreeing to a completely physiological third stage... He simply didn't want to have to hang around waiting for the placenta for what could be a very long time. Enormously respectful of his honesty, I then tentatively suggested we limit the time to two hours, which he agreed to.

" Two times I prepared birth plans, and two times my births proceeded as I wished. There were small differences, small unexpected eventualities, but nothing proceeded against my wishes, and after each birth I felt strong and satisfied. My midwives were very happy too, although somewhat astonished!

" Stating in my birth plan that I wanted no pain relief meant that none was offered to me while I was clearly in pain. This was important because I would have been more likely to have 'given in' if I had been offered things.

Finally, remember that in helping your client to prepare a care guide, you're collaborating for success. Both you and your client will be hoping for positive outcomes and this is one way of facilitating them.

A care guide will help you collaborate for success

# Questions for reflection...

### In the prenatal period:
1   How can you encourage women to take the idea of a care guide seriously?
2   How should you respond to unsolicited care guides? And how do you think your clients will expect you to respond? What kind of response do you think they would ideally like?
3   What should you say (or do) in cases where you feel that a woman's care guide is not realistic or is bound to result in problems during labor?

### When the woman goes into labor:
1   How can you encourage the woman to feel confident in your care?
2   How can you facilitate rapport, a sense of security and openness?
3   What can you say if you feel yourself deviating from a woman's care guide?
4   What do you need to do, in the light of protocols you need to work to, when using a care guide or deviating from it in any way?

To conclude on rather an ironic note, next we have two accounts which show how it should be done... but which ultimately also resulted in an unassisted birth, although for entirely different reasons from those in other accounts.

## Birthframe 69: Rebecca Wright—Empowering midwives

The first of the following births, experienced by Rebecca Wright, who wrote one of the forewords to this book, shows the importance of sensitive midwifery. Not only can it ease a woman's journey through pregnancy and birth, it can also make the transition into parenthood smoother and a great deal less stressful. In this case care was provided by an independent midwife in the UK. (Since Rebecca is American she did not automatically use the state-funded British NHS (National Health Service).) It is surely possible to work out systems so that the same high standard of continuity of care is also possible in maternity units.

The second birth account, which is a description of Rebecca's third experience of giving birth, demonstrates how a woman can become more and more comfortable about giving birth. Given that this birth was so smooth and fast, it can also be considered safe... After all, it is generally when labors are delayed that risks increase because the unborn baby is more likely to get distressed, or have difficulty getting itself born. In this case, the laboring woman had a known midwife, who was the partner of the midwife who had attended Rebecca in her first two labors.

When I conceived my daughter, I was in the UK on a temporary visa and was not eligible for the NHS though I had private cover that would pay for NHS care. The medical system was a bit of a mystery to me and, anxious as I was to start my prenatal care, I was dismayed to be passed off from one person to another because no one knew what to do with me. (I have learned since that they should have offered me care and then billed me.) Finally, I started looking for an alternative, and found an independent midwife, Chris Warren. She explained to me how the NHS worked, that I would have different midwives for prenatal care, birth care and postpartum care. Because continuity of care was so important to me, I decided to book with her and go for a home birth even though the idea of it made me a little nervous at first.

I was dismayed to be passed off from one person to another...

It was fabulous having Chris and her partner Michelle look after me in pregnancy. Their visits were always enjoyable and reassuring and, over time, I became very comfortable with them.

The day before my due date, I woke up when my water went at about 1:00AM. I was so excited!—so I woke my husband and asked him to get up with me to set up the birth pool. Contractions began very slowly as we set up the pool and as soon as we were done my husband went back to bed. I tried to rest, but was too excited to sleep. I called the midwives about 7AM to let them know what was going on. They were calm, and asked me just to call back in an hour or so with an update. I went for a walk with my husband then, after speaking with the midwives, had a bath and just rested, enjoying my contractions and the breaks between them, looking forward to my baby's arrival.

By lunchtime the contractions were stronger, but I was still hungry and able to eat, so I did. Soon after, things intensified. I asked my husband to call the midwives and ask them to come, but to tell them they didn't need to hurry. I got into the pool for a bit, but was soon out and up by myself in the bathroom, where I shut myself in for some privacy and to isolate myself from my husband's well-meant questions about what I needed and how I was doing. Soon the contractions were so strong, with virtually no break in between. I couldn't move and I lost my sense of time. The midwives arrived about 4:30PM, having been held up in traffic. Chris came up to me quietly, without interrupting me or asking any unnecessary questions. She placed a reassuring hand on my shoulder and quietly asked if she could monitor the baby in between contractions. I nodded my assent, and she checked the baby's heart.

I was amazed that Chris seemed to know instinctively what I was thinking and what I needed. She responded to me with minimal talking. She knew when I began to lose my concentration in the face of contractions without a break in between, and she helped me to regain my focus on the space between as the one contraction ended and the next began. When she saw that I was getting exhausted, she asked if I would like to get back into the pool. As soon as I did, something changed. The contractions became even more powerful and I could feel the baby descending rapidly. She was born a few minutes later, at 6PM. Chris caught her and handed her to me immediately. I birthed the placenta a few minutes later, easily, and with gratitude for Chris's statement that it wouldn't hurt as it had no bones!

After we sat and celebrated together for awhile, Chris helped me to shower and then tucked me and baby Hannah into bed. She visited every day for a few days, then every other, gradually spreading out her visits over the first 6 weeks. She was an amazing support and inspiration—and it's no wonder I decided I would have her for all my babies!

I remember being very surprised that she didn't ask to examine me internally or tell me when to push as all the books I'd read had told me my midwife would tell me when it was time and how to push. Instead, she stayed quiet and let me follow my body. I am so grateful to her for supporting me in this way, and with such respect for me, my body and my baby.

## And now on to Rebecca's third experience of birth...

This was an amazing and strange birthing experience, starting with five weeks of pre-labor, truly—where nearly every day I thought he might come that night, but then didn't.

So at 43 weeks, although I knew the day was getting closer, I didn't give too much thought to contractions all day Monday. I woke up at 11PM with my water broken. I called the midwife and went back to bed. Then I had an hour of stronger contractions, so I called the midwife again (just to update her) and started to get things ready for the birth as I waited for Robert, my husband, to come from Durham.

I called the midwife again and started to get things ready

Once all was set, I went to bed and went to sleep until the next morning, when William (my toddler) woke me up. I was still having contractions, strong ones, but well-spaced, so I went about my morning routine. Took Hannah to the neighhbors' at 7:45. Came back to read a book to William. Went upstairs, telling Robert to make me something to eat and call the midwife to come, but not to hurry. Went to the bedroom, tried to rest, had a massive contraction which I realized was the start of the second stage. Called to Robert to phone the midwife to come NOW and fill the pool. Went into the bathroom and had the baby there myself sometime around 9AM. I'm proud to say I caught him myself while standing!! Chris, my midwife, arrived around 10AM and helped me get settled in. My only regret is that I missed out on the birthing pool—but Hannah and William had a blast in it later.

I am really pleased to have experienced such a private, optimal birth for this last baby. We finished it off with a lotus birth—not cutting his cord, but letting him 'release' the placenta himself when he was ready. This turned out to be early Friday morning (after his birth on Tuesday). It was lovely, and there was no smell, mess or icky-ness.

# Pause to reflect...
## ...about time and the timing of support in labor

1   How can you ensure that women call you early enough in labor?

2   How can you ensure that women don't call you too early?!

3   How can you be sure of the progress a woman is making when communicating with her on the phone?

4   To what extent do you think you can communicate destructive, unhelpful (fear-inducing) anxiety when you're speaking over the phone to a woman who's in early labor?

5   How can you usefully spend your time when you're with a woman who's taking a long time to work towards second stage?

6   What if you don't arrive in time? How should you behave (apart from doing what's required of you in either medical or legal terms)? In other words, how can you behave in very human terms?

7   To what extent do you worry about women giving birth alone? To what extent do you think women's increasing confidence with birth can help to facilitate their labors and also optimize safety factors?

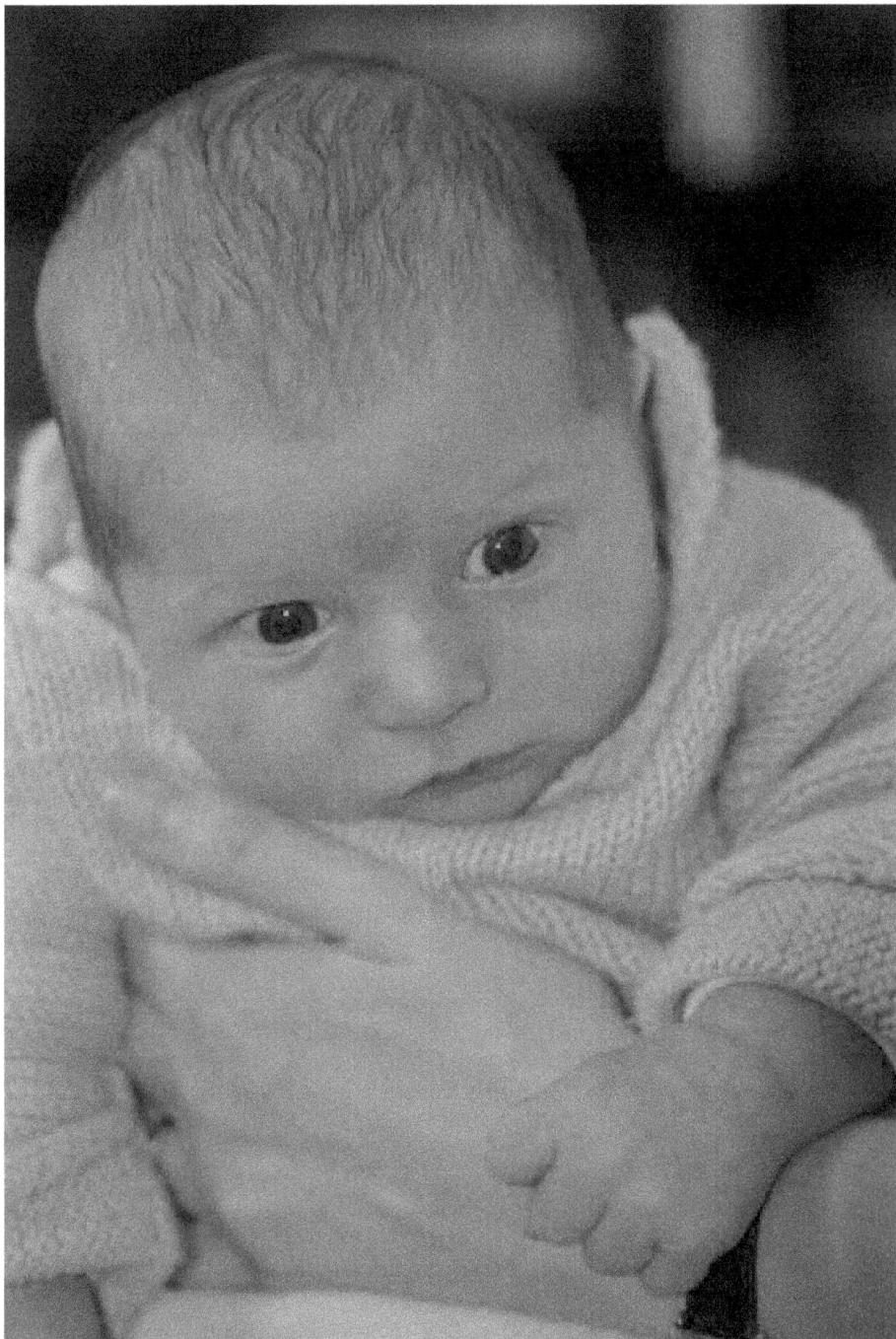

*Gabriel, Rebecca's third baby, very soon after his birth. Notice how alert he is!*

## Requests for home birth

Even if you agree that circumstances in the US are much better than those in other parts of the world, you may still not be very happy about the idea of supporting a woman who comes to you asking to have a home birth.[61] You may even have been quite shocked by some of the accounts earlier on in this book, if you have not already attended home births yourself. But what kind of risks are we dealing with here?

Actually, nowadays, homes are usually such sanitary places and hospitals are harboring so many infections that the idea of giving birth at home is not a romantic one. If you feel disinclined to believe this, consider that in 2003 the *Reader's Digest* reported that an estimated 100,000 people in the UK were picking up an infection in the hospital each year. In the same article they stated that hospital-acquired infections killed more than 5,000 patients and that they contributed to the deaths of some 15,000 more. Are these kinds of statistics possible in the US? In any case, is it realistically possible for hospitals to become places which are 'cleaner' than houses, given that their main purpose is to serve people who are sick? Even in maternity wards, bugs can collect on catheters, intravenous lines, computers, watches, drapes and male doctors' ties—not to mention hands which are not washed thoroughly.[62]

A visit to one of Britain's 'living museums' serves as a good reminder of how much things have changed over the last century. At the Beamish Living Museum in Co Durham, it's possible to stroll around houses which are as they would have been in 1913 in the UK, which is not dissimilar to how they must have been in the US around that time too, in terms of sanitation and facilities. Not only were surfaces all difficult to clean and cleaning products simpler and less effective than those available today, fabrics were more challenging to launder, rooms couldn't be heated as quickly or easily, and water had to be pumped. Drugs, expertise and technology were also not readily available and of course, most of what was used at that time hadn't been subjected to randomized controlled trials. (There was an even bigger mix of tested and untested treatments than exists today!) All in all, women giving birth at home in the early 1900s—just a hundred years ago—were in a completely different situation to women giving birth in the first decade or two of the 21st century. Modern women not only have better homes, they also have quick access to telephones, roads are better, more vehicles are available (and they're more reliable), medical facilities are more developed and numerous drugs and technologies can be used to treat a range of conditions quickly and efficiently. Even caregivers are better, because they now have better training and arguably work in much better conditions than caregivers working in homes and hospitals only a few decades ago. (How would you like to work without plastics, cars, telephones, central heating, piped water and electricity?!)

We not only have better homes, we also have quicker access to phones, transport, medical facilities, drugs and expertise

Thinking back to Africa and other poorer countries (e.g. in Asia), given the open drains, the poor sanitation, the slums, the water shortages or floods, and the inadequate health facilities, which are usually inaccessible to many women in any case, we need to realize we're in a very different situation in the developed world. Our homes and infrastructure are entirely different.[63]

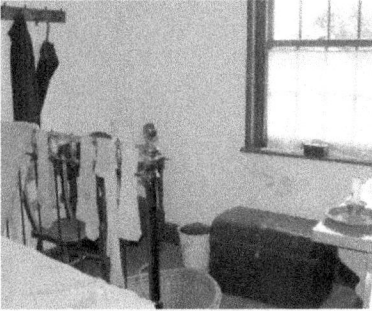

*Homes in Britain and the US have improved dramatically over the last few decades. The photos in the left column are of the Beamish Museum in County Durham and those on the right show a contemporary home in the US... We've obviously come a long way.*

The reminiscences of a midwife, Carol Walton, serve as a reminder of how different people's homes could be only a few decades ago... Over to Carol:

Some of the home circumstances were quite appalling and some of the home births I attended were in the most awful places. You just wouldn't want to wash your hands because the facilities were so bad. I remember in one apartment, they had done their best to make the bedroom clean and tidy, but you couldn't get into the kitchen for unwashed dishes and clothes all over the place. The husband then said to me that he would have made me a cup of tea, but he wasn't expecting me... His wife was nine months pregnant but he wasn't expecting a midwife to call! I was actually quite grateful because I didn't want to drink out of any of their cups, but I wouldn't have wanted to appear rude.

I was shocked by some of the housing districts I had to visit and the very poor circumstances and living conditions people lived in. One evening I once came across a child sitting in the gutter, eating out of the trashcan. Families lived in one room up flights of stone stairs, with one bathroom for three apartments, halfway down. There was one room with a basin, which did duty for the potatoes, the laundry (of course everybody had cotton diapers) and the baby's bathtub. We would be horrified now at some of the home conditions in which people had home births. A lot of the births I attended were in big families and I assisted many women who already had eight, nine or 10 children. Believe it or not, this was in the UK in the 1960s!

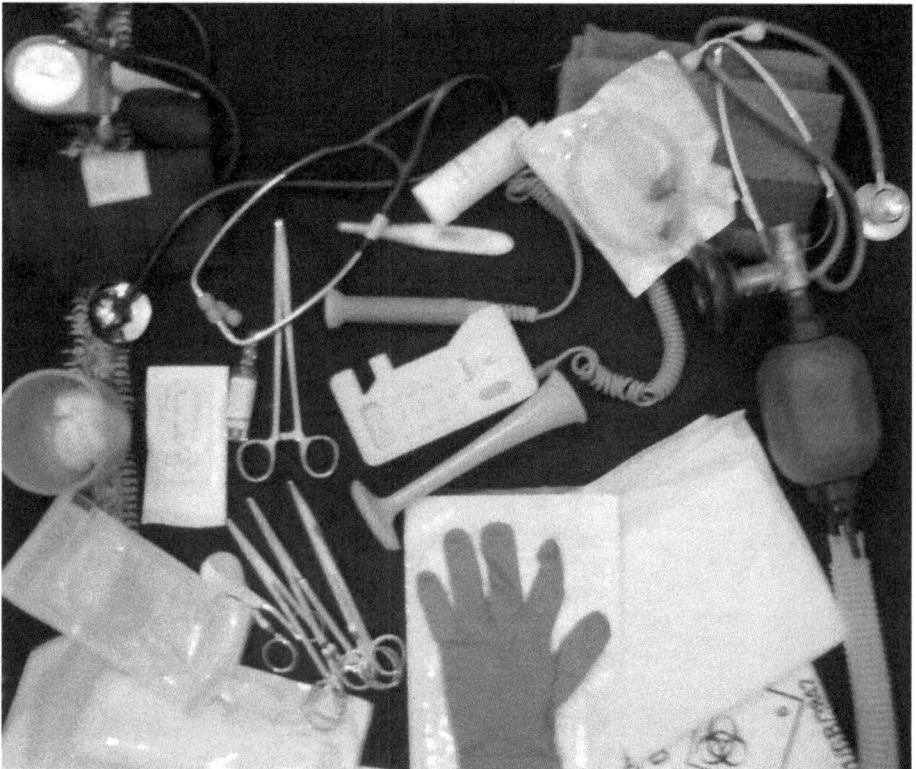

*As you know, very little equipment is needed to attend a woman at home...*

What kind of homes do your clients have? How easy would it be for them to get to hospital, in the case of an emergency?

*Pottering around at home helps many women relax in the early stages of labor and later on it might also be the most comfortable place to be. How does this environment compare with that offered by local hospital maternity units and birthing centers?*

## Birthframe 70: Justine Rowan—Home vs hospital decision

The following account by Justine, who we first met a few birthframes ago, may give you an idea of the thought processes some women might go through when they make decisions about where they want to give birth. Here, Justine explains why she persevered with the idea of staying at home and even giving birth there, despite the fact that it was actually illegal for her to do so!

### I decided that a healthy birth was best achieved at home

My first child, Hana, was born at home... I was always planning for a healthy birth, and very early on I decided that a healthy birth was best achieved at home. Several factors resulted in this strong conviction.

Firstly, personal inclination. I am a believer in a holistic approach to health and the fact that ill health has mental and emotional dimensions which Western medical practices do not always address appropriately or adequately. As part of this philosophy, I see birth as a natural process requiring a sense of security and the ability to relax, requirements not easily met in hospital. A hospital full of sick people and the equipment and training for intervention are not conducive to the healthy birth process. Hospital and doctors (rather than midwives) just seem to me a far from ideal set up for childbirth.

Secondly, I was lucky to have the experience of my sister, who having had one section followed by two home births was an excellent healthy birth example. She showed me first-hand how seriously traumatizing childbirth could be, and also how empowering it could be. Her journey from what she felt was an unnecessary cesarean to home birth, which she shared with me, made me think deeply about all these childbirth issues even before I was pregnant.

As a result, well before I was pregnant, I read widely about childbirth, particularly books about healthy birth. I was fascinated with birth stories and was astounded at how clear and easy it was to chart 'things going wrong' in a hospital delivery; the so-called 'cascade of intervention'. The first point of this cascade was often the mother not dilating or 'progressing' because she did not feel sufficiently secure or relaxed, even though she might not even have been fully aware of this or able to articulate it. I just knew I would not be relaxed in a hospital where strangers, however nice, were coming in and out and where I was not in my own comfortable environment.

Thirdly, the risk element seemed just as high, if not higher with a hospital birth. Even if you have enlightened midwives, there is still an element of risk. You don't know these people. Someone possibly giving off nervous vibes, a chance comment about how dilation is slow, etc has panicked many first-time mothers. You are on their territory and I would find it difficult to feel in control enough.

### The risk element seemed just as high, if not higher with a hospital birth

I do believe the instinct of the mother is so important in labor and her lead should be followed. When I was in labor myself I trusted my body implicitly to do what was necessary. It is difficult to have this trust for someone else, and hospital staff just see the need to keep things moving at their pace. In any case, there are risks to the baby with induction, all the pain-relieving drugs, with the supine position of the mother, with forceps and ventouse, with a too speedy labor and removal of the baby, etc, etc. Of course, there are risks with a home birth; there is risk with any birth. However, all the research shows that for a low-risk mother, a home birth is just as safe, if not better in terms of morbidity and mortality levels. This safety factor seems to be what most people have difficulty getting their heads around, but all the research is there. I suspect the real problem that many people have with home birth (and there is a problem—many people act as if you should be committed to an asylum) is the fear of taking responsibility for something we're trained in our society to abrogate control of.

So, based on the research that I had done on the subject, I was absolutely convinced that the best way to give birth for me, and the safest way for mother and baby, was an active birth with no drugs and almost no intervention, unless there were complications, which are less likely with a healthy birth anyway. And I was not confident that this kind of birth could be achieved in a hospital, and if it was, I felt it would be lucky. I wasn't prepared to take that risk. I don't think it matters which hospital you go to, as they are all geared to a 'medicalized' birth. So to have a healthy birth, you have to fight for it to an extent—a major distraction you don't need during labor. I would want to manage the labor differently from how the medical staff attending me would want. I wouldn't want to be hooked up to a monitor. I wouldn't want any vaginal exams. I knew I didn't want to lie down. I didn't want an episiotomy. I didn't want to be in a hospital.

In my case, I knew birth was going to be hard, and I felt it was only really possible for me at home and with total determination, and secure, away from any temptation of drugs. I am not brave with pain, and needed to keep telling myself that it was only for a certain number of hours and I was just going to have to tough it out for once in my life. I spent a lot of time thinking, planning and visualizing the birth, what I would do at each stage, what I could do to help myself. I kept very fit, walking an hour a day and doing yoga and meditation, with affirmations every day. These helped to make me feel confident and strong, rather than actually helping during labor when your body takes over anyway. I was so, so determined that this was how it was best for me personally to give birth—and therefore this was how I would do it—that I didn't allow myself to think of any other option, apart from telling myself that if things were not going well, I would, of course, accept the need to go to hospital and get help.

Finally, I think it is important to note that my husband was fully involved in my decision-making process. He listened to hours and hours of 'birthtalk' and he agreed with the home birth idea fully and was very supportive. I do think that this is an area where the woman must sort all the issues out for herself and decide what is best for her, and then explain and enlist the support of her husband, if there is one. After all, the fear of the husband often influences decision-making to the detriment of the birth giver!

## Birthframe 71: Clare O'Ryan—Disturbance or not at home

The next account reminds us that home may not always be the best place to be. After all, we're aiming for an undisturbed birth, aren't we? As this account shows, *how* things are done at home is just as important as where a birth takes place. Disturbing the natural processes by disturbing the woman's sense of security, privacy and confidence can have dramatic consequences whether the woman is at home or in hospital. Of course, as this account illustrates, continuity of care is sometimes crucial.

For the birth of my sixth child, Finn, I had decided on a home delivery for a combination of reasons. The previous five births had been straightforward and progressively quicker and on a practical level it meant not having to worry about where the other children went. The main reason, however, was that by No. 6 I had gained the confidence in and awareness of my body and its instincts to want to be left to labor on my own, with as little interference as possible. The birth went beautifully, with the help of some very supportive midwives, who had given me wonderful continuity of care from the start of pregnancy. This is something I'm sure gave me added reassurance and confidence during the birth. Finn is now a very happy and easy-going 4-year-old and I am sure that the peaceful atmosphere into which he was born is responsible.

> On a practical level, having a home birth meant not having to worry about where the other children went

When I discovered I was expecting No. 7 (Joseph) I decided to opt for a home delivery again. The pregnancy proceeded uneventfully and despite the baby lying in a slightly awkward posterior position in the last few weeks, everything signaled another hopefully quick and straightforward birth. 10 days overdue, and after a false alarm (contractions caused by the baby turning into an anterior position), labor began at 6:00 in the morning. As before, I kept upright and moving, walking around the house, and dealing with the contractions as well as the children's demands for breakfast.

By 8:00AM the 'on-call' midwife arrived. On examination I was almost fully dilated and she prepared for the imminent delivery. At this point the 'day shift' community midwives (plus student) arrived to take over. I was surprised, given the stage of labor, that the on-call team didn't stay to see the birth through to delivery. So at transition I had six people—midwives, students and my husband—in the bedroom, all filling in charts and forms or talking in corners and, most disconcertingly, giving each other glances of concern and borderline panic whilst telling me how well I was doing. I paced around them trying to deal with the contractions and not get in their way. Other visitors into the busy bedroom included two of the older children, who were getting ready for school and needed to know where their gym kits were. By now, I was beginning to feel the first sensation of pressure from the baby's head and was told by one midwife to push if I wanted to.

> I was told by one midwife to push if I wanted to

An examination revealed that the cervix had swollen, preventing the baby from descending

When, half an hour or so later there was still no sign of the baby's head, an examination revealed that the cervix had swollen, preventing the baby from descending. I was told I would have to allow the swelling to reduce by resisting the pushing urge, possibly for up to two hours. Fighting against my body's overwhelming instinctive urge to push out the baby and the powerful contractions doing exactly that became—after 20 minutes or so of trying—an impossibility. An epidural was suggested to reduce the pushing urge and help with the pain, but this would mean transfering to the hospital to have the epidural administered. By now, the pain of resisting pushing was excruciating even with gas and air, which I had gratefully used for the previous 20 minutes. I agreed, and the midwives called an ambulance, which rushed me into the hospital. My husband followed the ambulance in our car and just as he arrived to join me in the delivery room Joseph's head was born with a huge contraction over which I had no control. His shoulder then got slightly stuck but with the midwife pushing me into an almost yoga-like position—bringing my feet up towards my shoulders, while I lay on my back—he slipped out and, despite his shock entry into the world, was fine and didn't need any resuscitation.

Being distracted by so much activity at a crucial stage in the labor and losing control may have had an adverse effect

Although the birth was not what I had expected, it worked out for the best. Strangely, the noisiest and busiest part of the delivery was at home. It is possible that being distracted by so much activity at a crucial stage in the labor and losing the control that up to then I had had, and had always maintained during previous labors, had an adverse effect. Being told to push before my body was ready may have caused the swelling and the resulting battle with my body, and loss of control over what it was doing led to extreme pain and complications. However, being alone with Joseph in the delivery room after his birth was calm and peaceful. We had a few hours alone— uninterrupted, quiet and warm. I lay back and held him and we were both awake but able to recover from the trauma of the birth and rest together, something I'm sure I would not have been able to do at home with Joseph's brothers and sisters desperate to see and hold him.

Sometimes, hospital births can be peaceful and home births quite chaotic! I think this experience taught me that where the birth takes place is less important than how the mother is treated. Too much well-meaning advice, monitoring, admin, instruction and activity can interfere with the mother's control of the labor and result in more pain and unnecessary intervention, whether it occurs at home or in hospital. Being left to follow her natural instincts and urges in a calm atmosphere can only be beneficial for the mother and child.

Sometimes hospital can be peaceful and home chaotic!

## Birthframe 72: Joanne Whistler—Giving birth on a boat

This last account about place of birth addresses some wider issues... After telling us about her own experience of birth, healthcare manager Joanne Whistler reflects on the space in which she labored—which happened to be a boat. Her account and her comments raise interesting questions about optimality in terms of stress levels, expectations, space and territory, and also colors. Various people have considered these issues, but of course no final conclusions have been reached. Here's Joanne's account...

Early in my pregnancy I decided to have a home birth with an independent midwife. The place I felt safest was at home and it felt like the right place to deliver my baby. I'd originally been undecided about whether to have regular or independent care, but when I discovered that under regular hospital care it could be any two of 20 midwives who would attend me my mind was made up. My independent midwife, Debs, was fantastic. My unhurried appointments with her allowed us to discuss all sorts of issues in detail and—based on the information she gave me—I changed my mind on several things, whilst never feeling pressured into a particular course of action. For example, I'd assumed that I'd have a managed third stage, but on hearing that it made lots of women sick and wasn't necessarily safer I decided on a physiological third stage.

During my pregnancy I was very busy at work and, looking back, I didn't take seriously the need to reduce my work commitments and stress levels. My partner Paul (now husband) was working ridiculous hours and I didn't really get the support I needed. I remember calling my mom in floods of tears because I was too tired to cook myself dinner, Paul was working, and my low blood sugar had sent my mood plummeting.

At 32 weeks I had been down to a work meeting downtown. I'd taken the subway back and dizziness and strong Braxton-Hicks contractions had forced me to sit on my suitcase with my head as low as my baby belly allowed for a long time, before a kind police officer got me a drink of water from station staff. When I finally reached home my Braxton Hicks contractions were every two minutes. I expected them to die away once I got off my feet, but they didn't and I started to panic. Paul arrived home and took me to the hospital, where I was monitored, given a shot to mature the baby's lungs and my cervix was swabbed. The latter was very traumatic and painful. It turned out that I wasn't in early labor as I'd feared, but I was kept in for two days for observation. After that wake-up call I reassessed my priorities and stopped stressing about work—my baby was more important. I also felt hugely relieved that I had decided on a home birth. Although the circumstances of my hospital visit were intrinsically more distressing than laboring there at term would have been, I knew I would never feel safe in the hospital. (I'm not talking about 'objectively' here. I mean I knew I wouldn't have a deep down feeling of physical and emotional security.)

The rest of my pregnancy was uneventful. My due date came and went, and went, and went, and I started getting Paul to drive over as many road bumps as we could find. Having read about induction I felt there were too many disadvantages to intervention and resigned myself to waiting. I also decided not to try any strong alternative methods to start labor, such as acupressure. I thought that doing what had got me into this position in the first place couldn't do any harm though. About two hours afterwards (12:30AM) I woke up in a wet patch: my hindwater had broken. I tried to get back to sleep, but I was too excited and soon contractions started.

At first they were very mild, like period pain, but they gradually increased in intensity and I got up and circled my hips sitting on a birthing ball for a while. It was very cosy and relaxing in the dark, with the red glow of the heater close by. At about 3:00AM I roused Paul as I wanted some company and also felt that the TENS machine might help. By 6:00 in the morning or so I wanted to phone Debs for reassurance. Paul persuaded me not to ask her to come out, but I talked to her and said I'd call her later when I needed her.

I used the visualizations on a Hypnobirthing® CD to cope with the intensifying contractions, but after a couple of hours the CD stopped being helpful because I couldn't concentrate through the contractions. I then definitely felt that I needed Debs. By the time she arrived I was on all fours on the floor, singing a long loud low note with every contraction. I asked to use the gas and air which was very enjoyable. It made me feel 'high' and a bit giggly. During each contraction I was kneeling on all fours on my scrumpled-up bathrobe, and sitting back in between. My bathrobe felt like my nest. Its familiar smell and the way it was positioned became very important to me, and I was outraged when Paul rearranged it to make it more comfortable. When I talked about the strength of this feeling afterwards, he reminded me that I'd stopped using it during contractions shortly afterwards. He was right, but someone else interfering with my nest 'spoiled' it for me as a safe place with the *result* that I stopped using it.

For the vast majority of the time both Paul and Debs supported me quietly and without fuss, reassuring me when I needed it and being unobtrusive. Paul and I did have one falling-out though. During a contraction he made the big mistake of walking close to my head. I felt threatened and snarled at him. He didn't do it again.

Around the time of transition I had the strangest feeling of a split consciousness. Part of me was completely absorbed in the physicality of birthing, and part of me was rationally saying: "Ooh, curling your toes. Could be in transition!" I kept tweaking my pelvic floor to see if I could feel the head descending and it was frustrating that it didn't seem to be doing so.

Transition felt long and tiring, but when the pushing urge came it was overwhelming. In fact, I don't think describing it as an urge does it justice. The closest thing I can compare it to is vomiting—I really didn't feel as though I had any choice about pushing or not. My body was doing it involuntarily. The pushing stage took awhile, about two hours. It was probably partly because I spent a large portion of this period with my butt in the air and my head on the ground. I just didn't feel I could cope with the intensity of the contractions that came when I was upright (my frequent visits to the bathroom were distressingly painful). Eventually, Debs suggested that I lean over the birthing ball to let gravity help, and by this time I felt ready to be more upright—I somehow felt that the time had come. Around this time the gas and air ran out. Once we'd established that I was too close to the end for anyone to be able to go and get the back-up cylinder from the car, I just knew I had to do it, and that I *could* do it. Eventually my baby started moving down. It felt as though my pelvis was dislocating, a very scary sensation, and I said "Help!" repeatedly. Debs stayed silent but Paul asked what was wrong. I said how I felt and Debs reassured me. My baby descended slowly, allowing everything to stretch. When his head had crowned Paul said, ""e'll be able to see his ears soon." Unfortunately, what I heard was: "His ears are stuck. He'll never come out..." but strangely this didn't worry me as I was fully occupied with pushing. Then he was out, and I was holding him against my belly and crying out: "My baby! My baby!" The cord was uncomfortably short (does this explain the features of my second stage?), so once the cord had stopped pulsing Paul cut it. He held our newborn son skin-to-skin whilst I delivered the placenta. I felt quite trembly and my legs were stiff and threatening to cramp as I got up onto the bed. Once I was snuggled up in bed with my beautiful son, nothing mattered but the joy and love I felt.

## Epilogue

I had only a tiny nick in my perineum, despite my boy's head circumference being on the 91st percentile, but I did have a bad graze which was distressingly painful in the first couple of weeks after the birth. I also found establishing breastfeeding difficult due to first engorgement, then repeated bouts of thrush. Despite these difficulties, I found that my wonderful birth experience gave me a bedrock of confidence that I could do anything, overcome any obstacles. In the first weeks of my son's life I didn't feel 'in love' with him, but rather a fierce protectiveness like a tigress for her cub. In some ways we were still the same person—his distress was my distress, his need for food was my own hunger. Once I started to see him as a separate person, my love for him blossomed. He's 3½ now and almost finished with breastfeeding, but I still look back with pleasure to his birth: a beautiful beginning to our life as a family.

## Reflections...

Was my head down/butt up position because the cord was short, and my baby needed to rearrange things before he could come out? Or because the use of gas and air and/or TENS blocked one/some of the birth hormones? I don't have regrets about using TENS or gas and air because I feel great about the birth I had, and the feelings and beliefs about birth I had at the time led me to make those decisions. That's not to say that I won't do things differently next time now that my feelings and understanding of birth have changed and developed!

How does color of her surroundings affect a birthing woman? Might warm, muted colors feel safer (womb-like, cave-like), or are familiarity, personal preference or the cultural meaning of colors more important? I felt very strongly positive about the colors of the room I birthed in (dark red/orange fabrics, wooden walls), but I can't separate out the familiarity aspect from the other elements which I've suggested might be important.

I birthed on a boat, in a fairly confined space. Clearly no mammal would want to birth in an aircraft hangar, but is there a point where a very small room feels too small? Presumably if you have one or more caregivers present this introduces a personal space factor (see my snarling episode above), and might significantly decrease the size of room that feels safe and comfortable to a birthing woman. What effect does furniture and equipment have, particularly in terms of height? I hated anything looming over me when I was laboring on the floor. Is floor area or ceiling height more important in determining how a laboring woman feels about the room? How consistent are these effects? Are they too variable to use in planning delivery rooms, or are there design principles that could be widely applied?

Clearly no mammal would want to birth in an aircraft hangar, but is there a point where a very small room feels too small?

*Decisions about birthplace may well be affected by practical issues, such as the easy availability of transport*

# Intrapartum care

Of course, your own institution will have its own protocols and guidelines, probably including a decision tree for watchful waiting, and these should all be informed by official ACOG/ACNM guidelines. However, since there is enormous scope for variation it's important to consider what's important and also what's *possible,* so as to optimize in-labor care.[64]

## The importance of non-disturbance

I get the impression many people think that nothing much constitutes a disturbance when it comes to birth. All kinds of people are invited along to watch a labor and birth, cameras and even camcorders are used, and medical personnel come in and out, frequently with no respect for the woman's privacy. People freely talk to the laboring woman, unaware that this might have a profoundly negative effect. Midwives and obstetricians interrupt the laboring woman whenever they perform vaginal exams or listen to the fetal heartbeat and they sometimes suggest hooking her up to an electronic fetal monitor, which usually limits her movements.[65] Many caregivers issue commands or make suggestions, they offer drugs and other options for pain relief and if they actually administer any they completely change the woman's internal chemistry. They touch the woman and usually catch the baby, they offer comments and information without being asked (e.g. on the baby's gender) and they often interrupt early mother-baby interactions. All these things are done without any real concern that disturbance is a possibility. Insensitivity on top of—or instead of—all this disturbance can also result in poorer outcomes, because a mother's negative reactions can severely affect bonding.[66]

Although at times there are excellent safety reasons for stepping in, sometimes caregivers, relatives or friends disturb things out of impatience. Sometimes their speech or actions are prompted by worry or fear; sometimes caregivers are not really following protocols at all, but doing things out of habit. At times it may even be a desire to control a seemingly wild experience which prompts someone to cause a disturbance. And, of course, very often the processes of labor and birth are disturbed by attempts to alleviate pain. The idea that a woman is entirely capable of giving birth without anyone else's help is shocking to many people... and the idea that it's *possible* to help prompts many people to action.[67]

However, as I've stressed before, I really do believe that the processes of birth are so delicate that many things can disturb a laboring woman and consequently make her labor slower and more dangerous.[68] Just as one small jolt changes the pattern in a kaleidoscope or one mistake at work might have enormous repercussions, affecting numerous people, even one inappropriate action, comment, intervention or therapy can have a dramatic effect on a birth.

One inappropriate comment or intervention can have an effect

If you personally question this, please remember the link between emotions and hormonal production... It may cause a dramatic chain of negative developments. The link between speech and the 'wrong'—non-instinctive—state of mind explains why even simple one-word interventions can be very disruptive. Furthermore, disturbance is particularly likely because of the sexual nature of birth. As you may know, even a mistimed comment or wrong movement during a sexual experience can take a person 'out' of the mood. If a person outside the experience should also come along... well that might *really* change outcomes.[69]

States of mind are very important and things seem to proceed most smoothly if the laboring woman is allowed to drift off into a different, totally absorbed state of mind. To do this, a feeling of privacy and security is necessary. It's worth noting that first of all, as a caregiver, *you* are the person who can create both a sense of privacy and security for the laboring woman. You have enormous power in this respect, even if it may seem rather Zen-like! (It's not exactly inaction but *considered* action or refraining from action from a standpoint of wisdom.) It's also worth noting that in not disturbing the laboring woman you're not handing over all control to her. As one experienced prenatal teacher (with first-hand experience of physiological birth) pointed out to me, it's not a question of handing control to the laboring woman, it's a question of *not controlling her.* As you know, while she's in labor and giving birth physiologically, she's going to seem well and truly out of control—totally wild!—so the issue of control seems a pretty irrelevant one really.

Here, in an adapted extract from *Birth and Breastfeeding* (Clairview Books 2007), Michel Odent explains why our minds need to remain undisturbed and how it is that our bodies tap into their instinctive knowledge about giving birth:

The activity of the primitive brain prevails during the process of birth. We share this primitive or archaic brain with all the mammals. It is old also in the sense that it reaches maturity very early on in our lives, at the age when we are still dependent on our mothers. It cannot be dissociated from the hormonal system and the immune system, with which it forms a complex network. This network itself represents the adaptive systems involved in what we commonly call 'health'. The archaic brain, which governs the emotions and instincts, can also be looked on as a gland releasing the hormones necessary for the process of birth, inducing efficient uterine contractions, and protecting against pain as well.

The process of birth is all the easier when the other brain, the new brain, takes a backseat. This new brain—the neocortex, whose huge development is the main feature of human beings—does not reach maturity before adulthood. Its activity during the process of birth only hinders the activity of the old brain. All inhibitions come from the neocortex during a delivery (and in any other event of the sexual life, as well). That is why, in a very spontaneous birth according to the method of the mammals, there is a stage when the woman seems to be cut off from our world, as if on her way to another planet. This changing level of consciousness is obviously related to a lesser degree of control by the new brain. Then the mother-to-be is freed from any sort of inhibition. She dares to scream out; to open her sphincters; to forget about what she has learned, what is cultural, even what is decent. That is why the best way to make a birth longer, more difficult, more painful (and more dangerous) is to stimulate the neocortex where all the inhibitions originate. The laboring woman needs to be protected from any sort of neocortical stimulation.

The neocortex can be stimulated by light, or by having to listen to people talking logically and rationally, or by being surrounded by people who behave like observers. A feeling of privacy, on the other hand, accompanies a reduction in neocortical control. The need for privacy, along with the need to feel secure, is a basic mammalian need in the period surrounding birth.

## Birthframe 73: Jenny Sanderson—Four undisturbed births

To illustrate how this non-disturbance might work in practice, here is an account by a first-time mother from London, Jenny Sanderson. Her account provides some clues as to how the physiological processes can be facilitated effectively. As you will see, Jenny started out planning a hospital birth and she only started considering a home birth when she learned about typical hospital procedures. Laboring undisturbed with sensitive caregivers in attendance, Jenny experienced four optimal births.

When I was pregnant with my first child, and in a state of almost total ignorance about childbirth, my husband Tim and I attended Active Birth prenatal classes. I retained very little of the information we were given but will never forget the assertiveness role-play we did. It seemed that a hospital birth would involve countless interventions that had to be resisted—and I knew that I would not be able to resist them.

Then a friend told me about a lecture by Michel Odent that she had attended and suggested I phone him; he lived in London, and delivered babies at home. So as not to disappoint her, rather than anything else, I called and made an appointment. My appointment took place at his home, during early pregnancy. Over about one-and-a-half to two hours we discussed my work, education, childhood, Tim's work, where we lived, leisure activities, etc. He asked if I knew anything about my own birth—I didn't—and he was interested to know about my mother's experience of stillbirth; he felt it would not be a good idea for her to be present for my labor. We went over my pregnancy to date and he listened to the baby's heartbeat and felt its position.

It was all very straightforward—my pregnancy was normal, he was available around the time I was due and, almost before I realized it, we had arranged for him to deliver me at home.

My Active Birth teacher, surprisingly, was concerned. How would I cope just with my husband, Tim, as a labor companion? Did I know that Michel would pretty well leave us to it? This didn't seem like a problem to me compared with the treatment I seemed certain to get and the battles we'd have to fight in the hospital.

About a month before the EDD Michel visited Tim and me at home, mainly to meet Tim, see the house and our heating arrangements and where the bedrooms and bathroom were. He again listened to the baby's heartbeat and felt its position.

A week before the EDD, Michel sent me a letter, with a copy to my doctor and the hospital. It confirmed his intention to attend me at home during labor.

*Did I know that Michel would pretty well leave us to it?*

## Here's Jenny's account of the birth itself:

On the morning of my EDD I didn't feel too good, not very well, a bit queasy—but not bad enough to cancel friends who were coming to lunch—an arrangement deliberately made for this date, on the assumption that the first baby would be 'late'. I ate a normal breakfast.

> On my EDD friends were coming to lunch... This was an arrangement deliberately made for this date

Our friends arrived about 11:00AM. Soon after, I began to feel that I didn't want to sit still and went round the garden and up and down the house. We called Michel; I spoke to him but he wasn't anxious, especially when he heard about the breakfast. By lunchtime, I didn't want to be sociable or to eat anything and went upstairs. Tim phoned Michel again and everyone had lunch, leaving by early afternoon just as Michel arrived. He saw that I was not in 'hard labor', felt the baby's heartbeat and pronounced everything to be normal.

During the afternoon and early evening I spent some of my time walking round the bedroom but mostly in the bathtub or on the toilet. Later on I found leaning against the towel rail useful but I didn't want to use Tim for support and the one time I tried lying down on some cushions I felt stranded and found the contractions harder to manage. I spent a good deal of time on the toilet, though I'm sure my bowels were long since empty. During this time Michel listened to the baby's heartbeat several times with his Doppler machine and confirmed that the mucous plug had been ejected into the bathtub. He spent most of the time upstairs in the spare room with a book and occasionally talking to Tim.

> Michel spent most of the time upstairs in the spare room with a book and occasionally talking to Tim.

Shortly after 8PM Michel could hear that my breathing had turned to grunting and suggested that I move out of the bathtub into the bedroom. Tim supported me for two or three contractions before Rebecca was born at 8:25, by candlelight.

Michel laid her on a towel and used his mucus extractor.

[This was simply the funnel of a hand-held stethoscope. The procedure is explained in detail on p107 of *Birth Reborn* (Souvenir Press 1984).] Then Rebecca and I lay down on some cushions. She didn't want to suck but didn't cry much either. Michel lay on our bed for half an hour or so; Tim made hot drinks. At about 9PM I delivered the placenta into a hastily found casserole dish; we did not eat it! Michel weighed Rebecca (8lb) and did the necessary paperwork before leaving us to a leisurely meal.

The following morning he returned and we phoned the hospital, my doctor and the community midwives. Both Michel and the midwife visited for most of the following 10 days.

## Here's Jenny's account of her second labor:

About three days after my second baby was due I went out to eat with Rebecca (my first daughter), experiencing occasional indigestion-like twinges. By the time Tim came home I thought I was probably in labor but didn't mention it until about 6:30 or 7:00PM. We put Bec to bed and I phoned Michel at around 8:30PM; this time he said he'd come immediately. Tim and I then had dinner, though I ate only moderately. After Michel arrived we had hot drinks. Tim and I went for a short walk and when I returned I went at once to the bathtub, where I stayed for most of the rest of the labor.

From about 10:15PM contractions were getting very strong. At approx. 10:40PM I thought I needed to go to the bathroom, but after straining for a bit, I reached down and felt the head.

Rosamund was born all in one go with the next contraction, at 10:45PM; Michel caught her as she came, and laid her on a towel.

I lay in the bathroom with the baby for half an hour or so and then squatted to deliver the placenta with no assistance. Michel checked me, weighed Ros ($7\frac{1}{2}$lb) and did his paperwork before leaving us together.

This labor was certainly the shortest; I was out visiting friends at about 5PM when I felt the first early contractions and Ros was born just over five hours later. There obviously was a second stage but it was very short and I didn't need to do any strenuous pushing as I did for the other three. But it was quite a shock for the baby to be born so quickly.

*The Sanderson children enjoying a bedtime story with their father*

## And a comment about her third and fourth labors...

For me there were overwhelming advantages in having home births and I went on to have two more (which, unfortunately, Michel was unable to attend) with an excellent midwife. [These were two more normal, healthy births.]

*Jenny with two of her children*

## Birthframe 74: Clare Winter—Midwifery records

It's sometimes difficult to imagine how the very woolly term 'sensitivity' teams up with the hard realities of clinical practice. Here's Jenny's midwife's account of one of the normal, healthy, physiological births she attended. We can clearly see that the midwife's focus is on numbers and safety... but she does seem to have a relatively hands-off approach too, which is likely to make an optimal birth more probable—in terms of both safety and satisfaction. It's interesting that in this account we can glean nothing at all about how the midwife behaved with the woman. While you are reading, try to imagine how much or how little was said, imagine where the midwife stood or sat while the woman was in labor, what—if anything—she suggested, apart from the change of position, and what her concerns might have been at each stage in this labor. How might your own records differ? Could or should anything be added to records?

### Summary of labor

First stage: 6:45AM, 12 hrs 15 mins

Second stage: 7:00PM, 31 mins

Time of birth: 7:31PM

Third stage: 14 mins

Weight: 8lbs 4.5oz. Length: 52cm. Head circum. 36cm

### Notes

4:30PM Jenny bleeped me. She has had mild contractions since 6.45AM. She would like me to come as she feels things might start happening.

5:30PM arrived. Jenny seems very relaxed, experiencing weak irregular contractions 1:6-4 mins for 30 secs. Abdominal palpation. Long ceph LOP-L 3/5 engaged. FHHR 148. Doesn't look as though she is in established labor but I will stay awhile to see what happens. Suggested all fours to encourage baby to go LOL.

6:30PM. Contractions seem a bit better 1:4-5 mins for 30-40 secs. Jenny in the bathroom on all fours. FHHR 140.

6:50PM. SROM FHHR 150. Heavy show, clear liquor draining.

7:00PM. Contractions hotting up, suddenly more pressure in Jenny's pelvis. She is kneeling in the bathtub. Tim is washing bath water over her back. FHHR 130.

7:15PM. Strong contractions 1:2-3 mins 30-40 secs, some strong pushes. Jenny now out of bath kneeling over the bath. Tim giving pressure on her lower back.

7:31PM. Good pushes, strong contractions, head descends slowly. Kneeling over the bathtub, as the head was born lifted leg. With brilliant control head born gently. Cord round neck; gently pulled over head, body born. Baby passed meconium at birth.

7:40PM. Cord stopped pulsing, Jenny clamps and cuts the cord. Tim is holding their new daughter, tears of joy!

7:45PM. Jenny squats over plastic bowl and pushes out the placenta. Placenta ragged? Little piece missing. Membranes complete. 300 ml TBL. Perineum intact.

8:30PM. Jenny and the baby clean and resting in bed. The little girl has fed well at the breast. Uterus firm.

Jenny experienced a lot of discomfort during 2nd stage and the birth of the head and body of the baby. This was probably due to the fact that when Juliet was born, she had her right arm up at her neck. This made her elbow stick out so that the rotation of the head and body was more uncomfortable due to the extra bulk of the arm and elbow. Jenny would have felt the discomfort of the arm rotating with the body.

*A recent photo of the Sanderson girls, out in their back yard*

## Disturbance through speech... or facilitation?

We've already seen in various birthframes how interruptions from other people or simply having too many people around can appear to disturb a labor. Let's recap a few situations in which talking proved to be problematic in some way...

### Deborah Jackson (Birthframe 22)

In this case labor was taking longer than the mother expected. In her account she says "I kept moving and visualizing but unfortunately the midwife also kept talking and I didn't have the guts to ask her to keep quiet. I found it impossible to 'let go' while she chatted to me and I felt obliged to smile politely and respond between contractions."

### Liz Woolley (Birthframe 49)

During this labor, the way the midwife spoke to the laboring woman actually seemed to have a positive effect. This is what we think of midwifery as being all about... The mother was feeling doubtful and the midwife reminded her that being scared would make the birth harder. (These words had such a powerful effect that the mother actually *decided* not to be scared and then went on to have a breech baby vaginally. On a more negative note, did you also notice how the midwife had earlier on told the mother that 'the next part (second stage) would be very hard work'? What purpose did that comment serve except to create fear and worry? Is it possible that the midwife actually created the problem that she then 'solved'—through speech?

### Nuala OSullivan (Birthframe 85)

You haven't read this one yet, if you're reading through this book from beginning to end, but in this account you'll see another mention of speech. In this case the laboring woman comments that while it's okay for the laboring woman to scream blue murder, everyone else must keep their mouths shut and not even make the noise of turning pages in a book! In fact, this woman's screams in her first labor were so spine-chilling that the local police knocked on her door. Nevertheless, this woman felt that a double-standard was needed.

### Anonymous (Birthframe 52)

In this case, as you may remember, the mother was furious by the way in which the midwife dismissed her right to consent (or refuse) a treatment. The woman reports: "She did not ask me if she could [clip a monitor to the baby's scalp] and was actually laughing when I asked her what she was doing. She made a comment about me being 'away with the fairies anyhow.'"

I actually exchanged a few emails with this mother since the reason she contacted me in the first place was because she felt that she'd been treated poorly during labor. While I could only agree that this midwife's attitude (and speech) was awful—and in conflict a woman's right to choose, or refuse, care— I did tell her that I suspected that her midwife must have been in a blind panic at that stage because placental abruption may have been a concern. It was interesting and perplexing that the mother hadn't even realized that she was in such danger. Perhaps this is another place where communication went wrong... prenatally? Or was it a good thing that the mom-to-be wasn't aware of the risk?!

# Questions for reflection...

1   What kinds of responses do women give you when you speak to them when they're in labor? Do you think you should follow up and keep conversation going? In which cases should you minimize talking? Why?
2   What kinds of things do women spontaneously say? How do you respond?
3   How might your own speech habits and tendencies affect women's labors?
4   Have you ever been able to facilitate a woman's labor through speech?
5   Do you think there's ever a role for humor when a woman's in labor?
6   Do you ever *not* act as an advocate for a woman because of judgments you are making about her real needs in labor? If so, how valid is this in terms of the role you are supposed to be playing in professional terms?
7   How can a caregiver not create fear, while still giving moms-to-be the information they need? One woman told me how her request for an enema had been refused. It seemed, from her account, that she was in transition and an enema probably wasn't what she needed. But do you think it's right that her request was ignored? How can a caregiver be an advocate but also protect the woman from unnecessary disturbance? Would this be a time to communicate sensitively? Or would the speech stimulate the neocortex at this crucial time and disturb the birth, compromising its safety?
8   Can you think of any way in which women could ensure they got the level of 'support' (either verbal or physical) or non-disturbance they might prefer, perhaps using a standard intrapartum preference form prenatally? Would they need to be educated about their possible real needs too?
9   What would your own preferences be if you were in labor yourself?

Whatever your answers and comments... perhaps the most important thing is to recognize the enormous *power* of speech in all its forms, and also the power of silence, and use them both very wisely, remembering the importance of not disturbing the neocortex during labor, birth and postpartum.

## Disturbance vs negligence

It's important to note that I'm not advocating a completely laissez-faire attitude. As we know, things *do* sometimes go wrong during labor and birth and things can be done to prevent maternal or fetal mortality or morbidity. That's where the decision tree comes in (either on paper, or in your head), along with your years of experience, which will help you interpret it, and your sensitivity, which will help you relate to the laboring woman entirely silently and without disturbance. In births which feature only watchful waiting this mostly constitutes a complete lack of disturbance, not a situation in which there is a lack of care—where a woman's labor is disrupted, neglected or mismanaged.

## Birthframe 75: Sylvie Donna—Hearing about neglected birth

I remember once speaking to a very bitter German woman who'd given birth in Japan. Not speaking any Japanese, she'd nevertheless communicated her wish to have a natural labor. After three days in labor, she gave birth to a brain-damaged baby. I listened carefully to what she said and it was only after much reflection that I realized that there were various things that were odd about her account. Certain comments she'd made indicated clearly that her birth had not been at all 'natural'. For example, she'd mentioned in passing that she'd had to drag her IV, on its stand, to the public phone, and had said what a nuisance this had been. This must have meant either that she was being denied all food and drink and had therefore been put on a glucose IV, or—a more likely scenario—that she was being administered pitocin in order to accelerate or induce her labor. It's very sad that she insisted on continuing in this situation for a full three days, before finally giving birth. She'd clearly had interventions, so the natural processes had been disturbed. Her labor was then neglected at her own insistence.

Perhaps this is not such an uncommon scenario, not because of the communication difficulties which were obviously a factor in this birth, but also because lack of understanding on the part of women or caregivers as to what really constitutes disturbance. Very often, caregivers will talk about or suggest procedures which disturb the laboring woman, without realizing that they are doing anything disruptive. The laboring woman will then insist on continuing in a disturbed situation because of a misguided view about natural birth. This is sad because really we have a choice between leaving nature alone completely, or disturbing it and having a very 'managed' labor. Unless we are extremely lucky, there is usually no middle road.[70]

*Anyone who's ever let children play in the bathtub knows the difference between disturbance and negligence...How would you differentiate the two?*

## The prevalence of accidental disturbance

I've often come across cases where women have told me they wanted a natural birth, but it didn't work out. After listening to them I usually found it easy to pinpoint the disturbance which occurred in their labor and in almost all the cases I've come across it was a disturbance which wasn't at all necessary from a safety point of view. I have been amazed by the predictability of this element in 'failed' natural births.

What has saddened me is people's apparent unawareness of disturbance. Again and again, the assumption that nothing can possibly 'disturb' birthing processes emerges. In one article I read in a popular women's magazine, a woman expressed disappointment and sadness at not having the natural birth she'd wished for *after being induced.* What did she expect? Induction is an enormous intervention! In the same way, I've been repeatedly amazed that women who've had epidurals have then expected the natural processes to proceed smoothly. After any major intervention, such as induction or the use of drugs for short-term pain relief, medical management is essential for the sake of safety. In his book *Birth Reborn* (Souvenir Press 1994) Michel Odent commented that the more medicine gets involved with childbirth, the more complex and difficult everything becomes.

Modern prenatal clinics and hospitals don't usually provide an ideal environment for lack of disturbance. Many medical professionals intervene out of a sense of caution or a fear of litigation... Perhaps they feel they need to 'manage' the process of pregnancy, labor and birth; perhaps they are afraid of the processes; perhaps they simply lack faith in them; or perhaps they feel they need to be seen to be 'doing something'. Maybe health care providers offer drugs or treatments because they mistakenly believe they will prevent, relieve or eradicate all pain. Some even claim their main motivation in operating as they do is a fear of litigation: by tracking everything (irrespective of whether it causes disturbance) they at least have extensive records of the trouble they took, which they feel is helpful in a court case, even if the truth is that it made the birth less safe.[71] And to make matters worse, relatives tend to come in with their camcorders... Are you yourself guilty of any of these types of behavior?

It's no wonder that the phenomenon of the 'cascade of interventions' has become so well-known. Very often what started out seeming 'helpful' ends up being a trigger for another problem and another intervention. By definition, disturbing things means changing them in some way, and this is particularly the case with birth.

Does all this surprise you? It's true that sometimes things do proceed towards a happy conclusion despite any number of disturbances... but usually they don't. We can only be sure women will produce all the right hormones at the right moment if we leave the processes of pregnancy, labor and birth undisturbed and if we create a safe, private environment for each process.

## Birthframe 76: Anonymous—Cascade of interventions

The following birth story is an unhappy example of interventions which were perhaps unnecessary, which the mother could have refused... but didn't.

'A woman's intuition' is something that's often mentioned, but usually dismissed. In this case, the woman in question had a strong feeling that her upcoming birth needed to be natural. Unfortunately, though, she did not realize that in order for this to be possible she would have to say 'No' at one or many points in her pregnancy and labor. Not refusing one seemingly small intervention—allowing her water bag to be broken—led to a cascade of interventions because her babies clearly were not ready to be born. In a way this woman did have the 'natural' birth she wanted, but it was one which made a mockery of what nature really has to offer. The writer of this birth story has submitted it and approved it (including this blurb and my commentary on the next page) because she wants her story to help other women. Incidentally, when I spoke to her over the phone she said she was glad she avoided a cesarean, even though she did not have quite the vaginal birth she had hoped for. [Names have been changed here, but all other details are as in the original account.]

I gave birth to twin boys, Jamie and Sean, on April 24, 1997.

Ever since I found out I was pregnant with them I always told my husband I was going to have them the most natural way possible, no matter what.

I went for a prenatal check-up at 38 weeks and they told me I was 4cm dilated. They told me I had probably been in labor for a week since my mom and sister took me on a 10-mile walk the weekend before, which gave me backache.

They sent me up to the labor ward to monitor the babies. Around 5PM, they decided to break my water. I was adamant they were not going to intervene but for the babies' sake they had to. For some reason, I was not having any contractions. They found Jamie was the wrong way round, not breech but as the doctors put it 'face-to-pubis' and stuck! He was not budging. All I could think of was not having a section. Under no circumstances was anyone going to make me have these babies unnaturally, after all it may be the only chance I have in life.

I was given an episiotomy but still Jamie was not budging. Awhile later they took me to the operating room. I told them I was not having a section so they decided to give me a spinal injection and as I was hooked up to a contraction monitor because I couldn't feel the contractions, I was told to push when I had them. Jamie came out with forceps at 4:39AM and then Sean followed at 5:15PM, after forceps and suction. I guess that was as natural as they could have made it for me.

I did not start to bond with them until they were about 12 months old. I had a difficult time after it all. The doctors told me it was probably because of the birth. Because of what my body went through, it just rejected the boys. I guess it makes sense really.

Let's consider the interventions which occurred:

- There was no need for the vaginal exam at the 38-week check-up. These 'internals' are now widely recognized as being unnecessary in most cases and of course they have the disadvantage of increasing the risk of infection.[72] How often do you carry out vaginal exams yourself?
- There was no need to break this woman's water. Of course, as soon as an amniotomy was performed, the risk of infection rose again and the woman was under pressure to give birth. How often do you do ARMs and why?
- The electronic fetal monitoring, made necessary by the amniotomy, was unhelpful for two reasons. Firstly, the very fact of monitoring in this way must have created anxiety in the laboring woman, whose state of mind must be safeguarded at all times. It's hardly surprising her contractions stopped. Secondly, it would have compromised the babies' oxygen supply because women who are being monitored are almost always asked to remain immobile in a supine or leaning-back position. The fetal distress which the equipment is aiming to detect actually creates it! What's more, a supine or lying-back position would have done nothing to help the babies get in a good position for birth. Leaning forward positions and moving around would have been much more helpful from this point of view.[73] How much do you use EFM yourself and why in each case? Do you think you overuse or underuse this technology? Who is it popular with, do you think, and why? What alternative forms of monitoring could be used which would cause less disturbance to the laboring woman?
- The episiotomy clearly did nothing to help the birth of the first baby, so constituted another unnecessary intervention and would have caused the woman additional, unnecessary distress, not to mention pain or discomfort at best, for weeks or months after the birth.
- The spinal would no doubt have relieved the woman's pain but the anesthesia would have made it difficult, if not impossible, for the woman to perceive what was happening within her as her babies descended the so-called 'birth canal'. (In this case, it really would have lost its suppleness and organic properties.) Given that this woman was so keen to give birth naturally and spinals and epidurals, etc are well-known to increase the cesarean rate, precisely because of the lack of sensation that results, this woman might well have refused the spinal if she had not felt so disempowered during her labor and if she had realized the possible consequences.[74]

Left to her own devices, this woman probably would have gone into established labor spontaneously hours or even days after her 38-week prenatal appointment. Do you agree? What would you have said to this woman prenatally if you'd heard she wanted to have a 'natural birth'? If a cascade of interventions had taken place while this woman was in your care (as above), how would you have spoken to her postpartum? How would you help her improve her experience another time? Do you think it's likely she will ever have any more children?

   My first baby was a planned hospital birth. I expected to just turn up in labor, have the baby and come home. I did not really take on board the 'cascade of intervention' which can and does happen in hospitals to upset the natural balance of a labor that is going well.

   At some point in transition or second stage, the second midwife arrived—this is normal procedure, just in case both mother and baby need attention after the birth. Next, the contractions became more spaced out, and more painful.

   My husband took some video immediately after the birth. I stopped him filming the birth, even though we'd planned to, because it felt like an intrusion.

   I was put under a lot of pressure to deliver the placenta after Kiz's birth. They had their arbitrary time limit of an hour and were planning on pulling on the cord if I hadn't managed to push it out when I did. I'm sure the fact that they had me sitting on a bed pan on a chair (extremely uncomfortable) meant that it took longer than it would otherwise. I had to lean on my hands to keep all my weight off the bed pan, which meant I couldn't hold Kizzy to nurse her. She suckled for the first time—held in position by the midwife—just before I pushed the placenta out.

*Hospitals can be disturbing places, by definition, since they exist primarily to deal with things which are going wrong, not right. How can places be made as accommodating as possible for women in labor, from admission onwards? What needs to change? By the way, this is Pithiviers hospital (in 2009), where Michel Odent used to work.*

# The reality of non-disturbance

## Birthframe 77: Michel Odent—Birthing like a mammal

In case you're blasé about the possibility of a labor being disturbed, let's remember two forms of disturbance which have become commonplace in our society: cameras and husbands. We need to be very aware of the potential of these two very human kinds of disturbance and remember that we basically have the same needs as other mammals. Here's an account from Michel...

A week ago on Sunday, I was at a conference in California. I was a keynote speaker. My topic was provocative. It was why and how we should dehumanize childbirth. There were two organizers, two midwives, Iona and Laura. Iona was pregnant and due quite soon—this month—and expecting her first baby. Laura, who herself had three children, was at the same time Iona's partner as a practicing midwife, and also her midwife for this birth.

Iona introduced me on Saturday at about 11:30AM and while I was speaking some people noticed that she was often touching her back. Then at about 3PM she had a rupture of membranes at the conference. So she went back home, which was about 20 minutes from there, with Laura. Luckily there was a third midwife involved in the organization of the conference!

In the end, real labor started in the middle of the night and Iona gave birth on Sunday morning exactly a week ago. And I just heard before leaving (because I left on Sunday afternoon) that she had a baby boy. But I called her yesterday to find out more. She told me an interesting story.

She told me there was a time when Iona—when the baby was not far away—said, "Do something! I can't do it. I can't do it!" But Laura told her, "You can do it. You're a mammal!" [Hearty laughter.] Because of my lecture they had changed their approach. Originally, Iona had planned to give birth in front of some television cameras but then she heard me talking about privacy and cameras. I'd also said that it was important to be careful about the presence of the baby's father. So in the end she had much more privacy than she'd expected to have. It was her first baby and it was wonderful. And Laura had said two times, "You can do it! You're a mammal!"

Despite our previous discussion of principles, I can imagine you're thinking, 'Hang on a minute! Surely this doesn't apply to women who are categorized high risk?!' I asked Michel about this to check:

I have read that you have said the higher risk the woman the more important it is that she remains undisturbed during labor. The conventional approach seems to be the opposite—i.e. women who are high risk are told they will be monitored very closely. Do you think it is important that all women are left to labor undisturbed, or would you make exceptions in certain cases? If so, in which cases would you make an exception?

This is the basis of the art of midwifery: to know what is happening without disturbing. I wrote chapters about that, particularly in my book *Birth and Breastfeeding* (Clairview Books 2007).

So I assume you're saying that this can be achieved if (1) the caregiver observes without giving the laboring woman the feeling of being observed, if (2) he or she uses a Doppler to monitor the fetal heartbeat—or nothing if conditions seem to be good, if (3) the caregiver makes sure there is no disturbance from anyone else between first and second stage and then again between second and third stage... and if (4) the caregiver stays out of the way at these times too, i.e. throughout the length of the natural process—for example, by remaining in an adjoining room while the mother-to-be labors and gives birth alone. Have I forgotten anything?

Do you think it's always best if the mother-to-be gives birth alone, without the help or support of a midwife or other attendant?

The key word is privacy. It does not mean loneliness.

The key word is privacy. It does not mean loneliness. Privacy is compatible with the presence of somebody who does not behave like an observer or a guide.

Note, by the way, that I mention the use of the Doppler here. The reason is actually quite simple if a little surprising. In the last section I was discussing the use of ultrasound in non-essential cases. I chose to ask my own midwives to use a fetoscope throughout all of my pregnancies precisely so as to avoid ultrasound and because the fetoscope represented an equally effective means of checking the fetal heartbeat after 18 weeks or so of pregnancy.

During labor the priority is not to disturb the laboring woman

During labor the situation changes, of course, because the priority is to disturb the mom-to-be as little as possible. Instead of auscultating while chatting, it's helpful if you can either use watchful waiting (e.g. listening to the sounds the laboring woman makes and observing her movements), or aim to auscultate extremely discretely, without your client even noticing. Of course, since most midwives find it hard to auscultate with a fetoscope in any position apart from lying down, they will almost inevitably cause a disturbance if they use the fetoscope during a woman's labor. This is where the Doppler comes into its own because it can not only be used discretely, it can even be used silently (with earphones, with the speaker switched off)—and some models will even operate underwater.

When I first started discussing this and other issues with Michel Odent, this was one of the two issues we disagreed on. (The other was optimal fetal positioning, which he refuses to accept simply because there's no research evidence to prove it yet.) After hearing him explain the advantages, though, and after considering that the baby's exposure to ultrasound would be minimal and at a time when he or she is fully developed, I relented! So if I were to have another baby, I'd probably ask you to use a Doppler—but only minimally, i.e. if you felt, from watchful waiting and listening, that there might be a problem.

To be very specific and clear, I see non-disturbance as meaning the following:

- Don't talk to a woman while she's in labor, while she's giving birth or before the placenta has been born. If your client addresses you make your responses absolutely minimal or even ignore her comments, only responding nonverbally, aiming to be reassuring. If you need to talk to your colleagues, move well out of earshot so the laboring, birthing or postpartum woman can't hear what you're saying.[75]

*Don't talk to a woman while she's in labor, while she's giving birth or before the placenta has been born*

- Stop anyone else from talking to the woman too.
- Leave the woman (and baby) alone, without making her feel abandoned. (Quietly make your presence felt, while not intruding.) Of course, if your client specifically requests something (e.g. candlelight, a bowl, a hot bath, or some specific music)[76] silently meet her request. Ignore any request which you know to be unhelpful (e.g. administering an enema, performing 'yet another' vaginal exam or breaking her water), remembering that some requests will be made out of a lack of confidence. Your role is to reassure the woman and give her a quiet feeling of protection and confidence. Words are not necessary to achieve this, of course—silence and a reassuring smile are more effective than words. Remember she will find her way if allowed to 'go to another planet'. Her courage is likely to return after a few moments. (Of course, you can only behave this way if the woman has requested an optimal physiological birth in her prenatal check-ups or in her care guide.)

*Leave the woman (and baby) alone, without making her feel abandoned*

- Create a feeling of privacy and ensure there are no cameras or camcorders around until well after the birth. Nonverbally shoo people out of the woman's way!
- For safety reasons, don't approach the woman just as she's given birth (unless you've caught the baby). Leave her completely undisturbed (still no talking!) until after the placenta's been born. Don't let anyone enter the room at this point either, because that would also be a disturbance. Don't ask any questions or make any comments and *definitely* don't look yourself to find out the baby's gender. Let the woman do that herself in her own time. Only step in to help you if you observe that either the woman or the baby is experiencing difficulty. The initial Apgar scoring can easily be calculated by distant observation. (Of course, as at other times, this non-disturbance at this stage relates to safety because disturbance would dramatically increase the risk of postpartum hemorrhage, which used to be a major cause of maternal death.)[77] I've heard Michel Odent talking about a midwife needing to behave like a cat: it knows what's going on but it's discrete!

## Birthframe 78: Sarah Buckley—Understanding 'undisturbed'

Here, Sarah Buckley, author of *Gentle Birth, Gentle Mothering* (Celestial Arts, 2009), discusses the idea of non-disturbance further.

The term 'undisturbed birth' came to have great meaning for me when I gave birth to my fourth baby, Maia, at home, with only my family—including my medically qualified husband—present. This beautiful experience awakened me anew to the ecstasy of birth, and I realized that the process of birth can be very simple, if we avoid disturbing it. Comparing this birth to my three previous midwife-assisted home births, and to home and hospital births that I had attended, I saw also how ingrained is our habit of disturbance, and that our need to 'do something' so often becomes self-fulfilling in the birth room.

I realized that birth is also very complex, and that the process is exquisitely sensitive to outside influences. The parallels between making love and giving birth became very clear to me, not only in terms of passion and love, but also because we need essentially the same conditions for both experiences—to feel private, safe and unobserved. Yet the conditions that we provide for birthing women are almost diametrically opposed to these. No wonder giving birth is so difficult for most women today.

I realized that birth is exquisitely sensitive to outside influences

I came to realize first-hand that anything that disturbs a laboring woman's sense of safety and privacy will disrupt the birth process. This definition covers most of modern obstetrics, which has created an entire industry around the observation and monitoring of pregnant and birthing women. Some of the techniques used are painful or uncomfortable, most involve some transgression of bodily and/or social boundaries and almost all are performed by people who are essentially strangers to the woman herself.

All of these factors are disruptive to pregnant and birthing women. Underlying these procedures, is a deep distrust of women's bodies, and of the natural processes of gestation and birth, and this attitude in itself has a strong nocebo (fear-inducing), or noxious effect.[78]

On top of this, is another obstetric layer devoted to correcting the 'dysfunctional labor' that such disruption is likely to produce. The resulting distortion of the process of birth—what we might call 'disturbed birth'—has come to be what women expect when they have a baby and perhaps, in a strange circularity, it works. Under this model, women are almost certain to 'need' the interventions that the medical model promotes and to come away grateful to be 'saved', no matter how difficult or traumatic their experience.

These disturbances are counterproductive for midwives also. When a midwife's time and focus is taken up with monitoring and recording, she is less able to be 'with women' as the guardian of natural birth. When her intuitive skills and simple ways of knowing have been buried in service to the system, more and more invasive procedures will be needed to get information that, in other times, a midwife's heart and hands would have illuminated. And when a woman misses out on the joy and ecstasy of birth, so does her midwife, which will influence her expectations of birth, as well as her job satisfaction.

Our bodies have their own wisdom, and our innate system of birth, refined over 100,000 generations, is not so easily overpowered. This system—what I am calling undisturbed birth—has the evolutionary stamp of approval, not only because it is safe and efficient for the vast majority of mothers and babies, but also because it incorporates our hormonal blueprint for ecstasy in birth.

When birth is undisturbed, our birthing hormones can take us into ecstasy— outside (ec) our usual state (stasis)—so that we enter motherhood awakened and transformed. This is not just a good feeling; the post-birth hormones that suffuse the brains of a new mother and her baby also catalyse profound neurological or 'brain' changes. These changes give the new mother personal empowerment, physical strength and an intuitive sense of her baby's needs and prepare both partners for the pleasurable mutual dependency that will ensure a mother's care and protection, and her baby's survival.[79]

Undisturbed birth represents the smoothest hormonal orchestration of the birth process and the easiest transition

Undisturbed birth, then, represents the smoothest hormonal orchestration of the birth process and therefore the easiest transition possible—physiologically, hormonally, psychologically and emotionally—from pregnancy and birth to new motherhood and lactation for each woman. When a mother's hormonal orchestration is undisturbed, her baby's safety is also enhanced, not only during labor and delivery but also in the critical transition from intra- to extra-uterine life. Furthermore, the optimal expression of a woman's 'motherhood hormones' will ensure that her growing child is well nurtured, adding another layer of evolutionary 'fitness' to the process of undisturbed birth.

Undisturbed birth does not mean unsupported birth, though. Some anthropologists believe that human females have sought assistance in birth since we began to walk on two legs. The change in our pelvic shape that accompanied our upright stance added uniquely complex twists and turns to our babies' journeys during birth, making assistance more necessary than for other mammals.[80] It does mean having supporters who we have specifically chosen as our familiar and loving companions; who are confident in our abilities, and who will intervene as little and as gently as possible.

Undisturbed birth does not mean painless birth either. Giving birth is a huge event, physically and psychologically, and makes demands on the body which are hormonally equivalent to endurance athletics.[81] But when a woman feels confident in her body, well supported, and able to express herself without inhibition, any painful feelings can become just one part of the process and something that she can respond to instinctively from her own resources using, for example, breath, sound and movement.

Finally, we must recognize that having an undisturbed birth will not guarantee an easy birth. There are many layers, both individual and cultural, that can impede us at birth. But when we approach birth with the intention of minimal disturbance, we are optimizing the functioning of our birth hormones. This, coupled with our unparalleled levels of hygiene and nutrition, gives us a better chance of an easy and safe birth than any of our foremothers, from whom we have also inherited, through natural selection, the female anatomy and physiology that births most easily and efficiently.

Minimal disturbance, optimizes the functioning of hormones

*Sarah J Buckley with her daughter Maia, aged 3*

## Encouraging women to use upright positions

Of course, moving around during labor and using upright positions during childbirth have not always been encouraged or even permitted by health professionals. For the past three decades, Janet Balaskas has been doing an enormous amount to educate both women and health professionals of their usefulness. As well as founding the Active Birth Centre in London, she has also written several books on the subject. (At www.activebirthcentre.com you can see the original manifesto of her philosophy.)

### Birthframe 79: Janet Balaskas—Discovering upright labor

Here Janet explains how she first came to realize the importance of physical freedom during labor and birth and how it became her mission to help other women understand this.

Moving around during labor and using upright positions during childbirth have not always been encouraged or even permitted

During the late 70s and 80s, when my children were born, the majority of women labored and gave birth in a highly medicalized hospital environment and almost always in bed and semi-reclining. This was the peak of the hi-tech era when obstetricians launched the concept of 'actively managed birth'—one in which labor was induced at a convenient time, the baby electronically monitored continuously and the mother wired up to an epidural. Large obstetric units were the officially recommended birthplace for all women, whether they needed obstetric care or not. Not surprisingly, outcomes were not good—intervention and cesarean rates soared. No one seemed to know anything about birth physiology. Even 'natural' birth education at the time was geared to teaching coping strategies and breathing techniques for lying-down labors.

This horrified me. Surely, we can't be the only mammals who are incapable of giving birth by ourselves, I thought. I felt sure we must have a similar instinctive capacity to reproduce our species as other mammals do. While I was in labor with my second child I decided to get up from my semi-reclining position and to my astonishment I found that moving around helped my labor to progress more quickly.

I decided to do some research and look into the history of childbirth in different cultures prior to the advent of obstetrics. Availing myself of the amazing libraries in London, I soon discovered that there were many paintings, drawings, carvings and sculptures of birthing women from every continent in the world, some dating back to Neolithic times, some from Europe and as recent as the 17th and 18th centuries. I was struck by the fact that in all of them laboring and birthing women were depicted in upright positions, whether standing, kneeling or squatting. None were lying on their back! I then decided to take a look at existing ethnic cultures not yet influenced by Western obstetrics. Indeed, they too had rich traditions of birthing wisdom going back generations, which involved upright positions, the presence of other women and a generally secluded and private birthing environment.

My next step was to revisit the anatomy books. It was then that it dawned on me that the female pelvis was uniquely shaped for giving birth in upright positions. In a 'Eureka' moment, I realized that women are anatomically designed to be free to move instinctively during labor. This is surely the way, I thought, to make birth easier and safer for the majority of women, reserving the safety net of obstetrics for those who need it. Later, through further research, I confirmed that inevitably, when a woman chooses the positions which are most comfortable, her body aligns with the Earth's gravitational pull. This makes the contractions of the uterus more efficient and widens the pelvic diameters, making more space for the baby's descent. There is a better blood supply to the uterus and placenta and, not surprisingly, outcomes are much better, with a much greater chance of a vaginal birth and a huge reduction in the risk of complications. By moving, a mother can actually help her baby to be born.

I called this 'active birth'—a deliberate reinvention of the phrase 'active management of labor'. Freedom to move transforms a woman in labor from a passive patient to an active birth giver. No longer disempowered on her back like a stranded beetle, she regains control of her body and recovers her instinctive potential and power to give birth. Again, I experienced the reality of this when I gave birth to my third and fourth babies.

By being upright and free to move, without intervention and disturbance, the woman also reclaims the possibility of generating her own natural hormones. The birth process is governed by the maternal hormones of love and ecstasy, which mother and baby share. An active birth is not only safer and more practical, it also ensures that at the time of birth, babies are bathed in high levels of the hormones which foster love and healthy attachment. Having experienced this myself and witnessed it in thousands of women I have taught, it remains my mission in life to teach this to women and midwives and to continue to challenge the narrow-minded obstetric thinking and practice, which is not grounded in an understanding of birth physiology.

*Janet Balaskas, founder of the Active Birth Centre in London*

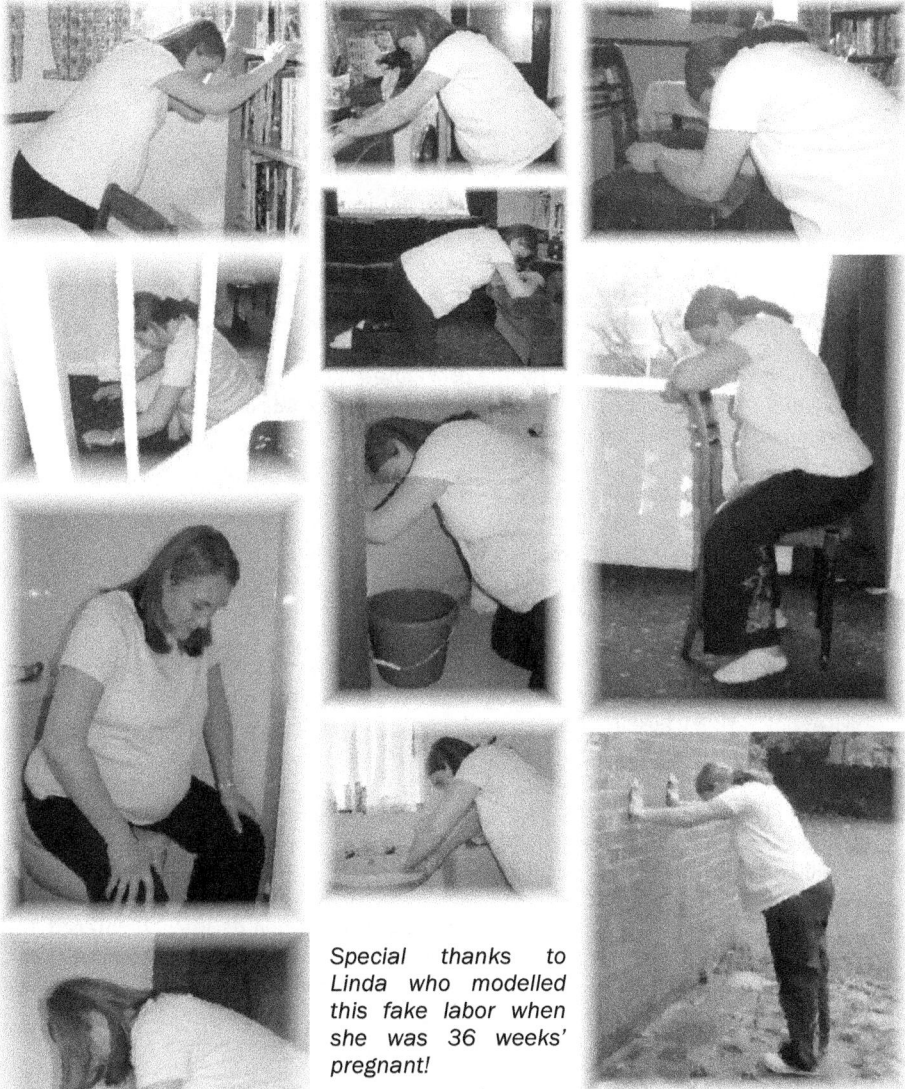

Special thanks to Linda who modelled this fake labor when she was 36 weeks' pregnant!

## Being upright helps mother and baby

During watchful waiting when active labor is progressing well you will probably observe your client adopting a range of leaning forwards positions, as depicted here. These positions are ideal for the baby because they ensure there is no pressure on the vena cava, so fetal distress is much less likely. Moving around also helps the baby to get into a good position for the birth.[82]

# Using water

What if your client asks for a water birth or at the very least to use the birthing pool? Here are some comments from women on their experience of using water in labor:

66 I tried the bathtub but found it too limiting in terms of movement.

66 I found the bathtub perfectly adequate. I leant forward during contractions then slumped down for a dreamy, sleepy rest between each one. After an hour or so like that, I suddenly wanted to get out and a few minutes later, I gave birth.

66 I got into the water and instantly the pain halved. It was beautiful. Mom stroked my head, and Dave my back. The lights were low, the music was on, the water was a giant woman-god warmly holding me together.

66 Get a birth pool if you can. They're wonderful.

66 I was a little scared that I would have no official pain relief once in the pool. But I didn't need it—the water was amazing. I could just move around and get into really good positions whenever I had a contraction.

66 My husband started to fill the pool, as I'd chosen a water birth. It was so hard for me to get downstairs at this point to the pool in our living room. I just didn't want to move, but with my husband's encouragement I did, and felt so much better for being in water. It just felt the most natural place for me to be.

66 Water is fantastic pain relief. Stepping into the birth pool felt like a miracle. Instantly the pain disappeared. No, it was still there, but suddenly it was manageable. I could have stayed in the pool all day.

66 Personally I found no need for any analgesic intervention as when I entered the water the pain I was feeling was substantially reduced. I wanted to labor, and possibly give birth, in a birthing pool because I was worried about suffering perineal trauma and, although there is conflicting evidence on the subject, I felt sure that the water would work as an aid in softening my perineum, thereby reducing tearing.

66 Before the second stage started there was a complete lull in the contractions. The lull felt like about 20-30 minutes and in this time I just dozed in the pool. When the contractions started up again I wanted to push and my second daughter was born within minutes under the water.

66 Baby was born in the water, and I gently brought her to the surface with the midwife's help. She was soft and velvety, purple, wonderful-smelling and crying. I sat there holding and feeling and looking for ages, and then was moved to check she was a girl after all, only to find she had a willie! "It's a boy," I was the one to announce.

Warm water really can help your clients open up to give birth
and many women report it really relieves the pain

*Here's Michel Odent, fully-clothed, in the first custom-designed birthing pool in 1982.
After having the idea to use pools in birthing rooms (perhaps inspired by practices
he'd heard about in Russia?), Michel bought a wading pool for the Pithiviers hospital—
in rue de la Couronne! It soon became popular with women during labor and/or birth.*

But what about your point of view? Is it worthwhile taking the trouble that birthing pools must involve? Will a bathtub do just as well? Here are some different perspectives on water birth, in its various forms.[83] First, an extract from one of Michel's books...

From *Birth and Breastfeeding* (Clairview Books 2007):

We had observed that immersion in a pool full of warm water was an effective way of facilitating the phase of dilation of the cervix. So long as the mother-to-be does not get into the bathtub until the onset of hard contractions in the middle of this phase, we learned to expect that once immersed she would be fully dilated quite quickly—after around perhaps an hour, or an hour and a half for a first baby. Although the contractions are apparently less intense and less painful in the water, the mother can feel that they are nevertheless more efficient. But when the baby is not far away, there comes a time when some mothers feel that the contractions are not working effectively any longer. After a series of five or six or seven contractions, there may be no further progress; and then many women feel the need to get out of the pool. As soon as they leave the warm water and return to the cooler atmosphere of the room, a puzzling phenomenon often occurs. It is as if a kind of reflex were triggered by the difference of temperature and, after a few huge contractions, the baby is born on the floor by the pool. This is a fetus ejection reflex.

In *Birth Reborn* (Souvenir Books 1994) Michel also mentions that at Pithiviers it was never the practice to insist that the laboring woman remain in the water for the birth itself—or even to encourage her to do so. They also used to prefer it if women got out of the water before the third stage (if the birth had taken place under water) so there was no risk of water entering their bloodstream via the open vessels in the womb, which could cause a life-threatening embolism. Focusing on the needs of newborn babies, Michel emphasizes that what they need is warmth and the feeling of their new moms' gentle touch. Here are some more comments and anecdotes from Michel to explain his views...

Personally I don't use the term 'water birth', which implies that the baby is born in water. Birth under water is possible but it should not be the goal. The main objective is to reduce the need for drugs. The best indication for the use of the birthing pool is when the first stage is long, difficult and very painful, in spite of perfect conditions of privacy. When the first stage is straightforward, it may be risky to change the environment. When a woman appears to be having difficulties, a birthing pool can actually be used to check whether or not a cesarean is necessary. It is particularly useful because a decision can be reached before fetal distress has occurred. I call this the 'birthing pool test'. Carrying out this test is very simple: when the woman is in hard labor, she is immersed in water at body temperature for approximately 90 minutes. Usually, within this time period, something spectacular happens. (The period of time is approximate, of course.) By spectacular I mean, for example, if she enters the bath at 5cm she reaches full dilation, or if she enters at 3cm she reaches 7cm. If after an hour and a half in water nothing like that happens, if you can see no difference in dilation, it means something is wrong, that there is some kind of obstacle. So the best course of action is to do a cesarean section—waiting is not at all helpful in this case.

The main objective in using water is to reduce the need for drugs

In case you're new to birthing pools and reluctant to use them, here's a summary of information I've found which is geared towards increasing the optimality of outcomes. I'm including this here only because many midwives seem wary of using birthing pools, perhaps because they have had bad experiences for all the wrong reasons. *Not* allowing women to use water seems a shame as it is one form of natural pain relief which seems very helpful, which facilitates birth (without disturbing it), provided a few guidelines are followed...

- Since women tend to need water suddenly, it's worth setting up a birthing pool in early labor, so that nobody's panicking at the last minute.
- Of course, extra hot water needs to be available in order to keep the water in the pool at an even, safe temperature.
- Putting folded towels on the floor of the pool is helpful because it gives laboring women a greater sense of security and stability.

Putting folded towels on the floor of the pool is helpful because it gives laboring women a greater sense of security.

- Stop a woman from entering a birthing pool (or even a bathtub) before she is 5cm dilated. If she enters the water early, the hot water is likely to *delay* her labor, not help it progress, which is obviously not at all helpful. Of course, this may well be the reason why some midwives have a poor view of birthing pools. Many midwives advise women to have a bath when they phone in to announce they're in labor, but this is not helpful, since the chances are the woman will be far less than 5cm dilated. As you probably know, if the woman is having contractions thick and fast (perhaps one minute apart), it's likely she's sufficiently dilated. (It's not even necessary to do a vaginal exam to confirm this—the very frequent, very strong contractions will be enough to provide you with this information.)

Stop a woman from entering a birthing pool (or even a bathtub) before she is 5cm dilated

- If you see a laboring woman leaning back, gently encourage her to lean forward. One way of doing this is to silently take both her hands and pull her towards you, so that she ends up leaning against the side of the pool, rather than propped up against it. Of course, the idea is to avoid compression of the vena cava and subsequent fetal distress.
- Silently monitor the water temperature because excessive heat can be dangerous for an unborn baby. (Of course, if you're working within a maternity unit, there will be guidelines on this.)
- If a birthing pool is not available, consider preparing a bath for your client, because this could be equally beneficial. Even if she only gets into the tub for a few moments, the hot water could have the same effect of hot compresses on her perineum and might decrease the risk of tearing.[84]

## Birthframe 80: Kathy Kleere—Appreciating water in birth

This account shows clearly how helpful water can be for a woman, particularly if she's nervous about the second stage, or if she wants to avoid using drugs.

Joshua was my second baby so I felt more at ease with my body in pregnancy and really loved being pregnant. I had not had any drugs with my first labor (too late) but had a very long second stage and that was a huge concern with the second labor looming. I had been to active birth/yoga classes during this pregnancy and felt much better prepared this time round. About a month before I was due I started reading and listening to stories of water births. When a woman from our yoga class had a water birth and came back with her baby I was convinced I wanted to try it. So I booked myself an appointment with the midwife in charge of the water birth unit and had it written in my notes. They only had one pool at the time so I was told if it was in use I wouldn't be able to use it. My luck was in! One week after my due date my water broke at home as I got out of bed after a nap. I called the hospital and they said to come in, so I also phoned and told my husband to come home. My mom-in-law was staying so as my first son slept in his crib, we went off to the hospital. The contractions started once we got there—about every four minutes—but I was able to breathe through them. We had to wait awhile for the room to be prepared but I was fine. About 40 minutes later I was examined and told I was 4-5 cm dilated! Wow, what a feeling that was—I was so proud of myself for getting that far already! My husband put on the TENS machine and they prepared the pool. About an hour or so later it was time to get in and I was a little scared that I would have no official pain relief once in the pool. But I didn't need it—the water was amazing. I could just move around and get into really good positions whenever I had a contraction.

A couple of hours later the midwife said she thought I was in transition as my breathing had changed. (Up until then she had just monitored the baby's heartbeat with a Doppler in the water.) At this stage I had all the classic signs of transition—shaking, crying, feeling like I couldn't do it—but my midwife and husband encouraged me all the way and Nikki (the midwife) told me to push whenever I felt ready. I remember being fearful, afraid it would take as long as last time but once I concentrated it all happened quite quickly—obviously being upright helped this time. Once Joshua was born Nikki slipped the cord over his head as it had been around his neck and as I sat back she handed him to me to bring him up to the surface. He didn't cry, just opened his eyes and looked around curiously. I felt like superwoman—amazed, empowered and overjoyed that birth could be such a wonderful experience! I got out for the third stage and had a small tear which didn't require suturing. Joshua was a very placid baby, but they say second babies usually are.

The whole birth experience has led me on a whole new path in my life... Training to become a prenatal teacher, I learned things about myself I didn't know and my journey continues every day. I have just qualified as a baby massage instructor and am now teaching new moms, as well as being a prenatal teacher. My teaching is a direct result of my wonderful empowering experience of birth.

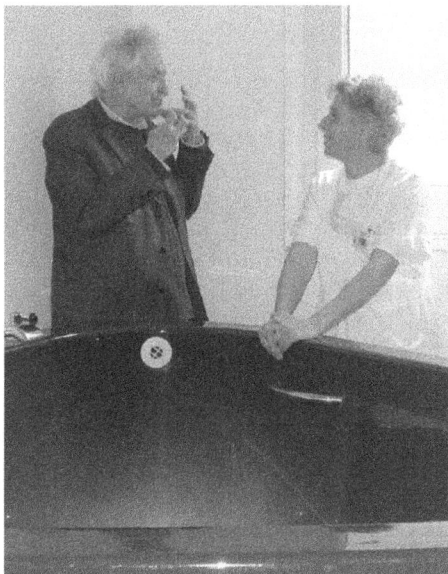

*Michel reminiscing with a midwife at the current birthing pool in Pithiviers, where he first introduced the use of wading pools in the 1970s, followed by specially-designed birthing pools*

*Michel catching a baby as it is born under water*

Birth underwater is possible but it should not be the goal

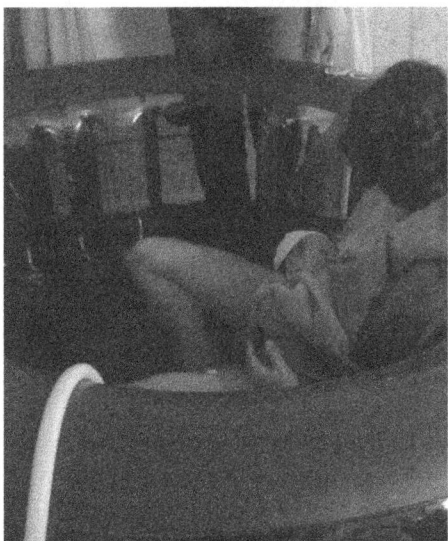

*Moments after a home water birth. Here the baby's having her first feed.*

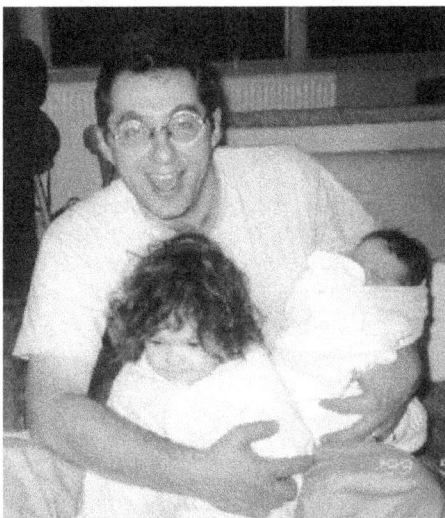

*Of course, water births can be fun for all the family too! Here the Jacoby family gets dry after Jennifer Jacoby's home water birth (see Birthframe 39)*

## Birthframe 81: Angela Horn—A water birth in detail

Here's another fairly typical water birth, with a very well-informed mother—Angela Horn, who set up the UK Homebirth website at www.homebirth.org.uk. Many women hand over responsibility for their pregnancy and birth to healthcare professionals but this account shows how important the woman's own role is in the whole process of having babies. Through her careful preparation she was able to maximize her chances of having a water birth, and the final experience was clearly and deservedly very rewarding. It's debatable whether or not her midwife's advice on pushing—or not pushing—was appropriate... some questions are not easy to answer. At least she did not have the discomfort of sutures to endure after the baby was born. Instead, she was free to enjoy her new baby, back as her normal self, in an undrugged state of mind, and free from any side-effects.

### Early labor

I awoke at around 6:15AM on January 6—my baby's due date, according to my own calculations. When I got up, a small trickle of amniotic fluid ran down my leg. It was watery and clear with white flecks of vernix—not what I expected. For some reason I thought amniotic fluid would be thick, so I spent about half an hour reading pregnancy books to check. I was having contractions every four to five minutes from the time I woke up.

I tried to go back to sleep, but found lying down very uncomfortable so I told Graham that he wouldn't be going to work that day, and got up. I tried to eat, thinking that I would need energy to sustain me through labor, but I really didn't feel like eating.

*I thought this was pre-labor and that it would probably go on for a day or so as this was my first baby*

I felt fine and the contractions were no real problem, still coming every four to five minutes or so. They were like the Braxton Hicks contractions I'd had throughout pregnancy, which had been becoming stronger recently. I thought this was pre-labor and that it would probably go on for a day or so as this was my first baby. From stories I'd heard of other natural labors for first babies it was not uncommon for contractions at five-minute intervals to indicate that there was still a long, long time to go, so I didn't think that this was anything to get too excited about. When a contraction came I leaned forward, standing up, rocked my hips and focused on breathing out slowly and calmly. I'd practiced this so often in Active Birth classes that it was second nature. I tried a few contractions sitting, leaning forward on a chair, but found this really uncomfortable. (I felt too restless and needed to be on the move.) So after that I took them standing up, and later kneeling or on all fours. At around 8:30AM I phoned the community midwife.

*I'd practiced this so often, it was second nature*

I wonder, in retrospect, if I had been in pre-labor for awhile. I had been getting very strong Braxton Hicks contractions for months, but in the last week or so they had got stronger, and I would often have to stop what I was doing because of them. The day before I went into labor (January 5), I did an hour-long aqua-aerobics class. I felt strong contractions whenever I jumped up and down, so I put plenty of energy into the exercises! The aqua-aerobics teacher was pointing at me accusingly, saying: "I can see you're having contractions in my pool! We don't want a water birth *here,* thank you very much!" I think the contractions had been coming every 20 minutes or so for awhile but since they only felt strong enough to interrupt my activities if I was doing something, like exercising or walking briskly, I just considered them to be practice for the real thing. Anyway, on January 6 it definitely was the real thing.

I used these early hours of labor to do all the jobs that I wouldn't want to do later—checking my pets and phoning friends. I felt contractions coming on long before they required my full concentration, so I could say "I need to go now" and finish the calls. The activity took my mind off the contractions and I didn't find them difficult to deal with. Some were more intense than others—for the tougher ones I dropped onto all fours and rocked my hips backwards and forwards. When the midwife phoned back I was feeling inexplicably emotional and didn't want to talk to her—I actually cried a little bit, but wasn't really distressed. I wasn't scared at all, just excited. I put the tears down to my hormones, and the emotional patch only lasted about 10 minutes.

## The midwife arrives

At around 9:30AM the midwife arrived. It was Norma, who I'd met a couple of times before. She was supportive, confident, friendly and very hands-off. We talked and she observed me for awhile. Then she offered to do an internal. I had discussed with the midwives previously that I would prefer not to have any internals in labor unless necessary and Norma made it clear that it was entirely up to me whether I wanted to have one or not, but reassured me that she would be gentle and would stop if I wanted, so I was happy for her to go ahead.

I had discussed with the midwives previously that I would prefer not to have any internals in la

I thought that it couldn't be established labor, because it wasn't difficult to cope with, so I was surprised to be told that my cervix was 4-5cm dilated, and that the baby was likely to be born in the early afternoon. (It was about 10AM at this point.) Norma could feel the bag of water bulging in front of the baby's head, so the fluid I'd seen earlier must have been a hindwater leak. As the baby's head descended it was preventing any more fluid from leaking away and the forewater was still cushioning its head. The head was 2/5 palpable, i.e. 3/5 engaged; before labor it had been 3/5 palpable for the past six weeks.

The head was 2/5 palpable, i.e. 3/5 engaged

## Labor can be fun!

I asked Graham to start assembling the birth pool, and found that I needed to devote more effort to dealing with the contractions. I put on some music, turned the lights down and turned on the lava lamp; the Christmas tree was still up, complete with lights, and the whole effect was lovely. I have very fond memories of grooving along to my favorite music, feeling perfectly normal in between contractions and rocking on all fours during them, singing along with the music. Graham needed some help with the pool, so I didn't get to 'groove' for very long.

Norma just chatted and watched me, checking the baby's heart rate and my blood pressure regularly. We were talking about animals, and Norma said "Well, you've seen animals give birth. You know what to do. Just do whatever feels natural to you."

Helping to sort out the pool was a great distraction. Graham and I chatted and worked and it could have been any normal day—except that every few minutes I had to excuse myself and deal with a contraction. Our three cats found the pool very interesting and stood on the side watching the water. One of them fell in! I wasn't worried about the hygiene aspect as the pool came complete with a non-chlorine water sterilizing kit and filter.

The contractions got closer and more intense, and by 11:30AM I definitely wanted the pool right away, so I got in while it was still filling. This was a good move, as I could direct the hose on my back and belly during contractions.

Music was playing all the time. Rock and pop songs with rousing choruses (turned up loud during contractions) which I could sing along to were my main form of pain relief. I didn't need Demerol—Jimi Hendrix and Abba were quite effective and had few side-effects! I am normally very self-conscious about singing when anyone can hear me because I've been told so often that I have an absolutely dreadful singing voice. However, I was completely uninhibited about singing when in labor and also about making noises later on.

## How the pool helped

The pool was a great help. Although it didn't take the pain away, it was warm and soothing and I certainly found it easier to cope with contractions *in* the pool than *out*. The support of the water made it very easy for me to switch quickly between positions—I had to be on all fours for contractions but between them I was floating, squatting, sitting or standing. Squatting positions were far easier to hold than on land. Sometimes I went completely under the water during or between contractions. It was very relaxing to feel surrounded by warm water and to feel quite alone in this peaceful place. The water helped me to focus on breathing out slowly too. Breathing out under water, or with just my mouth under the water, I could blow bubbles and watch and hear the breath, as opposed to just feeling it. It was a rhythmic process and a good distraction during contractions—duck under the water, breathe out slowly, surface and breathe in and then duck under again.

During some moderate contractions I tried wearing a snorkel and staying underwater for the whole contraction. This was quite fun at first, but having the mask on my face got a bit annoying later. Sometimes I kept my head above the water and sang or made other sounds as I breathed out. When things got really intense I hummed and vocalized in other ways, again as practiced in the Active Birth classes. When I could have cried out in pain, I hummed "Ommmmmm" during the out-breath. It's not that I attribute any special qualities to that particular word, but it was a convenient sound to make, and it helped a lot. I thought that crying out would not be helpful... It would make me more tense and might make those around me tense.

I put some Lavender and clary sage essential oils in the water. Regardless of whether they help labor, they smell nice and it was relaxing to be surrounded by fragrance rising from the surface of the warm pool.

## Water temperature

The water temperature was around 34 degrees centigrade for the first stage—any hotter and I think I'd have felt sick. During the second stage I would have liked it to be warmer, but I was too preoccupied to ask or to do anything about it. Next time I would try to remember.

Norma recorded the water temperature at intervals, and asked if I was comfortable. Characteristically, she was focusing upon what would make me feel good, rather than trying to tell me what temperature the pool should be. Don't let anyone try to tell you what temperature the pool should be; as long as you are not overheating during labor (which would make you very uncomfortable) and your baby is not hanging around in cold water after birth, all that matters is that the temperature is right for the mother. I got out of the pool a few times to go to the bathroom—I knew that pee was sterile so it was okay to go in the pool and, let's face it, a cat had already fallen in there so I could hardly get fussy about water hygiene! However, I also knew that it was good to move around as much as possible and that the process of climbing in and out of the pool involved major leg-raises and hip-hitching that would help the baby to move down. Contractions outside the water were far harder to cope with.

## The hard work starts..

As my labor progressed I felt sharp pains in my hips and the all-fours position became vital to take the pressure off my sacrum. Graham rubbed my back during contractions and kept me supplied with drinks of hot water with honey and lemon juice. I needed to have Graham there... I missed him when he left the room. He was very calm—we all were—and very involved in the labor. In fact, I had far more physical contact with Graham than I did with the midwife.

Norma seemed instinctively to know that I wanted to be touched as little as possible and that I did not need her to 'deliver' the baby but just to check that the baby and I were both well, while I got on with the job of giving birth.

During what was probably transition the contractions were very intense and close together and overwhelming, but I felt dreamy. I still felt the pain, yet seemed able to fit a lot of rest into the minute between contractions. My body's natural pain-managing mechanisms were working well. Although dreamy, I was still rational, still myself. I felt quite romantic during this 'dreamy' time and kept looking at Graham and thinking soppy things, like how much I loved him and how wonderful it was that I was having our baby.

Suddenly, I felt wide awake and perfectly lucid so I asked Norma if I was in transition. She said that normally people got very ratty in transition—I'd not been at all ratty so far. (In fact, I was probably more polite than I am usually.) Norma offered to do an internal if I wanted to find out but I thought that either I was or I wasn't in transition and an internal wouldn't change the fact, so why bother?

Anyway, it turned out that I must have been because, during a very hard contraction, I felt the water break—fluid gushed out under pressure, and Norma said the cervix was probably totally drawn up.

## Nearly there...

At one point I decided to get out of the pool and walk around to help the baby descend. Having told everyone that I was going to get out for awhile, I stood up, then a contraction came. My precise words were "Stuff this for a game of soldiers!" I went straight back in the water, and didn't get out again until the baby was born!

When I felt the urge to push, things became primal. Low, bellowing, moose-like noises came out as I pushed. It was like vomiting—huge convulsions over which I had minimal control.

Norma said that when the head crowned, I should push it out however felt natural, and bring the baby to the surface in my own way. Crowning seemed to be a long way off when she said this. I poked a finger inside and the head felt like a lump of spongy meat. I wasn't sure that it was a head at all but what else could it be? I remembered my prenatal class teacher saying that the head would not feel like a head at this stage because of molding, and was reassured.

The head crowned around 15 minutes later. The burning sensation was intense, and I thought "Aha! This is what they mean by the 'ring of fire'!" but it didn't matter because I knew the baby would be born very soon. I was kneeling upright with my knees wide apart, so gravity was able to do a lot to stop the head receding. I remember instinctively making small movements to give the baby more space to come out. At this stage I was very alert and was giving a running commentary on what the baby's head was doing because there was no way anyone outside the pool could see what was going on.

I realized that I could make voluntary pushes between contractions and that they really did move things down. I knew that voluntary pushing was A Bad Thing but, hey, it could still get results! "Norma, I can push in between contractions. Shall I?" "No!" came the reply. "There's no point in pushing in between contractions. Just wait for the contractions to do the work." Huh—no official sanction. I pushed anyway when I thought she wasn't looking.

With a huge push and a contraction the baby's head was born

With a huge push and a contraction the baby's head was born. I felt around his neck to check that the cord wasn't around it. The baby's body was firmly clamped inside— there was no way this one was going to just slither out. I waited for the next contraction. (It felt like at least 10 minutes but according to my notes it was only one minute.) And then I thought "Sod it!" and gave a huge shove anyway, but there was probably a contraction there too. The baby's shoulders, arms and chest were born but he was *still* stuck fast again at the tummy. (If I'd been on my back this might have been a case of shoulder dystocia, by the way.) I thought, "Stuff this. I'm not waiting for another contraction" so I grabbed hold of the baby under the arms, gave a big push and pulled the rest of him out.

I thought "Sod it!" and gave a shove anyway

I brought the baby to the surface. He gave a huge yell and turned bright red, and I was amazed to see that it was a boy—Lee. We had been convinced that it was a girl. Then I noticed the midwives saying things like "Look at the size of that!" and, ominously, directed at the baby: "I dread to think what you've done to that perineum". He was big and strong, very solid, covered in vernix and obviously vigorous.

I climbed out of the pool, holding him close to me, and sat on the bed next to Graham. The baby was covered in a towel, but with his skin bare next to mine. I offered him my breast. He wasn't very interested in sucking but he made an attempt. It was a great relief to sit down after nine hours on my feet and knees. I felt perfectly normal, if a little tired. I certainly didn't experience the post-delivery euphoria that many women have, even though it was a great birth in perfect surroundings. I felt very business-like about it and, being English, I just wanted a cup of tea! I cuddled Lee and comforted him—he didn't cry for long, just briefly at the initial shock of being born—and we waited for the cord to stop pulsating so that he had his natural quota of blood. After about 15 minutes there was no detectable pulse in the cord so Graham clamped and cut it. No one had touched our son apart from us for these first minutes of his life. No one handled him roughly, stuck tubes in his nose or mouth to suction them, put him down on cold surfaces, or subjected him to any 'standard procedures'. They weren't necessary. All he needed was a cuddle and the warmth of his mother's body.

## The third stage

I had a natural third stage but the placenta took an hour to turn up. Norma wasn't worried about the time as there was just one large gush of blood and clots, then very little blood loss. I think the placenta was sitting there, waiting for me to push it out. I could have done with more direction here but both midwives were busy with the baby and paperwork. I did feel slightly abandoned but I guess I should have asked for help. Lee was in excellent shape and was huge: 4.25kg, 9lb 6oz.

Lee was in excellent shape and was huge: 4.25kg (9lb 6oz)

I had a second-degree tear straight down the midline,
which didn't need to be sutured as it wasn't bleeding

I had a second-degree tear straight down the midline, which didn't need to be sutured as it wasn't bleeding. Perhaps if I'd resisted that voluntary pushing I could have avoided the tear. Norma commented that being in the pool would have helped to minimize tearing. (Like having a hot compress *in situ*, it provides warm counter-pressure against the perineum). I lost at least 500ml (just over 1 pint) of blood. That was what made it to the measuring jug but there was plenty that landed elsewhere. I was healthy, my blood pressure was fine, I didn't feel faint or even unwell and blood tests three days later showed that I wasn't anemic.

I didn't feel faint or even unwell and blood tests
three days later showed that I wasn't anemic

When everything was over I felt absolutely fine—tired, but not exhausted, and very happy. That night the three of us slept together, Lee curled up against me in the most secure and natural way for a baby to sleep.

## Conclusions

I would do it all again just the same way... tomorrow! It took nine hours to deliver Lee, including a second stage of about 1 hour 15 minutes. It might seem that I was lucky but it took a full 13 months of careful preparation—exercise and research—to help luck produce the goods! Of course, there are some eventualities that you cannot prepare for but there are others which you *can* do something about. Perhaps preparing for those things I *could* do something about stopped me worrying about the factors I *couldn't* control.

If there had been some complication and transfer to hospital had become necessary, it would have helped me to know that I'd done just about everything I could to have a normal labor and birth and that if intervention was necessary in spite of all that preparation and my best efforts, then I would be very glad that it was available.

The midwifery care I received was excellent. Throughout my pregnancy the community midwives I saw were all supportive of homebirth and seemed confident and well informed on the subject.

During the labor, although I found that I did not want to be fussed over, it gave me confidence to see that Norma was so calm and relaxed. The impression I got from her throughout was that she was perfectly sure that this labor was going to progress smoothly and I felt that if she was so confident, with all her experience, then I certainly didn't have anything to worry about. Having said that, if there had been a problem then I have no doubt that she would have been well prepared to deal with it—and I would have trusted her and accepted her recommendations. Thank you, Norma!

Water is fantastic pain relief. Stepping into the pool felt like a miracle. The pain was still there but now it was manageable.

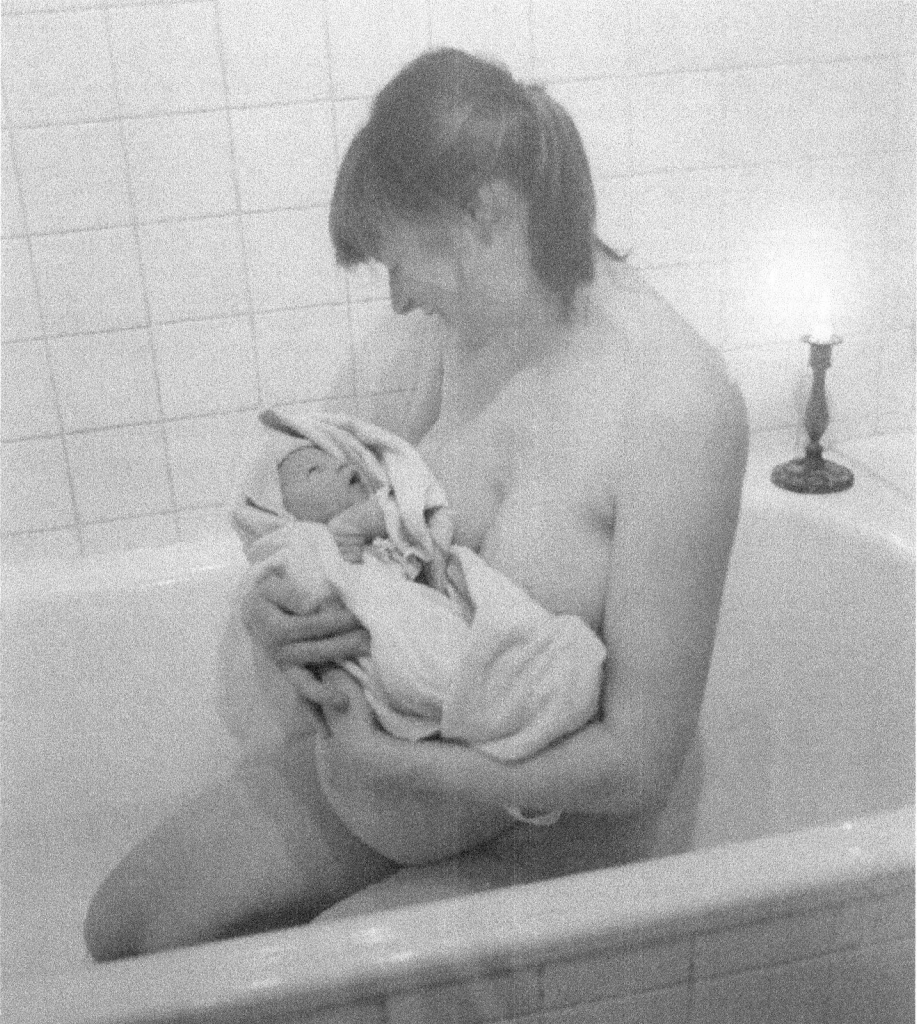

*Instead of a pool, some women are very happy to just use a bathtub, either at home or in the hospital. With bathtubs the same principles and 'rules' apply as with water birth... which may make you reconsider advising women in early labor to have a bath.*

I would do it all again just the same way... tomorrow!

# Prolonged labor / Failure to progress

## What if labor goes on and on and on?

Here's some more from Michel on how he uses water to actually *test* whether labor is really obstructed or not.

When a woman appears to be having difficulties, a birthing pool can actually be used to check whether or not a cesarean is necessary. It is particularly useful because a decision can be reached before fetal distress has occurred. I call this the 'birthing pool test'.

Carrying out this test is very simple: when the woman is in hard labor, she is immersed in water at body temperature for approximately 90 minutes. Usually, within this time period, something spectacular happens. (The period of time is approximate, of course.) By spectacular I mean, for example, if she enters the bathtub at 5cm she reaches full dilation, or if she enters at 3cm she reaches 7cm. If after an hour and a half in water nothing like that happens, if you can see no difference in dilation, it means something is wrong, that there is some kind of obstacle. So the best course of action is to do a cesarean section—waiting is not at all helpful in this case.

## Birthframe 82: Michel Odent—Water to speed things up

Michel told me two interesting stories to illustrate the 'birthing pool test'.

A woman was on a boat and I was with Liliana, the doula. It was obviously a big baby and we first went there early in the morning.

Around 2 or 3 o'clock in the afternoon she was in hard labor but just stayed at 4cm dilation. I suggested she get in the birthing pool. (Amazingly enough, they had rented a birthing pool for the boat!) She then spent two hours in the pool but after these two hours she was still only 4cm dilated. I said, "You know, you need to go to the hospital," because I was convinced she should have a cesarean section. I told Liliana to go with her because I could no longer be responsible for the birth.

At the hospital, the obstetrician put the woman on a pitocin IV and ruptured her membranes. However, in the end she had a cesarean.

Knowing what I knew about her labor I would have done a cesarean immediately, when she arrived at the hospital at around 7PM. What happened was that they tried everything and finally she had a cesarean section the day after, in the morning. The baby was okay but it might have been better for both the baby and the woman if the cesarean had been done earlier.

In the second case I was not there myself—it was Liliana who told me the story. It was at a hospital in London. The woman arrived in hard labor, even though she was only a couple of centimeters dilated. Eventually, she got into the birthing pool and no more progress was made so Liliana told them, "I think she needs a cesarean section." She probably mentioned my name but they didn't believe her in any case. Then, a senior obstetrician came along and she said it was a brow presentation—which is completely incompatible with the vaginal route. So without having any other way of diagnosing a problem, apart from through 'the birthing pool test', Liliana said it should be cesarean section. The senior obstetrician gave a reason for the diagnosis... but the result was still a cesarean section.

## Birthframe 83: Sylvie Donna—A stop-start labor for five days

At times, with or without the use of a birthing pool, it really does seem clear that labor is obstructed. Sometimes, as the following account suggests, there may well be psychological reasons for this. In these cases of course it's possible that talking to the woman in labor might actually be very helpful. (It's the exception to the rule, if you like.) Have you read the many accounts in Ina May Gaskin's book *Spiritual Midwifery* (Book Publishing Company 2002—originally published in 1975) where talking to the mom-to-be or getting her to laugh was often the solution? In the following account, as in some of the births Ina May describes in her book, the solution would perhaps have been to get the woman talking to her partner... although in the following case it was probably far too late by the time the woman was actually in labor.

While I was researching this book I got chatting to an old friend, who said she thought natural birth was all very well as long as it didn't last for five days, as had happened in her case! Five nights actually.

Her story was initially strange and perplexing: when she was 36 weeks' pregnant, for five nights in a row, she experienced strong contractions but each morning, as the sun rose, her contractions came to a complete halt. Eventually, her labor was augmented and her baby pulled out with forceps.

Since I know this woman quite well, I asked her if she'd mind if I asked some questions. Very quickly, I ventured to suggest why she might have had this stop-start labor... My friend immediately agreed with my interpretation. First of all, it was an unplanned pregnancy in a very well-established relationship where it had been agreed there would be no children. Her partner had reacted badly to the news of the pregnancy and had made no secret of his reluctance to go ahead with it. Apparently, even while she was in labor, this woman's partner had complained about her having a baby! Secondly, this woman had conceived just three months after the death of her mother, who she had been very close to; throughout her labor she said she had longed to have her mother's support. With so much emotional baggage, it's hardly surprising this woman couldn't relax into labor. And given her partner's strong negative feelings, it was as if she felt she needed to have the baby in secret, in the dead of night, as it were. But on some level of her mind, she was not even allowing herself to have this baby.

This story does have a happy ending. The child is now a healthy and happy 8-year-old. After being looked after by her mother for the first year, her father became her main carer because this is what worked out best financially—although her mother continued to nurse her until she was 3½ years old. Despite the enormous change this new baby made to her parents' lifestyle, she is now much loved by both parents and they all make a wonderful family.

# Pause to reflect...
## ...about conflict/anxiety prenatally and intrapartum

1 What would you do if you suspected conflict between a woman and her partner?

2 What methods could you use during prenatal check-ups or classes to draw out potential psychological problems?

3 How could such problems be resolved?

4 What would you do to support a woman who seems nervous during labor?

5 To what extent do you try to establish a rapport with a woman's partner?

6 What do you do when a partner seems especially nervous about a birth?

7 Have you ever encountered a situation where another person in the birthing room seems to be upsetting the laboring woman?

8 How do mothers of laboring women behave, in your experience?

9 Could a mother's experience affect her daughter's experience of birth?

10 How can you best deal with problems between friends and family members... or do you think it's none of your business?

Do you have any other thoughts or questions?

For material relevant to these questions, see Birthframes 3, 5, 6, 7, 15, 16, 17, 18, 19, 20, 21, 24, 25, 26, 27, 28, 30, 31, 32, 33, 34, 35, 37, 39, 42, 43, 44, 45, 46, 48, 49, 54, 57, 58, 59, 60, 61, 62, 63, 65, 67, 68, 69, 70, 73, 75, 78, 80, 82, 83, 85, 86, 87, 95, 96, 97 and 98.

## Birthframe 84: Elaine Batchelor—Longer than usual, but fine

There are actually times when bodies simply function a little more slowly than we would ideally like. How well do you personally cope with women who are at the edges of what is considered 'within the normal range'? How do you reach decisions? How much are you motivated by fear or by the prospect of litigation?

I was a midwife before I had any children and my first baby, Sophie, was born at home after 16 hours of normal but painful labor—there had been 24 hours of pre-labor, which I didn't count. I had the help of a supportive husband and a fab midwife.

Amy Rose is my middle daughter and she is 13 years old now. Being a second baby I thought that Amy would, of course, tumble out of me after three hours of manageable labor, but I was wrong.

*Being a second baby I thought that Amy would tumble out of me after three hours of manageable labor, but I was wrong*

I was three days past my due date and on my way to see the osteopath for treatment for lower back pain. I had a very strong contraction and then another, five minutes later, but that was all. On the osteopath's table the baby was writhing about like a bag of puppies and the poor guy was a bit taken aback at the sight of it. I didn't mention the contractions I had had as this may have made him even more nervous about treating me.

I got home and prepared dinner as usual and then Peter came home at 7:00PM as usual. At around 9 o'clock I had a show and contractions started every five minutes immediately—similar to my first labor. I had no idea what position my baby was in and it was before the days when I was so keen on optimal fetal positioning so I hadn't bothered to ask. By 11:00PM the contractions were very painful and every four minutes. (I thought I must be at least 8cm by now) Oh, how exciting—the baby would be born soon!

I woke Peter and said, "Peter, labor has started." He said, "Has the water broken?" I said, "No." He said, "Wake me when it does." At around midnight I was fed up on my own and I could not lie down so I made him get up and keep me company. We decided to call the midwife who came round and got things ready, etc and as there was no sign of the baby by 1:00AM I agreed to being examined. Oh joy... I was 2cm dilated. We plodded on through the night as I could not sleep. The midwife went away (she only lived round the corner) and Peter went back to bed. At around 8:00AM we all met up again, plus Sophie, aged 3½, and Grandma, aged 70—someone who was going to look after Sophie. I also tried to call a friend who was going to come and care for all of us as our doula, but her telephone was out of order.

During the day I ate, but vomited everything back. I walked and swayed and knelt and was on all fours and used the shower, which was horrible, by the way. At around lunch time I was pronounced 5cm dilated. My midwife suggested she break the water to speed things up and as I was exhausted I agreed. As it was, it was so hard to do that she never knew whether they went or not and no fluid actually came out at all—very odd! Because I had been so sure of my super quick and easy labor I had not thought it necessary to hire a TENS machine, or pool, or anything at all to help me in labor.

Not much happened for awhile after the water sac had been broken. Sophie and Grandma went off to the kindergarten Christmas party and came back with various little glittery things they'd made. They were quite sad not to find any babies at home, just mommy slogging it out with contractions every five minutes still.

The doctor called in. I didn't want to see him so requested he be kept outside the door. (You can do things like that in your own home.) Of course, this labor was going on much longer than usual for a second baby and in hospital they would have been pressurizing me to have a pitocin IV to speed it all up but my midwife was from the days when all births were at home unless it was life or death and she carried on supporting me with her quiet confidence, knowing that it would be fine.

*My midwife carried on supporting me with her quiet confidence, knowing it would be fine*

At one point at around 3:00PM I was in such agony that I secretly thought of going to the hospital for an epidural. But I was pronounced 8cm dilated, which made it all seem much better. Sophie came and went as she wished and was not at all disturbed by seeing or hearing me in labor, as opposed to Grandma who was finding it all quite hard going. Sophie came along and put a tiny Band-Aid on my belly to see if it would help. She also drew lots of pictures, lying on the bedroom floor.

At 6:00PM I was fully dilated and had started pushing spontaneously. 30 minutes later, just after Sophie had popped up onto the bed next to me, out came Amy Rose. The second stage was a doddle! Things I had not prepared Sophie for were that Amy was covered in vernix, which Sophie found most odd, and that the baby had no clothes on, at which she was very surprised. She also spent quite a bit of time calling her baby Jesus as she was born at Christmas and Sophie had just been to a Nativity that day.

After Amy was born I had more agony in the form of raging afterpains. I was in agony all over again! I was having a physiological third stage so should have been encouraged to change position and to push a bit with each afterpain but I wasn't, so it was taking awhile. Again, I didn't have the knowledge then that I have now. The doctor came along to check the baby over and was anxious to see the placenta out before he went off to a Christmas party. (I didn't know he had his wife waiting outside in the car as well.) He stepped in and started pushing and pulling on various bits of me and my cord until I could not bear it anymore so I shouted at him to "get off me and don't touch me again" and I pulled his hand off my abdomen. He backed off and the midwife encouraged me to push and the placenta came out just fine. In the end I think he prolonged the third stage with his impatience. After that he went away and everything settled down again. I had a bath and we all had dinner and it was all fine.

Why did it take so long? Because my baby was in a posterior position and I didn't know. Things were very different for the birth of No. 3! I had optimal fetal positioning, the pool, the doula and very direct instructions from everyone about what to do in third stage. [Optimal fetal positioning is basically about making sure you don't lean back and relax while you're pregnant—at any time! While you're in bed you need to lie on your left side, not on your back. Leaning forward helps the baby get into an anterior position, which means a shorter and less painful labor.] I had also changed to a new doctor as I didn't want a repeat of the brutal treatment I'd had with Amy's third stage.

Sophie was shocked that the midwife was clearly not going to take the baby away with her at any of her postpartum visits, but at age 16 years she has learned to live with it (just).

Questions none of us feel able to answer are: When did my labor start? Whatever happened to my membranes and the presumed fluid round my baby? There was never any sign of it before, during or after the birth! In hospital the presumption would have been that they went the day before (and that I had not noticed all that fluid pouring out of my vagina as I went about my business) and that my baby and I would need antibiotics (probably intravenous) to stop us dying of an overwhelming infection. But we were fine.

All in all, thank goodness I was at home with a midwife who had the confidence to see an unusual but normal labor. It saved us from a lot of medical intervention which could have had an adverse effect on the condition of me and my baby.

## Birthframe 85: Nuala OSullivan—Intervention at birth

Here's an account from Nuala OSullivan, the woman whose beautiful birthing photos you saw on pp 30, 37, 38 and 39, which portray the delicate balance needed between non-disturbance and vigilance. Although this was a difficult posterior labor, Nuala moved through the difficult moments in her safe but undisturbed environment and even managed to continue through to a face-to-pubis birth. (I know from my own first birth experience, just how difficult that can be.) Most importantly, although Michel Odent had been supremely focused on not disturbing Nuala while she was laboring and giving birth, he was clearly very much there when she needed him—to get her apparently lifeless baby breathing...

I booked Michel Odent for my second daughter's birth, having complete confidence in him and his approach to birth. I had first met Michel when I was being a birthing partner for a friend and watched her beautiful daughter emerging like a lithe Excalibur from the depths of the birthing pool. Michel had then helped me to deliver Ciara, respecting my laboring foibles. He not only accepted that I would soak anyone who came near me but also reassured the local beat police that in fact a child was being born and no one was being murdered despite the scary screams! His gentle consistency and faith in the laboring mother inspired confidence and allowed me to deliver as I had not believed I could.

Ciara had been 17 days late so when my due date of New Year's came and went, and all the women in my yoga class delivered before me, I wasn't unusually perturbed. On Thursday, January 14 I made the journey to kindergarten with my toddler dancing forwards and back to keep pace with my slow gait. I was expecting that the 17 hours of Ciara's birth would be divided by three for this baby. Contractions were steady but only 10 minutes apart so I still felt safe and anyway knew people in the shops and houses all the way to kindergarten . Her key worker and other parents were all excited for me and still maintained their enthusiasm when a day later we had only progressed to five minutes apart as we made the journey even more slowly.

On Friday afternoon we walked to Coram's Fields with the children and I sat on a seesaw with another parent to try to encourage contractions along. By the evening labor had become established.

I bathed Ciara walking around our tiny bathroom as she sang to me. Once she was asleep I called Michel, who came to see me, reminding me to call at any time.

My dear friend, Fee, arrived and sat with me. With night and our relaxed energy there seemed to be a lull and, although they never stopped, the contractions didn't seem as powerful as they had been. All night we sat up awake but contractions weren't developing. We looked at the birthing space with the empty birthing pool dominating my bedroom. Located over a cement archway the bedroom was deemed the best room for taking the weight of a filled pool. I had scented oils, the same ones I'd used for Ciara and candles at the ready—remembering how light sensitive I'd been the last time and how I'd craved the cocoon of my room. The freezer had frozen lemon ice cubes for sipping later and there was plenty of cooled water in the fridge.

I was getting worried by the slow development of contractions but by 6AM on Saturday contractions had re-established themselves more earnestly. Sally arrived early on Saturday followed by Danuta, my homeopath, and Jill Furmanovsky who had taken such precious pictures of Ciara's birth. Michel dropped in on his way for a swim, untroubled by what felt like established labor to me. I felt partially put out that my 'serious' labor didn't warrant any further attention and partially reassured that he was off for a swim, so delivery wasn't imminent. In between contractions I dressed Ciara and she went off for the day with friends. The day passed in a blur. I liked the sounds of the women around me as they talked, prepared meals and got on with other things somewhere on the periphery of my consciousness. I did not engage well with the group by this stage, didn't want people near me. All my social senses dulled, my need for attention became muted. Jill came and gave me a massage at the base of my spine as I had been rocking and could not get comfortable. It was welcome and comforting, and her gentle energy was calming. Mostly, I felt the need to be alone—to have them all somewhere near me, but as background rather than with me.

Michel seemed to materialize as stages progressed and then to fade into the background again. Fee came and sat quietly in the room, when she wasn't making hot drinks for everyone. Sally cared for Ciara and timed contractions from the kitchen, attuned to my sounds. Danuta checked on me and monitored changes in mood and energy but only sporadically as Michel was keen for me to be left in peace. He instructed that no one should make eye contact with me or disturb me unnecessarily. His protection of the birthing room was useful to me as I took myself out to deal with the contractions. I became acutely sensitive not only to light but even to whispering sounds from the other room. Nothing was expected of anyone, so each found their own role.

I felt sick. The waves of nausea were welcome as they meant that my body was starting to work and would bring my baby to me. Contractions escalated and I was aware of how tired I had become after two sleepless nights so I began using self-hypnosis. As the contraction began I took myself out and when it released I came back in again. I begged to go into the pool and was heartened when I saw them filling it. It paced me through the next few contractions and gave me something new to focus on.

The light faded and Ciara arrived home. Sally bathed her and I kept going in and out to change her story tape over—I was driven by an odd duty to be there for Ciara, even whilst laboring.

Because of Ciara, I think, there was no screaming this time and the contractions waned when I shifted attention from them to her. Ciara's birth had been loudly vociferous. As a trained soprano I had released each contraction with treble notes, as round and powerful as the labor itself. My neighbors knew I was laboring and were unconcerned but a passing pair of police officers called in to ensure that no murderous assault was in progress. Michel had protected me from prying eyes and the presence of extraneous visitors. Although the police wanted to see me, to set their fears to rest, Michel reassured them that all was as it should be and that I was having a baby. They returned the next day to ascertain that a baby really had been born with the male officer declaring that the sound had been 'bloodcurdling' and the female officer resolving never to have a baby after what she'd heard! Yet Ciara's birth had been beautiful to me. Sally lay down with Ciara, and I could hear her telling Ciara about the night she was born and answering Ciara's questions until they both drifted off to sleep.

The landscape of pain established. If I had been told two hours or four hours more it would have helped me to put a shape and sense to it but there was no way of predicting. I was 5cm and in the pool. Contractions took me out and when they abated I floated. I kept telling myself I could do 'just that much' again. I drank and sipped ice cubes but couldn't communicate with anyone. I was glad they were there somewhere nearby, but not near me. I tried counting, moving, dancing, going in and out of the water until I was too tired to move from the warm water.

Michel came and monitored me and told me the baby was fine and coping well. That was all I needed to hear. I could keep going if the baby was coping, so this time I stood still to allow monitoring and didn't drench anyone. I could feel my leg wriggling, ready to move away. I found the enforced sedentary pose intrusive but I brought all my reasoning to bear on keeping still for long enough for him to take a good reading. Then he moved away again, quietly leaving me to my own space, the journey of my baby and me, unfettered.

Contractions kept challenging me. Just as I got used to one level a new one opened up, demanding my full concentration. Sounds became excruciatingly magnified as all my senses heightened. I shouted to them in the kitchen to be quiet even though my own sounds were louder and they had only been whispering. Still, all complied with the unreasonable request.

My water exploded like a water bomb in the pool. Sudden, shocking. Danuta was topping up the pool and took the opportunity to ask me how I felt... I told her: "Everything irritates me". "Nox Vomica," she muttered, and went off to find a remedy. I heard the book pages flipping from the kitchen as she studied. It was all too noisy. The pages were like flapping sails to my ears.

I heard the book pages flipping from the kitchen
as she studied. It was all way too noisy

Within an hour of taking the remedy I went from 5cm to 10cm dilation and delivered. The last hour I found exhausting. The baby was aligned along my spine and not my front as Ciara had been. The pool no longer comforted me and I wanted to get out. I felt hot, cold and unable to find the center of the contraction anymore. My drive towards the birth and meeting my new baby wavered and I felt emotional and hesitant for the first time in over nine months. My 43-week pregnancy was drawing to a close. From nowhere I started to panic. I believed that I would not be able to bring my baby into the world. Then I felt the baby wasn't capable of going on this arduous journey.

Even as I crossed the bridge of transition I knew that if I could only get through it I would be able to deliver successfully.

Pushing started then and confidence returned. Here was a part I understood. I had a role and something to do with each contraction. I felt myself organizing my body's responses and planning how I would greet the new contraction. I got out of the pool thinking that I needed to open my bowels and that once I had I could have my baby. Once in the bathroom I realized that the sensation was confused because my baby was on the base of my spine and was in fact crowning.

I called out to them: "She's crowning!" As I staggered back to the bedroom, I noted to myself that even I had started to adopt Ciara's 'little sister' notions. The different voices passing on what I'd said didn't irritate me this time

Michel materialized, allowing me to find my position and my own connection with the contraction. I leant my hands against the wall and pushed. A head emerged with the cord round its neck and an arm up beside it, making the top centile head even wider. Michel deftly slipped the cord over the head. Another contraction and the rest of the baby slithered out, caught by Michel as I collapsed in a heap.

Tired out but terrified.

Because there was no sound. No crying.

A little grey daughter with no muscle tone and no sound.

3.6kg, 59cm and inert.

Michel was cradling my baby and sucking out her nose and throat. I splashed water from the pool to baptize her, desperate to do something for her.

Then I heard the indignant sound of Isolda's own voice. Michel rewarded me by handing me my wailing daughter. Perfect.

Sally took her whilst the placenta was delivered. Despite having size 4 feet (related to the pelvis size?)[85] and an 8lb (3.6kg) baby I didn't need any sutures! Everyone came around me to admire and meet Isolda Lily, each loving face welcome to me now. I bathed with her and then Ciara woke to meet her 'little sister'.

Ciara is 17 and taking her high school diploma now, hoping to study for a degree in Midwifery in September. Isolda is 14 and 5' 6", preparing for school exams and aiming to be either CEO of ICI or a runway model—whichever happens first! In the meantime, she is curiously delighted with how she has turned out... which is refreshing for a teenager.

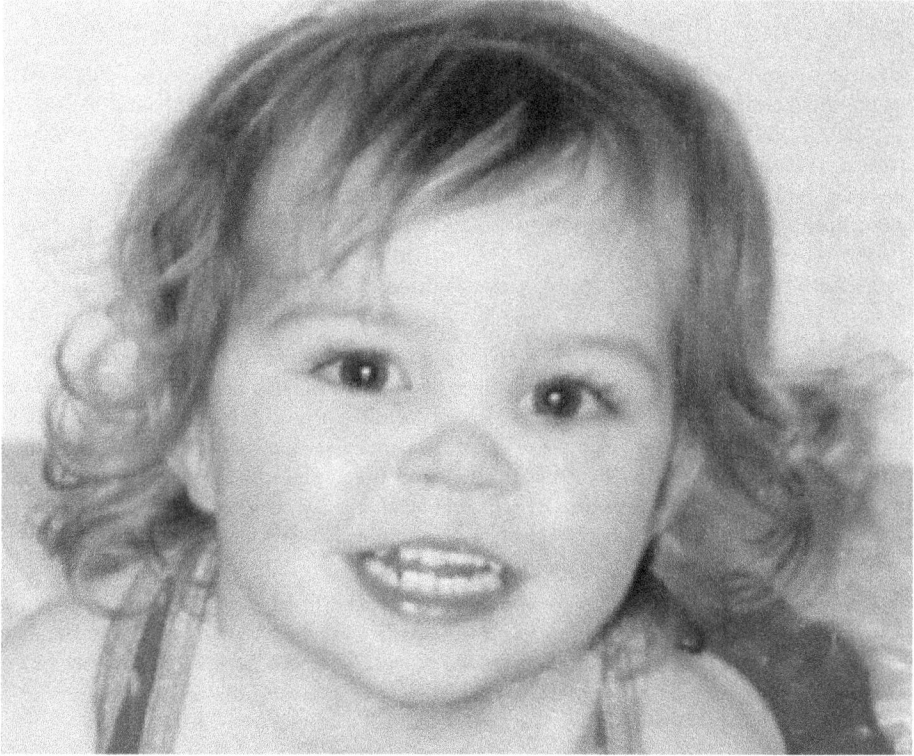

*Nuala's 'baby', Isolde, aged 14 months, enjoying a dip in the swimming pool*

## What about doulas?

While some women, like Nuala OSullivan, may want a lot of people around them—even if they want them to stay out of the way!—other women may choose to have a qualified doula with them in labor. What is this likely to mean for other caregivers, especially midwives, obstetricians and family doctors?[86]

## Questions for reflection...

### Assuming you're not a doula yourself:

1  What first-hand experience do you have of doulas?
2  Do you feel threatened by the idea of having a doula around?
3  What do you expect a doula to do... or not do?
4  Do you anticipate any problems in working with a doula?
5  Does your role as a caregiver change if there's a doula present at a birth?
6  In which ways is a doula different from any other birth attendant (e.g. a friend, mother or the father of the baby?)

## Birthframe 86: Sarah-Jane Forder—Doula acting as guard

In case you haven't had much contact with doulas, here are two accounts which give some insight into how a doula might behave and the role she might adopt. Of course, in this case the doula in question is Liliana Lammers, who we met in Birthframes 12 and 21. (This account was originally published in *LLL GB News*.) Note that other doulas may behave a little differently, because of guidelines provided by DONA and CAPPA (the US doula organizations), their personal view of the kind of care they need to provide, or the protocols they work within. Nevertheless, this approach does at least ensure non-disturbance!

I first heard the word 'doula' when about six months pregnant. Like most first-time mothers, I was hungrily searching for clues about childbirth in books and magazines— what would it really be like? Doulas, I was told by the childbirth expert, Sheila Kitzinger, are women who help other, less experienced women through birth, mainly by providing emotional and physical support and information. Research has shown, Kitzinger said, that the presence of a knowledgeable doula reduces the need for pain-relieving drugs, shortens labor, makes the birth an easier and happier experience, and results in fewer babies needing intensive care.

Wanting as I did a drug-free, non-intervention birth, it all sounded good to me, but I thought no more of it until a couple of weeks before my baby was due. The baby's head had not yet engaged and this, I was led to believe by the London hospital where I was booked to give birth, might be problematic. A local acquaintance put me in touch with Liliana, a highly experienced doula who lived only five minutes away.

A few days later I was sitting in her kitchen eating homemade soup. At the time, I don't think either of us knew that she would actually be present at the birth but she gave me reassurance. Liliana told me about her own experience of giving birth to four children and, above all, advised me to have faith in my body; simply to let it do what it was programmed to do, leaving the intellect well out of it.

I had no real concept of the wisdom of these words until about two weeks later when I went into labor. It was then that all the carefully laid plans for my husband Chris's involvement in the birth—offering lower-back massage, quiet encouragement and the sort of 'room service' you'd only ever expect in a five-star hotel—were summarily discarded. Once the contractions started forcefully to take over my body, I had no wish to communicate with anyone; I wanted to focus unreservedly on the baby.

Liliana had said that she would come over once labor was established, that she'd be able to help us judge the 'right' time to go into the hospital—I wanted to leave it until the last possible moment. When she turned up, just before midnight, I was on my hands and knees in the candlelit living room and contractions were coming quite strongly every three minutes. Liliana said hello and then almost immediately left the room. This was how she remained throughout the birth; a dreamlike comforting presence which I was vaguely aware of from time to time, but there was never any direct contact or intrusion. About an hour after her arrival, Liliana warned Chris that, if the baby was to be born in the hospital as planned, we should leave at once. She then generously offered to accompany us.

In many ways, Liliana's role—both at home and at the hospital—seemed to be that of a guard; discreetly, firmly she kept people (the midwife, Chris—the only others ever present in the room) away from me. [Research does in fact show that the sheer presence of a doula will tend to discourage attention from caregivers.[87] Nevertheless, neither DONA nor CAPPA recommend that doulas behave this way.] She was sympathetic too to the feelings of Chris, who, far from participating in the birth of his child, as laid out in the birth plan, was snarled at every time he came near. Towards the end of the labor, by which time I had retreated to crouch in the total darkness of the bathroom, Liliana stood sentinel at the door, reassuring Chris that I was perfectly fine. I knew I was fine, she knew I was fine; for him it must have been perfect agony.

My daughter Bea was born in a birthing pool approximately seven hours after labor had started. She was still encased in the membrane or caul, so her appearance as she slithered out between my legs was awesome and ghostly. I held her and we looked at one another for a long moment, then she rooted immediately for the breast. While I delivered the afterbirth, squatting on a table in the delivery room, Liliana rocked and sang to Bea in the half-light, welcoming her to the world.

I couldn't have wished for a better birth for me and my daughter, except had I known then what I know now I probably would have opted for a home birth. I had wanted no interventions; I had none. I had wanted no drugs; I had none. I had wanted a gentle, natural birth for my child; and this I believe I achieved (though Bea is really the only person who can vouch for this). And all this in a typical hospital setting.

The only other thing I'd do differently now would be to make quite sure that, in the days and weeks after the birth, I had a doula at hand to mother me. A doula's role is often described as 'mothering the mother'. While I certainly didn't want mothering during the birth, I could have done with it after I took my baby home. A doula will come in to see a new mother for a couple of hours a day up to six weeks after the birth, offering support at a time when many women are at their most vulnerable. If I were ever President for a day, I'd make this kind of essential care a provision of state.

## Birthframe 87: Natalie Meddings—Doula giving confidence

This account gives even more information about how a doula can best support a woman in labor.

It was an 11th-hour decision to have a doula at the birth of my second child. For eight months, my plan was to go to the local birth center, as I had for my first. And then as if by magic, a new trust arrived in me. One that told me I could manage at home. It really was a kind of magic. My head didn't convince me, or even my heart. Just as with my first pregnancy, an imaginary hand appeared—deep-down instinct I suppose—and I took hold of it.

*For eight months my plan was to go the birth center...*
*then, as if by magic, a new trust arrived in me*

As it turned out, that hand wasn't so imaginary. Two months before my due date, I attended a doula course with Liliana Lammers and Michel Odent. I found their ideas on birth exciting—at times breathtakingly so. I was especially fascinated by their belief in total privacy as being the key to a smooth birth and when Liliana helped me set up a home birth for myself, I knew she was the one I wanted by my side. As it turned out, she offered me more than that. This wonderfully calm and centered doula was more about and behind me, than beside me. Unseen, silent—but absolutely there.

I was over two weeks late and pressure was mounting to get things going. Forget induction. Liliana had convinced me that even gentler nudges, like stripping the membranes or reflexology were to miss the point. It must be the baby that gives the cue, she explained. "If the baby is ready, then the birth will go well." She urged me to feel for myself if everything was okay. Assured me that as the mother, I would know if something was wrong. I'd hang up the phone and feel a fresh energy. Something sure and strong and safe, guiding me. What Liliana was leading me to was my own instinct.

Finally, at 5AM on a dark November morning, I felt the first twinge. By 9AM, labor was really established and my husband Danny called Liliana. In true style, there was no urgency, no panic. She said she'd see to a couple of things, then cycle over mid-morning. Her ease was contagious. I did some cleaning, made some breakfast—calmly absorbing that most unabsorbable of notions. That at some point that day, I'd have my baby in my arms.

As soon as Liliana arrived, around midday, she made herself scarce—practicing absolutely what she and Michel preached. I remember wanting to offer her a hot drink, to make her comfortable, to talk—but she just shushed me and I closed my eyes. As I moved from room to room, from kneeling to standing and back again, I looked like someone alone. But Liliana was there all right. I could feel her unmistakable energy beaming through the walls. Could feel her listening—keeping an eye.

What I wasn't was being watched. I was totally private—and right inside myself as a result. Danny had made himself scarce, the house was silent, the room I'd somehow guided myself to, small and dark, and the world just fell away. I felt absolutely safe, wholly secure in my surroundings and with that in place, the chemicals just cued themselves up. I could almost feel the hormones firing, ratcheting up the pace—and my labor's progress.

A couple of times I asked Liliana a question: "The contractions pick up when I walk around, so should I keep walking?" She didn't reply. With a shrug, she simply handed the process back to me. Coaxed me back to myself.

After an hour, the pain accelerated and I needed a hand to hold. It was there. A silent squeeze. Liliana gave no encouragement, no commentary. Words would interrupt me, bring me back to thought when what I needed was this flow. My eyes were closed, but her support was surrounding. I could feel her focus on me—saw through half-shut eyes, that her own were shut too. It was as if she was moving through each contraction with me. So much birth assistance seems to tell the woman to turn away from her pain. But Liliana did the opposite. She helped me move to its center.

It suddenly felt right to get into the pool. As with every stage, she got me to follow instinct—her trust in me made me trust myself, like a circuit. The pain peaked and I practically pulled Liliana in with me. Just then, the doorbell rang—the midwife had finally appeared. "Oh, I'm not trained for water births," she told Liliana. "Don't worry, it's happening anyway," Liliana replied.

> As with every stage, my doula got me to follow instinct—
> her trust in me made me trust myself, like a circuit

It was almost a quarter before three and I'd begun to push. There was no cheering me on, no cautioning me to slow down and pant. "Your body knows how to get the baby out," I could remember Liliana saying on the doula course. And so it seemed. Two pushes, and she was there—my beautiful daughter, Pearl, had arrived.

On the subject of food, I was interested to come across some research which found that simple sugars can lower a person's pain threshold. (I'm afraid this was when I still had three under-5s, so I didn't note down the reference at the time. It will appear in a future edition.)[88] If this research is valid, it certainly doesn't seem to be well-known because laboring women are often encouraged (by family or friends) to 'keep up their energy' by having isotonic drinks (which contain high levels of glucose), chocolate or cake. If avoiding the use of analgesia and anesthesia is your client's aim, it would surely be helpful to be able to advise her as to best foods? Clearly, that would be another valuable area for research.

## Questions for reflection...

1   Do you ever give moms-to-be advice about eating and drinking during labor? If so, what is this advice based on?
2   If you have experience of giving birth yourself, what did you yourself feel like eating and drinking during labor?
3   How often have you observed women vomiting during labor? Do you see this as a problem or an inevitable part of the upheaval of labor?
4   What are the risks for the mom-to-be of becoming dehydrated?
5   How would you go about organizing research to test the effects of different types of food on labor and birth?
6   What do you usually offer women after the birth?
7   Do you think there might be any cultural considerations influencing women's choice of food and drink either during labor or postpartum?
8   Are you aware of any food taboos in different cultures? (They do exist!)
9   What kinds of food and drink do you think should ideally be available in a hospital environment, antepartum, intrapartum and postpartum?

*This is Rebecca Wright, the doula, nursing her own third, optimally-birthed baby!*

## What about the baby's father?

The question as to whether they should attend a woman in labor and what precisely they should do is a difficult one to answer. Sometimes women will come to you very sure of their views on this, while at other times they may well ask your advice. What are you going to say?

On the plus side, a laboring woman attended by her husband is likely to have continuous, one-on-one care from an intimate companion who is presumably supporting her all the way. On the negative side, there may be all kinds of problems involved in having a father-to-be around at the birth.

There may be all kinds of problems when a father is present

Here are a few reasons why:

1 Since a man, by definition, can never have given birth, he is unlikely to understand the laboring woman's real needs.

Since a man, by definition, can never have given birth, he is unlikely to understand the laboring woman's real needs

2 The couple's relationship might not be strong enough on many levels to withstand this kind of intimate endurance test. Will the man have the staying power? Will he be able to set aside resentments towards the woman? Will the woman feel entirely relaxed if their relationship isn't perfect? (I have heard a rumor that some couples have less-than-perfect relationships...)

3 If the man is anxious or fearful, his fear is likely to be conveyed to the woman and she is likely to become afraid and tense herself. Since adrenaline (the hormone associated with fear) is antagonistic to oxytocin (the hormone of love and relaxation), this will slow down and possibly even stop the woman's labor.

4 The man's presence may effectively disturb the woman—largely because it's extremely unlikely he will be capable of being an extremely unobtrusive, 'invisible' presence and it's impossible to imagine that he will not behave like an observer, which will inevitably slow down or stop his partner's labor.

5 When the man sees the woman performing a psycho-sexual act which is not... shall we say... entirely romantic, he may later no longer find her sexually attractive. Of course, poop is often a feature of a birth of any kind, and blood is also involved, so the father may react negatively.

When the man sees the woman performing a psycho-sexual act which is not... shall we say... entirely romantic, he may later no longer find her sexually attractive

Bearing these points in mind, it might be wise to ask women to consider how they would feel if their husbands saw them doing a poop on the floor, completely unintentionally, when they were about to give birth. Ask them if they would be happy for their husbands to see their genitals fully exposed, stretched and bloody. Mention the possibility of hemorrhoids too and ask women to consider whether having their husbands present is likely to affect their sex life in any way...

My own view is that husbands should not be in the birthing room, but they can certainly do other things nearby—as many birthframes in this book show.

It might be wise to ask women to consider how they feel

## Birthframe 88: Phil Anderton—What is the man's role?

As we see here, it's not necessarily easy for men to keep their women company at this quintessentially female time.

Please allow to me present my credentials: I am the father of four children, three of which were born without serious medical intervention but the last, being a breech baby, had to make its appearance via cesarean section. I attended all the births (three in the hospital, one at home) and therefore you might think I am eminently qualified to write something on the subject of childbirth—but you'd be wrong. I am no more capable of commenting on the pros and cons of the various forms of bringing babies into the world than a professional soccer player can offer advice about a good forward defensive stroke in baseball. This is simply because I am male and being male means I will never have the slightest idea of how it feels to give birth. Of course, once the line on the curious spatula thing purchased from the drugstore turned blue, I read books on the subject, I went to the prenatal classes more or less voluntarily and I did my best to empathize with my wife as she began to slowly swell. But no amount of lectures, movies and drinks with midwives could possibly convey the actual experience of labor to those of us with a Y chromosome. This put me in a very tricky situation.

My wife has very strong views about giving birth. She wanted everything to proceed with as little medical intervention as possible and without any painkilling drugs. I could relate to this at least; personally speaking I have always had a peculiar aversion to taking pills (not even an aspirin for a hangover). Covering my face with a mask has induced panic ever since a childhood trip to a drunken dentist and allowing anyone to stick a needle into my spinal column seems like asking for trouble. But it was not me who was going to give birth, it was not me who was going to have muscles and nerves stretched to breaking point and beyond. So when she asked me for opinions on everything from arnica and other homeopathic remedies to using a TENS machine, I felt like a complete fraud offering any advice at all. To compound the problem, there was my own ignorance. I thought a TENS machine was something to do with bowling alleys and one sip of the foul-tasting raspberry leaf tea made me feel glad I was a man. But if my sense of helpless confusion was bad during the pregnancy, things were about to get a lot worse.

I know everyone is different, but when my male friends say watching the birth of their children was one of the most rewarding experiences of their lives I can't help thinking they are either lying, complete sadists or both. How anyone can describe watching their loved ones suffer extreme pain as 'wonderful' is utterly beyond me— even when the compensation is holding your beautiful baby son or daughter in your arms. I guess the root of my problem was that there was absolutely nothing I could do during labor except to try not to get in the way. Of course, I could hold my wife's hand, offer her words of encouragement and mop her fevered brow as directed, but it all felt so inadequate. I mean how could I help my wife match her breathing with her contractions when I wasn't feeling the contractions?

How could I help my wife when I wasn't feeling contractions?

At the time, it struck me this is why most spectators at sporting events are men. The thousands of males who flock to pitches, tracks and courses every weekend are merely in training for the day when our other halves give birth. Coping with the frustration we feel about not being able to take that penalty kick, score runs or ride that winner in the 3.30 at the horse racing track is excellent practice for that crucial time in the labor ward, when all we can do is stand, watch and wait. Now don't get me wrong: the midwives and medical staff were absolutely marvellous. They did their best to make me feel included and involved but nothing they said or did could alter my belief that I had contributed nothing positive to the process. I suppose this feeling of helplessness is why some men feel the need to video the births of their children—something which, in my opinion, should be punished by a life sentence of watching endless repeats of *America's Funniest Home Videos*. Playing with electronic toys is what we guys do best and placing a lens between you and reality is an excellent way of coping with stress.

After the birth of No. 1, I felt like someone who had got too drunk at a party and made a thorough nuisance of themselves, so imagine my surprise when I was invited back, again, and again, and again! Did it get easier? Yes and no. Every birth is different but the risks remain the same. There is the worry that you are pushing your luck and that something may go horribly wrong. However, the other side of the coin is that the relief gets greater every time it goes right.

So did I do anything right? I suppose I must have done because I am still married to the mother of my four healthy children and she never, not even during our bitterest rows, refers to any mistakes I made during her confinements. If I did contribute anything positive it was, paradoxically, not what I did but what I didn't do. I did not insist she have a natural childbirth but once she had made her decision to do so, I did not insist she stick to it—cf. the medically necessary cesarean required to deliver No. 4. I did not allow my natural inclination to 'take charge' get the better of me; instead I let my wife and the midwives take the decisions and get on with it. The 'did not' I am most proud of, though, is that I did not squeal like a stuck pig when she dug her nails into my arm!

I was happy enough to do exactly as I was told but equally my wife knew she had my support whatever her decision. For example, No. 2 was three weeks late and the doctors were keen to induce the birth artificially. However, the baby was showing no signs of distress and my wife wanted to wait; if it had been me I would have wanted to be hooked up to every available machine in the hospital... but it wasn't me. I therefore shut up so my wife could listen to what her body was telling her. Our (or should I say her) reward was a bouncing, baby boy weighing an eye-watering 10lb 2oz (!) and delivered without any drugs or intervention.

Why, you may ask, if I found the births so distressing did I allow myself to be persuaded to attend? Some of you reading this may like to pick up on my earlier soccer analogy and ask why men spend hours in the wet and cold cheering on a bunch of overpaid, prima donnas, but I would prefer to sum up the whole experience in a single word. I bit the metaphorical bullet, swallowed my pride and allowed myself to be 'persuaded' into taking decisions against my male judgment for one reason and one reason only: it's called love.

However the woman's husband feels, it's important that she discuss her care guide with him. This will help him to understand her reasons and feelings on important issues and offer relevant support, with the woman's agreement.

If, for whatever reason, the woman isn't keen to have her husband around, he needs to understand why. Even if he very much wants to be at the birth, it's obviously best if he is sensitive to her wishes at this important time. If she wants her husband to be around but is concerned about how he'll behave, she should tell him about birth physiology—perhaps by showing him the companion volume to this book (for pregnant women): *Preparing for a Healthy Birth*.

     *I had initially had a fantasy of my husband being there for me—a time of deep intimacy! But we went to a talk with Michel Odent and someone brought up the question of men attending. Michel questioned the wisdom of it. My husband was relieved—he felt he didn't want to be the main attendant but hadn't felt confident to say so until Michel endorsed this. (My husband had a terrible experience with a birth in a previous relationship where, although he'd been a big support, the birth experience was about as bad as it can get and he was traumatized). When he said he didn't want to be there for me I was initially very disappointed and hurt. Then I came to terms with it, and, of course, it did turn out better for me to focus inwards.*

The idea that husbands are supportive and sensitive to moms-to-be may be a complete fallacy. I suppose it's not too surprising... with child No. 2, 3 or 4, men face increasing worries about finance, not to mention domestic chaos.

Despite the many chapters in pregnancy books helping men to gain an insight into their pregnant partner's feelings and moods, after traveling as a twosome through a first intriguing pregnancy, I've come to the conclusion from talking to other women that men merely tolerate a mom-to-be's presence a second or subsequent time. Instead of being understanding, supportive and sensitive, men tend to become bad-tempered. Pregnancy is then a lonely period for both husbands. The children, meanwhile, have to come to terms with this emotional climate and have to somehow work out what on earth's going on in Mommy and Daddy's relationship. They do still have one, don't they?

     *Although, to be honest, I hadn't really wanted my husband to be with me when I gave birth, I must admit he proved to be a wonderful support throughout my labor. He massaged my back for literally hours on end during my first labor and was always discretely positive and supportive of me. He told me afterwards that he felt offended about being ordered around so much by me during my first labor but he said he was prepared for this the second time round! Again, he proved to be the perfect assistant, perhaps because he really did want to be there.*

     *When I went into labor, attended only by my husband and a doctor, I had a sudden moment of panic. No one nearby had given birth and I felt a tremendous sense of loneliness—if only a woman could be nearby!*

66 I felt very strongly that I didn't want my husband to be there, but my husband decided he was going to be and the point was clearly non-negotiable. He took my preference for being alone as an implied rejection and considered it sufficient grounds to end our relationship. I put up with him in the end and I have to admit he was very helpful. I'm not at all sure it helped our relationship, though... It probably had a damaging effect in that respect, as part of me had suspected it might. Some things need to remain private... Strange how the same man will never let me into the bathroom while he's on the toilet!

66 My husband was amazing during the entire 36 hours that the birth day lasted, from walking with me during the day to making food in the evening and driving me around in the middle of the night. Neither of us has ever been so physically or emotionally exhausted in our entire lives. The whole thing brought us closer together, I have no doubt. He was an incredible support. I really feel as though we equally gave birth to our baby, especially since I didn't get to experience the delivery, because of the cesarean.

66 My husband cooked some sausages (which helped my ongoing nausea no end) and made a few phone calls. I wouldn't be surprised if he didn't switch his computer on and do a bit of work, but he's never told me that. Not because he didn't want to help, but more because I was doing okay on my own and really didn't want any interference.

66 I kept sending my husband out of the room, trying desperately to hold on to some of my dignity.

66 During second stage Adam had decided that he would put his hands down behind me to catch the baby; unfortunately he caught something entirely different! Apparently, his face was a picture but I was unaware of this until later; I was, however, very amused. A greater love hath no man!

66 I insisted on John being at my head-end as opposed to the 'business' end, as I'd heard of men who went off their husbands physically after seeing them give birth and I didn't want to risk that.

66 When a nurse suggested pushing my hemorrhoids back in, a few hours after I gave birth, my husband immediately said he'd like to come in and watch. Suddenly, I felt as if scales were falling from my eyes. I was not a circus sideshow and I immediately knew—and stated!—that it was not a good idea. It made me wonder about his motivation for watching the whole birth the day before and I remembered I'd read about men dressing up as women in 18th century France so that they could watch women giving birth. If I had a choice another time, I'd want this part of my life to be private. It's not for my man to see.

66 I think women should be allowed more privacy at such an instinctive time, when feeling uninhibited and unobserved, and tuning into the natural processes is so important.

Here's Michel's view, expressed in an article, entitled 'Is the participation of the father at birth dangerous?' which originally appeared in *Midwifery Today*.[89]

A century ago, when most babies were born at home, such a question would have been deemed irrelevant. At that time, everybody knew that childbirth is 'women's business'. The husband was given a practical task, such as spending hours boiling water, but he was not involved in the birth itself.

Today, the same question is still deemed irrelevant, even stupid. At the dawn of the 21st century, everybody knows about the importance of the active role of the father in the 'birth of a family'. Most women cannot even imagine giving birth without the participation of their 'husband'. We have heard countless wonderful stories of 'couples giving birth'. Fathers are welcome in the most conventional delivery rooms.

If it is commonplace to dodge the real question, in spite of a conceptual mutation, this means that the issues are complex. In order to interpret such sudden and radical changes in concepts and behavior, one must put them into their historical context. It is essential to recall that the intriguing phenomenon we are studying began unexpectedly in most industrialized countries in the 1960s. Then, a new generation of women felt the need to be assisted by the baby's father when giving birth. They started to express this new demand at the very time when births were more and more concentrated in larger and larger hospitals. From a historical viewpoint one cannot dissociate hospital birth and the participation of the baby's father. This was also the time when the family had a tendency to become smaller and commonly reduced to the nuclear family, so that in the daily life of many women the baby's father was the only familiar person.

Furthermore, the 1960s represent the time when the midwife became one of the members of a large medical team (in the countries where she had not completely disappeared). It is clear that the participation of the father was as an adaptation to unprecedented situations: it had not happened before in the history of mankind that women had to give birth in large hospitals among strangers; the nuclear family was unknown in any other culture and midwives had always been independent.

Those who have been active witnesses of such behavioral upheavals remember how quickly theoreticians established new doctrines. For example I heard around 1970 that the participation of the father will strengthen ties between the couple and that we should expect a decrease in the rate of divorces and separations. I also heard that the presence of the father, as a familiar person, should make the birth easier and that we should expect a decrease in the rate of cesarean sections.

The dawn of the 21st century represents, 30 years later, the beginning of another phase in the history of childbirth. The current turning point is related to the fast development of 'evidence based obstetrics' and 'evidence based midwifery'. One of the first effects of a scientific approach is to stimulate a new awareness of the importance of environmental factors in the perinatal period. For example, we learned from a series of prospective randomized controlled studies that an electronic environment tends to make the birth more difficult and has no other effects on statistics than to increase the rates of cesarean sections. Evidence based obstetrics is instrumental in the preparation for the 'post electronic age' in childbirth.

The current crisis, induced by evidence based practices, represents a unique opportunity to reconsider many theories and pre-conceived ideas and to take an inventory of the questions we must raise. Where the participation of the father at birth is concerned, we must raise at least three questions:

Where the participation of the father at the birth is concerned, we must raise at least three questions...

**First question: Does the participation of the father aid or hinder the birth?** Those who are old enough to remember what a birth can be like when there is nobody else around than an experienced, motherly and low profile midwife are inclined to formulate the question that way. Our objective is not to provide answers but to analyze the many reasons why it is such a complex issue.

There are many sorts of couples according to the duration of cohabitation, the degree of intimacy, and so forth. There are many sorts of men: some can keep a low profile while their partner is in labor; others tend to behave like observers, or like guides, whereas others are much more like protectors. At the very time when the laboring woman needs to reduce the activity of her intellect (of her neocortex) and 'to go to another planet', many men cannot stop being rational. Some look brave, but their release of high levels of adrenaline is contagious.

The double language of human beings appears as the main reason why the complexity of such issues is underestimated. There is a frequent conflict between the verbal language and the 'body language' of pregnant women. With words, most modern women are adamant that they need the participation of the baby's father while they give birth; but on the day of the birth the same women can express exactly the opposite in a nonverbal way. I remember a certain number of births that were going on slowly up to the time when the father was unexpectedly obliged to go out (for example to buy something urgently before the store closed). As soon as the man left, the laboring woman started to shout out, she went to the bathroom and the baby was born after a short series of powerful and irresistible contractions (what I call a 'fetus ejection reflex').

When raising such a question one must also take into account the particularities of the different stages of labor. It is often during the third stage that many men have a sudden need for activity, at the very time when the mother should have nothing else to do than to look at her baby's eyes and to feel the contact with her baby's skin in a warm place. At this time any distraction tends to inhibit the release of oxytocin and therefore interferes with the delivery of the placenta.

**Second question: Can the participation of the father at birth influence the sexual life of the couple afterward?** Through such a question we introduce the complex issue of sexual attraction. Sexual attraction is mysterious. Mystery has a role to play in inducing and cultivating sexual attraction. Once there were mother goddesses. At that time childbirth was enigmatic among the world of men. I had the opportunity in the past to talk about the birth of their baby with women who were themselves born at the end of the 19th century. They could not imagine being watched by their husband when giving birth: "And what about our sexual life afterward?" was their most common reaction.

Today I am amazed by the great number of couples who split up some years after a wonderful birth according to modern criteria. They remain good friends but they are not sexual partners any longer. It is as if the birth of the baby had reinforced their comradeship while sexual attraction was fading away.

**Third question: Can all men cope with the strong emotional reactions they may have while participating in the birth?** I am not thinking of a woman watching the TV while giving birth with an IV and an epidural, but of a woman relying on her own hormones. I would have never thought of raising such a question as long as I had only the experience of hospital birth. During the days following a hospital birth, nobody is wondering about the well being of the father. When visiting a family two or three days after a homebirth, I almost always found a happy and active mother taking care of her baby. I had a surprise when asking about the father. More often than not I heard that the father was in bed because he had a belly ache, or a backache, or grippe, or a toothache, or simply because he was 'drained', as a mother told me. When referring to my experience of homebirth, I am tempted to claim that male postpartum depression is more common than female postpartum depression, although it is not recognized as such. The concept of male postpartum depression is a reminder that many cultures have rituals whose effects are to channel the emotional reactions of the father. All these rituals belong to the framework of the 'couvade'. (Anthropologists use this term that originally means, in French, 'hatching'.) These rituals, whatever the local particularities, make the father busy while his wife is giving birth. The last example of couvade was the man spending long hours boiling water. I cannot help thinking of the case of young modern men who spend a long time rebuilding a rented transportable birthing pool: finally the baby is born before the pool is ready. Is it a revival of the couvade?

My only objective is to justify a series of questions by suggesting that the issues are much more complex than we commonly believe. It would be premature to offer clear-cut answers. Questions should precede doctrines.[88]

## Birthframe 89: David Newbound—A man as defender

There might be many reasons for wanting to have one's partner present, but the one presented here by a father as being paramount might come as a surprise.

My first child was born when I was 18, my last—fate permitting—when I was 44. I imagine men have been fathering children over that time period and much more for millions of years. Nothing unusual in it, then? Well, not in the time span, certainly, and with many men fathering children in their dotage, it's not even close to an extreme. What is unusual, now I come to think about it, is that my experience spans a revolution in a father's involvement in the birth process, starting as it did at the very time when men were first being tentatively allowed into the sacred labor ward.

Since then, my experience has moved with the times through discouragement of home births, the introduction (re-introduction!) of natural labor, and the latter-day prenatally-trained father oozing calm in the ward, sopping brows and monitoring birth plans (you wish!), while still filming the beautiful scene. Perhaps that's why Sylvie asked me to write a contribution, and I don't know what is about to appear. Much of it I haven't thought about recently.

At 18 (in the late 1960s), my wife was petrified when her water broke and we took our battered car to the hospital. The nurses spoke to her rather less well than a dog. (Actually a lot worse than a dog. Dogs at least cannot understand uncaring and cruel asides... "Yeah, she's just making a fuss. Look, she's not even one finger yet!") I sat, immature, scared and powerless. Having read all about fathers being welcomed, I was expecting someone might notice I was there, but I was surprised to be told to leave and, 25 minutes later, was told I had a new son and that I'd be able to go in in an hour or so. I remember throwing everything on my lap in the air. I had never felt ecstasy like it, but then at that stage I hadn't seen what my wife looked like either.

> I had never felt ecstasy like it... but than at that stage
> I hadn't seen what my wife looked like either...

Even with the distance of over 30 years, this is a traumatic memory. My wife and I had never felt so dominated, so insignificant and worthless. I have heard women be critical of male involvement in childbirth but in those days it was the nurses who didn't give a f\*\*\*. Perhaps fatherly attendance in the labor ward has one positive effect: it may keep medical staff more on their guard about common decency. I vowed to be present for every subsequent birth.

## Birthframe 90: Alan Low—Husband as supporter

To end on a positive note, this final birth story (by a father we've met before) shows just how supportive husbands can sometimes be. Indeed, the support this particular husband offered is probably the very best kind of support a pregnant and laboring woman can wish for... This man made sure his wife had caregivers she felt comfortable with by completely supporting her wishes, even under very difficult circumstances. Were you aware this kind of thing goes on in the background when families are calling you out to a home birth?! Would you be a favorite caregiver? If you were a 'disliked' caregiver and you saved a woman's life nevertheless, how would you speak to the woman afterwards?

We had returned late on the Saturday night from a friend's wedding reception and climbed into bed, with me still feeling the effects of an over-indulgence of food and alcohol. Sleep came upon me with ease until I was rudely awakened by a sharp dig in the ribs by Karen, my beloved, who rather nervously announced that she thought it had started. Being a loving, sensitive and supportive husband, my immediate response was to inquire: "Have you had a show yet? No?! Well go back to sleep and don't wake me until you have." Could this really have been Mr Sensitive talking or was it the alcohol and lack of sleep taking over? (That's my excuse and I'm sticking to it!) A slightly more aggressive dig in the ribs followed and was accompanied by a short, sharp volley of verbal abuse and threatening behavior. I was stung into action, so crawled out of bed and attempted to look sharp and alert, ready for anything, but in reality of course was a total wreck and looked it—no fooling anyone.

*I was stung into action, so crawled out of bed and attempted to look sharp and alert, ready for anything. In reality of course I was a total wreck and looked it—no fooling anyone.*

We agreed that we should keep busy, so picked up round the house, had a bath, played Scrabble (I won) and before we knew it 3:00AM turned into 7:00AM. It was now time to call Liz, who had helped and advised us so far and was our 'expert'. She had agreed to help out and had three children of her own, two of which had been born at home. I collected Liz around 9:00AM and returned to find that the contractions were regular enough to warrant calling the midwife.

*In the months leading up to the birth we had some difficulties (understatement) with 'The Establishment' regarding our desire for a home birth, with a few midwives vehemently opposed*

In the months leading up to the birth we had some difficulties (understatement) with 'The Establishment' regarding our desire for a home birth, with a few midwives vehemently opposed. We had expressed our discontent to the Head of Midwifery Services, who had agreed that the attending midwife would not be one of the dissenters. Unfortunately, on that fateful morn, No. 1 choice was ill, No. 2 had a day's leave, No. 3 was Head Dissenter (didn't even bother calling her, she would have been out on her broomstick anyway), but No. 4 answered. Great! Or so we thought. Much to our surprise, Karen was answered with a torrent of abuse for choosing a home birth and for also politely refusing the midwife's offer of having a student in attendance. She said she would be around soon, but was not happy. The effect of this event was to stop the contractions completely, but normal service resumed 20 minutes later.

We discussed and collectively agreed that this lady was not for us. Karen and Liz agreed that I would turn her away at the door while they hid—thanks. Time for my cool, calm approach towards problem-solving to surface. I took control and phoned the hospital to find out the alternatives and demand action. All they could suggest was to make the peace with midwife No. 4, as she was all that was available—time for Mr Cool to panic. "We can do it ourselves!" I proudly announced. "Boil some water and fetch towels!" Liz calmed the hysteria by informing us that this was illegal and that she would contact her ex-midwife and now friend, Jo, who was a practicing independent midwife. She agreed to come but would have to obtain the appropriate approvals from the state authorities first. This was to be a momentous obstacle but she managed to battle through okay and arrived on our doorstep at 1PM—phew! The next stop was to provide her with some assistance and back-up, courtesy of Jan, who we all knew and was an ex-nurse and current prenatal teacher.

*Jo agreed to come but would have to obtain approval first*

The first examination revealed, much to everyone's amazement, that Karen was 8cm dilated. Around 2:00PM the second stage commenced, a time Karen had been dreading, but Jo reassured her by saying that whilst it would probably be painful it shouldn't take too long as she was doing so well. 'Too long' to someone (like me) in intense pain would be 10 seconds.

'Not too long' lasted two hours for Karen, who was by this time starting to tire. Digby (the silly name we had christened the baby belly once the ultrasound had revealed our baby was to be a boy) started to crown, but a further 25 minutes passed. Jo and Jan constantly reassured Karen that 'one more push will do it' but none of us really believed it until suddenly, without warning, Digby shot out and was fantastically caught one-handed by Jo.

On closer examination Digby appeared to be woefully short of a few vital attributes that would make his position as the rightful Captain of the Los Angeles Galaxy soccer team difficult to attain. Digby's attributes were more suited for softball so I found myself wiping the tears of joy away, holding our beautiful daughter, while the third stage commenced. I looked at her and thought, "Just like her mother— late!" It made me want to weep!

Meanwhile, back at the action, Karen had finished the third stage relatively quickly but was hemorrhaging badly, was very pale and was starting to shake uncontrollably as she was going into shock. The scene started to resemble an out-take from *The Exorcist* but, fortunately, Jo and Jan quickly averted any danger and the bleeding subsided.

They quickly averted any danger and the bleeding subsided.
I dashed downstairs to crack the champagne...

At the first available moment I dashed downstairs to crack the champagne and returned to find the new mom feeding our daughter, which immediately restored the lump in my throat and dampness to my eyes. By 9:30PM, everyone had left and we were left alone in total amazement looking at the little bundle laid between us. Our first sleepless night was soon to be upon us...

*Here's a loving father with an older child. Fathers are clearly very important but what exactly do you conclude their role should be during pregnancy and the actual birth?*

# Children

## Birthframe 91: Cara Low (13 years old)—Being inspired

The next account was written by the 'baby' whose birth was described in the last birthframe. She witnessed her little sister's birth. Although she had mixed feelings about attending a birth (the birth of her little sister, in fact) this child—who was 7 years old at the time of the birth—was clearly surprised by what happened and it changed her views and expectations of childbirth.

I remember feeling warm all over. I had wanted a sister and now she was finally here. It was truly amazing. I fell asleep momentarily because I was so tired.

> It was truly amazing. I fell asleep because I was so tired.

Since then, I have learned a lot more about what happened that day, as at the age of 7 I didn't really take it all in. When I have seen DVDs at school, I have never seen an account of a home birth. From these DVDs I had been led to believe that there would be lots and lots of blood and screaming, but it was nothing like that. There had been surprisingly little blood and no screaming or signs of excruciating agony. The overall atmosphere was calm and pleasant and Marie made everyone feel relaxed and excited. We decided to give Christy the middle name of Marie for this wonderful midwife, who made the birth so much easier and relaxed. I am very grateful for the experience and, thanks to Marie, understood and saw all the things I'd seen on DVDs, in real life. I think it's a shame that so many people are unaware that home births are a possibility. I think that home birth awareness has grown since Christy's birth and it's still becoming more popular. If I have children I think that I will have a home birth and possibly in water too.

> I think it's a shame that so many people are unaware that home births are a possibility. If I have children I think that I will have a home birth and possibly in water too.

66 Very handy that my son woke up at just the right moment. But that's not really the point. What if yours doesn't? Children are very matter-of-fact about things. If you explain to them what you plan to do—have a baby right here in the bedroom—and what might happen—Mommy might make funny noises—then, chances are, if your children wake up at an inopportune moment, they probably won't be fazed in the least. Of course, you want to make sure you have back-up options, such as a neighbor who doesn't mind being woken up in the early hours. But in the event, chances are it won't be necessary. I rehearsed with my 2-year-old the grunting noises he might hear me making and explained that making a baby come out takes a lot of work. "You can help me practice," I told him, and we made pushing sounds together. Nina Klose

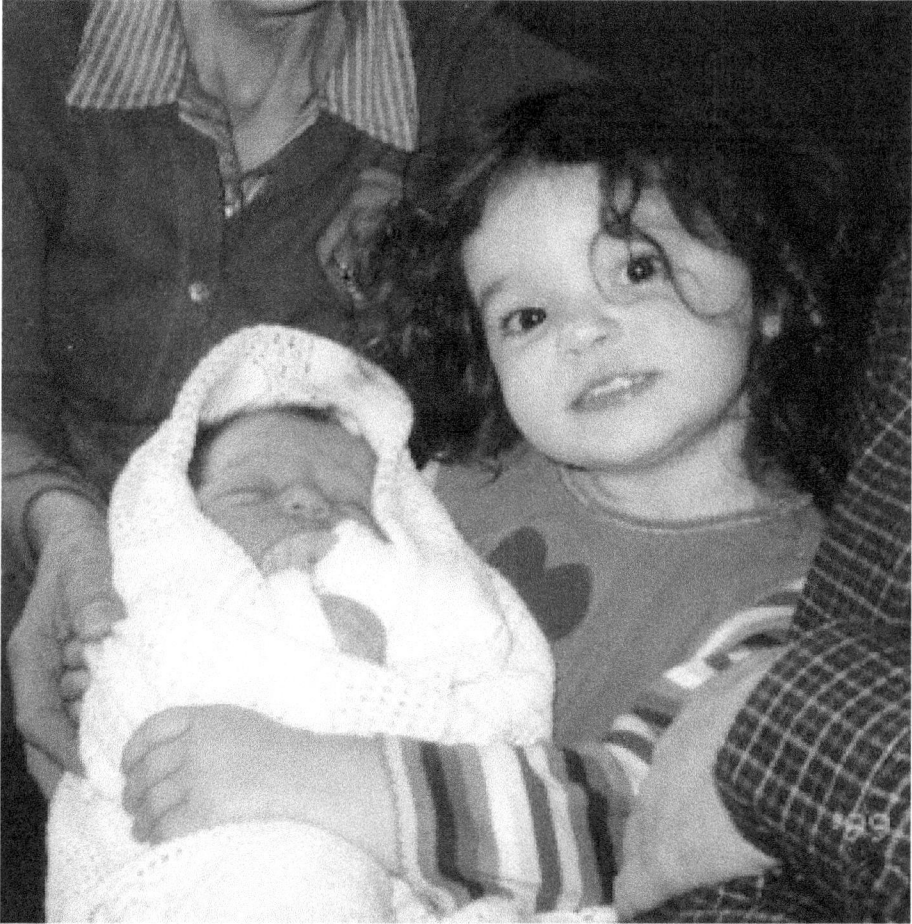

*A little girl who's delighted to be involved, this time with the help of the midwife*

66 It would be nice to be able to give a dreamy, romantic account, but I'm afraid my other children seemed pretty insensitive to my needs as I labored in front of them. Another time, I'd arrange for someone to be on call, to take them away somewhere to play so that I would be able to labor in peace and seclusion. Yes, I do love my children!—but I need to focus inward when I'm in labor and I can't do that when my kids are around.

It'd be nice to be able to give a dreamy, romantic account,
but I'm afraid my children seemed pretty insensitive

# The actual birth: optimal positions for late labor

*Women are often sick. Of course, leaning over a bucket is fine because it's another leaning forward position...*[90]

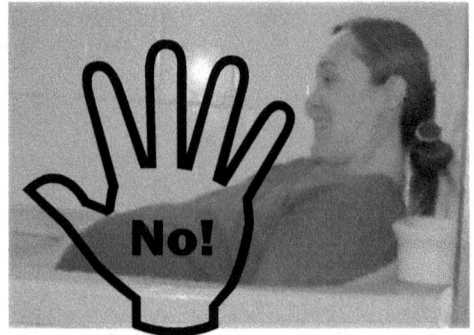

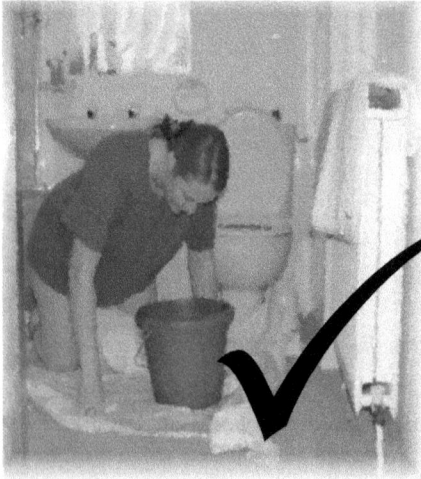

*As the birth approaches, it's nice to make sure there's something soft underneath the woman at all times, just in case the baby does actually choose to come out a little earlier than you expect! He or she really doesn't need to be caught, but a soft landing would make a nicer start to life. In the photo above note the presence of the waterproof tablecloth (which minimizes mess) and the old bathmat to kneel or stand on which stops the woman from sliding around uncomfortably.*

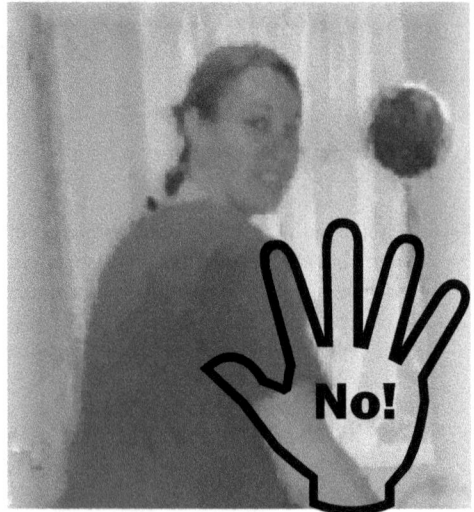

*By the way, remember not to talk to the laboring woman. She's supposed to be free to 'go to another planet', remember? 'Do not disturb' is the most important rule.*

# Optimal positions for the birth itself

*Women will spontaneously find a good position in which to give birth, when they are not being disturbed in any way.*

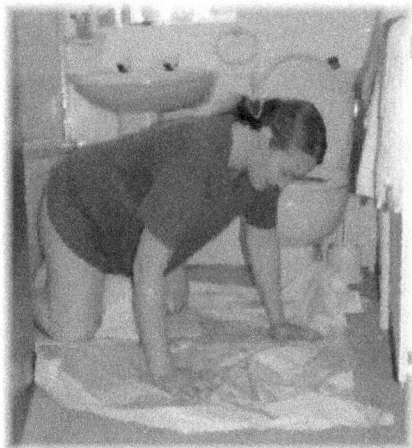

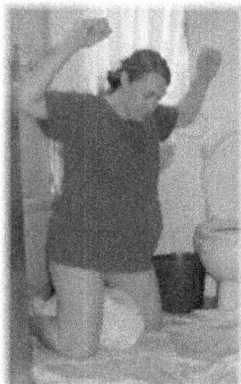

Upright positions will
facilitate descent

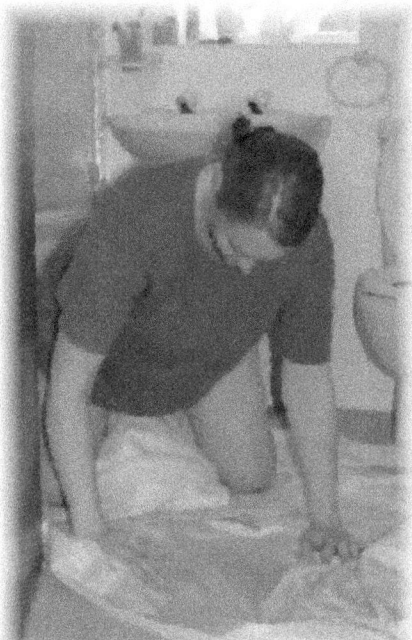

*This is my favorite position! Standing like this, I birthed my second and third babies.*

When undisturbed, women will spontaneously find a position

# Cord-cutting

Of course, it is now widely accepted that delayed cord-cutting is advantageous because it ensures that the newborn gets its last 'dose' of oxygen from the mother's blood before needing to access it on its own, through breathing.[91]

## Birthframe 92: Ashley Marshall—The realities of lotus birth

If a mom-to-be decides she doesn't want any cord-clamping, what's involved? Some clues are given in the following account, in which a doula—Ashley Marshall, whose photo you've already seen on pp 24 and 26—had what is called a 'lotus birth'.

Harper's birth at home was sweeter than I could have dreamed it. I labored easily and powerfully with loving support from my husband, doula, and midwife. When Harper's bag of water popped, birth energy rushed through me and soon his little head began to emerge. There was a slight case of dystocia and a tight umbilical cord but we were safe in the skilful hands of our midwife. Then, all of a sudden, there he was—pink, alert, and beautiful.

As I sat in awe of this perfect, wise being I rubbed the creamy coating of vernix into his delicate skin and knew we had made the right decision to give him a lotus birth. It was the most blissful beginning to follow the culmination of pregnancy and the sense of loss that often ensues.

Lotus birth is the process of leaving the baby's placenta attached via the umbilical cord until it falls off of its own accord. Its purpose is both physiological and spiritual. Physiologically, the baby receives 43% of the blood left in maternal/placental circulation.

Ever wonder why cord blood banking has become so popular? It is known that cord blood—or blood left in the cord and placenta that hasn't made its way to the baby—is useful in helping to fight childhood leukemia later in that child's life. Why not stop denying a baby this vital blood and allow it to pass to the baby after birth instead? Also, the placenta is the baby's life preserver before breathing is established. As long as the cord remains intact and while it still pulsates the baby is being oxygenated.

Long after the cord stops pulsating and the physical transference is complete, the spiritual transference has only begun. This quiet time is when the baby's aura, or spiritual presence, is being realized. The placenta originated from the same cells as the baby and because of this bond they are a genetic identical of one another. A lotus birth allows for a respectful goodbye to the baby's womb mate.

How does a parent care for an intact placenta? It was surprisingly easy. After Harper's birth my husband and midwife placed it in a colander and rinsed all of the blood clots out. Then we rubbed it with sea salt and sprinkled it with lavender flowers.

It was surprisingly easy to care to the intact placenta

Long after the cord stops pulsating and the physical transference is complete, the Harper's placenta wore a cloth diaper just like he did and we changed it daily. We swaddled the 'placenta package' right along with him so we were free to pick him up, nurse and cuddle without the fear of tugging at his navel.

On the third day after his birth, Harper let go of his placenta. He was content and whole and ready to be free. We said goodbye to the organ that had nourished and protected our son for his first nine months but we will see it again when we plant it at Harper's first Blessing Way ceremony. In case you don't know, a Blessing Way ceremony is a Native American ritual used to mark significant life passages in one's life, frequently held to commemorate a birth, a marriage, a death or a woman's journey into motherhood. It is a more spiritual ceremony than the traditional American baby shower, where the focus is on showering the mother with gifts for the baby. A Blessing Way ceremony honors the mother and helps her draw upon her own inner resources that she will need to later give birth. My husband and I decided to offer our children a Blessing Way ceremony to celebrate their first year of life as well as my first year postpartum. We also planted their placentas at this time. (They were kept in the freezer up till then.)

Harper is our second child to be born at home, but our first lotus birth. We now know that we would never have another baby without giving them a lotus birth. It was three very mindful days that enabled us to remain in that warped sense of time that follows birth. We were surrounded only by love and close family while Harper made his earthly transition. There was plenty of time for lively celebration later that week but I will always be grateful for those precious days when Harper was lotus-born.

# Questions for reflection...

1 If you usually cut and clamp the cord, when do you usually cut it and why?
2 How do you feel about not cutting the cord and waiting for it to break off naturally? What are your concerns?
3 What would you say to a woman who requests a lotus birth?
4 Would you feel confident about providing practical advice?
5 Would you attempt a lotus birth yourself?
6 Why do you think some mothers (and fathers) are so keen on the idea?
7 Do you accept there might be spiritual or emotional reasons as well as physical ones for delayed cord-cutting?
8 What do you think about other 'natural' approaches, such as eating the placenta? Can you think of any justification for it? Would you do it yourself?
9 Do you encourage new mothers and their husbands to look at the placenta?
10 Do you encourage new parents to bury the placenta in their garden? If not, what do you think of this practice? Why are placentas considered by many new parents to be important?

# Unexpected events

## Birthframe 93: Ruth Clark—Not going to plan, but also fine

Even when things go 'wrong' they don't necessarily need to go 'bad'... In my own pregnancies I became tired of meeting midwives who were full of doom and gloom. Please encourage your clients to embrace anything unexpected positively! I think both of the following accounts illustrate very well how *wrong* things can seem to go, while outcomes are still fine.

The first woman's labor took her by surprise. Although she briefly and reached for the TENS machine at one point, her labor was almost entirely natural. Even though it was not ideal that she should have so many interruptions between the birth of the baby and the birth of the placenta from a safety point of view, the third stage also went smoothly.

**The players:**

Me—Ruth

Adam—my husband

Ben—the star player

Eddy—our 2-year-old son

Sue—my midwife

Julie—my other midwife

Maggie—supporter for Ruth and Adam

Kerry—supporter for Eddy

Daniel—supporter for Eddy and also my brother

My first birth experience hadn't been as I had hoped. Due to a premature footling breech my plans for a home birth had been well and truly ruined. So this time round I was doubly determined to have my baby at home. I was delighted when I discovered the baby was cephalic (a perfect LOA) and ecstatic once we got past that magical 37 weeks, after which point my midwives would cover a home birth. In fact I was really excited about going into labor, no worries about anything... This was going to be good and everything that my last labor wasn't. My midwives were supportive and we got on really well.

I was woken on January 14 at 5AM by Eddy, who was crying. As I went in to see him I noticed a slight discomfort in my lower abdomen that was coming and going, I put this down to wind. After dealing with Eddy I went back to bed and listened to him singing and talking to himself until he went back to sleep at 6:15. All this time I was aware of the abdominal discomfort, only slight but enough to keep me awake—I'm a light sleeper.

The discomfort was only slight but enough to keep me awake

At 6:30 I decided that I needed to open my bowels. I thought I would feel better after this but I didn't. If anything, I felt slightly worse. I wondered if, as this was six days before my due date, these were contractions and decided that if they were they must be Braxton Hicks.

At 7AM I got up and fed the cats (all seven of them). The contractions were now coming about every 10 minutes and were getting stronger and longer but were easily bearable—I just breathed through them. I now also noticed a lower back pain but still did not think that I was in labor.

At 7:15AM I needed the bathroom again, this time it was much looser and I noticed a show. I wondered if this was the real thing as there were now four signs but I was not convinced.

At 7:30AM I thought that maybe I should wake Adam. I told him I thought that I might be in labor; he said okay and did I need him, and then he went back to sleep. I decided I'd better phone the other people who were due to come over. Kerry had just got home from working a night shift and she had taken a sleeping tablet and was about to go to bed; she was not going to miss this for the world and set out to walk across town to our house. Daniel was on his way to work so I couldn't get hold of him. Maggie did not answer her phone—it was broken—so I phoned her husband's cell phone; he was halfway across the country but I managed to get a message to Maggie via relatives who live nearby. I got hold of Daniel at 7:55AM, I told him there was no rush as he wanted to go elsewhere on the way.

By 8AM Adam was getting out of bed. I saw him naked at the top of the stairs, and through bleary eyes he told me he needed to go to the bathroom. I told him in no uncertain terms: "I need to go first." And once again I opened my bowels. My contractions were now very close together.

At 8:05AM, still sitting on the toilet, I had the first really intense contraction. It was very different to the previous ones, which had all been really low down and opening-out type contractions. This one was from the top of my uterus, a really strong pushing contraction. My first thought was that this shouldn't be happening yet... I couldn't possibly be at this part of my labor as I hadn't had the painful bit yet. But my body was pushing and there was nothing I could do that was going to stop it.

I hadn't had the painful bit yet... but my body was pushing

I then decided that I needed to do my hair (which is down to my waist), so I unbraided it and asked Adam for my hairbrush. I also decided that I needed my TENS machine on. It was far too late for this but it seemed the right thing to do at the time.

After two or three more contractions I put my hand down between my legs and could feel something sticking out! I lifted myself off the toilet seat and asked Adam what it was... it was the amniotic membranes bulging out and was about the size of a grapefruit. At this point I decided that I needed a midwife rather urgently. She was the one person that I had omitted to phone when I made my calls earlier so she didn't even know I was in labor.

Adam went downstairs and called the labor hotline. He asked to speak to Sue. The person on the other end told him that she was a community midwife and he would have to call their office. He came back up to me moaning about this and how useless they were. "Did you tell them that I am in labor?"... "Oh... no". He went back downstairs and tried again.

At 8:25AM Sue phoned back. She heard me shout out at the next contraction and suggested to Adam that he should be upstairs with me. I then shouted down that I could feel the baby's head. Sue said that she would phone back in two minutes once Adam had put the phone on the extension upstairs.

I instinctively kept my hand on the baby's head from when I first felt it. The next contraction came and the head moved out a bit and then back in again. Adam brought the phone upstairs and began to spread a groundsheet on the floor so that I could get off the toilet. As he finished doing this I had another contraction. With this one my water broke, out came the baby's head followed by his body. As I delivered my baby I lifted myself off the toilet seat, brought him up between my legs and cradled our second son in my arms as I sat back down again. Adam looked round and folded the groundsheet back up again. Then the phone rang, it was Maggie. Adam asked her the time, so we now knew our new son had been born at 8:30AM.

Almost immediately the phone rang again. It was Sue, expecting to talk Adam through the birth. He just said, "Listen!" She heard the cries and asked if we were okay. She also asked if we wanted to call an ambulance or just wait until she got there. As we were fine we said we would wait, and she said that she would get there as soon as possible. I offered baby my breast but he nuzzled at my nipple for a bit and then went to sleep.

At 8:35AM Eddy appeared in the bathroom doorway, he looked at me sitting on the toilet and said, "Baby!" and then came to join us. Adam got a big towel to wrap around baby and me to keep us warm and then decided that he'd better get some clothes on before everyone arrived; he'd had no time to dress up to this point and I had been wearing his bathrobe. I then asked Adam to take some photos while the cord was still intact. Kerry arrived first and took some photos of all four of us, she then looked after Eddy.

At 9:00AM Julie arrived. She had got dressed in the dark in a hurry and was wearing pink socks with her dark blue uniform and she had also been stopped by the police for speeding on her way to me. (They let her go as soon as she explained the situation.) As the cord had stopped pulsating, I was happy for her to clamp it and Adam cut it. Baby was then wrapped in another towel and Adam had his first cuddle while Julie helped me off the toilet and onto the floor. Adam then put baby on the floor so that I could lean against him while the placenta was born. While I was waiting for the placenta I noticed my window cleaner cleaning the bathroom window! At 9:10 I had a contraction and felt the placenta move down. Two minutes later I had another smaller one with which the placenta was born. Julie examined me and told me that I had sustained a small tear, which I decided not to have sutured. I stayed sitting on the floor while baby was checked and weighed. He was 7lb 12oz.

By this time Daniel, Maggie and Sue had arrived; Sue was later than she had hoped because she had skidded on ice and put her car into a ditch. At some point another midwife arrived but I have no idea who she was and she left once Sue got there. Maggie ran a bath for me and made drinks for everyone. Adam stood on our back doorstep to have a cigarette and shook a bit—poor man still hadn't been to the bathroom. I took baby into the bathtub with me where he had his first breastfeed for about 20 minutes. Eddy kept coming up to check on us and look at his new brother.

I was then dispatched to my bed even though I felt full of energy and not the least bit tired. I wanted to tell the whole world what a wonderful experience I'd just had; it was a total contrast to Eddy's birth. I was far too energetic and high to sleep, so Adam brought me the phone so that I could call all our friends and family to tell them the news.

It really was the most wonderful and empowering experience of my whole life. I thoroughly enjoyed my labor and birth and I wasn't the least bit worried or frightened about birthing my baby without a midwife in attendance. It was so natural and instinctive. All we needed now was a name for our son. Eddy offered Mr McGregor (he's a big Peter Rabbit fan)... Our baby was three days old when we decided to call him Ben.

Clearly, unassisted births happen occasionally for no good reason. In most cases, they proceed very smoothly because a fast and relatively easy birth is usually also safe. For this reason, when you arrive at someone's house, where the woman has given birth a little early, why not encourage her and be cheerful, while ensuring that all is safe, of course? I received one birth account when I was researching this book which described how negatively midwives had reacted in this situation. They changed what had been a very positive, uplifting experience into a negative one.

Of course, so as to avoid unassisted births from a safety point of view, it's a good idea if you encourage your clients to be well organized in terms of phone numbers, etc well before they go into labor. I think it's interesting and surprising how otherwise an well-organized woman can fall apart at the prospect of collecting all the relevant information and making sure it's easily and quickly accessible, wherever she may be when she happens to go into labor. As well as sticking a card up on the wall (perhaps for her husband to see), she should carry one in her purse, as I've suggested before, so it's accessible at all times.

## Birthframe 94: Sylvie Donna—Not as hoped, but also fine

My own third birth experience was not actually what I'd hoped for. All kinds of things 'went wrong' but there was nothing that couldn't be dealt with. It was true what I'd heard about it being impossible to 'plan' birth. But that didn't mean it was impossible to *prepare* for it... My preparation came to my rescue.

My preparation came to my rescue...

First of all I went into labor at rather an inconvenient time. Not only was my husband about to go off to work (where he was rather busy), I was also supposed to be taking my other two children along to a morning playgroup. I went ahead as planned, stopping in the street on my way to and from the playgroup, as each contraction made it impossible to walk for a few seconds. I hadn't felt comfortable with the idea of asking my new friends to help out with the children (since we'd only recently moved to the area). An hour or so after we got home, I eventually decided to phone my husband…

The next 'problem' was that it was broad daylight and we had no drapes in our bathroom. I'd had visions of a candlelit labor and birth because births always take place in the middle of the night, right? Oh well, I'd just have to carry on. Had to stay in the bathroom because I was vomiting, and somehow sitting on the toilet seemed the most comfortable position a lot of the time!

As I turned on the water to have a shower and run myself a bath I discovered there was no hot water. Aaaargh! What could have gone wrong? A cold bath was hardly likely to have the same effect… I decided to ask my husband if he could prepare me a bath using the kitchen kettle. ("Hot, but not too hot!")

The next minor blip was the midwives. I'd called them as late as possible because I wanted to avoid all possible disturbance. When they turned up, not only did I not know them at all, I also found them rather overly 'brusque' and business-like. No problem—they'd just need to be locked out of the bathroom! Michel had suggested this idea to me as a possibility over email only that morning.

Perhaps *because* of all these hiccups, I was not as calm and composed as I had hoped. I was even wrestling with some unresolved psychological issues… It's all very well for pregnancy books and magazines to tell us we need to discuss everything with our husband, but what if he's busy or doesn't want to talk? Now hardly seemed the time. I just continued, moment by moment, moving through each contraction as best I could. Somehow the time would pass… I gently talked to my baby to coax her out.

I gently talked to my baby to coax her out.

Another thing at the back of my mind was my lack of clarity on what to do with the placenta. Was I going to be really natural and have a lotus birth? (See Birthframe 92.) This question was answered for me very efficiently… The cord snapped as my new baby was born—she shot out with such enthusiasm. And no, I didn't manage to catch her. Landing on our thick-pile carpet covered by a plastic sheet seemed to do her no harm at all. (Giraffes apparently fall about eight feet when they're born!) The broken umbilical cord caused absolutely no problems—she breathed immediately and I myself lost only a small amount of blood. The placenta slipped out just then too.

Coming back to glitches in this labor and birth, the post-birth afterpains also took me a bit by surprise. My uterus seemed very intent on getting back to normal as quickly as humanly possible... Vomiting while nursing was not exactly pleasant, but then again the aftermath did only last a few hours in total and I had long breaks between each bout of afterpains. At least the midwives who visited me postpartum were impressed with my retracted uterus, not to mention my intact perineum. (I'd only sustained a few scratches to my vagina this time—no doubt caused by long newborn fingernails!) Beyond the afterpains in the first few hours after the birth, as before I had no postpartum pain and I was full of energy.

> Beyond the afterpains in the first few hours after the birth, as before I had no postpartum pain and I was full of energy

All in all, it had gone well, but NOT as expected! Even after the birth there were blips... First, we couldn't agree on a name and the matter was complicated by my wonderful mother-in-law trying to insist on her 'suggestion'. (That ended up being the middle name!) Then—just two days after the birth—I caught a bug off a visitor's baby. As I was chucking up for the umpteenth time that night I did at least console myself with the thought that my baby, Jumeira, would get my antibodies through my breastmilk. She was fine—and I was also back to normal a few hours later. After that I just had the usual problems of early motherhood—sleep deprivation, a messy house and a rather pudgy body.

A few days after the birth, I realized my faith in the natural processes had been reaffirmed. Our bodies and our babies know what to do. We simply need to have the courage not to intervene or disturb the natural processes. We need to observe and move through the sensations as we experience them, even if we perceive these as painful. And we need to hold on to that faith in the natural processes. We can and we must work through these times for the sake of our babies, our children, and the society into which they are born. And we're helping ourselves in doing so because postpartum our lives are likely to be much, much easier than they might otherwise be, after an alternative 'managed' birth using so-called 'pain relief'.

> We need to observe and move through the sensations

18 months later: Since Jumeira's birth I've experienced no pain, no discomfort—just a little tiredness now and then because of broken nights. The intense but brief pain I experienced during my third labor and birth was just that... outrageously strong but short-lived. Jumeira's on my lap, contentedly nursing as I type away at my computer. She's a lovely, lively, happy little girl.

> The intense pain was outrageously strong but short-lived

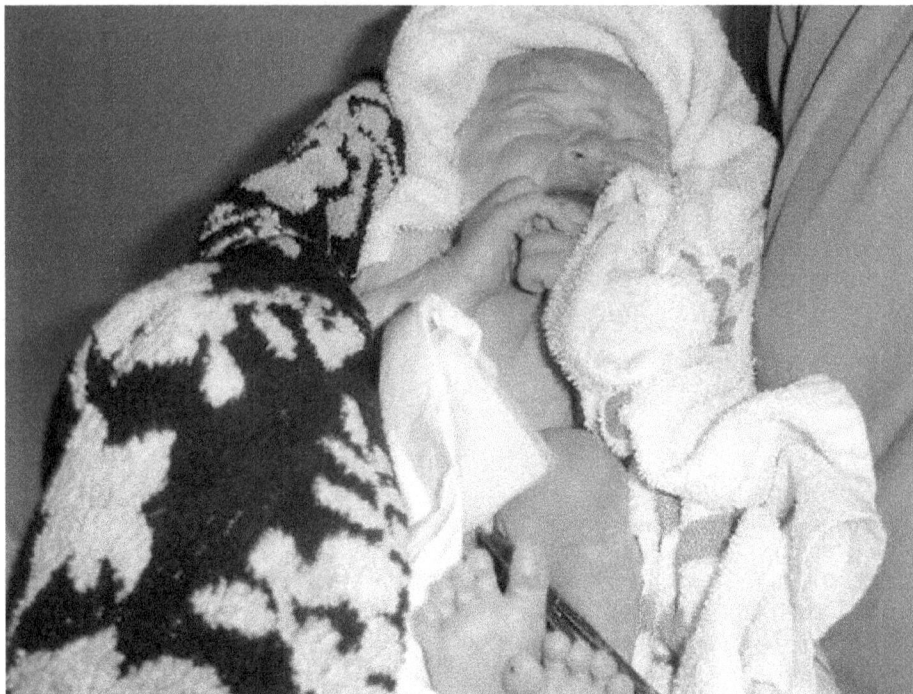

*Newborn Josh, Karen Low's last baby*

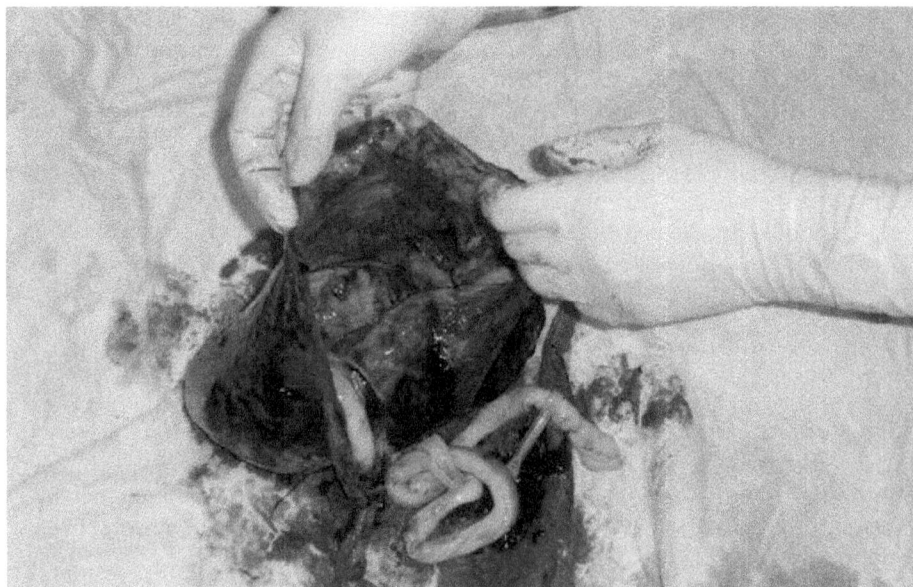

*A placenta*

# PPH

Sometimes, unfortunately, as you know, the placenta doesn't appear immediately after the baby and a postpartum hemorrhage takes place instead. The World Health Organization estimates that every year a staggering 14 million women have a postpartum hemorrhage and that around half a million die as a result each year.[92] It seems that the risk of PPH increases dramatically when women are disturbed between the second and third stages of labor. This is because their state of mind is irrevocably changed, which has dramatic physiological consequences. (It's that link between emotions and hormones again...) The following is an extract from one of Michel Odent's Primal Health newsletters.[93]

Over the years I have come to the conclusion that postpartum hemorrhages are almost always related to inappropriate interference. Postpartum hemorrhage would be extremely rare if a few simple rules were understood and observed. I am so convinced of the importance of these simple rules that I have two times agreed to attend a home birth, even though in each case I knew the woman's previous birth was followed by a manual removal of the placenta and a blood transfusion. These rules result in an approach which is in stark contrast to 'expectant' or so-called 'physiological' management used in randomized studies.

First, it is important to create the conditions for the 'fetus ejection reflex', which is a short series of irresistible contractions which allow no room for voluntary movements. If this is done, the need for privacy and the need to feel secure are met. The fetus ejection reflex typically occurs when there is nobody around but an experienced, motherly, silent and low-profile midwife sitting in a corner and, for example, knitting. Knitting—or a similar repetitive task—helps the midwife to maintain her own level of adrenaline as low as possible.

When conditions are physiological, at the very moment of birth most women tend to be upright, probably because of a momentary peak of adrenaline. They may be on their knees, or standing up and leaning on something. After an unmedicated delivery, it only takes a few seconds to hear and to see that the baby is in good shape. Then, in most cases, my first preoccupation is to warm the room. In the French hospital where I used to work, we just had to pull a string to switch on heating lamps. In the case of a planned home birth, instead of a written list of what to prepare, I focus on the need for a transportable heater that can be plugged in anywhere and at any time (including practical details, such as the need for an extension cord). When the heater is on it is possible, within a few seconds, to warm up blankets or towels and, if necessary, to cover the mother's and the baby's bodies. During the hour following birth women rarely complain that it is too hot. If the mother is shivering, it is not psychological: it means that the place is not warm enough.

From that time my main concern is that the mother is not distracted at all and does not feel observed. I want to make sure that she feels free to hold her baby, to look into his or her eyes and to smell him or her. It is easier to avoid disturbances if the light is kept dimmed and the telephone unplugged. I often invite the baby's father (or any other person who might be around) into another room to explain that this first interaction between mother and baby will never happen again and should not be disturbed. Many men have a tendency to break the sacredness of the atmosphere which ideally follows an undisturbed birth.

This first interaction will never happen again...

*During the first hour after the birth, I remain as silent as possible and keep a low profile. Either I sit down in a corner behind the mother and baby, or I disappear if there is an experienced doula present with personal experience of this situation.*

During the first hour after the birth, I remain as silent as possible and keep a low profile. Either I sit down in a corner behind the mother and baby, or I disappear if there is an experienced doula present with personal experience of this situation. Minutes after giving birth many mothers are no longer comfortable in an upright position. This is most likely the time when their level of adrenaline is decreasing and when women feel the contractions associated with the separation of the placenta. The birth attendant may have to hold the baby for some seconds, in order for the mother to find a comfortable position, almost always lying down on one side. After that there is no excuse for interfering with the interaction between mother and baby.

*For an hour I don't go anywhere near either the cord or the placenta. Clamping and cutting the cord before the delivery of the placenta is a dangerous distraction.*

For an hour I don't go anywhere near either the cord or the placenta. Clamping and cutting the cord before the delivery of the placenta is a dangerous distraction. Suggesting a position to the mother is another unneeded distraction. Her position is the consequence of her level of adrenaline. When her level of adrenaline is low and she feels the need to lie down, it would be unkind and unphysiological to suggest an upright position.

It is only when an hour has passed after the birth—if the placenta has not yet emerged—that I dare to disturb the mother in order to check that the placenta is at least separated from the uterus. With the mother on her back I press the abdominal wall just above the pubic bone with my fingertips: if the cord does not move, it means that the placenta is separated. In practice, the placenta is always either delivered or separated an hour after the birth, and bleeding is minimal, provided the third stage has not been 'managed'. I have never had to inject a uterotonic drug to control bleeding.

Such an attitude, based first on clinical observation, is based on physiological considerations. An easy delivery of the placenta with moderate blood loss implies that, immediately after the birth of the baby, a surge of oxytocin has been released. It is well-known that oxytocin release is highly dependant on environmental factors. It can be inhibited by adrenaline. This is more than empirical knowledge. A team from Sapporo, Japan (Saito *et al* 1991) studied the levels of adrenaline during the different phases of labor extensively by a non-invasive method (recording with a patch and analyzing the skin microvibration pattern of the palmar side of the hand) and confirmed the findings of a previous study in which adrenaline levels were measured through indwelling catheters (Lederman *et al* 1978). The Japanese team clearly demonstrated that postpartum hemorrhages are associated with high levels of adrenaline. The release of oxytocin can also be inhibited by the activity of the neocortex. After a physiological birth, the mother is still in a special state of consciousness, as if 'on another planet'. Her neocortex is still more or less at rest so my advice is: "Don't wake the mother up!" Once again, I must emphasize the need for privacy and silence.[94]

# Prematurity

How much do you know about kangaroo mother care? The following account will tell you more...

## Birthframe 95: Krisanne Collard—Kangaroo mother care

Here, one woman explains how she came to use a very natural approach to supplement her new daughter's technological support system.

In July of 1994, my husband and I moved to the little town of Canon City, Colorado. We had no family in the area—in fact, we had no family within a day's drive. I found out I was pregnant within two weeks of the move. We were ecstatic! We had tried to get pregnant for three years and suffered a miscarriage eight months before. We were finally going to have the baby we had dreamed about for so long.

I had my first prenatal appointment at the end of September when I was 11 weeks pregnant. The doctor asked me to provide a urine sample, confirmed I was pregnant and told me my due date would be March 4. He sent the rest of my sample off, so the lab could run all of the standard tests for new moms. He told me everything looked wonderful and I would be able to hear the heartbeat at my next check-up. I scheduled my next appointment for a month later, and my husband and I left the office overjoyed at the thought everything was fine. We went to Wal-Mart and looked at all of the pretty baby things, dreaming of what the nursery would look like.

The doctor called me three days after the appointment and asked to see me in his office later that day. He sat us both down and gave us the bad news. All of the standard tests he had run indicated that I no longer had a viable pregnancy. Our baby was probably no longer alive. He scheduled an ultrasound for Monday—just to be 'sure'. It was then Friday. All weekend we worried and panicked. By Monday afternoon, I had convinced myself that she was indeed dead—that we would have to try again. When I lay down on the table and told the technician why I was there, she had tears in her eyes as she prepared to take a look. For two agonizing minutes, all was quiet. Then, she screamed, "Oh my God! Look! Your baby's moving!" Sure enough, there was my little fighter, mad at the world—wriggling and trying to get comfortable. Her heartbeat was strong. My doctor never had an explanation for the incorrect test results.

For the next two months, my pregnancy went by the book. I had some morning sickness and started to get the 'pooch'. I felt her move for the first time on my birthday (November 14). We bought baby stuff and started to decorate the nursery. We were the happiest parents in the world!

But, on November 30, 24 weeks into my pregnancy, everything went black. I woke up with a dull ache in my lower back. I didn't think much of it till I started bleeding at about noon. I called my doctor and my husband, who rushed me to the local hospital. I was immediately hooked up to a fetal monitor and checked for dilation. I was already 5cm and her feet were in the birth canal. She was coming breech. Next thing I knew, I was being prepped for an immediate C-section. I was sobbing, screaming and confused. No one had time to talk to me—no one would let me know how the baby was doing. My husband and I were terrified!

When I woke up, I was in the recovery room. All I can remember was asking for a drink of water. I didn't ask about my baby because I thought she was dead. I just started crying—mourning her already. After about half an hour, another nurse came into the room. I asked her, "Was it a boy or a girl?" She smiled at me and said, "You have a beautiful little girl!" I was confused. She hadn't spoken in the past tense and she was happy. My daughter was still alive!

She hadn't spoken in the past tense and she was happy.
My daughter was still alive!

I was moved into my recovery room and my doctor came to talk to me. He told me that my daughter had been born weighing only 1lb, 12oz and only 13½ inches long. He said she came out kicking and screaming, angry at the world, but she was VERY sick. They were transporting her to a level three NICU (Neonatal Intensive Care Unit) at Memorial Hospital, an hour away, in Colorado Springs. He told me that there was only a 30% chance that she was going to make it through the night.

I didn't get to see her before they took her away. As I recovered from the C-section, I called the hospital she was in three times a day to check on her. She had been put on a special oscillating ventilator, I was told, and required maximum oxygen doses to keep her saturation up. They said she had gotten an infection from me and was very sick. I had a bladder infection I didn't know about (I have always been symptom-less), and she had come early because she was too sick to stay in my womb. They assured me over and over that she was a fighter and was going to be fine. They encouraged me to talk to her through the phone—but I felt silly because I hadn't even met her yet. Since we now felt she was going to make it, we decided on a name—Kaia Michele. A strong Norwegian name for a very brave little girl.

When I was released from my local hospital, four days after her birth, my husband and I immediately went to visit Kaia. When I walked into the NICU, nothing could have prepared me for what I saw. She was the size of two Barbie dolls stuck together. She had the blue bilirubin lights on her and tubes and wires everywhere. Her skin was translucent and her veins showed through her skin. I looked around, hoping to see 'my' baby in another bed. I looked back at Kaia. The nurse urged me closer and turned off the blue lights so I could take a closer look. I was terrified. The nurse urged me to touch her and talk to her—let her know I was there. I hesitated. She looked like she would break if I touched her. I started to talk to her, but it took 15 minutes to finally get up the nerve to touch her. When I did, her alarms went off. I jerked my hand back as the nurse came over. She had to rub her till her heart started again. I couldn't handle it—my baby had rejected me—I left the nursery in tears and went home.

The next night, as we were eating dinner, I burst into tears. My husband took my hand and led me to the couch. He put my shoes on and said, "Let's go—Mommy needs a baby fix." We drove the hour to Memorial Hospital. All the way there, I told my husband that THIS time I was going to be strong. I was going to count her toes and talk to her and let her know how loved she was. I was so excited!

But, when I got to her side and touched her, her alarms sounded again. The nurse had to start her heart yet again. I just sat there for the next 30 minutes and looked at her, crying. Why didn't my baby love me? I wondered. A nurse specialist named Theresa Kledzik (nicknamed 'The Whisper Lady') came over to talk to me. She asked me if I had held my baby yet. I just laughed and said that Kaia couldn't tolerate me touching her—I would kill her if I tried to hold her. Theresa just smiled at me and asked me to follow her into the other room. She gave me a gown, told me to take off my bra and shirt and put the gown on with the opening in the front. I was in shock as I changed.

When I walked back into the nursery, there were three nurses getting Kaia ready to be transferred to my chest. It took about 15 minutes for them to get her ready and I got more and more nervous every second. I changed my mind about 10 times, but the nurses didn't pay me any mind. I sat down in a strange-looking chair. (I learned later that it was specially designed to be used with babies on oscillating ventilators.) They laid Kaia on my chest, on her tummy, skin-to-skin with me. Her head was resting above my heart and her tiny feet were curled up in my hand. The tubes and wires were taped to my gown. I stared at the monitors as she wriggled into a comfortable position. I knew they were going to sound. I held my breath. She calmed down after about five minutes and stopped moving. I thought she was dead. I tried to make sense of the monitors. I called Theresa over and asked if I had killed her. She just laughed and said, "No, she's happy!" I looked down at Kaia and noticed the peaceful look on her face. I looked back up at Theresa and smiled. My daughter had just made me feel like a mom for the very first time!

I was able to hold Kaia every day for two hours. I couldn't get to the hospital fast enough every day to share that special time with her. Within a few days, I was able to transfer her to my chest all by myself, even with the tubes and wires everywhere. I was always disappointed when our time together came to an end. She responded very well to it, and her oxygen requirements lowered during every session. She never had any episodes where she stopped breathing (apneas) or periods where her heart stopped (bradyas) while I held her. She would just go into a deep, healing sleep and actually tolerated my touch when she was on the warming table. The future was looking very promising for her.

When she was two weeks old, she had trouble getting comfortable on my chest. I wound up putting her back after only half an hour. I spent another hour with her, then headed home. When I got there, there was a message telling me to immediately return to the hospital. My husband and I rushed back just in time to say goodbye as they rushed her to the operating room. They had no idea what was wrong, and no idea how long we would have to wait. Four hours later, the Neonatal Surgeon came out and told us she was going to be fine. She had got a small perforation at the end of her large intestines and the surgeon had made a colostomy. We got to see her within an hour. She had a huge bandage on her tummy and they had given her paralyzing drugs. She looked so little and sick. I cried as I held her hand for the next two hours. The nurses assured me I'd be able to hold her again in a few days. I missed her so much.

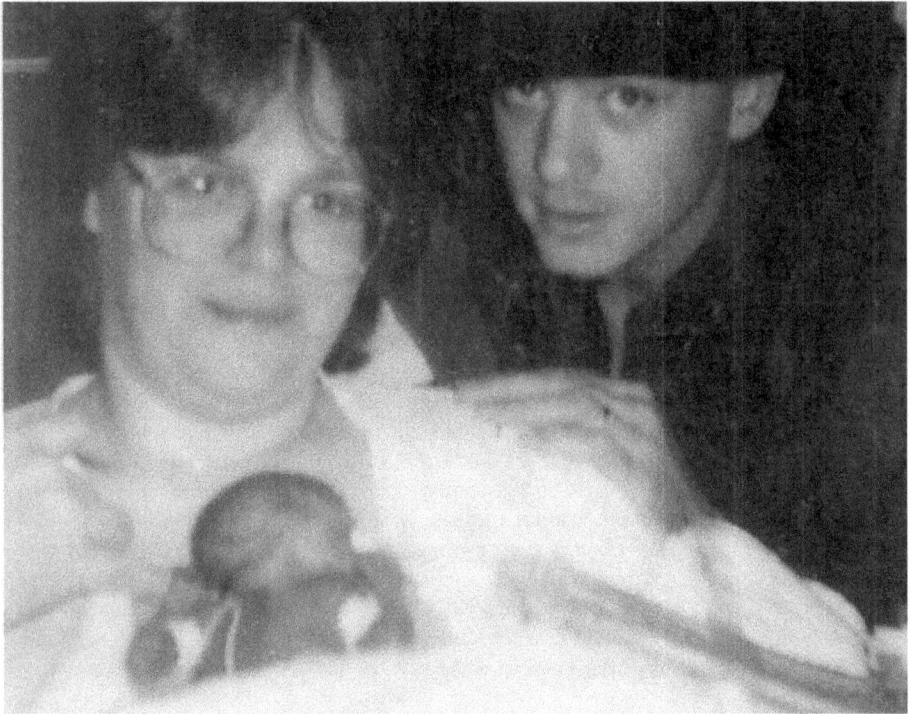

*First time kangarooing Kaia successfully (with non alarms), when she was 4 days old*

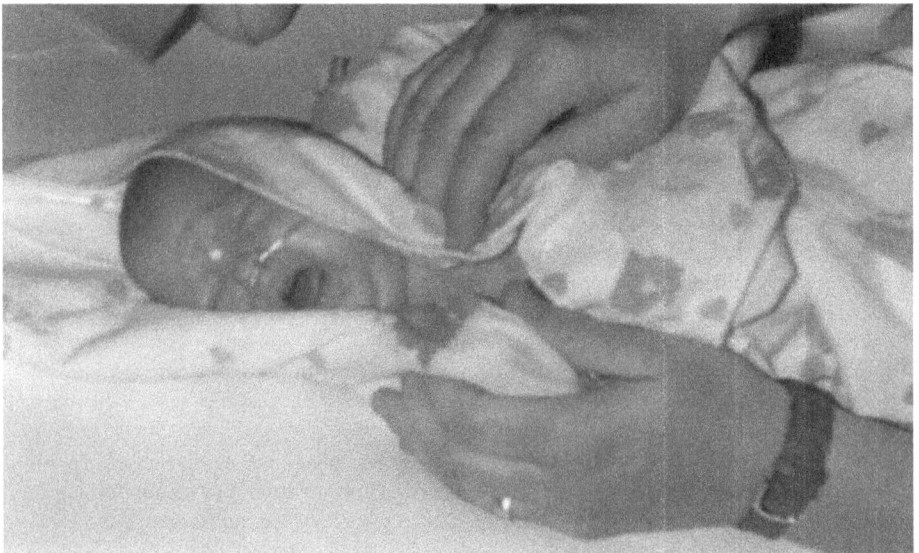

*Kaia at 3 months old, having her first bath at home*

I visited her the next morning. She was very uncomfortable and was requiring maximum pain medication doses. She wriggled and cried (silently because of the ventilator tube in her mouth). Finally, after an hour of seeing her suffer, the nurse said, "Hold your baby, she needs you!" I leapt at the chance and got her out of bed myself—making sure not to put too much pressure on her tummy. The nurse came over a few minutes later to check on us. She smiled. Kaia was actually lying on her newly operated-on tummy, sleeping peacefully. I held her for the full two hours and she didn't even need her next dose of pain medicine! I was on top of the world. I had gotten to hold and comfort my child!

Kaia required another surgery, at three weeks old, to modify her colostomy. She was moved to the Intermediate Nursery after two months, and came home on oxygen and a heart monitor a month later, on February 27, two weeks before her due date.

She went back into the NICU two weeks later for eye surgery to correct ROP (retinopathy of prematurity), which is common in preemies who have been on oxygen for long periods of time.

A month after that—she was 4½ months old by this time—she had a fundoplication [surgery in which the top of the stomach is wrapped around the esophagus, creating a valve] to correct severe reflux. We had to feed her through a tube (called a G-tube) for the next four months.

When she was 8½ months old, she had her final surgery to reconnect her colostomy and take out the feeding tube. She remained on oxygen and a heart monitor, continuously till she was a year old. This meant lugging around a 15lb oxygen tank and a 10lb heart monitor everywhere. People were very curious. In the grocery and department stores, I was given strange looks or stopped and asked 100 questions before I could continue my shopping. We didn't get out much because it was such a hassle.

That first year of Kaia's life was the most stressful year of my own life. There were so many things to do: doctor appointments, surgeries, therapy sessions, medications and colostomy bag changes. And so many milestones: first smiles, first words, first steps, first hugs! I was on a roller coaster that didn't seem to have an end in sight. I joined a local playgroup and an online preemie support group when Kaia was six months old. Both groups gave me so much love and support. They kept me from going insane. There are MANY premature baby support groups on the Internet and probably one in your own state or city. To find the one that's right for you, join several—don't limit yourself to just one. Every support group is different and unique. Your hospital will be able to tell you if there is a local support group in your area.

At our NICU reunion party in August 1995 Kaia was 9 months old. I ran into the nurse specialist who had encouraged me to hold my daughter the first time, Theresa Kledzik. We talked for almost an hour about 'kangaroo care'. That's what she called the skin-to-skin contact I had with Kaia. I asked her if all babies were able to be kangarooed and she told me that I had been very lucky. Most parents in other hospitals weren't able to kangaroo their babies as soon as I was. Most parents had to wait till their babies were 'stable', off the ventilator or 3lb before they were allowed to hold them. I was shocked! Kaia would have been over two months old if I had had to wait. I couldn't imagine having to wait that long to feel like a mom.

When I got home, I immediately told my online support group about what I had learned. Most wrote back asking what kangaroo care was and how they could do it. They had never heard of it before. I searched the Internet and found only one site that even mentioned kangaroo care. I was upset. How many other babies had to suffer without the loving touch of their parents? I emailed Theresa and she sent me all of the literature and research she had on the subject.

There was so much, it took days to go through it all. I was amazed at all I learned and angry that more hospitals didn't embrace kangaroo care. Here is just a glimpse of the amazing discoveries I made...

- Kangaroo care—originally called 'kangaroo mother care'—was developed in Bogotá, Colombia in July 1977 by neonatologists Edgar Rey and Hector Martinez. It received worldwide publicity in 1983 when the startling success of the approach became known. The mortality rate for premature babies in Bogotá had been 70% (39% in the US at that time) due to lack of power and reliable equipment (such as ventilators and incubators). After getting moms to carry their babies continuously in slings on their chests, the mortality rate decreased to 30%!

- Research has shown that moms and babies have a natural thermal synchrony. When Mom thinks her baby is getting too cool, her body heats up in response. When Mom thinks her baby is too hot, she cools down. Sorry, Dad, this only seems to happen with moms.

- The skin-to-skin contact helps in milk production and milk release. It is very important for mothers of premature babies to pump and store their milk for the time when the baby can start eating. The breastmilk of a mother who has given birth prematurely is tolerated by the baby's immature digestive system more easily than formula.

- Apnea (interruptions in normal breathing), bradycardia (severe heart rate decreases) and tachycardia (severe heart rate increases) decrease or disappear altogether because kangaroo care regulates breathing and stabilizes heart rates. These episodes are VERY common in premature babies and can easily lead to death or lack of oxygen to the brain, which could cause brain damage.

- Kangaroo care stimulates the baby to gain weight more rapidly and be discharged up to 50% sooner because there are greater periods of deep sleep that enable the preemie's body to heal faster.

- Kangaroo care can also be done with full-term babies who have colic. Colic is believed by many to be a baby's inability to transition from one sleep-state to another. Kangaroo care allows babies to enter into the deepest sleep state with ease. Most preemies that are kangarooed never suffer with colic. Mine didn't.

- Preemies who are kangarooed during treatments (such as 'heel sticks' to check the amount of oxygen in the blood—an hourly requirement for preemies on a ventilator) experience far less discomfort and require no extra pain medication during these painful procedures. Kaia didn't even seem to notice the treatments if she was being kangarooed.

After telling my online support group about what I had learned, I was encouraged to set up a website to educate the world about the wonders of kangaroo care. One of my online friends helped me design the website www.geocities.com/roopage and several preemie moms wrote stories about their first kangaroo care experience. I received so many wonderful stories, I decided to compile them all into a tiny booklet that I could send to NICUs and parents across the world. *Kangarooing Our Little Miracles* lets parents, nurses and doctors share the emotional joy kangaroo care brings to babies and parents. Most parents shared with me that by writing down their emotions, it helped with their healing process. I gave away over 200 booklets within the first year and received at least 15 emails a month from parents, neonatologists and nurses who wanted to implement kangaroo care in their hospitals. It feels wonderful to be an instigator.

In October 1996, I was asked to speak at the First International Congress of Kangaroo Care in Baltimore, MD. I was both flattered and nervous. I blew the minds of several nurses and neonatologists in the audience when I told them about my kangaroo care experience. They couldn't believe that I had been able to hold her on a ventilator (an oscillating one at that), or hold her when she was so unstable. They actually gasped when I told them about her reaction to being kangarooed after tummy surgery. They all cheered at the end of my speech when I told them that all babies deserve to feel their mother's and father's love and all parents deserve the opportunity to be parents, not visitors.

My 'preemie' is now 8 years old and going into 3rd grade. Where has the time gone? She is doing wonderfully in school and has no lasting effects of being born prematurely. She is sweet and kind and funny. She puts a smile on my face every day. The other day, when we were swimming at the local pool, another child asked her why her tummy was all messed up (she insisted on getting a bikini even though her massive scars showed). I held my breath, wondering if I should step in and help her explain. My eyes filled with tears as she said, proudly, "I was born the size of two Barbie dolls. I have these scars to show me how strong I had to be to live." I know that I helped give her that strength. I am so proud of her and thankful to Memorial Hospital of Colorado Springs for giving me the opportunity to be her mom!

We found out I was pregnant again in December 1995. My pregnancy was uncomplicated. My doctor checked me every two weeks for bladder infections. (I was treated for two during my pregnancy.) My doctor informed me that the prenatal pills I took with Kaia might have played a role in her early birth because they contained a high content of iodine, which I am highly allergic to. He switched me to a prenatal vitamin that contained no iodine. With Kaia, whenever I had missed one pill, I spotted. After missing two, I had Kaia. I am still very allergic to iodine and have to watch which multivitamins I take and keep salt out of my diet as much as possible. Even though I showed no signs of premature labor (i.e. contractions or bleeding), my doctor put me on terbulatine (used to stop contractions) at 28 weeks to be sure I wouldn't go into premature labor. I strongly encourage women who have ANY problems with their pregnancy to talk to their obstetrician. Don't just let it go. Even a small problem could snowball and mean life or death for your baby.

My mom teased me as I went through the summer months uncomfortable and VERY pregnant. Every time I would complain about how miserable I was, she would laugh and say, "YOU are the one who wants to see what it's like to carry a baby to term. You are going the full 40, young lady!"

My doctor left for his first vacation ever on August 1, a Friday, right after my 36-week appointment. He took me off the terbulatine and told me that I was NOT going to have her till he got back! I just laughed and told him that I wanted to have her that weekend. I did! I started having contractions within 24 hours of discontinuing the terbulatine and Katherine Elsie was born on August 4, Sunday, weighing a healthy 6lb, 15½oz by C-section. I was able to hold her within a couple of hours after she was born. I unwrapped her, counted her toes and fingers, and held her 'kangaroo care' style for hours as I slept.

She has been perfectly healthy—I was able to actually enjoy her first year. However, Katherine is a real free spirit. The tantrums that child threw! WOW! Kaia tried to imitate her sister when she was about 3 years old. She lay down on the floor, spread her arms wide and said, "Mom, I'm mad!" I just smiled down at her and said, "You forgot to scream." She got up, dusted off her pants and smiled at me. She never tried it again. I found out later that year that Kaia was short for Katherine in Norwegian. As they get older, I have noticed how ironic this is. They are as close as identical twins, but different as night and day. Just like their names!

The doctor who delivered Katherine suggested, after she was born, that I shouldn't have any more children. I was heartbroken AND relieved. My husband always wanted eight! He said that I hadn't healed well from my first C-section, and delivering Katherine had been very difficult with all of the scar tissue. It would be nearly impossible for him to deliver another baby safely. I was also told that I would never be able to deliver vaginally because I have a heart-shaped uterus (called a bicornuate uterus), which allows the baby only half the space to grow and no space in which to turn near the end of the pregnancy. (This had no affect on Kaia being born prematurely, incidentally.)

> Today, both of my children are healthy and happy.
> They are the lights of my life.

Today, both of my children are healthy and happy. They are the lights of my life. I know I wouldn't be the person I am today without the experiences of having them and raising them. They have taught me so much and made me stronger. I am forever grateful. Thank you, baby girls! I love you!

Note: There is some uncertainty as to when 'Kangaroo Mother Care' was first used. In some articles and websites it's given as being 1978. However, in the French translation of the first article by Edgar Rey, it is clearly given as July 1977.[95] It certainly seems to be a very useful practice, according to research.[96]

*Kaia, aged 8, with her sister Katherine, aged 7*

I asked Michel about prematurity and kangaroo mother care...

### What can a health professional do to prevent prematurity?

Recently, considerable research has focused on how useful antibiotics might be to prevent prematurity. A large multi-center randomized controlled trial involving 6,295 women did not support the use of antibiotics.[97] Another study concluded that the treatment of vaginal infection in early pregnancy does not decrease the incidence of preterm delivery.[98] Cerclage of the cervix, although widely used to reduce the risk of premature birth, has now also been called into question: research into this technique is inconsistent but has shown that the risk of postpartum fever is doubled as a result.[99] Medical interventions also do not reduce the risk of having a small-for-dates baby. Finally, even bed rest has been shown to be useless and even harmful.[100]

### When babies are indeed born prematurely, in what circumstances would you say it's appropriate to use so-called 'kangaroo mother care' with premature babies, instead of using conventional incubators?

In Bogotá [Columbia, South America], they have used kangaroo mother care with babies below 1000g. The prerequisite is that the baby can breathe without any assistance. In Pithiviers [the maternity hospital for which Michel was responsible near Paris, France] we could not be as audacious as in Bogotá. We used this method for babies weighing more than 1700 or 1800 grams (and breathing easily). If a woman wants to use kangaroo care, it is better to start in the birthing room, without interrupting at all the skin-to-skin contact. It is difficult without the enthusiastic cooperation of the health professionals.

> I was first able to kangaroo with Megan when she was 4 days old, the day after I was discharged. My husband and I hadn't expected to be able to hold her for a couple of weeks. We were absolutely thrilled. It was the first time since my pre-term labor began that I felt really happy. She was so fragile, so small. I recall her tiny hand lying on my chest. I could feel her body lying on me and it made my skin tingle all over. She seemed to weigh nothing at all. I never wanted it to end and I will never forget how amazing it was. I recall feeling as if I had been given a drug that made me feel more wonderful than I ever imagined. It was the first time I really felt like a mother.

*Nancy Redheffer, talking about her daughter Megan, who was born at 27 weeks 6 days*

I also consulted a midwife with lots of experience of Kangaroo Mother Care, Luisa Cescutti-Butler, who commented as follows...

Much has been written about the benefits of kangaroo mother care (KMC), which was introduced into neonatal care in response to high mortality and morbidity in resource poor countries.[101] The account by Krisanne Collard describes some of these benefits. Bergman, a highly respected expert has confirmed through much of his writings on KMC that to separate mother and baby is detrimental to the baby's future. Separation causes stress in newborn infants, and continued levels of raised stress hormones are toxic to the infant's developing brain.[102] KMC (also called 'skin-to-skin care') is one practice that will keep mother and baby together. Not only is this practice vital for term newborn babies but it's also of fundamental importance for premature infants.

In the traditional model of neonatal care premature babies are quickly removed from their mothers after the birth and transfered to a neonatal intensive care unit where they are housed in an incubator—a device intended to keep the baby warm and protect him/her from infection. In reality, though, this device only serves to keep the mother at arm's length from her newborn baby. Babies are only allowed out of incubators for KMC after guidelines have been written and enough staff are around to facilitate transfer from incubator to mother. If a baby becomes unstable when it's experiencing KMC it is usually swiftly removed from its mother and returned to the incubator until its condition stabilizes. The outcome is that many babies spend one to three hours undergoing KMC and 22-23 hours ensconced inside an incubator.

One neonatal unit in Sweden has turned this practice around and created a paradigm of care based on keeping mothers and babies together at all times. KMC is provided 24/7 and parents become the main carers, with neonatal staff providing support whenever necessary.[103] Numerous studies have demonstrated that premature infants show improved oxygenation and thermoregulation when experiencing skin-to-skin contact and the effects on the mother are far-reaching. She is able to start the process of attachment and this secures a positive future for the premature baby and may begin the process of healing any emotional problems which may have arisen as a result of becoming a mother prematurely.[104] I can put it no better than Bergman himself when he states: "Maternal-infant skin-to-skin contact from birth is the biological alternative to incubator care." This form of contact between all mothers and babies should be the norm and not the exception.

Recent research by the POPPY Project (Parents of Premature Babies Project) has highlighted that many neonatal units in the UK actively promote family centered care. Clearly, this is a very positive development.

### On the subject of prematurity generally, Luisa wrote the following:

Approximately 54,000 babies are born prematurely in England each year. [According to the NY Times, more than 540,000—10 times as many!—are born in the USA each year.] This represents 8.3% of the total number of live births.[105] There are a number of known risk factors associated with prematurity, such as high or low maternal age, obesity and smoking, but the single largest cause of spontaneous premature labor and birth remains unknown.[106] Where there is a known cause, such as pre-eclampsia, women may often be aware of the impending birth of their preterm baby and preparations prior to delivery may include a visit to the neonatal intensive care unit (NICU) to familiarize themselves with this new and sometimes unknown environment. However, for many women the birth of their premature baby is unexpected... They are not ready to become a mother yet[107] and they will have had little or no time to prepare for their baby's admission to a neonatal unit.[108]

During her pregnancy a woman may be experiencing many emotions about her unborn baby. She may be imagining what her baby looks like, what sex the child is and she will be preparing for her forthcoming role as a mother. Before meeting her new baby she is already acutely aware of him or her as she has been feeling her unborn baby's movements for many months and knows when her baby is asleep or active.[109] Unfortunately, for some women this dream of birthing a full-term, healthy baby is shattered when their babies are born too early.

Added to this stress is the far from ideal situation of learning to become a mother in an environment which is unknown and at times, frankly, daunting.[110] Women who have become mothers prematurely will often feel disappointed in themselves and useless as mothers because 'efficient healthcare practitioners' are caring for their vulnerable newborn infants.[111]

So what can we do as midwives? Research has shown that providing mothers and their partners with the opportunity to see and touch their infants in the delivery room or prior to transfer to the neonatal unit may help to reduce anxiety and uncertainty.[112] However, where this is not possible, perhaps due to the condition of the baby at birth, minimizing the delay between birth and the first contact may be helpful. At the very least, staff should be sensitive to the increased stress for mothers that may result from such delays. The following extracts from one of the women I interviewed for my Master's thesis highlights some of the issues when women are prevented from being reunited with their preterm infant for the first time.

**Case Study: "Justina"**

One of the mothers had birthed her baby late the previous evening and by 5:00AM the following morning was anxious to go to the neonatal unit, so she asked to do that. It was not until two hours later that she got to see her baby, though. When asked if there were any reasons for the delay, the woman responded:

"The midwife said I would disturb everyone that early. …. I had to wait and read a magazine and … I'd have to wait till a more reasonable hour."[113]

This delay clearly made her unhappy. She was finally given permission by the midwife to go and see her baby, but even then her visit was further delayed because the midwife first had to phone the neonatal unit to check that it was appropriate for the mother to come up.

"And then they said we'll have to phone and check, so they phoned and checked and said … 'Should we wheel you across?' and I said 'No thanks. It's okay, I'll walk.'"[114]

So frustrated was this woman at all the pitfalls in her way that in her haste to see her baby she refused an offer from the midwife to wheel her up to the unit in a wheelchair, despite having had a cesarean. Although she found it difficult and painful to walk, it did not matter—all she wanted was to see her baby.[115] Should midwives be calling NICU to obtain permission for a woman to see her baby, thus potentially adding to further delays? I don't believe it is strictly necessary and the midwife should take the woman to NICU as soon as the woman feels able to go. If it's not convenient, and by that I mean perhaps neonatal staff are undertaking a procedure on the woman's baby, the woman could wait in the parents' room or she could be part of the process if she wanted, until the episode of care has been completed.

Postpartum wards are busy environments where midwives are tasked with looking after an increasing number of women who are deemed high risk in the postpartum period because they've had a cesarean and require close observation. In some hospitals midwives are often taken away from the ward to make up for a shortfall of staff in the labor ward, thus depleting perhaps an already short-staffed area.[116] Many women find the postpartum stay an unrewarding experience. A study on women's views of postpartum care[117] discovered that women expected their hospital stay to be one of rest and recovery with support from midwives. In reality, time constraints, busy wards and increased work demands on postpartum midwives mean that women often felt unable to ask for help and advice so spent a great deal of time alone with their babies, feeling unsupported and wanting to go home.[118] If women who have experienced the birth of their baby at term find it difficult, imagine how much more challenging the postpartum environment is for those women who are separated from their baby because of his/her prematurity.

Imagine how much more challenging the environment is...

"Maternity staff are busy ... The second day that I wanted to see her, I had to wait *hours* for somebody to push me up there and I'm, like, I want to go and see my baby. Everybody else has got their baby. I want my baby! I had to wait and feel like a real pain in the butt till somebody could wheel me up there."[119]

Women who have birthed prematurely will now be exposed to unfamiliar people and procedures so if a midwife can guide them through the early days it will have a huge impact in terms of helping them to cope. Often women don't have any idea what a preterm baby may look like, let alone their own baby, as the following comment from "Beryl" suggests:

"I felt very emotional and terrified as I had not seen a baby that small before ... My first impression of her was as 'a tiny red thing'."[120]

I felt very emotional and terrified as I had not seen a baby that small before. My first impression of her was as 'a tiny red thing.'

It's clearly helpful to provide the mother with information before she sees her baby (e.g. a description of what her baby may look like and details on what equipment her baby is attached to) and to tell the mother not to be too scared if alarm bells go off. Honest, sensitive and open communication with mothers is necessary, because they may be very frightened and not know what to expect. If a student midwife is undergoing her placement on the NICU she can provide an important link between the postpartum ward and the NICU. The student could undertake the daily postpartum examinations required by a newly birthed premature mother in the early postpartum period and she could go down to the ward and update the woman on a regular basis. Better still, she could take the mother to the NICU and help to facilitate mother and baby attachment. Both midwife and student midwife have an important role in settling a new mother and in helping to orientate her to the unfamiliar and often hostile environment of a NICU. Doing this would go a long way towards helping a mother form a bond with her premature infant.

**Further suggestions for practice:**

- Encourage mothers to talk about their experience of premature birth, listening carefully and offering support when required.
- Introduce a scheme whereby midwives rotate through NICU because this will help inter-professional relationships. Rotation will increase midwives' knowledge and understanding of what happens on a NICU and this will in turn enable midwives to support women in their care.
- Encourage midwifery students on a NICU to do postpartum checks on the mother (if relevant) when they are assigned to care for a premature baby.
- Encourage parents to be their babies' primary carers and make sure any student midwives do the same.
- Support preterm breastfeeding and help mothers with pumping their breasts early on. Research has shown that mothers don't always sustain their lactation; when this happens, mothers can experience feelings of inadequacy and low self esteem.
- Try and give time to those mothers who don't have their baby on the postpartum ward, remembering that they are surrounded by mothers who have their baby with them.
- Create a position within your unit of neonatal/postpartum liaison midwife, or make sure this is recognized as a developmental opportunity.

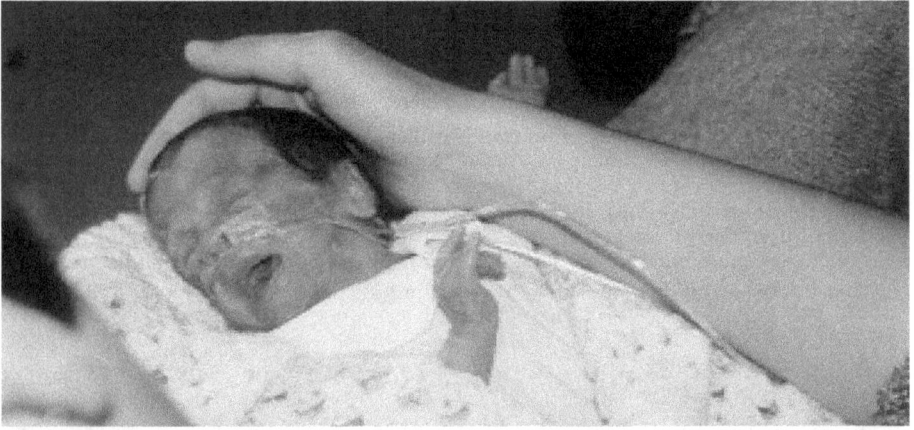

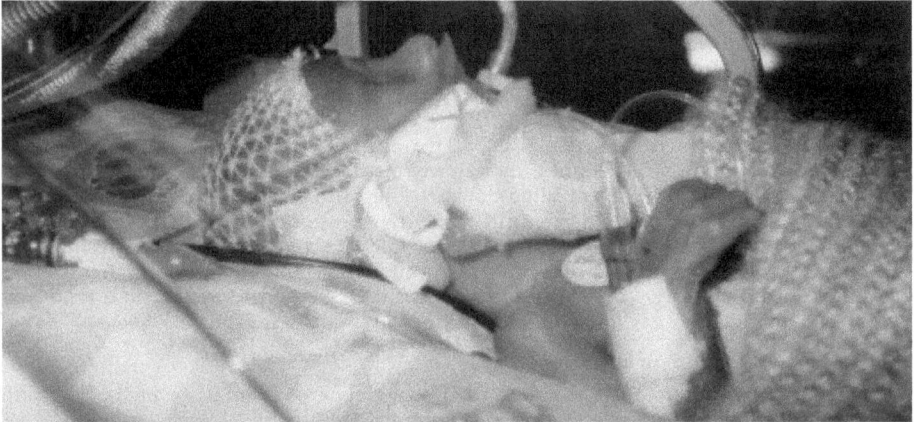

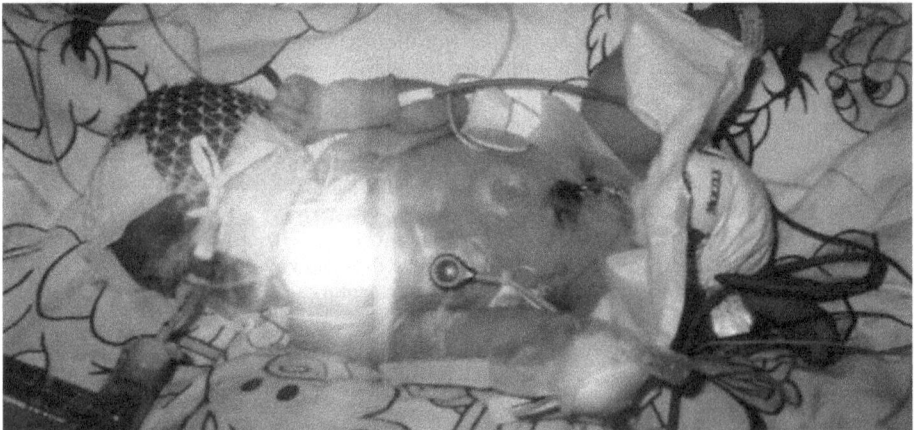

*Premature triplets, soon after their birth*

## What if the worse really comes to the worst?

### Birthframe 96: Monica Reid—Stillbirth followed by live births

This account is about a woman I met when she was 38 weeks pregnant with her third child—her name is Monica Reid. I was 36 weeks pregnant myself at the time and had no idea what significance '38 weeks' had for Monica. We became great friends almost immediately and within four weeks we had both given birth again.

One day when Monica came to visit me with her new baby (while our older children were at school and playgroup), she told me what had happened when she'd been 38 weeks pregnant with her first child. As our two babies sat or nursed on our laps, the story eventually came tumbling out. (She'd agreed to let me record it.) Monica said when we'd finished that this was the first time she'd told the story in such great detail since the birth.

To make the transcribed interview more readable—and also somewhat shorter!—I first edited the account to turn it into a monolog. Then, as I did with every contributor, I asked Monica to check through and rewrite (if necessary) before giving me permission to use it in this book. Monica has told her story because she would like to give other women who have similar difficult experiences the comfort that comes from knowing that we are never entirely alone in our experiences in life, however awful, and that, yes, there is a light at the end of the tunnel. Perhaps her account can also help improve practice.

---

It's funny because when people ask me how many children I have I always go to say actually I have three children but I always have to stop myself and say I have two children because I don't always feel comfortable talking about my first child, who we named Marley for the legendary Bob Marley. He was actually called Marley Levi Mandela. We called him after Nelson Mandela, the great civil rights leader of South Africa and Levi after his dad. It's a very powerful name. It's just unfortunate that he wasn't powerful enough to stay on this Earth. On record at the hospital I'm a mother of three but I only have two living children.

Marley was born on August 11 at 7:48 in the evening. I remember everything, It was a very rainy night in 1994. The labor was awful. He was actually born on a Thursday, but I was started off on labor on the Tuesday. And it all started, if I go back to the Monday morning before, when I was 38 weeks pregnant. That weekend before I had him had just been a normal weekend. Looking forward to the baby, going through the baby's things for about the hundredth time. Looking at it all and picking up the little matinee jacket and putting it on my tummy and giggling and getting really excited and then getting scared because, oh God, I knew it was going to kill—all the pain and everything—and then getting excited again. We were going to do so much with this child...

We were going to do so much with this child...

---

Then on the Monday, I'd come down with very bad grippe and was very ill and I'd actually called the doctor. I said that I had a bad cold and felt that there was something that wasn't quite right. I didn't feel quite right within myself. I was actually told by my doctor at the time that I only had a cold and I was pregnant, not to worry about it. He said many women had colds while they were pregnant. But sometimes you know that something's not quite right within your inner self. It was like a little voice in the back of my head telling me that something wasn't right. It was a sudden feeling, an inner voice telling me, "Something's wrong." I don't know whether it was preparing me or something but I just knew that something wasn't quite right. I had a very bad cold but beyond that I just knew that there was something not right, but I just couldn't put my finger on it. So, the rest of the day I actually spent in bed just sipping on honey and lemon and not really eating much.

On the Tuesday morning I woke up feeling a little bit better—not a lot but a little bit better. In the afternoon I had my last prenatal check-up for the pregnancy. So I got up as usual and filled the bathtub. I thought it very strange when I prepared the bath that there was no response from Marley, because usually he loved the sound of running water, be it a bath or just turning the faucets on. I got into the bathtub anyway and started by putting water over my tummy for a reaction, for kicking, for some kind of movement, and I didn't get anything. And at the back of my mind I kept thinking, "Oh, something's not quite right" again, but I dismissed it and carried on with the rest of the day.

And then Levi couldn't make the afternoon appointment at the prenatal clinic so my mom came with me. And we walked there. It was a very nice day that day. It was very sunny. And I remember it all exactly. I was wearing a blue checked maternity dress with blue sandals and I carried my prenatal notes in my hand. And I remember walking up the hill with my mom and we got to the prenatal clinic. Sat and waited for my turn. And then when the woman called me through I knew the first thing she would do was weigh me. When she weighed me, she said, "Oh, you've lost some weight." And I said, "Well, I have had a bad cold." So she took me through to the other room to listen to the baby's heartbeat using a Doppler. And she couldn't pick anything up. I know it's a long time ago but it feels like yesterday. I actually said to her, "The baby's dead." It's like God was saying to me, "It's just not meant to be."

When I had walked in there I hadn't thought anything. I didn't think... I just kept thinking there's something, something's not right. Something's not right at all. I didn't say anything to my mom or anything. I said to the midwife, "The baby's... the baby's dead." She said, "Don't be silly. It's probably just... it's gone into a breech position or something." Because sometimes they can't get the heartbeat. So she called for an ambulance and she said, "Look, we're going to send you off to the hospital. We'll take you to the hospital and do you a proper ultrasound to see what's happening." So I waited. My mom's very religious and we sat and prayed that everything would be okay. I kept saying to my mom, "Oh, don't... Everything's going to be okay." I hoped everything was fine and said, "Mom, it's alright" but knowing, knowing that it wasn't okay.

She was new to the job and had never had to deal with a stillborn baby at all. I didn't know how to handle the situation.

So we got to the hospital and I remember the midwife, Mandy, was a really nice girl. She was new to the job and she'd never had to deal with a stillborn baby at all. I was lying down and they put me on an ultrasound and I looked at the ultrasound and he was just lying there as if... with his head to the side. They just said, "We're so sorry. There isn't a heartbeat. We're so sorry." And of course then I just started vomiting and being sick and I don't know why I was just being sick. I couldn't hold any body fluids down. It sounds disgusting but I think my body just went into shock. So the doctors came down and one doctor said to me: "You've got two options. We can either start your labor now, or you can go home and wait for the labor to start yourself", which horrified me. I'd never had a baby before and the thought of being at home when the labor started... and I could have been overdue. I could have been sitting there another four weeks, waiting for it to happen, with a child basically rotting inside. So I said, "Look, you're going to have to start the labor off."

So they took me upstairs and hooked me up onto an IV. It must seem so funny... Before I had Marley I was a smoker but as soon as I found out I was pregnant I gave up smoking. At that moment, I desperately wanted a cigarette just to feel calm. But I wouldn't have one because I kept thinking, "Oh, no, I'm pregnant so I can't. I still can't." I hate seeing moms-to-be smoking or drinking because I think, "You've got a life inside you and for nine months can't you just... Don't be so selfish. Just give up, just for nine months, just let that living thing come out."

They started putting all the IVs in me and explained the procedure of what would happen. They'd got a drug that would start the labor. That was on the Tuesday night at about 4 o'clock.

About 7 o'clock, Levi, my husband, came. He'd been working. I hadn't been able to get hold of him because at the time cell phones weren't in abundance as they are now. He was very distraught and we were both of us just sitting on the... it was horrible. We were both just sitting there. Sitting there in silence, wondering why, wondering what we had done. To go down this path of fertilizing an egg, for it to become a human being, and then for me to deliver a dead baby, which I was going to do. It wasn't going to be alive. It was going to be dead. And that's what we had to keep telling ourselves. That it wasn't going to be a proper birth at all.

That evening a lot of friends and family were sort of hearing what was going on so they were coming to see me. I was lying in the hospital wanting to be brave, saying, "I'm okay", with everybody coming round. But inside I was breaking up and wanting to just be left alone. I didn't know how to handle the situation, to be honest. It was very difficult having people coming in and saying, "What's happened?" and to sort of go through the story over and over again: "I don't know what's happened, all I know is I went for an ultrasound and there was no heartbeat so now I'm going to have to deliver a..."

The midwives were absolutely fantastic. But that night I was just getting off to sleep and a midwife came in and woke me up to take a sleeping pill. I remember that really made me laugh. That was the only time I had a really good laugh. Both Levi and I thought, "How ludicrous!" We were laughing. We felt really guilty, but it was so funny because I'd just got off to sleep and she came in and woke me up to say, "Here's a sleeping pill for you." I was still in labor but it was very, very slow. I was just getting to the stage of period pain which is bearable.

The Wednesday morning, the pain started to get worse and they said, "We're going to have to break your water because nothing's happening and it's too slow." So I had my water broken. It didn't work the first time so they had to try two times. It was just horrible because I'd always pictured in my mind that I wanted a natural labor. I didn't want all these people coming in and out and peering at me and hoisting my legs up all the time. I just wanted it to be a personal thing between me and Levi. I didn't want people coming in and disturbing us. Anyway, they broke my water and that Wednesday nothing happened at all. I was in a lot of pain. A tremendous amount of pain. And then I just felt that I was no longer in control. I was basically just a piece of meat and they told me what was going to happen and I'd have to say, 'Yes', type of thing. They said, "We're going to have to give you an epidural because you're in so much pain." I said, "But it's going to affect the baby" because I kept thinking, "This is going to be a miracle child. It's going to be in the papers: BABY PRONOUNCED DEAD BUT BORN ALIVE." But I was just kidding myself, I think. I didn't want the epidural because I kept thinking, "Oh, the drugs are going to harm the baby and he's going to come out all floppy." The midwife just said, "Well, Monica, he's dead anyway. The baby's dead anyway." And one minute I'd be crying, then I'd be cooperating with them, then I'd be saying, "No, I don't want you to do it."

So they finally took me down to another ward to be given this epidural. The guy who did the epidural was in so much of a rush to get home, because it was the end of his shift, he didn't do it properly. It came out, so two hours later I had to be taken back down again. There was no anesthesiologist at the time because it was about 10:30 at night so they had to call an anesthesiologist in from home, so I had to wait for that. I was in a tremendous amount of pain. The gas and air was making me very sick, it just didn't agree with me, so I had another epidural. So now I suffer from a lot of backache because of it, especially at period times.

Then I was taken back up onto the ward. In the meantime, my mom had been sitting outside, praying with my dad and there was an Asian woman whose daughter was in the next room. She'd said to my mom, "Why are you crying? It's such a happy time... having a grandchild." And my mother said, "My grandchild's dead" but the Asian woman couldn't understand. I don't know whether it was because of lack of the English language or just... Sometimes when you've got your own happiness to be told that a child has died is not as easy to accept as to be told an adult's died. You can sort of accept that. When an adult has died you can think, "Well okay, they've had a life and they've seen and done things." But he hadn't harmed anybody. He hadn't even got to see that the grass was green or the sky was blue or anything like that.

I felt that things were going a little bit faster than they had been. My mom had set up this prayer circle and people were constantly coming in and out of the room, praying for me and for Levi and for the baby. I wanted the prayers but at the same time I just wanted to be left alone. I wanted to deal with it on my own and not have anybody there, because I just thought it was such a private thing. If it wasn't people from the church coming in it was doctors and pediatricians, and they had a lot of student nurses and doctors in as well to sort of have a look. Not that I was a guinea pig or anything. It was just a case to have a look at. They'd never seen a woman delivering a stillborn baby before. Looking back on it now, if it ever happened again, and touch wood it never will, I'd just ask to please be left alone and to let me deliver alone. When the doctor had come and said, "Do you want to go home?" I was horrified with him, thinking, "How can you allow me to go home?!" But now, looking back it's what I'd do. I'd just wait for nature to take its course.

It's funny because when I gave birth it was just me and Levi in the room. I felt something really pushing, so I went with my feelings and I pushed. And he said, "His head's here. The baby's head's here." He called for a midwife and I asked that only Mandy, the original midwife, come in. I didn't want anybody else in the room. She came and she encouraged me to keep pushing. And he was born at 7:48 that Thursday night. I'd been induced at about 4:30 on the Tuesday afternoon so it had been a long time, mentally as well as physically, because it was my first birth. Because it was my first birth I wasn't aware of what was going to happen. I wasn't aware of anything. And to be told the baby was going to be dead anyway. It was just a shock and horror.

As soon as I'd delivered him it was as if they were waiting, it was like vultures waiting outside the door. Two doctors just walked in and said, "Right do you want an autopsy?" I said, 'No' because the thought of them cutting him up... I just said, "No." In hindsight, I should have said 'Yes' because it would have helped to establish what was wrong. I have no idea why... They don't know why either. They took some blood and some tissue cells and tried to grow them to see if anything abnormal would come back but nothing did, there was nothing... And they said to me, "Do you want him?" I feel so bad now because I was scared and I said, "No, give him to his dad."

They said it was a boy. I asked what color hair he had, if he had red hair but they said he hadn't. And they said, "Do you want him on your tummy?" and I said, "No, give him to his dad." I really regret that now, when I look back. I think, how could I have just rejected him? I wish I'd been the first to hold him. It was difficult. It was very difficult because he was cold, he was very cold and I was scared and Levi was trying to give him the breath of life. He was blue but Levi kept saying, "If I breathe in his mouth, he might just come alive." Then he gave him to me and I held him and kissed him and dressed him. I put a little white all-in-one on him and a diaper too. Silly, isn't it? I put a diaper on him and I kept thinking, "Just in case he does a poop." And I talked to him. Then they said they'd have to take him and put him in the mortuary. So they took him off me but I called them back because I hadn't put a hat on his head. I wanted to put a hat on him because he was cold. Silly, isn't it? And that was it. They put him in the mortuary and...

As for me, I went to the special room they have for mothers who give birth to stillborn children. Actually it was a special apartment. It was beautiful. It was all pine, all rustic colors, really warm in there. There was a TV and a radio and a private nurse. I could stay in there as long as I wanted and I stayed in there for three nights. While I was there, there were three others on the ward who'd delivered stillborn children. The other two ladies already had children so they were quite happy to go home. I was scared to go home. I was very scared. I remember telling a very close friend of mine at the time to clear the apartment out of all the baby stuff. I didn't want the baby stuff there. I didn't want her to throw the stuff away or give it away, but I didn't want to see it. She very kindly kept it all at her house for two years, then I took it back. Once a year, around the time that I lost Marley, I would go through his things. I kept them and Noah got to wear them. He wore all the clothes—he's 3 now—and Isobel has worn them as well. [Noah and Isobel are the two children Monica had a few years later.] I kept trying to think back over what had happened. That weekend before, did I pick anything up awkwardly? Did I rush or did I have a hot sweat or a hot flush? I was trying to think back. Did I have any mucous in my knickers? Was there anything... There wasn't and I kept trying to think. Did I vomit? Did I have a headache? And then when you try to think and you start thinking too much, everything just becomes a muddle. That's how it all became in the end. It just became a muddle and I didn't know whether I was coming or going. So I don't know what happened.

When he was inside me, Marley was very active, especially at water time. When I was up in the bathtub or when I was swimming. I used to swim a lot as well. He was very active around that time. I was at university while I was pregnant and I was doing a drama course. We used to do a lot of dance and he used to become very alert to music. He used to love music and he used to sort of bounce about and kick. And Levi and I used to have a little game where he brought me a little pig and I used to put it on my tummy and the baby would kick the pig and we used to see the pig falling off. It was so much fun, amazing just to watch my stomach grow and to know that there was a little baby inside who was coming out.

And then the hospital priest came. He did the service for when he was buried.

It was just horrible because for the next five years I went to sleep praying that I wouldn't wake up sometimes. Especially, the first year because I kept thinking when I woke up I'd have to relive the nightmare of what I went through. I joined a counseling group of ladies who'd had stillborn children, but the thing is, out of everybody in the group, none of them had experienced it as a first-time mother. They'd all experienced it second or third time. And it's funny because that city has the highest rate of stillborn babies in Europe. They don't know whether it's because of the factories and the air or whatever. I think that's why the hospital's got this room set up.

*It was just horrible for the next five years I went to sleep praying that I wouldn't wake up sometimes. Especially the first year—I kept thinking I'd have to relive the nightmare...*

We were so scared to have another baby, in case it happened again, which—chances are—it wouldn't, which I've proved. I've got two now. I got pregnant five years later. It wasn't planned. I'd had a coil fitted immediately and it was only because the coil got infected that I had to have it taken out. If I hadn't had it taken out, I wouldn't have either of the children now.

But there was the horror of being told that I was pregnant. I was so scared. The pregnancy was fine, I could deal with the pregnancy, that wasn't a problem. It was just when I got to 38 weeks. When I got to 36 weeks I was a nightmare to live with. I was an absolute nightmare, constantly thinking something was wrong and phoning the doctors and going to the hospital. And then from 37 weeks I was literally at the hospital every day. They took ultrasounds to see that everything was fine and everything was fine. But they said because of what happened they were scared to let me go full term, so I would have to be started off again at 38 weeks. So, again, I was started off. Water sac was broken. Two hours later, I heard the cry. And when I heard the cry I knew that I'd done it. It was a very fast labor. Two hours from start to finish. And he was here. I was relieved. Relieved and... It's silly but it was wonderful just being able to talk to other mothers and saying, "Well, I've done it too."

A friend of mine phoned me a couple of weeks ago and her friend had gone through the same thing. I've never met her but I gave her my number and I've been talking to Bobby now for three or four months. She lost her baby in February. That was a stillbirth as well and she was a first-time mother. We've never met but we talk once a week even now. She's still in the early stages. She cries a lot and questions why. It's 1 in a 100 and it's almost always the boys who die. [Note: As you know, this statistic varies from country to country and time to time.] When we go to see him in the cemetery it's all boys. It's a special cemetery for babies and children and 9 out of 10 of the babies were boys. It's the odd little girl who gives up the fight. But they say that girls are fighters. If there's something wrong in the womb, the girls fight and they survive, but if it's a boy he'll give up. Boys don't fight. A friend of mine, her sister gave birth at 5 months. And we all expected that there was no way that she would live but she did. She fought and she lived.

Another friend of mine gave birth and everything was fine. She had the baby in the plastic thing they have at the hospital, you know the plastic thing next to your bed. She fell asleep and woke up and the baby had died. That was straight after, which is worse, because she'd held the baby and bonded and put him on the breast. I always say that I'm glad that he was taken then because if I'd built any type of bonding with him I honestly think I would have taken my own life as well, I really do. I know it sounds awful, but I do. But I say to Bobby, "Don't give up and... don't wait five years to have another baby. Don't. It's hard work, but you'd be depriving yourself." Noah's a hard little boy, he really is. He's 3 now. He tries my patience but he brings me so much laughter. When he comes up to me and says, "Oh, Mommy I like you. You're my friend." I think that's lovely. Don't deprive yourself because they're a blessing, they really are living angels. Definitely try again.

Photo © Colin Smith

*This isn't the woman Monica's talking about but she reminds me of the importance of never pre-judging people we meet. They sometimes have painful memories...*

It's ever so funny... I met an old woman yesterday. I was taking Isobel round the park and she stopped me and she said, "Oh. How many babies have you got?" And again, I went to say three but I said, "Two".

And she said to me, "Oh you don't want to be having anymore. You should get yourself sterilized." And I just thought, "That is the most horrible thing you could ever say to somebody!" And I said, "Why do you say that?" She said, "They're horrible things, children! You spend all your life looking after them and you get no thanks back, you know!" I said, "You do. Even if it's a smile or a hug. You get it back." What a cantankerous woman! It turned out, though, when I got talking to her that this old woman had lost two babies—a girl and a boy. Both had been stillborn. The boy at 6 months and the girl at 4 months. They'd both died because she'd taken the thalidomide drug while she was pregnant. So she had no living children. Maybe it was that that had made her so bitter. It was her way of coping, I suppose.

If it was to ever happen to anybody reading this, don't let it put you off having other children. Now I just look at it as a chapter in my life. It's something that I went through and it's something horrible that I went through but I can help Bobby who's going through it now. It's just hard sometimes.

Note: Just to let you know... In June 2003 Bobby had a baby boy too. He's doing fine. I also met up with Monica again a couple of years after she gave me this account and there she was with not just two, but three children. She's since had another beautiful daughter. They're all doing fine.

*Monica with Noah (almost 4 years old) and Isobel (1 year), who was born in June 2002*

# Pause to reflect...

## ... about miscarriage, stillbirth and neonatal death

1 Considering the birthframe you've just read, which aspects of the midwifery care were helpful to Monica and which aspects didn't seem so positive?

2 In very general terms, what do you think is the best way of helping women who experience stillbirth?

3 Do you know of any cultural differences which might affect the way in which you give care to different women?

4 How do miscarriage, spontaneous abortion, stillbirth and neonatal death differ in terms of parental reactions and the care you provide?

5 In what way do you think fathers' reactions might be different from mothers' reactions? What about the reactions of other family members?

6 What other support services do you know of, which might be able to offer help when you are unable to give it—e.g. because of time constraints?

7 What experience do you have of stillbirth? If you have any direct or indirect experience, has your attitude and approach changed since having that experience?

8 How can you best support a mom-to-be prenatally when she has previously experienced either repeated miscarriages or a stillbirth?

9 Would you offer this kind of woman different intrapartum care, in terms of monitoring, intervention or emotional support?

10 What support do you think might be helpful postpartum for women who have live births after a previous stillbirth or neonatal death?

Do you have any other thoughts or questions?

## Birthframe 97: Nina Klose—Contemplating birth and death
Here, Nina reflects on death in pregnancy...

I'm not religious in any formal Christian sense, and there's a lot in Christianity or organized religion that I don't agree with. But when you start talking about pregnancy and birth, you endlessly come up against the utter wonder of so many miracles. To be pregnant is to inhabit a divine state. How is it that I am blessed with such a gift, to carry another spirit?

While pregnant one must face the knowledge that not all pregnancies culminate in a birth. And not all births culminate in a new life. If the Life Spirit brings death in this pregnancy, I must be strong enough to accept it. Accepting pregnancy means accepting potential death, the baby's or even one's own. It is a huge—and potentially painful—gift to be given this risk.

I can hardly imagine what it would be like to experience the loss of a pregnancy through miscarriage or the loss of a child through stillbirth. Even healthy pregnancies contain huge losses. No one ever talks about the mourning we go through as mothers. I fret over my new, fat, just-pregnant self. Does that sound petty? But we are so conditioned to look at ourselves in the mirror, to try to be thin, pretty, desirable. I must relinquish my usual self for the childbearing year and more.

Then I grieve for the end of the pregnancy—for no longer carrying the babe under my heart; and for no longer being a Queen among Women, Bearer of Life, but only a tired, overtaxed new mother. I grieve for the changes in the family—the loss of a bond with my husband through exhaustion and the demands of the new baby. I grieve for the loss of the older children's places as youngest child, or only boy, or only girl. I mourn the passing of the huge drama of birth. I mourn that I can't return and try it again some other way. If the birth didn't go how I longed for, I will mourn the loss forever. Birth is a huge gift, but so painful.

*Sometimes life is far from perfect... In those cases optimality must be seen in terms of relationships and love and constructive planning for a better tomorrow.*

## NOTES & REFERENCES

1    The potential impact of prenatal care on birth outcome was confirmed in the following study:

  • Foster J. Innovative practice in birth education. In M. Nolan and J. Foster (eds) *Birth and Parenting Skills: New Directions in Antenatal Education.* Elsevier Science, 2005

2    Helping women create positive visualizations is not such a strange idea... After all, athletes often prepare for important events (such as the Olympics) in a similar way. There are even psychologists who specialize in helping athletes and other performers, such as musicians... Heba Zaphiriou-Zarifi, who we meet in Birthframe 99, is a therapist who uses harmonization, dream work and body therapy to help women work through psychological problems in preparation for birth. If you're in the London area and one of your clients would like to arrange some of this kind of therapy, she can contact Heba personally via Fresh Heart by emailing info@freshheartpublishing.co.uk.

3    Stress in general is something to be taken seriously indeed, not only during pregnancy but even before conception. Studies by Khashan, *et al* (2008 and 2009) found that maternal exposure to severe life events, particularly in the six months before pregnancy, may increase the risk of preterm and very preterm birth; and that mothers exposed to severe life events before conception or during pregnancy have babies with a significantly lower birthweight. (Note, their conclusion does not say that this *may* happen, but that it does.) Another study by a slightly different team of researchers (Khashan, *et al,* 2008) found that babies of pregnant women exposed to stress during the first trimester were at greater risk of having schizophrenia later on. However, as with other areas of research, results overall are not entirely consistent. See:

  • Khashan AS, McNamee R, Abel KM, Mortensen PB, Kenny LC, Pedersen MG, Webb RT, Baker PN. Rates of preterm birth following antenatal maternal exposure to severe life events: a population-based cohort study. *Human Reproduction,* 2009, Feb;24(2):429-37. Epub 2008 Dec 3

  • Khashan AS, McNamee R, Abel KM, Pedersen MG, Webb RT, Kenny LC, Mortensen PB, Baker PN. Reduced infant birthweight consequent upon maternal exposure to severe life events. *Psychosomatic Medicine,* 2008, Jul;70(6):688-94

  • Khashan AS, Abel KM, McNamee R, Pedersen MG, Webb RT, Baker PN, Kenny LC, Mortensen PB. Higher risk of offspring schizophrenia following antenatal maternal exposure to severe adverse life events. *Archives of General Psychiatry.* 2008 Feb;65(2):146-52

  • Short SJ, Lubach GR, Karasin AI, Olsen CW, Styner M, Knickmeyer RC, Gilmore JH, Coe CL. Maternal influenz infection during pregancy impacts postnatal brain development in the rhesus monkey. *Biological Psychiatry.* 2010 Jan 13. (Epublication ahead of print.)

  • Selten JP, Frissen A, Lensvelt-Mulders G, Morgan VA. Schizophrenia and 1957 pandemic of influenza: meta-analysis. *Schizophrenia Bulletin.* 2009 Dec 3. (Epublication ahead of print.)

- Brown AS, Vinogradov S, Kremen WS, Poole JH, Deicken RF, Penner JD, McKeague IW, Kochetkova A, Kern D, Schaefer CA. Prenatal exposure to maternal infection and executive dysfunction in adult schizophrenia. *American Journal of Psychiatry.* 2009 Jun;166(6):683-90. Epub 2009 Apr 15
- Boksa P. Maternal infection during pregnancy and schizophrenia. *Journal of Psychiatry & Neuroscience.* 2008 May;33(3):183-5
- Byrne M, Agerbo E, Bennedsen B, Eaton WW, Mortensen PB. Obstetric conditions and risk of first admission with schizophrenia: a Danish national register based study. *Schizophrenia Research.* 2007 Dec;97(1-3):51-9. Epub 2007 Aug 31.

4   Elective cesareans involve lower levels of endorphins (Facchinetti, 1990), catecholamines (Jones, 1982), and prolactin (Heasman, 1997), which could affect both the new mother's and the newborn baby's experience postpartum. See:

- Facchinetti F, Garuti G, Petraglia F, Mercantini F, Genazzani AR. Changes in beta-endorphin in fetal membranes and placenta in normal and pathological pregnancies. *Acta Obstetrica et Gynecologica Scandinavica,* 1990, 69(7-8): 603-7
- Jones CM 3rd, Greiss FC Jr. The effect of labor on maternal and fetal circulating catecholamines. *American Journal of Obstetrics & Gynecology,* 1982, Sep 15; 144(2):149-53
- Heasman L, Spencer JA, Symonds ME. Plasma prolactin concentrations after caesarean section or vaginal delivery. *Archives of Disease in Childhood, Fetal & Neonatal Edition,* 1997, Nov;77(3):F237-8

A study by Zanardo, *et al* (2001) found the amount of beta-endorphin in the colstral milk of mothers who'd given birth vaginally was significantly higher than colostrum levels of mothers who had a cesarean section. Having beta-endorphins in milk is thought to be important as these opiate-like substances make the newborn baby 'addicted' to its mother's milk. Of course, this fact alone suggests a VBAC is more 'optimal' than a repeat cesarean, since bonding and breastfeeding are so crucial to the baby's survival and well-being. See:

- Zanardo V, Nicolussi S, Giacomin C, Faggian D, Favaro F, Plebani M. Labor pain effects on colostral milk beta endorphin concentrations of lactating mothers. *Biology of the Neonate* 2001, 79(2):79-86)
- DiMatteo MR, Morton S, *et al.* Cesarean Childbirth and Psychosocial Outcomes— DiMatteo MR, Morton S, *et al.* Cesarean Childbirth and Psychosocial Outcomes— A Meta-Analysis. *Health Psychology,* 1996, 15(4):303-14

5   The importance of continuity of care has been confirmed by various studies. A study by Haggerty, *et al* (2005) emphasized the importance of different types of continuity—informational, management and relational. See:

- Haggerty J, Reid R, Freeman G, Starfield B, Adair C, McKendry R. Continuity of care: a multidisciplinary review. *British Medical Journal,* 2005, 327:1219-1221

The first of the following studies emphasized the importance of working in small team of midwives (numbering at the most six), while the other studies concluded that continuity was most important during labor, birth and postpartum (and perhaps not such a concern prenatally):

- Flint C. *Midwifery Teams and Caseloads.* Butterworth-Heinemann, 1993
- Flynn A, Hollins K, Lynch P. Ambulation in labour. *British Medical Journal,* 1978, 2(6137):591-593
- Green J, Curtis P, Price H, Renfrew M. *Continuing to Care: The Organization of Midwifery Services in the UK A Structured Review of the Evidence.* Books for Midwives Press, 1998
- Walsh D. An ethnographic study of women's experience of partnership caseload midwifery practice: the professional as friend. *Midwifery,* 1999, 15(3):165-176
- Page L, McCourt C, Beake S, Hewison J. Clinical interventions and outcomes of one-to-one midwifery practice. *Journal of Public Health Medicine,* 1999, 21(3): 243-248
- North Staffordshire Changing Childbirth Research Team. A randomised study of midwifery caseload care and traditional 'shared-care'. *Midwifery,* 2000, 16(4):295-302

6    In fact, as you may know, there are other reasons why a woman might want to avoid amniocentesis. In a commentary from *Birth Reborn* Michel Odent mentions that some studies suggest a higher incidence of respiratory difficulties for the newborn baby and that others also link amniocentesis carried out in the second trimester of pregnancy with orthopedic malformations. His recommendation is to suggest that moms-to-be consider their risk for Down's syndrome (Trisomy 21) and other abnormalities in a positive framework. In other words, instead of thinking about the 1 in 94 women aged 40 who are at risk of having a child with Down's syndrome, they should think about the 93 out of 94 40-year-olds who have a perfectly normal baby. Finally, Michel reminds us that the widely-believed increased risks of fetal abnormalities for women over 35 may be misguided if the abnormalities are in fact connected to an increased exposure to radiation, with increased age, rather than age itself.

7    See the following articles:

- Odent M. The Nocebo effect in prenatal care. *Primal Heath Research Newsletter,* 1994, 2(2)
- Odent M. Back to the Nocebo effect. *Primal Heath Research Newsletter,* 1995, 5(4)
- Odent M. Antenatal scare. *Primal Heath Research Newsletter,* 2000, 7(4)

8    There are usually clear-cut views amongst medical professionals as to whether a test is 'invasive' or not. It seems to me, though, that the boundaries are blurred. Why should we call a test (such as an ultrasound) non-invasive when we cannot be sure of its effects, especially since we know that some occur (even if not negative)? In fact, we can guess that an ultrasound definitely has an effect on the growing fetus (so is certainly invasive in that sense) because many research studies have detected changes which appear to have resulted from the repeated use of ultrasounds. And what about the psychological impact of each individual test? Somehow this should also be the focus of research. Whatever your view on individual tests, you might want to explain to women the risk of disturbing their pregnancy if they have any test beyond a blood or urine test.

9    Birlholz J,& Stephens JC. *American Journal of Roentology.* Fetal Movement

Patterns: A Possible Means of Defining Neurological Development Milestones *In Utero*, 1978, Vol 130 (pp 537-540)

10   Meire HB. The safety of diagnostic ultrasound. (Commentary) *British Journal of Obstetrics & Gynaecology*, 1987, Vol 94, (pp 1121-1122)

11   This was Prof. Nicholas Fisk at Queen Charlotte's Hospital in London in 1996.

12   Thomas P, Golding J, Peters TJ. Delayed antenatal care: does it affect pregnancy outcome? *Social Science & Medicine*, 1991, 32:715-23

13   Douglas KA, Redman CW. Eclampsia in the United Kingdom. *British Medical Journal*, 1994, 309:1395-400

14   See the following articles:

  • Symonds EM. Aetiology of pre-eclampsia: a review. *J R Soc Med*, 1980, 73:871-5

  • Naeye EM. Maternal blood pressure and fetal growth. *American Journal of Obstetrics & Gynecology*, 1981, 141:780-7

  • Kilpatrick S. Unlike pre-eclampsia, gestational hypertension is not associated with increased neonatal and maternal morbidity except abruption. SPO abstracts, *American Journal of Obstetrics & Gynecology*, 1995, 419:376

  • Curtis S, *et al.* Pregnancy effects of non-proteinuric gestational hypertension. SPO Abstracts, *American Journal of Obstetrics & Gynecology*, 1995, 418:376

15   The jury is still out on whether ultrasound is wholly helpful or wholly or partly harmful (either potentially or actually). The review of 16 studies which had all been published since 1990, conducted in 2008 by Chaimay, *et al* does not present convincing reassurance. After all, while the researchers concluded that ultrasound has no adverse effect on child development outcomes (when given during pregnancy) they also note that "all studies demonstrated that ultrasound examinations during pregnancy increased the risk of undesirable developmental outcomes". It seems the researchers' reasons for dismissing the importance of these possible undesirable outcomes were a) effects were considered 'minimal' and b) the studies were criticized on methodological grounds. (The studies which were reviewed by these researchers were 13 randomized controlled trials, one cohort study, and two case-control studies—but they were all deficient in some respect, apparently.) Oddly, perhaps, even though the abstract for this review includes the statement: "Presently, it is not clear whether [ultrasound] has a negative effect on the health and development of children" the conclusion is that ultrasound is safe, after all. Surely more research is needed, particularly since ultrasound has become so common? Surely, too, if ultrasound might only provide the *hope* of helpful information, while other methods are also available, shouldn't we prefer those other methods until we can be reassured that scans are safe?... if indeed they are. See:

  • Chaimay B, Woradet S. Does prenatal ultrasound exposure influence the development of children? Asia Pacific Journal of Public Health, 2008, Oct; 20 Suppl: 31-8

16   See the following articles:

  • Brown VA, *et al.* The value of antenatal cardiotocography in the management of high-risk pregnancy : a randomised controlled trial. *British Journal of Obstetrics*

& Gynaecology, 1982, 89:716-22

- Flynn AM, et al. A randomized controlled trial of non-stress antepartum cardiotocography. British Journal of Obstetrics & Gynaecology, 1982, 89:427-33
- Haverkamp AD, et al. A controlled trial of the differential effects of intrapartum monitoring. American Journal of Obstetrics & Gynecology, 1976, 126:470-76
- Haverkamp AD, et al. The evaluation of continuous fetal heart rate monitoring in high risk pregnancy. American Journal of Obstetrics & Gynecology, 1976, 125: 310-20
- Kelso IM, et al. An assessment of continuous fetal heart rate monitoring in labor. American Journal of Obstetrics & Gynecology, 1978, 131:526-32
- Kidd LC, et al. Non-stress antenatal cardiotocography—a prospective randomized clinical trial. British Journal of Obstetrics & Gynaecology, 1985, 92:1156-59
- Leveno KJ, et al. A prospective comparison of selective and universal electronic fetal monitoring in 34,995 pregnancies. New England Journal of Medicine, 1986, 315:615-19
- Lumley JC, Wood C, et al. A randomized trial of weekly cardiotocography in high-risk obstetric patients. British Journal of Obstetrics & Gynaecology, 1983, 90: 1018-26
- McDonald D, Chalmers I, et al. The Dublin randomised controlled trial of intrapartum fetal heart rate monitoring. American Journal of Obstetrics & Gynecology, 1985, 152:524-39
- Prentice A, Lind T. Fetal heart rate monitoring during labor—too frequent intervention, too little benefit. Lancet, 1987, 2:1375-77
- Sky K, et al. Effects of electronic fetal heart rate monitoring, as compared with periodic auscultation, on the neurological development of premature infants. New England Journal of Medicine, 1990, (March 1):588-93
- Wood C. A controlled trial of fetal heart rate monitoring in low-risk obstetric population. American Journal of Obstetrics & Gynecology, 1981, 141:527-34
- Impey L, Reynolds M, et al. Admission cardiotocography: a randomized controlled trial. Lancet, 2003, 361:465-70

17    Ewigman BG, Crane JP, Frigoletto FD, et al. Effect of prenatal ultrasound screening on perinatal outcome. RADIUS study group. New England Journal of Medicine, 1993, Vol 329, No 12 (pp 821-7)

18    Bucher HC, Schmidt JG. Does routine ultrasound scanning improve outcome in pregnancy? Meta-analysis of various outcome measures. British Medical Journal, 1993, 307:13-7

19    See the following articles:

- Larson T, Falck Larson J, et al. Detection of small-for-gestational-age fetuses by ultrasound screening in a high risk population: a randomized controlled study. British Journal of Obstetrics & Gynaecology, 1992, 99:469-74
- Secher NJ, Kern Hansen P, et al. A randomized study of fetal abdominal diameter and fetal weight estimation for detection of light-for-gestation infants in low-risk pregnancy. British Journal of Obstetrics & Gynaecology, 1987, 94:105-9

20    Johnstone FD, Prescott RJ, et al. Clinical and ultrasound prediction of macrosomia

in diabetic pregnancy. *British Journal of Obstetrics & Gynaecology*, 1996, 103:747-54

21　De Crespigny L, Dredge R. *Which Tests for my Unborn Baby?* Oxford University Press, 1996

22　Wagner M. Ultrasound; More harm than good? *Mothering* magazine, Winter 1995

23　De Crespigny L, Dredge R. *Which Tests for my Unborn Baby?* Oxford University Press, 1996

24　Oakley A. The history of ultrasonography in obstetrics. *Birth*, 1986, Vol 13,No 1, 8-13

25　Saari-Kemppainen A, Karjalainen O, Ylostalo P, *et al*. Ultrasound screening and perinatal mortality: controlled trial of systematic one-stage screening in pregnancy. The Helsinki ultrasound trial. *Lancet*, 1990, Vol 336, No 8712 (pp 387-391)

26　See the following for commentaries on this:

　　• Olsen O, *et al*. Routine ultrasound dating has not been shown to be more accurate than the calendar method. *British Journal of Obstetrics & Gynaecology*, 1997, Vol 104, No 11, pp 1221-2

　　• Kieler H, Axelsson O, Nilsson S, Waldenstrom U. Comparison of ultrasonic measurement of biparietal diameter and last menstrual period as a predictor of day of delivery in women with regular 28 day cycles. *Acta Obstetrica et* Kieler H, Axelsson O, Nilsson S, Waldenstrom U. Comparison of ultrasonic measurement of biparietal diameter and last menstrual period as a predictor of day of delivery in women with regular 28 day cycles. *Acta Obstetrica et Gynecologica Scandinavica*, 1993, Vol 75, No 5, pp 347-9

27　See the following articles:

　　• Ewigman BG, Crane JP, Frigoletto FD, *et al*. Effect of prenatal ultrasound screening on perinatal outcome. RADIUS study group. *New England Journal of Medicine*, 1993, Vol 329, No 12, pp 821-7

　　• Luck CA. Value of routine ultrasound scanning at 19 weeks: a four year study of 8849 deliveries. *British Medical Journal*, 1992, Vol 34, No 6840, pp 1474-8

28　Chan FY. Limitations of Ultrasound. Paper presented at Perinatal Society of Australia & New Zealand. 1st Annual Congress, Freemantle, 1997

29　See the following articles:

　　• Ewigman BG, Crane JP, Frigoletto FD, *et al*. Effect of prenatal ultrasound screening on perinatal outcome. RADIUS study group. *New England Journal of Medicine*. 1993 Vol 329, No 12, pp 821-7

　　• Luck CA. Value of routine ultrasound scanning at 19 weeks: a four year study of 8849 deliveries. *British Medical Journal*, 1992, Vol 34, No 6840, pp 1474-8

30　Brand IR, Kaminopetros P, Cave M, *et al*. Specificity of antenatal ultrasound in the Yorkshire region: a prospective study of 2261 ultrasound detected anomalies. *British Journal of Obstetrics & Gynaecology*, 1994. Vol 101, No 5, pp 392-397

31　Sparling JW, Seeds JW, Farran, D. C. The relationship of obstetric ultrasound to parent and infant behavior. *Obstetrics & Gynecology* 1988, Vol 72, No 6, pp 902-7

32　In 1975 a study of ultrasounds on unborn babies using Doppler ultrasound was published in the *British Medical Journal*. The researchers didn't tell the mothers whether the ultrasound machine was switched on, but when it was the fetuses were found to move about much more. See:

- David H, Weaver JB, Pearson JF. Doppler Ultrasound and Fetal Activity. *British Medical Journal*, 1975, Apr 12; 2(5962):62-4

33   Jumping Babies, *Aims Quarterly Journal,* 1993, Vol 5/7, pp 15-17

34   American Institute of Ultrasound Medicine Bioeffects Report 1988. *Journal of Ultrasound Medicine*, 1988, Sept 7S1-S38

35   American Institute of Ultrasound Medicine Bioeffects Report. *Journal of Ultrasound Medicine*, 1988, 7S1-S38, Sept

36   Liebeskind D, Bases R, Elequin F, *et al*. Diagnostic ultrasound: effects on the DNA and growth patterns of animal cells. *Radiology*, 1979, Vol 131, No 1, pp 177-184

37   Ellisman MH, Palmer DE, Andre MP. Diagnostic levels of ultrasound may disrupt myelination. *Experimental Neurology*, 1987, Vol 98, No 1, pp 78-92

38   Brennan P, *et al*. Shadow of doubt. *New Scientist,* 1999, 12 June, p 23

39   Testart J, Thebalt A, Souderis E, Frydman R. Premature ovulation after ovarian ultrasonography. *British Journal of Obstetrics & Gynaecology*, 1982, Vol 89, No 9, pp 694-700

40   See the following articles:

- Lorenz RP, Comstock CH, Bottoms SF, Marx SR. Randomised prospective trial comparing ultrasonography and pelvic examination for preterm labor surveillance. *American Journal of Obstetrics & Gynecology*, 1990, Vol 162, No 6, pp 1603-1610

- Saari-Kemppainen A, Karjalainen O, Ylostalo P, *et al*. Ultrasound screening and perinatal mortality: controlled trial of systematic one-stage screening in pregnancy. The Helsinki ultrasound trial. *Lancet*, 1990, Vol 336, No 8712, pp 387-391

41   See the following articles:

- Newnham J, Evans SF, Michael CA, *et al*. Effects of frequent ultrasound during pregnancy: a randomised controlled trial. *Lancet*, 1993, Vol 342, No 8876, 887-91

- Geerts JGM, Brand E, Theron B. Routine obstetric ultrasound in South Africa: cost and effect on perinatal outcome—a prospective randomised controlled trial. *British Journal of Obstetrics & Gynaecology*, 1996, Vol 103, pp 501-507

42   See the following articles:

- Thacker SB. Quality of controlled clinical trials. The case of imaging ultrasound in obstetrics: a review. British Journal of Obstetrics & Gynaecology, 1985, Vol 92, No 5, pp 437-444

- Newnham JP, *et al*. Doppler flow velocity wave form analysis in high risk pregnancies: a randomised controlled trial. *British Journal of Obstetrics & Gynaecology*, 1991, Vol 98, No 10, pp 956-963

43   Davies J, *et al*. Randomised controlled trial of doppler ultrasound screening of placental perfusion in pregnancy. *Lancet*, 1992, 340:1299-1303

44   Stark CR, Orleans M, Havercamp AD, *et al*. Short and long term risks after

exposure to diagnostic ultrasound in utero. *Obstetrics & Gynecology*, 1984, Vol 63, pp 194-200

45   Campbell JD, et al. Case-control study of prenatal ultrasonography in children with delayed speech. *Canadian Medical Association Journal*, 1993, Vol 149, No 10, pp 1435-1440

46   See the following articles:

- Salvesen KA, Vatten LJ, Eik-nes SH, *et al.* Routine ultrasonography in utero and subsequent handedness and neurological development. *British Medical Journal*, 1993, Vol 307, No 6897, pp 159-64

- Kieler H, Ahlsten G, Haguland B, *et al.* Routine ultrasound screening in pregnancy and the children's subsequent neorological development. *Obstetrics & Gynecology*, 1998, Vol 915 (pt 1), pp 750-6

- Salvesen KA, Ein-nes SH, *et al.* Ultrasound during pregnancy and subsequent childhood non-right handedness- a meta-analysis. *Ultrasound Obstetrics & Gynecology*, 1999, 13(4)241-6

- Kieler H, Cnattingius S, Haglund B, *et al.* Sinistrality—a side-effect of prenatal sonography: A comparative study of young men. *Epidemiology*, 2001, 12(6):618-23

47   See the following articles:

- Odent M. Where does handedness come from? *Primal Health Research Quarterly*, 1998, Vol 6, No 1

- Kieler H, Cnattingius S, Haglund B, *et al.* Sinistrality—a side-effect of prenatal sonography: A comparative study of young men. *Epidemiology*, 2001, 12 (6):618-23

48   Newnham J, Evans SF, Michael CA, *et al.* Effects of frequent ultrasound during pregnancy: a randomised controlled trial. *Lancet*, 1993, Vol 342, No 8876, pp 887-91

49   Michel Odent's overall conclusion is that ultrasounds should only be performed if the information they might provide would *change* decisions made about the pregnancy—in terms of treatment given by the caregiver, or action taken by the pregnant woman. If you would like to read some other expert comments, find former World Health Organization Director Marsden Wagner's comments at www.midwiferytoday.com/articles/ultrasoundwagner.asp or comments from Beverley Beech (honourary chair of the Association for Improvements in the Maternity Services (AIMS)) at www.midwiferytoday.com/articles/ultrasound.asp

50   Steer P, Alam MA, Wadsworth J, Welch A. Relation between maternal haemoglobin concentration and birth weight in different ethnic groups. *British Medical Journal*, 1995, 310:489-91

51   This is explained in the following article:

- Steer P, Alam MA, Wadsworth J, Welch A. Relation between maternal haemoglobin concentration and birth weight in different ethnic groups. *British Medical Journal*, 1995;310:489-91

52   Valberg LS. Effects of iron, tin, and copper on zinc absorption in humans. *American Journal of Clinical Nutrition*, 1984, 40:536-41

53 Rayman MP, Barlis J, et al. Abnormal iron parameters in the pregnancy syndrome preeclampsia. *American Journal of Obstetrics & Gynecology*, 2002, 187(2):412-8

54 Wen SW, Liu S, Kramer MS, et al. Impact of prenatal glucose screening on the diagnosis of gestational diabetes and on pregnancy outcomes. *American Journal of Epidemiology*, 2000, 152(11):1009-14

55 Michel writes more about his views on prenatal care at www.wombecology.com/newreasons.html

56 Hannah ME, Hannah WJ, et al. Planned caesarean section versus planned vaginal birth for breech presentation at term: a randomised multicentre trial. *Lancet*, 2000, 356:1375-83

57 See Coyle ME, Smith CA, Peat B. Cephalic version by moxibustion for breech presentation. *Cochrane Database of Systematic Reviews*, 2005, Issue 2. Art. No CD003928. DOI: 10.1002/14651858.CD003928.pub2

58 The following systematic literature review failed to identify any research basis for vaginal examinations to assess the progress of labor:

· Devane D. Sexuality and midwifery. *British Journal of Midwifery,* 1996, 4(8):413-

The following researchers also concluded that vaginal examinations were not necessary for facilitating physiological birth:

· Chalmers I, Kierse M, Neilson J. *A Guide to Effective Care in Pregnancy and Childbirth*. Oxford University Press, 1989

The practice of routinely conducting vaginal examinations was shown, in the following study, to be a ritual which legitimized intrusion into the very private birth place during the second stage of labor:

· Bergstrom L, Roberts J, Skillman L, Seidel J. "You'll feel me touching you, sweetie": Vaginal examinations during the second stage of labour. *Birth,* 1992, 19(1):10-18

Also see:

· Stewart M. "I'm just going to wash you down." Sanitizing the vaginal examination. *Journal of Advanced Nursing*, 2005, 51(6):587-594

· Warren C. Invaders of privacy. *Midwifery Matters*, 1999, 81:8-9

59 In fact, systematic reviews of research trials of hypnosis (Leslie, et al, 2007; Cyna, et al, 2004; Huntley, 2004) found that hypnosis reduced the need for analgesia, it reduced the level of pain experienced and shortened labor. Both Cyna (2004) and Smith (2003) found that hypnosis reduced the need for augmentation with oxytocin and that the incidence of spontaneous births increased. The CIMS review (the first in the list of references) also noted that no study reported an adverse effect. A couple of articles in *The Practising Midwife* also looked at relevant topics. One recommended hypnosis as a way of reducing the pain of labor and the other one looked at the use of NLP (neuro-linguistic programming) as a means of helping women prepare for labor. Incidentally, the CIMS review also looked at massage and hydrotherapy. See:

· Leslie MS, Romano A, Woolley D. Step 7: Educates Staff in Nondrug Methods of Pain Relief and Does Not Promote Use of Analgesic, Anesthetic Drugs: The Coalition for Improving Maternity Services. *Journal of Perinatal Education,* 2007, Winter;16(Suppl 1):65S-73S

- Cyna, A, McAuliffe, G. and Andrew, M. Hypnosis for pain relief in labour and childbirth: a systematic review. *British Journal of Anaesthesia,* 2004, 93(4):505-511

- Huntley AL, Coon JT, Ernst E. Complementary and alternative medicine for labor pain: a systematic review. *American Journal of Obstetrics & Gynecology,* 2004, Jul; 191(1):36-44

- Mottershead, N. Hypnosis: removing the labour from birth. *The Practising Midwife,* 2006, 9(3):26-29

- Spencer, S. Giving birth on the beach: hypnosis and psychology? *The Practising Midwife,* 2005, 8(1):27-29

- Smith CA, Collins CT, Cyna AM, Crowther CA. Complementary and alternative therapies for pain management in labour. Cochrane Database Systematic Review. 2003, (2):CD003521

60   The following researcher discovered that low-intervention birth brings healing to some women who've previously had a traumatic labor:

- Milan M. Childbirth as healing: three women's experience of independent midwife care. *Complementary Therapies in Nursing & Midwifery,* 2003, 9:140-146

If you are at a loss as to what to do to support women who've experienced sexual abuse, there are some pointers for midwives at www.midwiferytoday.com/articles/creatingspace.asp

You might also find it helpful to look at the following book, which is intended primarily for caregivers:

- Simkin P, Klaus P. *When Survivors Give Birth.* Classic Day Publishing 2009 (3rd edition)

61   In her book about birth in different cultures, Sheila Kitzinger provides evidence for the notion that place of birth has usually been highly significant. For this reason, it's probably very important for a woman to be able to choose where she gives birth. See:

- Kitzinger S. *Rediscovering Birth.* Little, Brown & Company, 2000

To consider the relative safety of home birth over hospital birth, in general, around the world, see the following, noting that studies generally draw very favorable conclusions, based on data collected and analyzed:

- Gyte G, Dodwell M, Newburn M, Sandall J, Macfarlane A, Bewley S. No rising trend in home birth mortality. *BJOG: International Journal of Obstetrics & Gynaecology,* 2009, Nov; 116(12):1686-7

- Groenendaal F. Homebirth: as safe as hospital? *BJOG: International Journal of Obstetrics & Gynaecology,* 2009, Nov; 116(12): 1686-5; author reply 1685-6

- Hutton EK, Reitsma AH, Kaufman K. Oucomes associated with planned home and planned hospital births in low-risk women attended by midwives in Ontario, Canada, 2003-2006: a retrospective cohort study. *Birth,* 2009 Sep;36(3):180-9

- Janssen PA, Saxell L, Page LA, Klein MC, Liston RM, Lee SK. Outcomes of planned home birth with registered midwife versus planned hospital birth with midwife or physician. *Canadian Medical Association Journal,* 2009 Sep 15;181

(6-7):377-83. Epub 2009 Aug 31

Olsen O, Jewell M. Home versus hospital birth. *The Cochrane Database of Systematic Reviews,* Issue 3, 2006

- Johnson K, Daviss BA. Outcomes *of* planned home births with certified professional midwives: large prospective study in North America. *British Medical Journal,* 2005, 330(7505):1416-1418
- Chamberlain G, Wraight A, Crowley P. Home births. Report of the 1994 confidential enquiry by the National Birthday Trust Fund, 1997, Parthenon (pp 107-113)
- Wiegers TA, Keirse MJNC, van der Zee J, Berghs GAH. Outcome of planned home and planned hospital births in low risk pregnancies: prospective study in midwifery practices in the Netherlands. *British Medical Journal,* 1996, 313:1309-1313 (23 November)
- Tew M. *Safer Childbirth? A Critical History of Maternity Care.* Chapman & Hail, 1998
- Campbell R. Place of birth reconsidered. In Alexander J, Levy V and Roth C (eds) *Midwifery Practice: Core Topics 2.* Macmillan, 1997
- Abel S, Kearns RA. Birth Places: A Geographical Perspective on Planned Home Birth in New Zealand, *Social Science & Medicine,* 1997, Vol 33,No 7,825-34

In the study by Hutton, *et al* (2009) the conclusion was: "Midwives who were integrated into the health care system with good access to emergency services, consultation, and transfer of care provided care resulting in favorable outcomes for women planning both home or hospital births." In the study by Janssen, *et al* (2009), the conclusion was: "Planned home birth attended by a registered midwife was associated with very low and comparable rates of perinatal death and reduced rates of obstetric interventions and other adverse perinatal outcomes compared with planned hospital birth attended by a midwife or physician."

In the American study by Johnson and Daviss (2005), the conclusion was: "Planned home birth for low risk women in North America using certified professional midwives was associated with lower rates of medical intervention but similar intrapartum and neonatal mortality to that of low risk hospital births in the United States." Despite all these positive research conclusions, home birth is very seriously under threat in some countries where it is legal now and many caregivers in all countries still tend to try and undermine it. Perhaps their motivation is to do with convenience, rather than safety or client satisfaction. Sorry, but I refuse to call a pregnant woman a 'patient'... but is the use of this word to describe a woman in labor also a clue to why some caregivers prefer births to take place in hospital? Could it be to do with control? Or is it just that so many caregivers and pregnant women cannot imagine birth without 'pain relief', which clearly needs to be managed in hospital settings, simply because it increases risk... See:

- Starr L. Legislation may drive homebirths underground. *Australian Nursing Journal,* 2009, Aug;17(2):31

More reasons and rationales for homebirth are presented at the Midwifery Today website, along with some research statistics. See: www.midwiferytoday.com/articles/homebirthuk.asp and also www.midwiferytoday.com/articles/homebirthissues.asp

62 As well as these problems relating to infection risk, some studies have also found that the ethos of hospitals (which tends to be hierarchical, institutional and medically-led), has the effect of making women feel like 'patients' and of making birth seem abnormal and unhealthy. Of course, this is likely to make unnecessary cascades of interventions more likely. The studies are as follows:

- Hunt S, Symonds A. *The Social Meaning of Midwifery.* Macmillan, 1995
- Machin D, Scamell M. The experience of labour: using ethnography to explore the irresistible nature of the bio-medical metaphor during labour. *Midwifery,* 1997, 13: 78-84
- Kirkham M. The culture of midwifery in the National Health Service in England. *Journal of Advanced Nursing,* 1999, 30:732-739
- Ball L, Curtis P, Kirkham M. *Why Do Midwives Leave?* Royal College of Midwives, 2002
- Stapleton H, Kirkham M, Thomas G, Curtis P. Midwives in the middle: balance and vulnerability. *British Journal of Midwifery,* 2002, 10(10):607-611

63 Many studies, such as the following, conclude that women resort to pain relief less often when they are outside large hospitals (e.g. at home or in a birth center) and also when they have continuity of care (i.e. one midwife right the way through their labor and birth):

- Olsen O. Meta-analysis of the safety of home birth. *Birth,* 1997, 24(1):4-13
- Walsh D, Downe S. Outcomes of free-standing, midwifery-led birth centres: a structured review of the evidence. *Birth,* 2004, 31(3):222-229
- Hodnett E, Downe S, Edwards N, Walsh D. Home-like versus conventional birth settings (Cochrane Review). In: *The Cochrane Library,* Issue 2. Chichester: John Wiley & Sons Ltd, 2006
- Waldenstrom U, Turnbull D. A systematic review comparing continuity of midwifery care with standard maternity services. *British Journal of Obstetrics & Gynaecology,* 1998, 105:1160-1170
- Wraight A, Ball J, Seccombe I, Stock J. *Mapping Team Midwifery: A Report to the Department of Health.* Brighton: Institute of Manpower Studies, University of Sussex, 1993
- Harvey S, Jarrell J, Brant R, Stainton C, Rach D. A randomised controlled trial of nurse/midwifery care. *Birth,* 1996, 23:128-135
- Page L, McCourt C, Beake S, Hewison J. Clinical interventions and outcomes of one-to-one midwifery practice. *Journal of Public Health Medicine,* 1999, 21 (3): 243-248

In case women in your care really do need to be near emergency facilities, it might be worthwhile helping them to explore all their options, particularly as regards birth centers and other midwifery-led units. After all, several papers have concluded that giving birth in a birth center (rather than a hospital) can help women avoid unnecessary interventions. At recent conferences I have also met or heard about many midwives who are setting up birth centers which sound amazingly well designed. A birth center might, therefore be a good option for a woman who is not confident about arranging a home birth. See the following papers:

- Tracy S, Sullivan E, Dahlen H, Black D. Does size matter? A population-based study of birth in lower volume maternity hospitals for low risk women. *BJOG:An International Journal of Obstetrics & Gynaecology,* 2005, 113:86-96
- Jackson D, Lang J, Swartz W, Ganiats T, Fullerton J. Outcomes, safety and resource utilization in a collaborative care birth centre program compared with traditional physician-based perinatal care. *American Journal of Public Health,* 2003, 93:999-1006
- Walsh D. Subverting assembly-line birth: childbirth in a free-standing birth centre. *Social Science & Medicine,* 2006, 62(6):1330-1340
- Stewart M, McCandlish R, Henderson J. *Report of a Structured Review of Birth Centre Outcomes.* Oxford: NPEU, 2004
- Walsh D, Downe S. Outcomes of free-standing, midwifery-led birth centres: a structured review of the evidence. *Birth,* 2004, 31(3):222-229
- Reddy K, Reginald P, Spring J, Nunn L, Mishra N. A free-standing low-risk maternity unit in the United Kingdom: does it have a role? *Journal of Obstetrics & Gynaecology,* 2004, 24(4):360-366
- Green J, Coupland B, Kitzinger J. *Great Expectations: A Prospective Study of Women's Expectations and Experiences of Childbirth.* Cambridge: Child Care & Development Group, 1998
- Hodnett E, Downe S, Edwards N, Walsh D. Home-like versus conventional birth settings (Cochrane Review). In: *The Cochrane Library,* Issue 2. John Wiley & Sons Ltd, 2006

The following studies also mention that less pharmacological analgesia is used in settings which are *not* large hospitals (i.e. in people's homes, in free-standing birth centers or in integrated birth centers):

- Olsen O. Meta-analysis of the safety of home birth. *Birth,* 1997, 24(1): 4-13
- Walsh D, Downe S. Outcomes of free-standing, midwifery-led birth centres: a structured review of the evidence. *Birth,* 2004, 31(3):222-229
- Hodnett E, Downe S, Edwards N, Walsh D. Home-like versus conventional birth settings (Cochrane Review). In: *The Cochrane Library,* Issue 2. John Wiley & Sons Ltd, 2006

Wherever women give birth, but particularly when they are in an institutional environment, it's important to remember that a birthing room does not need a bed and indeed that a bed might encourage women to adopt unhelpful positions during labor and birth. During my very first labor in 1997, when my obstetrician asked me what I thought of my hospital room and I said, 'Er... it looks a bit like a bed with walls around it'—he immediately arranged for the bed to be taken out! (This was one of many reasons he was a wonderful obstetrician.) The researchers Spiby, *et al* (2003) recommend that hospital beds be removed from birth spaces, or at the very least pushed against a wall so that women have more space. See:

- Spiby H, Slade P, Escott D, Henderson B, Fraser R. Selected coping strategies in labour: an investigation of women's experiences. *Birth,* 2003, 30:189-194

64 Some researchers have pointed out that a midwife's diagnosis of labor in hospital is not a simple, one-sided clinical judgement. Whether or not a diagnosis is made depends on many other institutional constraints. See:

- Burvill S. Midwifery diagnosis of labour onset. *British Journal of Midwifery*, 2002, 10(10):600-605
- Cheyne H, Dowding D, Hundley V. Making the diagnosis of labour: midwives' diagnostic judgement and management decisions. *Journal of Advanced Nursing*, 2006, 53(6):625-635

The importance of a midwife's attitudes and behavior is emphasized in the following paper:

- Downe S, McCourt C. From being to becoming: reconstructing childbirth knowledges. In Downe S (ed) *Normal Childbirth: Evidence and Debate*. Churchill Livingstone, 2004

The helpfulness of a 'working with pain' attitude, rather than a 'pain relief' orientation was emphasized in the following paper:

- Leap N, Anderson T. The role of pain in normal birth and the empowerment of women. In Downe S, McCourt C (eds) *Normal Childbirth: Evidence and Debate*. Churchill Livingstone, 2004 (pp 25-40)

65 The following systematic literature review failed to identify any research basis for vaginal examinations to assess the progress of labor:

- Devane D. Sexuality and midwifery. *British Journal of Midwifery,* 1996, 4(8):413-20

The following researchers also concluded that vaginal examinations were not necessary for facilitating physiological birth:

- Chalmers I, Kierse M, Neilson J. *A Guide to Effective Care in Pregnancy and Childbirth*. Oxford University Press, 1989

The practice of routinely conducting vaginal examinations was shown, in the following study, to be a ritual which legitimized intrusion into the very private birth place during the second stage of labor:

- Bergstrom L, Roberts J, Skillman L, Seidel J. "You'll feel me touching you, sweetie": Vaginal examinations during the second stage of labour. *Birth,* 1992, 19(1):10-18

Also see:

- Stewart M. "I'm just going to wash you down." Sanitizing the vaginal examination. *Journal of Advanced Nursing,* 2005, 51(6):587-594
- Warren C. Invaders of privacy. *Midwifery Matters,* 1999, 81:8-9

66 Grol and Grimshaw (2003) have written about doctors' impulse to act. Other writers have written about the way in which midwives tend to be dominated by obstetricians, when they work closely with them (Donnison 1998, Coombs and Ersser 2004). See:

- Grol R, Grimshaw J. From best evidence to best practice: effective implementation of change in patient's care. *Lancet,* 2003, 362:1225-1230
- Donnison J. *Midwives and Medical Men: A History of the Struggle for the Control of Childbirth*. Historical Publications, 1988
- Coombs M, Ersser S. Medoca; hegemony in decision-making—a barrier to interdisciplinary working in intensive care. *Journal of Advanced Nursing,* 2004, 46(3):245-252

67  Relevant issues are explored in the following articles:

- Rooks J. Low-Intervention Maternity Care. *Journal of Family Practice*, Vol 31,No 2, 125-7
- Tew M. Do Obstetric Intranatal Interventions Make Births Safer? *British Journal of Obstetrics & Gynaecology*, 1986, Vol 93,659-74
- Wagner M. *Pursuing the Birth Machine: The Search for Appropriate Birth Technology.* Sydney: ACE Graphics, 1994
- Wood LAC. Obstetric Retrospect, *Journal of the Royal College of General Practitioners*, 1981, Vol 31,80-90

68  Michel discusses this further in the following article:

- Odent M. New reasons and new ways to study birth physiology. International *Journal of Gynecology & Obstetrics*, 2001, 75:S39-S45

69  Michel Odent describes this process of transformation in the following article:

- Odent M. New reasons and new ways to study birth physiology. *International Journal of Gynecology & Obstetrics*, 2001, 75:S39-S45

70  In case you think that synthetic oxytocin (i.e. a drip of pitocin during labor) is a solution for one of your moms-to-be, please think again. Michel makes some brief comments on this obstetric 'solution' at www.wombecology.com/oxytocin.html. The most important thing to note is that synthetic oxytocin cannot cross from the body back to the brain because of a 'blood-brain barrier'. This means that when it's injected *into the body* it cannot enter the brain and function as the 'hormone of love'. The implication of this is that any synthetic form of oxytocin might impact not only the mother's ability to feel loving towards her newborn baby, but also her baby's ability to love. (Michel has written about this at length in *The Scientification of Love*, Free Association Books, 1999.) All in all, it's much better if conditions can be created so that the laboring and birthing woman's body spontaneously produces oxytocin because loving feelings then also appear as a bonus!

71  On this topic, you may be interested to see a DVD called *The Business of Being Born*, which explores possible financial and business motivations for having an extremely interventionist medical set-up for childbirth. You can find out more at www.thebusinessofbeingborn.com

72  The following systematic literature review failed to identify any research basis for vaginal examinations to assess the progress of labor:

- Devane D. Sexuality and midwifery. *British Journal of Midwifery*, 1996, 4(8):413-20

The following researchers also concluded that vaginal examinations were not necessary for facilitating physiological birth:

- Chalmers I, Kierse M, Neilson J. *A Guide to Effective Care in Pregnancy and Childbirth.* Oxford University Press, 1989

The practice of routinely conducting vaginal examinations was shown, in the following study, to be a ritual which legitimized intrusion into the very private birth place during the second stage of labor:

- Bergstrom L, Roberts J, Skillman L, Seidel J. "You'll feel me touching you, sweetie": Vaginal examinations during the second stage of labour. *Birth*, 1992, 19(1):10-18

Also see:

- Stewart M. "I'm just going to wash you down." Sanitizing the vaginal examination. *Journal of Advanced Nursing,* 2005, 51(6):587-594
- Warren C. Invaders of privacy. *Midwifery Matters,* 1999, 81:8-9

73 According to Dunn (1991), the instruction to caregivers to have women lying down was first printed in a textbook by Mauriceau in 1678. (Actually, I wonder if it's actually 1668—because copies of a book by him about this with that publication date is still available today!) According to Boyle (2000) the supine position became mandatory with the invention of forceps, which—it was thought, perhaps—might need to be used at any time. Hugh Chamberlain, who translated Mauriceau's book (from the French), belonged to the family who'd invented the forceps... and he tried to sell his secret to Mauriceau, so they obviously communicated. Later on, with the introduction of various forms of drug-based pain relief, women would inevitably end up lying down. The researcher Caldeyro-Barcia (*et al*) revealed the disadvantages of the supine position for labor and birth in 1979, especially for the baby. See:

- Dunn P. Francois Mauriceau (1637-1709) and maternal posture for parturition. *Midirs,* 1991, 66:78-79
- Boyle M. Childbirth in bed: the historical perspective. *The Practising Midwife,* 2000, 3(11):21-24
- Mauriceau F. *The Diseases of Women with Child and in Child-Bed: as also, the best means of helping them in natural and unnatural labours.* London, 1668—if you want to buy a copy, and can afford the price tag of £380 or £600! By 1727, this book was already in its sixth English edition, so it must have been popular
- Caldeyro-Barcia R. Influence of maternal bearing down efforts during second stage on fetal well-being. *Birth & Family Journal,* 1979, 6(i):7-15
- Caldeyro-Barcia R, Giussi G, Storch E. The influence of maternal bearing down efforts and their effects on fetal heart rate, oxygenation and acid base balance. *Journal of Perinatal Medicine,* 1979, 9:63-67

74 Lieberman R, Davidson K, *et al.* Changes in fetal position during labor and their association with epidural analgesia. *Obstetrics & Gynecology,* 2005, 105(5 Pt 1): 974-82

75 Michel Odent explains in detail why it's important that you make sure that no one (including you!) talks to the woman while she's in labor at www.wombecology.com/physiological.html

76 Some research has demonstrated that music helps to reduce anxiety and/or helps women cope with labor and stress, or give them a greater feeling of being in control. See:

- Spintge R. Some neuro-endocrinological effects of so-called anxiolytic music. *International Journal of Neurology,* 1989, 19/20:186-196
- Browning C. Using music during childbirth. *Birth,* 2000, 27(4):272-276
- Browning C. Music therapy in childbirth: research in practice. *Music Therapy Perceptions,* 2001, 19(2):74-81

77 Michel Odent explains some additional reasons why it's vital not to disturb the new mother in the first hour after the birth at www.midwiferytoday.com/articles/firsthour.asp

78   See the following articles:

  • Odent M. The Nocebo effect in prenatal care. *Primal Heath Research Newsletter*, 1994, 2(2)

  • Odent M. Back to the Nocebo effect. *Primal Heath Research Newsletter*, 1995, 5(4)

  • Odent M. Antenatal scare. *Primal Heath Research Newsletter*, 2000, 7(4)

79   To read more about these processes see:

  • Pearce JC, *Evolution's End: Reclaiming the Potential of Our Intelligence*. Harper, 1995 (pp 178-179)

  • Odent M. Orgasmic states, Ecstatic states and Mystical emotions. In: The *Scientification of Love*. Free Association Books, 1999 (pp 75-79)

80   Rosenberg KR, Trevathan WR. The Evolution of Human Birth. *Scientific American*, November 2001 (pp 60-64)

81   Goland RS, *et al*, Biologically Active Corticotrophin-releasing Hormone in Maternal and Fetal Plasma during Pregnancy, *American Journal of Obstetrics & Gynecology*, 1984, 159:884-890

82   The following studies noted various positive effects for the laboring woman of moving around in labor and using upright positions, including a reduction in the need for pharmacological analgesia, an increased sense of control and increased satisfaction with birth:

  • Simkin P, O'Hara M. Non-pharmacological relief of pain during labour: systematic review of five methods. *American Journal of Obstetrics & Gynaecology*, 2002, 186:S131-159

  • Spiby H, Slade P, Escott D, Henderson B, Fraser R. Selected coping strategies in labour: an investigation of women's experiences. *Birth*, 2003, 30:189-194

  • Albers L, Sedler K, Bedrick E, Teaf D, Peralta P. Midwifery care measures in the second stage of labour and reduction of genital tract trauma at birth: a randomised controlled trial. *Journal of Midwifery & Women's Health*, 2005, 50:365-372

83   For general research on water births, see:

  • Geissbuehler V, Stein S, Eberhard J. Waterbirth compared with landbirths: an observational study of nine years. *Journal of Perinatal Medicine*, 2004, 32 (4):308-314

The following research study established that hydrotherapy (i.e. the use of water) was far superior to augmentation with syntocinon when first-time mothers in labor experienced a 'prolonged' labor:

  • Cluett E, Pickering R, Getliffe K. Randomised controlled trial of labouring in water compared with standard of augmentation for management of dystocia in first stage of labour. *British Medical Journal*, 2004, 328:314

Another study, as follows, concluded that more women who immersed themselves in warm water used no analgesia during labor than women who labored on a bed:

  • Eberhard J, Stein S, Geissbuelher R. Experiences of pain and analgesia with water and land births. *Journal of Psychosomatic Obstetrics & Gynaecology*, 2005, 26 (2): 127-133

The following researchers found that women who labored in water had a high sense of control:

- Hall S, Holloway M. Staying in control: women's experiences of labour in water. *Midwifery*, 1998, 14(1):30-36

Finally, the following study concluded that after waterbirths more women had an intact perineum:

- Geissbuehler V, Stein S, Eberhard J. Waterbirth compared with landbirths: an observational study of nine years. *Journal of Perinatal Medicine*, 2004, 32 (4):308-314

84  Hot compresses are another possibility. One research study has confirmed the usefulness of warm packs applied to the perineum (during the second stage) by a caregiver. It's possible that if the laboring woman gets into warm water just before the birth this may have a similarly helpful effect, particularly since waterbirths have also been shown to result in more intact perineums. See:

- Dahlen H. The perineal warm pack trial. Abstract presented at the International Congress of Midwives, Brisbane, 2005
- Geissbuehler V, Stein S, Eberhard J. Waterbirth compared with landbirths: an observational study of nine years. *Journal of Perinatal Medicine*, 2004, 32 (4): 308-314

85  For a discussion of whether or not shoe size and height are related to higher risk see:

- Prasad M, Al-Taher H. Maternal height and labour outcome. *Journal of Obstetrics & Gynaecology*, 2002, 22(5):513-515

86  The following systematic review of nine randomized controlled trials concluded that continuous support during labor reduced the rates of cesarean sections, pharmacological analgesia, assisted vaginal birth, low Apgar scores and labor length and also helped women achieve more enjoyable births:

- Hodnett ED. Continuity of caregivers for care during pregnancy and childbirth (Cochrane Review). In: *The Cochrane Library*, Issue 2. John Wiley & Sons Ltd, 2006

The following study also supported the idea of woman having a 'known' support person during labor, although the conclusion was that a known *untrained* layperson provided the most effective care (so maybe you don't need to worry about how well-trained a doula is!):

- Rosen P. Supporting women in labour: analysis of different types of caregivers. *Journal of Midwifery & Women's Health*, 2004, 49(1):24-31

This strange conclusion was explained in the following study, which analyzed women's stress response:

- Taylor S, Klein L, Lewis B, Gruenewald T, Gurung R, Updegraff J. Biobehavioural responses to stress in females: tend-and-befriend, not fight-*or-flight*. *Psychological Review*, 2000, 107(3):411-429
- Bertsch TD, Nagashima-Whalen L, Dykeman S, Kennell JH, McGrath S. Labor support by first-time fathers: direct observations with a comparison to experienced doulas. *Journal of Psychosomatics in Obstetrics & Gynaecology*, 1990, 11:251-260

87 Bertsch TD, Nagashima-Whalen L, Dykeman S, Kennell JH, McGrath S. Labor support by first-time fathers: direct observations with a comparison to experienced doulas. *Journal of Psychosomatics in Obstetrics and Gynaecology.* 1990, 11:251-260.

88 Unfortunately, despite long searching, I really cannot as yet locate the reference for this study. (At one point I lost data from a whole computer hard disk and although I'd backed up most of it a few bits and pieces escaped, I'm afraid.) However, the following study into glucose levels in diabetics did find that subjects' pain thresholds decreased significantly on the days (in the study) when they had hyperglycemia (i.e. an excess of glucose in the blood):

- Thye-Rønn P, Sindrup SH, Arendt-Nielsen L, Brennum J, Hother-Nielsen O, Beck-Nielsen H. Effect of short-term hyperglycemia per se on nociceptive and non-nociceptive thresholds. Department of Endocrinology, Odense University Hospital, Denmark. *Pain,* 1994, Jan;56(1):43-9

In the following study, dextrose limited the duration of analgesia:

- Gage JC, D'Angelo R, Miller R, Eisenach JC. *Does dextrose affect analgesia or the side effects of intrathecal sufentanil?* Department of Anesthesia, Bowman Gray School of Medicine, Wake Forest University, Winston-Salem, North Carolina 27157-1009, USA. jgage@bgsm.edu. Anesth Analg, 1997, Oct;85(4):826-30

89 Copyright © 1999 Midwifery Today Inc. All rights reserved. Reprinted with permission from *Midwifery Today,* Issue 51. www.midwiferytoday.com / PH + 1 541 344 7438. You may also want to read Michel Odent's comments on what he sees as the 'masculinization' of birth over the last few centuries—because having a partner present at the birth is part of this. His full comments on this issue can be found at www.wombecology.com/masculinisation.html

90 Research has shown that upright positions for birth mean a shorter second stage, a much smaller risk of a need for episiotomy, forceps (or ventouse), less pain, easier 'pushing' and fewer fetal heart abnormalities. All in all, being upright seems a very good idea! See:

- De Jonge A, Teunissen T, Lagro Janssen A. Supine position compared to other positions during the second stage of labour: a meta-analytic review. *Journal of Psychosomatic Obstetrics & Gynaecology,* 2004, 25:35-45

- Gupta J, Hofmeyr G. Position for women during second stage of labour (Cochrane Review). In: *The Cochrane Library,* Issue 4. John Wiley & Sons Ltd, 2006

- Johnstone F, Aboelmagd M, Harouny A. Maternal position in the second stage of labour and fetal acid base status. *British Journal of Obstetrics & Gynaecology,* 1987, 94(8):753-757

- Chalk A. Pushing in the second stage of labour: Part 1. *British Journal of Midwifery,* 2004, 12(8):502-508

The only outcome which was found to be better when lying down was blood loss, although it is possible this was because of differences in ease of estimation. (It did not affect outcomes.) However, more generally, research has not yet confirmed that upright positions will help women avoid tearing. Nevertheless, the fact that the coccyx can move back when the woman is upright and thereby increase the size of the vaginal opening supports the idea that an upright position is better from the point of view of tearing. Research which has so far been carried out has been inconclusive and has only noted that a) a side-lying position may be slightly better than upright positions, b) that full squats did result in more perineal trauma, and—overall—that c) all-fours, standing, kneeling and semi-recumbent positions were all similar as regards the woman ending up with an intact perineum. However, it must be noted that research carried out will have looked mostly at women who were using some kind of drug-based pain relief—because it is used so frequently. No studies have as yet been conducted exclusively on women who have birthed without any drugs in their system, who are also undisturbed but supported by a sensitive midwife. For the research carried out so far see:

- Shorten A, Donsante J, Shorten B. Birth position, accoucheur and perineal outcomes: informing women about choices for vaginal birth. *Birth,* 2002, 29 (1):18-27
- Eason E, Labrecque M, Wells G, Feldman P. Preventing perineal trauma during childbirth: a systematic review. *Obstetrics & Gynecology,* 2000, Mar; 95(3):464-71
- Soong B, Barnes M. Maternal position at midwife-attended birth and perineal trauma: is there an association? *Birth,* 2005, 3:164-169

91 See the following articles:

- Kinmond S, *et al.* Umbilical Cord Clamping and Preterm Infants: a Randomized Trial. *British Medical Journal,* 1993, 306:172-175
- Pisacane A. Neonatal prevention of iron deficiency, *British Medical Journal,* 1996, 312:136-7

Physiologically, the baby receives the some 43% of the blood left in maternal/placental circulation. See the following articles:

- Gunther M. The transfer of blood between the baby and the placenta in the minutes after birth. *Lancet* 1957, I:1277-1280
- Wardrop CA, Holland BM. The roles and vital importance of placental blood to the newborn infant. *Journal of Perinatal Medicine,* 1995, 23(1-2):139-43

Cord blood—or blood left in the cord and placenta that hasn't made its way to the baby—is even useful in helping to fight childhood leukemia later in that child's life. For more about this, see:

- Steinbrook R. The cord-blood-bank controversies. *New England Journal of Medicine,* 2004, 351 (22):2255-7

92 AbouZahr C. Global burden of maternal death and disability. *British Medical Bulletin,* 2003, 67:1-11

93 Vol 12, No 2

94 See the following:

- Saito M, Sano T, Satohisa E. Plasma catecholamines and microvibration as labour progresses. *Shinshin-Thaku,* 1991, 31: 381-89. (Also presented at the Ninth International Congress of *Psychosomatic Obstetrics & Gynaecology.* Amsterdam 28; 1 May 1989 (Free communication No 502)
- Lederman RP, Lederman E, Work B *et al.* The relationship of maternal anxiety, plasma catecholamines and plasma cortisol to progress in progress in labour. *American Journal of Obstetrics & Gynecology,* 1978, 132:495-500

95 Rey Sanabria E. A l'autre bout du monde accueillir le prématuré. In Les cahiers du nouveau-né, No 6. Un enfant prématurément. Stock, 1983

96    In a typical study (Feldman, *et al,* 2002), which compared preemies which had had kangaroo care with preemies who hadn't, results of kangaroo care were seen to be quite dramatic. "After KC, interactions were more positive at 37 weeks' GA [gestational age]: mothers showed more positive affect, touch, and adaptation to infant cues, and infants showed more alertness and less gaze aversion. Mothers reported less depression and perceived infants as less abnormal. At 3 months, mothers and fathers of KC [kangaroo care] infants were more sensitive and provided a better home environment. At 6 months, KC mothers were more sensitive and infants scored higher on the Bayley Mental Developmental Index..." The researchers concluded by saying: "KC had a significant positive impact on the infant's perceptual-cognitive and motor development and on the parenting process." For this and other studies on kangaroo care, see:

- Feldman R, Eidelman AI, Sirota L, Weller A. Comparison of skin-to-skin (kangaroo) and traditional care: parenting outcomes and preterm infant development. *Pediatrics,* 2002. Jul;110(1 Pt 1):16-26

- Rey, ES, Martinez, HG (1983). Manejo racional del nino prematuro. In *Curso de Medicina Fetal.* Universidad Nacional, Bogota

- Charpak, N, Ruiz-Pelaez, JG, Figueroa de Calume Z (1996) Current knowledge of kangaroo mother intervention. *Curr Opin Pediatr.* (8:108-112)

- Doyle, LW (1997). Kangaroo mother care. *Lancet* 350:1721-1722

There is an article on kangaroo care at www.midwiferytoday.com/articles/timely.asp

97    Kenyon SL, Taylor DJ, Tarnow-Mordi W. Broad spectrum antibiotics for spontaneous preterm labour: the ORACLE II randomized trial. *Lancet,* 2001, 357:989-94

98    Guise JM, Mahon SM, *et al.* Screening for bacterial vaginosis in pregnancy. *American Journal of Preventive Medicine,* 2001, 20 (suppl 3):62-72

99    MRC/RCOG Working party on cervical cerclage. Final report of the Medical Research Council/Royal College of Obstetricians & Gynaecologists multicentre randomized trial of cervical cerclage. *British Journal of Obstetrics & Gynaecology,* 1993, 100: 516-23

100   Enkin MW, Keirse MJNC, Neilson J, Crowther C, Duley L, Hodnett E, Hofmeyr J. *A Guide to Effective Care in Pregnancy & Childbirth.* 3rd edition. Oxford University Press, 2000

101   See:   Hall, D. and Kirsten, G., 2008. Kangaroo Mother Care – a review. Transfusion Medicine, 18, 77-82.

- Feldman, R., 2004. Mother-Infant Skin-to-Skin Contact (Kangaroo Care). Theoretical, Clinical, and Empirical Aspects. Infants and Young Children, 17 (2), 145-161

102   Bergman, N.J. 2001. Kangaroo Mother Care and skin-to-skin contact as determinants of breastfeeding success

103   Cescutti-Butler, L.D.  2009. Eliciting parental views regarding early discharge to home care for premature infants. Infant, 5 (11), 23-27

104   See: Bergman, N.J. 2001. Kangaroo Mother Care and skin-to-skin contact as determinants of breastfeeding success.

- Stern, N.B. 1999. Motherhood: The Emotional Awakening. J Pediatr Health Care, May/June, 13, S8 – S11

105 Bliss, 2008. Premature babies – definitions and statistics. Available from: http:www.bliss.org.uk/

106 Tucker J, McGuire W, 2005. ABC of preterm birth: Epidemiology of preterm birth. STUDENTBMJ, April, 13, 146-148

107 Flacking R, Ewald U, Starrin B. 2007. "I wanted to do a good job": experiences of 'becoming a mother' and breastfeeding in mothers of very preterm infants after discharge from a neonatal unit. *Soc Sci Med* 64, 2405-16

108 Flacking R, Ewald U, Hedberg Nyqvist K, Starrin B. 2006. Trustful bonds: a key to "becoming a mother" and to reciprocal breastfeeding. Stories of mothers of very preterm infants at a neonatal unit. *Soc Sci Med*; 62: 70-80

109 Hepper, P, 2005. Unravelling our beginnings. *The Psychologist*, August. 18 (8)

110 See: Fenwick J, Lupton D. 2001. Mothers of infants in neonatal nurseries had challenges in establishing feelings of being a good mother. *Soc Sci Med1*; 53: 1011- 21.
- Heerman J.A., Wilson M.E., Wilhelm P.A. 2005. Mothers in the NICU: outside to partner. *Pediatric Nursing*; 31(3): 176-200

111 Stern N.B. 1999. Motherhood: The Emotional Awakening. *J Pediatr Health Care*, May/June, 13, S8 – S11

112 Shields-Poe D & Pinelli J. (1997) Variables associated with parental stress in neonatal intensive care units. *Neonatal Network*, February 16(1), 29–37

113 Cescutti-Butler, LD. Galvin, K. 2003. Parents' perceptions of staff competency in a neonatal intensive care unit. *Journal of Clinical Nursing*; 12: 752–761

114 Cescutti-Butler LD. Galvin K, 2003. Parents' perceptions of staff competency in a neonatal intensive care unit. Journal of Clinical Nursing; 12: 752–761

115 Cescutti-Butler LD, Galvin K, 2003. Parents' perceptions of staff competency in a neonatal intensive care unit. *Journal of Clinical Nursing*; 12: 752–761

116 See: Duddridge E, 2001. What are the advantages of transitional care for neonates? *British Journal of Midwifery*, February, 9 (2), 92-99.
- Wray J, 2006. Postnatal care: Is it based on ritual or a purpose? A reflective account. *British Journal of Midwifery*, September, 14 (9), 520-526

117 Beake S, Mccourt C, Bick D. 2005. Women's views of hospital and community-based postnatal care: The good, the bad and the indifferent. . *The Royal College of Midwives, Evidence Based Midwifery*, 3 (2), 6

118 See: Wray J and Davies, L. 2007. What women want from postnatal care? Midwives, March, 10 (3), 131.
- Wray J. 2006. Postnatal care: Is it based on ritual or a purpose? A reflective account. British Journal of Midwifery, September, 14 (9), 520-526

119 Cescutti-Butler L and Galvin K. 'Parents' perceptions of staff competency in a neonatal intensive care unit.' *Journal of Clinical Nursing* 2003; 12: 752–761

120 Cescutti-Butler LD. 2009. Eliciting parental views regarding early discharge to home care for premature infants. Infant, 5 (11), 23-27

# Conclusion:

## How can we move toward optimality?

We've been through so many reflections and so many birthframes, I hope it hasn't all left you feeling rather confused. In these last few pages I would like to remind you of a few key principles relating to optimality (based on my own conclusions) and make some commentaries on evaluating research generally.

## Ground rules for ensuring optimality

Firstly, I hope you will agree there seem to be some ground rules for care which promotes optimality, which most caregivers (midwives, their managers, doulas, etc) usually agree with. I would briefly summarize these as follows:

♥ **Continuity of care** There needs to be as much continuity of care as possible. This means that one woman receives care from just one, named midwife (who she is happy with) throughout the prenatal period and this same midwife then attends the woman in labor and right through until after the birth and postpartum. Other caregivers provide additional support.[1]

♥ **Agreement on care** Both the pregnant woman and her caregiver(s) need to be in agreement about the care given. In order to achieve this, time needs to be spent prenatally talking about possibilities (relevant to each woman's particular situation) and possible approaches. It is best if all decisions are confirmed in a care guide. Using a care guide simply ensures clarity of communication and full agreement. It's best, from the point of view of achieving this, if the main caregiver responsible for any particular woman's care can also sign the document and keep her (or his) own copy.

♥ **Backup arrangements** Since life doesn't always go to plan, it's obviously important for caregiver and mother-to-be to confirm what will happen if things don't go according to the care guide. This may mean deciding who will stand in for the main midwife, if the 'named' midwife isn't available at the time (because of some eventuality). It may mean considering which hospital she might transfer to in the case of a home birth and it would mean thinking through how she would physically get there. (This might be a particularly important consideration in some parts of the country!) It might also mean considering any other points which are particularly important to the woman... but note the next section in this respect. Mostly, it will mean having basic equipment (and expertise) available to ensure that care can be given effectively, if needed. However, it will also mean keeping this available in a very discrete way because *showing* a pregnant, laboring and birthing woman emergency equipment is not going to inspire her to believe that all will go well. Most women will want to be reassured that the equipment is there (or they will assume it is), but some won't. *All* of them will need to be able to forget all about this aspect of care if they are to visualize and experience an optimally optimal birth.

♥ **Lack of negative thinking** Even though life isn't perfect, it's important that the pregnant woman be allowed and indeed encouraged to build a positive expectation of birth.[2] If she is constantly asked to imagine what she would want if all kinds of awful things happened, this process is likely to be disturbed or disrupted completely. As a result, although you, as the caregiver, may need to think through certain eventualities, you need to be very careful about how you talk about them, you need to consider if it's necessary to, and you need to consider with everything you say whether you're building up fear in the pregnant woman, or a joyful sense of expectation. It's not for nothing that top sports people talk about visualizing their success. Business managers also talk about focusing on strengths (not weaknesses) if they are to be successful because it's by building on strength that we optimize chances of a good outcome. By contrast, the likely outcome if we focus time on weakness is a mediocre outcome. (I realize you're worried about all kinds of safety issues, but please remember the close link between emotional responses and hormone production and the importance of hormones in orchestrating a safe birth.)

♥ **Lack of disturbance** Many of the birthframes we've seen have revealed the importance of disturbance. Cascades of interventions, triggered by just one tiny first intervention (such as stripping the membraines, or asking a woman to lie down) have resulted in an enormous increase in the cesarean rate and a rise in the number of women who are either dissatisfied, depressed or incensed after giving birth. Cascades of hormones, on the other hand, leave women feeling empowered and energized, and well-prepared for motherhood.

♥ **Support and security** Creating an environment which protects the laboring and birthing woman from disturbance is difficult to do for all kinds of reasons but even this is not enough. As a caregiver, you also need to create an environment which somehow feels supportive and secure, from your client's point of view. (And remember, please, that she is not a 'patient'! She is simply a woman going through a healthy physical and psycho-sexual process.) How you personally make each woman feel supported and secure will depend not only on your personality, but also on hers, but whoever is involved your behaviour will need to be more silent than talkative, because—as we've seen—speech can be one of the biggest and most destructive sources of disturbance during labor and birth. Nonverbal communication needs to be sensitive and subtle, inspiring confidence, joy and perseverance, rather than fear, tension and worry. Again, this is to do with the hormonal cascade which should be taking place if you are being successful...

♥ **Handing over 'ownership' of the experience** Here I'm basically talking about allowing the woman to experience the pregnancy and birth herself, as she chooses to do, in her own way. You mustn't take that away from her through the way you speak or behave. In other words, invite her to share with you, don't assume you already know everything...

Beyond this basic foundation of care, I think there are also a few other factors that determine whether or not the care a woman finally receives is optimal, either from the point of view of safety or from a psychological point of view.

♥ **Her preparation prenatally** On one level it's impossible to 'prepare' to give birth. However, on another it's essential and a lack of preparation can mean a woman experiences something which is far from optimal, when it comes to childbirth. So what do I mean by 'preparation'? Here I'm basically talking about the chapters in the companion book for pregnant women called *Preparing for a Healthy Birth.* These can be summarized as follows:

   ◆ Understanding what birth is like in a healthy body ◆ Considering 'risk' ◆ Understanding the hormones of birth and why disturbance can be an issue ◆ Understanding what pregnancy and birth might be like from the growing baby's point of view ◆ Making decisions about prenatal care, birthplace and birth attendants ◆ Dealing with minor or major physical problems during pregnancy ◆ Dealing with psychological issues and visualizing a good birth ◆

♥ **The sensitivity of caregivers at all stages** As we've seen in various birthframes, even when everything appears to go well in terms of safety and outcomes in a particular birth, women are sometimes left feeling far from happy afterwards and there are all kinds of problems which go undocumented in the official medical records. These problems may relate to general emotions (anger at how midwives spoke, for example) or to medical problems (such as later sexual difficulties) which women are too embarrassed to report after the birth. It's interesting, but saddening, that births which appear well in statistics can leave such a negative taste in the mother's mouth, particularly given the impact this could potentially have on bonding, parental relationships, and new—or renewed—motherhood.

♥ **The woman's experience postpartum** Since the birth of any baby (the first or the fifth) is an unheaval for a woman, and since women increasingly have less experience of baby care by the time they give birth (since the birth rate is decreasing and more women are working), postpartum care is crucial to overall optimality. Visits after the birth are very difficult things to undertake for all kinds of reasons but somehow the associated problems need to be addressed. A few of these are as follows... The woman may feel she is being 'tested' when she is visited; she may not know what the midwife (or health visitor) is expecting her to do or say (so as to be categorized as 'normal); she may not feel enough rapport with the caregiver to open up in a relaxed and honest way; she may feel harassed because of the new baby's demands; she may feel in conflict about the birth, which may be partly (or fully) because of the midwife's behavior then; she may feel a whole host of emotions which she barely understands, which prevent her from wanting to express any particular thoughts or feelings. If you have an awareness of these issues it should help a woman to have the best possible care after the birth, so that she is left feeling she has had a birth which was as good as it could possibly have been. After all, that's the nature of optimality.

# Pause to reflect...
## ...about establishing systems for ongoing optimality

1 Do you agree with the 'ground rules' for optimality, as described here? Would you add any?

2 Do these ground rules apply to all caregivers involved with birth, or only to some?

3 Given the need for sensitive communication with the mother-to-be and birthing woman, how can caregivers liaise most effectively?

4 What kinds of in-service training supports optimality, in your view? Should specific groups of caregivers attend workshops independently of each other, or together? What are the pros and cons of each approach? How can each type of caregiver participate?

5 How can women be prepared prenatally for optimality? Do you 'believe' in preparation or are you one of the caregivers who claims that no preparation is possible? If so, what—in your experience—typically happens to a woman who is completely unprepared for labor and birth when she's actually experiencing it? Are outcomes in those cases determined solely by care given or by other factors too?

6 Are there any cases you can think of where sensitivity might be perceived in different ways by different women—perhaps from different socio-economic groups, or from different subcultures?

7 If you are a parent yourself, what were your own experiences intrapartum in terms of optimality, what were those of your partner, and how did you feel postpartum about the care you'd received?

8 From your viewpoint as a citizen, not a caregiver, what changes would you like to see in the maternity system?

9 How can women be better supported postpartum?

10 How can women be encouraged to provide useful feedback postpartum?

Do you have any other thoughts, suggestions, concerns or questions?

## Birthframe 98: Beth Dubois

Beth made some general comments about women's choices around birth...

At a certain point in my second labor, I suddenly stood up, leaned forward in a semi-squat, felt extreme determination and inner confidence, then suddenly pushed the baby out in one great push with a corresponding yell. My husband Joel said it was amazing... There was no baby visible, then suddenly there was a slight bulge and then a whole baby! Sylvie describes that in this type of birth, by fetus ejection reflex, there is typically very little blood loss and that was exactly what my midwife was surprised to report. My birth experiences have been profound and life-changing. My babies were born healthy and alert. I want the experience of physiological birth, and all the benefits that go along with it, to be available to women and babies everywhere.

Works such as this book are vital because the majority of women (as well as men and healthcare providers) don't understand or truly know about the option and incredible value of having an undisturbed birth. It is my hope that the power and transformative nature of physiological birth become known to our society and that women who wish for a physiological birth have the right support, care, birth attendants and birth settings to make it possible.

Having said that, I also recognize that every woman and every birth is unique and women may have medical or personal reasons that lead them to choose or need interventions during labor and birth. I trust that women make the choices that are best for them, to the best of their ability, in each moment. I had a friend who we strongly suspected had an abuse background. She, however, did not seem to be aware of it and did not seem open to delving into this part of her life. She planned a home birth but it didn't work out for her, perhaps due to unresolved issues she held in her mind and body. After laboring very painfully at home, she decided to go to hospital and have an epidural and in the common cascade of interventions ended up having a cesarean. She was physically and emotionally relieved when she got the epidural and was sure that was the route she'd go if she had another child. For her, it was great that she planned the home birth because the individualized midwifery care during the pregnancy and the increased sense of control she felt around the plan of having a home birth supported her. Another friend of mine, Sallie, told me that she was really terrified of giving birth, that she had a recollection of many past lives of having died in childbirth. When she learned her son was breech and that the consultant recommended a planned cesarean birth, she was truly relieved; she felt it was the right thing for her.

So while I feel passionately about the value of physiological childbirth and consider individualized midwifery care and home birth the gold standard, I also think it's important for women to have the right to choose what they feel comfortable with. Who can truly say what is optimal for anyone else? And most of us who have powerful unmedicated births have a difficult time during some part of the birth, even if it turns out to be a beneficial experience overall, or perhaps it's even the difficulty we go through that makes it such a beneficial experience. I have heard it called a rite of passage and in a traditional rite of passage, there is typically adversity that is overcome.

So, on the one hand, I wish more women had access to the type of birth I had and on the other hand, it may not be the right thing for everyone. For survivors of sexual abuse (or some other kind of abuse), it could be a breakthrough healing experience or, in some cases, if the woman doesn't feel well supported or has flashbacks, or what-have-you, it could be traumatic, actually. My midwife, Valerie, told me that she has known women who had cesareans who had had previous trauma around surgery and, for them, having their beautiful baby born surgically, i.e. having something so wonderful take place in that same region, was a healing experience.

Sylvie's work is invaluable because she is giving words and form to the experiences women have and other women wish to have, helping bring this reality further into concrete existence. After my daughter's birth, I had a lingering concern that I had pushed her out too aggressively or perhaps angrily, and felt remorse. However, from reading Sylvie's book, I realized with awe that, in fact, I had experienced the fetus ejection reflex. I realized that the emotions coursing through me, the posture I naturally assumed and the birth with just one push, were healthy, innate, and characteristic of a woman experiencing a physiological birth. What power they contain! What a relief for me to understand this! Since we live in a culture that is unfamiliar with physiological birth, even those of us who are fortunate enough to have such a birth may misunderstand it, not realizing how healthy, instinctive, and powerful it is. In this way, Sylvie's work is a guiding light for all of us.

# Questions for reflection...

1 Do you agree that undisturbed physiological birth with midwifery care is the gold standard? For which categories of women are cesareans better?
2 In cases where women choose (prenatally) to have a drug-free, physiological birth without any interventions if at all possible, what do you say to the pregnant woman? What could you say instead?
3 Given the apparent potential healing and empowering aspects of physiological birth, is it in your clients' best interests to learn about it? To what extent do you feel you need to educate women in your care?
4 What kinds of women do you think could be healed by other kinds of births?
5 How can you effectively differentiate (prenatally) between women who've had a cesarean, who would benefit from a positive cesarean experience for their second birth and women who would benefit more from a VBAC?
6 When women come to you prenatally saying they want to have as much pain relief as possible during labor, do you think there's any point in encouraging them to find out about undisturbed, unmedicated birth?
7 Do you think there's any value in encouraging pregnant women to learn about the differences between different types of unmedicated births—e.g. an unmedicated birth which was insensitive and disturbed and one which was caring and undisturbed and fulfilling?
8 Do you think there's any value in encouraging women to read birthframes?
9 Do you think it's helpful for women to write their own birthframe?

### Birthframe 99: Heba Zaphiriou-Zarifi

To remind us that women's bodies do in fact usually function extremely efficiently, here's a reflective account from another woman who gave birth before any midwives arrived. I'm including it here because I think it's so important to remember that although optimality involves a birth with a professional in attendance, those very professionals must never think themselves indispensible to the normal, healthy processes. I think this account also reminds us of the importance of a pregnant woman's attitude towards her own pregnancy and birth. When Heba (a movement therapist and Jungian psychotherapist in training) realized that she might give birth before the midwives arrived for her planned home birth she decided not to panic. Instead, she really did embrace the unexpected. Of course, this is what we must always do because one thing almost everyone can agree on is that ideals and reality do not always come close... but this doesn't mean we can't make things as good as they can possibly be.

My three daughters were born at home and yet every pregnancy was unique and every birth was a new and different experience. Each birth is influenced to some extent by the way the previous one occurred, it contains it as well as goes beyond it. It is also grounded in our maternal lineage with its roots embedded in the timeless, archetypal level of our nature, both personal and universal. Birth and death touch each other, like two sides of the same coin. A letting go, of some sort, is necessary for the birth to occur. Something has to surrender for the new life to emerge. The safety and sacredness of the 'chosen' time and space are a pre-requisite for a fulfilling birth experience. Partners involved, as well as the wider community, will be affected by that powerful experience, be it a hospital or home birth, which will mark and shape patterns of the child's life. The pattern of one's own birth is symbolic of all other transformative life events.

Perhaps one needs to reconnect consciously with the forgotten memory of one's own birth experience or with a major event in the family. Memories are stored in the many different layers of the body and might emerge as the contractions squeeze them out in waves. Like the ebb and flow of the tide, contractions bring to the shore what was trapped at the bottom of the sea, what was held in the tissues of the body. So I believe that whatever knots we can undo will help open a wider path for the birthing process to follow its natural course.

I had thoroughly prepared myself for a home birth the first time, and the second time too, by reading books on home birth, by doing my yoga and singing. Walking in nature was a delight, breathing and listening to birdsong. Harmonization was a fantastic tool too. (This is a very deep relaxation massage, which puts energies in harmony.) I also had at hand some homeopathic and herbal medicine. Every woman will find in herself all the treasure of knowledge she needs in order to face her pregnancy, birth and mothering experience. Empowering herself and making informed choices is of the essence.

As planned, I contacted my midwife at our nearby hospital while contractions were in progress. No response. The lines were constantly busy. I tried many times until I realized that the hospital's lines must be dysfunctional. There was no time to waste. Nothing can stop life when it is coming forth. So I decided to go with it. I filled my bathtub, breathing, sounding. The flow of the water was magical. I lit a candle and got in the bathtub. My husband suggested calling the ambulance and asking them to collect our midwife from the hospital. I was very calm... Having had a happy first home birth, there was no reason for me to be disturbed nor alarmed by this one. With total concentration and presence I surrendered to the process. And my baby came rushing out with such power, bursting out with life. I shall never forget her first intake of breath. The transition from water to air is such a significant transition it was a unique and unforgettable moment, something I might have missed had we been in a hospital environment. Hearing her first breath entering her body was a gift beyond words, a most gratifying spiritual experience. I trusted my instinctive and spiritual ability to give birth— something that all women have in them as an inborn knowledge, waiting to be revealed. In other words, I trusted the wisdom of my body and the circumstances for the birth of my daughter somehow became a manifestation of that basic trust.

I know from my experience as a therapist that the actual birth experience becomes a metaphor for other creative activities involved in one's journey. It is also a metaphor for the deep psychological transformation so necessary in order to find one's wholeness. No wonder women fall into depression and their energy scatters when their initiation from maidenhood to motherhood is tampered with. Other people's need to control them is a substitute for controlling their own fear of this archetypal energy of death and rebirth. Or is it an unconscious jealousy that makes the obstetricians want to take centre stage, when it is truly the woman's realm?

Scared men and women who are unable to surrender to the unknown, who are afraid of their own vulnerability tend to medicalize pregnancy, birth and even breastfeeding. Controlling a woman and her baby in their mutual birth experience can leave a scar so deep in the woman's and baby's psychological bodies that it may take many more painful wounds to find the root and meaning of it all. Dis-abling a mother and her baby when entering their heroic journey into the unknown is a psychological crime committed by many.

Going through a transitional, liminal birthing experience with tremendous vulnerability, a maiden sheds her skin in order to enter her new identity as a young mother. Giving birth is a struggle to be overcome by mother and baby and yet it enables a woman to take on the body of a mother, along with the energy necessary to fill this role. For nine months, her body has been the home of another unique individual and her psyche a land for her baby's development. By giving birth, she is being born unto herself, an experience that will change her for ever.

Going through a transitional, liminal birthing experience with tremendous vulnerability, a maiden sheds her skin in order to enter her new identity as a young mother

The actual birth experience is a metaphor for the deep, psychological transformation necessary to find wholeness

Above and below: *Sophia, who 'came rushing out with such power'*

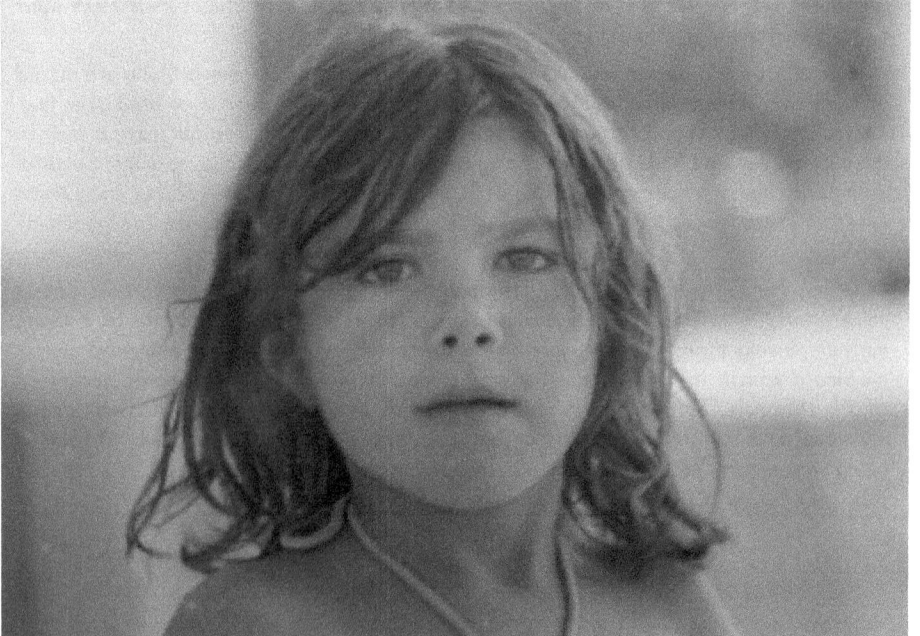

# Pause to reflect...

## ...about maternal evaluation and transformation

1   To what extent do you agree with Heba that birth is an important process in a woman's life?

2   Do you agree that the experience of birth is important for the baby too?

3   To what extent is the experience of a birth important to a new father?

4   Why do most women never seem to forget their experience of giving birth—assuming, of course, that they didn't have scopolamine (a drug which induces forgetfulness, which was given routinely a few decades ago)?! Why is it they can recount the story of their births in detail years later? How accurate are their accounts, in your opinion? Are caregivers accounts more or less accurate?

5   Do you agree that every new birth is unique? To what extent are there similarities between births? To what extent are differences important?

6   How can women deal with disappointment when their births don't go as they hoped they might?

7   Do you think it might be worthwhile keeping statistics on maternal satisfaction postpartum, tying data in with obstetric experience intrapartum and even prenatal preparation and details about care provided (e.g. continuity or not, midwife or obstetrician, etc)?

8   In your view, does birth change the way a woman behaves in the rest of her life? Does it change the extent to which she is assertive or confident about herself? Can it ever affect her relationship with her partner? Can new births heal the scars left by old births? If so, why?

9   Why is it that a certain number of women never recover from their experience of giving birth?

10  If you have given birth yourself, how did the experience change you? If you swam in the metaphorical waters of optimal, undisturbed, physiological birth, without any drugs or interventions, what effect did it have on you, if any? If you haven't experienced this kind of birth, would you want to? Would you wish it on other people? Why?

## Research and the need to evaluate it

You probably still have many unanswered questions... This is to be expected because optimality is not something which can easily and universally be defined. One way of exploring answers is to consider the evidence and then base practice on it. But how can we become experts at evaluating research? After all, if it turns out that research is poorly set up or analyzed (for whatever reason), the conclusions would be invalid and we would be taking misguided decisions about the care we provide. At the same time, we need to become used to evaluating data in qualitative terms, as well as quantitative terms because—as this book has tried to demonstrate—an enormous amount can be learned from qualitative data (e.g. case studies), which a more quantitative approach might miss altogether. (As you will remember, Soo Downe reminded us of the value of qualitative research in her Foreword.) With these points in mind, here are some guidelines for evaluating research...

### Consider the research design

However good a study is in other respects, if a research project has been poorly designed, none of its conclusions will have any validity. If, for example, a study were to look at women smoking in pregnancy and if this study concluded that smoking 'caused' fetal growth restriction, this conclusion would be invalid if it just so happened that all the women (or even many of them) were also heavy drinkers. We simply wouldn't know whether it was the smoking or the alcohol causing the growth restriction![3] Indeed, if a study has been poorly designed the cause of any outcome observed (such as growth restriction) could be due to something *other* than that identified in the conclusion.

At the risk of challenging your assumptions, I will mention the kinds studies which I believe may suffer from this problem:

- A breech trial, comparing vaginal births with elective cesareans[4]
- Studies conducted to establish the cause(s) of SIDS (neonatal death).[5]

In a breech trial, caregivers who are 'delivering' babies vaginally might not be experienced or skilled in breech birth. There is also a strong likelihood that they may not feel either confident or happy to be in the group of caregivers who are required to 'deliver' babies vaginally... and any fear could well skew results (because of the link between emotions and outcomes).

In studies looking into co-sleeping and SIDS it's very difficult for researchers to adequately take into account problems such as alcoholism (which is reportedly particularly widespread amongst Maoris) so inaccurate self-reporting might be a problem. (Imagine a couple who are in the habit of sleeping next to their baby. Imagine that the baby one day dies... Would the parents admit to a researcher that they'd been drinking that particular night? Unlikely, possibly.)

These or other studies may be problematic in terms of design in other ways. For example, a study of co-sleeping might only be valid if babies are considered according to whether they were breast– or bottle-fed and according to whether they were born in unmedicated births, or with drugs in their systems.

## Consider the researcher(s)

I'm being perfectly serious. Find out who the researchers are (e.g. where they work, what their job is and what their history is) and find out who is funding the research project. If a research study on anesthesia is carried out by an anesthesiologist and the project was funded by a pharmaceutical company a certain bias might come into the study... An obstetrician researching breech birth in a private maternity care system (i.e. virtually anywhere outside the UK) might not be motivated to conclude that breech births should all be vaginal births... because he or she would suddenly lose a lot of fees for cesareans.

This may sound enormously cynical. I'm sure it does. I know there are lots of very ethical people out there... But aren't we all a little bit too subjective for our own good? Is it really possible for a human being, attempting to evaluate things 'objectively', using time-bound, relative perception *within* the very system he or she is researching—to make objective evaluations? How can a researcher avoid bringing his or her personal experience to a situation? Is it possible at all?

Another aspect of this same issue is whether or not a researcher is likely to be able to set up a research study adequately, given his or her own first-hand experience. For example, how can a researcher who has never experienced an unmedicated, undisturbed physiological birth possibly ensure that conditions are appropriate for other unmedicated, undisturbed physiological births? If I were to set up a golf tournament to test out different golf balls, unless I consulted golfers who really had experience of the thing I was hoping to research, I really wouldn't be capable of checking the golf balls appropriately... simply because I know virtually nothing about golf. (I was actually the person responsible for organizing golf tournaments for Mitsubishi Electric in London for 14 months and used to regularly meet up with golfers and buy golf prizes... but I still don't have a clue!) Is it not possible that a researcher comparing two approaches to birth might make naïve mistakes—which any woman who's *really* been swimming might immediately understand?

## Consider the statistics

Any statistician—e.g. an accountant!—will tell you how statistics and numbers can be manipulated to present 'information' in different ways. So if you want to really evaluate statistics, you'll need to learn about them—about correlation, standard deviation, variable analysis, etc.

However little knowledge you have when it comes to statistical analyses, never forget the importance of a common-sense approach. If something sounds shocking (e.g. 30% of women eat chocolate every evening!) remind yourself that 70% don't. (No, I'm afraid that's not a real statistic!) And look at the language used to express conclusions. Consider, for example, the difference between that last statement (that 30% eat chocolate every evening) with the statement that 30% *may* eat chocolate in the evenings. In other words, read between the lines.

## Consider the details of any research study

- Are you looking at a randomized controlled study or a cohort study? In a cohort study, how were subjects selected? What were the exclusion criteria?
- Was the study prospective (i.e. looking into the future) or retrospective (i.e. analyzing past data)? If it was the latter, note that it's impossible to go back to the past to check details and the conclusions can only ever be based on the records kept. Have you ever made a mistake when in an administrative position (OK, OK, I'm the only one who's done that) and have you ever seen errors in medical records? (I won't make any comments about the latter, but some could certainly be made... and I'm sure you may well have thoughts on that of your own. After any accident, would five different reports of the scene written by different parties in the accident be identical? What might influence errors or omissions or wilful misrepresentation?)
- How well was the data collected checked and cross-checked? Is the research study analyzing objective data (e.g. heights of women and weights of their babies at birth) or unverifiable (e.g. reports of how much chocolate and fruit were eaten in pregnancy or—even worse—how often masturbation or infidelity occurred during a marriage)? In other words, how likely is it that the raw data of the research project is reliable and indisputable?
- If questionnaires were used, were there follow-up interviews to double-check information or opinions? If questionnaires or interviews were used, how well were these designed? What form did questions take? (Were answers given using a scale of 1-5 or were respondents required to write answers out in words? Note that while scales are likely to produce simplistic results, which are wonderfully easy to 'process', they are prone to error—because respondents may get confused halfway through, forgetting whether 1 is 'good' or 'bad', and also because answers are inevitably simplifications or misrepresentations of the truth, which is usually much more complex. Note that even responses written out in words are open to misinterpretation and may be incomplete.
- If experiments were carried out are the results valid or were there confounding factors? We're back to those hypothetical pregnant women who smoked and drank, who ended up having smaller babies...
- If samples were taken (of people) how big were the sample sizes?
- Where was the research conducted? Are these places representative?
- Could there be any other variables influencing the results?
- Could there be another interpretation of the data collected?
- Does a research study pose more questions than it answers?

## Find out what other evidence is available

Search wherever you can—e.g. at www.pubmed.com or using the database of journals you have access to via your place of work.

# How can we make decisions with inadequate data?

In some cases there is no research at all on a given topic and in some cases there is only one study—or a handful with inconclusive results or poor research design—and we nevertheless need to form an opinion on an issue. After all, these opinions are what allow us to 'use our discretion' in a given situation. So what should you do? Here are a few ideas...

## Consider any evidence you have available

Evidence you may have may come from anywhere in the hierarchy of research. At the top, we might conventionally put randomized controlled trials (RCTs) and at the bottom case studies... but in reality a well-focused case study might be a lot more useful for informing practice reliably than a poorly designed and conducted RCT. As you may already know, you will find formal or informal case studies (perhaps in the form of birth accounts) in journals (e.g. *Cases Journal*), magazines (e.g. *Midwifery Today* or even popular pregnancy magazines) or on the Internet. The point in reading this potentially unreliable, poorly recorded data is to stimulate ideas and get a 'window' into other people's worlds. However, when reading anything apart from this book, be aware that birth stories may not have been double-checked with contributors after editing (so may contain many errors or misrepresentations) and they may omit certain very important details—which, in this book, authors were cross-questioned on during the editing process. A few contributors who'd had their accounts published in other places commented to me that previous versions had contained mistakes or misrepresented their experience, which they'd found frustrating—so they were pleased about the Fresh Heart editing and approval process.

## Don't disregard your knowledge of the world

There are some things which are 99% likely to be true, which it is almost impossible to actually 'prove' through research, either because it would be impossible to set up a research study, or unethical or unsafe to do so. For example, we know that walking out onto an interstate without looking is likely to result in death, but who's going to try it out (with a control group of people who look!) in order to find out?! We also know that water is probably important for the growing fetus, but which researcher is going to ask one group of women to stop drinking water (or any other liquids) during the third trimester so as to check? There are hundreds of similar examples so we will often need to simply trust in our knowledge of the world and our ability to be logical. Yes, it's true that our knowledge is constantly changing, but it's all we have at any one time—and we have to trust it to a certain extent so that we don't endanger women and babies unnecessarily with research studies which are likely to harm large numbers of clients. For some reason, some medical people or researchers have become diffident about using logic to make decisions on care. Why, though? It's useful stuff, so please use it!

## Set up your own study and write your own articles

30 years ago when I first started working as a manager—yes, I really am that old!—I never saw myself as being the kind of person who could also conduct research or write articles. It is only now that I have given birth and become an empowered woman—YES! Empowered!—that I can see all my missed opportunities. Look at your own work situation and consider what you can realistically do to further any area of research which you're personally interested in exploring. Then, discuss your ideas with your line manager(s) and get the appropriate approval and guidance before going ahead. The world needs to hear about your own work and ideas!

*The world needs to hear about your own work and ideas!*

## Ask your clients prenatally about their opinion and research

It's easy to forget how motivating it is being in a particular situation. When I was pregnant I suddenly became tremendously motivated to read up on pregnancy. When I was about to go and live in Japan, I suddenly acquired an interest in books about Japanese culture... Let's not forget this motivation any pregnant woman (who has not become a total patient) will have to research her own situation. In any case, whether or not your clients are well-informed, their opinion is really what counts since caregivers are really there to be advocates for women, not dictators of care.

## Become aware of your clients' needs

Each individual will make choices based not only on information but also based on their past experience and on what they know about themselves and their own personal limits and potential. If you make an effort to become aware of every client's world view and state of mind, you will be in a much stronger position to advise her and discussions about care are much less likely to turn into battles of will.

## Become self-aware

It's amazing how much can be discovered simply by increasing our awareness of situations. Watch yourself in action (in your head), listen to the words you use, observe your client's reactions as well as those of your colleagues... and experiment with new approaches. As you try out and evaluate different approaches, moving further and further outside your comfort zone, even if you are not actively learning you will at least be gathering ideas and inspiration for research projects or reflective articles, which you can then submit to journals.

*Watch yourself in action and observe your clients carefully.*
*Even if you don't learn anything, you'll get plenty of ideas.*

And, just in case you feel inclined to dismiss this last suggestion, please let me remind you of the history of medicine. Only a few decades ago there was a lot more freedom and experimentation in the field of midwifery and obstetrics than there is now. How and why do you think procedures were first carried out? Yes... that's right... it was usually just because someone, somewhere thought it might be a good idea. People weren't always following protocols and they weren't always thinking 'inside the box'. It's more fun outside the box!

## Ask your client for feedback on an ongoing basis and postpartum

This is perhaps the most difficult thing of all to do effectively. After all, how are you going to elicit some really honest and useful information?

When you're providing prenatal care, during those delicate early phases of a relationship when attempts are being made to build rapport, how likely is it that your clients will tell you when they don't understand what you're explaining or don't like what you're doing? How assertive and strong in herself does a woman have to be in order to disagree with her midwife or other caregivers? What differences will there be between the things she says to you and the things she says to her closest friends, her partner and her mother?

Intrapartum this problem will be even greater. At that point, assuming you've provided at least some level of continuity of care, you will have developed quite a strong relationship. What's more, at this stage your client will be dependent on you and your goodwill, not only for the sake of safety but also for the sake of her own happiness because a grouchy or offended caregiver is unlikely to be an inspiring and empowering one... In any case, during labor, your client may well feel especially vulnerable, so she's very unlikely to want to 'rock the boat'. It may be even more difficult for new or renewed mothers to express their wishes, preferences and regrets clearly not only because of your shared history by that stage and the new mother's need to create a positive mental memory of what happened during the birth—for the sake of her own peace of mind—but also because she will be in that very fragile state of new motherhood. Again, she is likely to very much appreciate any support you can offer (even if only during fairly short visits) and is unlikely to feel able to say anything which might be construed as criticism.

If you feel, on reading this, that I'm painting a very bleak picture, consider for a moment most people's expectations within our particular modern world. Is our world not rather Orwellian at times (remember that book *1984*?) in that we use double-speak to conceal our real meanings? Are we not sometimes too polite for our own good? Although we have reality TV and very open discussions on the radio and the Internet, are we not still very private individuals who are in desperate need of friendship and support? While we're prepared to open ourselves up when there's little risk (e.g. when using an upbeat anonymous user name on an Internet forum) do we not all have limits in some areas of our lives? But do we not open up for certain business surveys very willingly? Why?

Your challenge is to decide how it might be possible to get feedback. Let's be creative and responsible in our quest for optimality. It's a goal worth reaching not just for the sake of the development of midwifery and obstetrics... It's something we need to do for the sake of every single pregnant woman and baby. These two key individuals will experience the outcome of our actions very keenly and their gratitude for what goes right is well worth having.

# Pause to reflect...
## ...about evaluating research and achieving change

1 How careful would you be if you were asked to participate in a research project?
2 If you were interviewed, how honest would you be when asked about sensitive, extremely personal subjects?
3 How often, in an interview, would you interrupt the interviewer to add extra information which didn't fit into the researcher's categories?
4 If you've ever set up a research project yourself, how easy was it to design?
5 How would you go about finding subjects for your study?
6 How would you check that your data analysis was correct?
7 Are you capable of interpreting and writing up information on statistics (standard variables, etc)?
8 Do you ever read about research conclusions which seem difficult to either believe or accept, or do you accept all research results unquestioningly?
9 Do you think all research questions are researched equally thoroughly?
10 Which kinds of research projects are difficult to fund, in your opinion?
11 Why are some studies replicated, while others, which seem very interesting, are ignored or dismissed?
12 Why are some research studies, with questionable data (when you look at the original full write-up) immediately accepted by the medical community, while others, with much clearer results, are ignored or unnoticed by midwives and obstetricians?
13 Who decides within your own working context (or potential working context) whether or not protocols should be changed to take into account new research?
14 What are the procedures in your workplace when you disagree with protocols, based on your own research or analysis of available evidence?
15 What response are you likely to receive if you suggest any changes in protocols?
16 How could you successfully negotiate hospital and other hierarchies?
17 How can your relationship with other caregivers (e.g. obstetricians and family doctors) become more collaborative and mutually respectful?
18 How can you succeed in making other caregivers value your own, very important experience, evaluation of research data and personal opinions?

## Birthframe 100: Sarah Hobart

Finally, to remind us of an aspect of a lack of optimality which is perhaps an inevitable part of childbirth, here's one more account which shows that motherhood is not all idyllic blissful bonding... Here one woman recounts how she experienced and coped with a rather unexpected problem postpartum.

When I was blessed a few years back with a baby boy, I had a vision. My life would go on much as before, only with a tiny baby keeping me company, smiling and gurgling as I went through the routines of daily living. That vision was almost immediately replaced by a reality where I barely maintained my sanity while struggling to meet the needs of one who, while adorable, was a tyrant. Any time of the day and especially at night I answered his clarion calls for "Milk, milk, milk, and step on it!" In those early weeks I was oblivious to everything but mastering this new role of motherhood.

It was then that I began to notice that not only had my world changed, but also I had changed. Namely, there seemed to be a lot more of me. 24lb more, to be exact. I told myself it was all in my breasts, but that myth was dispelled after my husband pointed out that I was wearing maternity dungarees. Embarrassed, I packed them away, and was left with a wardrobe of sweatpants and tight T-shirts.

I told my doctor, "I think I have a glandular problem. All the other nursing mothers are losing weight, but I hardly eat anything and I haven't lost an ounce!" She smiled gently and said, "If you burn more calories than you take in, you'll lose weight." She didn't believe me! I left her office in a huff, grumbling about the unfairness of it all. When did I, a new mother, ever have time for a meal? Most nights it was all I could do to grab a spoon and a pint of ice cream while my little one was nestled at my breast. I tried to ignore the growing problem of the larger me by avoiding full-length mirrors and clothes in general, but my new pounds fairly shouted for attention. They stood out in all sorts of awkward bulges, as if they'd been lobbed at me from some distance away and stuck fast. My ankles were thick and my feet were plump, my knees had a double chin and my midsection, well, let's just say I got tired of people asking me when my next baby was due. My upper torso was dominated by 'The Milk Factory', but the novelty of being well endowed quickly wore off. I wanted my old body back and was willing to try anything to get it.

"I'll exercise," I declared. The local gym was advertising an aerobics class for new mothers. Babies were welcome. I signed up right away.

On class day, I struggled to get out of the house on time, lugging my son in a car carrier that seemed to weigh 100 pounds. I couldn't find parking nearby, so I walked a few blocks with the car carrier bashing into my shins. By the time I reached the gym, I was sweating profusely. Class had already started, led by a boyish woman with a body fat percentage of 3.0. She whipped the class through a routine of jumping, step climbing and marching that hurt every part of my body. The other women moved smoothly through the program while I focused on not looking like a fool, but I was always a step or three behind.

Everyone was kind. I was glad, though, that my son slept through the entire pathetic performance. We went home and I weighed myself. I had lost a pound.

I went to my second class a few days later. This time I managed to keep up somewhat, and even to throw some verve into my moves. As I leapt onto my step and threw my right foot out, I lost my balance and fell, knocking over the woman in front of me. Red-faced, I finished the class and went home to weigh myself. I had gained a pound. I decided aerobics was not my 'thing'. Instead I bought a fancy jogging stroller, intending to zip around the block a few times a week with my baby riding in style. I had been an enthusiastic runner before my pregnancy, thinking nothing of knocking off two or three miles every other day. It wouldn't take long to fall back into my old habits.

Everything went fine with my new program until I came to my first hill and then I fell apart. It may as well have been Mount Rainier for all the gasping, grunting, and perspiring that went on—and that was just the first 20 yards. I felt awkward not having my hands free and altogether discouraged with my outing. "My running days are over," I thought dispiritedly.

Right about then I decided to leave my fat where it was for awhile and tackle another problem that had been bothering me—loneliness. My old friends didn't seem to understand my new life very well and sometimes blanched when I whipped out a breast to feed my son. I hadn't really gotten around to making new friends. So, despite my innate shyness, I dragged myself to a La Leche League meeting.

It was astonishing to see so many women brought together by a common interest, cheerfully nursing their babies and discussing the changes motherhood had brought as if they actually relished their new lives. I really enjoyed my first meeting. I met someone there, a woman with a baby girl, who happened to live in my neighborhood. Before long we had arranged to push our strollers around the lake a couple of times a week. She was such good company and we had so much in common that I really looked forward to these outings. After a few weeks she suggested that we jog a little way each time we did our three-mile route, and it seemed like a good idea. The pace was never so strenuous that we couldn't keep talking. One day we were so engrossed in our subject (I think it was spit-up) that we jogged the whole three miles. It wasn't long after that that I put my sweatpants and my scales away.

My jogging and mothering buddy lives 600 miles away now... If I had any advice to give to other women in my shoes, it would be this: you are so much more than the sum of your parts. You're somebody's mother, for crying out loud! Life will never be the same, and neither will your body. It will be better. Miraculously, you're nourishing a baby in the best way possible, using only your body, so despite its obvious flaws, strange bulges and odd sagging areas it's amazing. Focus instead on building and nurturing relationships with your child, your partner and especially other mothers. Have a 'tribe' with which to hang out, share stories, gripe, go to the park, exercise, talk parenting or any other subject... it'll make all the difference. The first year of new motherhood is tough. Establish a support network. You'll be glad you did when you discover how tough the second year is... and the third, fourth, fifth, etc.

The first year of motherhood is tough

That would be my advice. As far as my own weight goes, the less I think about it, the happier I am! I don't look like a model but I feel great: fit, healthy and occasionally like Supermom. I didn't get my old body back... I got a more fun version with bigger boobs! I still exercise regularly and I'm still nursing—my second baby now!

> I don't look like a model but I feel great:
> fit, healthy and occasionally like Supermom

*Sarah Hobart (on the left) with her running buddy*

I'm sure the babies of the world also have an opinion on it all...

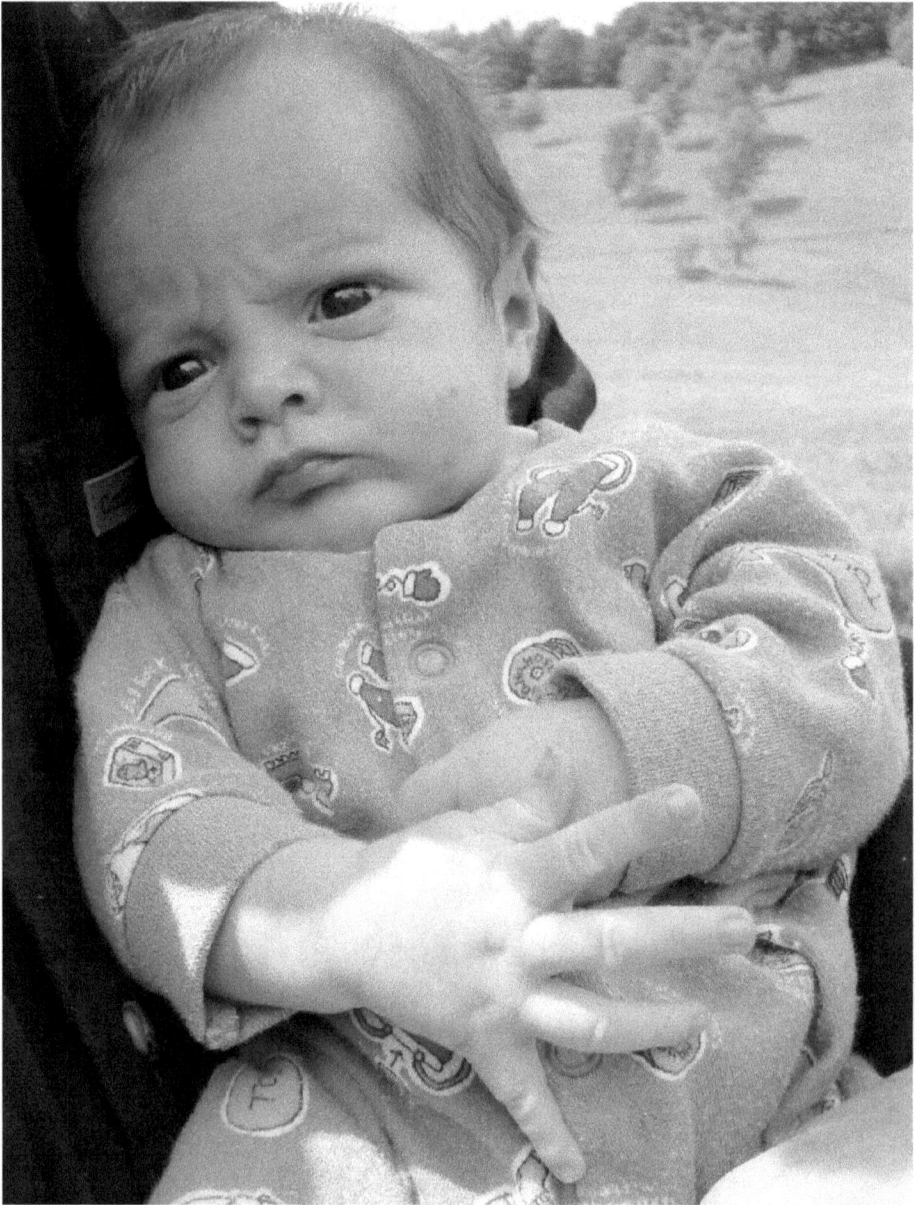

*One of the babies whose birth is described in this book—another optimal birth*

... and they're not the only ones who are likely to be opinionated

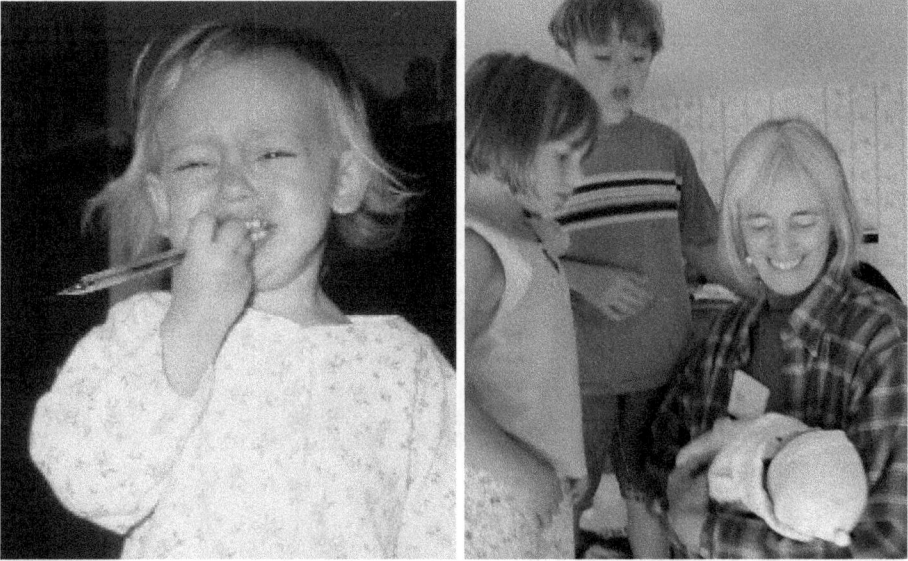

*Siblings may have one view (but not the same one all the time) and grandmothers may have a different perspective. Usually, of course, they aren't as 'disturbed' postpartum!*

*Perhaps others would comment on this mess. No, that's not a dead baby on the floor...*

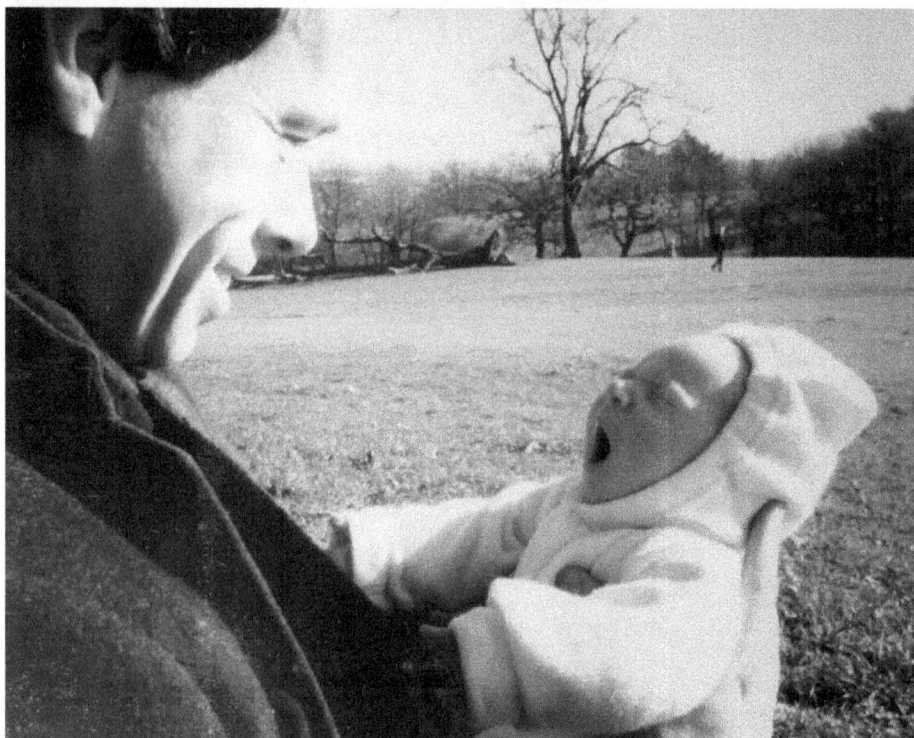

*Let's not forget Mom and Dad!*

# Pause to reflect...

## ...about the possible multi-dimensionality of birth

1   After all that you've read in this book, and after all that you've seen or experienced in real life (perhaps even when giving birth yourself), what is your current view of birth? Could this perhaps change in the future?

2   Do you agree, as some contributors have suggested that birth is not just a birth for the baby, but also a birth for the mother, in a practical, psychological or even spiritual sense?

3   Do you personally bring any religious or spiritual views to the births you attend? How aware of these are you when you're working and how do you feel about not being able to share your thoughts and feelings with clients? (You can either consider this alone, or with people you trust.)

4   Why do you think religions often talk about birth and death in their teachings? Are any of the teachings relevant to your consideration of what constitutes optimality? For example, when in Christianity Jesus talked about the need to be reborn, does this have a positive connotation or a neutral one? What assumptions underlie this story and do they have any relevance to your thoughts about optimality?

5   If you agree that birth at least constitutes change, do you feel that this element of birth always needs to be painful? And does change always need to involve confusion or some kind of realignment—practical or spiritual?

6   Why do you think most women expect birth to be painful and experience it in this way? Do you agree with Grantly Dick-Read's view on pain at all (i.e. that it results from fear)?

7   If you think they're possible, how could you facilitate painfree or orgasmic births, looking at the whole childbearing year, conception on?

8   Does the idea of facilitating a psycho-sexual experience which could be orgasmic and very private embarrass you at all? And do you feel uncomfortable dealing with women's confusion and tension? What would clarity and peace for *you* mean in very practical terms?

9   If you were pregnant yourself now... How would you like to give birth?

## NOTES & REFERENCES

1    Continuity of care is essential. A study by Haggerty, *et al* (2005) emphasized the importance of different types of continuity—informational, management and relational. See:

- Haggerty J, Reid R, Freeman G, Starfield B, Adair C, McKendry R. Continuity of care: a multidisciplinary review. *British Medical Journal*, 2005, 327:1219-1221

The first of the following studies emphasized the importance of working in small team of midwives (numbering at the most six), while the other studies concluded that continuity was most important during labor, birth and postpartum (and perhaps not such a concern prenatally):

- Flint C. *Midwifery Teams and Caseloads.* Butterworth-Heinemann, 1993
- Flynn A, Hollins K, Lynch P. Ambulation in labour. *British Medical Journal*, 1978, 2(6137):591-593
- Green J, Curtis P, Price H, Renfrew M. *Continuing to Care: The Organization of Midwifery Services in the UK A Structured Review of the Evidence.* Books for Midwives Press, 1998
- Walsh D. An ethnographic study of women's experience of partnership caseload midwifery practice: the professional as friend. *Midwifery,* 1999, 15(3):165-176
- Page L, McCourt C, Beake S, Hewison J. Clinical interventions and outcomes of one-to-one midwifery practice. *Journal of Public Health Medicine,* 1999, 21(3): 243-248
- North Staffordshire Changing Childbirth Research Team. A randomised study of midwifery caseload care and traditional 'shared-care'. *Midwifery,* 2000, 16(4):295-302

2    One study concluded that expectations shaped birth experience. See:

- Green, J, Coupland, B. and Kitzinger, J. *Great Expectations: A Prospective Study of Women's Expectations and Experiences of Childbirth.* Cambridge: Child Care and Development Group, 1998

3    Actually, as you probably know, research has already quite clearly confirmed that smoking in pregnancy is not advisable. In one study (Zammit, *et al,* 2009), pregnant women who smoked during pregnancy and drank more than 21 units of alcohol a week more often had babies who were at increased risk of suspected or definite psychotic symptoms. (Fathers smoking while their partners were pregnant was also found to be associated with increased risk.) Even if your clients generally drink less than this, please don't let them relax too much because another study (which draws tentative initial conclusions) (Sayal, *et al,* 2007) found that having more than one alcoholic drink per week during the first trimester of pregnancy was associated with childhood mental health problems in girls (behavioral and emotional) at 47 months (approx. 4 years old) and these problems apparently persisted—according to teacher reports—when checked when the child was almost 7 years old (81 months). So maybe it really is best for your pregnant women to abstain from alcohol altogether. See:

- Zammit S, Thomas K, Thompson A, Horwood J, Menezes P, Gunnell D, Hollis C,

Wolke D, Lewis G, Harrison G. Maternal tobacco, cannabis and alcohol use during pregnancy and risk of adolescent psychotic symptoms in offspring. *British Journal of Psychiatry*, 2009, Oct; 195(4):294-300

- Sayal K, Heron J, Golding J, Emond A. Prenatal alcohol exposure and gender differences in childhood mental health problems: a longtitudinal population-based study. *Pediatrics*, 2007, Feb;119(2):e426-34

4    One example would be the huge breech trial (Hannah, *et al*, 2000), which many caregivers found showed conclusively that a cesarean was the only way forward for breech births. I have heard this study criticized or dismissed many times by caregivers who are skilled in vaginal breech deliveries. In the study by Hannah, *et al*—which only looked at hospital births—caregivers may have had inadequate training and experience and, if this were the case, most would have been frightened of the possible consequences of a bad outcome. Of course, fearful caregivers are the last thing a laboring woman needs! Fear inculcates fear and inhibits the production of the hormones a woman needs in order to give birth successfully. For this reason, as well as because of their long experience of attending successful and safe breech births, many midwives are continuing to agree to attend vaginal breech births, despite this research data. See:

- Hannah M, Hannah J, Hewson S, Hodnett E, Saigal S, Willan AR. Planned caesarean section versus planned vaginal birth for breech presentation at term: a randomised multicentre trial. Term Breech Trial Collaborative Group. *Lancet*, 2000, 356(9239):1375-1383

5    For examples of recent co-sleeping studies, see the following articles:

- McIntosh CG, Tonkin SL, Gunn AJ. What is the mechanism of sudden infant deaths associated with co-sleeping? New Zealand Medical Journal. 2009 Dec 11;122(1307):69-75

- Obdeijn MC, Tonkin S, Mitchell EA. An audit of the sudden infant death syndrome prevention programme in the Auckland region. *New Zealand Medical Journal.* 1995 Mar 22;108(996):99-101

- Stewart AJ, Williams SM, Mitchell EA, Taylor BJ, Ford RP, Allen EM. Antenatal and intrapartum factors associated with sudden infant death syndrome in the New Zealand Cot Death Study. *Journal of Paediatrics & Child Health.* 1995 Oct;31 (5):473-8

- Ford RP, Mitchell EA, Taylor BJ. Well health care and the sudden infant death syndrome. *Journal of Paediatrics & Child Health.* 1994 Apr;30(2):140-3

- Lahr MB, Rosenberg KD, Lapidus JA. Bedsharing and maternal smoking in a population-based survey of new mothers. *Pediatrics.* 2005 Oct;116(4):e530-42

Note: The week-by-week easy-read guide to pregnancy, which follows was compiled from information in the following publications: *Myles Textbook for Midwives* by Bennett VR and Brown LK (eds). Churchill Livingstone, 1999; *Heart & Hands by* Davis 1997; *The National Childbirth Trust Book of Pregnancy & Birth; Birth & Parenthood* (Tucker 1996); *The Complete Book of Mother & Baby Care* (Fenwick 1995); *Babies Remember Birth* (Chamberlain 1988); *Babywatching* (Morris 1991); *Birthing from Within* (England and Horowitz 1998); *The Wish, the Wait, the Wonder* (Perry Johnston 1994); *Pregnancy & Birth* magazine, *Practical Parenting* magazine; *Pregnancy & Childbirth* (Kitzinger 1989) and the website www.pronatal.co.uk. I was interested and rather perplexed to note that often information was not consistent between these sources. Please email info@freshheartpublishing.co.uk if you would like to provide some corrections.

# Week-by-week easy use guide to pregnancy

This easy-read guide is intended to help you talk to pregnant women in such a way that they bond with their growing babies as well as possible, which might hopefully motivate them to do as much as possible for their babies' good. For this reason, instead of distancing terms such as 'fetus', more personal words such as 'baby' have been used. It's been written in such a way that you can just read out the texts to your clients, if you wish. I'm not assuming you're pregnant yourself! Although, who knows, you may be sometime soon...

## THE FIRST TRIMESTER

The first trimester is very important because this is when all the baby's organs and bones are being formed.

### — Weeks 1 & 2 —

During these first two weeks one or more ova (eggs) mature inside your ovaries. Meanwhile, the endometrium (the lining of the womb) builds up so as to provide a suitable place for the ovum to nestle in, after it's been fertilized. This will happen after the ovum is released from one of your ovaries and one of your partner's sperm comes up to meet it. (In previous months when no ovum was fertilized, the endometrium was eventually released in the form of a menstrual period.) As you probably realize, half of the developing baby's chromosomes (its genetic make-up) will be from you and the other half from your partner, but genes are only part of what influences what we're like.

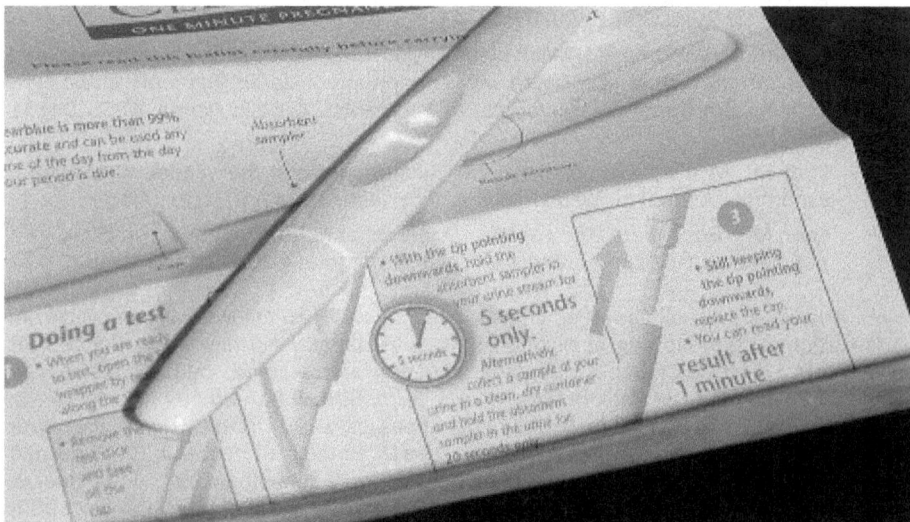

*As you may know, pregnancy tests are readily available at drugstores or superstores*

## — Week 3 —

Conception takes place either at the beginning of this week or later if you usually have a long cycle. (When women ovulate later, their due date is also later because it takes a baby around 266 days—or 38 weeks—to develop.) If only one baby is growing (instead of twins or triplets!), only one of a possible 200 million sperm fertilizes the ovum in this third week so all the sperm released into the vagina, when your partner had an orgasm inside you, were in competition with each other. It takes about 45 minutes for the sperm to reach the newly released ovum and some people say that lying down for half an hour or so after sex helps this process. It's helpful if you're already taking folic acid supplements by this stage because this is a nutrient that many people are deficient in.

During conception the pronucleus of the sperm which unites with the ovum is drawn into the ovum itself. Very quickly, the cell membrane around the ovum closes to seal out all the other sperm. This process is only successful 40% of the time when sperm and ovum meet so it's a really special month when you become pregnant. An incredible series of processes are taking place successfully.

Within 24 hours the newly fertilized cell, which is called a zygote, divides into two—a process called 'cleavage'!—and becomes a 'morula'. As the morula develops and fluid enters the mass of cells it becomes a 'blastocyst', which continues dividing so that after nine months it will consist of several hundred billion cells. The one fertilized cell transforms and differentiates itself to make a new person, thanks to the DNA inside the genes in the ovum and sperm.

Note: Of course, there may be variation in the experience of different mothers and babies, particularly when the mother has a serious health problem, such as heart or kidney disease. Caregivers may need to make some extra commentary in this case or in other special circumstances.

## — Week 4 —

On the 10th or 11th day after fertilization, having successfully traveled along one of your fallopian tubes, the blastocyst implants itself into the lining of your womb. From this point onwards your tiny baby is called an 'embryo'. It is nourished by tiny blood vessels called chorionic villi in the lining of your womb. These villi were produced by the embryo itself, even though it is still only the size of a pinhead.

The blastocyst implants itself into the lining of your womb

If you have conceived identical twins or triplets (or more!), by the end of this week the single blastocyst will have already divided up to form two, three or more separate embryos. If you are expecting non-identical twins or triplets, etc., more than one egg will have been released, fertilized and implanted two weeks ago.

## — Week 5 —

By the end of this week many women spontaneously realize they are pregnant and confirm this using one of the widely available home pregnancy tests. It's good if you confirm your pregnancy early on because knowing for sure may motivate you to do everything possible for the good of the little baby developing inside you. If a first test is negative but your period still hasn't started, you may still be pregnant. This may be due to low levels of hCG—human chorionic gonadotrophin, the hormone manufactured by the blastocyst, which the test tries to detect. Also, levels may be low if you ovulated late on in your cycle so it's a good idea to do another test in a few days' time and again after a week or so, if you still think you're pregnant.

Of course, taking at least 400mg of folic acid every day is particularly important this week and over the next few weeks because folic acid is crucial to ensure healthy development of the neural tube, which eventually becomes the spinal column and brain. It's also vital that you drink enough water. Requirements vary according to height, weight, activity levels, climate, etc, so it's not possible to make any precise recommendations, but do drink plenty of water, using your thirst as a guide. Remember to check bottled water for sodium levels, though, and only drink water when the label says the level is lower than 10mg per liter. (It's useful to know that levels in carbonated water tend to be dramatically higher, so it's probably best to mainly drink still water.)

Your baby is now an embryo and its brain already has two separate lobes. In appearance, the embryo is a bit like a grey, jelly-like, translucent tadpole, with a head and a tail. It's only a few millimeters long—perhaps as long as a grain of rice. It swims around in a miniature yolk sac and amniotic sac inside your womb, which protect it from harm.

Your future baby is already connected to you by a miniature umbilical cord and cells are beginning to differentiate so as to form your baby's skin, intestines, a primitive nervous system and bones.

## — Week 6 —

Several crucial developments take place this week. By the end of the week, your baby's umbilical cord, digestive tract, kidneys, liver, heart and blood-stream are already forming and beginning to function—the heart will start beating during this week—and all these organs and systems are helping your baby to get and use food from the rich lining of your womb. Your baby's mouth and jaw are also developing now and 10 discrete dental buds are growing at the top and bottom of the jaw. The four shallow pits which have already appeared on your little embryo's head will later develop into eyes and ears.

The most important development this week sounds unimpressive but is actually crucial. A groove appears down your baby's body; this will soon form the spinal cord and the brain. Of course, the folic acid which is recommended up until approximately 12 weeks after conception ensures that this process takes place smoothly. It is essential to ensure that this neural tube develops successfully because it consists of 125,000 cells. This is amazing when you consider that the embryo is still only the size of a pea.

## — Week 7 —

Your baby is still developing at a phenomenal rate. To get an idea just how fast the rate is, consider that while the embryo's limb buds will have developed by the 30th day after conception, by the 31st day (which is in the middle of this week) these same buds will have become subdivided into hands, arms and shoulders. At this point—on the 31st day— your baby's brain will be 25% bigger than it was just two days before.

By the end of this week, your baby is the size of a small grape, growing at a rate of 1mm a day. If we could see it, we might think its head looked big in relation to its body, but this is normal. Already a face is forming, with tightly closed eyelids over dark-looking eyes. Your baby also has rudimentary arms and legs and is making its first movements. At the end of each limb small indentations, which will later become fingers and toes, are already discernible. Bone cells are also appearing.

The development of your baby's lungs, intestines, liver and kidneys continues and your baby's sex organs are beginning to form. The pancreas and thyroid are already in place. Your baby's heart, which recently started beating, is currently beginning to form its four separate chambers.

Your baby's sex organs are beginning to form

The nervous system, including your baby's spinal cord and brain, is almost completely formed. In about a week's time (41 days after conception) the complex structure of your baby's mature brain will already be in evidence, albeit in miniature. Your baby is becoming so complex, that it will soon be known not as an embryo, but by a different term!

Note: These week-by-week notes were compiled from various sources. Much of this information must have been discovered incidentally, in the course of doing other research and much of it will have been discovered as a result of ultrasounds or X-rays.

## — Week 8 —

By the end of this week (six weeks after conception), your baby technically starts being called a 'fetus'—a word, which means 'little one'. It is no longer dependent on a yolk sac for nourishment, so this has already disappeared.

Your little one's face is continuing to form: its nose is now pointed and already has nostrils; its eyes and ears are growing; and the two sides of its jaw will have joined together to make a mouth. Inside its mouth, there is a tiny tongue and taste buds will begin to appear on it in two weeks' time. (The palate will also begin to form at that time.) Since the inner parts of the ears are developing your growing baby will soon have a sense of balance. This is useful because it is already moving around a lot in its sac, even though you won't be able to feel this yet.

Your little one's tiny skeleton is now fully formed out of soft cartilage; this will later develop into bone. All your baby's internal organs are in evidence, even though they are not yet fully developed or in their final positions. The arms and legs have grown longer in the last week and if only it were possible to take a peek, we would be able to discern shoulders, elbows, hips and knees. Even fingers and toes are beginning to form, although they are joined by webs of skin for the time being.

There is evidence that your baby's brain is working by this stage, as electrical activity can now be measured. By the end of this week your baby will be around 2½cm long.

## — Week 9 —

Your baby's limbs are continuing to develop rapidly and its fingers and toes are becoming more clearly defined.

Movements which were initially merely floating are gradually becoming a form of exercise for newly forming muscles. Movements will increasingly synchronize with your own movements as your baby's vestibular system (which is concerned with gravity and balance in space) starts to develop. Periods of activity will alternate with periods of rest from now on until the birth, when a pattern of nighttime rest and daytime activity will eventually become established. Your baby's brain, which coordinates all this movement, is continuing to develop: by now the brainstem (the lower portion of the brain) will be fully developed and the midbrain and forebrain (a little higher up) will start to expand. The wrinkles on the outer surface of the forebrain (the cerebral cortex), which are so characteristic of human beings and which allow us to accommodate vastly more brain cells than other species, are now also beginning to appear. Oxygen is being pumped round your baby's tiny body in red blood cells, which are already being produced by the liver. Beta-endorphins are already detectable in your baby's blood, which suggests that your tiny baby can already experience pleasure.

Beta-endorphins are already detectable in the blood...

Your baby's face, eyes, nose, lips, tongue and the first signs of teeth and bone would be clearly visible to us with ultrasound, although this is probably not a good idea, because the development going on now is very delicate indeed, and mustn't be disturbed, unless there's a very good reason to do so. By the end of this week urogenital development is taking place, whether your baby is a girl or a boy. If your baby is to be a boy, a penis would now also probably be visible.

All this is happening in miniature because your baby still only weighs little more than a grape.

## — Week 10 —

Just eight weeks after your baby was conceived, tactile sensitivity has developed in the face. If your baby's cheeks were to be stroked gently with a hair, your growing baby would move its head away, bend its trunk and pelvis and extend its arms and shoulders enough to push the hair away. (Don't ask me how researchers know this, but they claim that they do.) Sensitivity on other areas of skin will develop gradually over the next few weeks.

Chest movements, which are now detectable to researchers, might well be practice exercises for breathing after birth. Inside your baby's chest, the heart has finished dividing into four chambers and each chamber is connected to the other ones by tiny valves.

The placenta, to which your baby is already attached, begins to produce progesterone around this time. Your tiny baby is still being nourished entirely by the so-called corpus luteum, although this will soon change.

Fingers, hands, wrists, toes and ankles are well-defined and your baby's face is even more recognizably human. Your baby is now roughly the size of a large strawberry.

## — Week 11 —

Either by now or in a few days' time (depending on when you conceived), all your little one's major body organs will have finished forming. Even miniature ovaries and testicles will have finished forming and your baby's circulation will be functioning properly. This all means the most critical period of development is over. For some reason, this means the risk of miscarriage also decreases sharply around this time.

The most critical period of development is over.
For some reason, this means the risk of miscarriage
also decreases sharply around this time.

66 The first time I found out I was pregnant, we were on an Easter holiday above the Arctic Circle in Finland. My first noticeable sign of pregnancy was terrible insomnia. I crept outdoors at 4:00AM, as a pink spring dawn broke over a wide, flat snowscape. I felt overwhelmed with the knowledge that I was not alone. Although still only a microscopic ball of cells, my future offspring was already with me. What would this dawn of a new life be like, for me and for this child?

Three months later, my husband and I took the ferry to Bruges for the weekend. As our wake churned behind us towards Dover's chalk cliffs, I thought about how this point in my life was a metaphysical departure, too, towards a time in my life when I would come to know the new human beings who would enter my life through my own womb.

I never stop wondering at the journey that begins with a particular event of copulation—perhaps specifically remembered, perhaps not—into a single cell, into 10 billion cells born months later as a fully-formed human baby. *Nina Klose*

## — Week 12 —

The tiny baby living inside you is beginning to look more human. Its head is becoming more rounded, even though it would still look rather large in proportion to its body. Your baby's eyelids are now formed, but are closed over the eyes, and your baby's external ears even have earlobes.

The tiny baby living inside you is beginning to look more human

By the end of this week your growing baby is getting all its nourishment via the placenta, which should now be fully functioning. The blood vessels in the umbilical cord carry food and oxygen to your tiny baby and then take carbon dioxide back out again. Your little one can now open and shut its mouth and swallow; it uses this new skill to continuously take in amniotic fluid. Doing this will help its lungs develop, so that they are ready for breathing, as soon as he or she is born.

The kidneys are now beginning to function, which means that your growing baby will urinate into the amniotic fluid. This is nothing to worry about—your womb is constantly replacing this fluid so that your baby's environment stays fresh. Amazing as it might seem, a complete change of amniotic fluid is effected once every 24 hours. (This rate will even gradually increase over the next few weeks.) Any waste products in the urine from your growing baby are automatically dealt with by your baby's own kidneys; other waste products which do not come out in your baby's urine are stored in its own intestines, ready to be excreted soon after the birth as meconium, which will be your newborn baby's first 'poop'. This early poop is different from normal poops in that its tar-like consistency actually acts to block your baby's bowel before birth; only if a baby becomes distressed while in the womb is this secreted into the amniotic fluid. Mostly, that doesn't happen because your baby adapts well to living inside you, even if you experience a range of emotions.

Your baby's muscles, having further developed over the last couple of weeks, now allow for much more vigorous movement. In experiments, babies have been observed rolling from side to side, extending and then flexing their backs and necks, waving their arms and kicking their legs—straight into the side wall of the amniotic sac! The usually graceful and apparently voluntary and spontaneous movements your baby now makes are thought by some to be an early example of initiative and self-expression. Such creative gymnastics seem to demonstrate quite sophisticated brain activity, which only a few decades ago would have been considered impossible. Movements of the face—frowns, pursing the lips, opening and closing the mouth—all seem to indicate an emotional response to whatever's going on, given that the pituitary started producing pleasurable beta-endorphins a few weeks ago.

Babies in the womb at this age who have had their genitals stroked under experimental conditions (accidentally, of course) have shown a clear response, which suggests a high level of sensitivity. Small, well-coordinated movements of tiny fingers and toes (which are both growing miniature fingernails and toenails) suggest further sensitivity and emotional response.

Your own baby now weighs about 14g and is perhaps 7 or 8cm long. Obviously, a lot more growth and development are needed before this little being can survive independently, outside the safety of your womb. At least your baby is now a little less susceptible to harm from infection or chemicals circulating around your own body.

*Find out about the foods you should avoid altogether or be careful about. Food is extremely important and it's vital that you eat and drink healthily.*

## THE SECOND TRIMESTER

When you enter this second period of development your baby will be much more of a being in its own right and its support system, the placenta, should be functioning effectively, thanks to the production of progesterone. This means that if you suffered from nausea or vomiting in the first trimester, it's likely to disappear now, or at least fade very soon.

### — Week 13 —

Your baby now begins to do more and more things which seem more recognizably 'human'. It might suck its thumb, for example, extend its fingers, or even yawn when it's tired. Tiredness is perhaps understandable given that your baby's periods of movement are extremely frequent, with rest periods extending no longer than 15 minutes until around Week 20. Your baby's neck is also longer, which makes it look much more human.

Being completely formed, your baby is now focusing on growing (at an even faster rate than before), so good nutrition with plenty of protein is extremely important.

Your baby's skin sensitivity is increasing to the point where even the palms of its hands would respond to strokes; its arms and legs would certainly be sensitive to hair strokes. Some primitive reflexes are also either already present or will develop over the next four weeks.

Some people believe that personality is also beginning to develop at this stage because at least two research projects have concluded that exposure to influenza during the second trimester (between the third and seventh months of pregnancy) results in increased susceptibility to schizophrenia, so it might be especially important for you to avoid people who are snuffly over the next few months.

### — Week 14 —

Almost three months since conception, your growing baby is now approximately 9cm long and will definitely be receiving all its nourishment from its placenta, which some young children have referred to as a kind of friend in the womb. The placenta looks like a large, liver-like organ and it will be born separately, just after your baby (This is called the third stage of labor.)

Your baby's eyes have now moved away from the side of the head to the front and your baby's ears have moved up from the neck to their proper position on the head, but they're still fused shut for the time being. Your baby has eyebrows now, as well as a small amount of hair on its head. The basic parts of the spinal cord and brain are in place, and your baby's respiratory tract, which starts at the nose and branches again and again on the way to the lungs, is now ready for your growing baby's first breathing movements.

As for other movements, your baby may now start moving its arms and legs rhythmically. Increased skin sensitivity means that if it were possible to stroke your baby very gently on the cheek, we would see it turn its head and start searching for something, probably a nipple to suckle from!

The next four weeks are a period of rapid skeletal development. We know this because some researchers have tracked it on X-rays.

## — Week 15 —

It is in this week that your baby's sex organs will mature. Its nose will also be better formed and its head and eyebrow hair will gradually start to become coarser. If your baby has a gene for black hair, the pigment cells of the hair follicles are now beginning to produce black pigment.

It seems likely that taste buds are also functioning by now. We know for certain that no essential changes take place in taste receptors after this point, except for the fact that they multiply and spread themselves more widely. This means that your growing baby will be having taste experiences based on what you eat and drink yourself from now on up until the birth. It's possible that these experiences may affect your little one's preferences in childhood and possibly also beyond. Based on the contents of amniotic fluid which has been tested, we know that your baby will be tasting glucose, fructose, lactic, pyruvic and citric acid, fatty acids, phospholipids, creatinine, urea, uric acid, amino acids, polypeptides, proteins and salts, amongst other things. It doesn't sound very appetizing but it certainly indicates that your baby is having a wide range of gustatory experience.

As well as sucking its thumb, your baby may also suck its fingers, hands and toes and hold its own umbilical cord by this stage.

## — Week 16 —

By now, your baby will weigh about 75g and will be about 11½cm long. If we could look at it, we would see transparent skin, with fine networks of blood vessels underneath.

Hard bones are beginning to develop and your baby's legs will have become longer. Both arms and legs will now have joints. Your baby already has well-formed fingernails and toenails and its fingers would already produce a unique fingerprint. It can now coordinate all its movements and it will be moving around energetically, although you probably won't be able to feel it moving around just yet. (It won't be long before you do—maybe two to four weeks.)

Even though your baby's ears are not yet fully formed, your baby is already responding to sound, probably by feeling vibrations through its skin. Its eyes are now open and quite expressive. During invasive procedures, expressions such as squinting and sneering have been filmed at this stage of pregnancy.

Researchers who observed these expressions felt certain they represented meaningful reactions to what was happening.

Your baby's eyebrows and eyelashes are growing and your baby will already have fine downy hair ('lanugo') on both its face and body. This lanugo usually disappears by the time of the birth, but it can still be seen on premature babies. (A baby is called 'premature' if it's born before 37 completed weeks of pregnancy are over. Most babies—as you probably know—are born after about 40 weeks of gestation. In reality, most are born slightly before or after this.)

Although all the major organs are now fully developed, your baby will continue to grow rapidly this month and in the rest of your pregnancy so that it will be capable of independent life. Already, it will be making breathing movements as practice movements for after the birth.

You will be able to choose whether or not to have an ultrasound at this stage in your pregnancy, but you may first like to consider whether this is really necessary, after you've found out more about this technology.

## — Week 17 —

By now, your baby will measure about 15cm, which is approximately the length of a pen. It will weigh about 175g. These are landmark measurements because for the first time, your baby weighs more than its support system, the placenta. Your baby is still becoming increasingly sensitive... fetal abdomens and buttocks have responded to hair strokes in experiments at this stage and different responses have been recorded when babies were exposed to different kinds of music: Beethoven, Brahms and hard-rock music made them restless, while Vivaldi and Mozart calmed them down. Perhaps you should bear this in mind when deciding which music to listen to, although of course your own tastes in music are also important—so don't worry too much about this!

## — Week 18 —

Your baby may now be as long as 20cm and you may soon notice it moving around inside you. By the end of this week, your caregiver should be able to hear your baby's heartbeat with an ordinary fetal stethoscope (a Pinard), i.e. with a device which doesn't use ultrasound (unlike the Doppler).

Your baby will be drinking amniotic fluid through a perfectly formed mouth and lips. This is a practice exercise so that breathing will take place without undue effort or fatigue after your baby is born. Sometimes this breathing exercise will make your baby develop hiccups, which you might notice as a jerking of your abdomen. Your baby will also be practicing sucking, maybe using its thumb, which is obviously preparation for feeding after birth. Your growing baby may also be making other sounds quite voluntarily by now, which is perhaps the first step in language learning.

Bones are still continuing to form rapidly and your baby's nasal septum will now have fused with the palate.

## — Week 19 —

This week, buds for permanent teeth are forming behind those which have already formed for the milk teeth. Your baby's sense of touch is also still developing: your baby would now respond if it was touched almost anywhere on its body.

Your growing baby will now be making stronger and better coordinated movements, which might even include back flips, rolls and little punches. These gymnastics are only possible because its nervous system is now much more sophisticated.

By the way, it's probably not a good idea to practice for labor in any way. Nowadays it's been well-established that hormones (which are affected by emotions) pass across the placenta—so your own nervousness, tension, pain or fear would probably be shared with your developing baby if you were to do this.

## — Week 20 —

Your baby has been growing fast, so by now its body will have reached its correct relative proportions. Your baby is now about half as tall as it will be at birth; it will only weigh about 340g, though—about the weight of a grapefruit. Ear development is continuing, with myelin insulation taking place in the ear nerve (the cochlear nerve). This no doubt means that your baby is hearing even more sounds—those of your body as well as a muffled version of any you hear yourself. Your voice and that of your partner are likely to be a particular focus...

The parts of the system which allow a person to register head and body motion and the pull of gravity are all fully grown by now in your little one. This means that your baby will probably already be aware of its own movements. For some reason it is common for babies to have a quiet period at this stage, in terms of bodily movements, so don't worry if you can't feel it moving round quite as much this week.

Hair could probably be seen on your baby's head, if only you could take a look, and sebum from sebaceous glands mixes with skin cells so as to begin to form 'vernix caseosa'. (This is the greasy covering over the lanugo which protects your little one's skin while it is in your womb.)

Your baby can now also grip firmly with its hands and would reach out for a source of light (such as a torch beam panned slowly across your abdomen), even though its eyes are still fused shut.

It's possible that from now on protective substances may pass through your blood, through the placenta to your baby, so as to help your baby resist disease in the first few weeks of its life.

Your baby will probably already be aware of its own movements

## — Week 21 —

Your baby now weighs about 450g. Hair growth is a prominent feature of the next four weeks. It is also over this period that your baby's permanent teeth buds finish forming and that it will develop so-called 'brown fat', which is an important source of heat and energy for the newborn. It will also begin to make crying motions over the next few weeks and will continue to practice sucking.

You should by now feel very definite strong kicks sometimes high up in your tummy and sometimes low down near your pubic hair.

## — Week 22 —

At 22 weeks, your baby is about 30cm long but it will still be very red and wrinkled in appearance. The gradually increasing levels of fat on its body should be noticeable to you through your increasingly rounded form!

You may by now be able to feel the differences between different parts of your baby's body as it kicks, jumps and turns around. Sometimes you may feel a hand, sometimes a foot, its head or its buttocks. These movements may actually usually take place when you yourself are trying to rest because your baby is likely to be at its most active then. (Perhaps it finds the movement of your body soothing and sleep-inducing.) Your partner will also be able to become more aware of his baby this week because if he puts his ear to your abdomen, he should hear a distinct heartbeat.

Over the next four weeks, most organs will become capable of functioning. By this early stage of gestation, your baby will also have started renewing its own skin cells. Your growing baby's eyelids and eyebrows will be well-formed, even though the eyes will still be fused shut, and your baby's face will be even more expressive than it was before. In experiments, puckering of the lips, scowling and muscle tension around the eyes have all been associated with audible crying at this stage of pregnancy. (The sounds can actually be heard by the mother and researchers under certain experimental conditions.) These appropriate facial expressions are interesting in that they suggest that your baby has by now developed clear links between body and brain.

Also, at this point, just 20 weeks after conception, research has revealed that hearing and memory are becoming increasingly acute. Researchers have shown that newborn babies remember voices and music which they heard in the womb at this stage of pregnancy. This means that any lullaby you sing and any music you play from now on may have a noticeably calming effect on your newborn later on. (This could be a big advantage if ever you have trouble getting your new baby to sleep!) Your baby will also be calmed by your partner's voice if it's heard it a lot during pregnancy, so it's a good idea to encourage your man to speak to your growing baby—and for you to do the same, of course.

Newborn babies remember voices and music heard in the womb

## — Week 23 —

Your baby's arms and legs are now well developed. The fine hair (the lanugo) which covers its body is beginning to darken. You may find that Braxton Hicks contractions (painless tightenings of the womb which began around Week 6), now become more pronounced. This means that your baby is regularly getting practice hugs and massage sessions! If you make love with your partner, your body's own response may also have an effect on your baby. Research focusing on this period of pregnancy has demonstrated that an unborn baby's heart rate either shoots up or slows down when the mother has an orgasm.

## — Week 24 —

Most countries consider a baby legally viable by the end of this week, which means its birth would need to be registered if it should happen to take place prematurely. Your baby will now probably weigh over half a kilo and will be about 33cm long. Imagine it curled up in your womb, cushioned by the bag of water that surrounds your baby and totally dependent on the placenta for food and oxygen, as well as the disposal of its waste products. Everything you ingest (food, drink, drugs, other chemicals, air, smoke, etc.) still crosses the placenta and is shared with your baby, so you need to continue to be aware of this.

The amniotic fluid is now changed every four hours, so it's a good idea for you to drink lots and lots of water. (Don't go overboard—but be aware of this.) Amniotic fluid is important because it provides important liquid for your baby's digestion (enabling it to wee) and also regulates your baby's temperature and protects it from infection or any sudden bumps you may experience.

By now your baby will actually look and behave much the same as a baby at birth. It is continuing to make breathing movements (practicing for life outside the womb) and it coughs spontaneously whenever it needs to. It continues to kick and punch and even turn somersaults, so as to build up its muscles for life outside the womb. It might also sometimes make a fist, which indicates that it is developing its grasping reflex, which would be useful if we human beings still carried our babies round on our backs—without back-carriers!—like chimpanzees. However, your baby is unlike a newborn in that its eyes are still a bit bulgy (because of its thin face) and they are still sealed shut.

Your baby is making great progress in other respects too: its skin is getting thicker; its sweat glands (so important for temperature control) are forming in the skin; and its responses to its environment are becoming even more obvious. Awareness of sounds may have increased to such an extent that any sudden noise makes it jump and it is likely to react to sounds it finds unpleasant by moving in different ways. Loud music is likely to wake it up, if it's asleep. Your baby may continue to suck its thumb in an apparent attempt to comfort itself.

Over the next four weeks, the part of your baby's brain concerned with personality and intelligence is becoming much more complex so we can guess that its personality may be developing over this period too. Nevertheless, your baby is not considered legally viable in some countries because your baby's lungs are not yet sufficiently developed for independent survival outside your womb. Premature baby care is developing and some remarkable babies already survive.

## Resting while pregnant...

Be careful about how you rest from now on. Although no research has as yet confirmed that there is anything behind the thing called 'optimal fetal positioning', common sense and personal accounts confirm that it's worth taking seriously—as well as the fact that more and more women nowadays are having posterior labors, which are usually more difficult for the woman. Another reason to avoid lying back is because if you do lie on your back, or even lean back, your baby is likely to be deprived of oxygen. This is because the vena cava, which takes blood from the placenta and therefore your baby, will be compressed.

All this means it's probably best to avoid lying down on your back or leaning back in any sitting position, including in the bath, at any time outside your prenatal appointments. You can still have a bath, but you need to sit up and lean forward—or you can stick to showers while you're pregnant. In any case, your belly should not be exposed to very hot water because that can be very dangerous for your baby. In other words, *never* 'lie back and relax'!

When you're in a car, or sitting on a seat which tips back, consider using a cushion (under your butt, at the base of your spine) or use a triangular cushion or piece of foam, to keep yourself from tipping back. Triangular foam cushions (which can be bought from maternity shops or by mail order) are fantastic for putting under your baby belly at night. They'll help you get comfortable. Also consider using a kneeling chair—to help your baby get into the best possible position for a straightforward labor.

When you do lie down, for example to go to sleep, lie on your left side. Lying on the *left* side is much, much better for your baby than you lying on your right side (from the point of view of it getting enough oxygen). As well as optimizing oxygen flow, lying on your left side also lowers your risk of getting high blood pressure. Lying on your left side, you can read, sleep, read to children (they're usually very happy to climb on top), chat to friends... or even watch TV. Stick to gentle comedy or romantic movies, though, because frightening or hilarious movies might cause you to produce excessive amounts of hormones, which might have an impact on your baby or start labor off. Again, this is not based on research, but on anecdotal evidence... quite a few women have reported going into labor after over-strenuous laughter or something noisy or frightening happening nearby.

Beyond an awareness of these positions, just enjoy yourself, of course!

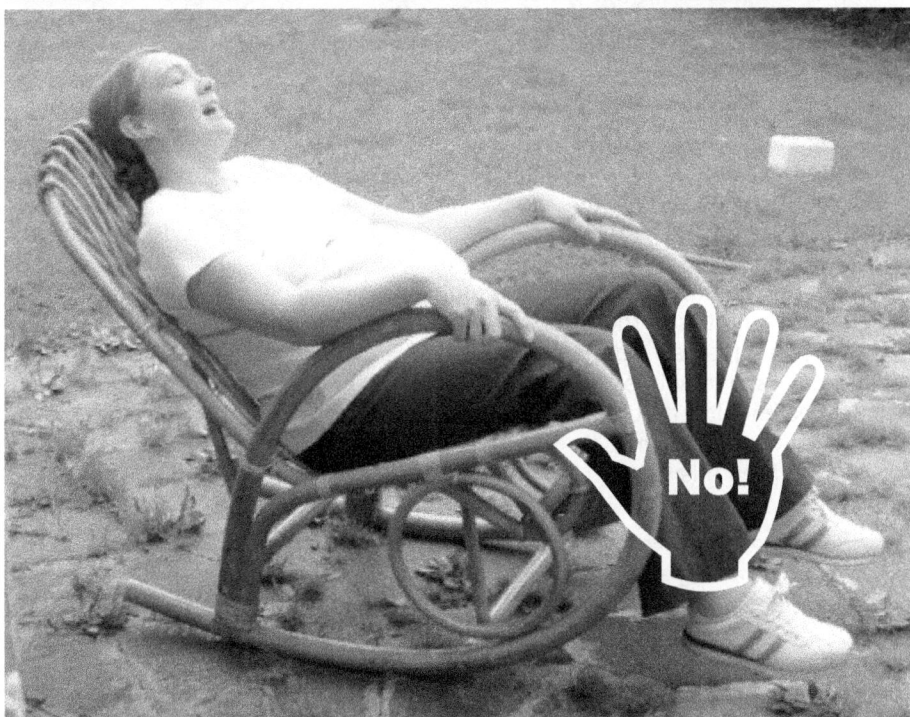

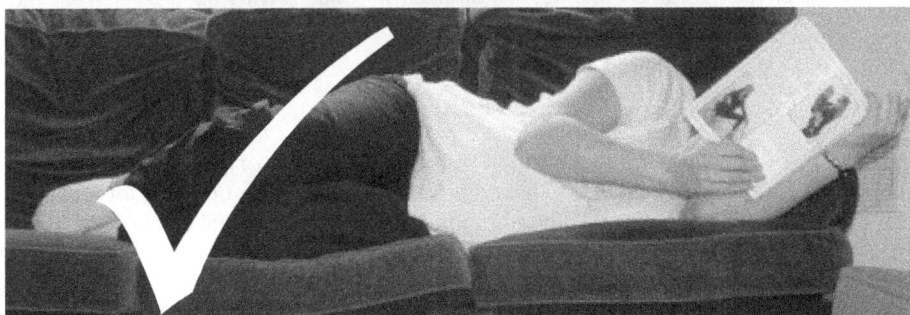

*Don't 'lie back and relax!' Lying on your left side is better.*

## — Week 25 —

Over the next month, your growing baby's eyes begin to open and shut, it gets significantly longer and puts on a substantial amount of weight. Its ears are now structurally complete so it can hear your voice clearly. This means you should perhaps be aware of the way you talk about your baby while it's living inside you, and afterwards too. (Some surprising perinatal memories have come out under hypnosis.) Remember that your baby is a person to be treated with respect and sensitivity. If you question this, consider how easily babies cry when a person is around who seems unsympathetic to them.

Your baby's reactions to loud noises may still be quite dramatic: you will simply feel a sudden jolt as your baby moves. Some sounds will be less alarming to your baby because it will be used to hearing many of them, e.g. the continuous sloshing and squelching of your stomach and bowels. Apparently, noises from these places peak at 85 decibels, which is really quite loud. (One researcher found it was the volume which his vacuum cleaner made at his ears when he was vacuuming a carpeted floor.) Sounds which are heard at around 55 decibels include the constant rumble of your blood in the arteries which supply your womb and the placenta, which move in synchrony with your heart. (To give you an idea how loud this is, normal speech is usually about 60 decibels.) As a result of your baby's development in hearing ability, some experts believe it's a good idea for you to sing to your baby from now on, as well as talk to him or her. On a more physical level, your baby's bone centers begin to harden this week.

66 I felt absolutely great, energized with the life force running through me—sounds weird, I know but it really was like that. I'd never felt better. My skin, nails and hair were all glossy and smooth. A lot of my hair fell out and I had to grow it longer in an attempt to cover up the bald patches (it grew back a year later). For the first time in my life I loved my body, it was doing things automatically and I felt impelled to feed it lots of things I don't like eating—meat, milk, oatmeal, custard (all foods my child loves). It did feel a bit like hosting an alien who was driving me around from the inside like I was a car.

## — Week 26 —

Your baby is now growing at a rate of 1cm a week and every extra centimeter is making it stronger. By now, it will weigh almost a kilo. The ongoing increase in length means that your baby's positioning in your womb needs to change. It will find itself either the right way up or upside down in the womb, depending on how it flexes or extends its knees. These movements must require an enormous amount of brain-to-body coordination and it's interesting that they won't be possible outside your womb until two to three weeks after the birth.

Other changes are taking place this week. Firstly, your baby's skin, which was previously paper-thin, is now becoming thicker and opaque. Secondly, your baby's eyelashes are lengthening. Thirdly, the number of tastebuds it has is increasing. Research from as far back as 1937 shows that if something sweet is introduced into the amniotic fluid, a baby will swallow more rapidly. This of course means that you need to consider carefully whether or not both you and your baby really need that chocolate bar or ice cream, before you eat them... Remember the importance of a good diet and avoid soda and coffee.

By the end of this week, your baby's chance of survival outside your womb would be much higher, although every extra week spent inside you is helping to prepare your baby for the outside world. Nature did not intend babies to be born at this early stage, but if your baby were to be born prematurely, you could perhaps use kangaroo mother care. (See the Birthframe 95 for more on this.)

## — Week 27 —

Your baby is still growing very fast and will probably weigh just over a kilo by now. Its eyes are now open and will have blue irises (slate grey, in fact—the color of muscle), but this color may change after the birth. Your baby's nostrils are open and your baby is practicing breathing.

If you want to try and visualize your baby, include a very wrinkled skin, coated in creamy vernix. The wrinkling is no doubt due to the watery nature of your baby's present home. Its creamy white covering is there for its own protection. It's a bit like staying in the bathtub, covered in protective, nourishing, moisturizing cream so as to avoid having waterlogged skin!

Your baby's heart will now be beating at a rate of 120-140 beats per minute, which is double the speed of an adult's heart rate.

## — Week 28 —

Most countries consider a baby to be legally viable by the end of this week. Pediatricians are optimistic about a premature baby's survival at this stage—some even put it at 95%.

Assuming that you are going to carry your baby to term (i.e. 40 weeks), you are now well on your way to having a full-term birth as long as you avoid licorice, shocks and an inadequate diet. (Surprisingly, perhaps, that's what research suggests.) The third trimester (which starts next week) is your 'home run', so to speak. Enjoy the last few months of your pregnancy so that your baby has a happy emotional environment to swim around in. Research suggests this is important, so it's good if you avoid all stress at this stage.

Your baby's length and weight will have increased even more. However, the variety of birth weights and lengths make measurements increasingly difficult to generalize from now on, so they won't be mentioned any more in these weekly guides. The ever-increasing size of your baby means that it is nearing the end of the time when it can lie stretched out. Perhaps because of an increasing lack of space, the volume of amniotic fluid in your womb will reach a peak between now and Week 32. As a result of this increase in size and a decrease in the amount of liquid to move about in, your baby will find it more and more of a challenge to find a position that feels comfortable over the next few weeks... so you will not be the only one who is shifting and twisting about in an effort to find the ideal position for a peaceful sleep! (Have you bought yourself a triangular cushion? It could really make a difference to your sleep.)

Your baby is definitely listening to your speech patterns very carefully now. A study in 1975 which analyzed the cries of premature babies born at 28 weeks clearly showed that babies were copying their mothers' basic speech patterns. Obviously, this requires a sophisticated level of filtering on the part of your baby. As well as screening out your body's squelching and gurgling, your baby will also be filtering out louder ambient sounds in order to focus on your voice. Obviously, this is another important stage in language development.

Your baby will also be appreciating tastes more, as it gets used to differentiating between whatever you eat. Research has shown that babies can now respond to sweet, sour and bitter tastes. Your growing baby will actually have more tastebuds at this stage in its development than it will have at birth, so its palate really might be quite discerning.

Difficult as it might be to believe, your baby will probably also start having another range of new experiences around this time: sexual feelings. Quite by accident, some American researchers observed male babies having an erection while analyzing a series of sonograms; in fact, six babies at about this stage of gestation were observed in this state! These erections doubtless prove that the appropriate nerve pathways are working by this time and they probably also indicate that unborn babies experience some of the feelings that adult men experience along with an erection. We can guess this because all six babies who were having erections were also thumb-sucking—which is almost certainly a pleasure-seeking pastime.

On a more general level, the thinking part of your baby's brain has now become much bigger and more complex. By now, it is possible for preborns to feel pain and they respond in much the same way as full-term babies. Muscles in your baby's body are also becoming even stronger, which you will notice in the form of assertive kicks from within! Your partner will almost certainly be able to feel your baby move too, if he puts his hand on your belly. You may both be able to watch the shape of a foot or a bottom travel across your middle, as your baby changes position.

Muscles in your baby's body are becoming stronger. Your baby is also laying down fat in preparation for the birth.

Finally, from this time onwards, your growing baby will be laying down fat underneath its skin in preparation for after his or her birth.

# THE THIRD TRIMESTER

Different people mark this trimester as beginning at different points. Assuming it begins at the beginning of Week 29, you are now entering the last phase of your baby's development. The main focus now is on growth and lung maturation, so your baby can breathe well when it's born. Growth is very important over the next few months because heavier babies tend to be stronger and healthier. That explains why it really is vital for you to eat well and drink sensible drinks over the next 12 weeks or so—even if you do feel huge!

It is vital for you to eat well over the next 12 weeks

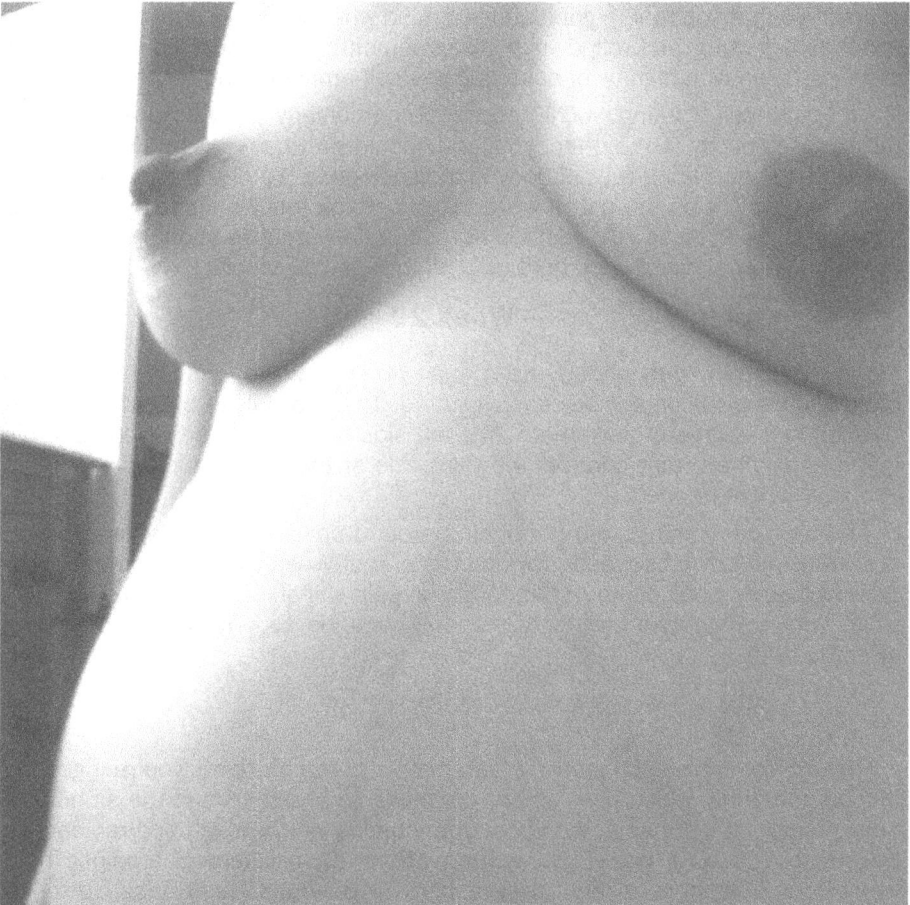

*How do you feel about your changing body? Try to appreciate the wonder of your body and be confident that it knows what it's doing! You'll be able to regain your figure at some point after the birth, with diet and exercise.*

You need to be careful about your stress levels

You also need to be careful about your stress levels. Many mothers seem to make a connection between a stressful third trimester and a difficult birth. But while some women find it stressful having to continue working, others have reported stress caused from boredom and having too much time to think. While a house move is too stressful for some women, for others it signals a joyous new beginning.

The key is probably self-awareness: whatever you do, you should not cause yourself too much *negative* stress. To do this—whatever your approach to rest, work and leisure activities—the key is to try to maintain harmony around yourself so that you're in a good state of mind when your due date approaches. Instead of getting embroiled in a disagreement, you should consider either stepping down with dignity, or asserting yourself amicably and respectfully.

If you're still working, you'll need to be particularly careful you don't allow the stresses of the job to affect you. You should do what you can in a methodical way, remembering that nobody is indispensable. You've got a much more important project on the go! However, if something really is upsetting you, you should get it sorted so your mind is at rest. Self-awareness should be followed by action. Many women work right up until they go into labor.

## — Week 29 —

By now your baby may already have quite an impressive head of hair and its brain will have become more sophisticated. The two halves of its brain have developed a degree of asymmetry, the left side being stronger. Of course, this is the hemisphere that controls the right side of the body, which is why most people are right-handed.

A lot of other things are happening around now... The fat deposits which are being laid down between now and the end of Week 32 are smoothing your growing baby's body contours, although its skin will still be covered with thick white vernix. Rhythmic breathing motions continue as practice for breathing after the birth.

A lot of things are happening around this time...

Your baby is also still growing fast, thanks to the nutrients you're ingesting in food and drink, which pass across the placenta and thanks to the amount of amniotic fluid it's drinking. We know from studies which have used radioactive tracers that growing babies drink from 15 to 40 milliliters of amniotic fluid every hour from now on until they're born. This nourishment adds up to 40 calories a day when a baby is swallowing normally and is supplemented by the nourishment provided via the placenta. Large, well-nourished babies swallow at a higher rate than small, grossly malnourished babies. This is why you need to avoid smoking, alcohol, junk food and bitter tastes, such as coffee. After all, studies have shown that when a bitter-tasting substance is injected into the amniotic fluid, babies suddenly stop drinking it. In the same studies, when saccharine was injected, some babies even doubled their rate of swallowing. No, this does not mean you can eat loads of cakes and candy! Your baby needs you to provide a balanced diet. You should persuade your baby to swallow fast by providing it with lots of good quality, tasty food. That way, it will get into a good pattern of growth in the last few weeks before its birth.

Your baby needs you to provide a balanced diet so it can get into a good pattern of growth in the last few weeks before its birth

## — Week 30 —

Your baby's facial features will now be well-developed. Other aspects of its appearance are also continuing to change this week: the lanugo (the fine layer of hair on its skin) is disappearing from its face and its skin is becoming paler and less wrinkled. This is mainly because of the fat which is continuing to build up underneath your baby's skin, which will help keep it warm after it's born. As well as storing fat, your growing baby is now also storing up iron reserves for after the birth. Finally, male babies' testes will descend into the scrotum sometime this week.

Your baby's facial features will now be well-developed. Other aspects of its appearance are also continuing to change.

Your baby's experience of the physical world is probably becoming much more intense. A startle reaction would already be recordable to researchers. It's also very likely that your baby will be aware of any Braxton Hicks contractions you are spontaneously having. Perhaps the experience of these intermittent tightenings in the womb is what gives us a liking for cuddles in later life—who knows? Your baby may also be becoming increasingly aware of its restricted environment. Sometimes children recall their time in the womb to their mothers (usually before the age of 2), saying something like: "It got very crowded in there at the end." (It's obviously not possible to ascertain whether these are 'real' memories or only imagined ones, but some young children really do give the impression of being able to remember their prebirth experience.) Perhaps the strong kicks you now feel are kicks of frustration as well as attempts at exercise! The constant wriggling you will be aware of is almost certainly a sign that your baby is trying to get comfortable.

Your baby's experience is probably becoming more intense. You yourself are likely to experience Braxton Hicks contractions.

## Getting your baby well-positioned...

As we've already noted, although there's no official proof, it does seem to be a good idea for you to help your baby get into a good position for the birth. Midwives Jean Sutton and Pauline Scott have written about this at length in their book *Optimal Foetal Positioning* (published by Birth Concepts in New Zealand in 1996). As they explain, not only is an anterior labor likely to be easier (because you will get breaks between contractions), it's also far less likely to result in a cesarean. It is surely possible that the dramatic increase in posterior labors in recent years is due to our contemporary lifestyle. Our ancestors were more likely to be on their hands and knees, scrubbing the kitchen floor, than lazing back on the couch, watching TV and reading a pregnancy book!

Kneel down as much as possible, for 10-15 minutes at a time, at least three times a day, especially in the last six weeks of your pregnancy. Make sure that when you do this you keep your back straight. (You can even arch your back slightly, but you mustn't let it sag.) Think of things you can do in this position. You can pick things up off the floor, vacuum clean using the hand attachment, wash the floor, help a child with a jigsaw... the list goes on!

When you are standing up, imagine yourself bouncing from the knees. This will help you to carry your weight well and avoid strain to you back. Your baby is also likely to get into and stay in a good position for the birth if you do this.

Your baby is likely to stay anterior if you avoid leaning back

A helpful reading position is kneeling down by the couch and reading a book or magazine propped up. As well as requiring relatively little effort, this also involves leaning forward. This is a great position for reading to a toddler.

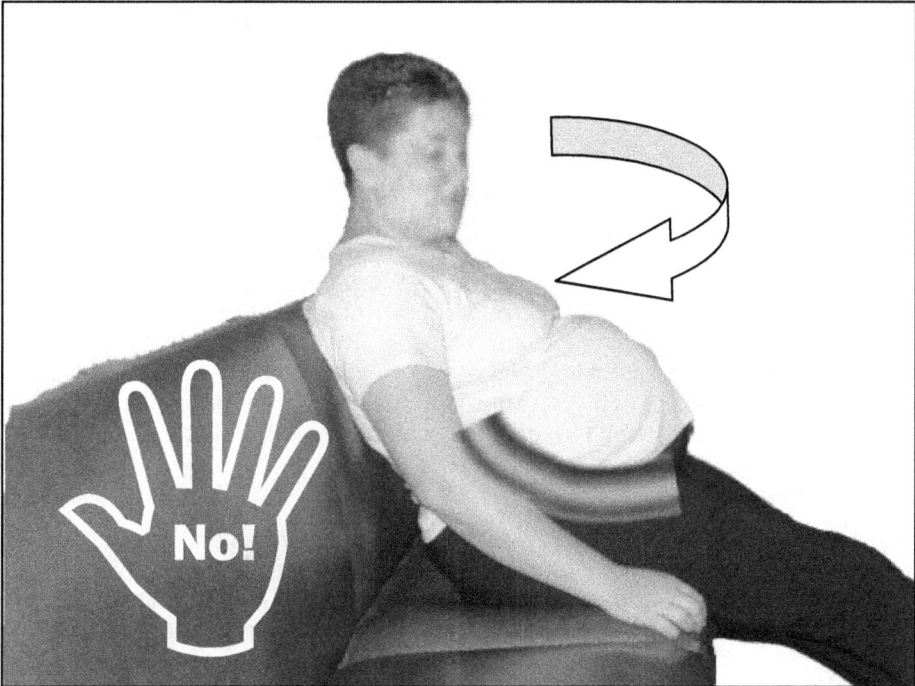

Looking at the photo above, if you imagine the striped line as your baby's back, it's easy to see how lying back is not helpful in the last few weeks of pregnancy. Your baby's back will tend to flip down because of the pull of gravity. This would result in a posterior labor, which would probably be much more painful.

In a more upright sitting position, your baby's back is more likely to stay round your front. With any type of chair, it's best to sit on the edge. It really is worth the effort!

It's easy to see how lying back is not helpful

Finally, when you are standing up, imagine yourself bouncing from the knees. This will help you to carry your weight well and avoid strain to your back. Your baby is also likely to get into and stay in a good position for the birth if you do this.

By the way, this model's face has been blurred at her request.

### — Week 31 —

Little by little, your growing baby will be getting plumper. It should now have gained around 50g of fat, which represents 3.5% of its total body weight. By the time it's born, body fat will account for 15% of its total body weight. (The percentage of body fat to body weight for an average-sized woman is around 27%.) The additional accumulated fat means it would no longer be possible to see the blood vessels beneath your baby's skin—if only we could take a peek. The shadowy images which represent your baby's growing bones would also no longer be visible now.

At birth, body fat accounts for 15% of total body weight

By this stage in your pregnancy you may occasionally feel very breathless, but your baby will certainly be getting plenty of oxygen from you, thanks to the flow of blood through the placenta, along the umbilical cord. Of course, this oxygen is essential for your baby's survival and brain development so you need to make sure you do nothing to compromise it. This means you should continue to avoid smokers (and smoking) and not lie on your back unless I ask you to at a prenatal check-up! (We'll prop your back up, so there's no pressure on the vena cava, and I can check the fetal heartbeat while you're sitting up...)

### — Week 32 —

By the end of this week your baby will be completely formed and its body will be in perfect proportion, taking into account the fact that the proportions will be that of a baby, i.e. large head and skinny limbs! It will be building up its immune system, taking antibodies from you so as to be able to fight off disease and infection after it's born. It will also have beautifully formed miniature eyelashes and eyebrows, so it must look very cute!

Your baby is continuing to gain weight... In fact, from now on until the time when it is born, it will gain at a rate of about 250g a week. This means it's more important than ever that you continue to eat well.

By now, your baby will have curled up into the well-known fetal position. Its head will probably already be pointing downwards, simply because the head is the heaviest part and gravity is likely to draw it downwards. If this isn't the case, don't worry. Before 32 weeks, as you know, 50% of babies will be in a breech position but as these babies' heads become heavier than their butts, they usually spontaneously turn round this week, or at least by the 36th week of pregnancy. Even if a baby stays breech, a normal optimal birth is still possible, providing your caregiver is happy to attend a vaginal breech birth.

Even if a baby stays breech a normal birth is still possible

## — Week 33 —

As your baby continues to become chubbier, its skin smoothes out even more. If it has turned to a head down position by now, it will probably stay like that now until its birth. After all, it no longer has the luxury of being able to perform somersaults—it's simply too cramped inside your womb now!

It will be going through some important development during the next few weeks. Its lungs will begin to produce surfactant (a substance similar to detergent), which will help them to expand and withstand pressure. It will also prevent your baby's lungs from collapsing at birth. Its brain will also develop some new abilities. Some people claim that a baby in the womb at this stage of gestation can be taught to recognize a nursery rhyme or simple piece of music and respond to it after the birth, as long as it's played daily for a month. Mozart and Vivaldi were the composers babies responded to best. Don't attempt any artificial exercises like these yourself, though. Doing what you spontaneously want to do is probably the best thing for your baby, who will benefit most from having a contented and healthy mother and an uncontrived environment. Babies can have interesting tastes in music!

## — Week 34 —

This week your growing baby is becoming more rounded. The lanugo, the fine covering of hair all over its body, begins to disappear and your growing baby's skin will be getting pinker. Its ear cartilage is still soft, though, and the so-called plantar creases—creases on the soles of the feet—are still visible. Over the next four weeks, the hair on your baby's head will get longer and your baby's nails will grow long enough to reach the tips of its fingers.

The movements you will now be feeling will probably be much gentler than a few weeks ago. The increasing lack of space which we've already mentioned makes those vigorous kicks and punches you felt before a near impossibility. You shouldn't worry about this apparent decrease in movement but should nevertheless continue to be aware of how and when your baby is moving about. If there is a change of pattern or a long absence of any movement, you should contact me so that I can check your baby's okay, using my fetal stethoscope.

You should avoid any potentially stressful over-stimulating events, e.g. carnivals, theater performances, violent movies or sports events. Loud music, overly prolonged and hearty laughter and energetic dancing are also things you should be careful of. Of course, this is because of the close link between emotional and hormonal states and the possibility that upsetting the balance could trigger labor. If you're at a concert or play (or other event) and suddenly you feel your baby moving around vigorously in response to some of the sound effects or music, you should leave immediately or at least move to a seat at the back of the hall or theater. Your baby might be truly upset!

## Optimal fetal positioning...

Are you being careful how you sit and lie down? From now on until the birth, it's a good idea to get down on your hands and knees three times a day, especially if your baby is now lying in a posterior position. It's worth it for an easier labor! Use a chart like this to persuade you to track you're doing this enough:

| 10-15 mins on hands and knees | Mon 10/3 | Tue 10/4 | Wed 10/5 | Thu 10/6 | Fri 10/7 | Sat 10/8 | Sun 10/9 |
|---|---|---|---|---|---|---|---|
| 1st time | ✓ | ✓ | ✓ | ✓ | ✓ | | |
| 2nd time | ✓ | ✓ | ✓ | ✓ | ✓ | | |
| 3rd time | ✓ | ✓ | ✓ | ✓ | | | |

*A tracking chart for ensuring that you really do get down on your hands and knees!*

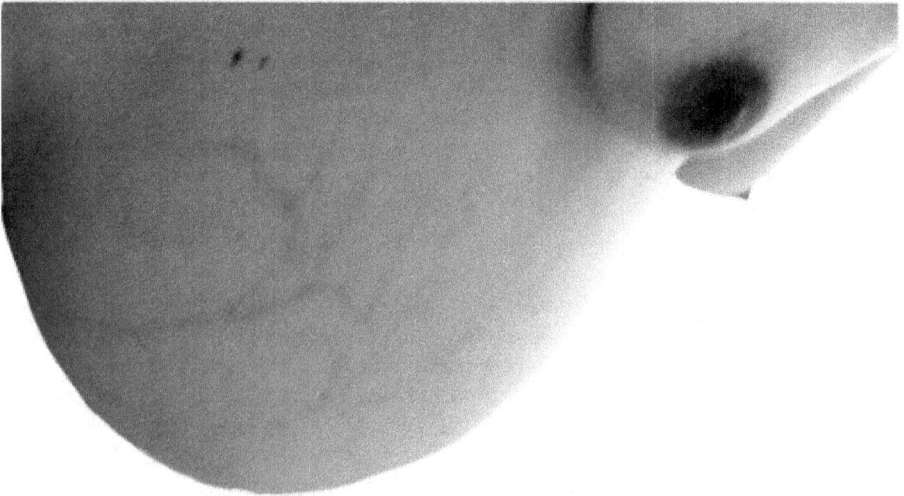

*How's your baby belly? Make sure it's pointing down more often than up... if you want to make your labor as straightforward as possible. A posterior labor is much harder.*

## — Week 35 —

If, for whatever reason, your baby were to be born now—anything from 34 weeks and two days' gestation onwards—it would have an excellent chance of survival.

During its last few weeks in the womb your baby is now doing an impressive amount of peeing... a remarkable 600ml every day. The pee initially goes into the amniotic fluid, of course, and then the waste products are filtered through the placenta into your own bloodstream. Eventually, your baby's waste products are dealt with by your kidneys, along with your own waste. Since your kidneys need to function very efficiently in order to deal with all this waste, it's very important for you to continue to drink plenty of good quality drinking water.

As well as peeing, your baby will be doing a lot of sucking practice now. If it is sucking its thumb or its hands (which is likely), it may even already have some sucking blisters! Developing a strong suck now and in the weeks to come is, of course, extremely important because it's necessary for successful breastfeeding.

As your baby sucks and manoeuvres carefully inside you, it will carry on trying to decipher sounds coming from outside its little world and will continue responding to tastes and changes in light.

## — Week 36 —

If you are one of the 1 in 80 moms carrying twins I will probably have noticed this by now when I have felt your belly. Perhaps it's a good idea to consider your options carefully because many mothers have managed to carry their twins to term and even have them vaginally. Even if you are carrying one baby, you may be interested to know that twins at this stage of gestation have been filmed hugging, stroking and patting things in the womb. This may mean that growing babies may have emotional needs at this stage of development, so it might be a good idea to tenderly stroke your expanding belly, if you aren't already doing so! You never know, it may well be reassuring for your baby living inside you. Remember too that you can talk to your baby with the expectation that it will be listening, and possibly even understanding... who knows?

In normal circumstances, you can imagine your baby gaining about 14g of fat a day at this stage. This even rises to about 28g a day during the last four weeks of gestation. Your baby is piling on the pounds (or grams, actually) so as to be able to cope with the lower temperatures outside the womb, after its birth. You should keep eating substantial quantities of good, wholesome food to help this process along.

Your baby is almost fully mature now, even in terms of lung maturation. (Obviously, this is important because your baby must be able to breathe when it's born). It may have quite long hair—it may be up to 5cm long. If you're carrying a boy, his testicles will probably already have descended at this stage.

Ideally your baby's position should be LOA, which means head-down and with its back round at your own front, on the left side of your belly. If your baby is in a different position, I'll explain the possible implications of that. If your baby is in another position (ROA, LOP, ROP or one of the breech positions) consider how you are using your body during the day and at night.

For ideal positioning, it's probably best for you to sleep on your left side and to avoid leaning back while you're awake. This will facilitate an anterior, rather than a posterior position, and will result in LOA, rather than ROA. The reason to take note of your baby's position is so as to optimize the chances of the birth being easy... and safe, of course. With anterior positioning (LOA or ROA) you will get 100% painfree breaks between contractions, which may in any case not be painful. (Different women experience them differently.) With posterior positioning a woman is likely to have no breaks at all—just varying degrees of intensity—which can be *extremely* hard work! If a baby who is initially posterior doesn't 'turn' to an anterior position by the time the woman reaches the second stage of labor (i.e. the birth), the birth itself ends up being much more painful, so this is another reason to take the baby's position seriously now. If your baby is anterior, the second stage of labor—the birth—can be completely painfree, apart from the famous 'ring of fire' everyone talks about. If you're anxious about all this, remember that many women still go on to have optimal births, even when their babies are not ideally positioned.

## — Week 37 —

For some midwives this is a landmark week because you normally need complete 37 weeks of pregnancy if you want to have a home birth because of the risks associated with prematurity. (Is your due date accurate? It might be a good idea to recheck this around about now.)

Your growing baby's vernix (the thick creamy coating covering its body) is disappearing now and your baby may have almost reached its birth weight, whatever that's going to be. You can imagine your baby with a thicker neck and with eyelids that open and close easily. Your baby will continue to rehearse breathing movements—even though no air will be going into its lungs as yet— and the repeated hiccups you may be aware of coming from your belly are ongoing proof of this. (It's a sign that amniotic fluid has passed into your baby's trachea.)

If your baby is lying in a breech position, it's still possible that your baby may turn round spontaneously before the birth so you may still be able to have a head-down birth. You may also want a vaginal breech birth.

If this is your first baby, and its head is pointing downwards, its head will probably drop into your pelvis very soon, ready for the birth. (We say it's 'engaged' when this happens.) If it's your second or subsequent baby it will probably descend just as you're going into labor.

At the end of this week, your baby is generally considered to be 'term'. A baby is considered term when you are between 37 and 42 weeks' gestation, i.e. when it's between 259 and 294 days since the first day of your last period—your LMP.

## — Week 38 —

Your growing baby's contours are now very well-rounded and its skull is firm. Its size will be even closer to its birthweight.

Your baby is now spending as much as 60% of its time asleep and its sleep is now falling into clearly detectable patterns. Now it even dreams like an adult. We know this because of the rapid eye movement (REM) associated with the dreaming state, which has been observed in babies in the womb at this stage of gestation. (REM sleep has been recorded from as early as 23 weeks' gestation.) Since it is thought that dreaming encourages brain development, we can guess that your baby is continuing to develop in terms of emotions and general intelligence. Measurement of brain waves increasingly shows more organization, steadier activity, and greater synchrony between the left and right halves of the brain. Research has also shown that your baby is by now uttering a much more definite cry (only audible in research conditions). Since crying is an essential means of communication for babies, this isn't bad news. If a baby couldn't cry at birth, how would it be able to get its mother's attention? Of course, it needs to fit in a little crying practice before the day of its birth.

Your baby needs to fit in a little crying practice before its birth

## — Week 39 —

Your baby's rate of growth is slowing down, which is just as well really because there is very little room left in your womb. Phenomenal development has taken place over the last 37 weeks (from conception onwards, remembering that your baby only really started growing from Week 3). The original fertilized cell has become a well-organized bundle of 200 million cells and these cells now weigh approximately six billion times more than the original fertilized egg!

The amniotic fluid around your baby is being renewed every three hours, so you need to keep drinking water. If your baby remains undistressed, the amniotic fluid will remain clear because the waste products which aren't released in its wee will remain in its bowel. The sticky, greeny-black substance which results from the storage of these waste products, which as we've already mentioned is called meconium, is a mixture of excretions from your baby's alimentary glands, bile pigment, lanugo and cells from its bowel wall. When this original gestational waste has been cleared in the first few diapers, your newborn baby's poops will become an interesting and pleasant-smelling orange color if you nurse him or her. (Bottle-fed babies' poops are browner and have a stronger odor.)

Whether or not your baby is ready to be born this week or next will depend not only on individual variation but also on the precise date your baby was conceived. (Do you know precisely when you ovulated in the month you conceived?) I would encourage you to resist attempting to hurry your baby along at this stage... Nature can cope very well without our interference; in fact, disturbance usually results in worse outcomes, not better ones. You will go into labor and give birth spontaneously when your baby is sufficiently mature and when your body has prepared precisely the correct balance of hormones to make the birth possible. (Postmaturity is a rare phenomenon and is perhaps worried about too much.)

Enjoy these last few days of life without a tiny baby to care for. Babies are much easier to look after when they're inside us! You should relax, walk, dance, paint, write, sing, bake cakes, contemplate the beauty of nature... If you're happy and contented, your baby is likely to be relaxed and happy too—because it's sharing your hormonal environment. Virtually everything crosses the placenta, of course, including emotions. True intimacy.

Having said all that, while focusing on your own well-being you should also spend some time preparing for your baby's arrival. Do you have everything you need... soft towels, receiving blankets, baby clothes?

## — Week 40 —

If you haven't already done so, you may now be nearing the time of your labor. However, since only 3% of babies arrive on their due date you may still have a fair wait. 80% of babies are born within 14 days of their due date, either before or after. The rest are born either earlier or later than 42 weeks—perhaps because they were conceived at the end of a very long menstrual cycle.

If your baby has now been growing inside you for a full 38 weeks, it will certainly be as heavy as it is going to be at the time of its birth, and also as long. Its fingernails may be so sharp that it scratches itself with them.

Most of your baby's lanugo (fine hair) and vernix will have disappeared by now, but at birth a little lanugo may still be seen over its shoulders, back, arms and legs and some traces of vernix (cream) may remain in skin folds.

The temporal lobe of your baby's brain—the part to which the ear sends its data—will now be fully myelinated. Other parts of its brain and nervous system will only be partially insulated at birth, so it's clear that your baby's hearing is given full priority. In this respect, it's worth remembering that under hypnosis in later life, people can often recall comments which are made around the time of their birth which have affected them in some way.

Your baby's nose is also fairly well-developed now, meaning that it will display a decided preference for certain smells as soon as it's born. It's interesting that a newborn will express these preferences even with no practice or experience. The smell of your own breasts—and armpits, actually!—will be particularly attractive to the newborn baby. So whatever you do, you shouldn't worry about body odor just now!

You shouldn't worry about body odor just now!

The part of your baby which is to be born first—its 'presenting part'—will probably be in the lower segment of your womb now, pressing through the already softened, partially opened cervix. If this baby is a second or subsequent baby later, it is likely to engage at the onset of labor. Obviously, the timing of engagement varies.

If indeed your baby is ready to be born, it will send a hormonal signal to your body calling for an end to the pregnancy. (This is a signal scientists would like to know more about!) Your body will respond by releasing oxytocin (the 'hormone of love') to make your womb contract and your cervix open up. The irregular and then regular contractions which follow are an inevitable part of labor—so you should try and view them very positively. They are there to help your baby exit your womb, descend through your pelvis and enter this world...

Of course, it's a dramatic business coming into this world from your baby's point of view. The journey itself involves traveling down a long, dark and unfamiliar passage. You shouldn't worry about your baby getting damaged, though. Its body will now be very flexible, thanks to fluid intervertebral disks and joints which can fold neatly in tight places. Even the head plates which guard your baby's brain will yield to pressure by overlapping, which may result in a pointed head or 'pixie' look at birth. (Remember that these plates will gradually reposition themselves again.) Your baby may well be shocked at birth... the reassuring hug of the womb will be gone, temperatures will no longer be constant, and supplies of food and drink won't be on tap any more. It will need plenty of tender love and care. How will all this affect your own approach to care at the birth?

66 I knew that the way in which due dates are calculated in England is about two weeks shorter than elsewhere in Europe, and medics get twitchy and start wanting to induce births here, when in Sweden they wouldn't consider it at the same point. Plus first babies are routinely late—probably because the calculations are wrong. My dates could not be accurate as I'd not known I was pregnant in the first place, but anxieties always seem to grow in the last two months. My way of avoiding them was to keep working and keep to a minimum contact with the world of pregnancy.

## — Week 41 —

It's worth noting, if you are still pregnant, that only a baby born after 41 weeks and 6 days (293 days) is technically overdue because human gestation can be anything from 37-42 weeks. However, following typical protocols, I will probably recommend I monitor you more closely from now on. Despite this, given the wide variation please don't worry! Your body really does know what to do.

Your body really does know what to do at each stage

Reconsider the precise date of conception if you haven't already done so, taking into account the length of your normal monthly cycle and of course also your sexual activity around the time you think you conceived. If your cycle is irregular, this could explain things, of course... How many days late has your period been in the past? The birth could be the same number of days late. For example, a 33-day cycle would mean you could only expect your baby five days after the due date given in charts. Also, was your partner away on business at any time in the month you got pregnant? You could only have become pregnant when you made love! Think through dates but still be confident in your body's abilities...

Think through dates carefully and be confident in your body!

At this potentially worrying time, remember just how amazing your body is, managing to conceive a baby and get it grown this far. Of course it knows precisely how to get it born too! It's a process which usually takes place with no help... The kind of monitoring I'm going to do, which is called 'watchful waiting', should involve an absolute minimum of disturbance, so that the physiological processes can take place smoothly. Tune into your baby and talk to it too. You will soon know if the 'it' is a girl or a boy.

While you're waiting, it's a good idea if you keep thinking about your baby, imagining what he or she might be like. It's not possible to calculate how heavy or tall he or she will be, no matter how large or compact you are, but here are a few interesting facts about babies' weights and heights around the time of their birth...

- A typical birth weight is 3.5 kilos. (That's 7½ pounds, if metric means nothing to you).
- Only 5% of babies weigh less than 2.5 kilos or more than 4.25 kilos.
- If your baby's a boy, it's likely he will weigh about 200 grams (½lb) more than a girl would.
- Whatever the newborn's weight, it will decrease in the first three days after the birth. This is nothing to worry about—it really is perfectly normal as the new baby adjusts to life outside your body.
- By about 10 days after the birth, your baby should have regained his or her birth weight and the long-term process of growth will have re-established itself.
- Your newborn's height will probably be just under a third of that of a typical adult, i.e. 51cm (or 20in).
- 95% of babies fall within the range of 46-56cm.

Here are a few more interesting facts for you to consider, to keep your mind off your wait...
- A newborn's body is made up of approximately 70% water, 16% fat, 11% protein and 1% carbohydrate.

- Despite the fact that your new baby's heart will weigh less than 30g at birth, your baby's pulse rate will be about 180 beats per minute during the actual birth and it will then average 140 beats per minute in the first few weeks. During the first year of your baby's life, this rate will gradually decrease to 115 beats per minute.
- Your new baby's eyesight will already be fairly well-developed at birth: your baby will be able to focus over the distance from your breast to your face and it's likely that he or she will be attracted to your face, rather than to less complex, less curvaceous, less mobile, inanimate objects.
- At birth, your baby will be ideally predisposed to bond with you, provided of course there are no drugs in his or her system. If you were to have any drugs in your system that would also be a disappointment for your baby, because your face would be less responsive. After all, research into the behavior of babies just after birth has revealed that babies typically spend up to an hour staring intently at their mother's face, if given the chance, before they fall asleep for the first time outside the womb. This is a first exchange which you should look forward to and cherish and it's one of the reasons why it's best for you to have a really healthy birth, i.e. a birth which involves an absolute minimum of intervention with no drugs. Many things will help you through any pain you might experience, as long as you're not disturbed in labor and you feel safe with your caregivers. Moving around, using water and a birthing ball may all help. Our aim is to make your baby's birth as gentle and as trauma-free as possible so that you and your baby can start off your relationship well.

## — Week 42 —

You really shouldn't worry. You probably ovulated later than you think. Perhaps that particular month when you conceived was a strange one in terms of your normal cycle. Your baby is probably developing as he or she should. He or she is probably soon going to start the birth process. Beyond carrying out non-invasive checks to confirm your baby is still okay, as long as you're not having any potentially dangerous symptoms (such as bleeding), I will be very confident in your body's abilities. Don't worry about induction. Too many inductions prove to be unnecessary and unsafe, not to mention traumatic for the women giving birth. Remember that only 1% of babies who are thought prenatally to be 'postmature' actually turn out to be so when they're born (and even then symptoms are not usually life-threatening). The other 99% show no signs of postmaturity and clearly weren't yet ready to be born.

You can even monitor your baby's movements yourself. Keep a kick chart, if you want to... this may provide me with some useful information. Remember that the process of labor and birth often begins very suddenly after a seemingly interminable wait! Strange and wonderful things are happening within your body. Continue to trust your baby's ability to get itself born, while continuing to talk to him or her, perhaps asking him or her to hurry up! You're going to meet up properly soon.

## — Week 43 —

If you're still thinking about week-by-week changes in pregnancy, note that female babies spend on average one day longer inside their mother's bodies than males. And white babies on average spend five days longer inside the womb than black babies. Apparently, these differences are purely racial and have nothing to do with individual size variations, wealth or even poverty. Until now, nobody has been able to explain why these variations occur.

In any case, if you are still pregnant now, you probably just miscalculated your LMP (last menstrual period), or maybe you ovulated very late in the month you got pregnant. If you are worried about not going into labor, focus on this feeling and try and work out why there's a delay... Are there any psychological issues you need to resolve before you give birth? Do you need to make any other preparations? Are you happy with the arrangements you've made with birth attendants, friends and family? Is there anything you haven't told me?

Also consider whether you really, intuitively, feel there's something wrong. If you feel pretty confident that everything's basically okay with both yourself and your baby, consider what's been happening lately... Has there been any diarrhea? Any loss of appetite? Any increased sexual desire? Any impatience or a deep need to tidy up and organize things? Of course, these are all signs that labor is very close indeed, so you should relax if the answer is 'Yes!'

When you do go into labor, as you will, continue to imagine what your baby might be experiencing alongside you. This is the beginning of your life together.

## Questions to think about and discuss...

1  Are you worried? If so, what are you most worried about? Is there anything you've seen in real life or on TV that's made you worry particularly?
2  Are you feeling anxious about some aspect of the birth?
3  Have you ever seen anyone else give birth in real life? What was it like?
4  What kind of birth are you hoping to have? What have you done to prepare for this? What do you feel you haven't done, which you should have done?
5  How do you feel about the place you're registered to give birth? Would you rather be giving birth somewhere else? Have you been thinking about giving birth at home? If you're registered for a home birth, would you rather be in hospital or in a birthing center? What aspect of your birth place is worrying you particularly and what could you change to make things seem better?
6  How do you feel about your caregivers? Is there anything you need to discuss?
7  Have you written a care guide and got your caregivers' agreement on it? Are there any points in it you're worried about?
8  What aspects of the birth and afterwards are you looking forward to?

*Liliana Lammers moments after giving birth physiologically
(see Birthframe 21)*

# Bibliography

Arms S. *Immaculate Deception II*. Celestial Arts 1994. ISBN: 978 0 890876 33 6.

Auel JM. *Clan of the Cave Bear*. Hodder Paperbacks 2006. ISBN: 978 0 340839 89 8.

Block J. *Pushed: The Painful Truth about Childbirth and Modern Maternity Care*. Da Capo 2007. ISBN: 978 0 738210 73 5.

Blumfield W. *Life After Birth*. Element Books 1992. ISBN: 978 1 852303 51 8.

Bryson B. *Bill Bryson's African Diary*. Doubleday 2002. ISBN: 978 0 385605 14 4.

Buckley S. *Gentle Birth, Gentle Mothering: The wisdom and science of gentle choices in pregnancy, birth, and parenting*. One Moon Press 2005. Available from www.gentlebirthgentlemothering.com.

Caldwell Sorel N. *Ever Since Eve*. Oxford University Press 1984. ISBN: 978 0 195034 60 8.

Campbell S and Lees C. *Obstetrics by Ten Teachers*. Hodder Arnold 2000. ISBN: 978 0 340719 86 2.

Chamberlain D. *Babies Remember Birth*. Ballantine Books 1990. ISBN: 978 3 45364 11 1.

De Beauvoir S. *The Second Sex*. Jonathan Cape 2009. ISBN: 978 0 224078 59 7

Diamant A. *The Red Tent*. Pan 2002. ISBN: 978 0 330487 96 2.

Donna S. *Preparing for a Healthy Birth*. Fresh Heart 2010. ISBN: 978 1 906619 01 5.

Downe S. *Normal Childbirth: Evidence and Debate*. Churchill Livingstone 2004. ISBN: 978 0 443073 85 4.

Enkin M, *et al. A Guide to Effective Care in Pregnancy and Childbirth*. Oxford University Press 2000. ISBN: 978 0 192631 73 2.

Gaskin I M. *Spiritual Midwifery*. Book Publishing Company 2002. ISBN: 978 1 570671 04 3.

Jadad AR and M Enkin. *Randomized Controlled Trials: Questions, Answers and Musings*. Blackwell 2007. ISBN: 978 1 405132 66 4.

Kitzinger S. *Birth Crisis*. Routledge 2006. ISBN: 978 0 415372 66 4.

Kitzinger S. *New Pregnancy and Childbirth: Choices and Challenges* (Dorling Kindersley 2003). ISBN: 978 0 751364 38 5.

Kitzinger S. *Ourselves as Mothers: Universal Experience of Motherhood*. Doubleday 1992. ISBN: 978 0 385403 20 7.

Leboyer F. *Birth Without Violence*. Inner Traditions Bear and Company 2002. ISBN: 978 0 892819 83 6.

McKay A. *The Birth House*. Harper Perennial 2006. ISBN: 978 0 00 723330 4.

Metland D (ed.) *The Complete Book of Pregnancy*. HarperCollins 2000. ISBN: 978 0 004140 99 5.

Myles M. *Myles Textbook for Midwives*. Churchill Livingstone 1999. ISBN: 978 0 443055 86 7.

Odent M and Vincent-Priya J. *Birth Traditions and Modern Pregnancy Care*. Element Books 1992. ISBN: 978 1 852303 21 1.

Odent M. *Birth and Breastfeeding*. Clairview Books 2003. ISBN: 978 1 902636 48 1.

Odent M. *Birth Reborn*. Souvenir Press 1994. ISBN: 978 0 285631 94 6.

Odent M. *Primal Health*, Clairview Books 2007. ISBN: 978 1 905570 08 9.

Odent M. *The Caesarean*. Free Association Books 2004. ISBN: 978 1 853437 18 2.

Odent M. *The Scientification of Love*. Free Association Books 1999. ISBN: 978 1 853434 76 1.

Odent M. *Water and Sexuality*. Penguin (Canada) 1990. ISBN: 978 0 140191 94 3.

Rank O. *The Trauma of Birth*. Routledge 1999. ISBN: 978 0 415211 04 8.

Rix J. *Is there sex after childbirth?* Thorsons 1995. ISBN: 978 0 722529 57 7.

Staff DA. *Our Bodies, Ourselves*. Touchstone Books 2005. ISBN: 978 0 844672793.

Stanway P. *Breast is Best*. Pan Books 2005. ISBN: 978 0 330436 30 4.

Sutton J and Scott P. *Optimal Foetal Positioning*. Birth Concepts 1996. (No ISBN available.)

Torgus J (ed.) *The Womanly Art of Breastfeeding*. Plume Books 2004. ISBN: 978 0 452285 80 4.

Wagner M. *Born in the USA: How a broken maternity system must be fixed to put women and children first*. University of California Press 2006. ISBN: 978 0 520245 96 9.

Walker A. *Possessing the Secret of Joy*. Vintage 1993. ISBN: 978 0 099224 11 2.

Wesson N. *Home Birth: A Practical Guide*. Pinter & Martin 2007. ISBN: 978 1 905177 06 6.

# Recommended reading

Most of the following books are readily available. If you have trouble finding any, you can do a search on the website www.bookfinder.com. Alternatively, contact the publisher directly or order any book via a bookshop, using its ISBN number.

Armstrong P and Feldman S. *A Wise Birth*. Pinter & Martin 2007. ISBN: 978 1 905177 03 5.

Balaskas J and Sieveking A. *Easy Exercises for Pregnancy*. Frances Lincoln Publishers 1997. ISBN: 978 0 711210 48 6.

Bauer I. *Diaper Free! The Gentle Wisdom of Natural Infant Hygiene*. Natural Wisdom Press 2001. ISBN: 978 0 452287 77 8.

Block J. *Pushed: The Painful Truth about Childbirth and Modern Maternity Care*. Da Capo Lifelong 2007. ISBN: 978 0 738210 73 5.

Castro M. *Mother and Baby*. Pan Books 1996. ISBN: 978 0 330349 25 3.

Chamberlain D. *Babies Remember Birth*. Ballantine Books 1990. ISBN: 978 3 45364 11 1.

Cohen N and Estner L. *Silent Knife: Caesarean Prevention and Vaginal Birth After Caesarean*. Greenwood Press 1983. ISBN: 978 0 897890 27 4.

Davis-Floyd R, Barclay L, Daviss BA, Tritten J. *Birth Models That Work*. University of California Press 2009. ISBN: 978 0 520 25891 4.

Donna S. *Preparing for a Healthy Birth*. Fresh Heart 2010. ISBN: 978 1 906619 01 5.

Downe S (ed.) *Normal Childbirth: Evidence and Debate*. Churchill Livingstone 2004. ISBN: 978 0 443073 85 4.

Enkin M, *et al*. *A Guide to Effective Care in Pregnancy and Childbirth*. Oxford University Press 2000. ISBN: 978 0 192631 73 2.

Fahy K, Foureur M, Hastie C (eds). *Birth Territory and Midwifery Guardianship: Theory for Practice, Education and Research*. Butterworth Heinemann 2008. ISBN: 978 0 7506 88 70 3.

Frye, A. *Understanding Diagnostic Tests in the Childbearing Year*. Labrys Press, 2007. ISBN: 978 1 891145 56 8.

Jackson D. *Three in a Bed: The Benefits of Sleeping with Your Baby*. Bloomsbury 2003 [first published in 1989]. ISBN: 978 0 747565 75 8.

Kitzinger S. *Birth Crisis*. Routledge 2006. ISBN: 978 0 415372 66 4.

Leboyer F. *Birth Without Violence*. Inner Traditions Bear and Company 2002 [first published in 1974]. ISBN: 978 0 892819 83 6.

McKenna J. *Sleeping with Your Baby: A Parent's Guide to Cosleeping*. Platypus Media 2007. ISBN: 978 1 930775 34 3.

Odent M and J Vincent-Priya. *Birth Traditions and Modern Pregnancy Care*. Element Books 1992. ISBN: 978 1 852303 21 1.

Odent M. *Birth and Breastfeeding.* Clairview Books 2003 [first published in 1992]. ISBN: 978 1 902636 48 1.

Odent M. *Birth Reborn.* Souvenir Press 1994 [first published in 1984]. ISBN: 978 0 285631 94 6.

Odent M. *Entering the World: The De-medicalization of Childbirth.* Marion Boyars Books 1984. ISBN: 978 0 714528 00 5.

Odent M. *The Caesarean.* Free Association Books 2004. ISBN: 978 1 853437 18 2.

Odent M. *The Farmer and the Obstetrician.* Free Association Books 2002. ISBN: 978 1 853435 65 2.

Odent M. *The Scientification of Love.* Free Association Books 1999. ISBN: 978 1 853434 76 1.

Raynor M, Marshall J, Sullivan A (eds). *Decision-Making in Midwifery Prac*tice. Churchill Livingstone 2005. ISBN: 978 0 443 07384 7.

Reid L. *Midwifery: Freedom To Practise? An international Exploration of Midwifery Practice.* Churchill Livingstone 2007. ISBN: 978 0 443 10312 4.

Rix J. *Is There Sex After Childbirth?* HarperCollins 1995. ISBN: 978 0 722529 57 0.

Schmid, V. *Birth Pain: Explaining Sensations, Exploring Possibilities.* Fresh Heart 2011. ISBN: 978 1 906619 14 5.

Sears W and M. *The Attachment Parenting Book.* Imported Little, Brown USA Titles 2001. ISBN: 978 0 316778 09 1.

Simkin P, Klaus P. *When Survivors Give Birth.* Classic Day Publishing 2004. ISBN: 978 1 59404 022 1.

Stanway P. *Breast is Best.* Pan Books 2005. ISBN: 978 0 330436 30 4.

Sutton J and Scott P. *Optimal Foetal Positioning.* Birth Concepts 1996.

Tellack P. *In The Womb.* National Geographic 2006. ISBN: 978 1 426 20003 8.

Torgus J (ed.). *The Womanly Art of Breastfeeding.* Plume Books 2004. ISBN: 978 0 452285 80 4.

Vadeboncoeur, H. *Birthing Normally After a Caesaeran or Two.* Fresh Heart, 2011. ISBN: 978 1 906619 15 2.

Wagner M. *Born in the USA: How a broken maternity system must be fixed to put women and children first.* University of California Press 2006. ISBN: 978 0 520245 96 9.

Walsh D. *Evidence-Based Care for Normal Labour and Birth: A guide for midwives.* Routledge 2007. ISBN: 978 0 415 41891 1.

Wesson N. *Home Birth: A Practical Guide.* Pinter & Martin 2006. ISBN: 978 1 905177 06 6.

Yntema S. *Vegetarian Pregnancy: Definitive Nutritional Guide.* McBooks Press 2004. ISBN: 978 0 935526 21 9.

# Useful websites

♥ American Association of Birth Centers (AABC)—www.birthcenters.org

♥ American College of Nurse-Midwives (ACNM)—www.acnm.org

♥ Attachment Parenting International (API)—www.attachmentparenting.org

♥ Birth Works ®—www.birthworks.org

♥ Birth Works ®—www.birthworks.org

♥ Childbirth & Postpartum Professional Organization (CAPPA)—www.cappa.net

♥ Childbirth Connection—www.childbirthconnection.org

♥ Citizens for Midwifery—www.cfmidwifery.org

♥ Coalition for Improving Maternity Services—www.motherfriendly.org

♥ DONA International (Doulas of North America)—www.dona.com

♥ Giving Birth Naturally —www.givingbirthnaturally.com

♥ Hypnobirthing—www.hypnobirthing.com

♥ International Cesarean Awareness Network (ICAN)—www.ican-online.org

♥ La Leche League—www.lalelecheleague.org or www.llli.org

♥ Lamaze International—www.lamaze.org

♥ Midwifery Today catalogue—www.birthmarket.com

♥ Midwifery Today—www.midwiferytoday.com

♥ Midwives Alliance of North America (MANA)—www.mana.org

♥ *Mothering* magazine—www.mothering.com

♥ Our Bodies, Ourselves—www.ourbodiesourselves.org

♥ Postpartum Support International—www.postpartum.net

♥ Primal Health Research Databank—www.primalhealthresearch.com

♥ Sidelines—www.sidelines.org

♥ Understanding Birth Better—www.understandingbirthbetter.com

♥ Waterbirth International—www.waterbirth.org

Photo © Virgilio Ponce

*Let's help couples get where they need to be, so they can get enough information!*

# Your own contacts...

*One of Nina Klose's optimally-birthed babies in a field of pumpkins!*

# Birthframes index

| No. | Name | Features | Page |
|-----|------|----------|------|
| 1 | Maria Shanahan | Special dreams of conception and birth | 15 |
| 2 | Sylvie Donna | First encounter with maternity care | 49 |
| 3 | Sylvie Donna | Fetus ejection reflex with Michel Odent | 52 |
| 4 | Sylvie Donna | Contemplations about midwifery care | 58 |
| 5 | Maria Shanahan | Optimal birth despite fears after first birth | 63 |
| 6 | Beth Dubois | Difference in approaches of two midwives | 70 |
| 7 | Rachel Urbach | Undisturbed birth experienced by a primipara | 73 |
| 8 | Anonymous | Intervention because of high risk of PPH | 74 |
| 9 | Tina C from the UK | Management of persistent hyperemesis | 75 |
| 10 | Jo Siebert | Management of obstetric cholestasis | 79 |
| 11 | Jill Furmanovsky | Optimality through in-labor intervention | 80 |
| 12 | Liliana Lammers | An in-labor cesarean with footling breech | 84 |
| 13 | Anonymous | Vaginal twin birth with supportive care | 87 |
| 14 | Debbie Brindley | Midwife arranging for minimal disturbance | 87 |
| 15 | Steve & Olga Mellor | Reflection on differing attitudes to birth | 90 |
| 16 | Nina Klose | Emotional memories of a cesarean | 103 |
| 17 | Anonymous | The emotional and physical aftermath | 104 |
| 18 | Anonymous | Postpartum experiences after cesareans | 105 |
| 19 | Anonymous | A positive experience of a cesarean | 106 |
| 20 | Jan Tritten | A midwife's motivation for qualifying | 145 |
| 21 | Liliana Lammers | A first disturbed birth, then three better ones | 166 |
| 22 | Deborah Jackson | Experiencing smooth and disturbed births | 167 |
| 23 | Caroline Turner | Woman regretting using the British 'Entonox' | 180 |
| 24 | Sue Pakes | Experiencing birth without pain | 187 |
| 25 | Gemma Shepherd | Psyching up for a bearable birth | 188 |
| 26 | Anonymous | Considering the role of pain in birth... | 191 |
| 27 | Anonymous | Anger about treatment | 202 |
| 28 | Christina Mansi | Intervention/non-intervention after SROM | 203 |
| 29 | Sylvie Donna | Intervention refused, after research/advice | 209 |
| 30 | June Worcester | Ineffective anesthesia and trauma | 215 |
| 31 | Fiona Lucy Stoppard | The experience of 'going to another planet' | 218 |
| 32 | Pauline Farrance | Positive birth after very managed first birth | 219 |
| 33 | Anonymous | Disempowering prenatal appointment | 276 |
| 34 | "Jackie" | Nocebo effect through prenatal comment | 278 |
| 35 | Elise Hansen | Prenatal reassurance and empowerment | 279 |

| No. | Name | Features | Page |
|-----|------|----------|------|
| 36 | 'Sarah' | Pregnant woman unaware of high risk | 280 |
| 37 | Pat T from the UK | Poor communication and its emotional effect | 281 |
| 38 | Karen Low | Disbelief resulting in unassisted birth | 288 |
| 39 | Jennifer Jacoby | Appreciating relationships with caregivers | 300 |
| 40 | Anonymous | Woman motivated to have a healthy body | 309 |
| 41 | Sarah Cave | Woman motivated to help her baby in labor | 312 |
| 42 | Marie-Claude | A real contrast postpartum after a VBAC | 316 |
| 43 | Céline | A feeling of empowerment after a VBAC | 316 |
| 44 | Anonymous | A failed VBAC, despite excellent support | 317 |
| 45 | Michel Odent | A successful VBAC | 319 |
| 46 | Nina Klose | A woman's perspective of a successful VBAC | 319 |
| 47 | Nina Klose | Woman reflecting on the ultrasounds she had | 338 |
| 48 | Esther Culpin | Vaginal breech in optimal conditions | 342 |
| 49 | Liz Woolley | Hospital breech birth with diffident mother | 346 |
| 50 | Anonymous | Primipara unaware of high risk | 348 |
| 51 | Anonymous | Woman under pressure to accept drugs | 349 |
| 52 | Anonymous | Induction and intervention without permission | 351 |
| 53 | Anonymous | Lack of privacy with close monitoring | 353 |
| 54 | Anonymous | A triplet birth with poor relationships | 356 |
| 55 | Mave Denyer | A vaginal triplet birth with minimal intervention | 358 |
| 56 | Janet Hanton | Another vaginal triplet birth, with ultrasounds | 364 |
| 57 | Gaia Pollini | A twin mother avoiding other people's fears | 368 |
| 58 | Nicolette Lawson | Using hypnotherapy to prepare for birth | 379 |
| 59 | Beth Dubois | Birth and breastfeeding after sexual abuse | 384 |
| 60 | Sylvie Donna | Motivation for writing—feeling protective | 396 |
| 61 | Beth Dubois | Struggling beyond midwife's help to strength | 401 |
| 62 | Helen Arundell | Pregnancy diary of daily niggles and 'treatment' | 408 |
| 63 | Anonymous | Secret plan to give birth unassisted | 411 |
| 64 | Laura Shanley | Giving birth unassisted by choice | 415 |
| 65 | Justine Rowan | Freebirth because of legal restrictions | 417 |
| 66 | Katya Korochantseva | Freebirth because midwives were unavailable | 421 |
| 67 | Katya S from Moscow | Freebirth because of uncaring midwives | 423 |
| 68 | Sonia Winter | Midwife commenting on illegal homebirth | 425 |
| 69 | Rebecca Wright | Midwife inspiring a belief in ability to give birth | 430 |
| 70 | Justine Rowan | Deciding on homebirth with self-awareness | 438 |
| 71 | Clare O'Ryan | Lack of continuity of care with difficult results | 440 |
| 72 | Joanne Whistler | Reflecting on optimal places to give birth | 442 |
| 73 | Jenny Sanderson | Non-disturbance in practice with Michel Odent | 448 |

| No. | Name | Features | Page |
|---|---|---|---|
| 74 | Clare Winter | Midwife's records with undisturbed birth | 452 |
| 75 | Sylvie Donna | Commentary on neglected birth | 456 |
| 76 | Anonymous | A cascade of interventions, after ARM | 458 |
| 77 | Michel Odent | The birthing pool test in practice | 461 |
| 78 | Sarah Buckley | Understanding undisturbed birth | 464 |
| 79 | Janet Balaskas | Discovering the value of upright positions | 466 |
| 80 | Kathy Kleere | An empowering water birth | 474 |
| 81 | Angela Horn | Water birth in practice | 476 |
| 82 | Michel Odent | What privacy means in practice... | 484 |
| 83 | Sylvie Donna | Stop-start labor with psychological issues | 485 |
| 84 | Elaine Batchelor | Slow progress, without adverse outcomes | 487 |
| 85 | Nuala OSullivan | Physiological birth with intervention at birth | 489 |
| 86 | Sarah-Jane Forder | Doula facilitates a smooth hospital birth | 494 |
| 87 | Natalie Meddings | Doula helps a woman 'go to another planet' | 495 |
| 88 | Phil Anderton | A man's experience of labor and birth | 500 |
| 89 | David Newbound | A father acting as 'guard' and 'defender' | 506 |
| 90 | Alan Low | Father as supporter, child being inspired | 507 |
| 91 | Cara Low | Child attending birth and getting inspired | 510 |
| 92 | Ashley Marshall | Lotus birth (i.e. no cord-cutting postpartum) | 514 |
| 93 | Ruth Clark | Unexpected events with good outcomes | 516 |
| 94 | Sylvie Donna | It all 'went wrong' but it was also optimal | 519 |
| 95 | Krisanne Collard | Premature birth and kangaroo mother care | 525 |
| 96 | Monica Reid | An unexplained stillbirth | 539 |
| 97 | Nina Klose | Reflections on birth, death and loss | 549 |
| 98 | Beth Dubois | Reflecting on what optimality really is | 576 |
| 99 | Heba Zaphiriou-Zarifi | Reflecting on how birth transforms a woman | 578 |
| 100 | Sarah Hobart | Adjusting to life after giving birth | 589 |

If you would like to contribute other birthframes, comments or research for possible inclusion in a future edition of this book, or other books, please do not hesitate to contact the publisher by emailing info@freshheartpublishing.co.uk. If any of your clients are interested in contributing their own experiences, please reassure them that they will have the opportunity to check how their accounts will be presented before publication and of course they would be able to choose to be named or anonymous. Even though it's vital that practices are research-based, we also need to recognize that certain areas can never be researched with randomized controlled trials (RCTs) or other types of research. For this reason, we need to acknowledge the value of first-hand accounts, which can give us enormous insight, not to mention inspiration and ideas. So email about your own experiences of attending women in labor... or even of giving birth yourself!

# Reflection index

As you will realize, the reflection questions ('Pause to reflect', etc) are not designed to be considered alone. They are positioned within the text, between birthframes and discussions about research and issues surrounding birth so as to stimulate further thought on key topics. However, this index might help you to negotiate your way around!

| Topic | Page |
|---|---|
| Conception and prenatal care | 19 |
| Labor itself and intrapartum care | 29 |
| Physiological birth and ways of facilitating it | 36 |
| Key issues in intrapartum care | 42 |
| Women's hope, expectations and realities | 67 |
| Feelings, anxiety and communication | 68 |
| Disturbing and upsetting women in labor | 73 |
| Choice and control (of parents and caregivers) | 90 |
| The history of midwifery and obstetrics | 151 |
| Evidence-based practice, changing protocols and disturbance | 165 |
| Pain, pain relief and the side– and/or after-effects of drugs used | 170 |
| Fast and/or painfree births | 191 |
| Perineal damage and sexual problems | 212 |
| Complaints about ineffective analgesia or anesthesia | 214 |
| Feelings around the time of a birth (your own and the mother's) | 226 |
| Optimality and how to justify it and explore it | 230 |
| Training, experience, research, the law, power, protocols, | 263 |
| personalities, optimal care at all stages of the process | -236 |
| The need (or not) to prepare the woman for birth | 272 |
| Communication amongst caregivers and with the mom-to-be | 300 |
| Information (giving it or not) and respect (for the woman's choices) | 305 |
| Induction and postmaturity | 306 |
| Offering pain relief or not | 312 |
| Perceptions of births | 316 |
| Vaginal birth after cesarean (VBAC) | 327 |
| Ultrasound | 333 |
| Testing for anemia and suspected iron deficiency | 340 |
| Information, decision-making and outcomes | 370 |

| Topic | Page |
|---|---|
| Mortality statistics and intervention | 377 |
| Self-disclosure in prenatal appointments (abuse, bad experiences) | 384 |
| Motivations prompting choices and action | 400 |
| Struggles (psychological or perhaps spiritual) and control | 401 |
| Your relationships with moms-to-be | 407 |
| Freebirth and control issues | 416 |
| Preparing birth plans / care guides with women and applying them | 429 |
| Time and the timing of support in labor | 432 |
| Communicating with women in labor (or not) and giving support | 455 |
| Conflict and/or anxiety prenatally and intrapartum | 486 |
| Doulas working prenatally and intrapartum | 493 |
| Eating and drinking in labor | 497 |
| Cord-cutting, lotus birth (or not) and dealing with the placenta | 515 |
| Miscarriage, stillbirth and neonatal death | 548 |
| Establishing systems for ongoing optimality | 575 |
| Dealing with clients prenatally and assessing their needs | 577 |
| Maternal evaluation and transformation | 581 |
| Evaluating research and achieving change | 588 |
| The possible multi-dimensionality of birth | 595 |
| Discussing hopes and fears about birth with moms-to-be | 632 |

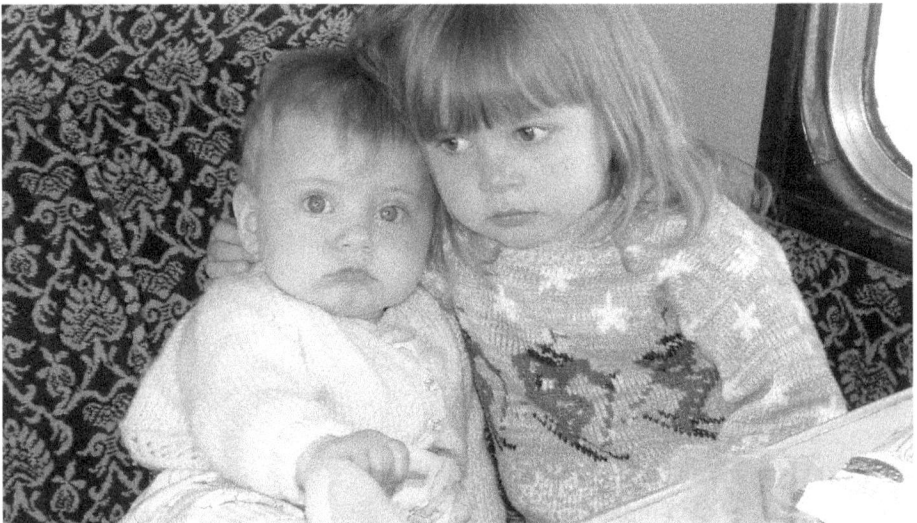

*I wonder what they think?*

*What experience of birth do people walk away with? What kind of parents can they be?*

# General index

This index has been provided primarily to help students with assignments on particular topics. For example, under the entries for 'control' and 'monitoring' you will find all cases where these topics feature in this book. Birthframes are mentioned so that it is possible for you to quickly see if relevant case studies are available relating to a given topic. (If mentions of a particular topic within a birthframe are only minor, however, the birthframe is not listed separately. This explains why there are sometimes double listings.) In other cases there may be double entries, particularly for notes and references when these are re-listed for the sake of easy access. Even minor mentions of topics are included since these may be relevant or thought-provoking when considering specific issues but bold numbers indicate more significant entries.

active birth 51, 63, 119-120, 138, 276, 439, 448, **466-468**, 474, 476, 479, 638
acupuncture 12, 158, 171, 183, **185-6, 251-2**, 300, 345
adrenaline 14, 31, 35, 148, 205, 278, 279, 282, 298, 332, 499, 505, 523, 524
Africa 154, 198, **210, 372-375**, 435, 539
after-effects—see 'side effects'
afterpains 44, 45, 60, 488, 521
alertness **7-9**, 12, 22, 34, 35, 40, 41-42, **44-45**, 51, 55-57, 59, 100, 115-117, **149**, 165, **171**, 182, 193, 208, 223, **228**, 234, 247, 250, 254, 291, 311, 378, 420, 433, 480, 507, 514, 544, **570**, 576
Alexander Technique 641
amniocentesis 154, 287, **328, 329, 552**
amniotic fluid 22, 100, 148, 203, 204, 209, 270, 329, 333, 346, 403, 476, **604, 607**, 608, **611**, 615, 616, **618-9, 625**, 626, **627**
amniotomy 156, 241, **459**
amphetamines 177, 195
analgesia 3, **7-9, 14**, 35, 50, 100, 108, **111-2, 113-114, 115-7, 120-1**, 158, 170, **171, 177, 178-181**, 182, 209, **243, 244, 245, 246, 249, 250, 260, 261**, 274, 497, **558-9, 562, 566, 567, 568**—also see specific drugs by name
anemia 22, 164, 204, **340**
anesthesia 3, 4, 9, 14, 35, **99, 100, 102**, 108, 110, 111-2, 114, 116, 121, **150-153**, 155, 156, 158, 159, 170, **171, 182, 193, 195**, 209, **214-216, 229**, 274, **314**, 318, 423, **459, 497, 583**—also see 'epidurals' and 'spinals'
anorexia 142, **195**
anterior position 26, **244**, 440, 488, **620-1**, 626
Apgar score 44, **121, 127-8**, 317, 410, **567-8**
ARM—see 'amniotomy'

arnica 118, 181, **183**, 500

aromatherapy 158, 183, **184**, **251**

asthma 136, **194**, 256

augmentation of labor 11, 12, **23**, **89**, **97**, **152-3**, **155**, **156**, 162, **173**, 203, **206**, 209, **237-8**, 241, 251, **297**, 301, 308, 351-2, 398, 485, **558-9**, **564**, **566-7**

auscultation **134-5**, 258, 259, **298**, **333**, **553-4**—also see 'fetoscope' and 'Doppler'

author's pregnancies and births—see Birthframes 2, 3, 4, 29, 60, 75 and 94

autism 142, **195**, 256-7

baby carriers 611, **641**

backache labor—see 'posterior position'

Balaskas, Janet 80, 282, 284, **466-468**, 636, 663

barbiturates **171**, 177, **195**

baths 32, 58, 105, **150**, **157**, 168, 169, **184**, 189, **192**, 205, 281, 287, 322, 323, 349, 358, 392, 409, 412, 415, 418, 420, 421, 422, 430, 449, 450, 452, 453, **456**, 463, **470**, **472**, **473**, **483**, **484**, 488, 507, **512**, 519, 520, 528, 540, **544**, 579, **612**—also see 'waterbirth'

birth 22-29, **30-43**—also see all birthframes

birthframes xiii—also see Birthframes index on pages 642-644

birthing pool—see 'waterbirth'

birthing pool test 472, 484

birth plans—called 'care guides' in this book—50, 51, 58, 88, 107, 168, 286, **297**, **312**, **349-351**, **407**, **427-429**, 495, **502**, 506, **572**, **632**, **663**

birth stories—see Birthframes index on pages 642-644

bleeding 45, 74, 76, **121-123**, 186, 189, **247-248**, 280, 281, **306**, 348-355, 368, 409, 421, 482, 509, **524**, 525, 531, **631**

blood pressure 7, **12**, 55, 71, 77, 89, 98, 100, 101, 103, 115, **124**, **172**, **177**, **182**, 220, 222, 224, **270**, 276, **332**, **341**, 380, 478, 482, **553**, **612**

bonding **8-9**, **12**, 41, 44, 83, 93, **96**, **97**, **115-7**, **149**, **165**, 171, **192**, **193**, 218, **228**, 233, **274**, 356-8, **446**, **458-9**, 514, **537**, 545, 549, **551**, **571**, **574**, 589, **598**, 631, **661**—also see Birthframes 2, 17, 32, 54, 57, 59, 76, 90, 92, 95, 96, 97 and 100

Braxton Hicks contractions 168, 287, 301, 408, 410, 442, 476, 477, 517, **611**, **619**

breastfeeding **6-9**, **12**, 31, 33, **35**, **40**, **41**, **44-5**, 55, **56**, 57, 60, 65, 69, 71, 83, 101, 106, 107, 108, **113-4**, 115-6, **123-4**, 125-6, 127, 136, 139, 140, 145, **147**, **148**, **149**, **159**, 162, 166, 168, **171**, **172**, **175**, **176**, **177**, **182**, **193**, **198**, **200**, 204, 205, 208, 218, 220, 223, 226, **228**, 232-3, 242, **245**, **246**, **247-8**, 255, 276, 291, **312**, 316, 324, 325, **326**, 349, 351, 354, 357, 367, 369, **381**, 384-394, 411, 412, 419, 422, 444, 453, 460, 481, 485, 495, 518, 519, 521, **530**, **537**, 539, 545, **551**, **570**, **571**, 579, **582**, **589-591**, 590, 591, 625, **627**, **628**, **631**, **637**, **638-9**, 659, **661**

breathing difficulties **34**, 41, **44**, 75, 99, **100**, 101, 159, 161, **173**, **176**, **177**, **182**, **192**, **193**, **195**, **216**, **256**, 291, 324, **329**, 356, 357, 367, **489**, **514**, 520, **527**, **530**, **534**, 543, **579**, 603, 604, 606, **608**, 611, 615, 617, 618, **625**, 626
breathing in labor xii, **55**, 81, 82, **121-3**, 129-131, **158**, 181, 189, 190, 222, **274**, 287, 291, **319**, 321, 322, 323, 349, 366, **378**, 388, 403, 412, 417, 418, 419, **449**, **465**, **467**, 474, 476, 478, 479, 500, 517, 579
breech position 30, 62, 129, 139-140, 180, 181, 185, 235, 241, 268, 285, 286, **315**, **335**, **342**, 369, 411, 454, 458, 500, 516, 526, 540, 558, 576, **582**, **583**, **597**, **622**, **626**—also see Birthframes 12, 3549, 49 and 64
camcorders 446, 457, 463
cameras **24**, **446**, **461**, 463—also see Birthframe 77
care guides **297**, **427-429**, 502, **572**, **632**, 663—also see 'birth plans'
cascade of interventions **35**, 73, **126**, **164**, 202, **206**, **240**, 372, 380-1, **416**, **457**, **458-9**, 576—also see Birthframes 27, 28, 32, 37, 60 and 76
catching the baby 31, 33, 34, **40**, 55, 60, 63, 65, 280, 350, 397, 415, 425, 431, 432, 446, 450, **463**, 475, 492, 503, 508, **512**, 520
cesareans xiii, **8**, 11, 13, **35**, 50, 51, 54, 65, 73, 79, 82, **86**, 92, **96-110**, **116**, 125, 126, **127-8**, 129, 134-141, **149**, **153**, **154**, **155**, **156**, **158**, **160**, **163**, **164**, 171, 172, **174**, **181**, 182, **191**, **194**, **195**, 198, **202**, 203, **206**, **208**, **214**, 215, 220, 232, 234, 236, 238, 240, 242, 243, 244, 245, 249, 250, 253, 255, 256, 259, 268, 274, 278, 279, 281, **285**, 287, **314-327**, 333, 338, **342**, **344**, **345**, 346, 353, 356, 358, 359, 360, 364, 365, 366, 368, **372**, **381**, **416**, 438, 458, 459, 467, **472**, **484**, 500, 501, 503, 504, 526, 532, 536, **551**, 558, 567, 573, 576, **577**, **582**, **583**, **597**, **620**, **636**, **637**, **638**, **641**, 659, 660, **661**, 663—also see Birthframes 8, 12, 16, 17, 18, 19, 26, 27, 28, 30, 34, 36, 37, 54 and 82
Chamberlain, David **192**, 560, 565, 597, 636
chorionic villus sampling 154, 328
commanded pushing 50, **121**, 129-131, 162, **174**, **250**—also see 'pushing'
communication—see 'talking in pregnancy', 'talking in pregnancy and birth' and 'talking postpartum'
complementary therapies **12**, **14**, 158, 171, 180, **183-6**, **251**, **252-3**, 408, 410, 559—also see Birthframe 62
conception **14-19**, 58, 63, 64, 90, **118**, 160, 170, 194, **234**, 277, **306**, 320, 330, 338, 339, 368, 369, 375, 402, 430, 485, 525, 545, 550-1, 595, **598-600**, 627, **628**, **630**, **631**, 632, **639**, 658, 659—also see Birthframe 1
continuity of care 60, **126**, **133**, **134**, **142**, **163**, **164**, **237**, **239**, **242-3**, **314**, 320, 430, **440-1**, **551**, **561**, **567**, **572**, **581**, 587, **596**
continuous monitoring—see 'EFM' and 'monitoring'

contractions xii, **11, 14, 23-4**, 25, 26, 27, **30**, 31, 32, 33, 53, 54, 58, 64, 70,
    71, 72, 73, 74, 80, 81, 82, 84, **89, 148, 149, 150, 152, 156, 157**,
    167, 168, 169, **173, 174**, 175, **178**, 180, 181, **185**, 186, 188, 189,
    190, **193**, 203, 205, 206, 215, 219, 221, 222, 223, **241**, 280, 287,
    289, 290, 291, **296, 298**, 300, 301, 302, 310, 312, 317, 318, 319,
    320, 321, 322, 323, 325, 343, 346, 347, 348, 350, 352, **353**, 354,
    359, 365, 366, 380, 381, 385, 388, 389, 390, 402, 403, 404, 411,
    412, **415**, 416, 417, 418, 419, 421, 422, 423, 424, 430, 431, 432,
    440, 441, 442, 443, 447, 449, 450, 452, 453, **454, 458, 459, 460,
    467**, 470, **472, 473**, 474, 476, 477, 478, 479, 480, 481, 485, 487,
    488, 489, 490, 491, 492, 494, 496, 500, **505**, 507, 508, 517, 518,
    520, **523-4**, 531, 532, 578, 579, 611, 619, **620, 626**, 629—also see all
    other birthframes
control ii, viii, 8, **9**, 25, 31, **34, 36, 41, 42**, 55, 60, **62**, 63, 76, 79, 81, **90, 102,
    105**, 106, **111, 112, 115-7**, 120, **121-3, 138, 147, 152, 154, 156,
    157**, 169, **170, 172, 175, 176, 182**, 188, **200, 209**, 211, **218**, 224,
    **228**, 239, 245, 249, **254**, 260, 261, **265**, 269, **271**, 284, 288, 300,
    318, 323, 352, 357, 381, 388, 393, **401**, 412, **416**, 438, 439, 441,
    **446-8**, 453, 467, 480, 482, 508, 509, 524, 542, **560**, 563, **565, 566-
    7, 576, 579**, 611, 618—also see Birthframes 1, 2, 3, 4, 6, 14, 15, 16,
    17, 18, 19, 25, 27, 28, 29, 30, 32, 33, 34, 37, 39, 40, 41, 42, 43, 44,
    47, 48, 52, 53, 56, 57, 58, 59, 61, 62, 63, 64, 65, 66, 67, 69, 70, 71,
    75, 76, 77, 78, 79, 80, 81, 83, 85, 86, 87, 88, 89, 90, 92, 94, 97, 98,
    99, 100
cord **9, 41, 42**, 45, **54**, 65, 84, **116**, 181, 190, **255-6**, 305, 324, 343, 344,
    **345**, 405, 412, 421, 444, 453, 460, 474, 481, 488, 492, 518, 520,
    **524, 600, 604, 607, 622**
cord clamping—see 'cutting the cord'
cord prolapse 84, 181, 344
cutting the cord **35, 42, 43, 45**, 59, 65, **89**, 94, **100, 125-6**, 166, **173**, 193,
    223, 291, 292, 343-4, 391, 412, 420, 421, 432, 444, 453, 481, **514-
    5**, 518, **524, 569**
delivery rooms 3, **51**, 87, **88**, 103, **153**, 187, **225, 248**, 311, 312, 359, 441,
    **445, 495**, 504, 536
Demerol 8, 11, 12, 50, **113-4**, 116, **153**, 168, **171, 176-7, 182**, 187, **246,
    248, 251**, 350, 359, 478
depression **7-9, 71, 97, 115-7, 199-209, 210**, 216, 217, 229, 243, 357, **378,
    380, 506, 570, 573, 579, 639**—also see 'postpartum scenarios'
developing countries 113, 123, **196, 198, 228**, 332, 340, **372, 375, 377**
diagnosis of labor 23, 53, 141-2, 562-3—also see 'onset of labor' and 'false
    labor'
diamorphine 8, **12, 113-4**, 116, **171, 176-7**, 178, **182**, 251, **297**
disturbance xii, 3, 7, **11-2**, 14, 24, 25, 26, 31, 32, 34, **35, 36**, 38, 40, 41, 44,
    49, 51, 52-6, 58-60, 63, 65, 66, 68, **73**, 165, 84, **101, 108-110**, 113,
    **117-8, 123-5**, 132, 134, **146-9**, 150, 156, 158, 159, 160, **162-9, 170**,
    173, 182, 183, 184, **186, 196, 198**, 208, 209, 218, **228-9**, 233, 240,

245, 246, **252**, **261**, **263**, **264**, 274, 278, **297**, **308-9**, 314, **319**, 325, 333, **342-4**, 350, 352, 353, 375, 378, **413**, 414-5, 417, **440**, **446-466**, **468**, 488, 489, 490, **499**, **513**, 520, 521, **523-4**, 536, 542, **552**, **566**, 569, **573**, **574**, 576, **577**, 579, **581**, **583**, 593, 603, **628**, **630**, **631**, 658, 659—also see 'going to another planet' and 'watchful waiting', and Birthframes 2, 3, 4, 5, 7, 11, 14, 21, 22, 28, 31, 32, 34, 37, 48, 54, 56, 57, 59, 61, 62, 64, 69, 70, 71, 73, 74, 75, 76, 77, 78, 79, 81, 82, 83, 84, 85, 86, 87, 89, 92, 93, 94 and 99.

Donna, Sylvie—see 'author's pregnancies and births'

Doppler xii, **28**, **42**, 70, **87-9**, **127-8**, **154**, **294**, **333**, **334**, **335**, 349, 449, 462, 474, 540, **608**—also see 'ultrasound'

doulas x, **xii**, 2, 166, **243**, 253, **265**, 275, 280, 316, 319, **413**, 427, **428**, 484, 487, 488, **493-7**, 514, 524, **567-8**, 572, **639**, 660

Down's syndrome 329, **330-2**, **336**, 339, 552, **639**

drug addiction **9**, **115-7**, **176**, **177**, **178**, **182**, **195**, 246, 247, 248, 257

drugs—see names of individual drugs or related procedures (e.g. Demerol, epidurals, ergometrine, etc)

due dates 22, 27, 50, 51, 64, 84, 92, 93, 168, 180, 187, 188, 209, 274, **306-7**, 310, 320, 330, 338, 343, 346, 348, 365, 368, 369, 379, 380, 386, 402, 409, 410, 411, 421, 422, 430, 440, 442, 448, 449, 450, 461, 474, 476, 487, 489, 494, 496, 517, 525, 529, 541, 589, **599**, **618**, **626**, **628-632**—also see 'ultrasound'

dystocia—see 'failure to progress' or 'shoulder dystocia'

eclampsia 330, 332, **553**

ectopic pregnancy 74, 97, 139, **640**

EFM (electronic fetal monitoring) **6**, 12, **50**, **86**, 88, 108, **127-8**, **132**, **134-5**, **143**, **153**, **157**, **158**, **162**, **163**, **164**, 182, **206**, 238, 240, 243, 249, 258, 259, **333**, **351**, **459**, 554—also see 'monitoring', 'fetal scalp monitoring' and 'ultrasound'

elderly primagravidae 49, 73, **308-313**, 332—also see Birthframes 2, 7, 40 and 41

elective cesarean 75, 84, **96-110**, **136**, **140**, **141**, 285, **314**, 342, **416**, 500, **551**, **582**—also see 'cesareans'

emergency cesareans 11, 50, 74, 102, 106, 136, 139, **171**, 281, 287, **314**, 315, 319, 333, 368—also see 'cesareans' and Birthframes 8, 12, 16, 17, 18, 19, 27, 28, 30, 34, 36, 37, 42, 43, 44, 45, 46, 57 and 95

emotions **2**, **7**, **14**, **19**, 20, **36**, 46, 71, 90, **102**, 104, 108, **117**, 138, **148**, **164**, 168, **192**, 193, **194**, **199-227**, 232, **240**, 269, **270**, **272**, 278, 281, 282, 285, **295**, **298**, 300, **308**, 318, **332**, 346, 348, 351, 353, 356, 358, 367, 368, 384, 391, 392, 393, 394, 397, **400**, 401, 403, 408, 424, **438**, 442, **446-8**, **465**, 477, **485**, 492, 494, **502**, 503, **505-6**, **515**, **523**, 531, 535, 537, **548**, **566**, 571, **573**, **574**, 576, 577, **582**, 596, 604, 605, 609, 615, 623, 625, 627, 628, **638**, **661**—also see Birthframes 1. 2. 5. 16. 17. 20, 21, 27, 28, 30, 32, 33, 34, 37, 39, 42,

43, 44, 52, 57, 58, 59, 61, 64, 65, 67, 78, 83, 86, 87, 88, 89, 90, 94, 95, 97, 98, 99, 100

endorphins 2, 12, 88, **117-8**, **138**, **141**, **147**, **149**, **160**, **173**, 179, 182, **200**, 228, **247**, **256**, 348, **551**, **602-3**, **605**

enemas 32, 81, **150**, **156**, 421, **455**, 463

Entonox 11, 12, 51, 63, **114**, **171**, **177-181**, **182**, 187, **195**, 220, **247**, **248**, 290, 291, 305, 309, 312, 349, 350, 352, 359, 366, 441, 443, **444**, **542**

epidurals **8-9**, **11-12**, 50, 51, **74**, 79, 82-3, **88**, **95**, 98, 100-1, 103, **111-2**, **113-4**, **115-7**, **119**, 122, 131, 142-3, **153**, **156**, **159**, **160**, 166, 168, **171**, **172-6**, **177**, **182**, 203, **233-4**, **243-5**, **249-250**, **255-6**, **257-8**, **260**, **261**, 305, 317, 320, 324, 338, 347, 356, 380, 404, 441, **457**, **459**, 467, 488, **505-6**, **542**, **565**, **576**, **661**—also see 'anesthesia'

episiotomy **50**, **120-1**, **128-9**, **138**, **152**, **153**, **154**, 156, 158, 162, **163**, 165, 166, **174**, **235-6**, 241-2, **261**, 302, 305, 319, 380, 423, 439, 458, 459, **568-9**

ergometrine **89**, 114, 124-5, **152**

eye contact **8**. 31, **33**, 46, 55, 93, **115-7**, **149**, **158**, **167**, **193**, 203, 204, 206, 223, 257, **276**, **294**, 301, **316**, 317, 323, 324, 348, 352, 357, 359, 365, 366, **372**, 419, 420, 470, 474, 480, 490, 495, **505**, 509, 518, 519, **523**, 525, 526, 527, 529, 543, 549

face-to-pubis births 49, 489

failure to progress 11, **108-110**, **142**, **244**, 314, **484-493**, 514, 566-7

false labor **22-3**, 25, 26—also see 'failure to progress'

false negatives and positives 135, **330**, 334, 335, 336, 338

female circumcision **154**, 198, **372**, **376**

fetal distress 9, 73, **104**, **108-110**, 116, **128-9**, 160, **172**, **173**, **174**, **235-6**, **237-8**, **241-2**, 257, **329**, **333**, 405, **430**, **459**, **469**, 472, **473**, 484, 501, **604**, **627**

fetal monitoring—see 'fetal scalp monitoring', 'Doppler' and 'fetoscope'

fetal scalp monitoring 109, 158, 172, 351, 352, 454

fetal stethoscope—see 'fetoscope'

fetoscope **42**, 54, **87-8**, 127-8, 269, **294**, **333**, **462**, 608

fetus ejection reflex **31**, **38**, 52, 55, 57, **59**, 60, 112, **121**, 143, 148, 165, 173, 179, 319, 472, 505, **523**, 577, 658—also see Birthframes 3, 45 and 98

first stage **22-29**, 27, 121, 129, 241, 246, **344**, 452, **472**, 479, 566

first trimester **20**, 75, **118**, 183, **234**, **270**, 402, **550**, **596-7**, **598-605**, 606

fistula 198, **375**

folic acid 118, **183**, **270**, 310, 599, **600**, **601**

footling breech 84, 279, 344, 415, 516—also see Birthframes 12, 64 and 93

forceps 6, 11, 63, 82, 99, **110**, 120, **121**, **127-8**, 130, **136-8**, 145, **150**, **152**, **154**, 155, 156, 158, 162, **163**, 165, 166, **174**, **193**, **194-5**, **208**, 234-

5, 249, **260-1**, **305**, 380, 439, 458, 485, **565**, **568-9**—also see
Birthframe 21
freebirth 413-425, 427—also see 'unassisted birth'
Friedman curve 23, 141
gas and air—see 'Entonox'
Gaskin, Ina May **111**, 119, 188, 189, **240**, 485
gestational diabetes 162, **164**, 270, **341**, 558
glucose 147, **156**, 341, 456, **497**, 558, 568, 607
going to another planet 73, **173**, **182**, 184, **185**, 218, **447**, **463**, **505**, 512,
524—also see Birthframes 3, 7, 21, 24, 25, 31, 32, 69, 72, 73, 74, 78,
86, 87, 90, 92 and 99
hemoglobin levels 22, 125-6, 304, 411, 557
hemorrhage—see 'PPH'
herbs 32, **118**, **171**, **183-4**, 310, 345, 421, 500, 578, **640**
high risk pregnancies and labors 3, **68**, 74, **87**, 97, 128, 134-5, 235, 237,
258, 274, **280**, **308-327**, **342-377**, **461-3**, 536, 553, 554, 556, 663
home birth 14, 41, 51, **52-6**, **58-60**, **63-5**, **73**, **80-3**, **84-5**, **87-9**, **90-5**, **104**,
125, **126**, 134, **142**, **145-6**, **152**, **165**, **166**, **167-9**, **177**, **178**, 186,
187, 188, 189, 191, **203**, 204, 205, 220, 221, 224, 225, 226, 231,
**237**, **242-3**, **268-9**, 275, 280, 283, 285, 286, 288-292, 300, **315**, 317,
318, 319, 342-4, 348, 349, 359, 364, **366**, **372**, **376-7**, 380, 384,
385, 386, 388, **390**, **402**, 403, 405, 411-2, 414, **417-420**, 421-2, 423-
5, **425**, **428**, 430-2, **434-445**, **448-451**, **460**, 461, **464-6**, 475, **476-
483**, **487-9**, **489-492**, 495, **495-7**, 500, **504-6**, **506**, 507-9, 510, **514-
5**, **516-9**, 519-521, **523**, **559-560**, **561-2**, **572**, **576**, **578-580**, **626**,
**632**, **637**, **638**, **639**, **659**, 660—also see Birthframes 3, 4, 5, 7, 11, 12,
14, 15, 20, 21, 22, 24, 25, 26, 28, 32, 28, 29, 44, 45, 48, 58, 61, 63,
65, 66, 67, 69, 70, 71, 72, 73, 77, 78, 81, 84, 85, 87, 90, 91, 92, 93,
94, 99
homeopathy **12**, **118**, **171**, **183**, 408-410, 422, 490, 500, 578, **640**
hormones **14**, **20**, **22**, **24**, **34**, **35**, **40**, **44**, **55**, **56**, **83**, **96**, **108**, **109**, 113, **117**,
**123**, **126**, **132**, **134**, 136, **138**, **140**, **147-9**, **159**, **160**, **164**, 174, **177**,
**196**, **200**, 205, **208**, **209**, 228, 232, 233, **240**, 246, **247**, **270**, 278,
279, **298**, **308**, 324, 343, 351, 386, 391, **398**, 444, **447**, **457**, **465**,
466, **468**, 477, 496, **499**, **505**, **523**, **534**, **564**, 566, **573**, **574**, **597**,
600, **609**, 612, **628**, **629**
hospital birth viii, 15, **25**, 27, 28, **36**, **49-51**, 58, **62**, **63-4**, 71, **73**, 74-5, 75-8,
79, 80-3, 84-5, **87**, 88, 90, **93**, 94, **95**, **102**, 104, **106-8**, **121-3**, **126**,
**127-8**, 130, **131**, **133**, **134**, 140, **143**, **150-161**, 162, **163**, **165**, **166**,
167-9, 176, 179, 180-1, 186, 187-8, **188-190**, **191**, 195, **196**, 198,
**200**, **203-6**, **207**, 209, 215-6, 219, **220-5**, **226**, **228**, 229, 237, **240**,
242-3, 253, **254**, 258, **268-9**, **272**, **275**, 280-1, 283, 285, 286, 288,
**289**, 296, **300**, 301, **302**, **305**, 309-311, 312-3, **314-5**, **316**, 317-8,
**319**, **320**, 324, 328, 332, **338**, 341, 342, 343, **344**, 346-8, 348-9,

349-351, 351-3, 353-5, 356-8, 358-363, 364-7, **377**, **379**, **380-1**, 384-394, 410, 411, **416**, **417-420**, 421-2, 423, **425**, **434**, **437**, 438-9, 440-1, **442**, **448**, **457**, **460**, **464**, **467**, 471, 474, 482, 483, **484**, 488, 489, 494-5, **497**, 500, **504-5**, 506, **523**, 525-532, 534, **536**, 539-547, **559-560**, **561-2**, **562-3**, 571, **572**, **576**, 578, **597**, **632**, **659**, **660**, **662**, 663—also see Birthframes 2, 5, 7, 8, 9, 10, 11, 12, 13, 14, 15, 15, 19, 21, 22, 23, 24, 25, 28, 30, 32, 36, 37, 39, 40, 41, 42, 44, 45, 49, 50, 51, 52, 53, 54, 55, 56, 59, 62, 65, 67, 68, 70, 71, 73, 80, 82, 86, 88, 89, 95, 96

hyperemesis 75-78

hypertension—see 'blood pressure'

hypnosis 73, 169, 192, 288, **379-382**, 443, 490, **558-9**, 614, 628

incontinence 105, **121-3**, **130-1**, **136-8**, **173**, **175**, 209, **210**, **211**, **212**, **234-5**, **249**, **260-1**, 375

induction **12**, 23, 27, 51, 65, 78, **97**, **114**, **150**, **152**, **153**, 155, **156**, 161, 162, **163**, 168, 184, 186, **193**, **195**, **206**, 209, 219, 220 238, **280-1**, 287, 288, **289**, **301**, **306**, **308**, **315**, **320**, 330, **333**, 338, **350-1**, **351-3**, **368**, 380, 385, 386, 351, 306, 410, 439, 442, 456, **457**, 467, 494, 501, 543, **629**, 631, **661**

infection 46, **96**, **98**, **121**, **129-130**, **147**, **159**, 173, 174, 175, **182**, 203, **205**, **206**, **209**, **225**, 234, 285, **289**, **314**, 380, **434**, **459**, 489, 526, 531, **534**, 535, 545, 550-1, 561, **605**, **611**, **622**

infertility 206, 259, 358

infibulation 154, 372, 376

internal examinations—see 'vaginal examinations'

intervention ii, **xiii**, **3**, **4**, **6**, **11**, **12**, **13**, **14**, **22**, **25**, **30**, **31**, **35**, **41**, **42**, **44**, 46, **49-51**, **52**, **56**, **58**, **62**, 63-5, **67**, **68**, **70-73**, 73, **74-85**, **86-110**, **118**, 119, **121**, **124**, **126**, 128, 129, **131**, **133-4**, 138, 141, **142-3**, 146, **150-161**, **162-169**, **170-186**, 188, **193-8**, **199-209**, **210-217**, **218-227**, **228-230**, 235, **236**, 237, 239, **240**, 242, 243, **253-4**, **254**, **257-9**, **260**, **264**, **270-1**, 274, 283, 286, 289, **297**, **308-9**, 309, 313, **314**, 320, **327**, **332**, 347, 348, **350**, **351**, **356**, 364, **371**, **372**, **377**, **413**, **416**, 425, 438, 439, 441, 443, **446-8**, 456, **457**, 458-9, 460, 464-6, 467, **468**, 470, 482, **489-493**, 494, 495, 500, 501, **521**, **534**, **548**, 552, **554**, **559-560**, **561**, **564**, 570, **573**, 576, **577**, **581**, 596, **631**, 660, **661**, **663**—also see Birthframes 2, 3, 4, 5, 6, 8, 9, 10, 11, 12, 16, 17, 18, 19, 20, 21, 22, 23, 27, 28, 29, 30, 31, 32, 34, 36, 37, 42, 43, 44, 45, 48, 50, 51, 52, 53, 54, 62, 68, 70, 71, 75, 76, 77, 78, 83, 84, 85, 92, 95, 96, 97

invasive tests —also see 'prenatal tests'

iron supplements 340, 341

kangaroo care 159, 353, **525-235**, **570**, 615

Kitzinger, Sheila 46, 60, 120, **132**, 150, 203, 282, 284, 299, 494, 559-560, 562, 596, 597, 636, **663**

labor **22-29**—also see all birthframes
Lammers, Liliana xi, 85, 179, 197, **319**, 633—also see Birthframes 12, 21, 82, 86 and 87
Leboyer, Frederick 192, 254-5, 279, 636
legal issues viii, **14**, 51, **58**, 94, 132, 156, 167, **264**, 320, 329, **408**, 417, 425, 432, 508, **560**, 611, 612, 615
listening **67**, 80, 84, **215**, **226**, 263, 268, 276-7, **294**, 288-293, **298-9**, 349-351, 362, **378-9**, 419, **446-8**, **455**, 461, 478, **502**, 505, 507, 516, 518, 595
lithotomy position 88. **157**, 162, **163**
lochia 46, 100, 101
lotus birth **259**, 432, **515**, 520—also see Birthframe 92
low blood pressure—see 'blood pressure'
low-lying placenta—see 'placenta previa'
mammals 147, 160, 202, 337, 445, **447-8**, 461, 465, 467—also see Birthframe 77
massage 27, 72, 171, **186**, 223, **232-3**, **243**, **252-3**, **261**, 310, 312, 385, 404, 421, 422, 474, 490, 494, 502, **558-9**, 578, 611
maternal mortality 3, 50, 108, 139, **151**, **198**, 238, 249, 330, **375**, **377**, 455, 569
miscarriage **19**, 75, 77, 80, **118**, 180, 184, 189, **328**, **337**, 525, **548**, 549, **603**, **640**
monitoring 3, **12**, **25**, **40**, **42**, **59**, 74, 78, 79, 83, **88**, 101, 103, **108-9**, **127-8**, **131**, **134-5**, **143**, **153**, **157**, **166**, **172**, **182**, 203, 206, **214**, 235, **243**, 255, 258-9, **266**, 270, 288, 289, **294**, **296**, **298**, 301, 302, 305, 308, 317, 329, **332-341**, 345, 348, 350, **351-3**, **353-4**, 356, 380, 388, 410, 431, 439, 441, 442, 446, 454, 458, 459, **461-6**, 467, **473**, 474, 490, 491, 506, 525, 527, 529, 548, **553-4**, **629**, **630**, **631**—also see 'watchful waiting'
mortality rate—see 'maternal mortality' and 'neonatal death'
MS (multiple sclerosis) 254
multiple birth—see 'twins' and 'triplets'
negligence 132, 455-6—also see 'watchful waiting' and 'disturbance'
neonatal death **377**, **548**, 560, 582, **641**
new natural 12
nitrous oxide—see 'Entonox'
nocebo effect **67**, **164**, 239-240, **278**, **328**, **464**, 552, 566—also see Birthframes 34 and 78
noise in labor **7**, 25, 51, 54, 64, **68**, **157**, 198, 222, 322, 323, 353, 359, 362, 412, 454, 478, 480, **491**, **489**, 510
nuchal translucency testing 154, 368
number of people at the birth 28, **40**, **49**, **63**, **68**, **82**, **88**, 94, 166, 167, **184**, **222**, 305, 354, 365, 366, **372**, 421, 423, **440**, **446**, **448**, **454**, **463**, **486**, 490, 493, **495**, 517, 541-3

nursing—see 'breastfeeding'

obstetric cholestasis 79

Odent, Michel x, 7, 15, 23, 31, 32, **52-6**, **62**, 63-5, 67, 73, 80, 82-3, **84**, **96**, 99, **109-110**, 112, **117-8**, **121**, **123**, **131**, **132**, **133-5**, **138**, **141**, **143**, 148, **160**, 164, 166, 168, **170**, 173, **179**, **184**, **185**, 189, **194-5**, **196**, 197, **203,-6**, **209**, 219, 221-4, **231**, 232, **236**, **239**, **245**, **254**, **256**, 256-7, 269, 284, **299**, **319**, **328**, **330**, **333**, **334**, **340**, **341**, **342-4**, **372**, **398**, 411, **416**, **447-8**, 448-451, **457**, 460, **461-3**, **471**, **472**, **475**, **484**, 489-493, 496, **499**, **502**, **504-6**, **520**, **523-4**, **534**, **552**, **557**, **558**, **564**, **565**, **566**, **568**, **658**, **659**, **663**—also see Birthframes 3, 5, 7, 12, 21, 31, 32, 45, 48, 73, 77, 82, 85

old natural 11-12

older mothers—see 'elderly primigravidae'

Omega-3 supplements 118

onset of labor 22-3, 25-28, 29—also see 'false labor', 'SRM' and 'contractions'

opiates **12**, 35, 101, **115**, 148, **150**, **176-7**, **178-9**, **182**, 195, 246, **247**, **248**, **256**, **257**, **551**

optimal fetal positioning 244, 248, 253, 462, **466-8**, **469**, 487, **488**, **513**, **612**, **620-1**, **624**, 637

oxytocin (or 'oxytocics') **12**, **14**, **24**, 25, **31**, **35**, **89**, **96**, **113**, 125, **126-7**, **136**, 140, **147**, **148-9**, **152**, **153**, 155, **165**, **177**, **196**, **200**, 205, 220, 232-3, **237-8**, 241, 246, **247**, 249, **278**, **279**, **297**, **298**, **315**, **332**, **351**, **398**, 484, 488, **499**, **505**, **524**, **558**, **564**, **629**

pain xii, **12**, **24-29**, 32, 33, **35**, **44-5**, 49, 51, 53, 55, 56, 60, **62**, 64, 65, **68**, 72-3, **73**, 74, 77, 78, 80, 81, 82, 90, **95**, 100, **102**, 104, 105, 107, **111-118**, 123, 128, **136-8**, **145**, **147**, **148**, 158, 159, **160**, 164, 167, 187-190, **191**, **193**, **200**, 203, 204, **208**, **209**, 215-6, 217, 221, 222, **228**, **229**, 233, **235**, 241-2, **243**, **244**, **245**, 246, 247, 249, 251-3, **253-4**, 256, **261**, **271**, 280-1, 283, 289, 290, **296**, **301**, 310, 312, **314**, **316**, **318**, **319-326**, 348, 349, 352, 354, 356, 359, 369, 370, **378**, **379**, 380, 385, 388, 390, 404, 405, 409, 412, 417-420, 424, 425, 429, 439, 441, 442, 443, 444, **446**, **447-8**, **459**, 460, 464, 465, **470-1**, **472**, 476-482, **483**, 487, 488, **491**, 496, **497**, 500, 508, 517, 521, 529, 530, 536, 537, 539-547, **549**, 551, **558-9**, **563**, **566**, 567, **568**, 576, 579, **595**, 609, **616**, **621**, **626**, 631, **636**, **661**—also see Birthframes 24, 25, 26, 46, 65, 80, 81 and 96

pain relief x, **3**, **4**, **7**, 11, **12**, 38, **42**, 54, 64, **68**, 73, 79, **95**, 101, **102**, **108**, 109, **112-118**, 120, **126**, 134, **148**, **150-3**, **154-6**, **158**, **161**, 164, **165**, **170-186**, **191**, 193, **208**, **209**, **229**, **230**, 237, 240-1, 242, 245, 246, 247, 248, 250, 251-3, 254, 261, **263**, **272**, 287, **288**, 305, **308**, 309, **312**, **327**, **343**, **348**, 349, 380, 381, 404, 429, 439, **446**, **457**, **470-1**, **473**, **474**, 478, **483**, **494**, 500, 521, 530, **558-9**, **560**, **561**, 563, **565**, 567, **568**, 569, **577**, **661**

palpation 54, **87**, 94, **270**, 298, 332, **333**, **335**, **353**, 369, 452 photographs—see 'camera', 'camcorder' and 'privacy'

physiological processes 3, 4, **7, 11, 12, 14-48, 52-7**, 79, 110, **113, 117-8,
132, 146-9**, 150, **160, 162-9, 182, 183**, 184, 185, **207**, 228, **240,
272**, 274, **308, 309, 345**, 410, 448, **630**, 659

physiological third stage—see 'third stage'

pitocin **89, 155, 156, 173**, 203, **206, 238, 297**, 301, 351-2, 398, **564, 566**

placenta 20, **22**, 32, **35, 38**, 45-6, 55, 59, 63, 64, 65, **74-5**, 79, 83, 89, 93,
94, **97**, 100, **121, 123-5, 125-6, 129-131**, 139, 141, **147, 148, 149,
152, 153**, 155, 168, 181, 186, 190, 198, 220, 223, **244**, 248, **250**,
253, **270, 280-1**, 291, 292, 321, **324, 335, 340**, 344, 349, 353, 354,
366, 368, 372, 391, 405, 411, 412, **420**, 422, 424, **429**, 431, 432,
444, 449, 450, 453, 455, 460, **463**, 467, 481, 488, 492, **505, 514-5**,
516, 518, 520, **522, 523-4**, 551, 556, **569, 603, 604, 606, 608, 609,
611, 612, 614, 618, 622, 625, 628**

placenta previa **74-5, 97**, 139, **335**

positions (for labor or birth) 466-468, 469, 512, 513

posterior position 26, 49, 54, 112, 174, **179, 186, 244, 248-9**, 253, 425,
440, 488, 489, **612, 620-1, 624**, 626—also see Birthframes 2, 71 and
85

postmaturity 202, 306, 410, 628, 631

postpartum depression 71, **97, 199-209**, 210, 216, 217, 229, **243**, 253, 357,
**378, 380, 506**, 570, 573, 579, 639

postpartum scenarios 35-48—also see all birthframes!

PPH (postpartum hemorrhage) 38, 59, 63, 70, 74-5, **96, 123-6, 149, 151,
152**, 181, **198**, 268, 292, **315, 340, 375**, 429, **463**, 509, **523-4**—also
see Birthframes 8 and 74

pre-eclampsia xiv, 195, 270, 296, 298, 332, 340, 535, **553**, 558

pregnancy **19**, 20-22, **598-632**

premature labor—see 'prematurity' and 'kangaroo care'

prematurity —also see 'kangaroo care'

prenatal care **19**, 20, 87, **152**, 194, 268, **270-2**, 274, **276-287**, 318, **330,
332-333, 340-1**, 372, 385, **407-8**, 411, 430, **550**, 553, **558**, 574, **587-
8**, 659, 662

prenatal checks **154, 340-341**

prenatal classes 81, **119, 200**, 220, **257-8, 272**, 356, 448, 500, **640**

prenatal tests 274, 282, **328, 330**, 368

primal health **115, 135-6, 170, 194-5**, 557, **640**, 659

Primal Health Research Database 170, 194, 195

primiparas (first-time moms) 73, 82, **114**, 129, **133**, 135, 137, 138, **141, 143,
156**, 235, 236, 242, 245, 260, 261, **266**, 268, **278**, 280, 283, **298**,
309, 312, 332, **438**, 448, **494**, 544, 545, **566-7**, 660—also see
Birthframes 7, 34, 36, 40, 41, 70, 73 and 86

privacy 15, 28, **36**, 40, **56**, **72**, 80, **82**, **88**, **109**, **110**, **133**, **134**, 143, **179**, **208**, 226, **254**, **308**, 314, **319**, 342, 344, 350, 353, **484**—also see 'disturbance' and Birthframes 45, 48, 53 and 82

protocols 4, **25**, **34**, **59**, **80**, **86**, 98, 126, **131**, 140, **141-2**, **153**, **160**, **164**, **165**, **172**, 205, **206**, **207**, 263, **265-6**, 288, **300**, **328-341**, 423, **446**, **586-7**, **588**

psychological difficulties **12**, 56, 63-66, 69, 96, **97**, 103-4, 104, 105, **109**, 111, 141, **155**, 160, **164**, 165, 166, 169, **192**, 202, **206**, 215-6, **228**, 229, 270, **275**, 276-7, 278, 281-7, **308**, **314**, **315**, **321**, 332, 333, **356**, 356-8, 364-7, 368-370, **378-406**, **407-426**, **427-433**, **465**, 485, **486**, **520**, **523**, 539-547, 549, **550**, **552**, 567, **574-5**, 578-9, 589-591, **595**, **632**—also see Birthframes 5, 16, 17, 22, 27, 30, 33, 34, 37, 54, 47, 48, 49, 56, 61, 64, 83, 96, 97, 99 and 100

pushing xii, 23, 27, **31-3**, **34**, **50**, **55**, 63, 65, **72-3**, 82-3, 93, 94, **120**, **121-3**, 129-131, **137**, 145, 148, **156**, **157**, **158**, **162**, 166, 169, **174**, 175, **178**, 180, 181, 187, 190, 204, 223, **250**, **260-1**, 291, 316, 317, 319-325, 347, 348, 349, **350**, **351**, 352, 354, 366, 369, **384**, 389, **390**, 401, 404, 405, 412, 416, 419, 421, 422, 425, 431, 440, 441, 443, 444, 450, 452, 453, 458, 460, 470, 474, 476, 480-2, 488, 492, **497**, 508, 510, 517, 543, **568-9**, 576, 577—also see 'commanded pushing' and Birthframes 6 and 46

randomized controlled trials **xiii**, 6, **49**, **97**, **118**, **160**, 236, 251-2, **259**, 334, 434, 567, **585**, **644**, 663—also see 'safety', 'watchful waiting' and all notes and references after the Intro, 'What', 'Why' and 'How' sections.

Rank, Otto 192

raspberry leaf tea 118, 183, 310, 500

rebirthing 368, 640

reflexology **252**, 496

ring of fire **30**, 33, 55, 405, 419, 480, 626

risk assessment 134, **306-327**, **342-377**, 661—also see 'safety'

safety v, **xii**, **xiii**, xiv, **2**, **3**, **4**, **6**, 11, **14**, **23**, **25**, 31, **35**, **38**, **40**, **41**, 51, **52**, 55, **56**, 59, **60**, 65, **70**, **87**, **96**, **97**, **108-110**, 118, **123-5**, 126, **134**, 139, 143, **149**, **150**, **152**, 154, **160**, **161**, 162, **165**, **173**, 179, 189, **200**, 202, 205, **208**, **226**, 228, **236-7**, **240**, 242-3, **246**, **250**, 251, 255, **259**, **263**, **266**, 268, 291, 294, **296**, **298**, **306**, **308-9**, **334**, **335**, **337**, **338**, **342**, 343, 345, **350**, **356**, 364, 366, **370**, 373, **374**, 375, 382, 384, 386, 388, 389, 392, **408**, **411**, **413**, 419, 430, **432**, **439**, 442, 443, 444, 445, 446, **452**, **457**, **463**, **464-5**, 467-8, 473, 489, 496, 514, **516**, **519**, 532, **553**, **559-560**, **561-2**, 564, **573**, **574-5**, 578, **585**, **587**, **597**, 605, 626, 631, 657, **659**—also see Birthframes 2, 3, 4, 6, 7, 8, 10, 12, 13, 14, 15, 21, 28, 29, 32, 33, 34, 35, 36, 38, 39, 45, 47, 48, 49, 50, 51, 52, 53, 54, 55, 56, 57, 59, 62, 63, 64, 65, 66, 67, 68, 69, 70, 71, 72, 74, 75, 76, 77, 78, 79, 80, 81, 82, 83, 84, 85, 86, 87, 89, 90, 92, 93, 94, 95, 97, 98 and 99

sanitation 110, **372-3**, 374, 435—also see 'developing countries'

schizophrenia **195, 234-235, 257, 550-551, 606**

screening tests—see 'prenatal tests'

second stage **12, 25, 30-34**, 105, **119-123**, 124, **128-134, 143, 174**, 204, **235-6, 240-242, 249-250**, 260, **315**, 366, 390, **432**, 444, 450, **452-453, 454, 460**, 462, **470**, 474, 479, 482, 488, 503, 508, **558, 565, 566, 567, 568**, 626—also see all birthframes

second trimester **20-22, 118**, 552, **606-616**

sepsis 151, 152, **375**

sex 8, 17, **29, 60 ,96**, 102, **111-3, 118, 128-129**, 133, 136-8, **175, 200, 211-2, 234-6**, 237, 241-2, 270, 287, **331**, 410, **446-8, 499, 505, 573, 574, 599**, 630, **632**, 637

sexual abuse 69, **384-399, 559, 576-577, 595**

SGA—see 'small for gestational age'

shiatsu 11, 12, **158, 183, 186, 252**

shoulder dystocia 341, **481**

side-effects **7-9, 35, 75, 108, 115-117, 165, 170-182, 193, 198, 214-7, 242**, 247, 287, **328**, 382, 476, **663**

skin-to-skin contact 38, 100, 101, **127, 149, 204, 232-3**, 316, **358**, 390, 527, 529, **530**, 534, **535, 570-1**—also see 'kangaroo care' and 'bonding'

slow labor—see 'failure to progress', 'disturbance' and 'number of people at the birth'

small for gestational age 237, **333**, 554

spinal **171, 249, 250**, 458, **459, 500**—also see 'epidural'

SROM (spontaneous rupture of the membranes) **203-6**, 289, **452-3**

stillbirth 275, 448, **539-549, 641**

Stoppard, Miriam 203

stress **9, 111**, 116, **135, 139, 145**, 147, **150, 158, 174**, 180, **193, 234**, 257, **278**, 281, 286, 291, **298, 311, 332, 352**, 420, **430, 442-5, 501**, 529, **534-8, 550-1, 565, 567-8, 571**, 615, **618, 623**

stress incontinence **121-3, 130, 136-8**, 209, **234-5, 260-1**

stripping membranes (or cervix) 156, 203, **206**, 409, **496, 573**

sucking reflex 81, 175, 179, 193

suicide 142, 195

supplements **118**, 162, 183, **270, 340-1**, 599

sutures 45, 63, 70, **71**, 82, 83, **100, 105**, 107, **128-9, 178**, 180, 181, 187, **211, 216, 247-8**, 302, 311, **315**, 316, 318, 325, 348, 405, 410, **420**, 421, 474, 476, 482, 518, **640**

syntometrine **46**, 63, **89, 123-5**, 155, 292, 366

talking in labor and birth 25, **26, 35, 52-57, 59-60**, 72, 84, 123, **133-134**, 143, 167-9, **215**, 236, 301, 322, **349-351**, 388, 390, 403, **416**, 419, **431, 440**, 443, **446-8**, 449, **454-6, 463**, 477, **485-6**, 489-493, **496**, 507, **512**, 518, **520, 565, 573** talking in pregnancy **67**, 77, **90**, **92**, **146**, 268, **270**, **276-299**, 301, 356, 359, 362, **369**, **370**, **378-9**, 396, 397, 421, **439**, 461, 478, **502**, 520, 531, **572**, **573**, **598-633**, **640**

talking postpartum **2**, 34, **62-3**, 94, **100-1**, 178, 187, **199-201**, **202**, **204**, **210**, 211, **212**, **215-7**, 223, **225**, **226**, 229, **230**, **316**, **362**, 390, 419, **443**, **463**, 502, 505, **526-7**, 529, **537**, 539, 543, **545**, 546, **549**, **590-1**, **640**, **633**—also see Birthframe index on pages 642-4

tandem feeding 412

tearing 12, 35, 45, 55, 65, 71, 112, **121**, **128-9**, 130, 136, **137-8**, 165, **175**, **178**, 181, 205, 211, 235-6, 241-2, 247-8, **261**, 302, 319, 324, 325, 360, 369, 390, 405, 419, 422, 423, 470, **473**, 474, 482, 518, 568-9

TENS **12**, 63, 79, **158**, **171**, **183-5**, **251**, 287, 300, 309, 443, 444, 474, 487, 500, 516, 517

tests 17, 44, 78, 106, 147, **154**, 159, **160**, 162, **164**, 220, 269, 271, 274, 282, 288, **328-332**, 333, 336, 338, 339, **340-1**, 346, 368, 411, **472**, 482, **484**, 499,525, **552**, 555, **574**, 598, 600, 607, **661**—also see 'prenatal tests' and 'prenatal checks'

third stage 32, 35, 38, 42, 45, 50, 63, 88, 89, **113-4**, **120-1**, **123-5**, **125-6**, 150, 153, 155, 168, **242**, 291, 344, 375, 429, 442, 452, 462, 472, 474, 481, 488, **505-6**, 509, 516, **523-4**, 606—also see 'placenta', 'physiological third stage' and all birthframes

third trimester **22**, **118**, **183**, 585, 615, **617-632**

transition 31, 42, 169, 180, 190, 322, **390**, 412, 440, 443, 455, 460, 474, 480, 492

trial of labor 285-6, 319, **566**

triplets **86**, 198, **358**, **538**, 599, 600—also see Birthframes 54, 55 and 56

twins 25, 27, 73, 74, 89, 268, 315, 364, 362, 372, 532, 599, 600, 625, 641—also see Birthframes 10, 13, 14, 15, 23, 56, 57, 62 and 76

ultrasound **87-89**, **136-138**, **152-153**, **154**, 162, **164**, **236-237**, **239**, **240**, 270, 277, 282, **287**, **294**, 329, **332-333**, **334-337**, 338, 339, 341, **462**, **552**, **553**, **554-557**, 603, **608**—also see 'ultrasounds'

ultrasounds (i.e. specific experiences)**74-78**, 79, **87-89**, 92, 94, **152-153**, **154**, **164**, 180, 189, **236-237**, **270**, **278**, 288, 320, **329**, **330**, **333**, **334-337**, **338**, 339, **341**, 350, 353, 356, 364, 366, 368, 411, 508, 540, 541, 545, **552**, **553**, **554**, **555**, **557**, 601, 608—also see 'ultrasound'

unassisted birth 58, 86, **288-293**, 408, **411-412**, **421-425**, 429, **430-433**, **516-519**—also see 'freebirth'

urine testing 76, 78, 220, **270**, **298**, **332**, 525, **552**

vaginal breech birth—see 'breech position'

vaginal examinations **42**, **59**, **131**, **133**, **143**, 174, **205**, 209, 305, 319, 350, 351, 356, 439, 446, **459**, **463**, **558**, **564-565**

VBAC 47, **97**, 105, 206, **314-327**, 346-348, 383, **551**, 577, **641**, 660

ventouse **121**, **127**, **138**, **152**, 156, **158**, 162, **163**, **234**, **261**, 316, 319, 380, 439, 525, 552, 568

Vincent-Priya, Jacqueline **150**, 636
visualization 17, **49**, 73, 80, 169, 189, **296**, 320, 360, 369, 391, 439, 443, 454, **550**, **572**, **573**, 574, 615
vomiting 7-9, 60, **75-78**, 89, 100, **115-117**, **123-125**, **176**, **177**, **182**, 406, 443, 480, 487, **497**, 521, 541, **606**
washing the baby **44**, 107, **193**, 436
watchful waiting **3**, **14**, **34**, 40, **42**, **59**, **131**, **132**, 206, 207, **218**, 289, **446**, **455**, **462**, 469, 630
water birth 15, 28, 63, 64, 87, 187, 189, 205, 275, 288, 321, 323, **470-473**, 474, **475**, **476-483**, **484**, 497, **566-567**
weighing the baby **93**, **103**, **193**, **197**, 291, **325**, 412, 420, 449, 450, 501, 518
X-rays **334**, 358, **371**, 601, 607

For research abstracts, go to www.pubmed.com

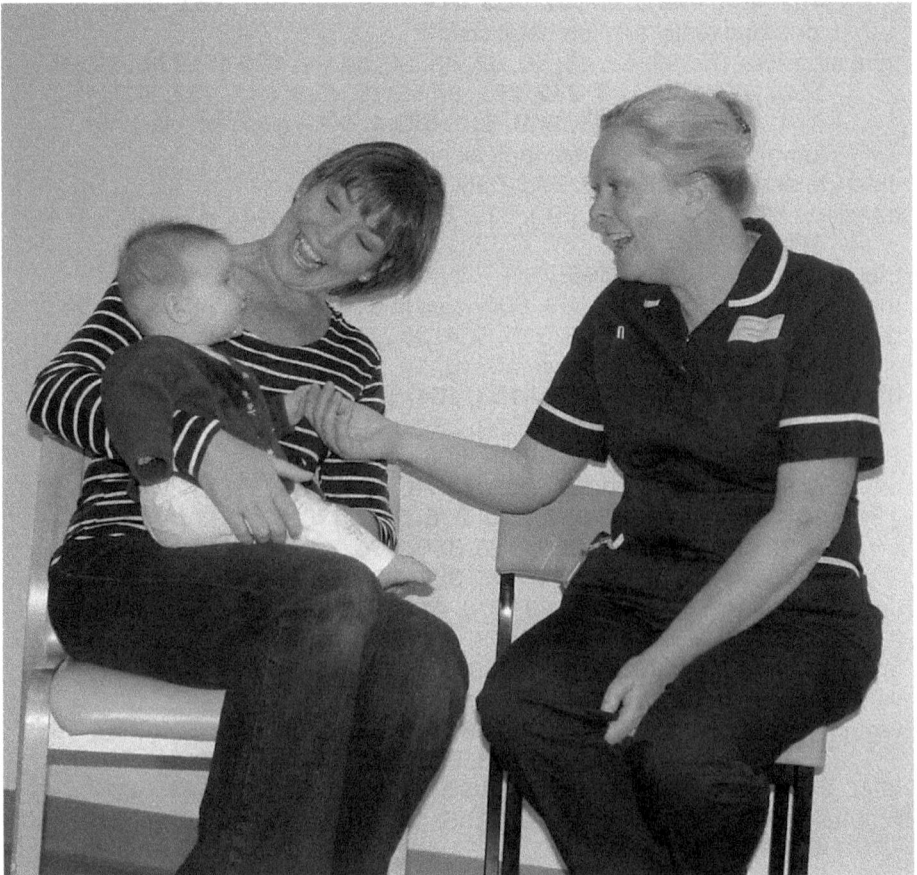

*Working towards optimality may be challenging on all kinds of levels...*

*... but let's not spoil our little ones' dreams. Let's make birth as beautiful, as safe, as happy, as fulfilling, as satisfying, as healthy... as optimal, in fact, as we possibly can—for this and future generations*

# Who is Sylvie Donna?

Before she started having children, Sylvie Donna worked in companies or taught English, mostly to working adults. She also trained teachers, or managed courses, departments—or a whole language center in one case. She taught or organized courses for middle and top-level managers (as well as clerical or research staff) in Europe, North Africa, South Asia, South East Asia, the Middle East and the Far East. Her first book—*Teach Business English* (Cambridge University Press 2000)—is a synthesis of her experience in this field. Since having children, Sylvie has worked from home, writing and editing, or marking Master's assignments. She still does some teaching and lecturing at Durham University in the UK, for MA, MSc, MBA and PhD students.

Sylvie started researching issues surrounding pregnancy and childbirth when she conceived her first child at the age of 37. (The reason she started checking things is explained in Birthframe 1.) Perhaps it was because of her long experience of working with managers and directors, that Sylvie found the confidence to challenge her caregivers when she disagreed with them! As she explains in this book, some of her other experiences in life (before she got pregnant) may also have influenced her and perhaps given her enough anger to assert herself and make sure she got the kind of care she felt she needed.

Anyway, Sylvie eventually found wonderful midwives and obstetricians and all went smoothly, thanks to her thorough preparation and a little bit of luck! A few days after the birth of her first baby, which was attended by an obstetrician, Sylvie read *Birth Reborn* by Michel Odent (Souvenir Press 1994) and it struck a note of recognition in her... For the first time she was reading about the kind of birth she'd experienced. She then felt fortunate to have Michel Odent agree to attend the birth of her second child. The experience of birthing with him in attendance and experiencing an authentic fetus ejection reflex (i.e. a completely undisturbed birth) then prompted her to think through more issues, which she was able to put to the test on a very modest scale when she gave birth to her third child, with state-registered midwives in attendance. The various things which went wrong, and right, with this third birth prompted further thought, which—interestingly enough—related to research she'd done for her own Master's, on the subject of language use and linguistic analysis.

Having had the idea to write a book to help other women and families, Sylvie eventually completed a book for moms-to-be (after many years of research). By chance, she was asked to help someone organize a conference stand at a midwifery conference a few days before it was published. As a result, she decided to go along to the conference and thought she might as well take a few of her brand new books along with her. To her astonishment, she found that midwives and midwifery managers were enormously interested in the research she'd done into physiological birth and why it did or didn't take place in particular situations... and she was soon encouraged to adapt the book for midwives. After a first, very rushed edition was produced for the UK this longer US edition was prepared, with academic references and reflections.

# Who is Michel Odent?

Born in 1930, Michel Odent initially qualified and worked as a general surgeon. He gradually became more and more interested in issues surrounding childbirth, after being put in charge of a government maternity hospital in Pithiviers, near Paris, France in the 1960s and '70s. While he was working there, he was responsible for around 15,000 births.

Coming new to the field and fortunate to have both a freshly-trained and an 'old-school' midwife to assist him, he soon realized that pregnancy and childbirth were not things with easy or clear-cut answers. This led him to develop various practices which he later checked out through extensive research, both within the field of obstetrics and on a cross-disciplinary basis. As the years progressed, he came to feel more and more that childbirth was at its safest when the healthy, physiological processes were left to take place undisturbed.

In the 1980s he moved to London in the UK, where he set up the Primal Health Research Centre and practiced as a homebirth midwife. His research has spanned topics such as preconceptual and prenatal care, nutrition in pregnancy, childbirth itself, breastfeeding and childhood innoculations.

Frequently interviewed on television, in radio programs and in the popular press, he has become known as the pioneer of the use of water during labor and homelike hospital birthing rooms.

He is the author of numerous scientific papers and 12 books, including *The Caesarean* (Free Association Books 2004), *The Farmer and the Obstetrician* (Free Association Books 2002), *The Scientification of Love* (Free Association Books 1999), *Birth and Breastfeeding* (Clairview Books 2007), *Primal Health* (Clairview Books 2007), and perhaps his most well-known title: *Birth Reborn* (Souvenir Press 1994). [Many of his commentaries are included in this book.]

## Michel's comments about Sylvie's writing...

After the birth of her second baby, Sylvie told me of her irresistible need to write a book about childbirth. My immediate tacit and skeptical reaction was: 'Yet another book about birth. If I had kept all the manuscripts and books that have been sent to me over the last 20 years, I would need to have an extension built onto my den!' After several conversations with Sylvie I started to change my mind. I realized that, thanks to her personal experience, Sylvie was aware of what very few people have understood. The turning point was on the day when she told me that *talking is an intervention*. Then I became convinced that her book should be published; I enthusiastically answered Sylvie's questions and occasionally added my comments.

There are many reasons why this book is special. One of them is that it is not a guide. Sylvie's aim is simply to inform and inspire. She has the power to achieve this goal because, after being pregnant in countries as diverse as Sri Lanka, the UK and Oman, and after giving birth in different environments, she has become a real expert in childbirth. She is immune to the countless received ideas that abound in magazines, newspapers and books. Often women who talk a lot about the birth of their babies are those who had a difficult birth, problems and so on. And women who had a very easy birth tend not to talk about that... This book is full of accounts we can learn from. Browse through this book and you'll absorb some authentic knowledge transmitted by an authentic expert.

# Who are the contributors?

*Elaine Batchelor, Jenny Sanderson and Nina Klose—three of over 200 contributors*

These are just three of the many women who contributed to this book. The woman on the left, Elaine Batchelor, is an independent midwife. She has contributed various comments and photographs, as well as advice—based not only on her experience of being a midwife, but also on her experience of giving birth herself. The woman in the middle, Jenny Sanderson, a wonderfully thorough editor, opted for a physiological birth at home, even as a first-time mother, after hearing about the numerous interventions which are common in hospitals and wondering how on earth she would be able to find the courage to resist them! She had an extremely straightforward and healthy birth then and also for her three subsequent children. Hearing about other women's less positive experiences, she later trained to be a prenatal teacher. Nina Klose, a Harvard graduate, had two home VBACs (vaginal births after cesarean) after experiencing a cascade of intervention the first time she gave birth in hospital. She has contributed a great deal of material and support.

Other contributors come from all walks of life... There's a birth story from a piano teacher, two from engineers, accounts by midwives, nurses and doulas, another from a professional photographer, yet another from a psychoanalyst, another from a shopkeeper... The list is long and varied. A few men also contributed their views, as did a couple of children. There are also plenty of contributions from other professionals from around the world.

Sylvie was able to attract so many different types of people by advertising for contributions in very different places, doing searches for people on the Internet, or by approaching people face-to-face when she met them in the street, at playgroups or at other events, such as midwifery conferences. She was also referred to many contributors by other contacts and more recently people have been attracted by the website. If you or your clients want to contribute to a future edition please email info@freshheartpublishing.co.uk.

# What inspired this book?

While she was pregnant with her own three children, the author and editor of this book, Sylvie Donna, kept on meeting women who'd had difficult experiences. She started trying to work out what could have gone wrong for these women. It seemed to her that too many women were being traumatized temporarily, if not for the rest of their lives, by childbirth. Sadly, the memory of giving birth was leaving these women feeling disappointed, alienated or betrayed. Their babies, partners and families were affected too.

For many people Sylvie met, pregnancy was more of an obstacle course of tests and worries, than a time of wonder and waiting. Somehow, amongst all the prenatal appointments, risk assessment and birthing pool hire, the baby-to-be got thrown out with the as yet non-existent bath water. And many women told Sylvie how the birth they'd planned went wrong in the end. From some of the women, who were the 'statistics' of care gone wrong, she heard horrendous stories of pain and trauma. Many women simply described their feelings of disempowerment as they were 'managed' through the maternity system. For others it was the breastfeeding or the bonding which didn't work out...

What was it, she wondered, that made things go wrong? Listening carefully to countless women, she started making connections between behavior in pregnancy and birth, and outcomes. She realized that things often start going wrong in pregnancy for no good reason, other than fear. She also discovered—through women's personal accounts—that drug-based pain relief often ended up causing more pain than it ever relieved, if postpartum pain was counted too.

While she was realizing these things, she also became increasingly aware that very few women see the chain of events which they set up for themselves by accepting or even requesting certain treatment while they're pregnant, in labor, giving birth and even afterwards. For example, how many women are aware that having an induction of labor increases their risk of having all kinds of other interventions? How many women would choose to have an epidural if they knew what it really involved and if they knew what consequences there might be for either herself or her newborn baby? How many women have found out about and thought through the potentially harmful effects of other forms of drug-based pain relief? How many women have been able to carefully compare postpartum scenarios after a vaginal birth and a cesarean, considering emotional, physical and practical aspects of the experience? Most importantly, Sylvie wondered, how many women know that a great deal of care offered prenatally and during labor flies in the face of research recommendations?

Given the recent trends to development evidence-based care, Sylvie felt optimistic about change in the future... Nevertheless, transition periods are not easy and both caregivers and moms-to-be need preparation if optimality is going to become a reality in maternity units around the country. But then again, she thought, what is optimality... and how on earth can we achieve it? A title and a book were born, for caregivers of all kinds, and medical students.

## Comments about the companion book for moms-to-be...

*... Birth: Countdown to Optimal,* now called *Preparing for a Healthy Birth*

### Betty-Anne Davies:

There are many books written about childbirth, each one valuable because each author adds a unique dimension to resonate with changing times. In academic circles, many hours are spent detailing important issues, but in so doing, the basics can get left behind. Sylvie has done a great job of bringing a resonant dimension, the literature, and common sense together in one book for parents who are as new at this process as are their newborns. She has created a prescription for healthy birth in a 10-step approach to keep parents focused on what is important in a potentially disastrous birth environment. The liberal use of anecdotes from other parents is reminiscent of the group prenatal care approach that has swept North America in its warm appeal because learning from other parents is often better than from professionals who don't understand well what the average parent needs.

Sylvie's book should ring a bell not only with the parents for whom it was intended, but also as a 10-step refresher course for professionals to remember what it is like to be in parents' shoes. She has gracefully unravelled the problems that health professionals and hospital policies have imposed on normal, healthy birth and provided us with a book that encapsulates anything you wanted to know about keeping birth normal and healthy, with some extras you may never have thought about.

### By the same author—books for your moms-to-be:

- ♥ *Surprising, Inspiring Birth!*
- ♥ *Preparing for a Healthy Birth* (American and British editions)
- ♥ *Better Pregnancy, Better Birth: A Week-by-Week Guide*

See the website for more info. All Fresh Heart books are available from either www.amazon.com or www.freshheartpublishing.com. Discounts are available on all Fresh Heart orders of five or more books. If you'd like to have more information, email Fresh Heart at sales@freshheartpublishing.co.uk.

Fresh Heart is currently preparing more titles on birth by other authors. If you have a book you would like to write yourself, please email a proposal, via the website.

## Sheila Kitzinger:

It is difficult to write a book about birth drawing on research, analyzing the effects and side-effects of interventions and also acknowledging that in the right setting and with loving, sensitive and unobtrusive support birth is a psycho-sexual process which can bring ecstasy. Sylvie has achieved this splendidly. She writes with energy and passion. Her book is rich with women's accounts of pregnancy, birth and after. Readers who do not relish childbirth may find it hard to take it all on board, but the tone, both of the women whom she quotes lavishly, and her own enthusiasm, is so compelling that many could be converted to a radically different view of birth. If they are brave enough to explore what she has to say, with another pregnancy birth may turn out to be a very much better experience. This is a book that can help its readers be adventurous. Breaking the barrier involves not only getting information on which to base choices, or of acquiring the knowledge to make a birth plan, but also getting our inner confidence to grow and blossom! An amazing book—with insight, scholarship and rich with vivid real-life experiences. An essential resource.

## Janet Balaskas:

This is a book that absolutely celebrates the normality of birth and the joy of welcoming a new life into family and community. It is completely authentic, full of personal stories and evidence that lots of families and the author herself have truly experienced... It is a truly scholarly work that weaves in the universal truths of birth physiology and research evidence while encouraging, inspiring and informing women. Professionals, as well as families, have so much to gain from its pages. This is a very important book for our times, when we sure have got this whole process of birth wrong. I heartily congratulate Sylvie for her passion and all the work that went into compiling and writing this book.

## Former Head of Midwifery for three large hospitals:

Brilliant, moving, challenging and educational. The stories about C-sections are particularly moving, especially where one woman describes a feeling of being burglarized! This is nicely balanced with a positive story of a C-section, which demonstrates that high risk women can have a positive birthing experience, even in a medicalized environment... The photos are also lovely. I would recommend this book to anyone interested in childbirth as I feel it gives more than the average clichéd information that most childbirth books offer.

## Comments from other readers:

  What every mom-to-be will want to know, put simply and clearly.

  Contains some really interesting, moving, informative and provocative material. The section on pain relief and interventions is very valuable and included info I'd not read in other books—very detailed and brilliantly researched.

  A wonderful and necessary piece of work

  This book should be compulsory reading on all medical and midwifery courses.

  Had a very late night as I could not put the book down!

  Reading this almost made me want to have another baby!

www.ingramcontent.com/pod-product-compliance
Lightning Source LLC
Chambersburg PA
CBHW072037020426
42334CB00017B/1299